Materials for Sustainable Energy Applications

Materials for Sustainable Energy Applications

Conversion, Storage, Transmission, and Consumption

edited by

Xavier Moya
David Muñoz-Rojas

PAN STANFORD PUBLISHING

Published by

Pan Stanford Publishing Pte. Ltd.
Penthouse Level, Suntec Tower 3
8 Temasek Boulevard
Singapore 038988

Email: editorial@panstanford.com
Web: www.panstanford.com

British Library Cataloguing-in-Publication Data
A catalogue record for this book is available from the British Library.

Materials for Sustainable Energy Applications: Conversion, Storage, Transmission, and Consumption

Copyright © 2016 Pan Stanford Publishing Pte. Ltd.

All rights reserved. This book, or parts thereof, may not be reproduced in any form or by any means, electronic or mechanical, including photocopying, recording or any information storage and retrieval system now known or to be invented, without written permission from the publisher.

For photocopying of material in this volume, please pay a copying fee through the Copyright Clearance Center, Inc., 222 Rosewood Drive, Danvers, MA 01923, USA. In this case permission to photocopy is not required from the publisher.

ISBN 978-981-4411-81-3 (Hardcover)
ISBN 978-981-4411-82-0 (eBook)

Printed in the USA

There are many things we could do to keep ourselves going: as the oil wells run dry. We have to keep our wits about us, though. We have to co-operate the whole world over and we have to work hard and *fast*.

—Isaac Asimov, 1982

About the Editors

Xavier Moya is a Royal Society University Research Fellow in the Department of Materials Science & Metallurgy at the University of Cambridge. He received his BSc and PhD in physics from the University of Barcelona in 2003 and 2008, respectively. He is interested in phase transitions in functional materials whose structural, magnetic, electrical, and thermal properties display strong coupling. His research focuses primarily on caloric materials for cooling applications and magnetoelectric materials for data storage. Dr. Moya has been a Fellow of Churchill College since 2014. He was awarded the 2009 Ramon Margalef Prize for his PhD work and the 2015 Young Researcher in Experimental Physics Prize by the Spanish Royal Society of Physics for his continuous work on multiferroic materials.

David Muñoz-Rojas received his degree in organic chemistry in 1999 and master's degree in chemical engineering (2000) from the Instituto Químico de Sarrià (IQS, Barcelona), obtaining the P. Salvador Gil, S.I. 2000 prize. He did his PhD in materials science (2004) at the Instituto de Ciencia de Materiales de Barcelona (CSIC-UAB). Thereafter, he worked as a postdoc at the Laboratoire de Réactivité et Chimie des Solides in Amiens (France), the Research Centre for Nanoscience and Nanotechnology in Barcelona, and the University of Cambridge (Device Materials Group). Dr. Muñoz-Rojas is currently a permanent CNRS researcher at the Laboratoire des Matériaux et du Génie Physique in Grenoble, France. His research focuses on using and developing cheap and scalable chemical approaches for the fabrication of novel functional materials for electronic and optoelectronic applications. In particular, he has pioneered the development of the novel spatial atomic layer deposition (SALD) technique for the deposition of active components for optoelectronic devices. He is currently further developing SALD to extend the possibilities and fields of application of this exciting technique.

Contents

Preface   xxiii

Part 1: Introduction

1. Energy in Transition   3
Pedro Gómez-Romero and David Muñoz-Rojas

    1.1   Introduction   4
    1.2   Materials for Energy   11
    1.3   How Far Ahead Is the Future?   17

Part 2: Energy Conversion

2. Materials for Photovoltaic Solar Cells   27
David Muñoz-Rojas, Hongjun Liu, João Resende, Daniel Bellet, Jean-Luc Deschanvres, Vincent Consonni, and Shanting Zhang

    2.1   The Physics of Solar Cells   27
        2.1.1   The p-n Junction   28
        2.1.2   Performance Assessment   32
        2.1.3   The Shockley–Queisser Limit   35
        2.1.4   Advanced Characterization   36
        2.1.5   Materials Requirement: The Ideal Solar Cell   38
    2.2   Types of Solar Cell   39
        2.2.1   Silicon   40
            2.2.1.1   Crystalline silicon solar cells   40
            2.2.1.2   Si heterojunction solar cells   41
        2.2.2   Gallium Arsenide   43
        2.2.3   Thin Film Technologies   44
            2.2.3.1   Copper indium gallium diselenide   44

		2.2.3.2	Cadmium telluride	45

		2.2.3.2	Cadmium telluride	45
		2.2.3.3	Amorphous Si	46
		2.2.3.4	Thin film Si, nano- and micro-Si	47
	2.2.4	Emerging Technologies		48
		2.2.4.1	Dye sensitized solar cells	48
		2.2.4.2	Organic and hybrid solar cells	49
		2.2.4.3	Inorganic and ultra-low-cost cells	50
		2.2.4.4	Quantum dots cells	54
		2.2.4.5	Hybrid perovskite cells	57
		2.2.4.6	Spectral conversion	60
		2.2.4.7	Nanostructured solar cells based on zinc oxide nanowire arrays	62
	2.2.5	Multijunction or Tandem Solar Cells		65
2.3	Transparent Conductive Materials			67
	2.3.1	N-Type TCOs		70
	2.3.2	P-Type TCOs		71
	2.3.3	TCM Based on Metallic Nanowires		72
	2.3.4	TCM Based on CNT and Graphene		73
2.4	Toward Low Cost, Fast and Scalable Processing			75

3. Low-Cost Electricity Production from Sunlight: Third-Generation Photovoltaics and the Dye-Sensitized Solar Cell — 93

Nadia Barbero and Frédéric Sauvage

3.1	Introduction		93
3.2	Basics of Organic Photovoltaics		95
3.3	Dye-Sensitized Solar Cell Principle		101
	3.3.1	Semiconductor and Electrolyte Development for Dye-Sensitized Solar Cells	103
	3.3.2	Dye Development with Molecular Engineering	108
	3.3.3	Liquid Electrolyte Development Based on Solvent and Solvent-Free Formulation for Stable Devices	123

		3.3.4	Current Understanding of Chemical/Photoelectrochemical Degradation Pathways	127
		3.3.5	Concluding Remarks	137

4. Thermoelectrics — 155
Damien Saurel

	4.1	Introduction			155
	4.2	Definition			156
	4.3	Applications of Thermoelectricity			157
		4.3.1	Temperature Sensing – Thermocouples		157
		4.3.2	Conversion		158
			4.3.2.1	Maximum efficiency	161
			4.3.2.2	Maximum power	163
			4.3.2.3	Thermoelectric modules	164
		4.3.3	Heat Pump		166
	4.4	Semiclassical Theory of Thermoelectricity in Solids			168
		4.4.1	Introduction		168
		4.4.2	Quasi-Free Electron Model		171
		4.4.3	Electron Conductivity		174
		4.4.4	Thermopower		177
		4.4.5	The Sommerfeld Expansion		178
		4.4.6	Electrons and Holes		182
		4.4.7	Multiband Contribution		183
		4.4.8	Thermal Conductivity		185
		4.4.9	Figure of Merit		187
			4.4.9.1	Optimum chemical potential and quality factor	187
			4.4.9.2	Treatment for a single band in the low temperature limit	189
			4.4.9.3	Optimum band gap	192
	4.5	Thermoelectric Materials			193
		4.5.1	Historical Overview		193
		4.5.2	Chalcogenides		196

		4.5.3	Silicon and Si–Ge Alloys	197
		4.5.4	PGEC Related Materials: Skutterudites, Clathrates, and Half-Heusler	198
		4.5.5	Oxides	200
		4.5.6	Other Materials	201
	4.6	Conclusion		201

5. Piezoelectric Conversion — 205

Steven R. Anton

	5.1	Introduction			206
		5.1.1	Vibration Energy Harvesting Concepts		206
		5.1.2	A Brief History of Piezoelectricity		209
	5.2	Principles of Piezoelectric Transduction			210
		5.2.1	Piezoelectric Transduction Phenomenon		210
		5.2.2	Piezoelectric Material and Transducer Types		212
		5.2.3	Mathematical Modeling of Piezoelectric Energy Harvesters		216
			5.2.3.1	Unimorph cantilever exact analytical solution	218
			5.2.3.2	Bimorph cantilever exact analytical solution for series connection of electrodes	221
			5.2.3.3	Bimorph cantilever exact analytical solution for parallel connection of electrodes	222
			5.2.3.4	Approximate distributed parameter solutions	223
			5.2.3.5	Summary	224
	5.3	Energy Conditioning Circuitry			224
		5.3.1	Rectification		225
		5.3.2	DC–DC Conversion		226
		5.3.3	Synchronous Extraction		228
		5.3.4	Impedance Matching		230
		5.3.5	Summary		232

5.4	Applications of Piezoelectric Energy Harvesting		232
	5.4.1	Self-Powered Sensing Systems	232
	5.4.2	Biological and Wearable Energy Harvesting	236
	5.4.3	Piezoelectric Harvesting in Microelectromechanical Systems	239
	5.4.4	Harvesting Fluid Flow Using Piezoelectric Transduction	242
		5.4.4.1 Harvesting of liquid flow	242
		5.4.4.2 Harvesting air flow using windmill-style harvesters	244
		5.4.4.3 Harvesting of air flow using flutter-style harvesters	245
	5.4.5	Summary	248
5.5	Current Research Thrusts		248
	5.5.1	Broadband and Nonlinear Harvesting	248
		5.5.1.1 Broadband piezoelectric energy harvesting	249
		5.5.1.2 Nonlinear piezoelectric energy harvesting	250
	5.5.2	Multifunctional Harvesting	251
	5.5.3	Multi-Source Energy Harvesting	253
	5.5.4	Novel Piezoelectric Materials	256
		5.5.4.1 Piezoelectric single crystals	256
		5.5.4.2 Piezoelectric nanocomposites	258
		5.5.4.3 Piezoelectret foams	260
		5.5.4.4 Lead-free piezoelectrics	262
5.6	Summary and Future Visions		263

6. Fuel cells   277

Jesús Canales-Vázquez and Juan Carlos Ruiz-Morales

6.1	Introduction		277
6.2	History		281
6.3	Types of Fuel Cells		284
	6.3.1	Alkaline Fuel Cells	286

		6.3.2	Polymer Electrolyte Membrane Fuel Cells	286
		6.3.3	Phosphoric Acid	288
		6.3.4	Molten Carbonate	288
		6.3.5	Solid Oxide Fuel Cells	289
	6.4	Thermodynamics		291
	6.5	Fuel Cell Efficiency		294
		6.5.1	Thermodynamic Efficiency	295
		6.5.2	Voltaic Efficiency	296
		6.5.3	Faradaic Efficiency	300
		6.5.4	Heat Efficiency	300
	6.6	Applications		302
		6.6.1	HyFLEET-CUTE	302
		6.6.2	UTSIRA	303
		6.6.3	Present and Future	304
		6.6.4	Last Trends in Fuel Cell Technology	304
			6.6.4.1 Alkaline fuel cells	304
			6.6.4.2 Phosphoric acid fuel cells	305
			6.6.4.3 Polymer electrolyte membrane fuel cells	305
			6.6.4.4 Molten carbonate fuel cells	306
			6.6.4.5 Solid oxide fuel cells	306

Part 3: Energy Storage

7. Batteries: Fundamentals and Materials Aspects — 313

Montse Casas-Cabanas and Jordi Cabana

	7.1	Introduction		313
		7.1.1	What Is a Battery?	313
		7.1.2	Materials Aspects	319
		7.1.3	Methods for Battery Testing	323
			7.1.3.1 Chronopotentiometry	323
			7.1.3.2 Chronoamperometry	324
			7.1.3.3 Cyclic voltammetry	325
			7.1.3.4 Electrochemical impedance spectroscopy	325

7.2	Rechargeable Battery Systems		325
	7.2.1	Lead Acid Batteries	325
	7.2.2	Alkaline Rechargeable Batteries	329
	7.2.3	Lithium Rechargeable Batteries	332
		7.2.3.1 From Li metal to Li-ion	332
		7.2.3.2 Negative electrodes	333
		7.2.3.3 Positive electrodes	336
		7.2.3.4 Electrolytes	339
		7.2.3.5 ... and back to Li metal	340
7.3	Beyond Li-Ion: From Single to Multivalent Ion Chemistries		342
7.4	Redox Flow Batteries		343

8. Environmentally Friendly Supercapacitors — 351

Ana Karina Cuentas-Gallegos, Daniella Pacheco-Catalán, and Margarita Miranda-Hernández

8.1	Introduction		351
8.2	Energy Storage Devices		352
8.3	Supercapacitors Background		357
8.4	Charge Storage Mechanisms		360
	8.4.1	The Electric Double Layer	360
		8.4.1.1 Helmholtz model	361
		8.4.1.2 Gouy–Chapman model	362
		8.4.1.3 Stern and modern models	363
	8.4.2	Pseudocapacitance Mechanism	365
		8.4.2.1 Redox reactions	366
		8.4.2.2 Ion electrosorption	368
		8.4.2.3 Intercalation	369
8.5	Classification		370
	8.5.1	Charge Storage Mechanism	372
		8.5.1.1 Electric double layer capacitors	373
		8.5.1.2 Pseudocapacitors	387
		8.5.1.3 Conducting organic polymers	388
		8.5.1.4 Transition metal oxides	394

		8.5.1.5	Hybrid supercapacitors	400
		8.5.1.6	Functionalized carbons	400
		8.5.1.7	Nanocomposites and/or hybrid materials	404
		8.5.1.8	Asymmetric assembly	410
	8.5.2	Electrolyte		421
		8.5.2.1	Organic electrolytes	422
		8.5.2.2	Ionic liquids	422
		8.5.2.3	Polymeric electrolytes	423
		8.5.2.4	Aqueous electrolytes	424
8.6	Designing High-Performance Environmentally Friendly Supercapacitors			425
8.7	Characterization			430
	8.7.1	Electrode Fabrication		430
	8.7.2	Electrode Material Characterization		433
	8.7.3	Cell Characterization		437
		8.7.3.1	Cyclic voltammetry	437
		8.7.3.2	Galvanostatic measurements	440
8.8	Future Perspectives			445

9. Power-to-Fuel and Artificial Photosynthesis for Chemical Energy Storage — 493

Albert Tarancón, Cristian Fábrega, Alex Morata, Marc Torrell, and Teresa Andreu

9.1	Energy Storage in Current and Future Energy Scenarios			494
	9.1.1	Energy Storage Systems		495
	9.1.2	Chemical Energy Storage		497
	9.1.3	Synthetic Fuels Production		500
		9.1.3.1	Fischer–Tropsch synthesis	500
		9.1.3.2	Sabatier reaction	501
	9.1.4	Efficiency of Converting Chemical Energy into Electricity		502
		9.1.4.1	Chemical-to-mechanical-to-electrical conversion: Heat engines	502

		9.1.4.2	Direct chemical-to-electrical conversion: Fuel cells	504
	9.1.5	One Possible Sustainable Generation/Storage/Consumption Cycle		505
9.2	Power to Fuel			508
	9.2.1	General Aspects of Electrolytic Cells		509
		9.2.1.1	Fundamentals of electrolysis	509
		9.2.1.2	Temperature and pressure effects on electrolysis	511
		9.2.1.3	Types of electrolysers according to the electrolyte	515
		9.2.1.4	Non-ideal electrolysers	517
		9.2.1.5	Cell efficiency	520
	9.2.2	Electrolysis of Water		520
		9.2.2.1	Low-temperature electrolysers	522
		9.2.2.2	High-temperature electrolysers	529
	9.2.3	Coelectrolysis of Water and Carbon Dioxide		536
		9.2.3.1	Low-temperature carbon dioxide electrolysis	536
		9.2.3.2	High-temperature co-electrolysis of steam andcarbon dioxide in SOECs	536
		9.2.3.3	Polygeneration in solid oxide fuel cells	539
9.3	Artificial Photosynthesis			540
	9.3.1	General Aspects of Artificial Photosynthesis		540
	9.3.2	Water Splitting		543
		9.3.2.1	Photolysis of water	543
		9.3.2.2	Photoelectrochemical water spltting	548
	9.3.3	Photoreduction of Carbon Dioxide		551
9.4	Concluding Remarks			555

10. Hydrogen Storage — 567
Raphaël Janot

- 10.1 Conventional Hydrogen Storages — 570
  - 10.1.1 Compressed Gas — 570
  - 10.1.2 Liquid Hydrogen — 573
- 10.2 Hydrogen Physisorption — 576
  - 10.2.1 Carbon Materials — 577
  - 10.2.2 Zeolites — 579
  - 10.2.3 Metal-Organic Frameworks — 581
- 10.3 Metal Hydrides — 584
  - 10.3.1 Elements — 587
  - 10.3.2 AB_5 Intermetallic Compounds — 589
  - 10.3.3 AB_2 Intermetallic Compounds — 590
  - 10.3.4 AB Intermetallic Compounds — 592
  - 10.3.5 A_2B Intermetallic Compounds — 593
  - 10.3.6 Solid Solutions — 594
- 10.4 Complex Hydrides — 595
  - 10.4.1 Borohydrides — 596
    - 10.4.1.1 Lithium borohydride: $LiBH_4$ — 596
    - 10.4.1.2 Magnesium borohydride — 600
  - 10.4.2 Alanates — 600
  - 10.4.3 Silanides — 603
- 10.5 Amides and Imides — 608
  - 10.5.1 Hydrogenation of Li_3N — 609
  - 10.5.2 The Li-Mg-N-H System — 612
  - 10.5.3 Other Li-Metal-N-H Systems — 614
- 10.6 Ammonia-Borane — 617
- 10.7 Conclusions — 622

Part 4: Energy Transmission and Consumption

11. Superconductors — 641
Stuart C. Wimbush

- 11.1 Introduction — 641

11.2	Fundamental Phenomenology of Superconductivity		642
	11.2.1	Origin of Lossless Current Transport	643
	11.2.2	Limitations on the Superconducting State	644
	11.2.3	Flux Penetration and Flux Pinning	645
11.3	Superconducting Materials for Application		648
	11.3.1	First Generation BSCCO Wires	648
	11.3.2	Second Generation RBCO Tapes	651
	11.3.3	MgB_2 Wires	653
	11.3.4	New Materials on the Horizon	655
11.4	Coated Conductor Fabrication		655
11.5	Superconductors for Energy Applications		659
	11.5.1	Superconducting Power Cables	660
	11.5.2	Superconducting Transformers	665
	11.5.3	Superconducting Generators	667
	11.5.4	Superconducting Energy Storage Devices	668
	11.5.5	Superconducting Fault Current Limiters	670
11.6	Superconductors for Transportation Applications		672
	11.6.1	Superconducting Motors	672
	11.6.2	Marine Propulsion Systems	673
	11.6.3	Magnetically Levitated Trains	674
	11.6.4	Electric Aircraft	676
	11.6.5	Personal Electric Vehicles	677
11.7	Paradigm-Shifting Energy Technologies		678
	11.7.1	Fusion Power	678
	11.7.2	Hydrogen Economy	680
	11.7.3	Room-Temperature Superconductivity	680
11.8	Other Applications of Superconductors		682
11.9	Cooling		683
11.10	Cost		685
11.11	Summary		687

12. Solid-State Lighting: An Approach to Energy-Efficient Illumination — 693

Mariano Perálvarez, Jorge Higuera, Wim Hertog, Óscar Motto, and Josep Carreras

- 12.1 Properties of Light — 693
  - 12.1.1 Introduction — 693
  - 12.1.2 The Visual System — 695
  - 12.1.3 The Chromaticity Diagram — 697
  - 12.1.4 Luminous Efficacy of Radiation — 698
  - 12.1.5 Colour Temperature — 699
  - 12.1.6 Colour Rendering Index — 699
  - 12.1.7 Spectrum and Quality of Light — 700
- 12.2 Light Sources — 702
  - 12.2.1 Introduction — 702
- 12.3 LED Physics — 705
  - 12.3.1 Semiconductors — 705
    - 12.3.1.1 Doping — 707
    - 12.3.1.2 p-n junctions — 708
    - 12.3.1.3 Direct and indirect band gaps — 709
    - 12.3.1.4 LED architecture — 710
    - 12.3.1.5 Manufacturing processes — 712
- 12.4 Light Emitting Diodes Based on III-V Junctions — 713
  - 12.4.1 Gallium Arsenide — 714
  - 12.4.2 Gallium Arsenide Phosphide and Gallium Phosphide — 714
  - 12.4.3 Aluminium Gallium Arsenide — 715
  - 12.4.4 Aluminium Gallium Indium Phosphide — 715
  - 12.4.5 Gallium Nitride and Indium Gallium Nitride — 716
  - 12.4.6 ZnSe — 717
  - 12.4.7 Materials for UV LEDs — 718
- 12.5 Organic Light Emitting Diodes — 718
- 12.6 White Light with LEDs — 721
  - 12.6.1 Wavelength Converters — 721

		12.6.1.1	Phosphors for LEDs	722
		12.6.1.2	Phosphor application methods	723
	12.6.2	Multichromatic LED Sources		724
		12.6.2.1	Dichromatic LEDs	724
		12.6.2.2	Trichromatic LEDs	724
		12.6.2.3	Polychromatic LEDs	725
		12.6.2.4	Spectral characteristics of multichromatic LED sources	726
12.7	New Approaches			727
	12.7.1	Silicon-Based Emitters		727
	12.7.2	Quantum Dots		728
12.8	LED Packaging			729
	12.8.1	Low-Power LED Packaging		730
	12.8.2	Mid- and High-Power LED Packaging		730
	12.8.3	Thermal Management		731
	12.8.4	Some Effects Related to Excessive Junction Temperature		733
	12.8.5	The Role of the Packaging on Light Extraction		734
12.9	LED Drivers			737
	12.9.1	Linear Constant-Current Drivers		738
	12.9.2	Switching Constant-Current LED Drivers		739
		12.9.2.1	Buck converters	740
		12.9.2.2	Boost converter	740
		12.9.2.3	Boost-buck converter	741
12.10	Lighting Control Systems and Applications			742
	12.10.1	Smart Lighting Control Systems		742
	12.10.2	Occupancy Sensors		743
		12.10.2.1	Illuminance sensors	744
		12.10.2.2	Colour sensors	745
		12.10.2.3	Spectral sensors	745
	12.10.3	Sustainable Energy-Efficient Applications for Smart Cities		746

		12.10.3.1	Smart outdoor urban lighting	746
		12.10.3.2	Visible light communications	747
		12.10.3.3	Adaptive LED lighting	748
		12.10.3.4	Indoor spectrally tunable LED luminaires	749

13. Solid-State Refrigeration Based on Caloric Effects   753
Seda Aksoy

13.1	Magnetocaloric Effect			754
	13.1.1	Theory of the MCE		754
	13.1.2	Magnetocaloric Materials		759
		13.1.2.1	Rare earth (lanthanide) elements and their compounds	759
		13.1.2.2	3d-transition metal compounds and manganites	761
	13.1.3	Refrigeration Technology		764
13.2	Mechanocaloric Effect			768
	13.2.1	Mechanocaloric Materials		769
	13.2.2	Mechanocaloric Refrigeration		771
13.3	Electrocaloric Effect			772
	13.3.1	Electrocaloric Materials		774
	13.3.2	Electrocaloric Refrigeration		775
13.4	Conclusion			778

Index   791

Preface

Materials are fundamental for us humans. Their importance is such that key stages of our civilization have been named after them, each new stage being brought about by a new material that revolutionized existing technologies. Early humans made most of their tools from flint during the Stone Age. The next stages of civilization, from the Copper Age, to the Bronze Age, to the Iron Age, represented a succession of stronger and stronger alloys that led to better tools. More recently, silicon permitted the extraordinary development of modern electronics that profoundly transformed the way we live and communicate. Arguably, the twentieth century was therefore the Age of Silicon, but we must not overlook the myriad of other modern materials that also helped revolutionize our lives. For example, carbon-fiber composites that are light and strong enabled us to fly affordably, and ceramics and metals that are biocompatible allowed us to rebuild ourselves.

With the turn of the century, there are difficult challenges ahead. According to current projections, the world population will reach eight billion by 2030 and will likely reach nine billion by 2050. Such a dramatic increase in population will lead, among other things, to a huge increase in energy demand worldwide. Meeting this ever-increasing demand represents without doubt one of the main challenges of the twenty-first century and will become more and more critical as the fossil fuels on which we rely to generate most of our energy start to run out. It is therefore vital to search for alternative energy sources that are renewable and to find new ways of using energy more efficiently. Any of such new technologies will most likely rely on new materials with outstanding properties, and so the twenty-first century will be perhaps eventually known as the Age of Materials for Energy.

In this context, the purpose of this book is to give a unified and comprehensive presentation of the materials that may underpin this so-needed energy revolution. After a general introduction

(Chapter 1), the book is divided into three blocks that describe materials for energy conversion (Chapters 2–6), energy storage (Chapters 7–10), and energy transmission and consumption (Chapters 11–13). Each chapter is self-contained and includes both fundamentals and latest research results. The book should therefore prove useful for undergraduate and graduate students and researchers working on sustainable energies.

This book would not have happened if it were not for the extraordinary work of all contributing authors, the thorough revision from a selected group of reviewers, and the continuous support from the whole team at Pan Stanford.

Xavier Moya
Cambridge, United Kingdom

David Muñoz-Rojas
Grenoble, France
March 2016

PART 1
INTRODUCTION

Chapter 1

Energy in Transition

Pedro Gómez-Romero[a] and David Muñoz-Rojas[b]

[a]*Catalan Institute of Nanoscience and Nanotechnology (ICN2), CSIC and The Barcelona Institute of Science and Technology, Campus UAB, Bellaterra, 08193 Barcelona, Spain*
[b]*Laboratoire des Matériaux et du Génie Physique (LMGP), University Grenoble-Alpes, CNRS, F-3800 Grenoble, France*

pedro.gomez@cin2.es

An introductory chapter is a good way to start a technical book. It provides a broad overview of the field, announces the general intention of the authors, describes the structure of the book, and outlines its contents. It should not be taken, however, as a mere summary or a highlight instrument—after all this is certainly not an executive summary. Instead, we believe this introduction could go beyond the conventional goals mentioned above and provide also an account of the reasons why it is worth working on energy materials, a vision of how research and development can contribute to the due re-evolution towards a sustainable model of generation, storage, distribution, management and consumption of energy and a hint of what are the major scientific, technical and social challenges ahead.

Materials for Sustainable Energy Applications: Conversion, Storage, Transmission, and Consumption
Edited by Xavier Moya and David Muñoz-Rojas
Copyright © 2016 Pan Stanford Publishing Pte. Ltd.
ISBN 978-981-4411-81-3 (Hardcover), 978-981-4411-82-0 (eBook)
www.panstanford.com

1.1 Introduction

On Wednesday December 5, 2012, at a press conference during the Fall Meeting of the American Geophysical Union in San Francisco, NASA and NOAA scientists unveiled the latest of a series of satellite views of Earth at night. The spectacular composite photograph was assembled from cloud-free shots acquired by the visible infrared imaging radiometer suite (VIIRS) installed at the Suomi NPP satellite. The VIIRS detects light in a range of wavelengths from green to near-infrared and uses filtering techniques to observe dim signals such as city lights, gas flares, auroras, wildfires or even reflected moonlight and is sensitive enough to detect the light from a single ship in the sea. This image is already a global icon, as it was the first "Earth at Night" released in 2004 and the daylight images that conform the *Blue Marble* project [1].

Figure 1.1   NASA-NOAA composite view of Earth at Night, 2012.

A mere glimpse at this image immediately conveys its many wonders. Our global nature, the asymmetry of our world, the tiny light from our own hometown, the cradle of civilizations like the Nile river, the darkness of human-free sanctuaries are sensed right away. But the most striking feature, in contrast with the daylight view of Earth, is the human footprint in the form of man-made lights.

It is not language, nor the use of tools, nor the social nature of our species. What makes us different from any other living species on Earth is our use of energy. We are the only ones making use of exosomatic energy, namely, that used to sustain activities external to our own biological metabolism. In our modern technological society this *social* energy has been estimated to be two orders of magnitude larger than the somatic energy necessary to keep our bodies alive. That means a consumption of ca. 200,000 Kilocalories per person and day [2].

A decade and a half ago, when the world was mesmerized by the Y2K Millennium bug, most of our global exosomatic energy was extracted from fossil fuels. Coal, oil, and natural gas accounted for a total 80–85% of it, a figure that fluctuates a bit depending on whether and how the biomass used by Third World countries is accounted for. Fifteen years later, Information and Communication Technologies have evolved so wildly that our forgotten fear to a mere change of formatting dates might seem childish. However, not much has changed concerning our sources of energy. Our fossil fuel share is still 80–85% depending on estimates of biomass consumption in the Third World, and coal consumption has even grown.

Figure 1.2 shows this distribution and the trend during a period of 10 years by including percent values for years 2012 and 2002. It is interesting to note that the share of each fossil fuel changed significantly in that decade, both concerning total primary energy and electricity generation, but in both cases the overall fossil fuels share remained practically unchanged.

The development of emerging countries, in particular China, is frequently claimed as one key factor contributing to our present situation and to the forecasting of ever-growing global energy consumption in years to come. Unfair as it is to blame a country with lower consumption of energy per capita than ours, the fact is that China stands as the first producer of coal in the world, accounting for almost half of the world production (and yet, still a net importer) [3]. This supports the claims that China has based its recent growth on burning coal and has boosted global fossil fuels consumption. But China is not the only culprit in our global energy status quo. The First World overconsumption model is at the heart of the present situation, with archaic sun in the form of fossil fuels feeding our wasteful society. Indeed, the initial perception of an advanced technological world that comes with the view of Earth at night in Fig. 1.1 quickly vanishes when we realize that the overall efficiency of making those shiny technological flares is just a single-digit figure close to 5%.

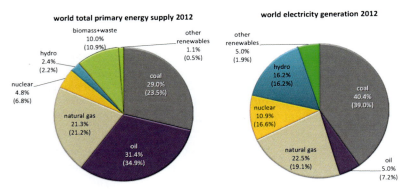

Figure 1.2 Left: world total primary energy supply in 2012. Distribution by sources (%). Right: world electricity in 2012 by sources. Figures in parentheses indicate the same values, from the same source for year 2002, for comparison (data from International Energy Agency (IEA), Key World Energy Statistics, 2014) [3]. It should be noted that in these graphs, peat and oil shale are aggregated with coal.

How have we come to develop such a wasteful technosphere is not easy to understand, but at the risk of oversimplifying we could mention a couple of technological revolutions which contributed significantly to massive energy consumption. The first came about with the conjunction of the Great Britain, coal, the Steam Machine and the railroad in the XIX century. The second one was boosted by the USA, oil, the internal combustion engine and roads during the XX century. Cheap energy was at the heart of both. Cheap energy that—now we know—comes with hidden (a.k.a. non-internalized) but very high environmental, social, and public health costs.

Our energy model is bound to change. As a matter of fact, it is already gradually changing, towards a new sustainable model that will change the way we generate but also the way we distribute, store, manage, and use energy. This categorical assertion is not only based on the growing evidence for our contribution to global warming but also on the self-evident fact that any non-renewable resource will get eventually depleted if its consumption is not phased-out. Oil, for instance, will be the first fossil we will stop burning, leading our cars to quit smoking and allowing our lungs to breathe cleaner air. The future of oil as a fuel is very clear; it will be P.O. (Phase Out) or P.O. (Peak Oil). Therefore, the question is not whether we will quit burning fossils as our predominant way to get energy but when will we.

Certain energy analysts with shortsighted vision and/or with vested interests have celebrated the advent of non-conventional oil and gas as a new era in which miracles like the USA becoming a net exporter of oil could take place. But let us be clear, the very fact that the same multinational companies that have exploited easy-to-get oil during the last century are now bothering to invest heavily in oceanic deep-water prospections and in fracking (at the expense of higher oil prices and higher environmental costs) is the proof that the fossil fuel era (in particular oil) is approaching its end.

The time has come, therefore, to set the basis for the transition to a sustainable energy model. A transition that will not take place instantly or automatically. A gradual transition that nonetheless will require of proactive actors, including citizens and entrepreneurs, scientists and engineers, financial and industrial corporations and, of course, last but not least, policy makers.

Governmental support to new or strategically important technologies is not new. Whether in the form of legislation or direct subsidies every heavy industrial sector has benefitted from very generous public help, from coal to nuclear, from electrification to the car industry. Why the big fuzz then concerning the subsidies to renewables or legislation penalizing CO_2 emissions?

Damage to the economy is frequently argued as a factor against CO_2 penalties, from fossil fuels burning to concrete manufacturing. However, this primary and simplistic accounting approach has been demolished by more rigorous analyses by economists like Stern [4]. Indeed, one of the main conclusions of the Stern review is that the overall costs of climate change will be equivalent to losing between 5% and 20% of the global gross domestic product (GDP) each year, now and forever. When compared with the 1–2% of the global GDP per year that would need to be invested to avoid the worst effects of climate change, it becomes clear that the benefits of strong, early action on climate change would outweigh the dreadful future costs. But even if we stay in the realm of economic and strategic orthodoxy there are very strong arguments in favor of an ordered shift towards low-carbon and renewable technologies. These will reduce the heavy burden of non- internalized costs, derived from our way to produce energy but accounted for in other spreadsheets: health problems derived from smoking cars, cleaning costs of environmental pollution, from black tides to nuclear, or oil-war costs. Renewables will also definitively reduce our energy supply dependency from producers based on unstable territories. Renewable industries and new manufacturing players closing the circle or circular economy in all industrial sectors will be the main actors of a much-needed new productive economy. Finally, early re-evolution towards sustainability will give a competitive advantage to the societies collectively adapting themselves to the inexorably forthcoming sustainable model. Note that we have written societies, not industries or corporations. This is so because the energy conundrum is so complex and tightly intertwined with our ways of life that only an integral and collective change will suffice.

Energy, Environment and Economy are three threaded E-word concepts strongly intertwined, which can push and pull each other in complex feedback cycles (Fig. 1.3). They can configure a vicious circle of wasteful overconsumption as witnessed during the 20th century. But they can also converge into a virtuous feedback circle leading to environmental, social and economic sustainability as it will become increasingly apparent in the 21st century.

Figure 1.3   The three threaded E's.

What primary energy will be feeding this virtuous feedback cycle in our present century is no secret. As a matter of fact, it will be primarily the same that led to our successive industrial revolutions in the 19th and 20th centuries: solar energy, of course. Except that instead of using million-year-old canned solar energy in the form of coal, oil or natural gas, we will eventually be just using contemporary solar energy. Directly or in the form of biomass, hydropower, and wind power, solar energy will dominate over non-solar (geothermal and nuclear) (See Fig. 1.4).

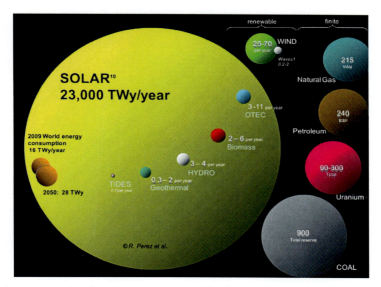

Figure 1.4 Graphical comparison of the current annual energy consumption of the world to (1) the known reserves of the finite fossil and nuclear resources and (2) to the yearly potential of the renewable alternatives. The volume of each sphere represents the total amount of energy recoverable from the finite reserves and the energy recoverable per year from renewable sources. Solar energy received by emerged continents only, assuming 65% losses by atmosphere and clouds. From reference [5].

The technologies for harnessing solar power are very diverse. Some have reached a maturity level that makes them stand very close to direct economic viability even without subsidies. That is the case of Thermal-solar electricity plants for instance. Other technologies are nonetheless in its infancy, like solar-driven photocatalytic reduction of CO_2. All of them will contribute to take advantage of the massive prevalence of solar energy on our blue planet.

Indeed, the solar system depicted in Fig. 1.4 is actually a graphical account of the energy available from the Sun (every year for the next 5000 million years) in comparison with estimated finite total reserves of various "primary" sources. It also depicts the annual global energy consumption by humans, presently standing at 16 TW· year and expected (by linear extrapolation) to reach 28 TW· year by 2050. The energy granted by our yearly solar bath, even after atmospheric filtering, is three orders of magnitude larger than our yearly energy needs and two orders of magnitude larger than that available from the total of estimated oil reserves. No matter how far off the latter could be, the energy we could ever get from all fossil fuels will never be comparable to the free energy provided by the sun every year.

With all this in mind, it is not difficult to forecast that solar energy, in all its forms, will have to be massively developed in order to cover our energy needs in a coming world of shrinking fossil power. Conversion of solar radiative power into electrical, chemical (and biochemical) or thermal energy will have to be mastered before we could abandon our fondness for cheap-n-wasteful energy "generation" methods.

But in addition to primary sources, we will have to worry about the setting up of a new model of energy use, especially in transportation. That is a sector overwhelmingly dominated by the burning of petroleum distillates (practically 100%, land, sea, and air) and it will take great efforts and long times to change. But it will change. We are slowly shifting from a society of *energy-gatherers* to one of *energy-farmers*. Thus, the get-it-and-burn-it approach we have been using to move our society (first with coal, then oil and gas) will be transformed to one making use of energy vectors. Biofuels, and electricity, and hydrogen and other synthetic fuels will have to be generated taking up more energy than they will later release (welcome to thermodynamics!). So we better make the best of these energy vectors.

Which takes us to the last-but-not-least topic of energy efficiency and conservation. This is a topic that goes way beyond switching off unnecessary lights and involves aspects like supply chains or life cycle assessment, and one in which we will also find surprising materials breakthroughs waiting to be made.

Indeed, in all the aspects of the energy cycle, namely, generation, storage, and use, materials science will play a critical role in years to come in order to improve the performance and lower the costs of existing materials. But research on new advanced materials will also be needed in order to produce the pending breakthroughs which will finally make clean energy technologies for a sustainable world competitive. In the following section, we will present an overview of the important projection of materials in the world of energy.

1.2   Materials for Energy

There was a time when aluminum was an advanced material, that is, one prepared and studied ahead of its applications. That was one and a half century ago. Jules Verne had just published his novel *From the Earth to the Moon* (1865) in which aluminum was the light metal chosen to build their capsule to the Moon. At the same time, Napoleon III, by then emperor of the French, was dreaming of aluminum as a strategic material for his army and commissioned the construction of the first aluminum factory. The problem then was that aluminum was more expensive than gold. The reason was not scarcity; after all it was then, as still is, the third most abundant element on Earth's crust. But the methods to extract it from its minerals followed the cumbersome and costly procedures established by its discoverers. Thus, the material was extremely expensive. Indeed, it was perfectly suited to make luxurious cutlery to be displayed at state banquets but never made it to the helmets of French lancers.

Only 20 years later, in 1886, two *creators of future* patented simultaneously similar methods to produce aluminum by electrolysis in molten cryolite. The French Heroult and the American Hall devised an inexpensive method for the production of aluminum and hence all the applications of aluminum we know today were made possible. They changed the history of aluminum and the history of civilization.

The story of aluminum is by no means an exceptional example from the old times. Today, as 150 years ago, we could easily find an example of a material that is still quite expensive, not because of its scarcity, but due to its fabrication process—a material that could benefit enormously from a less costly preparation and that would also change the history of civilization. Silicon is not the third, but the second most abundant element on Earth's crust. And at 99.999% purity, the fabrication of solar silicon for photovoltaic application has been burdened by high costs.

There are two main ways to lower the cost of an emerging technology. The first one rests on mass-production in order to benefit from what is known as *economy of scale*. The second one relies on the discovery and implementation of new materials or processes through Research + Development. Solar silicon has benefitted from both. Indeed, it has come a long way since the meager photovoltaic industry was fed by scratch spare silicon from microelectronics. But despite the continuous reduction in production costs in recent years, the need still remains to decrease even further the price of solar silicon production or, alternatively, to develop other photovoltaic materials intrinsically cheaper and with equivalent or superior performance. This topic will be discussed in Chapters 2 and 3, where various concepts and materials related to the direct conversion of light into electricity will be covered. It is precisely in the field of photovoltaics (PV) where a "new" material has been causing a huge stir in the last 3 years. Organometal halide perovskites, a hybrid material known for many years, was not introduced in the PV game until very recently. And the unique properties of these exotic materials have yielded a record efficiency increase from an initial 3–4% in 2009 to over 20% at present! Of course with many issues yet to be solved and understood (see Chapter 2). This example illustrates how new or "rediscovered" materials can have a huge influence on a particular energy application.

Photovoltaics are not one isolated example of a technology in need of materials developments. Notwithstanding the importance of process engineering for the optimization of energy plants and devices, the truth is that essentially all energy technologies could benefit most substantially from materials development. Whether through the introduction of advanced materials or by optimization of established ones, every single type of device, every technology can be improved by improving the materials that make it possible.

Figure 1.5 shows a schematic diagram of the many materials research areas and topics relevant to the development of energy. An integral development which should include new and improved "generation" technologies, but also an increasing contribution from storage technologies and a big "last but not least" radical new consideration of energy consumption as a key factor in our energy equation.

Let us, for once, begin discussing the latter side of the energy triangle, which we have generically labeled as *use* of energy. We should note that in this category we are including not only the final use of energy, which should be tamed through conservation and efficiency, but also technologies related to distribution and transport of energy, such as superconductivity, a topic which actually goes well beyond the naïve bid to build megametric transmission lines and involves efficiency issues when superconducting motors, generators and transformers will be put to work in place of conventional ones, as it will be discussed in Chapter 11.

But of course the portfolio of technologies under the "use" of energy is not limited to high advanced technologies like superconductivity which will have to be developed in conjunction with other enabling technologies (in that case cryogenics). Improvements in more traditional technologies like refrigeration could also provide great contributions to energy conservation strategies. Chapter 13 will provide an introduction to magneticaloric, electrocaloric, and mechanocaloric materials for energy efficient solid-state cooling technologies.

The last technology to be discussed in relation to energy efficiency will be that of materials for LEDs (in Chapter 12). It is indeed clear from the image in Fig. 1.1 that an efficient illumination is mandatory if we want to control our energy consumption.

Energy conservation and efficiency will be increasingly important within a sustainable energy model. Yet, its need and intrinsic importance could be rightly considered independent of the dominant energy model or generation technologies. That is not the case for energy storage.

Energy storage has been essentially absent from the equation of the conventional energy model dominant in the 20th Century. The perception of counting with an uninterrupted supply of cheap energy led to the consequent renounce to storage. Even if saving was preferred to wasting energy, wasting was most frequently preferred to storing.

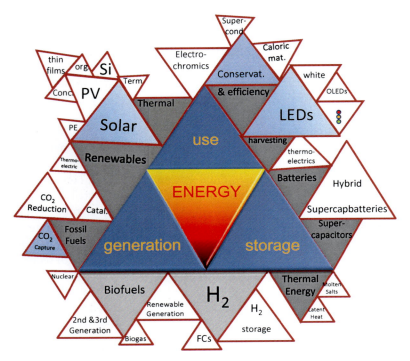

Figure 1.5 Schematic diagram of areas, research topics, and specific technologies under generation, storage, and final use of energy. All the technologies displayed are in great need of advancements in materials research.

Conversely, energy storage in all its modalities will play an increasingly important role as a consubstantial part of a sustainable model in which energy generation is neither cheap nor uninterrupted. Yet, the storage of large amounts of energy is an unresolved challenge. We are presently unable to store any amount of energy coming close to a 10% of our daily consumption in a cost-effective way whether in the form of electrical, chemical, or thermal energy. All three varieties will strongly benefit from materials developments, but we will center on the first one, which has reached by now a greater maturity level. Indeed, electrochemical energy storage (ECES) has come a long way from the heavy and contaminating lead-acid battery (introduced by Planté in 1859) to the last generation of rechargeable lithium-ion batteries, ruling now the kingdom of consumer electronics, and the new generation of supercapacitors [1, 2]. But our problem is that these technologies have been focused on small power applications related to power supply for consumer electronics (see Fig. 1.6). Yet, when it comes to high-power storage applications pumped hydro (and to a lesser extent compressed air) is presently the only technology with a capacity high enough to respond to our oversized collective needs of power.

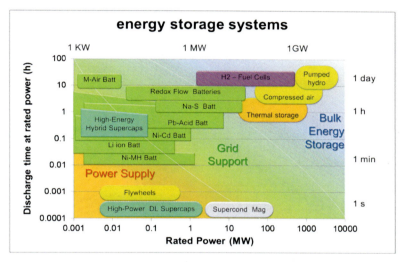

Figure 1.6 The collection of energy storage technologies is quite large and varied. From mechanical (yellow) or magnetic (grey) physical techniques to chemical (in purple), electrochemical (green) or electrophysical (cyan) methods. However, in the past only small power applications such as power supply for consumer electronics have been widely implemented. When it comes to bulk energy storage only pumped hydroelectric is presently practical. Yet, in order to implement a sustainable energy model, novel smart grid and bulk energy storage technologies will have to be developed.

On the other hand, ECES will predictably be a key player within the framework of distributed energy generation and storage networks. But before letting utility companies know about the good news of consumers turning into producers-storers-consumers (and discounting any support from policy makers), a few giant breakthroughs need to take place within the field of ECES itself. In short, ECES systems should be able to store more energy and do it at a much faster rate (i.e., higher power), all of it at a lower cost and topped with environmental friendliness. Batteries and supercapacitors as well as hybrid systems are quickly advancing in that direction [6] as it will be discussed in Chapters 7 and 8.

Hydrogen and fuel cells (FC) technologies could be considered as a particular way of energy storage (in combination with electrolyzers, for instance) or as a particular energy generation technology (with hydrogen as a distributed vector). In any event, this is one of the emerging technologies in greater need of materials. Pt-free catalysts for low-temperature FCs is the first strategic research avenue that comes to mind. Technologies that would allow for the use of other fuels (including for example bioethanol, or methane) at reduced temperatures would also help boost these technologies. There is also an important research focused in developing smaller, micro FCs that could eventually compete with batteries of capacitors for powering portable devices. The different FC technologies and materials involved, as well as the description of several current examples showing the successful application of FCs will be presented in Chapter 6. While hydrogen would of course be the preferred fuel in terms of energy density and (a priory) availability, one of the main issues faced by H_2-powered FCs is the storage and distribution of H_2. Here many types of materials and approaches have been and are being studied, but the matter is far from being satisfactorily solved, as described in Chapter 10.

Speaking of H_2, and fuels in general, wouldn't it be great if we tried and managed to imitate nature, i.e., using the sun energy to power us by transforming CO_2 into burnable fuels in a closed cycle? The idea is of course very appealing and obviously many research efforts are being dedicated to developing so-called "artificial leaves" or "artificial photosynthesis." Apart from being a CO_2 neutral approach, photosynthesizing our fuels will benefit from all the current infrastructures, making its implementation very straightforward. This exciting topic, and in particular the stringent materials challenges to be overcome, is covered in Chapter 9.

To close this brief overview of the chapters ahead, we will come back to physical methods like photovoltaics and other energy harvesting technologies. All of them rest on sophisticated processes leading to direct conversion of energy, from light to electricity in photovoltaics, mechanical to electrical in piezoelectricity or from thermal to electrical in thermoelectrics, which convert heat into electricity. Photons, excitons, phonons and, of course, electrons play their role in these elegant conversion processes. Yet, all these technologies are as elegant as untamed, lagging far behind their full potential. Photovoltaic technologies and materials will be discussed, as we have seen, in Chapters 2 and 3, whereas Chapter 4 will cover piezoelectricity, and thermoelectricity will be dealt with in Chapter 5. Never mind the low power of the resulting devices. Their strength lies in the strategic nature of their applications and in their cumulative effects upon extended use and integration in multi-device architectures or smart grids. Harvesting materials and devices could indeed close the intertwined circle of energy conversion, storage, and last but not least, conservation and efficiency.

1.3 How Far Ahead Is the Future?

Speaking of scientists and entrepreneurs, Tesla and Westinghouse are two names rightly associated to the use of alternating current (AC) for the transmission of electricity. They eventually won the commercial battle fought against the defenders of Direct Current (DC) distribution, most notably, Thomas A. Edison and J. P. Morgan in the late eighties of the 19th century. Back then DC current was not well suited to be transmitted through long distances; generation and consumption were expected to be in close proximity (within 1–2 Km), and despite the head start advantage of Edison and DC technologies of generation and use, AC distribution at high voltage was more practical and finally leaned the balance towards Westinghouse's interests. The consequences for our technological evolution were profound and lasting, going far beyond the proliferation of transmission towers. A system of remote centralized energy generation grew since then, which has led to large companies producing large amounts of energy, transported and distributed very long distances (frequently by public enterprises) to a very large number of consumers. This structure was not only justified by the feasibility of long-distance transmission of high-voltage AC electricity but also by the economy of scale, which was in principle favoring the centralized generation and consequent transmission and distribution of electricity over distribution of fuels to small local plants. Just imagine your coal supplier pouring the weekly load of carbon at your local residential building power plant. However, that economy of scale began to change in the late 20th century, when the grid itself and not the power plants were increasingly contributing to high costs and poor reliability of the supplied electricity. That, together with the realization that current (rather than archaic) sun could be our primary energy source, is leading to a redefinition of our grids, which are slowly making room to distributed (i.e., decentralized) generation and storage and to the conception of smart grids. These will integrate a variety of energy technologies with Information and Communication Technologies (ICTs) in ways that will strongly improve efficiency and reliability (Fig. 1.7).

A hybrid system with both centralized and distributed generation of electricity (Fig. 1.7) will not only contribute to a more robust network and more reliable energy supply. It could also contribute to lower overall costs. This surprising remark rests on two main contributions to direct costs reduction (leaving out further reductions in non-internalized costs): (i) Distributed generation-storage-use would reduce the need of new and expensive transmission-distribution infrastructures, and (ii) alternative energy generation reaching grid parity should directly contribute to a lower overall expense by the consumer. Grid parity takes place when an alternative energy source generates electricity at an overall cost matching the price of power from the electric grid.

In order to answer the question used as a title of this section, we could arguably state that the future of a given renewable technology will come when grid parity is achieved. This might come as a surprise to many people, but according to several independent sources [7, 8], photovoltaic generation has already reached grid parity in several countries, including Australia, Denmark, Germany, Italy, Portugal and Spain, as well as in a good number of states in the USA. Let us then conclude that the future is now.

Grid parity is not the end of the road, though. It accounts for overall costs over the whole life time of the technology considered. But in cases like photovoltaics, initial capital and financial expenses are substantially larger than deferred running expenses after full amortization (low-cost maintenance and free fuel strongly contribute to that asymmetry). Thus, even with grid parity, these technologies are hard to adopt by the common citizen. This means there is still the need for substantial reductions in costs. Even more clearly so if renewable generation is accompanied by electrical energy storage.

Energy storage will also be a sector of increasing importance and will also lean towards novel distributed models. Building huge batteries of batteries in order to level the electrical output of renewable plants does not make much sense, especially if a large number of distributed storage units primarily designed for other purposes could be used for load-leveling. That could be precisely the case with electric vehicles (EVs) if they were massively used as storage units in addition to mobility units. Of course, that distributed network would be quite complex to manage and has not even been considered feasible until the advent of ICTs. Now the so-called smart grids, integrating centralized and distributed generation, distributed storage and optimized consumption with sensors, actuators, and controlling ICTs, make that model possible (Fig. 1.7).

We are going to witness important changes in the way we use energy. Energy management is one aspect normally unnoticed by the citizen that might also change substantially. Indeed, our present model has traditionally included the management of the offer of energy as part of our energy cycle. Production is always adjusted to cover demand. This might change in the future and the so-called *demand side management* could come as a relatively new and increasingly important ingredient of a more sustainable and efficient energy model. In short, it would mean that the user and not only the production and distribution of energy should adapt in a flexible way in order to match production and consumption. This approach will especially encourage the consumer to use less energy during peak hours, or to move the time of energy use to off-peak times such as nighttime and weekends. The resulting reduction in energy consumption might not be too large, but this approach would reduce the need to over-dimension energy production to account safely for unusually large peak demand. So, smart microgrids will play a growing role in meeting local demand, enhancing reliability and efficiency and ensuring local control of electricity.

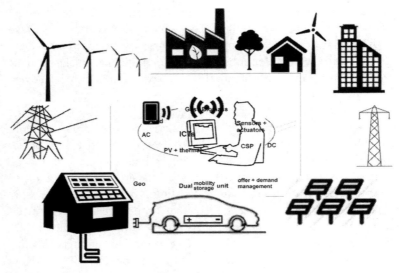

Figure 1.7 Schematic representation of a smart grid, integrating centralized and distributed generation of energy and incorporating distributed storage in the form of electric vehicles acting as dual mobility and storage units. Existing AC transmission would be kept and high-voltage DC (HVDC) could be implemented for special long-haul transmission. Photovoltaic (PV) and thermal-solar are integrated at home and concentrated solar power (CSP), wind, hydro and remote PV plants account for centralized renewable generation. Present natural gas (combined cycle) to be gradually replaced by biomass plants. Smart grids will rest on information and communication technologies (ICTs) and on novel artificial intelligence monitoring and control, integrating also demand-side management, all leading to a more efficient and robust grid.

If this trend towards local grids was not enough, Edison would be happy to know about another trendy change involving high voltage direct current (HVDC) transmission. What was unfeasible in Edison's times is practical now and, although HVDC transmission has some drawbacks (conversion equipment at the terminal stations are costly), the transmission line costs over long distances are lower than for AC (HVDC requires less conductor per unit distance than an AC line) and HVDC transmission losses are estimated to be smaller (ca. 3.5% per 1000 km). Thus, HVDC is increasingly considered for the transmission of large amounts of electricity through long distances.

So we see that just when a trend towards smaller grids comes into play, an apparently contrary trend towards long-haul transmission shows up in the form of HVDC. Maybe these two trends are not so incompatible and the latter could provide strong interconnections (including submarine) between networks of smaller smart grid networks in a future that is not so far ahead.

But the future of energy will not be marked exclusively by the evolution of the electricity vector. Bio vectors and chemical vectors could also play important roles. The case of biomass and biofuels is quite controversial. On the one hand, the primary perception of biomass as a source of non-fossil fuel that, in principle, could be sustainable led to the consideration of bioethanol as an alternative to gasoline. As a matter of fact, that is an alternative already implemented for a long time in Brazil, with bioethanol obtained from sugarcane and cars equipped with *flex* motors able to burn gasoline or ethanol. The first trend in the USA was to use corn for the production of bioethanol. This type of first-generation agrofuels has led to a great controversy based on their impact on food markets but also low overall yields and use of land and fresh water. A second generation of biofuels could rely on the use of cellulosic biomass for the production of bioethanol after enzymatic digestion followed by fermentation. These would not involve food crops but, if specific energy-crops are targeted, land and water would still be needed and the low yield of the overall process would compromise the sustainable nature of the whole technology. This consideration would change if instead of energy-oriented crops waste biomass would be used. Finally there is a third generation of biofuels presently under very active research that involves algae with high lipid contents targeted for the production of biodiesel. These would not need land or fresh water and therefore represent the highest hope for a viable biofuel vector. Aside from biofuels, we should always remember the need to use biomass, especially waste biomass, from food crops, wood chips, farming, or urban wastes to close the circle of sustainability by using them as energy resources.

Finally, chemical vectors represent the most variable scenario taking into account the many different fuels and technologies involved under this generic denomination. Maybe one of the most widely recognized chemical vectors is hydrogen. With as many intrinsic advantages as technological drawbacks, hydrogen has been the great long-term hope as a high-energy vector, but, as it has been mentioned above, major breakthroughs are still pending concerning its sustainable generation, efficient storage and viable use in low-cost (Pt-free) fuel cells working at reasonably low temperatures. Fuel Cells are also developed for other chemical fuels such as methanol, ethanol, and methane, all working at different temperatures and targeted for diverse niche applications. But in addition to these electrochemical vectors there is also a whole world of R+D dealing with a sustainable control of synthetic chemical fuels the development of which will also provide a clue of how close we could be to *the future*. In particular, there is a most challenging process which has defied scientists for centuries and which would represent the ultimate closure of the sustainability cycle. As briefly discussed above, artificial photosynthesis, effectively converting CO_2 (and H_2O) into useful high potential energy chemicals (not only for fuel) would represent the inverse process of our intensive CO_2—producing activities. Of course, we are talking about a sun-driven low-temperature process with a high envisaged efficiency that could reduce CO_2, not to noxious CO, but to methanol or methane. It could be said that is utopia. But converting utopia into knowledge is what science does best.

References

1. http://earthobservatory.nasa.gov/Features/BlueMarble/. Accessed September 22, 2014.
2. G. Tyler Miller Jr. (2000) *Living in the Environment*, 11th ed. Brooks/Cole.
3. International Energy Agency (IEA), Key World Energy Statistics, 2014.
4. Stern, N. (2006). "Summary of Conclusions". *Executive summary (short)* (PDF). Stern Review Report on the Economics of Climate Change (pre-publication edition). HM Treasury. Retrieved November 28, 2014.
5. R. Pérez and M. Pérez A fundamental look at energy reserves for the planet. The IEA SHC Solar Update. vol. 50 pp. 2–3

6. D. P. Dubal, O. Ayyad, V. Ruiz, and P. Gomez-Romero (2015) Hybrid energy storage. The merging of battery and supercapacitor chemistries. *Chem. Soc. Rev.*, **44**, 1777–1790.
7. K. Branker, M. J. M. Pathak, J. M. Pearce (2011). A review of solar photovoltaic levelized cost of electricity. *Renewable Sustainable Energy Rev.*, **15**, 4470– 4482.
8. J. Channell, T. Lam, S. Pourreza (2012). Shale & renewables: a symbiotic relationship. *Citi Research Equities*. September 12, 2012.

Part 2
Energy Conversion

Chapter 2

Materials for Photovoltaic Solar Cells

David Muñoz-Rojas, Hongjun Liu, João Resende, Daniel Bellet,
Jean-Luc Deschanvres, Vincent Consonni, and Shanting Zhang

Laboratoire des Matériaux et du Génie Physique,
University Grenoble-Alpes, CNRS, F-3800 Grenoble, France

david.munoz-rojas@grenoble-inp.fr

In this chapter, we provide a brief introduction to (i) solar cell basic physics together with a description of the common and advanced characterization and performance evaluation techniques and (ii) the main types of solar cell technologies being currently used and studied. The chapter is completed by a section dedicated to transparent conductive materials, a key component of photovoltaic solar cells. The last part gives an introduction to spatial atomic layer deposition (SALD), a recent variation of ALD allowing processing without vacuum and at high deposition rates and that has proved to be a very useful tool for fabricating solar cell components.

2.1 The Physics of Solar Cells

A short introduction to solar cell physics is presented next. Interested readers are directed to more comprehensive works, such as: J. Nelson, *The Physics of Solar Cells*, ICP, 2003; or: P. Würfel,

---

Materials for Sustainable Energy Applications: Conversion, Storage, Transmission, and Consumption
Edited by Xavier Moya and David Muñoz-Rojas
Copyright © 2016 Pan Stanford Publishing Pte. Ltd.
ISBN 978-981-4411-81-3 (Hardcover), 978-981-4411-82-0 (eBook)
www.panstanford.com

The Physics of Solar Cells: From Basic Principles to Advanced Concepts, Wiley-VCH, 2009.

2.1.1 The p-n Junction

A solar cell is a device capable of transforming light into electricity. Absorption of a photon causes the excitation of an electron in the absorbing material. In order to do work, i.e. to get the electrons to flow out of the cell generating a current and a voltage, an asymmetry must be present in the device. This asymmetry is, for instance, created by having a p-n junction formed by two semiconductors with opposite electrical behavior. As it is known, in p-type semiconductors, the majority carriers are holes, while in n-type semiconductors, electrons are the majority carriers. In the case of inorganic semiconductors (i.e. silicon), when n-type and p-type are in contact so that an interface is created, the excess electrons in the n-type material will diffuse into p-type material, and the other way around for holes from the p-type into the n-type. The consequence of this flow is the spontaneous generation of an electrical field near the interface, which opposes to a further diffusion of charges, until thermal equilibrium is reached [1]. This balancing space is called depletion region, which has a width W_0, as presented in Fig. 2.1. As we will illustrate below, W_0 depends on the carrier concentration of the semiconductors according to Eq. 2.1, and thus the n- and p-type materials can contribute differently to the depletion region (in Fig. 2.1, for instance, the p-type semiconductor contributes more to W_0 since it has a lower carrier concentration than the n-type semiconductor).

The depletion layer thickness can be affected by an imposed external bias, as shown in Fig. 2.2. When the applied external electrical field opposes the built-in potential, then it is called a forward bias, facilitating electron flow across the junction, as shown in Fig. 2.2a. When the external field is applied in the same direction of the built-in potential, it is reverse bias [1], as presented in Fig. 2.2b.

Control over the width and built-in potential V_{bi} can be achieved by several methods. By combining different semiconductor materials, the depletion zone and built-in potential can vary. For a specific material, in the absence of an external electrical bias, the depletion width is mainly dependent on carrier concentration and thus doping can modify the depletion region width. The built-in potential is associated with the bending of the band structure, which is a consequence of both materials having the same Fermi level once connected, as shown in Fig. 2.1. (In p-type materials, the Fermi level, E_{Fp}, is closer to the valence band, while for n-type ones, E_{Fn}, it is closer to the conduction band.)

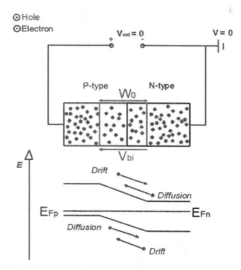

Figure 2.1 Band structure of an inorganic p-n junction with zero bias.

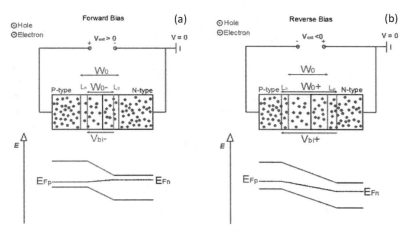

Figure 2.2 (a) Forward bias decreases depletion zone and built-in potential. (b) Reverse bias increases depletion zone and built-in potential.

Introducing a controlled amount of impurities in very pure semiconductors (doping) allows one to tune their electrical properties. Simple calculations show that in an intrinsic semiconductor, the Fermi level, E_{Fi}, is located in the middle of the band gap. By introducing dopants, donors in n-type semiconductors (such as phosphorus for silicon), the Fermi level raises toward the conduction band. Contrarily, introducing acceptors (such as boron for silicon) leads to a decrease of the Fermi level toward the valence band, thus obtaining a p-type doping.

To mathematically describe the depletion zone, charge neutrality and the Poisson equation are used [2]. For the sake of simplicity, the discussion is limited to a 1D model. Thus, equations are given as follows:

$$-\frac{d^2\psi}{dx^2} = \frac{dE}{dx} = \frac{\rho(x)}{\varepsilon_s}, \quad W_{Dp}N_A = W_{Dn}N_D, \tag{2.1}$$

where ψ is the potential, E is the electric field, ρ is the charge density, x is the distance from the junction, and ε_s is the dielectric constant. To simplify the model further, we assume an abrupt junction where both N_D and N_A are constant. A boundary condition assumes as well that at each end of the depletion zone there is no electrical field. Thus, the electrical profile can be solved by Eq. 2.1, and Eqs. 2.2 and 2.3 are obtained.

$$E_p(x) = -\frac{qN_A(x + W_{Dp})}{\varepsilon_p}, \quad \text{for} \quad -W_{Dp} \leq x \leq 0 \tag{2.2}$$

$$E_n(x) = -\frac{qN_D(W_{Dn} - x)}{\varepsilon_n}, \quad \text{for} \quad 0 \leq x \leq W_{Dn} \tag{2.3}$$

By integration of Eqs. 2.2 and 2.3, plus the boundary condition setting $\psi_p(W_{Dp}) = 0$, $\psi_p(0) = \psi_n(0)$, the potential for the p and n regions can be given, as shown in Eqs. 2.4 and 2.5.

$$\psi_p(x) = \frac{qN_A(x + W_{Dp})^2}{2\varepsilon_p} \quad \text{for} \quad -W_{Dp} \leq x \leq 0 \tag{2.4}$$

$$\psi_n(x) = \psi_p(0) + \frac{qN_D(W_{Dn} - x/2)x}{\varepsilon_n} \quad \text{for} \quad 0 \leq x \leq W_{Dn} \tag{2.5}$$

Thus, the total potential across the device, also called built in potential ψ_{bi}, would be the sum of ψ_p and ψ_n.

$$V_{bi} = \psi_{bi} = \psi_n(W_{Dn}) - \psi_p(-W_{Dp}) \tag{2.6}$$

With the charge neutrality equation, the depletion width on each side can be determined, as shown in Eq. 2.7.

$$W_0 = W_{nP} + W_{nD} = \sqrt{\frac{2V_{bi}\varepsilon_p\varepsilon_n}{q(\varepsilon_n N_D + \varepsilon_p N_A)}} \left(\sqrt{\frac{1}{N_A}} + \sqrt{\frac{1}{N_D}} \right) \tag{2.7}$$

In solar cells, i.e. when one or two semiconductors of the p-n junction have a gap suitable for visible light absorption, illumination creates electron–hole pairs in the depletion zone, a phenomenon called photovoltaic effect. Due to the built-in electrical field in this region, the holes will move to the p-type semiconductor, while the electrons move to the n-type material. If the circuit is open, the positive charges will accumulate in the p-type material and the negative ones in the n-type material, reducing the electric field in the depletion zone (Fig. 2.3a). However, if an external load is introduced in the system (i.e. the circuit is closed), the charges can move outside of the junction, generating an electrical current and increasing once again the electrical field in the depletion zone (Fig. 2.3b). Consequently, the energy produced in the solar cell is dependent on the rate of electron–hole generation, bias voltage and all the conduction processes inside and at the borders of the cell [3].

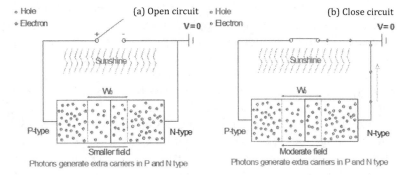

Figure 2.3 (a) Carrier movement in p-n junction under illumination with no load. (b) Carrier movement in p-n junction with load.

2.1.2 Performance Assessment

The performance of a photovoltaic solar cell can be evaluated using a variety of parameters [4]. One of the first and most important types of analysis is the power conversion efficiency (PCE), defined as the ratio of the power produced by the cell (P_{prod}) with regard to the power from the incident light on the cell (P_{inc}). It is given by

$$\eta = \frac{P_{prod}}{P_{inc}} \tag{2.8}$$

The values for the incident light power are usually standardized for light sources with an AM 1.5 G spectrum and a light flux of 100 mW/cm^2. This solar spectrum is a standard value taking into account, for example, the absorption by H_2O or CO_2 molecules in the atmosphere.

The produced power is the result of multiplying the obtained current for a particular applied voltage. For the calculation of the PCE, the maximum power is considered, as indicated in Eq. 2.9.

$$P_{prod} = I_i V_i, \tag{2.9}$$

where I_i and V_i are the electric current and potential values that give maximum power. Thus, *I* vs. *V* (*IV*) curves are measured to estimate the maximum power as well as other parameters. The *IV* curves are obtained by measuring the electrical current dependence on the bias voltage applied to the terminals of the cell, under illumination. When the applied voltage is zero, the current measured, I_{SC}, corresponds to the short circuit current. This current is described as the electric flow produced by the illumination of the cell (photocurrent) with no external loads, which strongly depends on the number of photons absorbed and the efficiency with which carriers are collected.

When the bias voltage increases, the electrical current drops, due to the increase of the recombination current, opposite to the photocurrent generated in the cell. When the net current is zero, the recombination and photocurrent are equal, and the voltage applied is called the open circuit voltage V_{OC}. This voltage depends on the properties of the p- and n-type materials and the contacts between the electrodes and the semiconductors on the cell. Figure 2.4a represents the variation of the produced power with the bias voltage, resulting from a typical *IV* curve (Fig. 2.4b).

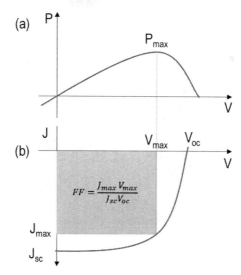

Figure 2.4 (a) Power variation with bias voltage. (b) *IV* curve of a solar cell.

Considering the two concepts, V_{OC} and I_{SC}, we can define the PCE based on these two parameters:

$$\eta = \frac{V_{OC}\, I_{SC} FF}{P_{inc}}, \qquad (2.10)$$

where FF is defined as fill factor:

$$FF = \frac{P_{max}}{I_{SC}\, V_{OC}} \qquad (2.11)$$

In Fig. 2.4b, the power produced by the solar cell is directly related to the area of the square whose edges are equal to V_{max} and I_{max}. Therefore the fill factor can then be graphically interpreted as the ratio between the maximum power ontained from the cell and the product of the short circuit current and open circuit voltage.

The shape of the *IV* curve, is strongly dependent on the different resistances of the solar cell. The shunt resistance, R_{Sh}, is mainly responsible for the leakage current troughout the cell. In this case, the resistance should be maximized to prevent recombination of electron–holes in the p-n juntion, reducing the leakage current in the cell. In the *IV* curve, this maximization is expressed by a smoother slope at lower bias voltages. Conversely, the series resistance, R_S, is the component of resistance regarding to the electrodes contacts, bulk resistance or other interfacial barriers (i.e. against the flow of electrons across the cell). The reduction of this type of resistance leads to an increase of electrical current, which increases the slope of the *IV* curve at bias voltages near V_{OC}, thus a more squared shape of the curve (i.e. closer to the ieal diode). The control of both resistances is fundamental to increase the final PCE of the cell.

In general, the equation that expresses the dependence of current and voltage can be defined as

$$I = I_{SC} - I_0 \left(\frac{e\,q(V - IR_S)}{nkT} - 1 \right) + \frac{V - IR_S}{R_{Sh}} \qquad (2.12)$$

where *n* is the non-ideality factor which can vary between 1 and 2.

In practice, because measurements are made illuminating a particular cell area, short-circuit current density (J_{SC}, i.e. I_{SC}/illuminated area) is commonly quoted when evaluating *IV* curves.

While the evaluation of *IV* curves under illumination is a basic evaluation of cell performance, dark *IV* curves (i.e. without any illumination) are also routinely performed. This measurement allows evaluating the ideality factor of the p-n junction in the absence of any light induced bias. The comparison of R_S and R_{Sh} for light and dark measurements also provides useful information [5].

Another basic means for evaluating the performance of a photovoltaic solar cell is measuring its external quantum efficiency (EQE, sometimes also quoted as incident photon-to-current efficiency (IPCE)). EQE yields the ratio of collected charges per photon incident on the device as a function of wavelength. EQE is analytically expressed by Eq. 2.13, where n_e is the number of electrons collected, n_{ph} is the number of incident photons (both per unit of time), h is Planck's constant, c is the speed of light, P_o is the incident optical power, q is the elementary charge, and λ is the wavelength.

$$\text{EQE}(\lambda) = \frac{n_e(\lambda)}{n_{ph}(\lambda)} = \frac{J_{SC}(\lambda)}{P_o(\lambda)} \frac{hc}{q\lambda} \qquad (2.13)$$

Thus an EQE of 100% would correspond to a hole–electron pair collected per each incident photon. In practice, EQEs are always smaller since EQE takes into account the total number of photons arriving to the device, regardless they are reflected, transmitted or absorbed. The EQE onset will in principle match the band gap of the absorbing materials in the cell (i.e. the corresponding wavelength). EQE varies for different wavelengths and thus it is a very useful parameter to evaluate the sensitivity of the solar cell along the spectrum. In the case of multiple absorber materials (with different band gaps) being used in a device, evaluation of the cell's EQE allows to elucidate which materials contribute to the collected photocurrent (i.e. EQE allows to correlate the solar cell's J_{SC} with particular spectrum regions). The relationship between J_{SC} and EQE (is given in Eq. 2.14, $\phi(\lambda)$ being the solar photon flux). If only the absorbed photons are taken into account (i.e. with the help of an integrating sphere in the set-up), then the internal quantum efficiency (IQE) can also be calculated. This is a more meaningful measure to evaluate the photoconversion efficiency of a cell. Because normally not all light is absorbed by a cell, IQEs are higher than EQEs.

$$J_{SC} = q \int_0^\infty \text{EQE}(\lambda) \phi(\lambda) d\lambda \qquad (2.14)$$

2.1.3 The Shockley–Queisser Limit

The Shockley–Queisser limit is a theoretical calculation of the maximum efficiency of a single p-n junction solar cell under 1.5 AM sunshine light, which is around 33%. This theoretical estimation is based on the assumption that one photon ($E > E_g$) can only generate one exciton (an electron–hole pair bonded by Coulomb interactions). It gives an important idea of how much solar energy can be converted into electricity. There are mainly two parts of energy losses taken into consideration in this estimation: irradiative recombination and spectrum losses.

The first limitation concerns the unavoidable blackbody radiation, which occurs for a solar cell at high temperature. This energy radiation increases with temperature and then lowers the efficiency. The second limitation is related to the fact that energy production is limited by the band gap, i.e. photons with energy lower that the band gap not contributing to the output power while photons with higher energy than the band gap generate "hot electrons" that rapidly thermalize and fall to the bottom of the conduction band realeasing part of their energy as heat (and not voltage).

While one might think that using very low band gap materials as the absorbing layer can cover larger ratio of the spectrum, a low gap results in a smaller built-in potential to drive the generated hole and electron pairs, thus resulting in a high J_{SC} but a low V_{OC} (i.e. low output power). An optimized band gap is therefore required, as shown in Fig. 2.5. To obtain the maximum converting efficiency of the sunshine for a single junction cell, the band gap should be around 1.1–1.5 eV.

To overcome the SQ limit, several strategies have been undertaken, such as cells based on quantum dots as absorber or multi-junction cells. These and other will be discussed in more detail later.

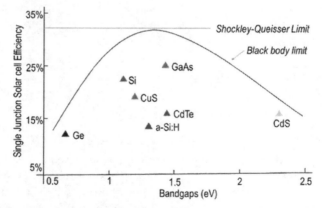

Figure 2.5 Different semiconductor materials. Single p-n junction efficiency and Shockley–Queisser limit [6].

2.1.4 Advanced Characterization

Other more advanced types of cell characterization can enlighten on how to improve the performance of the device and understand which phenomena are occurring and limiting cell efficiency, complementing the *IV* curve characterization and EQE of a solar cell. Theses advanced techniques can vary from AC electric performance, light absorption mechanisms, microstructural characterization, or even optical and optoelectronic characteristics.

On the electrical analysis, **electrochemical impedance spectroscopy** (EIS) can be used to probe different components of a cell. As a widely employed analysis technique in different areas of materials science, it is a quasi-static method that permits the study of electrical mechanisms in the different materials and interfaces of a solar cell [7]. This is possible since the different electrochemical processes inside the cell have different rates (i.e. time scales) and thus respond to different frequencies [7]. The interpretation of the results is achieved by finding an equivalent electrical circuit (EEC) to model each separated mechanism on the device. In previous research, EIS has been used to study the electric phenomenon of different types of solar cells [8, 9]. A variety of parameters can be modified in order to extract more information from the analysis and the experiment can be performed under illumination or in the dark, in open or short circuit mode and with different resistance loads.

Another spectroscopy technique, which allows the evaluation of defect creation in the cell, is **admittance spectroscopy**. In this case, the capacitance of a rectifying junction is studied by varying the current frequency and temperature [10]. The application of small AC bias voltage produces a response that can be evaluated in variable frequencies or temperature. The results can provide different characteristics, such as activation energy, density of states, and recombination rates [11].

Photovoltage and photocurrent decay measurements can be used to evaluate the carrier's lifetimes in the different materials of a cell structure. It allows measuring the carrier decay in a nondestructive way. However, the setup can result costly [12]. The method usually needs two ohmic contacts on the cell so that the photoconductivity can be sensed regarding to the voltage change [13].

The analysis of the transmission variations in a semi-transparent cell can be used to study the electron–hole creation and recombination processes. **Photo-induced absorption spectroscopy** uses a frequency-modulated monochromatic light beam to examine the total transmission. Slightly variations in the transmittance of the cell can be related with the kinetics of photo-generated species [14].

The detection of microstructural information, like dopants or defects at boundaries between different layers of materials, can be achieved by **infrared modulation spectroscopy**. This technique is based on the detection of optical transmittance differences in the cell with a varying sinusoidal reverse bias voltage [15]. The optical transmission changes can provide the charge states of dopants and defects near the interfaces due to the electrical potential modulation.

By using highly sensitive **photothermal spectroscopy**, analysis of optical behavior and thermal characteristics can be obtained. This technique is based on the rise of the sample temperature caused by absorbed light. The heating causes changes in the thermodynamic properties, such as pressure or density changes, that can indirectly affect the optical absorption of the specimen [16].

The use of advanced **imaging techniques** such as **STEM tomography** can enable a deep microstructural analysis of the different layers and materials. Consequently, a better understating of the cell structure can be presented. For instance, the connectivity and porosity of the nanoparticles in the photo-sensible component, or adhesion problems with the electrode can be observed using a 3D view of the different layers [10]. Moreover, a quantitative study of the charge percolation pathways in organic and hybrid solar cells can be performed by this tomography method.

2.1.5 Materials Requirement: The Ideal Solar Cell

An ideal semiconductor material for a photovoltaic cell should have a direct band gap of around 1.1 to 1.5 eV in order to provide maximum conversion efficiency. In addition, it should be easily doped into p- or n-type. From the economical point of view, the use of non-toxic earth-abundant elements is preferred. Finally, the required synthesis and processing steps should be scalable, low-cost, and environmentally friendly.

The combination of different technologies in the future of photovoltaic energy generation is fundamental to achieve the sustainability and success of solar cells. The adaptation, in terms of materials and processing, to each specific situation based on weather, global location or economical development of the place will determine the most suitable PV technology. This adaptability should also reduce the dependence on a specific compound or group of compounds, leading to a more sustainable development of photovoltaic energy generation.

2.2 Types of Solar Cell

Because there is a large number of potential absorber materials and combinations of p-type and n-type semiconductors, both organic and inorganic, different photovoltaic technologies have emerged. These range from the initial crystalline silicon (c-Si)-based solar cells to the exciting novel solar cells based on hybrid perovskite absorbers, going through thin film technologies, all-oxide devices, organic, and hybrid solar cells. The main characteristics, functioning, assets, and drawbacks of these different technologies are briefly described in the following sections. The reader is directed to the respective references if wishing to gain a deeper understanding on any of these technologies. We have followed here the classification of the different technologies made by the National Renewable Energies Laboratory (NREL). Figure 2.6 shows an efficiency chart for the different technologies (published and regularly updated on the NREL Web site).

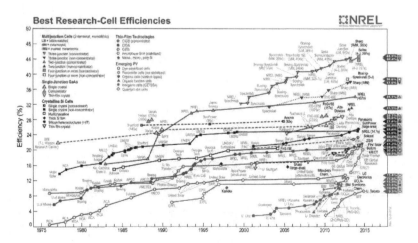

Figure 2.6 Efficiency of different types of solar cells (this plot is courtesy of the National Renewable Energy Laboratory, Golden, CO [17]).

2.2.1 Silicon

2.2.1.1 Crystalline silicon solar cells

At present, the most widely used solar cell in the market is based on crystalline Si [18]. Crystalline Si is a semiconductor material with an indirect band gap of 1.1 eV. There are generally two routes to obtaining single crystal bulk Si material. The well-known Czochralski method, (illustrated in Fig. 2.7) and RF partial heating.

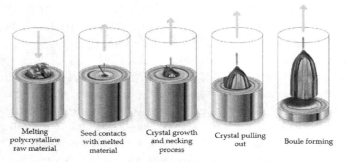

| Melting polycrystalline raw material | Seed contacts with melted material | Crystal growth and necking process | Crystal pulling out | Boule forming |

Figure 2.7 Single crystal Si by Czochralski process [19]. Image courtesy of Galaxy Compound Semiconductors, Inc.

In the former, the raw material is distilled and purified several times and then melted at around 1500 °C (melting temperature of Si is around 1400 °C), then a seed crystal is dipped into the melted liquid and crystal starts to grow around the seed. With a slow and stable pulling out speed, a big Si single crystal is obtained. After cutting and carefully polishing single crystal Si wafers of up to 450 mm can be made. In RF partial heating, a single crystal Si column can thus be obtained by partially heating a polycrystalline Si material under a moving hot coil. During this process, the polycrystalline Si will be melted and re-crystallized, yielding a large Si single crystal. Additionally, when the material is partially heated and melted, the impurities are carried from one end to another, thus a high-purity Si single crystal can be obtained. Nevertheless, due to surface tension limitation, the silicon single crystal produced in this way is not as large as the Czochralski process.

In addition to various mature fabrication processes yielding high-purity and high-quality Si wafer, Si is also very easily doped with other group elements to form p- and n-type material. This makes it straightforward to acquire homojunctions, which is not the case for other semiconductor materials. Also, with an optimum band gap of 1.1 eV, c-Si-based solar cells can reach reasonably high efficiency, with values up to 27.6% being reported [17]. However, c-Si has an indirect band gap, and thus photons need to be coupled with the phonon momentum of the lattice. This results in c-Si having a low absorption coefficient and thus thick layers (~100 μm) are needed to maximize absorption [18]. Such a thick layer requires very high-purity bulk material standard, meaning more strict purification process, and thus penalizing the cost/efficiency ratio of c-Si cells despite Si being a cheap and abundant element.

2.2.1.2 Si heterojunction solar cells

An alternative to conventional crystalline Si solar cells are silicon heterojunction solar cells [20]. HeteroJunction SC (HET) have demonstrated 24.7% power conversion efficiency (www.Panasonic.com), thus being a promising technology. The key point of this type of cells is the displacement of (ohmic) contacts, which are highly active toward recombination, from the crystalline surface by insertion of a film having a wide band gap. To reach the full device potential, the state density should be minimal at the heterointerface. Hydrogenated amorphous silicon (a-Si:H) films of only a few nanometers thick are used since their band gap is wider than that of c-Si and, when intrinsic, are able to reduce the c-Si surface state density by hydrogenation. The active element of HET is thus the stack formed by amorphous silicon layers (intrinsic for the good passivation at interfaces, doped to produce the electron and hole collectors on both sides), which are normally deposited by plasma-enhanced chemical vapor deposition (PECVD) on both sides of c-Si. Figure 2.8 shows the structure of a typical HET cell.

To further increase the device performance, research is focused on (i) the improvement of the amorphous silicon/crystalline silicon interfaces, i.e. to reduce the interface defect density, and (ii) the electrical contacts necessary to collect photogenerated charges. Point (ii) includes research on the metal grid and on transparent conductive layers. Such layers (transparent conductive oxides (TCOs)) are necessary to compensate the low conductivity of amorphous silicon, and thus to ensure an efficient lateral transport of charge carriers from crystalline silicon to the metal grid. In addition, they act as an antireflective coating, thus optimizing the amount of light reaching the cell. Thanks to the excellent passivation properties provided by the a-Si:H layers, HET cells can achieve very high efficiencies, while remaining a cost-effective technology (simple fabrication process, low temperatures and thus less energy consuming processes, possibility to use thinner c-Si wafer thus lowering Si consumption).

Figure 2.8 Structure of a standard HET cell.

TCO properties are critical for the functioning of HET solar cells: Their optical properties have to be finely tuned to ensure weak reflection and high transmission to the crystalline silicon absorber. In addition, their electrical properties, in particular their work function, has to be adjusted to favor a high band bending in crystalline silicon, yielding a high field effect needed to separate photogenerated charges. The extremely thin (<10 nm) amorphous layers also impose soft deposition processes to preserve interface and layer properties. Thus, the temperature for TCO processing should be low enough (<200 °C) to preserve the electrical properties of the active layers, mainly the amorphous state of silicon. Currently, HET cells are made using indium tin oxide (ITO) as TCO, which is sputtered. This means that the TCO and its deposition represents up to a 28% (*DnB NOR Markets Research 2009*) of the total cell cost. Thus alternative In-free TCO and cheap, scalable and low temperature deposition methods need to be developed to further increase the efficiency/cost ratio of HET cells and their stability.

2.2.2 Gallium Arsenide

As one of the most famous stars in the family of solar cell materials, gallium arsenide (GaAs)-based devices maintain the highest efficiency record for a single p-n junction device, reaching an astonishing 29.1% [21]. GaAs is a III-V group semiconductor material that has a direct band gap of around 1.4 eV, slightly larger than crystalline Si. With a much higher absorption coefficient than c-Si, 1 μm of GaAs is enough to replace 100 μm of c-Si. Nevertheless, the scarcity of those elements raises its price thus being one of the most costly materials in thin film solar cell technologies. Due to its direct band gap and good electrical properties, it can also be used in other applications such as light detectors, light emitting devices or integrated circuits.

In addition to price, GaAs has several other limitations. From the point of view of material fabrication, improper stoichiometry and inhomogeneity of element mixing are common problems. Importantly, GaAs is not environmentally friendly and is dangerous to human health, therefore demanding a strict safety control.

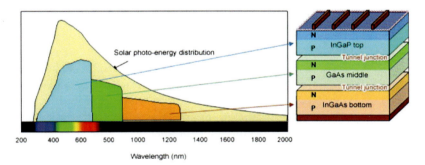

Figure 2.9 Wavelength distribution of solar photo-energy and wavelength sensitivity of triple-junction cell for the InGaP, GaAs and InGaAs parts of the structure [22]. Image credit: Semiconductor Today magazine/Sharp Corp. Please see: http://www.semiconductor-today.com/news_items/2012/DEC/SHARP_101212.html.

Even though the high price of GaAs is still affordable in large projects like solar cell panels for spaceships or vehicles, there is a lot of research to improve the efficiency. In October 2012, Sharp Corporation claimed a record 37.7% efficiency for non-concentrator solar cell by utilizing triple junction cell structure, as illustrated in Fig. 2.9 (see Section 2.2.5 for more details on multijunction cells).

In the triple junction structure shown in Fig. 2.9, three absorption layers with different band gaps are used, and each layer has a different absorbing window; by connecting the three layers with tunnel junctions, the absorption windows are stacked together. Thus, a much larger portion of the sunshine spectrum can be converted into electricity.

2.2.3 Thin Film Technologies

2.2.3.1 Copper indium gallium diselenide

The appearance of new Si free technologies in PV research has increased drastically in the last decades. One alternative has been the use of more absorbing materials allowing thinner layers to be used. Apart from GaAs, and due to its drawbacks, other thin film solar cell technologies are being developed. Copper indium gallium diselenide, a p-type semiconductor usually designed as CIGS, has been widely studied due to its promising band gap tunability [18]. While $CuInSe_2$ presents a band gap of 1.04 eV, lower than the ideal 1.4 eV for solar cells, incorporating gallium to substitute the indium leads to an increase of band gap up to 1.2 eV [3]. Moreover, the substitution of selenium by sulfur can be an alternative solution to obtain the ideal semiconductor characteristics. CIGS is commonly combined with n-type cadmium sulfide (CdS) to form the p-n junction. This has yielded solar cells with efficiencies higher than 20% (23.3% in the NREL plot), and high fill factors above 80% for cells deposited by three-stage co-evaporation [23].

One of the more relevant advantages of this technology is its reliability, since they do not present any degradation under illumination. Even more, CIGS solar cells surprisingly present an increase of V_{OC} and efficiency after they start operating (i.e. light soaking), which also means a higher sensibility to heat and humidity comparing with Si-based cells [24].

CIGS devices normally operate with light entering through the top layer and contacts and not the substrate (Fig. 2.10). The layered structure of these cells is deposited on a soda lime glass, covered with a molybdenum layer. The absorbing CIGS layer is then deposited followed by the CdS layer. The two layers differ drastically in thickness, since the first layer has usually 1.5 to 2 µm, while the n-type part is only 50 nm thick. Then a layer of ZnO is deposited, followed by top contacts that are made of highly conductive aluminum-doped ZnO layer (AZO) [21, 25].

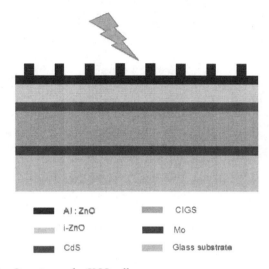

Figure 2.10 Structure of a CIGS cell.

2.2.3.2 Cadmium telluride

Still another thin film photovoltaic technology, cadmium telluride (CdTe)-based cells have been described as a viable low-cost alternative. CdTe cells are based on a hetero-junction between CdS and CdTe, which presents a band gap near the optimum value for solar light harvesting, as predicated by *P*. Rappaport in the 1950s [26]. The active layer is commonly deposited by close space sublimation (CSS), which yields the best results. However, other chemical and electrochemical methods such as screen-printing or electrodeposition are also used [27]. The scheme of a typical CdTe cell is shown in Fig. 2.11.

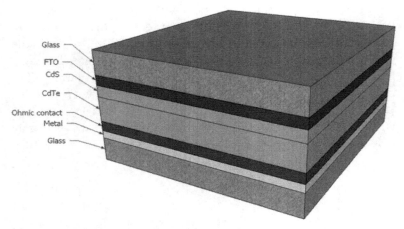

Figure 2.11 Structure of a CdTe solar cell.

In terms of efficiency, the record for a laboratory cell is 21.1% by First Solar, and large modules are typically produced in the 15% efficiency level. The lowest solar cell cost with this technique is around $0.50/Wp, according to First Solar, which is also the lowest price for all the PV technologies available [27].

Nevertheless, the main drawback of this technology lies on the materials. Cadmium is considered a hazard element, even if CdTe itself is a stable compound. Additionally, tellurium is a scarce element on earth, resulting in costs around $$10^5$–$10^6$/ton [28]. These issues account for the low popularity of CdTe in the PV industry.

2.2.3.3 Amorphous Si

The use of hydrogenated amorphous silicon has been intensively tested since the 1970s as a thin film solution for Si-based photovoltaic [29]. The indirect band gap in c-Si eliminates the possibility of adopting thin film technology; Therefore, to avoid this problem amorphous silicon is a suitable alternative. Thanks to its higher absorption coefficient, due to having a direct band gap, a-Si layers of only around 1 μm are required [30].

The direct band gap in a-Si is due to the presence of unsaturated dangling bonds, but these also act as charge traps and recombination sites, thus limiting the efficiency and increasing the degradation problems. The reduction of recombination losses can be achieved by creating a three layered structure, p-i-n, where an intrinsic thin layer of silicon is present in the middle of the p-n junction, working as absorbing film (see Fig. 2.12) [30]. The amorphous silicon cells use transparent conductive glass as subtract, where the layers are sequentially deposited by PECVD.

Figure 2.12 Structure of amorphous Si solar cells.

In the best case, using stable a-Si, the efficiency has only reached 13.4%, far behind other types of solar cell. This low efficiency is attributed to the Staebler–Wronski effect of light-induced degradation, causing a 25% efficiency drop before stabilization of the cell. Since the price of these cells is just slightly lower than crystalline Si cells, widespread use has been limited [25].

2.2.3.4 Thin film Si, nano- and micro-Si

The studies around Silicon continue in other forms than monocrystalline and amorphous silicon, with the appearance of other types based on granular films, with nano and micro sizes. An evolution from a-Si, microcrystalline silicon solar cells were created in the 1990s in the University of Neuchatel, presenting higher stability than amorphous cells. The deposition technique is similar as for a-Si but nucleation is promoted to start grain growth, originating a grain structure that can reach micro-sizes. The best cells are in the transition regime, where the space between grains is filled with amorphous phase. The efficiency of these cells are around 10%, for a 2 µm layer thickness, creating a maximum short-circuit current density of 24 mA/cm^2 [30].

2.2.4 Emerging Technologies

2.2.4.1 Dye sensitized solar cells

In the 1990s, Michael Grätzel invented the dye-sensitized solar cell (DSSC) concept, a low-cost and eco-friendly method inspired in natural photosynthesis [31]. The elementary structure of these cells contains three main materials: nanoparticles of a semiconductive oxide, a dye, and an electrolyte. The sensitizing dye is responsible for the absorption of the solar light by electron excitation from the HOMO (highest occupied molecular orbital) to the LUMO (lowest unoccupied molecular orbital). The electrons are then injected into the conduction band of the oxide where the dye is chemically fixed. The solid metal oxide structure has a mesoporous configuration to produce an extensive surface area for dye coverage, increasing the solar light absorption. Additionally, a high connectivity between nanoparticles leads to an optimized electron's conduction to an electrode, usually a conductive glass. Lastly, the dye is regenerated thanks to the electrolyte. The structure of this type of cell is shown in Fig. 2.13.

The semiconductor material is usually mesoporous TiO_2, since it has a wide band gap above 3 eV. Other oxides such as ZnO have been studied with less positive results [33]. Regarding dyes, the first dye used was called N3, based on ruthenium, but the most traditionally used one has been the N719, also based on the same element [33]. Since the beginning of DSSCs, iodide has been the most frequently used for the electrolyte since it presents a redox couple (I^{-3}/I^{-}), which enables an effective replacement of electrons in the dye and a good infiltration in the oxide structure when liquid [34]. Nevertheless, the need for higher efficiencies has pushed the development of novel materials and currently the most promising DSSC are based on Co electrolytes (with a record efficiency of 12.3%) and Zn porphyrines as absorbers [35].

Compared to silicon cells, the differences are mainly in the uptake of solar radiation. While in the case of silicon, the process of solar absorption and charge transport is made all in the same semiconductor material in the case of DSSC these processes are separated, where the sensitive dye is responsible for the absorption of light and the semiconductor responsible for the transport of electrons to the collector. As opposed to Si cells, DSSC is also an example of excitonic solar cell, where after photon absorption the excited electron and hole are electrostatically bound [36]. In view of its importance, DSSC are reviewed in full detail in the next chapter.

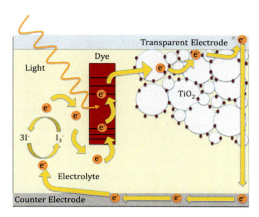

Figure 2.13 DSSC working scheme.

2.2.4.2 Organic and hybrid solar cells

Organic solar cells are devices that use organic semiconductors instead of conventional inorganic ones such as Si and GaAs. The organic semiconductors can be either polymers or small molecules and that affects their properties and how they can be deposited and processed. Because organic solar cells can be deposited by solution methods and at low temperature, thus compatible with plastic substrates, they are often called plastic solar cells. Another advantage of organic solar cells is that molecular engineering of the organic semiconductors provides a key resource to tune the semiconductor's properties such as band gap.

The main limitation of organic solar cells (apart from stability issues) is that the exciton formed upon photon absorption (electro-hole pair) has a relatively high binding energy of around 0.3–0.4 eV, caused by electrostatic interactions, and it is thus not readily split, as opposed to what happens in Si, for instance, where charges are immediately separated after absorption. Organic cells and other cells where excitons are present are thus also called excitonic solar cells [36]. In order to split, the exciton must find a driving force such as that present at the interface between a donnor and acceptor organic semiconductor. The low diffusion length of exictons in organic semiconductors (of the order of 10 nm) limited the initial efficiency of organic devices made of bilayers of p- and n-type organic semiconductors. Later, the discovery of the self-nanostructuration of the two semiconductors in what is known as a bulk heterojunction, resulted in much higher efficiencies being obtained since the interface area between the two organic semiconductors increased enormously. Efficiencies are still limited by the incompatibility between absorption coefficients of the organic materials and diffusion lengths of the separated charges. This issue is being addressed by introducing inorganic semiconductors, thus obtaining hybrid devices. In this sense, nanostructuring of the inorganic component has been a very active field of research in order to improve the efficiency of such devices [36, 38]. Organic solar cells is a hot topic and one of the most promising alternatives to Si solar cells [39]. Organic cells are covered in more detail in the next chapter.

2.2.4.3 Inorganic and ultra-low-cost cells

In order to have solar cells with a higher efficiency/cost ratio, the used materials should ideally be abundant and processable using low-cost and scalable techniques. In this context, there has been a recent drive toward the utilization of stable inorganic semiconductors based on earth abundant elements (EAE; mainly Si, Fe, Cu, O, S), and with the theoretical capacity of producing enough energy to cover the demand [40] (see Fig. 2.14).

Among the identified semiconductors, Cu_2O is being heavily investigated since Cu is abundant, non-toxic and theoretical maximum obtainable efficiencies for Cu_2O-based single junction cells are high (20%) [41]. Despite this promising theoretical values, efficiencies values of around 5.38% are the maximum that has been achieved so far for Cu_2O-based solar cells, thus still far from predicted efficiency values, and using energy intensive methods not suitable for large-scale production [42].

Cu_2O is an intrinsic p-type semiconductor with a direct band gap of around 2.1 eV. The difficulty in obtaining n-type Cu_2O is the main factor limiting the maximum efficiencies that can be obtained, since an efficient homojunction of the type n-Cu_2O/p-Cu_2O (equivalent to the Si homojuntions currently used in commercial solar cells) [43] has not been possible to make. Although there are some reports on n-type Cu_2O [44–46], the origin of the n-type character is not well established (with theoretical calculations even indicating that n-type Cu_2O is not possible without doping [47]) and the homojunction cells prepared with such materials presented very low efficiencies [43, 48].

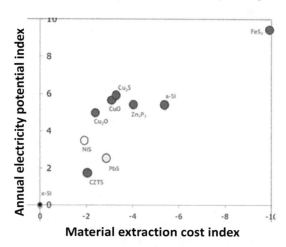

Figure 2.14 Annual electricity potential vs. materials extraction cost for several semiconductors. Those with better ratio than the currently used crystalline Si are located in the first quadrant (the one shown) [40]. Adapted with permission from Wadia, C., Alivisatos, P., and Kammen, D. M. *Environ. Sci. Technol.*, **43**, 2072 (2009). Copyright (2009) American Chemical Society.

For this reason, Cu_2O has been traditionally combined with an n-type semiconductor, typically ZnO. For this junction, a built-in bias of approximately 0.6 to 0.7 V is expected, which should approximate the achievable open-circuit voltage (V_{OC}) [49]. In a Cu_2O/ZnO solar cell, visible in Fig. 2.15, light is absorbed by Cu_2O and the generated electrons are injected into the ZnO conduction band and extracted, thanks to the high electron mobility of ZnO, through the contact toward the external circuit. Holes are thus transported through the Cu_2O layer to the external circuits as majority carriers, and electrons toward the junction as minority carriers. Up to now the maximum efficiencies obtained for this system have been accomplished for cells in which Cu_2O was synthesized by thermal oxidation at very high temperatures

(~1000 °C) [42]. This yields Cu_2O films with large crystals that present high charge mobilities, thus allowing the use of thicker layers and therefore a higher absorption of light. In spite of the attractive efficiencies obtained with thermally oxidized Cu_2O, this method is not desirable for mass production due to its elevated cost. In this respect, there has been an intensive study on ZnO/Cu_2O cells fabricated by electrodeposition. Although, the initial reports suffered from rather low V_{OC} values, engineering of the ZnO/Cu_2O and of the Cu_2O/contact interface has had a huge impact on efficiency, clearly showing that in addition to the bulk properties of the different semiconductors used, the interfaces (surface chemistry and morphology) play a key role on device efficiency [50, 51]. By engineering the interfaces, maximum efficiencies of 2.61% and 5% have been achieved for cells prepared by low temperature and high temperature approaches, respectively [42, 51].

Apart from a complete understanding and control of the interfaces (i.e. surface chemistry, micro/nanostructure), the other main problem of electrodeposited Cu_2O films is that they present incompatible collection vs. depletion and absorption width lengths (<1 µm vs. up to 3 µm, respectively) [52]. This results from a decay of the absorption coefficient with increasing light wavelength and the combination of a short electron transport length (mainly due to a small electron mobility in Cu_2O) and low hole concentration (see Fig. 2.15). These factors produce a recombination region at distances far (>1 µm) from the heterojunction interface in the Cu_2O where the minority carriers recombine before reaching the junction and hence do not contribute to the resulting current density [49, 52, 53]. Limited collection length can be addressed by nanostructuring the interface to reduce the required minority carrier (electrons) transport length [49]. In another approach the Cu_2O layer thickness was reduced to the charge collection length taking advantage of light confinement (i.e. absorption near the junction) produced by a polymer coating with a low refractive index [53]. While these two strategies resulted in an increase of the current density of about 15%, the devices displayed low Voc, due to inhibited formation of a complete depletion layer in the Cu_2O. Thus, in low-temperature, atmospherically processed ZnO/Cu_2O devices, the challenge remains to maximize both charge collection and high voltage output.

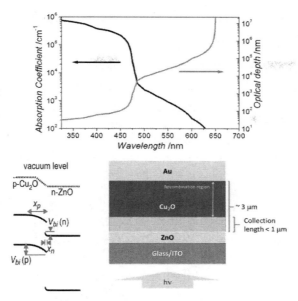

Figure 2.15 Top: Cu_2O absorption coefficient and optical depth as a function of light wavelength [49]; Bottom-left: Energy level diagram of Cu_2O–ZnO heterojunction with components of the built-in bias ($V_{bi}(p)$, $V_{bi}(n)$) and depletion layer thicknesses (x_p, x_n) [52]; Bottom-right: Schematic representation of a ZnO/Cu_2O cell showing the collection and recombination regions [53]. Top and bottom-left images kindly provided by Prof. K. Musselman.

Another material that offers a huge potential is $Cu_2ZnSn(S,Se)_4$, typically called CZTS. CZTS is indeed an a priori ideal material in view of its properties, specially the band gap, 1 to 1.5 eV, which can be tuned by controlling the ratio between S and Se. In addition, the abundance and non-toxicity of all the constituent elements of CZTS make it a scalable thin film technology, with the prospect of replacing CIGS cells. After many years of research on this technology with slow increase in efficiency, there is a regained interest in CZTS after IBM achieved a 10% efficiency in 2010. The record efficiency is currently above 12.5%, with J_{SC} values exceeding 80% of the theoretical maximum efficiencies. Conversely, the efficiency is still far from predicted values. This has been ascribed to three main general issues that still remain to be solved: a non-Ohmic back contact, a high degree of defects and disorder in the

CZTS film and, finally, a non-optimized interface between CZTS and the buffer layer [54].

2.2.4.4 Quantum dots cells

Decreasing materials down to the nanoscale has a strong impact on the physicochemical properties compared to bulk. According to the International Organization for standardization, nanoparticles correspond to particles having a size in the range roughly from 1 to 100 nm, at least in one direction. At this scale, the number of atoms at the surface is comparable to the whole volume.

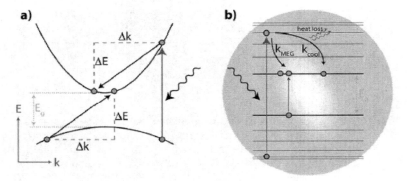

Figure 2.16 (a) Energy converting route in bulk semiconductor material. (b) Energy absorbing route in QD [55]. Reprinted with permission from Beard, M. C., Luther, J. M., Semonin, O. E., and Nozik, A. J. Third generation photovoltaics based on multiple exciton generation in quantum confined semiconductors. *Acc. Chem. Res.*, **46**, 1252–1260 (2012). Copyright (2013) American Chemical Society.

As we have mentioned above, in a single p-n junction solar cell device using bulk material the Shockley-Queisser limit limits efficiency to below 33%, due in part to thermalization of hot carriers generated by photons of higher energy than the band gap. Except under high incoming energy ($E_{photon} > 4E_g$), which is impossible in real working conditions, only one pair of exciton can be generated, the rest of the energy being lost to heat. As it is shown in Fig. 2.16a, in a bulk semiconductor material there are other three factors limiting efficiency. First, the crystal momentum must be conserved at the initial and final states, which sometimes leads to phonon generation upon photon absorption. Also, a limited density of states results in waste of the excess part of the photon energy. Finally, the difficulty to control defects in bulk material leads to higher recombination rate.

Conversely, in a nanometric semiconductor material higher density of states can be reached, thus more possibility to place extra energy excitons. Also, in such a small dimension, Bloch states are not eigenstates and thus no energy needs to be wasted toward conservation of momentum. Another crucial point in such confined system is the fact that due to small dimension, the exciton with extra energy can interact with others before relaxing to the lowest band edge. Thus, all together, those factors lead to a higher chance to obtain multi exciton generation ($E_{photon} > 2E_g$), this process is also called hot carrier impact ionization.

The number of excitons generated per incoming photon absorbed is named quantum yield (QY) and it is described in Eq. 2.15 [56]:

$$QY = 1 + \frac{k_{MEG}}{(k^{(1)}_{MEG} + k_{cool})} + \frac{k^{(1)}_{MEG} + k^{(2)}_{MEG}}{(k^{(1)}_{MEG} + k_{cool})(k^{(2)}_{MEG} + k_{cool})} + ... \quad (2.15)$$

where k_{MEG} means multiple exciton generation rate, $k_{MEG}(i)$ refers to the rate of creating $(i + 1)$ exciton from i hot excitons, k_{cool} is the cooling rate of excitons to the band edge (assumed to be the same in single and multiple excitation). Each extension term is valid only when energy conservation is met. Such as, when $E_{photon} > 3E_g$, higher order series will be extended (to $k^{(3)}_{MEG}$). The energy required for $QY > 1$ is called threshold voltage, hV_{th}. The extra energy could affect the k_{MEG}, which is described in Eq. 2.16:

$$k_{MEG} = Pk_{cool} \left(\frac{h\nu_{ex}}{h\nu_{th}}\right)^2 \quad (2.16)$$

where $h\nu_{ex}$ is defined as $h\nu_{ex} = h\nu - h\nu_{th}$, P as a parameter to evaluate the competition between MEG and cooling down process. When $h\nu = 2h\nu_{th}$, $\nu_{ex} = \nu_{th}$, thus $P = \frac{K^{(2)}_{MEG}}{K_{cool}}$. Thus, in this case, the relation

between threshold energy and parameter P can be described as
$$v_{th} = \left(2 + \frac{1}{P}\right)E_g = \left(2 + \frac{K_{cool}}{K_{MEG}^{(2)}}\right)E_g.$$

Now, if we suppose that for each extra exciton pair generated, ε_{eh} extra energy is needed, then $\varepsilon_{eh} = \Delta h\nu/\Delta QY$, $h\nu > h\nu_{th}$. On the other hand, for MEG, $h\nu_{th} = E_g + \varepsilon_{eh}$. Thus, v_{th} and the quantum yield (number of excitons generated) are related. It means that, ideally, for a material having bandgap $E_{g'}$ a photon energy $2E_g$ is enough to achieve MEG. But, in reality, extra energy ε_{eh} is always required, which depends on the material physical properties, size and volume. To further describe this phenomenon, a term named *MEG efficiency* is defined as ε_{MEG}, $\varepsilon_{MEG} = E_g/\varepsilon_{eh}$. Thus, also $h\nu_{th} = E_g(1 + 1/\varepsilon_{MEG})$.

Therefore both exciton generation and $h\nu_{th}$ will increase when k_{MEG}/k_{cool} is increased. Figure 2.17 shows how k_{MEG}, and thus P, increases when going from bulk to nanometric structures while threshold frequency v_{th} minimize close to ideal $2E_g$.

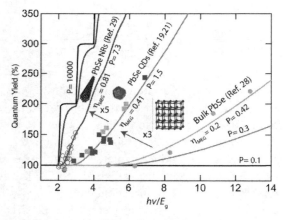

Figure 2.17 Improvement of MEG by changing dimension confinement (nanowire/nanorods, QD, bulk material). Reprinted with permission from Beard, M. C., et al. Comparing multiple exciton generation in quantum dots to impact ionization in bulk semiconductors: Implications for enhancement of solar energy conversion. *Nano Lett.*, **10**, 3019–3027 (2010). Copyright (2010) American Chemical Society.

Figure 2.18a shows a common QDs-based solar cell. In this case, ZnO NPs form a hetorojunction with PbS QDs. Electrons are collected through the ITO electrode while MoO$_3$ collects holes to the metal contact. The band structure and working mechanism is usually simplified to a conventional structure with one CB and VB. However, the potential distribution of those interfaces is not fully understood yet. Thanks to the MEG energy-conversion route mentioned above, the ideal energy harvesting efficiency can reach up to 45%.

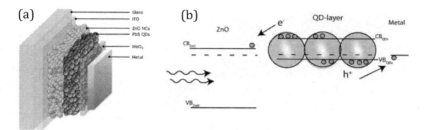

Figure 2.18 (a) Common structure of a QDs solar cell. (b) Approximated band strcuture of QDs layers. Reprinted with permission from Beard, M. C., Luther, J. M., Semonin, O. E., and Nozik, A. J. Third generation photovoltaics based on multiple exciton generation in quantum confined semiconductors. *Acc. Chem. Res.*, **46**, 1252–1260 (2012). Copyright (2013) American Chemical Society.

However, to reach expected efficiencies in this type of solar cell, several issues need to be tackled and solved, the main one being the interfaces and the QD layer. Interfaces are potential barriers for carriers, while the QD layer needs to facilitate the transport of generated excitons to the ZnO/QDs interface so that dissociation of photons takes place. Controlled growth kinetics during the QD synthesis (size and shape of QDs, distance between QD) is therefore a key step to enhance carrier transport and dissociation of excitons [36].

2.2.4.5 Hybrid perovskite cells

The liquid solvent presented in the best dye sensitized solar cells, combined with a moderate efficiency with a very slow asymptotic efficiency growth rate over the years, have been the key problems for the wide-spread application of DSSC. The quest for higher

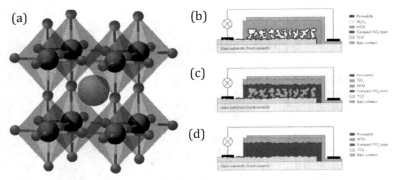

Figure 2.19 (a) Typical perovskite crystallographic structure. In the case of organohalide perovskites, the organic cation is located at the A position (green), while metal cations (gray) and halides (pink) are located in the B and X position, respectively; (b–d) evolution from a mesoscopic to a planar cell architecture. Adapted with permission from Grätzel, M. The light and shade of perovskite solar cells. *Nat. Mater.*, **13**, 838–842 (2014) [63].

efficiencies in solid state devices, i.e. without liquid electrolyte, has yielded a new type of solar cell in which the absorber material is a family of hybrid materials having the perovskite structure and with general formula $CH_3NH_3PbX_3$ (X = Cl, Br, I), generally combined with Spiro-OMeTAD, a solid state hole-transporting semiconductor traditionally used for solid state DSSC (Fig. 2.19a). The first results of cells based on hybrid perovskites appeared in 2009 showing a 3–4% efficiency, thus still far from traditional dye sensitized solar cells, and showing stability issues [57]. This technology is progressing impressively (exponentially!) and in just five years after the first report, the maximum efficiency rises to 20.1% with fill factor above 77% [58]. It is important to note though that these record efficiencies have been obtained for small area, laboratory devices. Since the theoretical maximum efficiency is 26% for a 1 μm layer of material [59], hybrid perovskite solar cells appear to offer more economical viability than other technologies, such as GaAs. Nevertheless, the stability and cost-effectiveness of hybrid perovskite solar cells is still an issue to be addressed [60]. Additionally, the use of lead in the structure is an obvious problem because of its toxicity and increasing legal restrictions. Hence, other elements such as tin have been studied to replace lead in the perovskite structure, with lower efficiencies being obtained [61]. The ability to reproduce high efficiencies for larger, module-size cells is also a challenge for this technology.

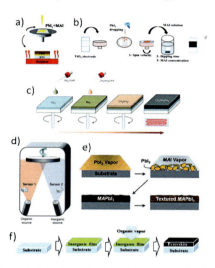

Figure 2.20 (a–c) Solution-based deposition methods: (a) One-step precursor deposition (OSPD). (b) Sequential deposition process (SDP). (c) Two-step spin-coating deposition (TSSD); (d–f) Vapor-based deposition methods: (d) Dual source vacuum deposition (DSVD). (e) Sequential vapor deposition (SVD) (f) Vapor-assisted solution process (VASP). Adapted with permission of The Royal Society of Chemistry from Shi, S., Li, Y., Li, X., and Wang, H. Advancements in all-solid-state hybrid solar cells based on organometal halide perovskites. *Mater. Horiz.*, **00**, 1–28 (2015).

Although initially considered to be a type of solid DSSC in which the classical dye was replaced by the hybrid perovskite, further research has shown that the mobility of charges in the HP is indeed so large that the use of the mesoporous TiO_2 is not necessary to collect the electrons, and that the obtained V_{OC} are indeed larger when other insulating materials such as SiO_2 are used [62]. Several cell architectures have been proposed (Fig. 2.19b), and so far the best efficiencies are obtained for cells having the thin-film planar structure. Efficiencies are, nevertheless, still debated since a hysteresis of the *IV* curves for different scan senses and speeds are observed and not yet fully understood [63].

In the early studies, the deposition of the perovskite layer started as a two-step spin coating process, where PbI_2 was deposited first and then followed by the CH_3NH_3I complex that reacts thermally with the deposited layer. The deposition techniques evolved to other possible opportunities, such as one-step spin coating, thermal evaporation and vapor-assisted solution process, which lead to the formation of layers with different grain structure, roughness or porosity [60, 64]. The different deposition methods can be classified in solution and vapor methods and are summarized in Fig. 2.20 [65].

2.2.4.6 Spectral conversion

One of the major limitations in PV devices is the fact that the different absorbing materials are tuned to a particular part of the visible spectrum, thus being very inefficient out of that particular wavelength range. There have been several strategies focused in maximizing the amount of visible light absorbed and converted by the cell. An approach to do that is based on modifying the PV device to better exploit the solar spectrum as in the case of tandem cells (see Section 2.2.5). An alternative solution consist in manipulating the solar spectrum through the use of spectral conversion layers in order to better match the band gap of the absorber material. After a first report in 1995 by F. Auzel [66], this approach has experience a surge since 2002 following the pioneering works published by M. Green and B. Richards [67]. As described in Fig. 2.21, spectral conversion covers two photonic processes:

- *Downconversion (DC)*, which consists in absorbing high-energy photons and then re-emitting a photon with wavelength just above the band gap (*down shifting*) or alternatively re-emitting two photons (*quantum cutting (QC)*, only for incident photons with energy higher than $2E_g$). These processes allow the reduction of the thermalization losses and are implemented in layers integrated on the front side of the PV cells.
- *Upconversion (UC)*, which consists in absorbing several low energy photons which are transmitted through the PV cell and then re-emitting photons with energy higher than E_g.

This process thus takes place in a layer integrated on the rear side.

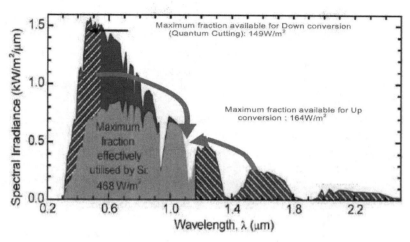

Figure 2.21 AM1.5G solar spectrum showing the fraction of terrestrial sunlight that is currently absorbed and effectively utilized by a thick crystalline silicon device, and the additional regions of the spectrum that can contribute to up- or down-conversion. Adapted with permission from Richards, B. S. *Sol. Energy Mater. Sol. Cells* 2006, **90**, 2329–2337.

A detailed presentation of each of the processes and the related materials challenges are reported in recent review papers [68–70]. As a first advantage, the UC or DC layers are passive photonic components which can be applied to existing PV devices. The effective potential of each process depends strongly on the solar cell type [68] and a lot of work has been developed in the last 10 years to demonstrate the predicted efficiency gain provided by spectral conversion [71–74].

Concerning the type of materials used, the spectral conversion is mainly based on compounds (typically oxides and halogenides) doped with rare earth ions, which cover a large number of possible transitions ranging from UV to NIR.

In the case of DC for c-Si PV cells, the most studied dopings are Pr, Er or Tb, which act as absorbers or activators, coupled with Yb to obtain an energy transfer to two Yb ions (a case of QC), and finally the emission following $^2F_{7/2} - {}^2F_{5/2}$ Yb transitions at 980 nm, so just above de Si band gap. Internal quantum efficiencies higher than 190% have been reported but, unfortunately, due to weak absorption of lanthanide 4f–4f forbidden (for trivalent lanthanides) transitions, the external quantum efficiencies are always very small and no proof of global efficiency improvement have been reported yet on an integrated solar cell. In order to improve the absorption step, application of photonic crystal structures as well as plasmonic phenomena are under study in parallel to the sensitization by a third ion with high absorption cross section such as Eu^{2+} or Ce^{3+}.

UC processes are more interesting for higher band gap cells than c-Si since $NaYF_4$ codoped with Yb and Er (the most efficient upconverter to date), absorbs strongly at 980 nm through Yb and allows efficient transfer toward the high excited levels of Er. Thus upconversion of 980 nm has been already demonstrated in DSSCs [75] and in Si:H solar cells [76]. Nevertheless, as UC is a nonlinear process requiring at least the combined absorption of two photons, UC quantum yield increases with the photon density [74]. This means that concentration of sunlight is necessary for achieving an efficiency improvement induced by UC. Another limitation that must be overcome by further work is related to the narrow spectral absorption of the lanthanides compared to the broad spectral range of sunlight.

2.2.4.7 Nanostructured solar cells based on zinc oxide nanowire arrays

Over the last decade, semiconductor nanowires (NWs) have emerged as promising building blocks for a large number of sensing, electronic, optoelectronic, and photovoltaic devices [77]. Their large surface to volume ratio (i.e. aspect ratio) at nanoscale dimensions offers a wide variety of outstanding physical phenomena such as stress relaxation by lateral surfaces, quantum and dielectric confinements, or electron transport, for instance [78]. In the field of photovolt aics, the assets of silicon microwires as well as III–V and zinc oxides NWs offer a great potential for reducing the amount of materials used and favoring light trapping as well as efficient charge carrier management (i.e. both separation and collection) [79, 80]. The arrays of NWs in radial heterostructures (Fig. 2.22a) can efficiently trap the light by exciting sophisticated optical processes such as Fabry–Perot resonances, optically guided modes and optically radiated modes (i.e. diffraction process), for instance [81, 82]. Charge carrier separation is also achieved over the short distance of the NW radius, favoring in turn its collection thanks to the high crystalline quality of NWs. In contrast, the arrays of NWs in axial heterostructures (Fig. 2.22b) enhance the absorption process through efficient anti-reflective properties and also benefit from the charge carrier collection through the high crystallinity of NWs. It is thus expected that NW-based solar cells can be a good alternative to standard planar layer-based solar cells especially for mobile devices, by drastically reducing their cost while maintaining fairly high photo-conversion efficiency. Photo-conversion efficiencies of 9% [83] and 13.8% [84] have, for instance, been reported for Si microwire- and InP NW-based solar cells, respectively.

Alternatively to Si and III–V materials, zinc oxide exhibits a lot of advantages as an abundant, nontoxic materials with a high electron mobility of about 200 cm^2/Vs [86]. It can also easily be grown as NWs by a large number of physical and chemical deposition techniques [87], including the low-cost, low-temperature and easily implemented chemical bath deposition process [88]. Since ZnO only absorbs the ultra-violet light owing to its wide band gap energy of 3.3 eV, the use of an absorbing materials is required and typically combined with ZnO NWs in the so-called core shell heterostructures, as depicted in the schematic of Fig. 2.23a. These heterostructures are typically composed of ZnO NWs as the core and the absorbing materials as the shell and deposited on transparent conductive oxides (TCO) coated glass substrates. An alternative way is to completely fill the ZnO NW arrays with the absorbing materials, as presented in the schematic of Fig. 2.23b. The incident photons go through TCO/glass substrates and are only absorbed by the absorbing materials, generating electron/hole pairs in their center. One very critical aspect is that the ZnO/absorbing materials interface should comply with a type II band alignment for favoring efficient charge carrier separation (i.e. electron injection into ZnO), as shown in Fig. 2.23b. ZnO NWs thus act as an electron transporting materials and the electrons are eventually collected by the TCOs. These heterostructures are filled by a p-type semiconductor acting as the hole transporting materials and generally involving CuSCN or electrolytes (i.e.

iodide/triiodide, sulfur/polysulfur...) with gold or platinum contacts. When ZnO NWs arrays are already filled by the absorbing material, Spiro-OMe TAD or direct metallic contacts can be used for instance. The integration of ZnO NWs into solar cells is particularly suited to the fabrication of extremely thin absorber solar cells (ETASCs), quantum dot-sensitized solar cells (QDSSCs), DSSCs, and hybrid solar cells. In all of these ZnO NW–based solar cells, only the absorbing material is changed: semiconducting extremely thin films or quantum dots (e.g. CdSe [89–91], CdS [92], CIS [93], PbS [94], Cu_2O [95, 96], ...), dyes [97–99], and organic compounds such as P3HT [100, 101] or organohalide perovskites [102, 103]. Until now, the photo-conversion efficiency of ETASCs and QDSSCs lies in the range of 1% to 5%. In contrast, while the photo-conversion efficiency of hybrid solar cells is low when using P3HT, the integration of organohalide perovskites has resulted in a photo-conversion efficiency exceeding 10% [103, 104]. In most of these heterostructures, the optical absorption is usually high, but several limitations are still challenging for electron/hole pair collection, including surface and interface recombination, doping control or hole transport by using relevant hole transporting materials for instance. Increasing efforts are thus being currently dedicated to overcoming and circumventing these limitations.

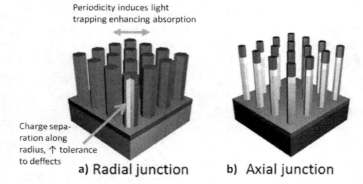

Figure 2.22 NW (a) radial (i.e. core shell) and (b) axial heterostructures. Adapted with permission from Yao, M., et al. GaAs nanowire array solar cells with axial p-i-n junctions. *Nano Lett.*, **14**, 3293–3303 (2014). Copyright 2014 American Chemical Society [85].

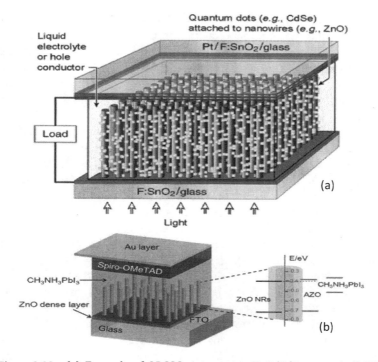

Figure 2.23 (a) Example of QDSSCs integrating ZnO/CdSe core shell NW heterostructures. Reprinted (adapted) with permission from Leschkies, K. S., et al. Photosensitization of ZnO nanowires with CdSe quantum dots for photovoltaic devices. *Nano Lett.*, **7**, 1793–1798 (2007) [90]. Copyright (2007) American Chemical Society. (b) Example of hybrid solar cells integrating ZnO NWs covered with halides perovskites. Reproduced from Dong, J., et al. Impressive enhancement in the cell performance of ZnO nanorod-based perovskite solar cells with Al-doped ZnO interfacial modification. *Chem. Commun.*, **50**, 13381–13384 (2014) [103]. With permission from The Royal Society of Chemistry.

2.2.5 Multijunction or Tandem Solar Cells

Generally speaking third generation solar cells approaches aim to tackle one or both of the major losses in solar cells, i.e. non-absorption of low energy photons and thermalization of high-energy photons [105]. An approach for limiting these losses has been the fabrication of tandem cells, which consist of several cells of different band gap stacked on top of each other. Each cell absorbs a different portion of the solar spectrum, with band gap decreasing in the direction of light incidence. Therefore each cell absorbs photons closer to its band gap and the resultant electron–hole pairs undergo less thermalization than they would do in a single band gap cell, finally resulting in an higher solar cell efficiency. The principle is schematically depicted in Fig. 2.24. Roughly speaking the open circuit voltage of the tandem cell is the sum of those of the individual slabs while the short circuit current is limited to the one obtained from the less efficient slab. Tandem cells allow a theoretical efficiency limit of 42.5% for a two-cell stack and 47.5% for a three-cell stack tandem cell, as opposed to the Shockley–Queisser limit of 31% for a single band gap cell [106]. A good example is the amorphous/nanocrystalline silicon heterojunction with intrinsic thin layer (HIT) cell, which has lately reached a conversion efficiency of 24.7% on thin Si wafer [107]. Solar cells with record efficiencies (about 45% [37]) are based on triple junctions using a germanium substrate for a better absorption in the red region of the solar spectrum [108]. Indeed III–V multijunction solar cells demonstrate the highest efficiency in photovoltaics: 40% under concentration was already obtained in 2007 [109]. Such high efficiency can be obtained from the optimization of band gaps of each layer as well as lattice constant of III–V and IV semiconductors.

An attractive idea would be to synthesize a nanostructured material with the appropriate photovoltaic properties, which would provide a promising route toward the realization of an "all-silicon" tandem solar cell. For this purpose Si nanoparticles embedded in an amorphous matrix (SiO_2, Si_3N_4, or SiC) are well suited since the quantum confinement of carriers allows band gap engineering [110]. Such nanoparticles can be fabricated, for instance, by alternate deposition of doped silicon-rich oxide and SiO_2 followed by a high temperature post-annealing [111]. Although this idea appears promising it is still not clear whether it is actually possible to have at the same time quantum confinement as well as carrier transport through the quantum particle network. Organic tandem photovoltaic cells have also attracted a lot of research efforts with efficiencies reaching 10%. The key limiting factor and necessary developments to be carried out in order to reach efficiencies of 15% have been reviewed in detail [112].

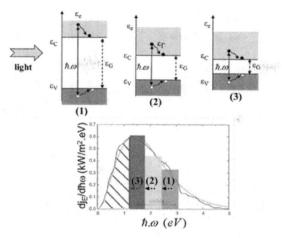

Figure 2.24 Schematic representation of a tandem cell. High energy photons are first absorbed in the top cell while photons of lower energy are absorbed in a lower cell. The overall voltage generated is then enhanced as the associated photovoltaic efficiency. A tandem solar cell can use quantum dots to engineer the band gap of the top cell and potentially also the lower cells for a better efficiency [105].

2.3 Transparent Conductive Materials

Transparent conductors, or transparent conductive materials (TCMs), constitute an extremely important part of modern technology and are used in applications such as coatings for defrosting windows [113], touch panels or as electrodes in solar cells [114, 115]. For PV, transparent electrodes should extract separated charge carriers from the absorbing layer while being as transparent as possible. A lot of research has been devoted in the past to the deposition, characterization and understanding of thin layers which exhibit both high transparency and high electrical conductivity [116–118]. Clearly the most studied materials so far have been TCOs [117, 119] such as ITO, fluorine-doped tin oxide (FTO) [120] or aluminum-doped zinc oxide (AZO) [114]. Indeed the most commonly used material so far is ITO, which also exhibits very good physical properties. However, the increasing need for transparent electrodes for different applications has led to an increase in the cost of In (a scarce element) and, additionally the fact that TCOs are by nature brittle, have both prompted the search for new emergent TCMs [121]. These upcoming materials are carbon nanotubes [122], graphene [123], or metallic nanowire networks [124].

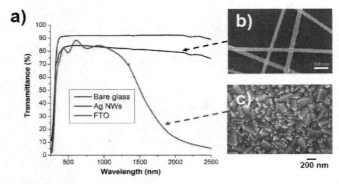

Figure 2.25 Optical total transmittance for a bare glass substrate (blue) and two transparent conductive materials in the UV-VIS-NIR region. A spray pyrolysis deposited FTO and Ag nanowire network deposited by spin coating exhibit a high transparency in the visible range. The sheet resistance is reported to be close to 11 ohms per square for both layers. The associated SEM images are presented with rather similar magnification. Ag nanowire network exhibits large areas for which no absorption by the TCM occurs while FTO layer is a homogeneous polycrystalline layer. Note that AgNW network is still very transparent in the near infra-red (NIR) region as opposed to TCOs (for which plasma effects occur decreasing the transparency in this spectral domain).

The most important requirements for TCMs are optical transparency (T) and sheet resistance (R_{sheet}). For the sake of comparison between different TCMs it is common to consider the following figure of merit (FOM) after Haacke [125]: FOM = T^{10}/R_{sheet}. While quoted requirements can vary, it is generally accepted that for solar cells T should be close or higher than 90% while R_{sheet} should be lower than 10 Ω/sq [126]. Solar cells appear to be the most stringent among all applications in terms of TCM requirement and the impact of TCM properties on photovoltaic efficient is significant (i.e. 10–25% of power loss) [126]. In order to get good figures of merit, a compromise between transparency and sheet resistance must be achieved. This is so since making the TCO layer thin to increase its transparence results in a higher resistivity.

However, other properties are important in addition to T and R_{sheet}. Mechanical flexibility is often required for large area production of flexible electrodes. This will be mandatory for the expected move toward flexible electronics. Haziness is also an advantage for solar cell applications (as opposed to displays): The haze factor is the ratio between the diffuse transmitted light intensity divided by the total transmitted intensity. Attempts are made for increasing the haze factor since it enhances the optical path of incident photons in the absorbing layer. This parameter can influence largely the solar cell efficiency, as briefly described below.

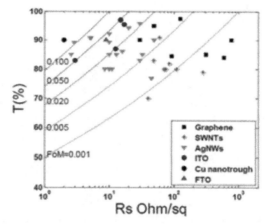

Figure 2.26 Transmittance (T) and sheet resistance (R_s) associated with different transparent conductive materials. The target is the top left area. The general trends are depicted: Graphene and carbon nanotubes (SWNTs) still exhibit larger sheet resistance compared to TCOs, Ag nanowire networks or Cu nanothrough. Iso-values of the Figure of Merit are also depicted as guides for the eyes.

Other properties related to interfaces also play important roles for an efficient integration of TCMs in solar cells [127]. Indeed a detailed control of the Fermi level position both in the bulk and at surfaces and interfaces of polycrystalline TCO is mandatory for a good understanding of the physical properties of TCOs. A very convenient tool for proving surface and interface properties is photoelectron spectroscopy, which enables performing in situ observations of energy band alignment at interfaces [127].

Depending on the type of solar cells, different TCMs should be considered and there is still a large room for scientists to develop and further understand TCMs. The main type of TCMs currently being evaluated are detailed next. The FOM of the different TCM is compared in Fig. 2.26, showing that n-type TCOs and Ag nanowire are currently the most promising TCMs.

2.3.1   N-Type TCOs

The best-known and most used TCM so far is tin-doped indium oxide ITO, a n-type oxide semiconductor. Despite its good figure of merit and properties, indium is likely to become a commodity in short supply in the near future, thus making ITO too expensive. Other n-type TCOs are zinc-oxide-based thin films (such as Al-doped Zn oxide: AZO [114, 128]) and FTO, which have received much attention in the last decades. As depicted in Fig. 2.25c, n-type TCOs are thin polycrystalline layers (i.e. generally les than 100 nm thick) and exhibit a low resistivity of a few 10^{-4} $\Omega \cdot$cm, associated with a high mobility of about several tens of $cm^2 \cdot V^{-1} \cdot s^{-1}$ and transmittance up to 90%. Efficient TCO thin films are generally heavily doped n-type materials with a charge carrier density larger than 10^{19} cm^{-3}. As a result, TCO thin films are polycrystalline degenerate semiconductors.

In order to improve solar cell efficiency, one parameter that needs to be optimized is the carrier mobility within the TCO. Four main electron scattering mechanisms have been reported to account for the electrical properties of TCO thin films: grain boundary scattering, phonon scattering, ionized impurity scattering and (namely for FTO) twin boundary scattering. Several studies have focused on the relationship between structural and electrical properties. An example on FTO can be found in ref. [120].

The main drawback of TCOs (in addition to indium scarcity for ITO) concerns their poor mechanical properties: When deposited on polymers, TCOs are ceramic and therefore are prone to cracking when bending is considered [129].

UV-VIS-NIR spectroscopy enables to measure the transmission of the TCO with two components, diffuse and specular, and thus evaluate the haze factor of a TCM film. Attempts are made to minimize the haze factor in flat-panel display (for a better spatial resolution) and to maximize it in a solar cell (to increase the optical path of the photons in the absorbing layer). The impacts on a solar cell efficiency can be drastic: For instance, Tsai et al. [130] showed that an impressive increase in the photo-conversion efficiency (from 8.18 to 10.1%) of dye-sensitized solar cells when using increasingly hazy FTO front transparent electrodes with haze factor from 2 to 17%. Recently ZnO-SnO$_2$:F thin film nanocomposites were fabricated by combining spin-coated ZnO nanoparticles (NPs) on glass with FTO thin films subsequently grown by spray pyrolysis [131]. Depending on ZnO NP surface coverage prior to the FTO growth step, the resulting nanocomposite exhibits an average haze factor enhancement from 0.4% to 64.2% while the sheet resistance only increases from 9.5 to 14.8 Ω/sq. The integration of such transparent and very hazy films in solar cells have thus interesting prospects.

2.3.2 P-Type TCOs

At present, transparent electrodes based on p-type oxides need yet to be optimized, their present performance being much lower than that of n-type oxide semiconductors.

For p-type oxides, carrier conduction path (valence band) is mainly formed from the oxygen p asymmetric orbitals, which severely limits the carrier mobility. So, p-type oxides have very low carrier mobility compared to their n-type counterparts. The development of robust p-type semiconductors remains a considerable challenge, where substantial material development is needed before high performance devices can be realized. The accomplishment of a low temperature process and reasonable device stability/reliability will also be critical, since a fine control of the transport properties of the materials obtained is necessary. This is so because different applications require different properties, i.e. while conventional passive applications (transparent electric conductors) require the maximum possible hole density for optimal conductivity, application in thin film transistors (TFTs) as channel layer requires a low hole density in order to modulate the channel conductance by an applied electric field. Currently, Cu-based oxides are the most promising candidates for p-type transparent conductors both because Cu(I)-based delafossites of the type $CuB^{III}O_2$ (B being Al^{+3}, Ga^{+3}, In^{+3}, Cr^{+3}, Fe^{+3}, Co^{+3}, Y^{+3}, La^{+3}, Sc^{+3}, etc.) show the best performance so far and because Cu is a nontoxic, earth abundant element. In addition, simpler binary Cu oxides (CuO, Cu_2O, and Cu_3O_4) offer many possibilities of study through processing and doping [132, 133].

2.3.3 TCM Based on Metallic Nanowires

The need for transparent conducting materials suitable for flexible applications has stimulated research on metallic nanowire (NW) networks [134] and especially on silver nanowires (AgNW) [135]. The main advantages of AgNW concern, on the one hand, the very high electrical conductivity of Ag and, on the other hand, its synthesis, especially by the polyol process, which is well mastered and can lead to large-scale production [136, 137]. But other metals are also being explored, such as Cu and Cu–Ni alloys. Randomly oriented metallic nanowires can be easily deposited on flexible or rigid substrates at low temperature and under atmospheric conditions, by spraying or spin coating for example. Finally, the malleable nature of metals in general makes the AgNW suitable for flexible applications.

Regarding the optimization of the optical and electrical properties of AgNW networks, several parameters are important: network density, NW dimensions, post-deposition treatment. First, finding an optimum nanowire density allowing a high transmission of light through the electrode coupled together with a high electrical conductivity over the whole cell surface is primary. Depending on the network density two regimes take place: For low density a percolation regime occurs, while for denser networks the bulk regime is considered [113, 138]. The other important step for an efficient metallic NW-based TCM is often an adapted post-deposition treatment that can lead to a drastic reduction of the network electrical resistance. Such treatment can be, for instance, thermal annealing [124], light-induced plasmonic nanowelding [139] or mechanical pressure [140].

Concerning the optical properties of AgNW-based TCM, transparency and light scattering properties are dependent on the wire dimensions [141]: A linear decrease of the transmittance versus AgNW network density is observed along with a linear increase of the haze factor [137, 141]. Indeed, AgNW transparent conducting paper-based electrodes can exhibit high optical haze factor which is again very beneficial for the integration into solar cells [136].

The integration of metallic NW networks as transparent electrodes in solar cells has already shown promising results. Preston et al. [136] demonstrated that AgNW networks appear cheaper, more flexible, more amenable to large-scale manufacturing, and exhibit at the same time better optoelectronic properties than other TCMs. Especially the large haze factor of AgNW networks constitutes a clear asset for solar cell integration.

A key issue in views of the actual application is the stability of metallic NW-based TCM. Mayousse et al. showed that a long term (months) humidity treatment may cause degradation of the AgNWs electrodes [142]. Solutions such as embedding with inorganic nanoparticles constitute a serious solution, among others, for improving the thermal stability and integration in solar cells [143]. Morgenstern has demonstrated that solution-processable AgNW films coated with ZnO nanoparticles can be used efficiently as transparent electrodes in organic devices [143]: The devices prepared using AgNW-based electrode exhibited similar device performance as when using ITO. Similar conclusions have also been obtained by Leem et al. [144]. Kang et al. concluded that AgNW network can even lead to enhanced power conversion efficiency (i.e. relative increase of 35%) thanks to surface plasmon resonance as well as waveguide effects [145] when compared with ITO-based devices. Moreover, this AgNW-based technology is compatible with transfer printing technique, flexible substrates and with roll-to-roll process and therefore with high-throughput fabrication.

2.3.4 TCM Based on CNT and Graphene (Including Hybrids with Nanowires and Other Materials)

Carbon nano tubes (CNTs) constitute another emerging nanoscale material along with AgNW networks. Its main physical characteristics concern, for instance, a very high carrier mobility within one individual CNT ($>10^5$ cm^{2}V^{-1}s^{-1}) and a current density as large as 10^9 A/cm^2 [146]. However, excitonic transitions in the visible range cause significant transmittance losses, making them less efficient compared with AgNW networks or TCOs [147] (Fig. 2.26). In spite of such interesting physical properties it still remains difficult to produce bulk quantity of CNTs with the suitable purity. However, much progress has been made lately and CNT still constitute a promising material for transparent electrodes. As for AgNW networks, random CNT networks exhibit a high transparency in the near infrared range. Interestingly, flexible transparent electrodes based on functionalized metallic single walled carbon nanotubes can be obtained with tuned work function (from 4.6 to 5.1 eV): This constitutes a valuable asset for an optimized integration in many optoelectronic devices including solar cells [148].

Since its experimental isolation in 2004, graphene has been the focus of much research due to, among others, its high charge mobility, mechanical strength, flexibility properties as well as transparency. Graphene can thus be considered as a rather promising next-generation conductive material for electrical and optical devices [123, 149]. The possibility to synthesize ultra-large-scale ($\approx$30 inch) graphene films using roll-to-roll transfer as well as chemical doping processes has paved the way for pratical applications on a large scale [150].

Barnes et al. [147] compared in detail the different conductive nanostructured networks with TCOs through their potential performances. Calculating the current losses due to absorption in TCMs appear to be a useful tool for choosing a transparent electrode for a solar cell. This analysis concluded that AgNW networks appear to reach more optimal physical properties (optical and electrical) than CNT networks.

Finally, hybrid nanostructures based on two-dimensional graphene and AgNW have been studied yielding interesting results: Sheet resistance of 33 Ω/sq with a transparency of 94% in the visible range associated with a very good flexibility and stretchability (100%) response have been observed [151].

Carbon nanotube networks have already been integrated into organic-inorganic metal halide perovskite solar cells. The $CH_3NH_3PbI_3$/CNTs solar cells were semi-transparent and showed photovoltaic output with dual side illuminations due to the transparency of the CNT electrode with an efficiency of up to 6.87% [152]. Thanks to the incorporation of a hole transporting material spiro-OMeTAD, the $CH_3NH_3PbI_3$/CNTs perovskite solar cell efficiency was even improved to 9.90% [152] showing good potential for integration of CNT networks in some solar cells.

2.4 Toward Low Cost, Fast and Scalable Processing

Atomic layer deposition (ALD) is a powerful deposition technique offering the deposition of high-quality films at low temperatures, very precise control over film thickness and highly conformal coatings even on high aspect ratio features. The burst of nanoscience and nanotechnology in the last decades has brought an associated boom of ALD popularity, since it allows the nanoengineering of surfaces with precise nanoscale control, and has indeed already been applied to many different fields [153–159], in particular energy [160–163]. Despite this increasing success, ALD faces two main drawbacks, which compromise both its generalized use in the laboratory and its implementation in large-scale, high-throughput production lines: Slow processing and usually processing in vacuum, thus making it complicated and expensive to scale up.

ALD is a particular case of CVD in which the reaction is restricted to the sample surface, thus being self-limited, through the sequential exposure of the sample to the reactants [164] (Fig. 2.27a). In this way, the metal precursors are supplied and react with the surface, ideally forming a monolayer. Excess precursor is then purged. The second precursor is then injected and reacts with the chemisorbed layer forming a monolayer of the desired material plus by-products that have to be purged along with the excess precursor. The cycle is then repeated the necessary number of times to obtain a very precise film thickness.

Although there have been several approaches to vaccum-free ALD processing, atmospheric pressure *Spatial* ALD (AP-SALD) has been the most popular in the last years due to its conceptual simplicity and the high deposition rates that can be achieved. The key idea was indeed patented in 1977 and 1983 by Suntola et al. [166, 167], and it simply consists of separating the precursors in space and not in time (Fig. 2.27b). In SALD therefore the different precursors are supplied constantly in between inert gas regions.

Films are then grown by alternatively exposing the substrate from one precursor region to the other going across the inert gas regions (Fig. 2.27b). Thus, the oscillation of the substrate from one precursor zone to the second one reproduces the classical ALD scheme. SALD allows much faster deposition rates, which are mainly limited by the precursor reaction kinetics allowing the full reaction to take place while the sample is within the precursor zones. From 2008 there has been a variety of SALD designs exploiting this idea [165, 168].

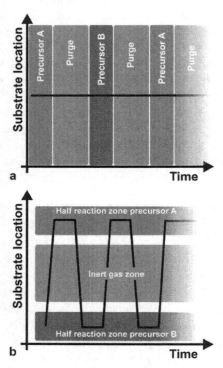

Figure 2.27 (a and b): Temporal vs. Spatial ALD. Copyright 2011, AIP Publishing LLC; Reprinted with permission from Poodt, P., et al. Spatial atomic layer deposition: A route towards further industrialization of atomic layer deposition. *J. Vac. Sci. Technol. A Vacuum Surf. Film*, **30**, 010802 (2012) [165].

One particularly advantageous design is the use of a short distance between the precursor delivery zone and the substrate (typically below 100 µm, Fig. 2.28a). In this "close proximity

SALD", the different gas regions are located along alternative channels where the gases flow parallel to each other. The close distance of the substrate prevents precursor mixing since the inert gas channel acts as an effective barrier.

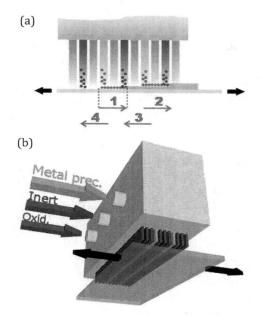

Figure 2.28 (a) Close-proximity SALD approach. The sample oscillates under the different flows reproducing the cycles in conventional ALD (numbered arrows); (b) Scheme of the deposition head designed by Kodak. Copyright 2013 John Wiley & Sons, Ltd. Adapted with permission from Muñoz-Rojas, D., et al. High-speed atmospheric atomic layer deposition of ultra thin amorphous TiO_2 blocking layers at 100 °C for inverted bulk heterojunction solar cells. *Prog. Photovoltaics Res. Appl.*, **21**, 393–400 (2013) [169].

The intrinsic advantages of ALD combined with the high throughput and atmospheric processing make AP-SALD a very attractive technique for the deposition of solar cell components. AP-SALD has indeed already shown its high potential as a deposition technique for new generation photovoltaics, allowing the fabrication of high-quality materials at high speeds and to perform fundamental studies through material doping [170]. We briefly describe two examples to illustrate the potential of AP-SALD.

In the first example AP-SALD was used to grow high-quality TiO_2 blocking layers for organic (poly(3-hexylthiophene-2,5-diyl):[6,6]-phenyl-C61-butyric acid methyl ester (P3HT:PCBM)) solar cells [169]. Dense, uniform thin TiO_2 films were grown at temperatures of only 100 °C in just 37 s (~20 nm/min growth rate). Incorporation of these films in P3HT-PCBM-based solar cells showed performances comparable with cells made using TiO_2 films deposited with much longer processing times and/or higher temperatures (SALD at 350 °C and spray pyrolysis at 450 °C). The high quality of the amorphous films obtained at 100 °C allowed the use of only 12 nm-thick blocking layers, thus compensating for the lower conductivity of amorphous TiO_2 with respect to crystalline TiO_2, and resulting in two orders of magnitude faster deposition than other low-temperature scalable methods.

In the second, AP-SALD was used to deposit Cu_2O. As been detailed above, ZnO/Cu_2O suffer from an incompatibility in terms of absorption and collection lengths. Also, the low hole concentration of electrodeposited Cu_2O imposes the use of thick Cu_2O films in order to have a full depletion layer. The depletion width in a semiconductor depends mainly on the carrier concentration and, therefore, synthesizing Cu_2O with a higher hole concentration would result in higher cell V_{ocs} for a same cell thickness and higher electric field within the Cu_2O layer [96]. The Cu_2O films obtained using AP-SALD presented mobilities from 1.5 to 5.5 $cm^2V^{-1}s^{-1}$, for deposition temperatures from 125 to 225 °C. The lowest mobility obtained for the SALD Cu_2O is already one order of magnitude higher than that of electrodeposited Cu_2O films. What is more important, all films, for temperatures from 150 °C, had a carrier concentration of around 10^{16} cm^{-3}, which is three orders of magnitude higher than for electrodeposited films (Fig. 2.29) [132].

Taking advantage of the high carrier concentration of the SALD Cu_2O films, a back surface field (BSF) cell with the following structure $Cu_2O_{SALD}/Cu_2O_{ED}/ZnO_{ED}$ (where ED stands for electrodeposited; Fig. 2.30) was implemented [96]. In a BSF cell, the top layer, with a high carrier concentration, dopes the layer underneath thus contributing to decrease the depletion width in the cell, and increasing the electric field within the electrodeposited Cu_2O layer. In this way, record current densities were obtained for a ZnO/Cu$_2$O cell deposited by low temperature, atmospheric methods, while using a Cu$_2$O layer of around 1 micron (as opposed to the standard 3 microns for fully electrodeposited devices). Apart from enhanced electric field, the use of a thinner Cu$_2$O layer and the nanoparticulated AP-SALD Cu$_2$O layer also contributed to a more effective collection of charges generated by photons of wavelengths above 450 nm (as shown in the EQE measurements, Fig. 2.30 right). The obtained V_{OC} was lower than expected, due to the nanometric nature of the AP-SALD Cu$_2$O films (from 50 to ~250 nm, see example in Fig. 2.29), which is believed to cause charge recombination.

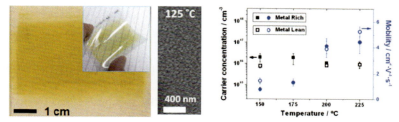

Figure 2.29 Left: Optical photograph of Cu$_2$O film deposited at 150 °C on glass. Inset: Cu$_2$O film on a 6.3 × 6.3 cm^2 PEN thin film deposited at 150 °C; Middle: SEM image of a film deposited at 125 °C; Right: Carrier concentration and mobility values for the films grown by SALD at 150, 175, 200 and 225 °C. Copyright 2012, American Institute of Physics. Licensed under Creative Commons Attribution 3.0 Unported. Reprinted from ref. 132.

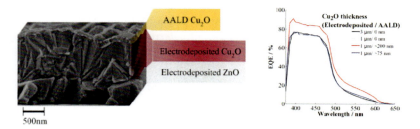

Figure 2.30 Left: Cross section SEM images of an SALD-enhanced cell; Right: EQE spectra of a thin (~75 nm) and thick (~200 nm) SALD enhanced cell (ZnO/Cu$_2$O$_{ED}$/Cu$_2$O$_{SALD}$) compared to a 1 μm-Cu$_2$O and 3 μm-Cu$_2$O fully-ED (ZnO/Cu$_2$O) cell. Copyright 2013 Wiley-VCH Verlag GmbH & Co. KGaA. Reproduced with permission from ref. 96.

After these promising initial results, further labs are becoming interested in AP-SALD, with more materials being deposited and studied, including TCOs. The coming years will surely show an increase in the number of publications concerning the development of further AP-SALD systems and their application to hot research fields, in particular optoelectronics.

References

1. Kittel, C., McEuen, P., and McEuen, P. *Introduction to Solid State Physics*, 8th ed. (Wiley New York, 1976).
2. Pierret, R. F. *Semiconductor Device Fundamentals* (Addison-Wesley Reading, MA, 1996).
3. Möller, H. J. *Semiconductors for Solar Cells* (Artech House Publishers, 1993).
4. Boyle, G. Renewable energy: Power for a sustainable future. *J. Energy Lit.*, **2**, 106–107 (1996).
5. Servaites, J. D., Ratner, M. A., and Marks, T. J. Organic solar cells: A new look at traditional models. *Energy Environ. Sci.*, **4**, 4410 (2011).
6. Green econometrics. What's Pushing Solar Energy Efficiency? (http://greenecon.net/what%E2%80%99s-pushing-solar-energy-efficiency/energy_economics.html).
7. Chang, B. Y., and Park, S. M. in *Annu. Rev. Anal. Chem.* (eds. Yeung, E. S., and Zare, R. N.) **3**, 207–229 (2010).
8. Wang, Q., Moser, J.-E., and Grätzel, M. Electrochemical impedance spectroscopic analysis of dye-sensitized solar cells. *J. Phys. Chem. B*, **109**, 14945–14953 (2005).
9. Fabregat-Santiago, F., Bisquert, J., Garcia-Belmonte, G., Boschloo, G., and Hagfeldt, A. Influence of electrolyte in transport and recombination in dye-sensitized solar cells studied by impedance spectroscopy. *Sol. Energy Mater. Sol. Cells*, **87**, 117–131 (2005).
10. Li, J. V., Crandall, R. S., Repins, I. L., Nardes, A. M., and Levi, D. H. Applications of admittance spectroscopy in photovoltaic devices beyond majority-carrier trapping defects. in *Photovolt. Spec. Conf. (PVSC), 2011 37th IEEE* 75–78 (2011). doi:10.1109/PVSC.2011.6185849.
11. Kavasoglu, A. S., and Bayhan, H. Admittance and impedance spectroscopy on Cu (In, Ga) Se$_2$ solar cells. *Turkish J. Phys.*, **27**, 529–536 (2004).
12. Pisarkiewicz, T. Photodecay method in investigation of materials and photovoltaic structures. *Opto Electron. Rev.*, **12**, 33–40 (2004).

13. Singh, S. N., Gandotra, R., Singh, P. K., and Chakravarty, B. C. Application of photoconductivity decay and photocurrent generation methods for determination of minority carrier lifetime in silicon. *Bull. Mater. Sci.*, **28**, 317–323 (2005).
14. Boschloo, G., and Hagfeldt, A. Photoinduced absorption spectroscopy of dye-sensitized nanostructured TiO_2. *Chem. Phys. Lett.*, **370**, 381–386 (2003).
15. Zhu, K., Schiff, E. A., and Ganguly, G. Infrared charge-modulation spectroscopy of defects in phosphorus doped amorphous silicon. in *MRS Proc.*, **715**, A5. 4 (Cambridge University Press, 2002).
16. Bialkowski, S. *Photothermal Spectroscopy Methods for Chemical Analysis.*, **134** (John Wiley & Sons, 1996).
17. NREL. Best Research-Cell Efficiencies (2015). Available at: http://www.nrel.gov/ncpv/images/efficiency_chart.jpg.
18. Fonash, S. *Solar Cell Device Physics* (Elsevier, 2012).
19. Galaxy Compound Semiconductors, I. Czochralski Crystal Growth (2015). Available at: http://www.iqep.com/galaxy/technology/crystal-growth/.
20. Muñoz, D., Desrues, T., and Ribeyron, P. J. a-Si:H/c-Si Heterojunction solar cells: A smart choice for high efficiency solar cells (2010).
21. Chopra, K. L., Paulson, P. D., and Dutta, V. Thin-film solar cells: An overview. *Prog. Photovoltaics*, **12**, 69–92 (2004).
22. Semiconductor Today. Sharp claims record 37.7% efficiency for non-concentrator solar cell (2012). Available at: http://www.semiconductor-today.com/news_items/2012/DEC/SHARP_101212.html.
23. Repins, I. et al., 19.9% efficient $ZnO/CdS/CuInGaSe_2$ solar cell with 81.2% fill factor. *Prog. Photovoltaics Res. Appl.*, **16**, 235–239 (2008).
24. Liu, Q. Optimization and characterization of transparent oxide layers for CIGS solar cells fabrication. University of Toledo. Theses and Dissertations. Paper 1305 (2007).
25. Wennerberg, J. Design and stability of Cu (In, Ga) Se_2-based solar cell modules. Uppsala University (2002).
26. Rappaport, P. The photovoltaic effect and its utilization. *Sol. Energy*, **3**, 8–18 (1959).
27. Dobson, K. D., Visoly-Fisher, I., Hodes, G., and Cahen, D. Stability of CdTe/CdS thin-film solar cells. *Sol. Energy Mater. Sol. Cells*, **62**, 295–325 (2000).
28. Andersson, B. A. Materials availability for large scale thin film photovoltaics. *Prog. Photovoltaics Res. Appl.*, **8**, 61–76 (2000).

29. Carlson, D. E., and Wronski, C. R. Amorphous silicon solar cell. *Appl. Phys. Lett.*, **28**, 671–673 (1976).

30. Aberle, A. G. Thin-film solar cells. *Thin Solid Films*, **517**, 4706–4710 (2009).

31. O'Regan, B., Grätzel, M. A low-cost, high-efficiency solar cell based on dye-sensitized colloidat TiO_2 films. *Nature*, **353**, 737–740 (1991).

32. Barbara, U. C. S., Department of Geography. Dye sensitize solar cells scheme. Available at: http://www.geog.ucsb.edu/img/news/2010/Dye_Sensitized_Solar_Cell_Scheme.png.

33. Gratzel, M. Dye-sensitized solar cells. *J. Photochem. Photobiol. C Photochem. Rev.*, **4**, 145–153 (2003).

34. Hardin, B. E., Snaith, H. J., and McGehee, M. D. The renaissance of dye-sensitized solar cells. *Nat. Photonics*, **6**, 162–169 (2012).

35. Ye, M. et al., Recent advances in dye-sensitized solar cells: from photoanodes, sensitizers and electrolytes to counter electrodes. *Mater. Today*, **18**, 155–162 (2014).

36. González-Valls, I., and Lira-Cantú, M. Vertically-aligned nanostructures of ZnO for excitonic solar cells: A review. *Energy Environ. Sci.*, **2**, 19–34 (2009).

37. Dimroth, F., Grave, M., Beutel, P., Fiedeler, U., Karcher, C., Tibbits, T. N. D., Oliva, E., Siefer, G., Schachtner, M., Wekkeli, A., Bett, A. W., Krause, R., Piccin, M., Blanc, N., Drazek, C., Guiot, E., Ghyselen, B., Salvetat, T., Tauzin, A., Signamarcheix, T., Dobrich, A., Hannappel, T., and Schwarzburg, K. Wafer bonded four-junction GaInP/GaAs//GaInAsP/GaInAs concentrator solar cells with 44.7% efficiency. *Prog. Photovoltaics Res. Appl.*, **22**(3), pp. 277–282 (2014).

38. Iza, D. C. et al., Nanostructured conformal hybrid solar cells: A promising architecture towards complete charge collection and light absorption. *Nanoscale Res. Lett.*, **8**, 359 (2013).

39. Brabec, C., Dyakonov, V., Parisi, J., and Sariciftci, N. S. *Organic Photovoltaics: Concepts and Realization* (Springer-Verlag, 2003).

40. Wadia, C., Alivisatos, A. P., and Kammen, D. M. Materials availability expands the opportunity for large-scale photovoltaics deployment. *Environ. Sci. Technol.*, **43**, 2072–2077 (2009).

41. Olsen, L. C., Addis, F. W., and Miller, W. Experimental and theoretical studies of Cu_2O solar cells. *Sol. Cells*, **7**, 247–279 (1982).

42. Minami, T., Nishi, Y., and Miyata, T. High-efficiency Cu_2O-based heterojunction solar cells fabricated using a Ga_2O_3 thin film as N-type layer. *Appl. Phys. Express*, **6**, 044101 (2013).

43. McShane, C. M., and Choi, K.-S. Junction studies on electrochemically fabricated p-n Cu_2O homojunction solar cells for efficiency enhancement. *Phys. Chem. Chem. Phys.*, **14**, 6112–6118 (2012).
44. Xiong, L., et al. p-Type and n-type Cu_2O semiconductor thin films: Controllable preparation by simple solvothermal method and photoelectrochemical properties. *Electrochim. Acta*, **56**, 2735–2739 (2011).
45. McShane, C. M., and Choi, K.-S. Photocurrent enhancement of n-type Cu_2O electrodes achieved by controlling dendritic branching growth. *J. Am. Chem. Soc.*, **131**, 2561–2569 (2009).
46. Fernando, C. A. N., et al. Investigation of n-type Cu_2O layers prepared by a low cost chemical method for use in photo-voltaic thin film solar cells. *Renew. Energy*, **26**, 521–529 (2002).
47. Scanlon, D. O., and Watson, G. W. Undoped n -Type Cu_2O: Fact or Fiction? *J. Phys. Chem. Lett.*, **1**, 2582–2585 (2010).
48. Jiang, T., et al. Photoelectrochemical and photovoltaic properties of p–n Cu_2O homojunction films and their photocatalytic performance. *J. Phys. Chem. C*, **117**, 4619–4624 (2013).
49. Musselman, K. P., et al. Strong efficiency improvements in ultra-low-cost inorganic nanowire solar cells. *Adv. Mater.*, **22**, E254–E258 (2010).
50. Gershon, T. S., et al. Improved fill factors in solution-processed ZnO/Cu_2O photovoltaics. *Thin Solid Films*, **536**, 280–285 (2013).
51. Lee, Y. S., et al. Ultrathin amorphous zinc-tin-oxide buffer layer for enhancing heterojunction interface quality in metal-oxide solar cells. *Energy Environ. Sci.*, **6**, 2112–2118 (2013).
52. Musselman, K. P., Marin, A., Schmidt-Mende, L., and MacManus-Driscoll, J. L. Incompatible length scales in nanostructured Cu_2O solar cells. *Adv. Funct. Mater.*, **22**, 2202–2208 (2012).
53. Gershon, T., Musselman, K. P., Marin, A., Friend, R. H., and Macmanus-driscoll, J. L. Solar Energy materials & solar cells thin-film ZnO/Cu_2O solar cells incorporating an organic buffer layer. *Sol. Energy Mater. Sol. Cells*, **96**, 148–154 (2012).
54. Gershon, T., et al. Understanding the relationship between $Cu_2ZnSn(S,Se)_4$ material properties and device performance. *MRS Commun.*, **4**, 159–170 (2014).
55. Beard, M. C., Luther, J. M., Semonin, O. E., and Nozik, A. J. Third generation photovoltaics based on multiple exciton generation in quantum confined semiconductors. *Acc. Chem. Res.*, **46**, 1252–1260 (2012).

56. Beard, M. C., et al. Comparing multiple exciton generation in quantum dots to impact ionization in bulk semiconductors: Implications for enhancement of solar energy conversion. *Nano Lett.*, **10**, 3019–3027 (2010).

57. Kojima, A., Teshima, K., Shirai, Y., and Miyasaka, T. Organometal halide perovskites as visible-light sensitizers for photovoltaic cells. *J. Am. Chem. Soc.*, **131**, 6050–6051 (2009).

58. Burschka, J., et al. Sequential deposition as a route to high-performance perovskite-sensitized solar cells. *Nature*, **499**, 316–9 (2013).

59. Yin, W.-J., Shi, T., and Yan, Y. Unique properties of halide perovskites as possible origins of the superior solar cell performance. *Adv. Mater.*, **26**, 4653–4658 (2014).

60. Jung, H. S., and Park, N. Perovskite solar cells: From materials to devices. *Small*, **11**, 10–25 (2015).

61. Ghanavi, S. *Organic-Inorganic Hybrid Perovskites as Light Absorbing/Hole Conducting Material in Solar Cells* (2013).

62. Lee, M. M., Teuscher, J., Miyasaka, T., Murakami, T. N., and Snaith, H. J. Efficient hybrid solar cells based on meso-superstructured organometal halide perovskites. *Science* (80-). 1–5 (2012).

63. Grätzel, M. The light and shade of perovskite solar cells. *Nat. Mater.*, **13**, 838–842 (2014).

64. Liu, M., Johnston, M. B., and Snaith, H. J. Efficient planar heterojunction perovskite solar cells by vapour deposition. *Nature*, **501**, 395–398 (2013).

65. Shi, S., Li, Y., Li, X., and Wang, H. Advancements in all-solid-state hybrid solar cells based on organometal halide perovskites. *Mater. Horiz.*, **00**, 1–28 (2015).

66. Gibart, P., Auzel, F., Guillaume, J., and Zahraman, K. Below band-gap IR response of substrate-free GaAs solar cells using two-photon up-conversion. *Jpn. J. Appl. Phys.*, **35**, 4401 (1996).

67. Trupke, T., Green, M. A., and Würfel, P. Improving solar cell efficiencies by up-conversion of sub-band-gap light. *J. Appl. Phys.*, **92**, 4117–4122 (2002).

68. Richards, B. S., Ivaturi, A., MacDougall, S. K. W., and Marques-Hueso, J. Up- and down-conversion materials for photovoltaic devices. *Proc. SPIE*, **8438**, 843802–843802–8 (2012).

69. Wang, H.-Q., Batentschuk, M., Osvet, A., Pinna, L., and Brabec, C. J. Rare-earth ion doped up-conversion materials for photovoltaic applications. *Adv. Mater.*, **23**, 2675–2680 (2011).

70. Lian, H., et al. Rare earth ions doped phosphors for improving efficiencies of solar cells. *Energy*, **57**, 270–283 (2013).
71. Badescu, V., and Badescu, A. M. Improved model for solar cells with up-conversion of low-energy photons. *Renew. Energy*, **34**, 1538–1544 (2009).
72. Atre, A. C., and Dionne, J. A. Realistic upconverter-enhanced solar cells with non-ideal absorption and recombination efficiencies. *J. Appl. Phys.*, **110**, 034505 (2011).
73. Van Sark, W. G., de Wild, J., Rath, J. K., Meijerink, A., and Schropp, R. E. Upconversion in solar cells. *Nanoscale Res. Lett.*, **8**, 81 (2013).
74. Fischer, S., et al. Upconverter materials and upconversion solar-cell devices: Simulation and characterization with broad solar spectrum illumination. *Proc. SPIE*, **8981**, 89810B–89810B–7 (2014).
75. Shan, G. B., and Demopoulos, G. P. Near-infrared sunlight harvesting in dye-sensitized solar cells via the insertion of an upconverter-TiO_2 nanocomposite layer. *Adv. Mater.*, **22**, 4373–4377 (2010).
76. De Wild, J., Meijerink, A., Rath, J. K., Van Sark, W. G. J. H. M., and Schropp, R. E. I. Towards upconversion for amorphous silicon solar cells. *Sol. Energy Mater. Sol. Cells*, **94**, 1919–1922 (2010).
77. Lieber, C. M., and Wang, Z. L. Functional nanowires. *MRS Bull.*, **32**, 99–108 (2007).
78. Law, M., Goldberger, J., and Yang, P. Semiconductor nanowires and nanotubes. *Annu. Rev. Mater. Res.*, **34**, 83–122 (2004).
79. Garnett, E. C., Brongersma, M. L., Cui, Y., and McGehee, M. D. Nanowire solar cells. *Annu. Rev. Mater. Res.*, **41**, 269–295 (2011).
80. Peng, K.-Q., and Lee, S.-T. Silicon nanowires for photovoltaic solar energy conversion. *Adv. Mater.*, **23**, 198–215 (2011).
81. Garnett, E., and Yang, P. Light trapping in silicon nanowire solar cells. *Nano Lett.*, **10**, 1082–1087 (2010).
82. Michallon, J., et al. Light trapping in ZnO nanowire arrays covered with an absorbing shell for solar cells. *Opt. Express*, **22**, A1174–A1189 (2014).
83. Kelzenberg, M. D., et al. High-performance Si microwire photovoltaics. *Energy Environ. Sci.*, **4**, 866–871 (2011).
84. Wallentin, J., et al. InP nanowire array solar cells achieving 13.8% efficiency by exceeding the ray optics limit. *Science*, **339**, 1057–1060 (2013).
85. Yao, M., et al. GaAs nanowire array solar cells with axial p-i-n junctions. *Nano Lett.*, **14**, 3293–303 (2014).

86. Özgür, Ü., et al. A comprehensive review of ZnO materials and devices. *J. Appl. Phys.*, **98**, 041301 (2005).

87. Yi, G.-C., Wang, C., and Park, W. Il. ZnO nanorods: Synthesis, characterization and applications. *Semicond. Sci. Technol.*, **20**, S22 (2005).

88. Xu, S., and Wang, Z. L. One-dimensional ZnO nanostructures: Solution growth and functional properties. *Nano Res.*, **4**, 1013–1098 (2011).

89. Lévy-Clément, C., Tena-Zaera, R., Ryan, M. A., Katty, A., and Hodes, G. CdSe-sensitized p-CuSCN/nanowire n-ZnO heterojunctions. *Adv. Mater.*, **17**, 1512–1515 (2005).

90. Leschkies, K. S., et al. Photosensitization of ZnO nanowires with CdSe quantum dots for photovoltaic devices. *Nano Lett.*, **7**, 1793–1798 (2007).

91. Xu, J., et al. Arrays of ZnO/Zn_xCd_{1-x}Se nanocables: Band gap engineering and photovoltaic applications. *Nano Lett.*, **11**, 4138–4143 (2011).

92. Tak, Y., Hong, S. J., Lee, J. S., and Yong, K. Fabrication of ZnO/CdS core/shell nanowire arrays for efficient solar energy conversion. *J. Mater. Chem.*, **19**, 5945–5951 (2009).

93. Krunks, M., et al. Extremely thin absorber layer solar cells on zinc oxide nanorods by chemical spray. *Sol. Energy Mater. Sol. Cells*, **94**, 1191–1195 (2010).

94. Jean, J., et al. ZnO nanowire arrays for enhanced photocurrent in PbS quantum dot solar cells. *Adv. Mater.*, **25**, 2790–2796 (2013).

95. Izaki, M., et al. Electrodeposited ZnO—nanowire/Cu_2O photovoltaic device with highly resistive ZnO intermediate layer. *ACS Appl. Mater. Interfaces*, **6**, 13461–13469 (2014).

96. Marin, A. T., et al. Novel atmospheric growth technique to improve both light absorption and charge collection in ZnO/Cu_2O thin film solar cells. *Adv. Funct. Mater.*, **23**, 3413–3419 (2013).

97. Law, M., Greene, L. E., Johnson, J. C., Saykally, R., and Yang, P. Nanowire dye-sensitized solar cells. *Nat. Mater.*, **4**, 455–459 (2005).

98. Baxter, J. B., and Aydil, E. S. Nanowire-based dye-sensitized solar cells. *Appl. Phys. Lett.*, **86**, 053114 (2005).

99. Puyoo, E., Rey, G., Appert, E., Consonni, V., and Bellet, D. Efficient dye-sensitized solar cells made from ZnO nanostructure composites. *J. Phys. Chem. C*, **116**, 18117–18123 (2012).

100. Greene, L. E., Law, M., Yuhas, B. D., and Yang, P. ZnO-TiO_2 core-shell nanorod/P3HT solar cells. *J. Phys. Chem. C Lett.*, **111**, 18451–18456 (2007).

101. Briseno, A. L., et al. Oligo- and polythiophene/ZnO hybrid nanowire solar cells. *Nano Lett.*, **10**, 334–340 (2010).
102. Bi, D., et al. Efficient and stable $CH_3NH_3PbI_3$-sensitized ZnO nanorod array solid-state solar cells. *Nanoscale*, **5**, 11686–11691 (2013).
103. Dong, J., et al. Impressive enhancement in the cell performance of ZnO nanorod-based perovskite solar cells with Al-doped ZnO interfacial modification. *Chem. Commun.*, **50**, 13381–13384 (2014).
104. Zhang, J., Barboux, P., and Pauporté, T. Electrochemical design of nanostructured ZnO charge carrier layers for efficient solid-state perovskite-sensitized solar cells. *Adv. Energy Mater.*, **4**, 1400932 (2014).
105. Conibeer, G., et al. Silicon nanostructures for third generation photovoltaic solar cells. *Thin Solid Films*, **512**, 654–662 (2006).
106. Green, M. A. *Third Generation Photovoltaics*, Springer-Verlag Berlin Heidelberg, New York (2006).
107. Taguchi, M., et al. 24.7% record efficiency HIT solar cell on thin silicon wafer. *IEEE J. Photovoltaics*, **4**, 96–99 (2014).
108. Wolden, C. A., et al. Photovoltaic manufacturing: Present status, future prospects, and research needs. *J. Vac. Sci. Technol. A*, **29**, 03080 (2011).
109. King, R. R., et al. 40% efficient metamorphic GaInPGaInAsGe multijunction solar cells. *Appl. Phys. Lett.*, **90**, 3–5 (2007).
110. Hao, X. J., et al. Effects of phosphorus doping on structural and optical properties of silicon nanocrystals in a SiO_2 matrix. *Thin Solid Films*, **517**, 5646–5652 (2009).
111. Huang, S., and Conibeer, G. Sputter-grown Si quantum dot nanostructures for tandem solar cells. *J. Phys. D. Appl. Phys.*, **46**, 024003 (2013).
112. Ameri, T., Li, N., and Brabec, C. J. Highly efficient organic tandem solar cells: A follow up review. *Energy Environ. Sci.*, **6**, 2390 (2013).
113. Sorel, S., Bellet, D., and Coleman, J. N. Relationship between material properties and transparent heater performance for both bulk-like and percolative nanostructured networks. *ACS Nano*, **8**, 4805–4814 (2014).
114. Ellmer, K., Klein, A., and Rech, B. (eds.). *Transparent Conductive Zinc Oxide: Basics and Applications in Thin Film Solar Cells*, Springer, Berlin [u.a.] (2008).
115. Fortunato, E., Ginley, D., Hosono, H., and Paine, D. C. Transparent conducting oxides for photovoltaics. *MRS Bull.*, **32**, 242–247 (2007).
116. Gordon, R. G. Criteria for choosing transparent conductors. *MRS Bull.*, **25**, 52–57 (2000).

117. Granqvist, C. G. Transparent conductors as solar energy materials: A panoramic review. *Sol. Energy Mater. Sol. Cells*, **91**, 1529–1598 (2007).

118. Ellmer, K. Past achievements and future challenges in the development of optically transparent electrodes. *Nat. Photonics*, **6**, 809–817 (2012).

119. Minami, T. Transparent conducting oxide semiconductors for transparent electrodes. *Semicond. Sci. Technol.*, **20**, S35–S44 (2005).

120. Rey, G., et al. Electron scattering mechanisms in fluorine-doped SnO_2 thin films. *J. Appl. Phys.*, **114**, 183713 (2013).

121. Hecht, D. S., Hu, L., and Irvin, G. Emerging transparent electrodes based on thin films of carbon nanotubes, graphene, and metallic nanostructures. *Adv. Mater.*, **23**, 1482–1513 (2011).

122. Yao, S., and Zhu, Y. Nanomaterial-enabled stretchable conductors: Strategies, materials and devices. *Adv. Mater.*, **27**, 1480–1511.

123. Jo, G., et al. The application of graphene as electrodes in electrical and optical devices. *Nanotechnology*, **23**, 112001 (2012).

124. Langley, D. P., et al. Metallic nanowire networks: Effects of thermal annealing on electrical resistance. *Nanoscale*, **6**, 13535–13543 (2014).

125. Haacke, G. New figure of merit for transparent conductors. *J. Appl. Phys.*, **47**, 4086–4089 (1976).

126. Rowell, M. W., and McGehee, M. D. Transparent electrode requirements for thin film solar cell modules. *Energy Environ. Sci.*, **4**, 131 (2011).

127. Klein, A. Transparent conducting oxides: Electronic structure-property relationship from photoelectron spectroscopy with in situ sample preparation. *J. Am. Ceram. Soc.*, **96**, 331–345 (2013).

128. Charpentier, C., Prod'Homme, P., and Roca I Cabarrocas, P. Microstructural, optical and electrical properties of annealed ZnO: Al thin films. *Thin Solid Films*, **531**, 424–429 (2013).

129. Muthukumar, A., Giusti, G., Jouvert, M., Consonni, V., and Bellet, D. Fluorine-doped SnO_2 thin films deposited on polymer substrate for flexible transparent electrodes. *Thin Solid Films*, **545**, 302–309 (2013).

130. Tsai, C. H., et al. Influences of textures in fluorine-doped tin oxide on characteristics of dye-sensitized solar cells. *Org. Electron. Phys. Mater. Appl.*, **12**, 2003–2011 (2011).

131. Giusti, G., Consonni, V., Puyoo, E., and Bellet, D. High performance ZnO-SnO_2: F nanocomposite transparent electrodes for energy applications. *ACS Appl. Mater. Interfaces*, **6**, 14096–14107 (2014).

132. Muñoz-Rojas, D., et al. Growth of 5 cm^{2}V^{-1}s^{-1} mobility, p-type copper (I) oxide (Cu$_2$O) films by fast atmospheric atomic layer deposition (AALD) at 225 °C and below. *AIP Adv.*, **2**, 042179 (2012).

133. Meyer, B. K., et al. Binary copper oxide semiconductors: From materials towards devices. *Phys. Status Solidi*, **249**, 1487–1509 (2012).

134. Langley, D., et al. Flexible transparent conductive materials based on silver nanowire networks: A review. *Nanotechnology*, **24**, 452001 (2013).

135. De, S., et al. Silver nanowire networks as flexible, transparent, conducting films: Extremely high DC to optical conductivity ratios. *ACS Nano*, **3**, 1767–1774 (2009).

136. Preston, C., Xu, Y., Han, X., Munday, J. N., and Hu, L. Optical haze of transparent and conductive silver nanowire films. *Nano Res.*, **6**, 461–468 (2013).

137. Araki, T., et al. Low haze transparent electrodes and highly conducting air dried films with ultra-long silver nanowires synthesized by one-step polyol method. *Nano Res.*, **7**, 236–245 (2014).

138. De, S., King, P. J., Lyons, P. E., Khan, U., and Coleman, J. N. Size effects and the problem with percolation in nanostructured transparent conductors. *ACS Nano*, **4**, 7064–7072 (2010).

139. Garnett, E. C., et al. Self-limited plasmonic welding of silver nanowire junctions. *Nat. Mater.*, **11**, 241–249 (2012).

140. Tokuno, T., et al. Fabrication of silver nanowire transparent electrodes at room temperature. *Nano Res.*, **4**, 1215–1222 (2011).

141. Bergin, S. M., et al. The effect of nanowire length and diameter on the properties of transparent, conducting nanowire films. *Nanoscale*, **4**, 1996 (2012).

142. Mayousse, C., Celle, C., Fraczkiewicz, A., and Simonato, J. Stability of silver nanowire based electrodes under environmental and electrical stresses. *Nanoscale*, **7**, 2107–2115 (2015).

143. Morgenstern, F. S. F., et al. Ag-nanowire films coated with ZnO nanoparticles as a transparent electrode for solar cells. *Appl. Phys. Lett.*, **99**, 48–50 (2011).

144. Leem, D.-S., et al. Efficient organic solar cells with solution-processed silver nanowire electrodes. *Adv. Mater.*, **23**, 4371–4375 (2011).

145. Kang, M. G., Xu, T., Park, H. J., Luo, X., and Guo, L. J. Efficiency enhancement of organic solar cells using transparent plasmonic Ag nanowire electrodes. *Adv. Mater.*, **22**, 4378–4383 (2010).

146. Hu, L., Hecht, D. S., and Grüner, G. Carbon nanotube thin films: Fabrication, properties, and applications. *Chem. Rev.*, **110**, 5790–5844 (2010).

147. Barnes, T. M., et al. Comparing the fundamental physics and device performance of transparent, conductive nanostructured networks with conventional transparent conducting oxides. *Adv. Energy Mater.*, **2**, 353–360 (2012).

148. Spadafora, E. J., et al. Work function tuning for flexible transparent electrodes based on functionalized metallic single walled carbon nanotubes. *Carbon N. Y.*, **50**, 3459–3464 (2012).

149. Bonaccorso, F., Sun, Z., Hasan, T., and Ferrari, A. C. Graphene photonics and optoelectronics. *Nat. Photonics*, **4**, 611–622 (2010).

150. Bae, S., Kim, S. J., Shin, D., Ahn, J.-H., and Hong, B. H. Towards industrial applications of graphene electrodes. *Phys. Scr.*, **T146**, 014024 (2012).

151. Lee, M.-S., et al. High-performance, transparent and stretchable electrodes using graphene-metal nanowire hybrid structures. *Nano Lett.*, **13**, 2814–2821 (2013).

152. Solar, P., et al. Laminated carbon nanotube networks for metal electrode-free. *ACS Nano*, **8**, 6797–6804 (2014).

153. (Erwin) Kessels, W. M. M., and Putkonen, M. Advanced process technologies: Plasma, direct-write, atmospheric pressure, and roll-to-roll ALD. *MRS Bull.*, **36**, 907–913 (2011).

154. Leskelä, M., and Ritala, M. Atomic layer deposition (ALD): From precursors to thin film structures. *Thin Solid Films*, **409**, 138–146 (2002).

155. Leskelä, M., and Ritala, M. Atomic layer deposition chemistry: Recent developments and future challenges. *Angew. Chem. Int. Ed.*, **42**, 5548–5554 (2003).

156. Puurunen, R. L. Surface chemistry of atomic layer deposition: A case study for the trimethylaluminum/water process. *J. Appl. Phys.*, **97**, 121301 (2005).

157. Clavel, G., Rauwel, E., Willinger, M.-G., and Pinna, N. Non-aqueous sol–gel routes applied to atomic layer deposition of oxides. *J. Mater. Chem.*, **19**, 454 (2009).

158. George, S. M. Atomic layer deposition: An overview. *Chem. Rev.*, **110**, 111–131 (2010).

159. Miikkulainen, V., Leskelä, M., Ritala, M., and Puurunen, R. L. Crystallinity of inorganic films grown by atomic layer deposition: Overview and general trends. *J. Appl. Phys.*, **113**, 021301 (2013).

160. Meng, X., Yang, X.-Q., and Sun, X. Emerging applications of atomic layer deposition for lithium-ion battery studies. *Adv. Mater.*, **24**, 3589–3615 (2012).

161. Cassir, M., Ringuedé, A., and Niinistö, L. Input of atomic layer deposition for solid oxide fuel cell applications. *J. Mater. Chem.*, **20**, 8987 (2010).

162. Bakke, J. R., Pickrahn, K. L., Brennan, T. P., and Bent, S. F. Nanoengineering and interfacial engineering of photovoltaics by atomic layer deposition. *Nanoscale*, **3**, 3482–3508 (2011).

163. Van Delft, J. A., Garcia-Alonso, D., and Kessels, W. M. M. Atomic layer deposition for photovoltaics: Applications and prospects for solar cell manufacturing. *Semicond. Sci. Technol.*, **27**, 074002 (2012).

164. Ritala, M., and Leskelä, M. in *Handbook of Thin Film Materials* (Nalwa, H. S., ed.), Academic Press, pp. 103–159 (2002).

165. Poodt, P., et al. Spatial atomic layer deposition: A route towards further industrialization of atomic layer deposition. *J. Vac. Sci. Technol. A Vacuum Surf. Film*, **30**, 010802 (2012).

166. Suntola, T. S., Pakkala, A. J., and Lindfors, S. G. Apparatus for performing growth of compound thin films. US Patent 4,389,973 (1983).

167. Suntola, T. S., and Antson, J. Method for producing compound thin films. US Patent 4,058,430 (1977).

168. Yersak, A. S., Lee, Y. C., Spencer, J. A., and Groner, M. D. Atmospheric pressure spatial atomic layer deposition web coating with in situ monitoring of film thickness. *J. Vac. Sci. Technol. Vacuum Surf. Film*, **32**, 01A130 (2014).

169. Muñoz-Rojas, D., et al. High-speed atmospheric atomic layer deposition of ultra thin amorphous TiO_2 blocking layers at 100 °C for inverted bulk heterojunction solar cells. *Prog. Photovoltaics Res. Appl.*, **21**, 393–400 (2013).

170. Muñoz-Rojas, D., and MacManus-Driscoll, J. Spatial atmospheric atomic layer deposition: A new laboratory and industrial tool for low-cost photovoltaics. *Mater. Horizons*, **1**, 314–320 (2014).

Chapter 3

Low-Cost Electricity Production from Sunlight: Third-Generation Photovoltaics and the Dye-Sensitized Solar Cell

Nadia Barbero[a] and Frédéric Sauvage[b,c]

[a]*University of Torino, Department of Chemistry and NIS Interdepartmental Centre, Via Pietro Giuria 7, 10125, Torino, Italy*
[b]*Laboratoire de Réactivité et Chimie des Solides, Université de Picardie Jules Verne, CNRS UMR 7314, 33 rue Saint Leu, 80039 Amiens, France*
[c]*Institut de Chimie de Picardie (ICP), CNRS FR 3085, 33 rue Saint Leu, 80039 Amiens, France*

frederic.sauvage@u-picardie.fr

3.1 Introduction

The photovoltaic effect was first discovered by the French scientist Antoine Cesar Becquerel in 1839. This effect was twice described at the Academie des Sciences in Paris by his son Alexandre Edmond [1]. They revealed an intriguing phenomenon corresponding to the generation of a current flowing when setting two platinum plates in contact with an aqueous acidic galvanic cell under illumination. To enable this new physical occurrence, one of the two electrode surfaces was modified beforehand by a halide vapor (iodide, chloride, and bromide). Depending on the experimental procedure

---

Materials for Sustainable Energy Applications: Conversion, Storage, Transmission, and Consumption
Edited by Xavier Moya and David Muñoz-Rojas
Copyright © 2016 Pan Stanford Publishing Pte. Ltd.
ISBN 978-981-4411-81-3 (Hardcover), 978-981-4411-82-0 (eBook)
www.panstanford.com

that varies this halide-based layer thickness, they found a photocurrent flowing in one or the other direction, which is the premise of an n- or p-type film depending upon the deposition conditions. For a long time, this starting point of the photovoltaic effect was only considered a scientific curiosity. A mechanistic explanation of this effect was given by A. Einstein in 1905 through his general theory. He earned his first Nobel prize in Physics in 1921 for the explanation of the photoelectric effect. Nobody envisioned at that time that the field of photonics would grow into a new area combining abundant fundamental and applied research from which the underlying technologies would become an important stake for the human being.

The starting point for PV applications goes back to the Cold War, when one considers perhaps the irrational race to become the first nation to land on the Moon. This is under the support of billions of dollars worldwide that Chapin Pearson and Fuller from Bell Laboratories pioneered the silicon p-n technology. This new architecture paved the way for the substantial enhancement in the light-to-electricity power conversion efficiency (PCE) compared to the selenium materials that were exhibiting at best *ca.* 1% efficiencies. The main objective was to make use of this new p-n technology to supply energy to space vehicles by relying on the extraterrestrial sun power. With its 1.1 eV indirect bandgap affording panchromatic visible light absorption, they first demonstrated 4.5% in 1954, 6% a few months later, and even 8% power conversion efficiency as reported in the original patent published in 1957 [2, 3]. This is one of the rare cases in which within the span of one year the technology has progressed from a scientific discovery to a real commercial product distributed by Western Electric (USA). They were the first to commercialize a PV device to a "large" public with small systems of 14 mW for $25 that exhibited 2% conversion efficiency ($1500 per W).

Photovoltaic research reached wide public awareness following the 1973 and 1979 petrol (and financial) crisis. Due to the increase in the price of crude petroleum and greater public awareness of the limitations of petroleum in terms of its cost and availability, there is a corresponding increase in research effort on alternative low-cost energy production and storage technologies. This has largely contributed not only to the progress of PV but also on electrochemical battery storage systems, which was stimulated by the discovery of alkali insertion compounds used in current lithium and lithium-ion technologies. The most promising low-cost PV technology is the third generation of PVs developed at the end of the 1980s. Organic photovoltaics (OPV) and hybrid organic/inorganic dye-sensitized solar cells (DSSCs) are fulfilling not only such low-cost requirements but also easy processing favorable for technological transfer and low environmental footprint. However, what history tends to forget is that the forward thinking Elliot Bermann already envisioned OPVs back in the late 1960s. He was supported by Exxon Company, which included the prospection for PV cutting-edge technology in their research program. They were able to rapidly cut the energy cost by five times within the first two years (from 100 to 20 \$/W_p).

This chapter is dedicated to the third generation PVs, including organic PVs (OPV) and dye-sensitized solar cells, with a particular emphasis on the latter as it probably has the most promise in passing successfully all the requirements of the IEC61646 standard protocol for accelerated ageing tests [4]. Therefore, an in-depth review will be provided to the reader on the current progress and knowledge regarding the stability of this technology. The chapter is composed of two parts: First, a short discussion offers the reader a brief overview of the OPV technology, including a description of how it functions and the challenges that remain to be overcome, and second, a more in-depth discussion related to the operation of the dye-sensitized solar cell at the dawn of its commercialization, progress in electrolyte stability, development of organic and organometallic sensitizers, and finally a current understanding of the chemical degradation in this technology.

3.2 Basics of Organic Photovoltaics

Organic photovoltaics (OPV) is a technology that converts sunlight into electricity by making use of an organic semiconducting junction between donor and acceptor entities. Research on OPV materials and devices has flourished in recent years due to their potential for offering low-cost solar energy conversion. Respectable solar conversion efficiencies over 10% have been achieved as a result of a deepened understanding of the fundamental photovoltaic processes in organic electronic materials and the development of tailored materials and device architectures. This clearly attracts tremendous commercial interests for further development and manufacturing. Indeed, organic semiconductors show great promise for photoconversion displays through their synthetic design capability, their low temperature processing similar to that applied to plastics, offering lightweight, flexible, easily manufacturable materials and low-cost solar cells.

Studies on the photovoltaic behavior of organic solids can be traced back to the 1950s and 60s [5–7] and the power conversion efficiencies of these early OPV devices were generally significantly below 0.1% [8]. The fundamental understanding of the chemical, electrical, and photonic properties of these materials were limited for much of this time.

However, Tang's organic donor–acceptor heterojunction (HJ) device is considered as the foundation for all efficient OPV devices [9]. Since then, the performance of OPV devices, including the efficiency and stability, has steadily improved due to new materials and novel device architectures, with efficiency now exceeding 10% for both polymer- and small molecule-based OPV devices [10, 11].

A typical planar OPV device consists of two photoactive materials sandwiched between two current collectors (Fig. 3.1) [12]. Generally, it consists of a transparent electrode, i.e., a conductive oxide such as indium tin oxide (ITO), two organic light-absorbing layers of opposing electronic properties, donor on the one hand and acceptor on the other hand, and a second electrode. A significant improvement of OPV device performance has been accomplished by introducing various OPV architectures, such as bulk-heterojunction (BHJ) and inverted device structures, and developing low bandgap conjugated polymers and innovative organic small molecules as donor materials [13].

The device architecture is particularly more challenging than other technologies based on inorganic (Si, CIGS, GaAs…) or hybrid materials (DSSC). One of the reasons comes from the low dielectric constant of the organic materials compared to their inorganic counterparts ($\varepsilon \approx 3.5$–5.5) [14–16], which hinders the process of photoconversion; in other words, the electron–hole pairs in an OPV are more strongly bounded, consequently reducing the charge separation lifetime compared to more conventional PV technologies.

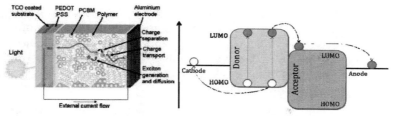

Figure 3.1 Organic bulk heterojunction photovoltaic device made of the classical PEDOT:PSS/PCBM donor/acceptor materials (Left), and the basic mechanism of the photoelectrochemical charge separation and collection processes (Right).

It was also realized in the original studies aiming at understanding the fundamental of OPVs that the photovoltaic behavior is in some ways fundamentally different from that of conventional technologies [8, 17, 18]. The production of photocurrent and photovoltage can be achieved in the absence of any junctions, either p-n or Schottky. Photocurrent action spectra indicated a process that is controlled by light absorbed closed to the surface of the absorber rather than in its bulk. This was latter ascribed correctly to the photoinduced creation of excitons, i.e., electrically neutral quasi-particles consisting of bound electron–hole pairs, favored again by the low dielectric constant of organic absorbers, and their subsequent diffusion towards the illuminated surface and its dissociation.

The mechanism of the photon-to-electron conversion process and recent progress in OPVs are excellently reviewed in literature [14, 19, 20] and can be summarized in four steps:

(1) Upon the absorption of an incident photon, an electron in the donor material undergoes photoinduced excitation from the highest occupied molecular orbital (HOMO) to the lowest unoccupied molecular orbital (LUMO), forming a strongly bounded Frenkel exciton (e^-) and hole (h^+). Thus, the organic material serves as an electron donor while a second material is needed to serve as an electron acceptor. The most widely used acceptor materials are fullerenes, which have electron affinities greater than those of polymers or small molecules.

(2) The formed excitons diffuse to the donor–acceptor (D–A) interfaces and separate into free holes (positive charge

carriers) and electrons (negative charge carriers). This phenomenon is typically referred to as the charge separation process. One requirement for approaching unity quantum efficiency is to have an excitonic diffusion length (L_D) as short as possible to prevent recombination to the ground state.

(3) Excitons at a D–A interface undergo a charge transfer (CT) process to form a CT exciton. The hole and electron remain in the donor and acceptor phases, respectively, held together through relatively weak coulombic attraction.

(4) The CT exciton dissociates into free holes and electrons, which are then transported through the donor and acceptor phases, respectively, to their respective electrodes (anode and cathode). This process is expressed in terms of the charge collection efficiency (η_{cc}), defined as the ratio between the number of carriers collected and the number of excitons that have undergone the CT process. A photovoltage is generated when the holes and electrons move to the corresponding electrodes by following either the donor or the acceptor phase (charge extraction).

Donor and acceptor materials in an OPV device need to have high molar extinction coefficients for light harvesting, high chemical and excited state stabilities, and good integration/compatibility of the donor and acceptor to form a perfect interface quality free of any cracks. Film morphologies exert a very strong impact on resulting PV characteristics. This latter plays a key role on the charge collection efficiency as a result from the low dielectric constant of the organic semi-conductors. Since the donor molecule plays a critical role as the absorber of solar photon flux, it should have a wide optical absorption to match the solar spectrum. Another basic requirement for an ideal donor/acceptor is a large hole/electron polaron mobility to maximize charge transport towards charge collection. This requirement calls for the creation of a continuous interpenetrating network with optimal domain sizes. In tandem, this demands very specific characterization tools to reveal thin film nanostructure, such as based on grazing x-ray scattering (e.g., GIWAXS and GISAXS, R-SoXS), small angle neutron scattering (SANS), and more general tools such as AFM, NMR and TEM. A careful and well-developed up-to-date review focused on the importance of the heterojunction nanostructure can be found in Ref. [21].

Donor materials are usually conjugated polymers based on electron push/pull molecular units in the conjugated main chain in order to maximize photon harvesting. Among them, the most used is P3HT due to its easy synthesis, high charge carrier mobility, and good processability. As an alternative to the polymer, solution-processed conjugated small molecule-based solar cells have attracted much attention in the last few years due to their reproducible fabrication protocols and a better understanding of structure–property relationships [22, 23]. A wide range of molecules has been explored and studied, such as oligothiophenes, phthalocyanine, merocyanine, squaraine, diketopyrrolopyrroles, borondipyrromethene, perylene diimides, fused acenes, and triphenylaminederivatives [24–27]. On the other hand, acceptor materials are generally fullerene derivatives, such as $PC_{61}BM$ and $PV_{71}BM$, or fullerenes blended with polymers. The main types of molecules are reported in Fig. 3.2.

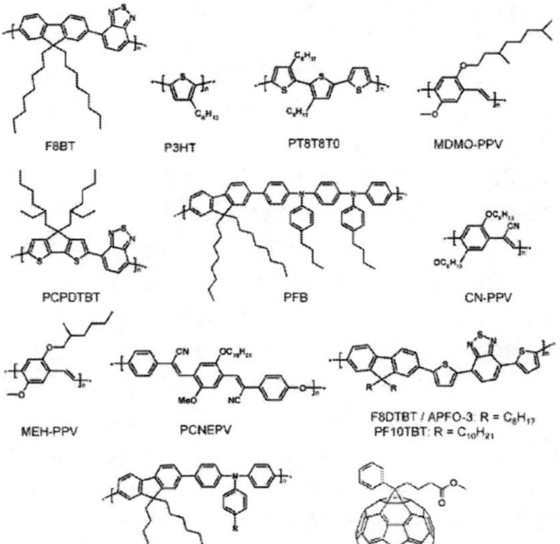

Figure 3.2 Chemical structures of the main types of small molecules and polymers integrated in OPV devices. Names for the small molecules and polymers are presented in the tabular form at the end of the chapter.

The origin of the photovoltage in OPV cells has been the subject of much discussions since values as high as 400 mV have been found, which can be generated in symmetrical OPV cells (i.e., small molecules and polymer sandwiched between the two current collectors) [19, 28]. The hidden asymmetry necessary to separate the charge carriers in this case derives from two factors: (i) illumination of only one side of the photoactive film (ii) the interfacial exciton dissociation process that preferentially injects only one carrier type into the electrode. In a donor–acceptor architecture, the quasi-Fermi-level description, although to some extent is simplistic, affords a good understanding of the photovoltage. In this case, it corresponds to the energy difference of the quasi-Fermi level between the acceptor and the donor material. The one-dimensional quasi-Fermi levels of electrons and holes, respectively, are expressed as

$$qE_{fn,A} = qE_{cb,A} - kT \ln (n_A(x)/N_{c,A})$$

$$qE_{fp,D} = qE_{vb,D} - kT \ln (p_D(x)/N_{v,D}),$$

where $n_A(x)$ and $p_D(x)$ are the electron and hole concentrations in the acceptor and donor, respectively, and $N_{c,A}$ and $N_{v,D}$ are the corresponding carrier densities.

With various optimization strategies developed in the last few years, the maximum power conversion efficiency of OPV devices has exceeded 10% in laboratory cells [29, 30] and 7–8.5% in modules [31]. The thermodynamic limit of PCE in OPVs was recently reported to be around 20–27% as stated by Forrest et al. and Janssen and Nelson [32, 33]. Aside from efficiency, another issue that should be tackled before their commercialization is the stability of OPV devices. Although OPV devices with a lifetime over seven years have been demonstrated [34], a more robust encapsulation technology and more detailed studies about the degradation mechanism are required to improve device stability. On the other hand, the unquestionable advantage of OPV technology is its ability to be made into large area and flexible solar modules by allowing roll-to-roll production. Additionally, manufacturing cost can be reduced for organic solar cells due to their lower cost compared to silicon-based materials and the ease of device manufacturing. Unfortunately, whereas encapsulation is a critical issue to avoid water ingress, the stability of OPV devices remain too poor to pass the IEC61646 performance standard, which is arguably the most challenging hurdle facing OPV's commercial adoption. (IEC61646 protocol: the main three independent steps are 80% power conversion efficiency retention after 1000 h storage at 60 °C under 100 mW/cm^2 light soaking, 80% retention after 1000 h at 85 °C/85% hygrometry in dark, 80% retention after 200 cycles of thermal shocking between −40 and 85 °C.)

3.3 Dye-Sensitized Solar Cell Principle

The working principle of the DSSC is vastly different compared to other technologies in the sense that polycrystalline materials of large bandgaps are used, high purity of the materials are not necessarily required, nanocrystalline semiconductors contain large amounts of point defects, the device relies on electrochemical processes, exciton separation and transport are separated processes, and that finally no internal electric fields lead to the charge separation. The original idea of providing light sensitivity to larger bandgap semiconductors has more than one hundred years of history. It was prompted by Prof. Vogel who pioneered dye-sensitized silver halide emulsions for practical use in argentic photographic films. This principle has been shifted from its primary scope to energy-related applications, in particular by Gerischer and Tributsch in 1968 and 1972, who used ZnO single crystals sensitized by chlorophyll [35–37]. A new impetus was given by Brian O'Regan and Michael Grätzel, who published the sensitization of a mesoscopic film of nanocrystalline anatase TiO_2 particles using red heteroleptic ruthenium(II) polypyridyl complexes to achieve a remarkable enhancement in the light harvesting properties of the photoanode [4].

Dye-sensitized solar cells offer the potential to combine a low solar energy conversion cost, facile processability and high photoelectrochemical stability of the system with the fact that Ru(II) polypyridyl complexes in association with TiO_2 or ZnO can enable turnover numbers above 10^7–10^8 for the electron injection from the excited dye state to the electron acceptor levels in the semiconductor, which in turn would guarantee more than 10 years lifetime of the panel [38]. Rapidly passing the threshold of 10%

PCE in 1993 [39], despite endless efforts on material development, cell efficiency has stagnated for a long time in the range of 11.0–11.5% [40–45] before scoring four new records subsequently, first 12.3 and 13.0% under AM 1.5G conditions (100 mW/cm^2) using organic dye molecules in association with a stronger redox cobalt polypyridyl complex [46, 47], and more recently, certified 14.1% and even certified 17.9% PCE achieved by means of a hybrid organic/inorganic lead halide perovskite absorber combined with a solid hole transporting material (HTM) [48, 49]. This black absorber combined with an HTM is actually paving the way for a new technology separately named as perovskite solar cells for which progress is remarkably fast (21% achieved in December 2015).

Besides the excellent performance in terms of light-to-electricity power conversion efficiency, which clearly meets the standard requirement for market introduction, the race for achieving the highest PCE outshined in many aspects the requirements for high stability and long lifetime for mass industrialization. The fewer number of publications in this area are evidence of this disparity. Yanagida et al. in 1980 already raised issues of dye stability in devices artificially mimicking the photosynthetic process such as dye-sensitized solar cells [50].

A schematic representation of the operating principle of the DSSC is given in Fig. 3.3. At the heart of the electrochemical system is a mesoscopic layer of anatase TiO_2 sintered together to provide efficient electronic conduction. The layer is typically composed of well-dispersed 20–30 nm size particles of anatase TiO_2 sheltered by a scattering layer of *ca.* 400 nm particles for which the role is to scatter unabsorbed photons. Attached to the surface of the nanocrystalline film is a monolayer of the light sensitive dye. Photo-excitation of the latter induces charge separation, ultra-fast femto/picosecond electron injection into the conduction band of the oxide and hole capture by the reductive redox species composing the electrolyte. This reaction, called regeneration, enables the recovery of the original oxidation state of the dye. For a sufficient concentration of reductive species, this reaction is sufficiently fast to prevent the recapture of the conduction band electron by the oxidized dye and as we will see in the following to prevent/limit dye degradation.

The role of the dye is then pivotal to the cell's operation. It manages the light absorption ability of the device, and in contact with TiO_2 and the electrolyte governs the charge separation properties at the molecular level. The main family of redox couples used in DSSCs is the two-electron system I_3^-/I^-. Subsequent to dye regeneration, tri-iodide diffuses towards the counter-electrode, collects two electrons from the external circuit in order to form back the iodide. The counter-electrode is typically supported by platinum nanoparticles to reach fill factors in the range of 65 to 75%. Platinum plays the role of an electrocatalyst to lower the charge transfer resistance associated with the energy barrier for I–I bond breaking.

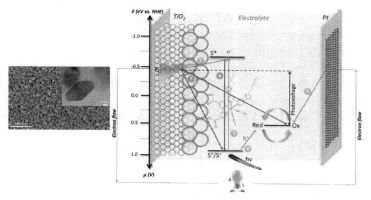

Figure 3.3 Operating principle of the dye-sensitized solar cell, including the favorable charge transfer pathways (in blue) and the loss pathways (in red). Also included is a top view scanning electron micrograph of a typical mesoporous TiO_2 film and a high-resolution transmission electron micrograph of a TiO_2 nanocrystal.

3.3.1 Semiconductor and Electrolyte Development for Dye-Sensitized Solar Cells

In the development of dye-sensitized solar cells, alternatives to anatase TiO_2 have been suggested. Investigations have primarily been devoted to other binary metal oxides materials. The most obvious has been to explore the two other principal polymorphs of TiO_2: brookite [51] and rutile, which the latter can enhance the scattering effect of the reflecting layer from its higher refractive index [52]. Alternatively, ZnO [53–72], SnO_2 [73–77], Nb_2O_5 [78–84] and In_2O_3 [85] have also been explored. Substituted ternary metal oxides have also been tailored at the nanoscale such as the two perovskites $SrTiO_3$ [86–88] and $BaSnO_3$ [89] or the spinel $ZnSn_2O_4$ [90, 91].

Although such effort to further improve the characteristics of the photoanode, in particular to enhance charge collection efficiency and tune the energy of the conduction band edge position, anatase TiO_2 remains so far the leading contender photoanode material for highly efficient devices. For these reasons, another area of research concentrates on anatase TiO_2 in optimizing the particle texture or by voluntarily introducing further point defects through aliovalent doping. This will be further discussed in the following of this chapter. The reduction of particle dimensionality offers faster electron transport when the photoanode is composed of vertically aligned nanowires or nanotubes [92–100]. But the use of TiO_2 beads leads to the most substantial improvement stemming from excellent particle interconnectivity, shorter mass transport pathways into the mesopores and a combination of light scattering ability and high surface area, thus offering excellent optical characteristics of the photoanode with enhanced light confining properties [44, 101–103]. When optimized beads are employed, remarkable efficiencies exceeding 11% (A.M. 1.5 G) can be obtained in combination with high molar extinction coefficient C101 or C106 ruthenium dyes.

The strategy to introduce point defects relies on the fact that Ti^{4+} adopts a $3d^0$ electronic configuration, which is recognized to have particular sensitivity to cationic and/or anionic doping. The literature on this topic as applied to DSSCs is increasingly extensive, although this approach had for a long time been viewed as undesirable as it may lead to an increase of the density of free carriers, which is equivalent to an increase in recombination. However, in the case of TiO_2, when close to thermodynamic synthetic methods are employed, charge compensation mechanisms take place by means of oxygen vacancy formation or oxygen intake in interstitial sites leading, regardless of the type of dopant, to the preserving of n-type characteristics and low carrier concentration [104]. We can enlist a series of hypervalent cations that have been successfully incorporated into the anatase crystal structure, namely W^{6+} [105], Nb^{5+} [106–109] and Ta^{5+}[110], subvalent cations like Zn^{2+} [111], Cr^{3+} [112], Fe^{3+} [113], Sc^{3+} [114], Y^{3+} [115] and Ga^{3+} [116], or isovalent cations such as Sn^{4+} [117] and Ce^{4+} [118]. The common observation is that hypervalent doping affects the energy of acceptor trap states. A careful control of dopant concentration can therefore adjust trap energetics, and thus in some cases promote ultra-fast charge injection. In contrast, subvalent dopants do not influence the energy nor the distribution of traps (Fig. 3.4) [115–116]. With the appropriate dopant type and concentration, an improvement in charge collection efficiency can be achieved, thus potentially contributing to higher power conversion efficiency, although the efficiencies reported to date remain below 10%.

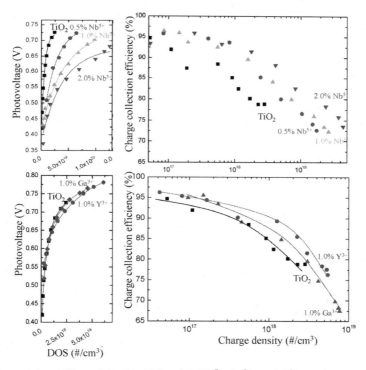

Figure 3.4 Effect of doping TiO_2 with Nb^{5+}, Ga^{3+}, and Y^{3+} on the energy and distribution of trap states, and charge collection efficiency.

As for the electrolyte, attempts to replace the iodine/iodide redox couple have been motivated by a need to reduce the important energy loss (ca. 600 mV) resulting from the energy alignment mismatch between the HOMO level of the dye and the chemical potential of iodine/iodide. It is also motivated by the strong corrosive character of this redox couple as well as its deep orange coloration limiting the conversion of the blue part of the solar spectrum. Figure 3.5a–c shows the actual redox molecules and hole transporting materials successfully introduced in DSSCs as a function of their redox potential. The closely related two-electron Br_3^-/Br^- redox couple has been proposed despite its high corrosive character [119–121]. In combination with carbazole-based sensitizers, Hagfeldt et al. achieved voltages beyond the threshold of 1 V. In particular, relevant for quantum dot-sensitized solar cells, the sulfide/polysulfide redox couple has been proposed [122] and an alternate disulfide/thiolate showed appealing features in combination with ruthenium or organic dyes [123–125]. Other one-electron redox systems have been investigated using metal organic complexes based on cobalt (II/III) [126–131], copper [132], iron [133], or nickel [134]. Fast redox couples based on nitroxide have been proposed by Grätzel et al. [135]. Finally, alternatives to redox molecules have been pioneered by Grätzel et al., using hole transporting materials based on p-type polymeric or organic materials leading to solid-state DSSC [136], the leading contender being the so-called Spiro-OMeTAD (2,2',7,7'-tetrakis(N, N-di-p-methoxyphenyl-amine)9,9'-spirobifluorene). Also listed in Fig. 3.5c, solid or liquid, amorphous or crystallized organic and inorganic hole transporting materials have been suggested even though they typically exhibit lower performances compared to Spiro-OMeTAD [137–140].

The output voltage is limited in principle by the absorption bandgap value of the dye, which lies in a broad range from 1.5 to 2.5 eV. The photovoltage generated corresponds to the energy difference between the chemical redox potential of the mediator in the electrolyte and the Fermi level position in the semiconductor placed under illumination. The integration of cobalt complexes have been beneficial by increasing significantly the photovoltage of champion devices, therefore breaking new records in power conversion efficiency. It is interesting to note that for the case of DSSCs in breaking records by means of photovoltage improvement is similar to single junction technology for which new records were systematically a result of photovoltage improvements.

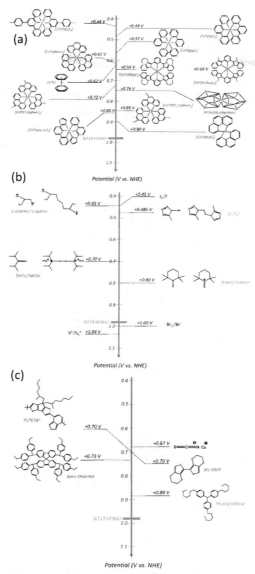

Figure 3.5 Potentials for the main redox complexes, redox molecules, and hole transporting materials used in dye-sensitized solar cells.

3.3.2 Dye Development with Molecular Engineering

The central role of the sensitizing dye molecule places demands on its chemical, electrochemical, and photoelectrochemical properties. The dye should obviously have chemical and photoelectrochemical stability. It should possess (i) a high molar extinction coefficient with panchromatic light absorption ability, (ii) perfect HOMO/LUMO energy alignment with respect to the redox couple (or alternatively the working function of the p-type transporting material) and the acceptor levels in the semiconductor.

The pioneering work at EPFL presented a series of ruthenium polypyridyl complexes, among them the ubiquitous N719 and Z907 dyes are still the benchmark sensitizers today [141]. After more than 20 years of research to develop the ideal sensitizer with high molar extinction coefficient and panchromatic response, two main types (besides perovksites) have been developed: the organometallic complexes based on ruthenium or osmium, and the organic chromophores.

In the following, some of the more pertinent classes of sensitizers for DSSCs and their latest results will be reviewed. Innovative new aspects of DSSC sensitization will also be briefly discussed.

Pyridine-Based Ruthenium Dyes

Among the photosensitizers used in DSSCs, ruthenium (Ru) complexes have been widely investigated and have shown some of the best photovoltaic properties. They are exhaustively reviewed in literature [142–146]. Recent advances as well as synthetic methods, properties and main applications of most relevant homoleptic and heteroleptic ruthenium complexes are described in detail in a minireview by Palomares's group [146] and what follows is a short summary of the molecular modifications, starting from N3, the first photosensitizer that is able to attain a solar-to-electric energy conversion efficiency of 10%. Subsequent modifications were introduced in order to overcome the limitations of N3 and to achieve better performance and panchromatic photosensitizers. Ru complex N3 is a homoleptic cis-$(SCN)_2$bis(2,2'-bipyridyl-4,4'-dicarboxylate)ruthenium(II) complex and exhibits a broad visible light absorption spectrum, an incident power-to-conversion efficiency (IPCE) spectrum up to 800 nm, and strong adsorption on the semiconductor surface due to binding with up to four carboxyl groups. Most of the following complexes have been designed around modifications to the N3 structure as the protoypical Ru complex (Fig. 3.6).

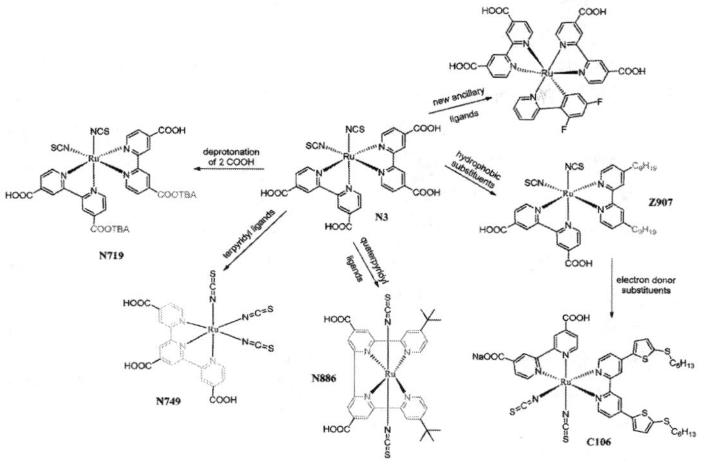

Figure 3.6 A brief summary of the structural modifications of Ru complexes from N3 up to panchromatic terpyridine (N749) and quaterpyridine (N866) complexes.

For instance, the well-known N719 Ru complex, which shows a 11.18% efficiency [147], is the deprotonated version of N3 where two carboxylic acid groups are replaced by the deprotonated carboxylate counterpart using tetrabutylammonium hydroxide. The efficiency of the solar cell is thus influenced by changing the protonation of the acid groups. The effect of protons carried by the sensitizer on cell performance has been reported by studying the photocurrent–voltage characteristics of a nanocrystalline TiO_2 cell sensitized with N3 (4 protons), N719 (2 protons), N3[TBA]$_3$ (1 proton) and N712 (zero proton) dyes measured under AM 1.5G sun [148].

To enhance the conversion efficiencies of Ru complexes, the N3 structure can be modified by keeping one of the 2,2′-bipyridyl-4,4′-dicarboxylate ligands and adjusting the ancillary 2,2′-bipyridyl ligand with different substituents (alkyl, alkoxyl, phenylene, etc.) [149–151]. This leads to an increase in the molar extinction coefficient, suppression of aggregation on the semiconductor, and optimization of the redox potential. The introduction of long dinonyl alkyl chains leads to the hydrophobic dye Z907 [151]. The advantages of this dye are higher thermal stability, lower solubility in polar solvent-based electrolytes, and an enhanced conversion efficiency under long-term thermal stress measurements owing to the lower dark current, namely reduction of iodine on the surface of the photoanode corresponding to the reduction pathway 7 in Fig. 3.3. This molecular engineering approach was applied to obtain other Ru complexes by varying the nature of the ligands, e.g., hydrophobic (in Z907) or electron donor groups [151], that provided complexes with greater conjugation, red-shift of absorption maxima and higher molar extinction coefficients. For example, a triarylamine moiety inserted on one bipyridyl ligand caused a 1000-fold retardation of the recombination dynamics in comparison with N719 [152].

Among electron-rich heteroaromatic conjugated bipyridine-based ruthenium sensitizers for efficient DSSCs, thiophene-based ancillary ligands significantly enhance performance, e.g., the short-circuit photocurrent can be raised to 19.7 mA/cm^2 with cell efficiencies reaching up to 11.4% with the so called "supersensitizer" C106 bearing a 2-(hexylthio)thiophene conjugated bipyridine [41]. These results paved the way for the synthesis of a series of efficient heteroleptic and homoleptic ruthenium sensitizers with new bipyridine ligands carrying donor heteroaromatic end groups.

It is worth noting that all these complexes show improved stability (due to the hydrophobic nature of the substituents on the bipyridyl ligand) and higher J_{SC} but lower V_{OC} compared with N719. However, N3 and N719 show low IPCEs in the red and near-infrared (NIR) regions of the electromagnetic spectrum. For this reason, the control of HOMO and LUMO levels is required to develop better red-absorbing dyes. In order to get such a property, another molecular engineering approach was undertaken: the bipyridyl ligands were replaced by a tricarboxylate terpyridyl ligand. The complex obtained is the N749, usually called "black dye" for its red shift in the metal-to-ligand charge transfer band (MLCT) [153]. An IPCE spectrum was obtained over the whole visible range with extended absorption into the near-IR region up to 920 nm, and very high J_{SC} (above 22 mA/cm^2) and a 10.4% conversion efficiency were achieved under AM 1.5 G and certified 11.1% by Sharp [40].

Following the results obtained with the black dye, further developments were realized by replacing the terpyridine by a quaterpyridyl ligand leading to the synthesis of N886 and N1044 [154, 155]. The synthesis of these structures were challenging and above all called for thorough purification of the complexes [156]. The obtained dyes show absorption bands across the entire visible and NIR regions of the spectrum with good conversion efficiencies.

Another aspect to take into account is that, for almost 20 years, ruthenium complexes have included thiocyanate ligands. Many attempts to replace the thiocyanate donor ligands were made, because, from the chemical stability point of view, the monodentate NCS group is believed to be the weakest part of the complex. Another motivation to replace the thiocyanate unit is to destabilize the t_{2g} orbitals of the ruthenium by introducing a donating unit stronger than the thiocyanate. However, these efforts have yielded to a large extent only limited success. A breakthrough was made by Grätzel's group in 2009 describing a novel thiocyanate-free cyclometalated ruthenium complex identified as YE [157]. The lowest energy MLCT band is found to red-shift by 25 nm and contains a new absorption band at 485 nm with a remarkably high molar extinction coefficient. This new band is characterized by an electronic transition between the HOMO, which has a sizable π-orbital contribution from the cyclometallated ligand, and the LUMO formed by a set of π^* orbitals localized on the bipyridine ligands that provide strong electronic coupling with acceptor states in the TiO_2. These features help to explain the high photocurrent of around 17 mA/cm^2 but also the remarkable photovoltage exceeding the threshold of 800 mV that leads to a 10.1% conversion efficiency.

This represents the start of the development of a novel generation of robust and panchromatic ruthenium complexes that is expected to further enhance DSSC performance in terms of stability and efficiency.

Last but not least, another feasible approach in tuning the spectral response of the dye is to modify the metal center by either the 5d metal osmium [158] or 3d copper [159] in which spin-forbidden transitions could play a role in extending the spectral response in the NIR region for photoelectrochemical applications.

However, these complexes show inferior overall efficiency but superior photochemical stability compared to the black dye.

Porphyrin Dyes

Porphyrins are large π-aromatic molecules and represent an important class of sensitizers for DSSC applications. They are in general photostable and exhibit high light-harvesting capabilities, which paves the way for thin film and low-cost PV devices.

In recent years, porphyrins have received significant attention as sensitizers in DSSCs owing to their suitable LUMO and HOMO energy levels and very strong absorption of the Soret band in the 400–450 nm region as well as the Q band in the 500–700 nm region. Moreover, they can be appropriately functionalized at the four meso or eight β positions at the periphery or at the metal center of the macrocycle, allowing a rational molecular design and synthesis in order to tune their photophysical and electrochemical properties.

At present, porphyrins have achieved a record efficiency of as high as 13% under standard AM 1.5G conditions without the requirement of any co-sensitization [160]. The history of porphyrins in DSSCs has shown that they have not always been a success. The ability of porphyrins to sensitize the TiO_2 semiconductor was demonstrated by Grätzel in 1987 [161]. A few years later in 1993, Kay and Grätzel first introduced porphyrin as a sensitizer in DSSCs, specifically using a copper-based porphyrin substituted with propionic acid as anchoring groups to achieve an overall efficiency of 2.6% (structure 1 in Fig. 3.7) [162]. Since then, numerous porphyrins have been designed for sensitizing DSSC without achieving better efficiencies compared with typical Ru dyes (around η = 4–5%). Actually, a common perception was that the best porphyrins had to be β-substituted and that there was no possibility of further improvement in this field.

The renaissance of porphyrins is due to Officer and Grätzel, who reported a series of meso-aryl-zinc porphyrins bearing a β-substituted butenylidene malonic acid group in which a remarkable efficiency of 7% was achieved [163]. The idea of designing a meso-substituted porphyrin was first introduced by Anderson [164] and Therien [165], and a complete history is well described in the review of Diau et al. [166]. The most common ones are the free-base and zinc derivatives of the meso-benzoic acid substituted porphyrin (TCPP) (structures 2 and 3 in Fig. 3.7), which have been used for the photosensitization of not only TiO_2 but also other wide-bandgap semiconductors like NiO, SnO_2, and ZnO. Different anchoring groups such as carboxylic acids, catechols, phosphonic acids, acetylacetones, sulfonic acids, and pyridines have been employed [163]. Carboxylic and phosphonic acid groups were found to be the most successful in providing fast electron injection [167].

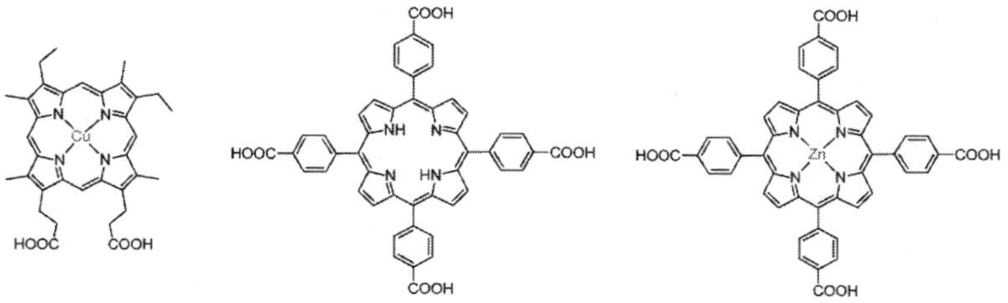

Figure 3.7 The molecular structures of copper-based porphyrins with propionic acid (1), and free-base (2) and zinc derivatives (3) of the meso-benzoic acid substituted porphyrins (TCPP).

However, porphyrin-based DSSC efficiencies remained much lower than ruthenium sensitizers [168]. Very recently, structural and theoretical investigations looking at push–pull type D–A porphyrins have yielded dramatically improved efficiencies of up to 13% [160]. The real breakthrough was introduced by YD2, a sensitizer having a diarylamine group attached to the porphyrin ring, which acts as an electron donor, and an ethynylbenzoic acid moiety, which serves as both an acceptor and an anchoring group (Fig. 3.8) [169]. The porphyrin chromophore itself constitutes the π bridge in this particular D-π-A structure. The IPCE spectrum exhibits a broad absorption from 400 to 750 nm with a maximum peak of over 90% at 675 nm, and this yields an overall power conversion efficiency of 11% under illumination with standard AM 1.5 G simulated sunlight. The advantage of these types of porphyrins comes from the directionality of the electronic distribution along the π-cloud of the porphyrin ring, which lowers the HOMO–LUMO gap as well as symmetry and thus relaxes the electronic transitions, resulting in an improved absorption efficiency in the visible region. A tailored variant of YD2, named YD2-o-C8, which incorporates two octyloxy groups in the ortho positions of each meso-phenyl ring, produces a marked improvement in the photo-induced charge separation in DSSCs using Co(II/III)tris(bipyridyl)–based redox electrolyte, reaching PCE values of 11.9% under standard AM 1.5 sunlight. The cosensitization of YD2-o-C8 with the previously prepared organic D-π-A dye, Y123, yielded an efficiency of 12.3% when used in conjunction with the Co(II/III)tris(bipyridyl)–based redox electrolyte [170].

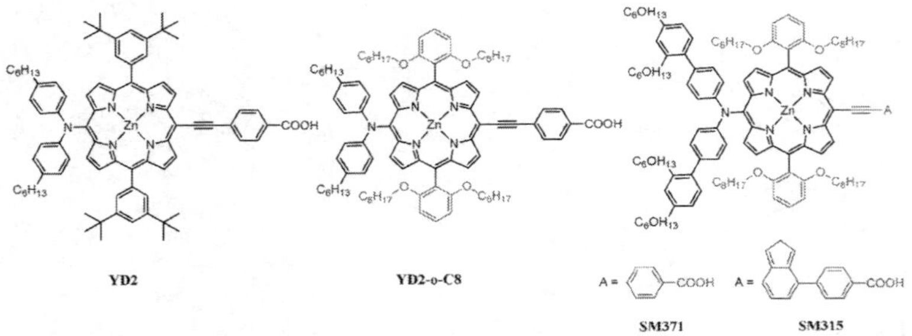

Figure 3.8 The molecular structures of YD2, YD2-o-C8, SM371, and SM315 porphyrin dyes.

There have been several excellent reviews published so far. Porphyrin-sensitized solar cells reported before 2009 are summarized by Imahori et al. [171], who extended the original review of Campbell et al. [172]. More recently, a systematic review of the progress of various kinds of porphyrins and their derivatives, as applied in DSSCs have been published by Diau et al. [166]. Focused on reports during the period of 2007 to 2012, the review discusses the correlation between molecular design and photovoltaic performance. Other reviews focused on porphyrins and where porphyrins are summarized along with other sensitizers for DSSCs are also available in literature [173–176].

The latest and best advancement was achieved very recently with the SM315 dye [160]. Grätzel's group redesigned the D-π-A structure of porphyrins to simultaneously maximize cobalt-electrolyte compatibility and improve the panchromaticity of the dye. Functionalization of the porphyrin core with the bulky bis(2′,4′-bis(hexyloxy)-[1,1′-biphenyl]-4-yl)amine donor and a 4-ethynylbenzoic acid yielded the green dye SM371, whose PCE (η = 12%) is slightly improved compared to the previously reported YD2-o-C8 (η = 11.9%). But it is the panchromatic sensitizer

SM315, obtained by the incorporation of the proquinoidal benzothiadiazole (BTD) unit into SM371 structure, that achieved a record 13.0% PCE at full sun illumination without the requirement of a co-sensitizer. SM315 is a panchromatic porphyrin sensitizer with a significant broadening of Soret and Q band absorbance features compared to SM371, yielding improved light harvesting in both the green (500–600 nm) and red (up to 800 nm) regions of the spectrum.

Since the most fruitful way to enhance J_{SC} is to harvest a broader region of the spectrum by using panchromatic dyes, some attempts have been made by functionalizing porphyrins at the meso-position in order to extend the π-conjugation. Other strategies were attempted by fusing porphyrins with a chromophore to make a π-elongated macrocycle or by combining two porphyrin moieties through a chemical bond.

Even if porphyrins show promise for enabling high efficiency DSSCs, many problems still remain. One problem is aggregation caused by their planar structure, which can significantly decrease electron injection efficiency. To overcome this, long hydrophobic alkyl chains have been introduced to protect the porphyrin core. Long and bulky chains can improve the packing density and reduce charge recombination between the injected electrons and the electrolyte.

Organic Dyes

All-organic dyes have shown potential as an effective sensitizer in DSSCs given their facile synthesis and purification by well-established techniques. There are some simple molecular design rules that can be followed to produce all-organic dyes with improved light-to-electric energy conversion [177]. By tuning dye absorption, electron injection, dye regeneration, and recombination, the performance of devices can be improved to deliver higher efficiencies. Although it is not possible to cover all the work that has been done in this area, we will focus our attention on a specific class of dyes, NIR dyes, which have recently drawn the most interest.

Among the metal-free organic dyes, the arylamine dyes, which have shown a validated efficiency of over 10.3%, are one of the promising candidates for making highly efficient DSSCs [178]. Other dyes that have also found interest are coumarins, perylenes, carbazoles, anthracenes, and indolines, and they are all excellently reviewed in the literature [179].

NIR Dyes

The need to move towards longer wavelengths in order to capture the more of the solar energy spectrum is pushing researchers to molecular engineer NIR dyes with higher photovoltaic performance and enhanced stability. Ideally, all the photons below a threshold wavelength of about 920 nm could be harvested and converted into electrical power [180]. However, NIR dyes present some disadvantages: susceptibility to aggregate, poor stability under thermal stress and light-soaking tests, and above all difficulty in obtaining dyes with the appropriate energy levels. In the following, a short summary of the main classes of NIR dyes will be discussed but the reader is also encouraged to refer to a recent review dedicated to the NIR sensitization in DSSCs [181].

(1) Squaraine dyes

Squaraine dyes are well known for their intense and sharp absorption in the visible/NIR region along with a strong emission in the long-wavelength region [175, 176, 182]. They have been widely applied in DSSCs to take advantage of their relative photo- and thermal stability.

Very recently, Han's group provided a summary on the development of squaraines in DSSCs, from the first investigation of their sensitization behavior to the latest results containing a detailed structure-performance relationship study [183, 184]. The first squaraine dyes used in DSSCs were limited to symmetrical dialkyl-substituted anilines or quaternarized indolenines, where the focus was on introducing different anchoring groups (COOH, OH, SO_3^-) for attaching to the TiO_2 surface [185]. Squaraine sensitization was therefore overshadowed for some years by other dyes appearing in literature that were able to reach more promising quantum and power conversion efficiencies. Yet at the same time, the notion that unsymmetrical structures are better performing than symmetrical squaraines was starting to spread, even if the efficiencies remained very low.

The first breakthrough came in 2007 when Yum et al. reported a respectable 4.5% efficiency with an unsymmetrical squaraine identified as SQ01 (Fig. 3.9). For the first time, the carboxylic acid function was directly attached to the conjugated π-system of the dye, which provided a strong electronic coupling with the conduction band of TiO_2. The asymmetric structure created by the octyl aliphatic side chain also hampered dye aggregation and limited self-quenching of the excited state.

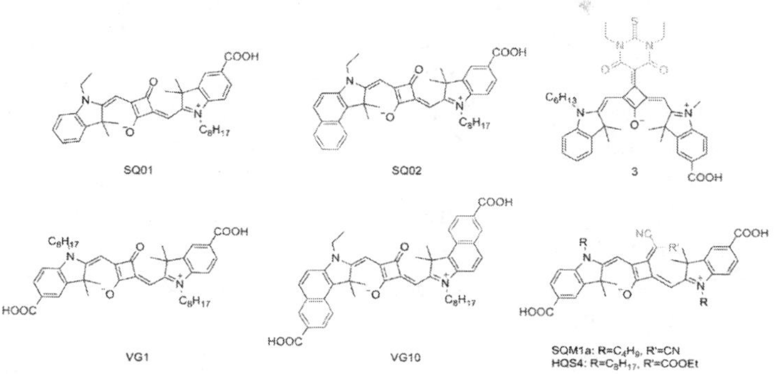

Figure 3.9 Selected examples of squaraine dyes showing the structural development from symmetrical to unsymmetrical and again to symmetrical structures.

At the same time, the need to increase the absorption towards the NIR region to further improve capture of more of the solar spectrum became increasingly evident. One strategy was to extend the conjugation as demonstrated by Burke et al., who replaced one indolium by a benzoindolium moiety, yielding an unsymmetrical squaraine SQ02 with extended absorption in the red region that achieved a 5.4% efficiency. Since then, a large number of structures with more extended absorption have been reported with very interesting results. JK216 [186], YR6 [187], JD10 [188] and VG10 [189] structures exhibit panchromatic light harvesting properties and cell efficiencies around 6–7%. However, when going towards the NIR region, key problems appear to be either a lower charge injection [190] or slower regeneration process, as has been described in a recent review [181].

However, even with the positive advances towards the NIR region, the problem of a narrow absorption range still remained. Beverina et al. proposed the functionalization of the central core with a diethylbarbiturate residue (dye 3 in Fig. 3.9) instead of an oxygen atom as the electron-withdrawing group to produce an additional high-energy absorption band, thereby broadening the squaraine's light harvesting property [191]. This approach was used very recently by other groups, which used a dicyanovinyl or a cyanoesther group that led to NIR squaraine dyes with decent performance (SQM1 and HQS4 in Fig. 3.9) [192, 193].

These latest dyes confirmed what has been stated by our group that against the preconceived idea that unsymmetrical squaraine structures are required, symmetrical squaraines in reality show superior optical properties and comparable performance in terms of light-to-electricity conversion efficiencies compared to their corresponding unsymmetrical counterparts [194]. Moreover, the synthesis of symmetrical dyes is significantly easier and the bi-dentate grafting mode significantly stabilizes the dye against desorption. Finally, we have also proved that going from unsymmetrical to symmetrical structures enables a red-shift of light harvesting by about 20 nm. VG10 dye remains a benchmark dye in DSSCs and if we consider the cost-to-cell efficiency ratio is, so far, probably the best-performing dye in the literature.

(2) Phthalocyanine Dyes

Phthalocyanines (Pcs) are structurally related to porphyrins but are even more conjugated. They are composed of 18 planar π-electron macroheterocycles with strong absorption in the far visible/NIR region. They have extremely high extinction coefficients over 100,000 $M^{-1}cm^{-1}$ even up to around 700 nm where ruthenium sensitizers start to tail off in absorbance [195]. The advantages of Pcs include their thermal and chemical stability as well as suitable redox properties for sensitization and dye regeneration by the electrolyte [196].

A very detailed and recent review on the development of Pcs structures in DSSCs has recently been published by the Torres group, who, along with Kimura and Mori, heavily contributed to the use of Pcs in DSSCs [197]. At present, the best-performing Pc is PCS18 with a 5.9% efficiency [198], but, as with the porphyrins, the history of Pcs was not always bright with success as a result of aggregation problems and poor electronic coupling between their LUMO orbitals and the TiO$_2$'s conduction band. A significant progress was made in 2007 with the design of PCH001 [199] and TT1 [200] (η = 3.05% and 3.5%, respectively) (Fig. 3.10) that introduced bulky groups in the heterocycle to enhance solubility and reduce aggregation. Moreover, attention was given to the number of carboxylic groups to improve charge transfer directionality of the excited state of these push-pull systems.

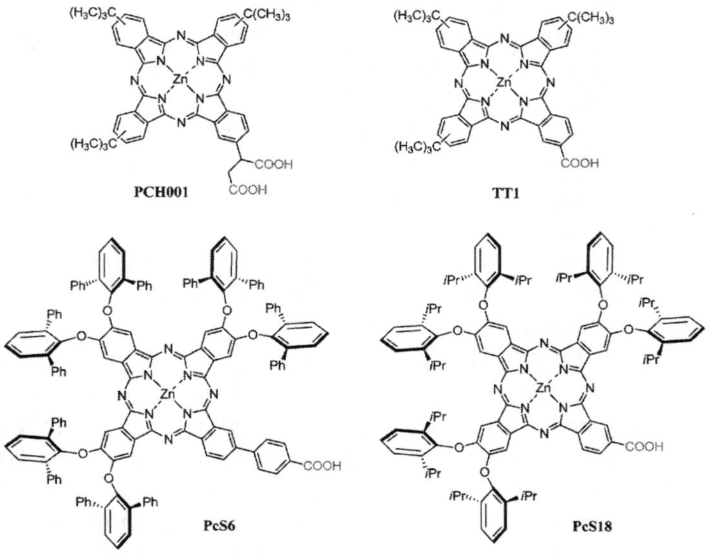

Figure 3.10 The molecular structure of PCH001, TT1, PcS6, and PcS18 phthalocyanine dyes.

The lastest development in Pcs dyes considered these design principles. For instance, PCS6 has good conversion performances (η = 4.6%) without the need of co-adsorbents unlike in the past [201]. Currently, the best-performing Pc is PcS18, which has been designed again with bulky end groups at the periphery of the dye.

(3) Cyanine Dyes

Cyanines belong to the class of polymethine dyes that are characterized by a high molar extinction coefficient, high solubility, tunability, and very strong light absorption in the visible and NIR regions. Originally, they were standard sensitizers used in silver halide photography. They have been recently applied as sensitizers in DSSCs but with relatively low efficiencies [202].

New Aspects in DSSC Sensitization

(1) Co-sensitization

It is clear from the discussion above that it is extremely difficult to design a single sensitizer with efficient absorption in the entire visible and NIR spectrum (up to 920 nm) and fulfilling all the requirements necessary to obtain a very efficient device in terms of performance and stability. However, it is possible to tailor and develop NIR-efficient sensitizers and to combine them with other dyes to obtain panchromatic systems. It is obvious that in order for mixtures of two or more sensitizers to harvest sunlight, each of these dyes has to be able to attach to the metal oxide surface to achieve panchromatic co-sensitization. Finally, the combination of dyes should complement each other in their absorption properties and should not interfere with their sensitization properties [203]. Generally, a co-sensitized solar cell has an extended spectral response across the whole visible domain, showing a greater photocurrent than if individual dyes were used, and therefore is expected to exhibit superior power conversion efficiency. The co-sensitization process can be carried out by various means and has been recently reviewed by Balasingam et al. [204]. Early work dealt with the preparation and application of what is usually referred to as a "cocktail," a mixture of two different dyes in a common solvent or the mixing of two dyes dissolved in different solvents and usually of different density. A second approach involved the sequential dipping of the electrode in two dye solutions.

Nazeeruddin and coworkers compared the simultaneous and sequential adsorption of JK2 and SQ1 [205]. Other systems include the dye-bilayered structure proposed by Hayase and coworkers [206], the multiple electron transfer processes on the co-sensitized TiO_2 working electrode by Al_2O_3 layers [207], and the very recent ultra-fast multiple co-sensitization method, which overcomes the limitations caused by having different partition coefficients and molar extinction coefficients [208]. Spitler and coworkers combined dicarboxylated cyanine dyes and achieved higher photocurrent in cells with the combination of three dyes vs. a single dye [209]. Noda and coworkers achieved as high as 11% conversion efficiency by combining the black dye with an organic dye (D131) [210]. Recently, the best examples of co-sensitization have been achieved by using porphyrins and organic dyes such as YD2-o-C8 and Y123, reaching a record efficiency of 12.3% [171].

When we employ more than two dyes in a working electrode, the final current densities are obviously expected to increase. However, the trend in open circuit voltage is more complicated. In most cases, the V_{oc} is lower than with the two dyes each acting alone. V_{oc} is closely related to the interfacial charge transfer rate at the interface between the dye layer and the electrolyte (i.e., injection and recombination) [211]. For all the co-sensitization systems investigated, only one example has been reported where the photovoltage has also been improved, and this was achieved by combining LD12 (V_{oc} = 0.711 V) and CD5 (V_{oc} = 0.689 V), which yielded a final V_{oc} of 0.736 V, a value that is unprecedented for co-sensitized porphyrin and organic-dye systems [212].

Co-sensitization is still at an early stage. This approach needs to be further investigated in order to better address the choice of the dyes and systems to be used.

(2) Quantum dots

Another option is the use of inorganic semiconducting quantum dots for which tuning the size of the dots alter its absorption energy from the NIR to visible range. Inorganic semiconductor quantum dots as applied to DSSCs are referred to as quantum dot sensitized solar cells (QDSSC) [213–215]. Semiconductor quantum dots (QDs), which absorb light in the visible region, such as CdS [216], CdSe [217], CdTe [218], PbS [219], InAs [220], and Zn_3P_2 [221], have been used as sensitizers in DSSCs. The main reasons QDs have received considerable attention are that the physical and optical properties of QDs can be controlled by changing the size/shape of the QDs and they can generate multiple excitons from the absorption of one photon through an impact ionization mechanism (multiple exciton generation or MEG effect).

Niitsoo et al. reported an efficiency of 2.8% under 1 sun illumination in a three-electrode configuration using chemical bath deposited CdS/CdSe on porous TiO_2 [222]. Diguna et al. reached a very similar efficiency but in a sealed two-electrode cell configuration [223]. In these cases, they employed a polysulfide as an electrolyte for the redox couple to regenerate the photoexcited holes in the QDs. A polysulfide-based electrolyte is preferred as iodine, which is the common component in DSSC electrolytes, rapidly photo–corrodes the QDs leading to their dissolution [224–225].

PbS QDs have attracted significant interest as sensitizers because they allow the extension of the absorption band towards the NIR part of the solar spectrum. PbS quantum dots show a minimum transition energy in the limit of large dots that is reflective of the bulk bandgap (~0.41eV). Nozik and coworkers reported PbS/PbSe colloidal quantum dots in Schottky solar cells with a remarkably high photocurrent (>21 mA/cm^2) [226]. Furthermore, Nazeeruddin and coworkers demonstrated PbS cells that involved PbS deposition by the successive ionic layer adsorption and reaction (SILAR) method and which made use of a cobalt redox mediator [Co(o-phen)$_3$]$^{3+/2+}$ instead of the polysulfide [227]. This enabled IPCE of over 50% and opened up possibilities to harvest light up to 900 nm and beyond.

A more advanced configuration was proposed by Grimes and coworkers who achieved high external quantum efficiencies with red photons by employing high surface area nanowire TiO_2 arrays and donor chromophores displaying high fluorescence quantum yields through Förster resonance energy transfer (FRET) [228]. The FRET mechanism has been contemplated as an alternative mechanism for charge separation and a way to improve exciton harvesting when close to the heterojunction interface. This effect was also made use of by Barolo and coworkers who fabricated a FRET cell using CdSe QDs and a symmetric squaraine dye (VG1) to efficiently collect photons from the visible to the NIR spectrum [229].

With this configuration, they obtained 47% enhancement in overall efficiency compared to QDs alone.

3.3.3 Liquid Electrolyte Development Based on Solvent and Solvent-Free Formulation for Stable Devices

Champion efficiencies reported above are systematically obtained in conjunction with an electrolyte based on a volatile solvent. Pure acetonitrile or an acetonitrile/valeronitrile solvent mixture is often preferred. Their intrinsic properties, in particular their volatility, are insufficiently stable to pass the accelerated IEC61646 protocol. The excessive vapor pressure exerted on the cell weakens the seals even at temperatures well below 60 °C. In addition, this might lead to undesirable chemical/electrochemical reactions in the cell. This volatility issue has been largely reduced by replacing acetonitrile with lower volatility solvents such as γ-butyrolactone (GBL), propylene carbonate (PC), propionitrile (PN), sulfolane, butyronitrile (BN), or 3-methoxypropionitrile (MPN). Their integration led to significant prolonging of device lifetime. It appears that the last two solvent alternatives show actually the best compromise between high efficiency and stability. The sulfolane-based electrolyte is promising for extreme conditions even though its high viscosity impedes high photocurrent production owing to mass transport limitations [230].

The electrolyte is not solely composed of iodide and iodine. It also contains a series of additives. The beneficial impact of these additives on cell performance is undeniable. However, their exact role/action in the complete cell remains speculative to some extent. Table 3.1 gives the benchmark composition of nitrile-based electrolytes used to obtain either high efficiency or good stability. Note that there are also solvent-free ionic liquid-based electrolytes which are composed of binary or ternary eutectics melts and include components and compositions different from that reported in Table 3.1. A review of these is provided by Grätzel et al. [231]. The two additives typically incorporated in electrolyte are the guanidium thiocyanate and a Lewis base, namely *tert*-butyl pyridine to achieve high efficiency, or either *N*-butylbenzimidazole or benzimidazole to achieve good stability. The utilization of a Lewis base in the electrolyte causes the deprotonation of the TiO_2 surface, shifting up the quasi-Fermi level and consequently the open circuit photovoltage. *N*-butylbenzimidazole (NBB) is preferred over *tert*-butylpyridine (TBP) regardless if MPN or BN is used as the solvent because of stability issues of the latter at 60 °C under light soaking or 85 °C in the dark (Fig. 3.11).

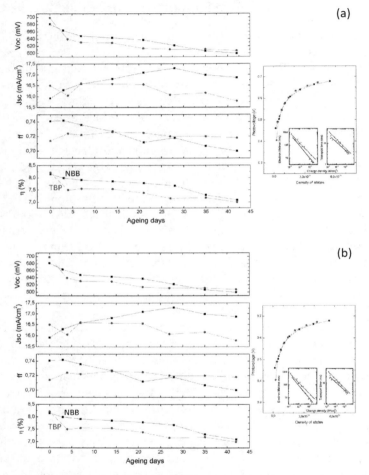

Figure 3.11 Comparison of cell characteristics during ageing at 60 °C/ 100 mW·cm^{-2} using an electrolyte containing TBP or NBB (composition: 1 M DMII, 0.1 M GuNCS, 0.15 M I$_2$, 0.5M NBB/ TBP) in (a) 3-MPN solvent and in (b) BN solvent. Included also is a comparison of the trap state distribution and energy as well as a comparison of electron lifetime and transport time as a function of charge density.

Table 3.1 Composition and structure of main liquid electrolyte components

Name	Acetonitrile	3-Methoxy-proprionitrile	1,3-Dialkylimi-dazolium iodide	Iodine	Tert-butylpyridine	N-butylbenzimidazole	Guanidinium thiocyanate
Structure	H₃C—C≡N	(structure)	(structure)	I—I	(structure)	(structure)	(structure)
High efficiency (concentration)	Solvent	—	1 M	0.03 M	0.5 M	—	0.1 M
Stable efficiency (concentration)	—	Solvent	1 M	0.15 M	—	0.5 M	0.1 M

The comparison of the two molecules suggests that they have similar strengths for deprotonation since the distribution of the sub-conduction band energy levels are not modified. The charge collection efficiency is also similar or slightly improved using the *tert*-butylpyridine, i.e., marginally higher electron lifetime compensated by slow electron transport (Fig. 3.11). The guanidinium thiocyanate has a more subtle function. In general, we often refer to the role of the guanidine cation going to the exposed surface of TiO_2 in between the dye molecules to explain the improvement in charge collection efficiency.

It is also clear that this additive is beneficial for cell stability, in particular, to obtain a high fill factor. An impedance study carried out on TCO-Pt/electrolyte/Pt-TCO symmetric cells has highlighted that the charge transfer resistance is strongly dependent on whether the electrolyte contains the guanidinium thiocyanate additive. Indeed, whereas the benchmark electrolyte containing 0.1 M of guanidinium thiocyanate shows a slight decrease in R_{ct} to *ca.* 1 $\Omega \cdot cm^2$ during ageing, this trend is completely the opposite for an electrolyte composition free of guanidinium thiocyanate, i.e., R_{ct} increases (Fig. 3.12). This trend combined with the visual observation of the symmetric cell during ageing which showed bleaching of the counter electrode in contact with the electrolyte, enables us to conclude that guanidinium thiocyanate also plays the role of a molecularly self-assembled protection layer on the Pt to prevent its dissolution through soluble PtI_6^{2-} complexes.

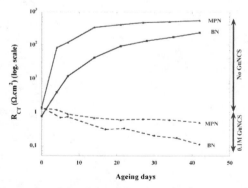

Figure 3.12 Evolution of the charge transfer resistance as a function ageing time at 60 °C/100 mW·cm^{-2} for a TCO-Pt/electrolyte/Pt-TCO symmetric cell configuration comparing BN and 3-MPN-based electrolytes with 0.1M GuNCS and without GuNCS.

In terms of stability performance, lab cells (<1 cm²) with 10% efficiency and PCE retention over 95% after 1000 h under 60 °C/100 mW/cm² light soaking actually represent the best values reported so far on lab devices. This has been achieved by using either BN [232] or alternatively MPN-based gel electrolyte [230]. Dyesol recently reported an HSS electrolyte that passed even a 95 °C/dark accelerated test for 1000 h in combination with the N719 dye with efficiencies in the range of 5% under standard illumination conditions (AM 1.5G) [230]. These are two conditions among the three of the IEC61646 protocol validated. Out of this protocol, stable performances were reported for the following: over 2000 h at 60 °C in the dark that maintained a 5.5% PCE [234]; 2.5 years in outdoor conditions without a specified module efficiency [235]; 2280 h at room temperature under 80 mW/cm² light soaking [236]; or more recently Dyesol achieved as long as 25600 h at 55–60 °C under continuous light soaking while preserving 4% efficiency [237]. Since the seminal publication by Grätzel, only very few articles have reported stability at higher temperatures, requiring as it seems, more viscous liquids to alleviate sealing issues, as for instance gel electrolytes or solvent-free ionic liquids [238–241]. Even with the very few highlighting reporting this achievement, still there are no publications of actually passing the challenging threshold of 1000 h at 85 °C in the dark with efficiencies above 5% [230, 242, 243]. On the one hand, these results emphasize the potential of DSSCs for larger scale applications, but on the other hand, they also point to the important gap between achieving champion efficiencies and stable devices for which closing this gap will require more focused effort on understanding the ageing mechanisms leading to cell failure. This will enable the design of more robust and better performing cell components (i.e., ligand, solvent, additive(s) in the electrolyte).

3.3.4 Current Understanding of Chemical/ Photoelectrochemical Degradation Pathways

Whereas single-crystalline and amorphous silicon modules can last up to 20 years, the stability and ageing behavior of emerging PV technologies are still a work in progress towards achieving at least a 10-year lifetime under working conditions. The ability to control the dynamics of charge transport and charge recombination in DSSCs is crucial for reaching high device performance. Nevertheless, the stability of each cell component taken separately and the understanding of all the chemical, electrochemical and photochemical reactions and interplays for the different material interfaces or components are key to bring further development on DSSC, not only to improve device stability under severe ageing conditions but also to close the gap between high efficiency and stable efficiency. A more complete understanding of the ageing mechanisms will pave the way to the design of more robust components. These new materials will give impetus to DSSC commercialization and broader adoption in the PV arena. Only a very few groups are actively working on this more fundamental domain. Besides the intrinsic sealing and permeability issues of the device, which have been shown to influence device stability, there are six distinct intrinsic features contributing to cell degradation as discussed below, which have been found in the literature (Fig. 3.13):

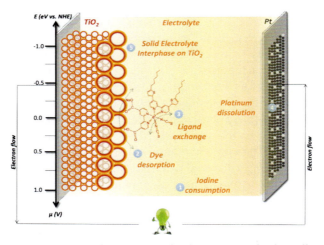

Figure 3.13 Known stability issues in the dye-sensitized solar cell.

(1) One of the most critical issues stems from the well-known iodine consumption in the electrolyte. This reaction takes place during ageing. It translates into a bleaching and loss of coloration of the electrolyte. This side reaction must be suppressed as it is an important source of failure since the iodine concentration controls the rate of tri-iodide mass

transport. Excessive depletion will limit the short-circuit current density of the device.

(2) Dye desorption at elevated temperatures (>60 °C). This possible event takes its origin from the competition between dye solubility in the electrolyte and the binding strength of the anchoring group on TiO_2. It can also be promoted by the rupture of the ruthenium hexacoordination. Water uptake in the electrolyte has also been proposed in assisting this dye desorption mechanism. The development of hydrophobic dyes is anticipated to go counteract this process.

(3) Lack of chemical and photochemical stability of the monodentate thiocyanate ligand in heteroleptic ruthenium (II) complexes, which tends to undergo substitution reactions with different external components.

(4) Platinum dissolution in electrolytes free of thiocyanate (cf. Fig. 3.12).

(5) Formation of a polymeric solid electrolyte interphase (SEI) on TiO_2 and sensitized-TiO_2 (see Fig. 3.14 inset).

(6) UV irradiation causes direct bandgap excitation of TiO_2 leading to conceivable dye or electrolyte component oxidation [234, 244]. Addition of MgI_2 or CaI_2 in the electrolyte was found to improve electrolyte tolerance to UV irradiation [235]. It is also typical to cover the glass photo anode with an anti-reflecting polymer coating, which also serves as a UV filter [245].

In the following, we will review in more detail the two most severe issues that need to be overcome, namely the iodine depletion in the electrolyte and the dye's chemical/photoelectrochemical stability.

Iodine consumption

Iodine consumption has been found by numerous groups, which have given different possible reasons to explain this depletion. There are eight hypotheses/observations, some of which are controversial, that have been proposed:

- Formation of IO_3^- induced by traces of water [246–250]. In this case, it is postulated that I_2 reacts with ingressed water or residual water in the electrolyte. It leads to the formation of the iodate anion (IO_3^-). As commented also by a few of these authors, the formation of IO_3^- has, however, never been really detected so far either by spectroscopic methods (UV-Vis, Raman) or by chromatography (LC/MS).

- I_2 reacts with the glass frit used in some technologies as a sealant [251]. The experimental evidence reported seems convincing, but this depletion is experienced regardless of the type of sealant utilized (Bynel or Surlyn polymers, glass frit). Therefore, the glass frit is most likely not the sole reason for this depletion though it will contribute to this consumption.
- Sublimation of molecular I_2 has been proposed by different groups [243, 252, 253]. This hypothesis has been contradicted by ex situ and in situ spectroscopic methods such as Raman spectroscopy on aged electrolytes and GC/MS or TGA/MS performed also on aged electrolytes or on electrolytes heated to 60 or 85 °C. The results suggest the absence of any spectroscopic bands or mass fragments, which could be attributed to iodine or iodide-based compounds even in trace amounts [250].
- Bandgap excitation of TiO_2 leading to the photoelectrochemical reduction of tri-iodide [254]. This reaction would be particularly unexpected knowing that TiO_2 is a strong n-type semiconductor. This reaction takes place faster without the dye than with the dye and therefore could not be attributed to a recombination procedure between iodine and electrons in the conduction band. Last but not least, it has also been pointed out that iodine consumption proceeds in the dark and is only thermally activated, as discussed below [251].
- The electrolyte bleaching is solvent dependent. This has been clearly highlighted from the work led by Hans Desilvestro et al. at Dyesol [230, 255]. The authors compared γ-butyrolactone, 3-methoxypropionitrile and a so-called nitrile-free solvent HSS. They have concluded that iodine depletion significantly slows down going along this series. They have also reported slower iodine consumption using tetraglime compared to 3-MPN [255].
- Iodine consumption is not an intrinsic reaction taking place in the electrolyte alone. It is believed to be triggered by the

surface of TiO_2, which plays a catalytic role in the electrolyte degradation [250]. The authors concluded that iodine depletion is exclusively activated by temperature while light has no action on this reaction, at least for the 3-MPN-based electrolyte (Fig. 3.14).

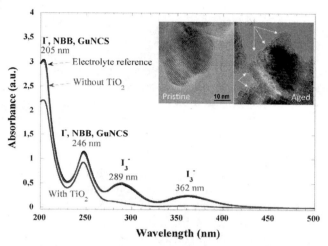

Figure 3.14 UV-Vis absorption spectrum of reference MPN-based electrolyte (in black), electrolyte aged alone for 500 h at 85 °C/dark (in blue), and aged in contact with TiO_2 (in red). The inset shows high-resolution transmission electron micrographs comparing the TiO_2 nanoparticles before and after ageing showing the formation of the SEI layer.

- Iodine reacts with electrolyte additives like 4-*tert*-butylpyridine (TBP) leading to an iodo-pyridinate complex or the thiocyanate ligand of the dye to form I_2NCS^- species in the particular case of the electrolyte free of any TBP [257].
- Formation of a solid electrolyte interphase (SEI) on the surface of TiO_2 in the dye monolayer pinholes (Fig. 3.12 inset) [250]. The authors highlighted that SEI formation traps or solvates iodine (and other electrolyte components), explaining its depletion from the bulk electrolyte. It is supported by UV-Vis and XPS spectroscopies, and combined with ToF-SIMS, the SEI is found to possess a highly complex composition and a high concentration of iodine/iodide-based species along with other degraded components of the

electrolyte. The current understanding is that this SEI layer is likely a result of the polymerization between acrylonitrile radicals to form a very cohesive polyacrylonitrile polymer. GC/MS experiments showed that 3-MPN in presence of TiO_2 thermally decomposes to form two highly volatile compounds: acrylonitrile (bp = 77 °C) and methanol (bp = 65 °C) [259]. The free radical polymerization of acrylonitrile is well-established to take place spontaneously by mild thermal activation in the range of 50 °C. The polymerization rate can be substantially faster in the presence of a metal halide catalyst [256]. The formation of these two degradation compounds can also give a more rational explanation to the electrolyte evaporation pointed out by many authors in the field which takes place suddenly during ageing; event typically due to poor cell sealing.

The literature often focuses on iodine consumption in the electrolyte as it is the chromatic component of the electrolyte with its characteristic strong purple color that is easy to track. Our group presented at the HOPV2014 conference that many other electrolyte components are in reality consumed as well. We determined by cyclic voltamperometry using a platinum microelectrode that about 50% of the iodide is depleted after 500 h ageing at 85 °C; FT-IR carried out on an aged electrolyte indicated that the thiocyanate from guanidinium thiocyanate is almost completely depleted; last but not least, *N*-butylbenzimidazole is also quantitatively consumed (*ca.* 50%) as shown by LC/MS [258]. These experiments were carried out by looking at the interface between bare TiO_2 and the electrolyte. When sensitized, these reactions are still occurring but at a slower rate. This was similarly found in the case of iodine consumption for which Lund et al. have made careful comparison between unsensitized TiO_2 and sensitized TiO_2, highlighting that the dye monolayer only slows down this reaction [231]. The MPN-based electrolyte stability is strongly affected by the surface of TiO_2, which tends to destabilize significantly the thermal stability of electrolyte components by about 50 °C (Fig. 3.15). The surface of TiO_2 also collaterally induces gas formation as aforementioned in this review [250, 258].

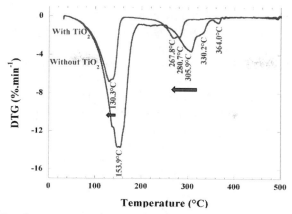

Figure 3.15 Comparison of TGA derivative curves as a function of temperature comparing the stability of intrinsic MPN-based electrolyte and in the presence of TiO$_2$.

Chemical/photochemical stability of benchmark ruthenium sensitizers

Here, the focus is on ruthenium polypyridyl complexes. Much less is known about the core stability of organic dyes. Toyoda et al. reported that the major degradation pathway in indoline-based D-π-A dye, namely the yellow D131 chromophore, is the decarboxylation of the cyano acrylic acid anchoring group [261]. This decarboxylation reaction contributes to dye desorption, lowering significantly the device performance whether the ageing is at 60 °C/1 sun or 85 °C/dark. This reaction has been attributed to the harmful action of iodine or amine components of the dye. This reaction can also occur to some degree with the family of "push-pull" D-π-A dyes, although some of the dyes in this family have been reported to be very stable under 60 °C/light and 85 °C/dark ageing conditions. Most of the organic dyes are known to have stability issues given their susceptibility in undergoing fast photo-oxidation when they are anchored onto TiO$_2$. Following this hypothesis, increased stability will thus require a fast recovery of the reduced form to prevent this side reaction from occurring.

The chemical/photo(electro)chemical durability of ruthenium polypyridyl-based sensitizers in contact with TiO$_2$ has been questioned well before the seminal paper of O'Regan and Grätzel in 1991 [262]. The underlying effect of the TiO_2 nanostructure on the ruthenium-complex stability has been discussed [4]. The authors presented photocurrent stability over 2 months subjected to continuous visible light illumination stress with less than 10% degradation, corresponding to a stability of over 5×10^6 turnovers. Similar results were reported on monomeric dye with no noticeable degradation [263].

These intriguing results are in relative contradiction with other reports on cis-Ru(bpy)$_2$(SCN)$_2$ (bpy = 2,2'bipyridine) and trimeric ruthenium dyes [264]. The lack of stability of N719 and Z907 has been reported in particular by Tributsch, Hagfeldt, and Lund. These authors were concerned with the vulnerability of the strong electron donor thiocyanate ligand, which can easily exchange with electrolyte components as a result of the antibonding character of these orbitals [265–266]. This ligand exchange reaction is not only triggered by light action (photolysis) but can also be activated by temperatures above 80 °C. The monodentate thiocyanate ligand can exchange with acetonitrile or 3-methoxypropionitrile as the solvent, TBP, residual water or even iodide [264–272]. The different exchange reactions reported are summarized in Fig. 3.16. This reactivity of the monodentate thiocyanate ligand was explained by Grünwald et al. on the basis of an incomplete reduction of the photooxidized dye by the I_3^-/I^- redox couple together with a certain lack of stability of the oxidized form of the complex [235]. This explanation was further confirmed by Kohle et al. who showed in solution that dye instability can also be prompted by too low a concentration of iodide, i.e., when the rate of dye regeneration is slower [273]. The results provided by Lund et al. suggest that again TiO_2 provides significant catalytic activity by accelerating this ligand exchange reaction by a factor of two.

It is also suggested that some degradation products of N719 and Z907 dyes are in equilibrium with each other [272]. The following for instance are three inter-related equilibrium reactions:

(1) [RuLL'(NCS)$_2$] + 3-MPN ↔ [RuLL'(NCS)(3-MPN)]$^+$ + NCS$^-$
(2) [RuLL'(NCS)(3-MPN)]$^+$ + 4-TBP ↔ [RuLL'(NCS)(4-TBP)]$^+$ + 3-MPN
(3) [RuLL'(NCS)$_2$] + 4-TBP ↔ [RuLL'(NCS)(4-TBP)]$^+$ + NCS$^-$

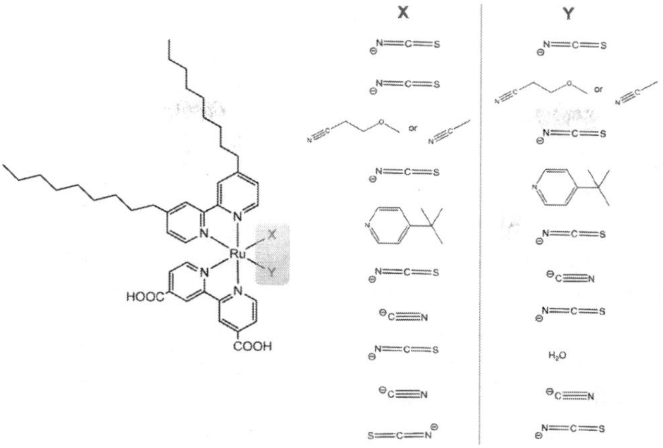

Figure 3.16 Resume of the different dye side-products characterized on either N719 or Z907Na after ageing (adapted from ref. 274 and 275).

The rate for this ligand exchange reaction can be lowered by a factor of two when a buffer of thiocyanate anions is included in the electrolyte (e.g., guanidinium thiocyanate) [268, 273, 274]. The degradation products based on TBP can be prevented by replacing it with N-butyl benzimidazole or the closely related benzimidazole, which enhances device stability. Such ligand exchange reactions will affect the optical MLCT (metal-to-ligand charge transfer) contribution, resulting in either a slight bathochromic or hypsochromic absorption shift [230]. The real implication of this in situ/in operando dye structure modification on practical cell characteristics remains unclear. Finally, again in the work published by Grünwald et al., the authors showed that the ruthenium hexacoordination can become fragmented leading to the desorption of the metal core unit while retaining the anchoring bipyridine group attached to the TiO_2 [263].

Besides these intrinsic degradation mechanisms highlighted to date, more external factors have also contributed to cell degradation, namely sealing conditions, substrate corrosion, and sealing issues. An assessment of these technical challenges is relatively difficult since they are rarely published. An interesting work reported by Lindquist et al. showed that it is important to consider the temperature of sealing as it will influence device efficiency [275]. This work compared N719 with the organic D5 sensitizer and found that N719 appeared more sensitive to thermal degradation than D5. Cell efficiency was shown to change from a maximum of *ca.* 4% to almost zero when cells were sealed at a sealing temperature exceeding 200 °C. They also highlighted by using IMPS/IMVS that the electron diffusion length, related to the charge collection efficiency, can reach as low as 10%.

The encapsulation of the cells should prevent the exchange of material between the inner parts of the cell and the ambient environment. For cell stability, it is crucial that the encapsulation creates a barrier against water and oxygen as they can be assimilated by the electrolyte and the mesoporous sensitized TiO_2 layer. However, water has also been reported to enhance the power conversion efficiency of DSSCs based on ionic liquids [276]. The proportion of water in the cell can drastically modify the proton concentration in the electrolyte. This can in turn affect energy potentials as protons tend to adsorb on the surface of TiO_2 leading to a downward shift in the energy of the Fermi level and conduction band edge [277]. The critical threshold of the amount of water to maintain high stability and improve cell efficiency is difficult to determine although it is an important factor for device stability. One difficulty in determining this threshold value is due to the interference of iodine in the DSSC electrolyte towards a Karl-Fischer titration. This threshold value is also expected to strongly depend not only on the type of dye used but also on the type of electrolyte.

Finally, the general photocatalytic properties of TiO_2 should be mentioned since it could trigger the formation hydrogen peroxide from oxygen and water, which can destructively oxidize organic compounds [278]. The production of H_2O_2 is initiated by the reduction of O_2 from conduction band electrons in TiO_2 leading to superoxide anion O_2^- that will react with water. It should also be stressed that because of the reactivity of the superoxide anion radical, it can also directly participate in internal chemical reactions with other organic cell components [279].

Related to the sealing of the cell is the conductive substrate (transparent conducting oxide [TCO]), which under certain conditions can lose its conduction properties. The indium-doped tin oxide (ITO) layer can irreversibly lose its electrical conductivity when voltages above 1.5 V are applied, voltages that can be reached in series modules [280, 281]. This may lead to issues, for instance, for flexible devices for which the substrate is typically made of ITO deposited on poly(ethylene naphthalate) (PEN) or poly(ethylene terephthalate) (PET).

3.3.5 Concluding Remarks

Since the seminal publication of Grätzel et al. describing the nanocrystalline dye-sensitized solar cell, the stability of DSSCs has been reported in hundreds of publications. The IEC61646 accelerated test has been to date passed by a few companies. The development of stable DSSCs has principally relied on knowledge and experience gathered by the community after more than 20 years of research. However, there is as yet very little understanding as a whole about the interrelated chemical reactions responsible for device ageing and the stress factors responsible for this degradation. Is UV, visible light or temperature the most critical stress parameter? The experiments gathered by our group on electrolyte stability pointed out that temperature appears to be the most critical parameter.

The field of DSSCs has been extremely prolific in terms of the photophysics and photoelectrochemistry, but also, as we have discussed in this chapter, very rich in terms of the organic chemistry and variety of molecules which have been developed following various guiding principles to improve on the original design (e.g., extension of conjugation, metal center replacement, introduction of bulky groups to reduce aggregation).

In practice, studies on device stability mainly report on change in cell characteristics as a function of time. Less often, IMVS-IMPS, EIS, or transient photovoltage/photocurrent decay techniques have been used to monitor the influence of ageing on charge transfer kinetics and the distribution and energy of surface trap states in nanocrystalline TiO_2. For further improvements, not only in obtaining more stable power conversion efficiency but also in prolonging DSSC lifetime to be competitive with silicon technology, there is a great need to establish the experiments and characterization techniques for a careful analysis of the degradation products that are formed during device ageing. In this area, only a very few groups are currently active in deciphering the complexities with these devices.

We also recognize that TiO_2 plays a major role in device stability. This goes particularly against the preconceived idea that the inorganic part of the cell is the most robust compared to the organic and organometallic compounds. TiO_2 displays in fact a catalytic role not only in dye degradation as it has been clearly demonstrated by the pioneering work of Lund's group but also in affecting electrolyte stability, in particular those based on alkyl and alkoxynitrile solvents for which we are actually scrutinizing. Understanding the multifaceted nature and chemistry interplay of DSSCs and developing new materials based on such knowledge will indisputably open up significant breakthroughs in stability for this artificial photosynthetic solar cell.

List of acronyms

Acronyms	Full names
F8BT	Poly[(9,9-di-*n*-octylfluorenyl-2,7-diyl)-*alt*-(benzo[2,1,3]thiadiazol-4,8-diyl)]
P3HT	Poly(3-hexylthiophene-2,5-diyl)
PT8T8T0	Poly(3-octylthiophene-2,5-diyl)
MDMO-PPV	Poly[2-methoxy-5-(3,7-dimethyloctyloxy)-1,4-phenylenevinylene]
PCPDTBT	Poly[2,6-(4,4-bis-(2-ethylhexyl)-4*H*-cyclopenta [2,1-*b*;3,4-*b'*]dithiophene)-*alt*-4,7(2,1,3-benzothiadiazole)]
PFB	Poly(9,9'-dioctylfluorene-co-bis(*N*,*N'*-(4,butylphenyl))bis(*N*,*N'*-phenyl-1,4-phenylene)diamine)
CN-PPV	Cyano-polyphenylene vinylene
MEH-PPV	Poly[2-methoxy-5-(2-ethylhexyloxy)-1,4-phenylenevinylene]
PCNEPV	Poly-[oxa-1,4-phenylene-(1-cyano-1,2-vinylene)-(2-methoxy-5-(3,7-dimethyloctyloxy)-1,4-phenylene)-1,2-(2-cyanovinylene)-1,4-phenylene
F8DTBT	Poly((2,7-(9,9-(di-n-octyl)fluorene)-*alt*-5,5-(4',7'-dithienyl-2',1',3'-benzothiadiazole
APFO-3	Poly[(9,9-dioctylfluorenyl-2,7-diyl)-*alt*-5,5-(4',7'-di-2-thienyl-2',1',3'-benzothiadiazole)]

PF10TBT	Poly[2,7-(9,9-didecylfluorene)-alt-5,5-(4,7-di-2-thienyl-2,1,3-benzothiadiazole)]
TFMO	Poly[9,9-dioctylfluorene-co-N-(4-methoxyphenyl)diphenylamine
TFB	Poly[(9,9-dioctylfluorenyl-2,7-diyl)-co-(4,4'-(N-(4-sec-butylphenyl)diphenylamine)]
PCBM	1-[3-(Methoxycarbonyl)propyl]-1-phenyl-[6.6]C_{61}

References

1. A. E. Becquerel, *Comptes Rendus des Séances Hebdomadaires*, 1839, **9**, 561–567.
2. D. M. Chapin, C. S. Fuller, G. L. Pearson, *J. Appl. Phys.*, 1954, **25**, 676–677.
3. D. M. Chapin, B. Ridge, C. S. Fuller, G. L. Pearson, *US Patent #2,780,765*, "Solar Energy Converting Apparatus."
4. B. O'Regan, M. Grätzel, *Nature*, 1991, **353**, 737–740.
5. I. Levin, C. E. White, *J. Chem. Phys.*, 1950, **18**, 417–426.
6. D. Kearns, M. Calvin, *J. Chem. Phys.*, 1958, **29**, 950–951.
7. H. Kallmann, M. Pope, *J. Chem. Phys.*, 1959, **30**, 585–586.
8. P. J. Reucro, K. Takahashi, H. Ullal, *J. Appl. Phys.*, 1975, **46**, 5218–5223.
9. C. W. Tang, *Appl. Phys. Lett.*, 1986, **48**, 183–185.
10. M. A. Green, K. Emery, Y. Hishikawa, W. Warta, E. D. Dunlop, *Prog. Photovoltaics*, 2013, **21**, 1–11.
11. G. Li, R. Zhu, Y. Yang, *Nat. Photonics*, 2012, **6**, 153–161.
12. B. Kippelen, J. L. Brédas, *Energy Environ. Sci.*, 2009, **2**, 251–261.
13. Y.-W. Su, S.-C. Lan, K.-H. Wei, *Mater. Today*, 2012, **15**, 554–562.
14. P. Peumans, A. Yakimov, S. R. Forrest, *J. Appl. Phys.*, 2003, **93**, 3693.
15. B. A. Gregg, *J. Phys. Chem. B*, 2003, **107**, 4688.
16. S. R. Forrest, *Nature*, 2004, **428**, 911.
17. M. Pope, C. E. Swenberg, *Electronic Processes in Organic Crystals and Polymers*, 2nd ed., Oxford University Press, New York 1999.
18. N. Geacintov, M. Pope, H. Kallmann, *J. Chem. Phys.*, 1966, **45**, 2639.
19. L. Dou, J. You, Z. Hong, Z. Xu, G. Li, R. A. Street, Y. Yang, *Adv. Mater.*, 2013, **25**, 6642–6671.

20. W. Cao, J. Xue, *Energy Environ. Sci.*, 2014, **7**, 2123–2144.
21. Y. Huang, E. J. Kramer, A. J. Heeger, G. C. Bazan, *Chem. Rev.*, 2014, **114**, 7006–7043.
22. B. Walker, C. Kim, T.-Q. Nguyen, *Chem. Mater.*, 2011, **23**, 470.
23. Y. Li, Q. Guo, Z. Li, J. Pei, W. Tian, *Energy Environ. Sci.*, 2010, **3**, 1427.
24. Y. Lin, Y. Li, X. Zhan, *Chem. Soc. Rev.*, 2012, **41**, 4245.
25. S. Shen, P. Jiang, C. He, J. Zhang, P. Shen, Y. Zhang, Y. Yi, Z. Zhang, Z. Li, Y. Li, *Chem. Mater.*, 2013, **25**, 2274.
26. M. Lloyd, J. Anthony, G. Malliaras, *Mater. Today*, 2007, 10, 34.
27. W. Tang, J. Hai, Y. Dai, Z. Huang, B. Lu, F. Yuan, J. Tang, F. Zhang, *Sol. Energy Mater. Sol. Cells*, 2010, 94, 1963.
28. B. A. Gregg, M. A. Fox, A. J. Bard, *J. Phys. Chem.*, 1990, 94, 1586.
29. R. F. Service, *Science*, 2011, 332, 293.
30. J. You, L. Dou, K. Yoshimura, T. Kato, K. Ohya, T. Moriarty, K. Emery, C.-C. Chen, J. Gao, G. Li, Y. Yang, *Nat. Commun.*, 2013, 4, 1446.
31. M. A. Green, K. Emery, Y. Hishikawa, W. Warta, E. D. Dunlop, *Prog. Photovoltaics*, 2013, 21, 827–837.
32. N. C. Giebink, G. P. Wiederrecht, M. R. Wasielewski, S. R. Forrest, *Phys. Rev. B*, 2011, 83, 195326.
33. R. A. Janssen, J. Nelson, *Adv. Mater.*, 2013, 25, 1847.
34. Q. Gan, F. J. Bartoli, Z. H. Kafafi, *Adv. Mater.*, 2013, 25, 2385–2396.
35. H. Gerischer, M. Michel-Beyerle, E. Rebentrost, H. Tributsch, *Electrochim. Acta*, 1968, 13(13), 1509–1515.
36. H. Tributsch, M. Calvin, *Photochem. Photobiol.*, 1971, 14(14), 95–112.
37. H. Tributsch, *Photochem. Photobiol.*, 1972, 16, 261.
38. M. Grätzel, *Inorg. Chem.*, 2005, 44, 6841–6851.
39. Md. K. Nazeeruddin, A. Kay, J. Rodicio, R. Humphry-Baker, E. Müller, P. Liska, N. Vlachopoulos, M. Grätzel, *J. Am. Chem. Soc.*, 1993, 115, 6382.
40. Y. Chiba, A. Islam, Y. Wanatabe, R. Komiya, N. Koide, L. Han, *Jpn. J. Appl. Phys.*, 2006, 45(25), L638–L640.
41. F. Gao, Y. Wang, D. Shi, J. Zhang, M. Wang, X. Jing, R. Humphry-Baker, P. Wang, S. M. Zakeeruddin, M. Grätzel, *J. Am. Chem. Soc.*, 2008, 130, 10720–10728.
42. F. Gao, Y. Wang, J. Zhang, D. Shi, M. Wang, R. Humphry-Baker, P. Wang, S. M. Zakeeruddin, M. Grätzel, *Chem. Commun.*, 2008, 2635–2637.

43. Y. Cao, Y. Bai, Q. Yu, Y. Cheng, S. Liu, D. Shi, F. Gao, P. Wang, *J. Phys. Chem. C,* 2009, 113(15), 6290–6297.
44. F. Sauvage, J.-D. Decoppet, M. Zhang, S. M. Zakeeruddin, P. Comte, Md. K. Nazeeruddin, P. Wang, M. Grätzel, *J. Am. Chem. Soc.,* 2011, 133(24), 9304–9310.
45. F. Sauvage, D. Chen, P. Comte, F. Huang, Y. B. Cheng, R. A. Caruso, M. Grätzel, *ACS Nano,* 2010, 4(8), 4420–4425.
46. A. Yella, H. W. Lee, H. N. Tsao, C. Yi, A. K. Chandiran, Md. K. Nazeeruddin, E. W. G. Diau, C. Y. Yeh, S. M. Zakeeruddin, M. Grätzel, *Science,* 2011, 334(6056), 629–634.
47. S. Mathew, A. Yella, P. Gao, R. Humphry-Baker, B. F. E. Curchod, N. Ashari-Astani, I. Tavernelli, U. Rothisberger, Md. K. Nazeruddin, M. Grätzel, *Nat. Chem.,* 2014, 6(3), 242–247.
48. J. Burschka, N. Pellet, S. J. Moon, R. Humphy-Baker, P. Gao, Md. K. Nazeeruddin, M. Grätzel, *Nature,* 2013, 499(7458), 316–319.
49. Sang Il Seok, *Hybrid Org. Photovoltaics,* 2014, 116.
50. M. Matsumura, S. Matsudaira, H. Tsubomura, M. Takata, H. Yanagida, *Ind. Eng. Chem. Prod. Res. Dev.,* 1980, 19 (3), 415–421.
51. C. Magne, S. Cassaignon, G. Lancel, T. Pauporte, *ChemPhysChem,* 2011, 12(13), 2461–2467.
52. H. J. Koo, J. Park, B. Yoo, K. Yoo, K. Kim, N. G. Park, *Inorg. Chim. Acta,* 2008, 361, 677–683.
53. K. Tennakone, G. R. R. Kumara, I. R. M. Kottegoda, V. S. P. Perera, *Chem. Commun.,* 1999, 15–16.
54. M. Quintana, T. Edvinsson, A. Hagfeldt, G. Boschloo, *J. Phys. Chem. C,* 2006, 111, 1035–1041.
55. J. Elias, M. Parlinska-Wojtan, R. Erni, F. Sauvage, C. Niederberger, J. Michler, L. Philippe, *Nano Energy,* 2012, 1, 742–750.
56. J. Fan, Y. Hao, A. Cabot, E. M. J. Johansson, G. Boschloo, A. Hagfeldt, *ACS Appl. Mater. Interfaces,* 2013, 5, 1902–1906.
57. R. L. Willis, C. Olson, B. O'Regan, T. Lutz, J. Nelson, J. R. Durrant, *J. Phys. Chem. B,* 2002, 106, 7605–7613.
58. K. Keis, J. Lindgren, S. E. Lindquist, A. Hagfeldt, *Langmuir,* 2000, 16, 4688–4694.
59. B. Liu, H. C. Zeng, *J. Am. Chem. Soc.,* 2003, 125, 4430–4431.
60. C. Bauer, G. Boschloo, E. Mukhtar, A. Hagfeldt, *J. Phys. Chem. B,* 2001, 105, 5585–5588.

61. P. Tiwana, P. Docampo, M. B. Johnston, H. J. Snaith, L. M. Herz, *ACS Nano*, 2011, 5, 5158–5166.
62. C. Y. Jiang, X. W. Sun, G. Q. Lo, D. L. Kwong, J. X. Wang, *Appl. Phys. Lett.*, 2007, 90, 263501.
63. Q. Zhang, C. S. Dandeneau, X. Zhou, G. Cao, *Adv. Mater.*, 2009, 21, 4087–4108.
64. K. Westermark, H. Rensmo, H. Siegbahn, K. Keis, A. Hagfeldt, L. Ojamäe, P. Persson, *J. Phys. Chem. B*, 2002, 106, 10102–10107.
65. R. Schölin, M. Quintana, E. M. J. Johansson, M. Hahlin, T. Marinado, A. Hagfeldt, H. Rensmo, *J. Phys. Chem. C*, 2011, 115, 19274–19279.
66. Q. Zhang, T. P. Chou, B. Russo, S. A. Jenekhe, G. Cao, *Angew. Chem.*, 2008, 120, 2436–2440.
67. Y. Shi, C. Zhu, L. Wang, C. Zhao, W. Li, K. K. Fung, T. Ma, A. Hagfeldt, N. Wang, *Chem. Mater.*, 2013, 25, 1000–1012.
68. A. B. F. Martinson, J. W. Elam, J. T. Hupp, M. J. Pellin, *Nano Lett.*, 2007, 7, 2183–2187.
69. H. Rensmo, K. Keis, H. Lindström, S. Södergren, A. Solbrand, A. Hagfeldt, S. E. Lindquist, L. N. Wang, M. Muhammed, *J. Phys. Chem. B*, 1997, 101, 2598–2601.
70. M. Law, L. E. Greene, J. C. Johnson, R. Saykally, P. Yang, *Nat. Mater.*, 2005, 4, 455–459.
71. J. B. Baxter, E. S. Aydil, *Appl. Phys. Lett.*, 2005, 86, 53114.
72. A. K. Chandiran, M. Abdi-Jalebi, Md. K. Nazeeruddin, M. Grätzel, *ACS Nano*, 2014, 8(3), 2261–2268.
73. S. Ferrere, A. Zaban, B. A. Gregg, *J. Phys. Chem. B*, 1997, 101, 4490–4493.
74. A. Kay, M. Grätzel, *Chem. Mater.*, 2002, 14, 2930.
75. E. N. Kumar, R. Jose, P. S. Archana, C. Vijila, M. M. Yusoff, S. Ramakrishna, *Energy Environ. Sci.*, 2012, 5, 5401.
75. Y. P. Y. P. Ariyasinghe, T. R. C. K. Wijayarathna, I. G. C. K. Kumara, I. P. L. Jayarathna, C. A. Thotawatthage, W. S. S. Gunathilake, G. K. R. Senadeera, V. P. S. Perera, *J. Photochem. Photobiol. A*, 2011, 217, 249.
76. H. J. Snaith, C. Ducati, *Nano Lett.*, 2010, 10, 1259–1265.
77. J. Z. Ou, R. A. Rani, M. H. Ham, M. R. Field, Y. Zhang, H. Zheng, P. Reece, S. Zhuiykov, S. Sriram, M. Bhaskaran, *ACS Nano*, 2012, 6, 4045–4053.
78. K. Sayama, H. Sugihara, H. Arakawa, *Chem. Mater.*, 1998, 10, 3825–3832.
79. R. Ghosh, M. K. Brennaman, T. Uher, M. R. Ok, E. T. Samulski, L. E. McNeil, T. J. Meyer, R. Lopez, *ACS Appl. Mater. Interfaces*, 2011, 3, 3929–3935.

80. A. Le Viet, R. Jose, M. V. Reddy, B. V. R. Chowdari, S. Ramakrishna, *J. Phys. Chem. C,* 2010, 114, 21795-21800.
81. S. G. Chen, S. Chappel, Y. Diamant, A. Zaban, *Chem. Mater.,* 2001, 13, 4629-4634.
82. K. Sayama, H. Suguhara, H. Arakawa, *Chem. Mater.,* 1998, 10, 3825.
83. P. Guo, M. A. Aegerter, *Thin Solid Films,* 1999, 351, 290 3825.
84. K. Hara, T. Horiguchi, T. Kinoshita, K. Sayama, H. Sugihara, H. Arakawa, *Sol. Energy Mater. Sol. Cells,* 2000, 64, 115.
85. S. Burnside, J. E. Moser, K. Brooks, M. Grätzel, D. Cahen, *J. Phys. Chem. B,* 1999, 103, 9328-9332.
86. R. Dabestani, A. J. Bard, A. Campion, M. A. Fox, T. E. Mallouk, S. E. Webber, J. M. White, *J. Phys. Chem.,* 1988, 92, 1872-1878.
87. S. Yang, H. Kou, J. Wang, H. Xue, H. Han, *J. Phys. Chem. C,* 2010, 114, 4245-4249.
88. S. S. Shin, J. S. Kim, J. H. Suk, K. D. Lee, D. W. Kim, J. H. Park, I. S. Cho, K. S. Hong, J. Y. Kim, *ACS Nano,* 2013, 7, 1027-1035.
89. B. Tan, E. Toman, Y. Li, Y. Wu, *J. Am. Chem. Soc.,* 2007, 129, 4162-4163.
90. S. H. Choi, D. Hwang, D. Y. Kim, Y. Kervella, P. Maldivi, S. Y. Jang, R. Demadrille, I. D. Kim *Adv. Funct. Mater.,* 2013, 23, 3146-3155.
91. K. Zhu, N. R. Neale, A. Miedaner, A. J. Frank, *Nano Lett.,* 2007, 7, 69-74.
92. H. E. Prakasam, K. Shankar, M. Paulose, O. K. Varghese, C. A. Grimes, *J. Phys. Chem. C,* 2007, 111, 7235-7241.
93. J. R. Jennings, A. Ghicov, L. M. Peter, P. Schmuki, A. B. Walker, *J. Am. Chem. Soc.,* 2008, 130, 13364-13372.
94. D. Kim, A. Ghicov, S. P. Albu, P. Schmuki, *J. Am. Chem. Soc.,* 2008, 130, 16454-16455.
95. G. K. Mor, S. Kim, M. Paulose, O. K. Varghese, K. Shankar, J. Basham, C. A. Grimes, *Nano Lett.,* 2009, 9, 4250-4257.
96. F. Sauvage, F. Di Fonzo, A. Li Bassi, C. S. Casari, V. Russo, G. Diviniti, C. Ducati, C. E. Bottani, P. Comte, M. Grätzel, *Nano Lett.,* 2010, 10(7), 2562-2567.
97. X. Feng, K. Zhu, A. J. Frank, C. A. Grimes, T. E. Mallouk, *Angew. Chem. Int. Ed.,* 2012, 51, 2727 -2730.
98. K. Zhu, S. R. Jang, A. J. Frank, *J. Phys. Chem. Lett.,* 2011, 2, 1070-1076.
99. K. Zhu, T. B. Vinzant, N. R. Neale, A. J. Frank, *Nano Lett.,* 2007, 7(12), 3739-3746.
100. D. H. Chen, F. Z. Huang, Y. B. Cheng, R. A. Caruso, *Adv. Mater.,* 2009, 21, 2206-2210.

101. D. H. Chen, L. Cao, F. Z. Huang, P. Imperia, Y. B. Cheng, R. A. Caruso, *J. Am. Chem. Soc.*, 2010, 132, 4438–4444.

102. A. R. Pascoe, D. Chen, F. Huang, N. W. Duffy, W. Noel, R. A. Caruso, Y. B. Cheng, *J. Phys. Chem. C*, 2014, Ahead of Print.

103. E. M. Hopper, F. Sauvage, A. K. Chandiran, M. Grätzel, K. R. Poeppelmeier, T. O. Mason, *J. Am. Cer. Soc.*, 2012, 95(10), 3192–3196.

104. X. Zhang, F. Liu, Q. L. Huang, G. Zhou, Z. S. Wang, *J. Phys. Chem. C*, 2011, 115(25), 12665–12671.

105. B. Mei, M. D. Sánchez, T. Reinecke, S. Kaluza, W. Xia, M. Muhler, *J. Mater. Chem.*, 2011, 21, 11781–11790.

106. N. Tsvetkov, L. Larina, O. Shevaleevskiy, B. T. Ahna, *J. Electrochem. Soc.*, 2011, 158 (11) B1281–B1285.

107. S. G. Kim, M. J. Ju, I. T. Choi, W. S. Choi, H. J. Choi, J. B. Baek, H. K. Kim, *RSC Adv.*, 2013, 3, 16380–16386.

108. A. K. Chandiran, F. Sauvage, M. Casas-Cabanas, P. Comte, M. Grätzel, *J. Phys. Chem. C*, 2010, 114(37), 15849–15856.

109. R. Ghosh, Y. Hara, L. Alibabaei, K. Hanson, S. Rangan, R. Bartynski, T. J. Meyer, R. Lopez, *ACS Appl. Mater. Interfaces*, 2012, 4, 4566–4570.

110. X. Zhang, S. T. Wang, Z. S. Wang, *Appl. Phys. Lett.*, 2011, 99, 113503.

111. Y. Xie, N. Huang, S. You, Y. Liu, B. Sebo, L. Liang, X. Fang, W. Liu, S. Guo, X. Z. Zhao, *J. Power Sources*, 2013, 224, 168–173.

112. N. A. Kyeremateng, V. Hornebecq, H. Martinez, P. Knauth, T. Djenizian, *ChemPhysChem*, 2012, 13(16), 3707–3713.

113. A. Latini, C. Cavallo, F. K. Aldibaja, D. Gozzi, D. Carta, A. Corrias, L. Lazzarini, G. Salviati, *J. Phys. Chem. C*, 2013, 117, 25276–25289.

114. M. Khana, W. Cao, *J. Mol. Cat. A: Chemical*, 2013, 376, 71–77.

115. A. K. Chandiran, F. Sauvage, E. Lioz, M. Grätzel, *J. Phys. Chem. C*, 2011, 115(18), 9232–9240.

116. Y. Duan, N. Fu, Q. Liu, Y. Fang, X. Zhou, J. Zhang, Y. Lin, *J. Phys. Chem. C*, 2012, 116, 8888–8893.

117. J. Zhang, W. Peng, Z. Chen, H. Chen, L. Han, *J. Phys. Chem. C*, 2012, 116 (36), 19182–19190.

118. Z. Wang, K. Sayama, H. Sugihara, *J. Phys. Chem. B*, 2005, 109, 22449.

119. C. Teng, X. Yang, C. Yuan, C. Li, R. Chen, H. Tian, S. Li, A. Hagfeldt, L. Sun, *Org. Lett.*, 2009, 11, 5542.

120. C. Teng, X. Yang, S. Li, M. Cheng, A. Hagfeldt, L. Z. Wu, L. Sun, *Chem. Eur. J.*, 2010, 16, 13127–13138.

121. L. Li, X. Yang, J. Gao, H. Tian, J. Zhao, A. Hagfeldt, L. Sun, *J. Am. Chem. Soc.*, 2011, 133, 8458.
122. M. Wang, N. Chamberland, L. Breau, J. E. Moser, R. Humphry-Baker, B. Marsan, S. M. Zakeeruddin, M. Grätzel, *Nat. Chem.*, 2010, 2, 385–389.
123. D. Li, H. Li, Y. Luo, K. Li, Q. Meng, M. Armand, L. Chen, *Adv. Funct. Mater.*, 2010, 20, 3358–3365.
124. M. Cheng, X. Yang, S. Li, X. Wang, L. Sun, *Energy Environ. Sci.*, 2012, 5, 6290.
125. H. Nusbaumer, *Alternative redox systems for the dye-sensitized solar cell*, PhD thesis, 2004, EPFL, Switzerland.
126. A. Yella, H. Lee, H. Tsao, C. Yi, A. Chandiran, Md. K. Nazeeruddin, E. Diau, C. Yeh, S. M. Zakeeruddin, M. Grätzel, *Science*, 2011, 334, 629.
127. S. M. Feldt, G. Wang, G. Boschloo, A. Hagfeldt, *J. Phys. Chem. C*, 2011, 115, 21500.
128. J. H. Yum, E. Baranoff, F. Kessler, T. Moehl, S. Ahmad, T. Bessho, A. Marchioro, E. Ghadiri, J. E. Moser, C. Yi, Md. K. Nazeeruddin, M. Grätzel, *Nat. Commun.*, 2012, 3, 361.
129. S. Feldt, E. Gibson, E. Gabrielsson, L. Sun, G. Boschloo, A. Hagfledt, *J. Am. Chem. Soc.*, 2010, 132, 16714.
130. M. Xu, M. Zhang, M. Pastore, R. Li, F. De Angelis, P. Wang, *Chem. Sci.*, 2012, 3, 976–983.
131. Y. Bai, Q. Yu, N. Cai, Y. Wang, M. Zhang, P. Wang, *Chem. Commun.*, 2011, 47, 4376.
132. T. Daeneke, T. Kwon, A. Holmes, N. Duffy, U. Bach, L. Spiccia, *Nat. Chem.*, 2011, 3, 211.
133. T. Li, A. M. Spokoyny, C. She, O. K. Farha, C. A. Mirkin, T. J. Marks, J. T. Hupp, *J. Am. Chem. Soc.*, 2010, 132, 4580.
134. Z. Zhang, P. Chen, T. N. Murakami, S. M. Zakeeruddin, M. Grätzel, *Adv. Funct. Mater.*, 2008, 18, 341.
135. U. Bach, D. Lupo, P. Comte, J. E. Moser, F. Weissortel, J. Salbeok, H. Spreitzer, M. Grätzel, *Nature*, 1998, 395(6702), 583–585.
136. E. Premalal, G. Kumara, R. Rajapakse, M. Shimomur, K. Murakami, A. Konno, *Chem. Commun.*, 2010, 46, 3360.
137. S. H. Im, C. Lim, J. Chang, Y. H. Lee, N. Maiti, H. Kim, Md. K. Nazeeruddin, M. Grätzel, S. Seok, *Nano Lett.*, 2011, 11, 4789.
138. X. Liu, Y. Cheng, L. Wang, L. Cai, B. Liu, *Phys. Chem. Chem. Phys.*, 2012, 14, 7098.

139. M. Wang, P. Chen, R. Humphry-Baker, S. M. Zakeeruddin, M. Grätzel, *ChemPhysChem,* 2009, 10, 290.

140. H. Snaith, S. M. Zakeeruddin, Q. Wang, P. Pechy, M. Grätzel, *Nano Lett.,* 2006, 6, 2000.

141. K. Kalyanasundaram, Md. K. Nazeeruddin, *Chem. Phys. Lett.,* 1992, 193, 292–297.

142. B. Happ, A. Winter, M. D. Hagera, U. S. Schubert, *Chem. Soc. Rev.,* 2012, 41, 2222–2255.

143. Y. Qin, Q. Peng, *Int. J. Photoenergy,* 2012, Article ID 291579.

144. B. Bozic-Weber, E. C. Constable, C. E. Housecroft, *Coord. Chem. Rev.,* 2013, 257, 3089–3106.

145. J.-F. Yina, M. Velayudham, D. Bhattacharya, H.-C. Lin, K.-L. Lu, *Coord. Chem. Rev.,* 2012, 256, 3008–3035.

146. A. Reynal, E. Palomares, *Eur. J. Inorg. Chem.,* 2011, 4509–4526.

147. Md. K. Nazeeruddin, F. De Angelis, S. Fantacci, A. Selloni, G. Viscardi, P. Liska, S. Ito, T. Bessho, M. Grätzel, *J. Am. Chem. Soc.,* 2005, 127 (48), 16835–16847.

148. Md. K. Nazeeruddin, S. M. Zakeeruddin, R. Humphry-Baker, M. Jirousek, P. Liska, N. Vlachopoulos, V. Shklover, C. H. Fischer, M. Grätzel, *Inorg Chem,* 1999, 38, 6298.

150. A. Abbotto, L. Bellotto, F. De Angelis, M. Grätzel, N. Manfredi, C. Marinzi, S. Fantacci, J.-H. Yum, Md. K. Nazeeruddin, *Chem. Commun.,* 2008, 5318–5320.

151. N. Hirata, J.-J. Lagref, E. J. Palomares, J. R. Durrant, M. K. Nazeeruddin, M. Grätzel, D. Di Censo, *Chem. Eur. J.,* 2004, 10, 595.

152. C.-Y. Chen, M. Wang, J.-Y. Li, N. Pootrakulchote, L. Alibabaei, C.-H. Ngocle, J.-D. Decoppet, J.-H. Tsai, C. Grätzel, C.-G. Wu, S. M. Zakeeruddin and M. Grätzel, *ACS Nano,* 2009, 3, 3103–3109.

153. Md. K. Nazeeruddin, P. Péchy, T. Renouard, S. M. Zakeeruddin, R. Humphry-Baker, P. Comte, P. Liska, L. Cevey, E. Costa, V. Shklover, L. Spiccia, G. B. Deacon, C. A. Bignozzi, M. Grätzel, *J. Am. Chem. Soc.,* 2001, 123, 1613.

154. C. Barolo, Md. K. Nazeeruddin, S. Fantacci, D. Di Censo, P. Comte, P. Liska, G. Viscardi, P. Quagliotto, F. De Angelis, S. Ito, M. Grätzel, *Inorg. Chem.,* 2006, 45 (12), 4642–4653.

155. A. Abbotto, F. Sauvage, C. Barolo, F. De Angelis, S. Fantacci, M. Grätzel, N. Manfredi, C. Marinzi, Md. K. Nazeeruddin, *Dalton Trans,* 2011, 40, 234.

156. C. Barolo, J.-H. Yum, E. Artuso, N. Barbero, D. Di Censo, M. G. Lobello, S. Fantacci, F. De Angelis, M. Grätzel, Md. K. Nazeeruddin, G. Viscardi, *Chem Sus Chem,* 2013, 6, 2170–2180.

157. T. Bessho, E. Yoneda, J.-H. Yum, M. Guglielmi, I. Tavernelli, H. Imai, U. Rothlisberger, Md. K. Nazeeruddin, M. Grätzel, *J. Am. Chem. Soc.,* 2009, 131 (16), 5930–5934.

159. T. Bessho, E. C. Constable, M. Grätzel, A. Hernandez Redondo, C. E. Housecroft, W. Kylberg, Md. K. Nazeeruddin, M. Neuburger and S. Schaffner, *Chem. Commun.,* 2008, 3717–3719.

160. S. Mathew, A. Yella, P. Gao, R. Humphry-Baker, B. F. E. Curchod, N. Ashari-Astani, I. Tavernelli, U. Rothlisberger, Md. K. Nazeeruddin, M. Grätzel, *Nat. Chem.,* 2014, 6, 242–247.

161. K. Kalyanasundaram, N. Vlachopoulos, V. Krishnan, A. Monnier, M. Grätzel, *J. Phys. Chem.,* 1987, 91, 2342.

162. A. Kay, M. Grätzel, *J. Phys. Chem.,* 1993, 97, 6272.

163. D. L. Officer, W. M. Campbell, K. W. Jolley, P. Wagner, K. Wagner, P. J. Walsh, K. C. Gordon, L. Schmidt-Mende, Md. K. Nazeeruddin, Q. Wang, M. Grätzel, *J. Phys. Chem. C,* 2007, 111, 11760.

164. T. E. O. Screen, K. B. Lawton, G. S. Wilson, N. Dolney, R. Ispasoiu, T. Goodson Iii, S. J. Martin, D. D. C. Bradley, H. L. Anderson, *J. Mater. Chem.,* 2001, 11, 312–320.

165. V. Lin, S. DiMagno, M. Therien, *Science,* 1994, 264, 1105–1111.

166. L.-L. Li and E. W.-G. Diau, *Chem. Soc. Rev.,* 2013, 42, 291.

167. M. K. Panda, K. Ladomenou, A. G. Coutsolelos, *Coord. Chem. Rev.,* 2012, 256, 2601–2627.

168. P. A. Angaridis, T. Lazarides, A. C. Coutsolelos, *Polyhedron,* 2014, 82(4), 19–32.

169. T. Bessho, S. M. Zakeeruddin, C.-Y. Yeh, E. W. G. Diau, M. Grätzel, *Angew. Chem. Int. Ed.,* 2010, 49, 6646–6649.

170. A. Yella, H.-W. Lee, H. Nok Tsao, C. Yi, A. K. Chandiran, Md. K. Nazeeruddin, E. W. G. Diau, C.-Y. Yeh, S. M. Zakeeruddin, M. Grätzel, *Science,* 2011, 334, 629–634.

171. H. Imahori, T. Umeyama, S. Ito, *Acc. Chem. Res.,* 2009, 42 (11), 1809–1818.

172. W. M. Campbell, A. K. Burrell, D. L. Officer, K. W. Jolley, *Coord. Chem. Rev.,* 2004, 248, 1363–1379.

173. M. K. Panda, K. Ladomenou, A. G. Coutsolelos, *Coord. Chem. Rev.,* 2012, 256, 2601–2627.

174. E. W. G. Diau, Lu-Lin Li, Porphyrin-Sensitized Solar Cells, *Handbook of Porphyrin Science*, 2014, vol. 28, 279–318.
175. A. Hagfeldt, G. Boschloo, L. Sun, L. Kloo, H. Pettersson, *Chem. Rev.*, 2010, **110**(11), 6595.
176. Y. Ooyama, Y. Harima, *Eur. J. Org. Chem.*, 2009, 2903–2934.
177. B.-Gi Kim, K. Chung, J. Kim, *Chem. Eur. J.*, 2013, 19, 5220–5230.
178. W. Zeng, Y. Cao, Y. Bai, Y. Wang, Y. Shi, M. Zhang, F. Wang, Y. Pan, P. Wang, *Chem. Mater.*, 2010, 22, 1915–1925.
179. R. K. Kanaparthi, J. Kandhadi, L. Giribabu, *Tetrahedron*, 2012, 68, 8383–8393.
180. A. De Vos, *Endoreversible Thermodynamics of Solar Energy Conversion*, Science Publishers, Oxford, 1992, p. 301.
181. J. Park, G. Viscardi, C. Barolo, N. Barbero, *Chimia*, 2013, 67, 129–135.
182. A. Mishra, M. K. R. Fischer, P. Bäuerle, *Angew. Chem. Int. Ed.*, 2009, 48, 2474–2499.
183. C. Qin, W.-Y. Wong, L. Han, *Chem. Asian J.*, 2013, 8, 1706–1719.
184. P. V. Kamat, S. Hotchandanit, M. de Lind, K. G. Thomas, S. Das, M. V. George, *J. Chem. Soc. Farad Trans.*, 1993, 89, 2397–2402.
185. S. Alex, U. Santhosh, S. Das, *J. Photochem. Photobiol. A Chem.*, 2005, 172, 63–71.
186. S. Paek, H. Choi, C. Kim, N. Cho, S. So, K. Song, Md. K. Nazeeruddin, J. Ko, *Chem. Commun.*, 2011, 47, 2874.
187. Y. Shi, R. B. M. Hill, J.-H. Yum, A. Dualeh, S. Barlow, M. Grätzel, S. R. Marder, Md. K. Nazeeruddin, *Angew. Chem. Int. Ed.*, 2011, 50, 6619.
188. J. H. Delcamp, Y. Shi, J.-H. Yum, T. Sajoto, E. Dell'Orto, S. Barlow, Md. K. Nazeeruddin, S. R. Marder, M. Grätzel, *Chem. Eur. J.*, 2013, 19, 1819.
189. J. Park, N. Barbero, J. Yoon, E. Dell'Orto, S. Galliano, R. Borrelli, J.-H. Yum, D. Di Censo, M. Grätzel, Md. K. Nazeeruddin, C. Barolo, G. Viscardi, *Phys. Chem. Chem. Phys.*, 2014, 16, 24173–24177.
190. S. Kuster, F. Sauvage, Md. K. Nazeeruddin, M. Grätzel, F. A. Nüesch, T. Geiger, *Dyes Pigments*, 2010, 87, 30–38.
191. L. Beverina, R. Ruffo, C. M. Mari, G. A. Pagani, M. Sassi, F. De Angelis, S. Fantacci, J.-H. Yum, M. Grätzel, Md. K. Nazeeruddin, *ChemSusChem*, 2009, 2, 621–624.
192. T. Maeda, S. Mineta, H. Fujiwara, H. Nakao, S. Yagi, H. Nakazumi, *J. Mater. Chem. A*, 2013, 1, 1303.
193. C. Qin, Y. Numata, S. Zhang, X. Yang, A. Islam, K. Zhang, Han Chen, L. Han, *Adv. Funct. Mater.*, 2014, 24, 3059–3066.

194. J. Park, C. Barolo, F. Sauvage, N. Barbero, C. Benzi, P. Quagliotto, S. Coluccia, D. Di Censo, M. Grätzel, Md. K. Nazeeruddin, G. Viscardi, *Chem. Commun.*, 2012, 48, 2782–2784.
195. G. de la Torre, C. G. Claessens, T. Torres, *Chem. Commun.*, 2007, 2000–2015.
196. Phthalocyanines as sensitizers in dye-sensitized solar cells. *Handbook of Porphyrin Science Schlettwein*, Derck., 2012, vol. 24, pp. 389.
197. M.-E. Ragoussi, M. Ince, T. Torres, *Eur. J. Org. Chem.*, 2013, 6475–6489.
198. M. Kimura, H. Nomoto, H. Suzuki, T. Ikeuchi, H. Matsuzaki, T. N. Murakami, A. Furube, N. Masaki, M. J. Griffith, S. Mori, *Chem. Eur. J.*, 2013, 19, 7496–7502.
199. Y. Reddy, L. Giribabu, C. Lyness, H. Snaith, C. Vijaykumar, M. Chandrasekharam, M. Lakshmikantam, J. H. Yum, K. Kalyanasundaram, M. Grätzel, Md. K. Nazeeruddin, *Angew. Chem. Int. Ed.*, 2007, 46, 373–376.
200. J.-J. Cid, J.-H. Yum, S.-R. Jang, Md. K. Nazeeruddin, E. Martínez-Ferrero, E. Palomares, J. Ko, M. Grätzel, T. Torres, *Angew. Chem.*, 2007, 119, 8510.
201. S. Mori, M. Nagata, Y. Nakahata, K. Yasuta, R. Goto, M. Kimura, M. Taya, *J. Am. Chem. Soc.*, 2010, 132, 4054–4055.
202. Y. Ooyama, Y. Harima, *Eur. J. Org. Chem.*, 2009, 2903–2934.
203. A. Mishra, M. K. R. Fischer, P. Bäuerle, *Angew. Chem. Int. Ed.*, 2009, 48, 2474–2499.
204. S. K. Balasingam, M. Lee, M. Gu Kang, Y. Jun, *Chem. Commun.*, 2013, 49, 1471.
205. J.-H. Yum, S.-R. Jang, P. Walter, T. Geiger, F. Nüesch, S. Kim, J. Ko, M. Grätzel, Md K. Nazeeruddin, *Chem. Commun.*, 2007, 4680.
206. F. Inakazu, Y. Noma, Y. Ogomi, S. Hayase, *Appl. Phys. Lett.*, 2008, 93, 093304.
207. J. N. Clifford, E. Palomares, Md. K. Nazeeruddin, R. Thampi, M. Grätzel, J. R. Durrant, *J. Am. Chem. Soc.*, 2004, 126, 5670.
208. P. J. Holliman, K. J. Al-Salihi, A. Connell, M. L. Davies, E. W. Jones, D. A. Worsley, *RSC Adv.*, 2014, 4, 2515.
209. A. Ehret, L. Stuhl, M. T. Spitler, *J. Phys. Chem. B,* 2001, 105, 9960.
210. R. Y. Ogura, S. Nakane, M. Morooka, M. Orihashi, Y. Suzuki, K. Noda, *Appl. Phys. Lett.*, 2009, 94, 073308.
211. J. N. Clifford, A. Forneli, H. Chen, T. Torres, S. Tan and E. Palomares, *J. Mater. Chem.*, 2011, 21, 1693.

212. C. M. Lan, H. P. Wu, T. Y. Pan, C. W. Chang, W. S. Chao, C. T. Chen, C. L. Wang, C. Y. Lin, d E. W. G. Diau, *Energy Environ. Sci.,* 2012, 5, 6460.
213. Y. Itzhaik, O. Niitsoo, M. Page, G. Hodes, *J. Phys. Chem. C,* 2009, 113, 4254.
214. C.-H. Chang, Y.-L. Lee, *Appl. Phys. Lett.,* 2007, 91, 053503.
215. R. D. Schaller, V. I. Klimov, *Phys. Rev. Lett.,* 2004, 92, 186601.
216. Y. H. Lee, S. H. Im, J. H. Rhee, J.-H. Lee, S. I. Seok, *ACS Appl. Mater. Interf.,* 2010, 2, 1648.
217. I. Robel, M. Kuno, P. V. Kamat, *J. Am. Chem. Soc.,* 2007, 129, 4136.
218. I. R. S. Aga, D. Jowhar, A. Ueda, Z. Pan, W. E. Collins, R. Mu, K. D. Singer, J. Shen, *Appl. Phys. Lett.,* 2007, 91, 232108.
219. S. Kruger, S. G. Hickey, S. Tscharntke, A. Eychmuller, *J. Phys. Chem. C,* 2011, 115, 13047.
220. P. Yu, K. Zhu, A. G. Norman, S. Ferrere, A. J. Frank, A. J. Nozik, *J. Phys. Chem. B,* 2006, 110, 25451.
221. M. Bhushan, A. Catalano, *Appl. Phys. Lett.,* 1981, 38, 39.
222. O. Niitsoo, S. K. Sarkar, C. Pejoux, S. Rühle, D. Cahen, G. Hodes, *J. Photochem. Photobiol. A,* 2006, 181, 306.
223. L. J. Diguna, Q. Shen, J. Kobayashi, T. Toyoda, *Appl. Phys. Lett.,* 2007, 91, 023116.
224. C. J. Liu, J. Olsen, D. R. Sounders, J. H. Wang, *J. Electrochem. Soc.,* 1981, 128, 1224.
225. Y. Ueno, H. Minoura, T. Nishikawa, T. Masayasu, *J. Electrochem. Soc.,* 1983, 130, 43.
226. J. M. Luther, M. Law, M. C. Beard, Q. Song, M. O. Reese, R. J. Ellingson, A. J. Nozik, *Nano Lett.,* 2008, 8, 3488.
227. H. J. Lee, P. Chen, S.-J. Moon, F. Sauvage, K. Sivula, T. Bessho, D. R. Gamelin, P. Comte, S. M. Zakeeruddin, S. Il Seok, M. Grätzel, Md. K. Nazeeruddin, *Langmuir,* 2009, 25, 7602.
228. K. Shankar, X. Feng, C. A. Grimes, *ACS Nano,* 2009, 3, 788.
229. L. Etgar, J. Park, C. Barolo, V. Lesnyak, S. K. Panda, P. Quagliotto, S. G. Hickey, Md. K. Nazeeruddin, A. Eychmüller, G. Viscardi, M. Grätzel, *RSC Adv.,* 2012, 2, 2748.
230. N. Jiang, T. Sumitomo, T. Lee, A. Pellaroque, O. Bellon, D. Milliken, H. Desilvestro, *Solar Energy Mater. Solar Cells,* 2013, 119, 36–50.
231. S. M. Zakeeruddin, M. Grätzel, *Adv. Funct. Mater.,* 2009, 19, 2187–2202.
232. F. Sauvage, S. Chhor, A. Marchioro, J. E. Moser, M. Grätzel, *J. Am. Chem. Soc.,* 2011, 133(33), 13103–13109.

233. Q. Yu, D. Zhou, Y. Shi, X. Si, Y. Wang, P. Wang, *Energy Environ. Sci.*, 2010, 3, 1722–1725.

234. A. Hinsch, J. M. Kroon, R. Kern, I. Uhlendorf, J. Holzbock, A. Meyer, J. Ferber, *Prog. Photovolt. Res. Appl.*, 2001, 9, 425–438.

235. N. Kato, Y. Takeda, K. Higuchi, A. Takeichi, E. Sudo, H. Tanaka, T. Motohiro, T. Sano, T. Toyoda, *Solar Energy Mater. Solar Cells*, 2009, 93, 893–897.

236. E. Leonardi, S. Penna, T. M. Brown, A. Di Carlo, A. Reale, *J. Non-Crystalline Solids*, 2010, 356, 2049–2052.

237. R. Harikisun, H. Desilvestro, *Solar Energy*, 2011, 85, 1179–1188.

238. P. Wang, S. M. Zakeeruddin, J.-E. Moser, Md. K. Nazeeruddin, T. Sekigushi, M. Grätzel, *Nat. Mater.*, 2003, 2, 402.

239. D. Kuang, P. Wang, S. Ito, S. M. Zakeeruddin, M. Grätzel, *J. Am. Chem. Soc.*, 2006, 128, 7732–7733.

240. D. Kuang, C. Klein, Z. Zhang, S. Ito, J.-E. Moser, S. M. Zakeeruddin, M. Grätzel, *Small*, 2007, 3, 2094–2102.

241. Z. Zhang, S. Ito, J.-E. Moser, S. M. Zakeeruddin, M. Grätzel, *Chem. Phys. Chem.*, 2009, 10, 1834–1838.

242. W. Kubo, T. Kitamura, K. Hanabusa, Y. Wada, S. Yanagida, *Chem. Commun.*, 2002, 374–375.

243. H. Matsui, K. Okada, T. Kitamura, N. Tanabe, *Solar Energy Mater. Solar Cells*, 2009, 93, 1110–1115.

244. P. M. Sommeling, M. Späth, H. J. P. Smith, N. J. Bakker, J. M. Kroon, *J. Photochem. Photobiol. A*, 2004, 164, 137–144.

245. H. Pettersson, T. Gruszecki, *Sol. Energy Mater. Sol. Cells*, 2001, 70, 203–212.

246. B. Macht, M. Turrion, A. Barkschat, P. Salvador, K. Ellmer, H. Tributsch, *Sol. Energy Mater. Sol. Cells*, 2002, 73, 163.

247. S. Mastroianni, A. Lembo, T. M. Brown, A. Reale, A. Di Carlo, *Chem. Phys. Chem*, 2012, 13(12), 2964–2975.

248. E. Figgemeier, A. Hagfeldt, *International J. of Photoenergy*, 2004, 6, 127–140.

249. A. Hagfeldt, G. Boschloo, L. Sun, L. Kloo, H. Pettersson, *Chem. Rev.*, 2010, 110, 6595–6663.

250. M. Flasque, A. Nguyen Van Nhien, J. Swiatowska, A. Seyeux, C. Davoisne, F. Sauvage, *ChemPhysChem*, 2014, 15(6), 1126–1137.

251. K. F. Jensen, W. Veurman, H. Brandt, K. Bialecka, S. Sarker, M. M. Rahman, C. Im, A. Hinsch, J. J. Lee, *27th European PV Solar Energy*

Conference and Exhibition, September 24–28, 2012, Frankfurt, Germany.

252. B. Muthuraaman, S. Murugesan, V. Mathew, S. Ganesan, B. J. Paul, J. Madhavan, P. Maruthamuthu, S. A. Suthanthiraraj, *Sol. Energy Mater. Sol. Cells*, 2008, 92, 1712–1717.

253. M. Gorlov, L. Kloo, *Dalton Trans.*, 2008, 2655–2666.

254. A. Hinsch, J. M. Kroon, R. Kern, I. Uhlendorf, J. Holzbock, A. Meyer, J. Ferber, *Prog. Photovolt. Res. Appl.*, 2001, 9, 425–438.

255. A. G. Kontos, T. Stergiopoulos, V. Likodimos, D. Milliken, H. Desilvestro, G. Tulloch, P. Falaras, *J. Phys. Chem. C*, 2013, 117, 8636–8646.

256. H. Greijer, J. Lindgren, A. Hagfeldt, *J. Phys. Chem. B*, 2001, 105, 6314–6320.

257. P. E. Hansen, P. T. Nguyen, J. Krake, J. Spanget-Larsen, T. Lund, *Spectrochim. Acta Mol. Biomol. Spectrosc.*, 2012, 98, 247–251.

258. M. Flasque, A. Nguyen Van Nhien, G. Gachot, J. Swiatowska, A. Seyeux, F. Sauvage, *6th International Conference on Hybrid and Organic Photovoltaics*, Lausanne, Switzerland, May 2014.

259. N. G. Gaylord, A. Takahashi, *J. Polym. Sci.*, 1968, 6(10), 743–748.

260. M. I. Asghar, K. Miettunen, S. Mastroianni, J. Halme, H. Vahlman, P. Lund, *Solar Energy*, 2012, 86, 331–338.

261. H. Tanaka, A. Takeichi, K. Higuchi, T. Motohiro, M. Takata, N. Hirota, J. Nakajima, T. Toyoda, *Solar Energy Mater. Solar Cells*, 2009, 93, 1143–1148.

262. S. Anderson, E. C. Constable, M. P. Dareedwards, J. B. Goodenough, A. Hamnett, K. R. Seddon, R. D. Wright, *Nature*, 1979, 280, 571.

263. Md. K. Nazeeruddin, A. Kay, I. Rodicio, R. Humphry-Baker, E. Muller, P. Liska, N. Vlachopoulos, M. Grätzel, *J. Am. Chem. Soc.*, 1993, 115, 6382.

264. R. T. H. Grünwald, H. Tributsch, *J. Phys. Chem. B*, 1997, 101, 2564.

265. H. Greijer, J. Lindgren, A. Hagfeldt, *J. Phys. Chem. B*, 2001, 105, 6314–6320.

266. P. E. Hansen, P. T. Nguyen, J. Krake, J. Spanget-Larsen, T. Lund, *Spectrochim. Acta Mol. Biomol. Spectrosc.*, 2012, 98, 247–251.

267. N. T. Hoang, N. T. P. Thoa, T. Lund, *Adv. Nat. Sci.*, 2009, 1(10), 51–58.

268. H. T. Nguyen, H. M. Ta, T. Lund, *Solar Energy Mater. Solar Cells*, 2007, 91, 1934–1942.

269. M. Thomalla, H. Tributsch, *C.R. Chim.*, 2006, 9, 659–666.

270. F. Nour-Mohammadi, H. T. Nguyen, G. Boschloo, T. Lund, *J. Photochem. Photo Biol. A: Chem.*, 2007, 187, 348–355.
271. P. T. Nguyen, R. Degn, H. T. Nguyen, T. Lund, *Solar Energy Mater. Solar Cells*, 2009, 93, 1939–1945.
272. P. T. Nguyen, A. R. Andersen, E. M. Skou, T. Lund, *Solar Energy Mater. Solar Cells*, 2010, 94, 1582–1590.
273. O. Kohle, M. Grätzel, A. F. Meyer, T. B. Meyer, *Adv. Mater.*, 1997, 9, 904.
274. M. Grätzel, *C.R. Chim.*, 2006, 9, 578.
275. K. Fredin, K. F. Anderson, N. W. Duffy, G. J. Wilson, C. J. Fell, D. P. Hagberg, L. Sun, U. Bach, S. E. Lindquist, *J. Phys. Chem. C*, 2009, 113, 18902–18906.
276. S. Mikoshiba, S. Murai, H. Sumino, T. Kado, D. Kosugi, S. Hayase, *Curr. Appl. Phys.*, 2005, 5, 152–158.
277. D. F. Watson, G. J. Meyer, *Annu. Rev. Phys. Chem.*, 2005, 56, 119–156.
278. M. R. Hoffmann, S. T. Martin, W. Y. Choi, D. W. Bahnemann, *Chem. Rev.*, 1995, 95, 69.
279. S. T. Martin, H. Herrmann, W. Y. Choi, M. R. Hoffmann, *J. Chem. Soc. Trans.*, 1994, 90, 3315.
280. C. C. Chen, X. Z. Li, W. H. Ma, J. C. Zhao, H. Hidaka, N. Serpone, *J. Phys. Chem. B*, 2002, 106, 318.
281. A. Kraft, H. Hennig, A. Herbst, K. H. Heckner, *J. Electroanal. Chem.*, 1994, 365, 191.

Chapter 4

Thermoelectrics

Damien Saurel

CIC energiGUNE, Parque Tecnológico de Álava,
Albert Einstein 48-ED. CIC 01510 Miñano (Alava), Spain
dsaurel@cicenergigune.com

4.1 Introduction

In 1821 T. J. Seebeck discovered that when a temperature gradient is established between the two junctions of a loop made of two different metals, a magnetic field is generated. Later on, in 1834, J. Peltier discovered that a temperature difference appears between the junctions of two different materials under the flow of an electrical current and in 1838 H. Lenz showed that heat is liberated or absorbed depending on the direction of the electrical current. In 1851, soon after starting establishing the first rules of thermodynamics, W. Thomson (also known as Lord Kelvin) demonstrated that the coupling between electrical current and heat flux through a material, the Thomson effect, is the origin of both Seebeck and Peltier effects. The magnetic field measured by Seebeck was actually the consequence of an electrical current generated by the temperature gradient. He also showed that Seebeck and

---

Materials for Sustainable Energy Applications: Conversion, Storage, Transmission, and Consumption
Edited by Xavier Moya and David Muñoz-Rojas
Copyright © 2016 Pan Stanford Publishing Pte. Ltd.
ISBN 978-981-4411-81-3 (Hardcover), 978-981-4411-82-0 (eBook)
www.panstanford.com

Peltier effects actually appear in each of the individual materials, the properties of a junction being the coupling of the individual effects. It was decades before was discovered what an electrical current is made of: electrons.

Nowadays this coupling between heat current and electric current appears as an appealing way to convert heat into electric power, even if a long way remains to be walked before reaching efficiencies that compete with best energy conversion technologies. Nevertheless, thermoelectricity has its own advantages that make the actual state of the art and near future perspectives attractive for specific applications, namely lower power applications and waste heat conversion.

4.2 Definition

These thermoelectric effects, Seebeck's, Peltier's, and Thomson's, originate in the coupling between electric current and heat current. First of all, Seebeck's discovery indicates that, in addition to Ohm's law, $\vec{J} = \sigma \vec{E}$ (which defines the electronic conductivity σ as the coupling term between the local electric current density $\vec{J}$ and the local electric field $\vec{E}$) in presence of a temperature gradient $\vec{\nabla} T$ one should consider a second contribution to the electric current density:

$$\vec{J} = \sigma(\vec{E} - S\vec{\nabla}T), \tag{4.1}$$

where S is the Seebeck coefficient. In a similar way, Peltier's experiment indicates that, in addition to the contribution of the thermal conductivity K one should consider a second contribution to the heat current density in presence of local electric field

$$\vec{Q} = -\kappa \vec{\nabla} T + \Pi \vec{J}, \tag{4.2}$$

where Π is the Peltier coefficient. Lord Kelvin demonstrated, and it has been further generalized to other systems by Onsager, that the coupling between electron current and heat current is totally reversible, which translates here into a direct relation between Seebeck and Peltier coefficients:

$$\Pi = TS \tag{4.3}$$

The three historical thermoelectric effects are actually manifestations of the same phenomena, the Thomson effect later designated "thermoelectric power." The convention within the scientific community is to use the Seebeck coefficient as the root expression of the thermoelectric power.

4.3 Applications of Thermoelectricity

4.3.1 Temperature Sensing – Thermocouples

This is the simplest and the most spread use of the thermoelectric power. If a metallic wire kept in open electrical circuit (i.e., no electric current) is thermally connected to two distinct temperature sources it will present, according to Eq. (4.4), an internal electric field $\vec{E}$ directly proportional to the internal temperature gradient:

$$\vec{E} = S\vec{\nabla}T \qquad (4.4)$$

This leads to a voltage difference ΔV proportional to the temperature difference ΔT between the extremities:

$$\Delta V = S\Delta T \qquad (4.5)$$

From a practical point of view, it is difficult to measure the voltage difference between the ends of a single thermoelectric wire without perturbing the temperature difference. In order to avoid this issue a couple of wires of different materials are used (see Fig. 4.1), thermally connected in parallel and electrically in series, allowing the measure of the generated voltage difference on the cold side. This couple of wires is called a thermocouple.

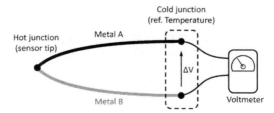

Figure 4.1 Thermocouple used as temperature sensor. The voltage is measured on the reference temperature junction. The cold junction must be in good thermal contact while electrically insulated.

Since the overall voltage difference generated on the reference side will depend on the difference of thermoelectric power of the two legs of the thermocouple, two distinct metals have to be used to get a non-zero output:

$$\nabla V = \int_{T_{COLD}}^{T_{HOT}} (S_A(T) - S_B(T))dT \tag{4.6}$$

The advantages of this technology for temperature sensing are low cost, simplicity, self-powered and high temperature range. It is used in a very broad range of applications, from industry, to science through consumer devices. The drawback is that metals are needed for mechanical reasons (long wires are needed to get enough distance between the reference temperature and the temperature under test), while their thermoelectric power is rather small, in the range of 10 to 100 µV/K. This limits the precision of the temperature sensing to about ±1 K, prohibiting the use of thermocouples for small temperature differences, unless using expensive high precision voltage meters and complex reference temperature shifting compensation.

4.3.2 Conversion

A thermocouple used as sensor is already working in conversion mode, i.e., generating electricity from a heat source. However, the electrical power output capability of thermocouples engineered for sensing is almost zero since only information matters (voltage difference) while the thermal flux and electrical current through the thermocouple have to be kept negligible to not perturb the temperature difference that needs to be estimated.

However, as Seebeck did in his historical experiment, it is also possible to use a thermocouple in order to generate electrical power from a heat source. The capability of a material to do so is commonly estimated through its dimensionless figure of merit:

$$ZT = \frac{S^2 \sigma}{\kappa} T, \tag{4.7}$$

where T is the average working temperature, S the thermoelectric power, σ the electrical conductivity and κ the thermal conductivity.

Note that even if the thermoelectric power has more weight in the figure of merit, a large electrical conductivity and a low thermal conductivity are also needed for large ZT in order to minimize both the internal electrical resistance of the material and the heat loss through thermal conduction.

The efficiency of a thermoelectric material in conversion mode can be defined as the ratio of the electrical power P_{OUT} generated by the thermal power Q_{IN} that penetrates the material from its hot end.

$$\text{eff} = \frac{P_{OUT}}{Q_{IN}} \tag{4.8}$$

The electric power output P_{OUT} generated by the Seebeck effect has the following expression:

$$P_{OUT} = I\Delta V \tag{4.9}$$

where ΔV is the voltage difference generated by the thermoelectric power S, I the electrical current and $\Delta T = T_H - T_C$ is the temperature difference between hot and cold sides.

The input heat power Q_{IN} that penetrates the hot side and the output heat power Q_{OUT} that dissipates from the cold side are derived from equation (4.2), to which the internal Joule heating $R_{in}I^2$ has to be added:

$$Q_{IN} = \kappa \frac{A}{l}\Delta T + T_H SI - \frac{1}{2}R_{in}I^2 \tag{4.10}$$

$$Q_{OUT} = \kappa \frac{A}{l}\Delta T + T_C SI + \frac{1}{2}R_{in}I^2 \tag{4.11}$$

where κ is the thermal conductivity and R_{in} is the internal resistance of the thermoelectric material, as can be seen in the power flow diagram of Fig. 4.2. If one suppose the material to be perfectly thermally insulated except on his ends, the Joule heating dissipation would be spitted equally between the hot end (considered negative) and the cold end (considered positive). One can see here that the electric power output P_{OUT} defined above is actually the difference between the heat input Q_{IN} and output Q_{OUT}, as seen in Fig. 4.2.

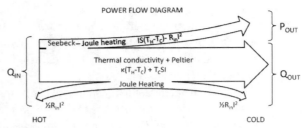

Figure 4.2 Power flow chart within a thermoelectric material working in conversion mode. The output electrical power P_{OUT} corresponds to the Seebeck effect's electromotive force minus the power lost through internal electrical resistance R_{in} by Joule heating. The latter is considered equally dissipated on both ends of the material.

Inserting (4.10) and (4.11) into (4.8) leads to an extended expression of the efficiency:

$$\text{eff} = \frac{I\Delta V}{\kappa \frac{A}{l}\Delta T + T_H SI - \frac{1}{2}R_{in}I^2} \quad (4.12)$$

where A and l are the materials cross section and length, respectively, allowing the conversion from heat flux (in $W \cdot m^{-2}$) to heat power (in W). The equivalent circuit of the thermoelectric material acting as voltage source is a voltage generator of electromotive force E in series with its internal resistance R_{in}, as seen in the schematics of Fig. 4.3 where the electrical current is driven into a resistive load R_{load}.

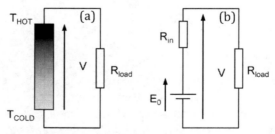

Figure 4.3 (a) Schematics of a thermoelectric material working in conversion mode and delivering its electrical power to a load resistance R_{load}. (b) Equivalent electrical circuit in which the thermoelectric material is represented as a voltage source of electromotive force E_0 and internal resistance R_{in}.

According to the equivalent circuit of Fig. 4.3b, the current I generated and the output power P_{OUT} are

$$I = \frac{\Delta V}{R_{in} + R_{load}} = \frac{S\Delta T}{R_{in}(1+M)} = \frac{A}{l}\frac{\sigma S\Delta T}{(1+M)} \quad (4.13)$$

$$P_{OUT} = I\Delta V = IE - R_{in}I^2 = \frac{A}{l}\sigma S^2(\Delta T)^2 \frac{M}{(1+M)^2}, \quad (4.14)$$

where $M = R_{load}/R_{in}$ is the load ratio and $\sigma = R_{in}^{-1}l/A$ the electrical conductivity. Replacing the expressions above into (4.12) leads to the following expression of the efficiency:

$$\text{eff} = \text{eff}_C \frac{MZT_H}{(1+M)^2 + (1+M)ZT_H - \frac{1}{2}Z\Delta T}, \quad (4.15)$$

where $Z = \sigma S^2/\kappa$ is the figure of merit in K^{-1} and the common factor $\text{eff}_C = \Delta T/T_H$ is the Carnot efficiency, e.g., the maximum efficiency one could expect from a heat machine.

4.3.2.1 Maximum efficiency

Expression (4.15) demonstrates that the efficiency eff depends on the ratio M of the load resistance over the internal resistance. The maximum efficiency eff_{max} can be calculated by solving

$$\left[\frac{\partial \text{eff}}{\partial M}\right]_{\text{eff}=\text{eff}_{max}} = 0, \quad (4.16)$$

which is verified if

$$M = M_{\text{eff}_{max}} = \sqrt{1+Z\overline{T}}, \quad (4.17)$$

where $\overline{T} = (T_H + T_C)/2$ is the average temperature of the material. The maximum efficiency that one could expect from a thermoelectric material in optimum conditions has then the following expression:

$$\text{eff}_{\max} = \text{eff}_{C} \frac{\sqrt{1+\overline{ZT}} - 1}{\sqrt{1+\overline{ZT}} + \frac{T_C}{T_H}} \qquad (4.18)$$

This is the maximum efficiency one could expect from a thermoelectric material whose load over internal resistances ratio respects equation (4.17). This can be achieved by controlling the material's geometrical factor A/l.

$$M = \frac{R_{\text{load}}}{R_{\text{in}}} = R_{\text{load}} \sigma \frac{A}{l}, \qquad (4.19)$$

where σ is the electrical conductivity of the material, A its cross section, and l its length.

The maximum efficiency $\eta_{\max}$ as function of hot temperature is represented in Fig. 4.4 for various values of the figure of merit $\overline{ZT}$. The efficiency of major conversion technologies has been also reported in Fig. 4.4 for comparison [1]. Commercially available thermoelectrics merely achieve $ZT = 1$ and even if they can reasonably be expected to reach $ZT \sim 1.5$–2 in the near future, their efficiencies remain significantly lower than state of the art heat conversion technologies. According to Vining [1], the interest of thermoelectrics is not in making the energy production more efficient, but rather in niche applications were the common technologies cannot apply satisfactorily: [1] portable electronics and other small power applications (efficiency of common technologies drops down when decreasing the power) [1], remote places far from the central grid (aerospace) and waste heat conversion. The major target for thermoelectric R&D nowadays is actually the conversion of waste heat, generated for instance by vehicle engines through exhaust pipes or chemical industry. Indeed, since the source of energy would be wasted anyway the efficiency of conversion is less critical as long as the overall gain compensates the investment. For instance, Vining and collaborators predicted in a near future the possibility to reduce by 10% the gas consumption of consumer cars using a thermoelectric generator to convert the heat generated by the exhaust pipe into electricity able to recharge the battery and hence reduce the load on the combustion engine [1].

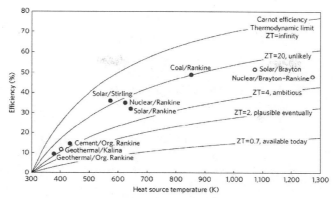

Figure 4.4 Maximum efficiency of thermoelectrics for various ZT values vs. heat source temperature (assuming cold end at 298 K), compared to state of the art conversion technologies. From Vining et al. [1] reprinted by permission from Macmillan Publishers Ltd: Nature Materials, copyright (2009).

4.3.2.2 Maximum power

The maximum electrical power output can be derived from the following condition:

$$\left.\frac{\partial P_{OUT}}{\partial M}\right]_{P_{OUT}=P_{OUT}^{max}} = 0, \qquad (4.20)$$

which is verified if the load resistance equals the internal resistance:

$$M = M_{P_{OUT}^{max}} = 1 \qquad (4.21)$$

One can see here that the condition leading to best power differs from the condition for best efficiency (equation 4.17). According to equation (4.17) the larger the figure of merit ZT the larger this difference is.

The maximum electrical power output would then be

$$P_{OUT^{max}} = \frac{1}{4}\frac{A}{l}\sigma S^2 (\Delta T)^2 \qquad (4.22)$$

where the material's dependent factor σS^2 is commonly called "power factor."

4.3.2.3 Thermoelectric modules

As for the metallic wires used as temperature sensor, if a single material is used as generator its efficiency will be greatly affected by the thermal short circuit induced by the electrical connections that will leak heat from the hot source to the cold source. In order to avoid this problem the solution is the same as for sensors: built thermocouples (see Fig. 4.5). Two major differences appear for the present application, though. First of all, due to their very low *ZT* metals are excluded. Rods of semiconductor ceramics are typically used. Second, the thermocouple will be made of materials with opposite thermopower sign in order to get an overall efficiency as close as possible to the theoretical efficiency of the single material. The internal resistances of the two legs will be in series, their thermal resistances in parallel, while their overall thermoelectric power is the difference of their individual thermopowers leading to an electromotive force defined by

$$E_0 = \int_{T_{COLD}}^{T_{HOT}} (S_A(T) - S_B(T))dT. \tag{4.23}$$

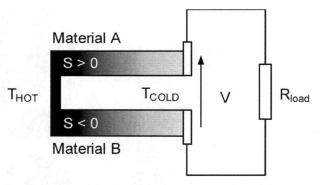

Figure 4.5 Schematic representation of a thermocouple working in conversion mode and powering a resistive load. In order to minimize efficiency loss the two legs of the thermocouple have to be made of two materials with opposite thermopowers.

Moreover, the absolute value of the thermopowers of the two "legs" needs to be as close as possible within the same temperature range. The engineering of a thermocouple for a given temperature of operation goes through a precise selection of the materials.

As can be seen in the examples shown in Fig. 4.6a,b the choices are still quite limited, and for some temperatures the best thermoelectrics do not have their counter parts of opposite sign.

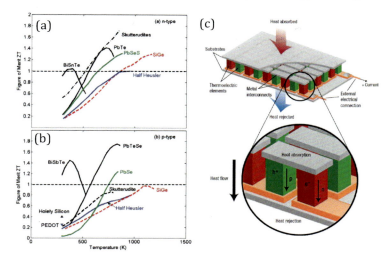

Figure 4.6 (a) and (b) Figure of merit ZT of state of the art of medium to large temperature thermoelectric materials displaying negative ("n-type", (a)) and positive ("p-type", (b)) thermopowers. Reproduced from Zebarjadi et al. with permission of The Royal Society of Chemistry [2]. (c) Schematics of a thermoelectric module constituted of thermocouples. From Snyder et al. [3], reprinted by permission from Macmillan Publishers Ltd: Nature Materials, copyright (2008).

Another important aspect of the design of thermoelectric generators is the output voltage. Best observed thermopowers are in the range of several 100 µV/K to about 1 mV/K, leading to a few 10 to 100 mV with 500–1000 K heat sources, which is way too small to have practical interest. In order to reach a significant voltage several couples will need to be connected in series. This implies a significant increase of the internal resistance of the generator, which is avoided by connecting thermocouples also in parallel. The tuning of the output voltage and internal resistance to reach maximum power or maximum efficiency for a target application will be thus performed by the geometrical dimensions of the legs, the selection of the couple of materials, and the series/parallel network of thermocouples. Such assembly of thermocouples is called a thermoelectric module, as the one shown in Fig. 4.6c.

Finally, apart for all what has been discussed above, it is important to keep in mind that the engineering of a thermocouple whose overall ZT actually reflects the ZT of the materials it is constituted of is not straightforward. Indeed, the previous efficiency equations suppose that all heat is absorbed or rejected at the end of the materials where the electrical connections extract the electrical power. In reality, there is heat dissipation in all surfaces of the material, which might decrease substantially its conversion efficiency. Moreover, since the modules will have to be subject to large temperature changes, all mechanical, thermal, and electrical bounds of the modules need to be perfectly stable to this constraint. Finally, all the thermal and electrical contacts add up to the overall internal thermal and electrical resistances of the module, and thus lower the efficiency. Especially challenging is the need of thermal contacts that are electrically insulating since best ceramics, e.g., alumina, are far from conducting heat as well as metals such as aluminum or copper.

In addition to a strong materials science research activity to discover more efficient thermoelectric materials, a strong effort in modules engineering is compulsory to get the maximum of their capabilities.

4.3.3 Heat Pump

Even if in the present global concern about sustainability in energy production and consumption the conversion of heat into electricity appears to be the main focus of thermoelectricians, the use of thermoelectric materials as heat pumps is worth mentioning since more effective heat pumps lead to lower energy consumption.

As demonstrated by Lord Kelvin, the thermoelectric effect is reversible: an electric current can induce a heat flow, the material acting as a heat pump. Indeed, in expression (4.2) of the heat flux the second term indicates that, in presence of a non-negligible thermopower, part of the heat flux is directly proportional to the electric current crossing the material, even in absence of a temperature gradient.

Following a comparable demonstration as in the case of conversion, one reaches the following expressions of the maximum efficiency of a thermoelectric material working as heat pump, for cooling (4.24) and heating (4.25).

$$\text{eff}_{\max}^{\text{cooling}} = \text{eff}_C^{\text{cooling}} \frac{\sqrt{1+Z\overline{T}} - \frac{T_C}{T_H}}{\sqrt{1+Z\overline{T}} + 1} \qquad (4.24)$$

$$\text{eff}_{\max}^{\text{heating}} = \text{eff}_C^{\text{heating}} \left[1 - 2\frac{\sqrt{1+Z\overline{T}} - 1}{ZT_C} \right], \qquad (4.25)$$

where $\text{eff}_C^{\text{cooling}}$ and $\text{eff}_C^{\text{heating}}$ are the Carnot efficiencies:

$$\text{eff}_C^{\text{cooling}} = \frac{T_C}{\nabla T} \qquad (4.26)$$

$$\text{eff}_C^{\text{heating}} = \frac{T_H}{\nabla T} \qquad (4.27)$$

The efficiencies of cooling and heating vs. temperature of heat absorption (T_H for heating and T_C for cooling) for various values of ZT and for heating mode and cooling mode are represented in Fig. 4.7. It appears clearly that the general trend is a higher efficiency for heating, while for lowest cold temperature the cooling efficiency comes negative. This is due to the Joule heating of the thermoelectric material, which adds up when in heating mode, while is counterproductive in cooling mode. Moreover, contrary to conversion that needs large temperature differences, for pumping heat the smaller the temperature difference the higher the efficiency.

If one could build thermoelectric heat pumps with overall ZT close to the materials ZT, which is typically around unity at room temperature for best thermoelectrics, one would be able to compete seriously with the common mechanical heat pumps. As discussed for conversion application, there are design issues to be solved (heat pump applications have their specific design requirements that will not be detailed here), and some of these materials are quite toxic or expensive. However, results are quite encouraging and allow imagining that, for instance, in a near future, silent domestic thermoelectric heat pumps will be available.

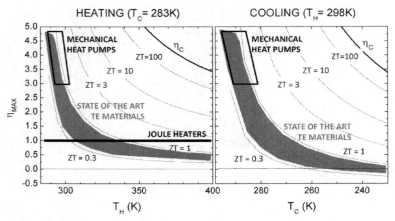

Figure 4.7 Efficiency vs. temperature of heat absorption (T_H for heating and T_C for cooling) for various values of ZT for heating mode (T_C = 283 K, e.g., 10 °C) and cooling mode (T_H = 298 K, e.g., 25 °C). The gray zones represent the state of the art of thermoelectric materials. The efficiency of electricity powered heating/cooling systems has also been reported for comparison (mechanical heat pumps and electric heaters based on Joule dissipation).

Moreover, while thermoelectric modules can be miniaturized if needed (like local CPU cooling in computers), mechanical heat pumps have a physical limit to miniaturization and an efficiency that drops drastically when reducing the power output [1].

4.4 Semiclassical Theory of Thermoelectricity in Solids

4.4.1 Introduction

When a large number of atoms are closely gathered to form a solid, their electron clouds overlap to form chemical bounds. As schematically represented in Fig. 4.8, rather than localized at the vicinity of a single nucleus occupying discrete energy levels, electrons in a solid present some degree of spatial delocalization which allows them to occupy continuums of allowed energies (bands), separated by leftover forbidden energy ranges (gaps).

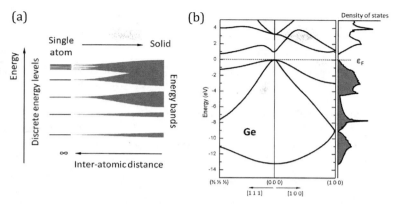

Figure 4.8 (a) Schematic representation of the effect of the interatomic distance on the electronic energy states. Single atoms present discrete energy states, while solids present energy bands separated by gaps of forbidden energies. (b) Example of band structure (germanium, redrawn from ref. [4]). Left: energy dispersion $\varepsilon(k)$ of the band structure in two characteristic crystallographic directions of the reciprocal space. Right: associated density of states $g(\varepsilon)$.

Rather than being all concentrated in the lower energy state, Pauli's exclusion principle forbids two electrons to occupy the same quantum state and hence imposes them to occupy gradually higher energy levels as lower energy levels come occupied. The energy states occupation follows the Fermi–Dirac statistics:

$$f(\varepsilon) = \frac{1}{1 + \exp\left(\dfrac{\varepsilon - \eta}{k_B T}\right)}, \tag{4.28}$$

where k_B is the Boltzmann constant and η the chemical potential. η corresponds to the energy an electron must hold in order to insert within the material, and its value directly depends on the density of electrons within the material, according to the following equation:

$$n = \int_{-\infty}^{+\infty} g(\varepsilon) f(\varepsilon) d\varepsilon, \tag{4.29}$$

where n is the total density of electrons (electrons per volume unit) and $g(\varepsilon)$ is the density of states, which corresponds to the density of electrons per energy unit assuming total occupancy (see Fig. 4.8b for the density of states of Ge)):

$$g(\varepsilon)=\frac{\partial n}{\partial \varepsilon} \tag{4.30}$$

As seen in Fig. 4.9c, at zero temperature the Fermi–Dirac statistics is a step function: the occupation probability is 1 below η and 0 above. In this special case, the chemical potential is called Fermi level, ε_F, and equation (4.29) simplifies as follows:

$$n=\int_{-\infty}^{\varepsilon_F} g(\varepsilon)d\varepsilon \tag{4.31}$$

For a given density of states, the larger the density of electrons the higher the chemical potential will be.

When temperature is increasing the occupation statistics is blurred a few $k_B T$ around η, presenting an S shape rather than the initial step function: There are empty energy states below η and electrons occupying energy states above (see Fig. 4.9b). According to Pauli's exclusion principle, electrons need an accessible (e.g., not occupied) energy state to move within the material, giving rise to electric current and heat flux. This means that the contribution of electrons of energy ε to the overall electronic properties will depend on the product $f(1-f)$, e.g., the statistics of occupancy multiplied by the statistics of vacancy, which is non-negligible only at vicinity of the chemical potential, as seen in Fig. 4.9a. Conversely, if the chemical potential falls in the middle of a gap of forbidden energies, the amount of electrons that can move within the material will be strongly reduced.

Depending on the band structure, three major kinds of electronic transport behavior are usually considered, represented schematically in Fig. 4.10: metals, for which the chemical potential stands in the middle of an energy band; insulators, for which the chemical potential stands within the gap; and semiconductors, for which the chemical potential is close to the a band edge.

The highest occupied energy band is usually called conduction band since its electrons, when occupied, are contributing to the electronic conductivity. The lower energy band is usually called valence bands since its electrons constitute the chemical bonds that hold the atoms together.

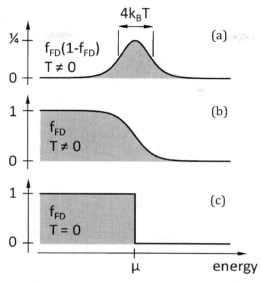

Figure 4.9 (b) and (c): Fermi–Dirac statistics at zero temperature (c) and non-zero temperature (b). (a): $f(1-f)$ deduced from (b).

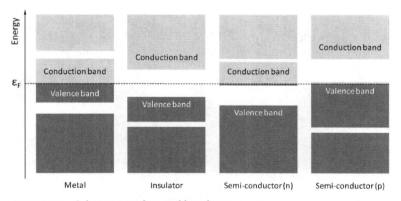

Figure 4.10 Schematics of typical band structures.

4.4.2 Quasi-Free Electron Model

A single free electron of mass m_e and speed v has the following energy:

$$\varepsilon_k = \frac{1}{2} m_e v_k^2 = \frac{\hbar^2 k^2}{2 m_e} \tag{4.32}$$

where v_k is the speed of an electron of momentum k:

$$v_k = \frac{\hbar k}{m_e} \qquad (4.33)$$

and $\hbar$ the reduced Planck constant. According to expressions (4.32) and (4.33) a Fermi momentum k_F and a Fermi speed v_F will be associated to the Fermi level ε_F:

$$\varepsilon_F = \frac{\hbar^2 k_F^2}{2m_e} \qquad (4.34)$$

$$v_F = \frac{\hbar k_F}{m_e} = \frac{2\varepsilon_F}{m_e} \qquad (4.35)$$

If these free electrons are now placed in a periodical lattice, there will be an energy dispersion relation originating at each point of the reciprocal space, as represented in 1D in Fig. 4.11.

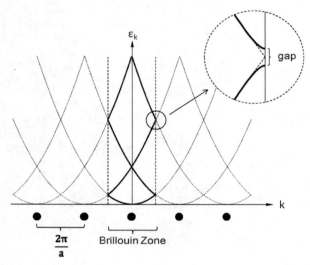

Figure 4.11 1D representation of the free electron energy dispersion in the reciprocal space for a periodic lattice of period a. Each point of the reciprocal space is the center of an energy dispersion curve. The Brillouin zone constitutes the base unit of the network of overlapping dispersion curves. Inset: when subject to a periodic potential (as in a real crystal, due to the attraction potential of nucleus), forbidden gaps open.

The total amount N of electrons filling the crystal is then enclosed within the so-called Fermi sphere of radius k_F centered in the Brillouin zone, i.e., the unit cell in k-space of the network of overlapping energy dispersion curves. The electron density of free electrons can thus be expressed as

$$n = \frac{N}{V} = \frac{1}{3\pi^2}k_F^3 = \frac{1}{3\pi^2}\left(\frac{2m_e}{\hbar^2}\right)^{3/2}\varepsilon_F^{3/2} \tag{4.36}$$

From the above equation, the density of states $g(E)$ can be calculated:

$$g(\varepsilon) = \frac{\partial n}{\partial \varepsilon} = \frac{\partial n}{\partial k}\frac{\partial k}{\partial \varepsilon} = \frac{1}{2\pi^2}\left(\frac{2m_e}{\hbar^2}\right)^{3/2} \tag{4.37}$$

When electrons are placed in a real crystal, they feel the periodic potential generated by the lattice of nucleus. This tends to open forbidden gaps in the intersection of the dispersion relations, as represented in the inset of Fig. 4.11. Moreover, while the band structure of Fig. 4.11 is quite simplistic, in a real crystal the symmetry leads to more complex free electron band structures, see for instance the case of germanium in Fig. 4.8b.

Although the quasi-free electron model is simplistic, it is commonly used for predicting and interpreting the conduction properties of real systems presenting electron delocalization, e.g., metals, semi-metals and extrinsic semiconductors. In order to take into account interactions with the lattice and possible electron–electron interactions, some adjustment of the free electron model has to be done by considering an effective electron mass m^*, which would be different from the free electron mass m_e. In k-space the effective mass is directly related to the curvature of the energy dispersion relation:

$$m^* = \frac{1}{\hbar^2}\left[\frac{\partial^2 \varepsilon}{\partial k^2}\right]^{-1} \tag{4.38}$$

According to (4.33) the speed of an electron of momentum k would then become

$$v_k = \frac{\hbar k}{m^*}. \tag{4.39}$$

Note that since within a crystal the effective mass depends on the crystallographic direction, it is actually a tensor.

4.4.3 Electron Conductivity

According to Ohm's law, the electrical conductivity σ is the coupling term between density of current $\vec{J}$ and electrostatic field $\vec{E}$:

$$\vec{J} = \sigma \vec{E} \tag{4.40}$$

As discussed previously, only partially occupied energies contribute to electron motion. In order to take this strong energy dependence into account $\vec{J}$ has to be written as the integral of its component in k-space:

$$\vec{J} = e \int_{-\infty}^{+\infty} f_k \vec{v}_k \cdot \frac{d\vec{k}}{4\pi^3} \tag{4.41}$$

where $\vec{v_k}$ is the speed of electrons of moment $\vec{k}$, f_k the probability of the state k to be occupied by an electron, i.e., the Fermi–Dirac distribution (4.28), and e the negative electron's charge.

Once the transitory period is over (during which polarization and current stabilize), equation (4.40) corresponds to a steady state with constant current density and electric field. The temperature, and hence the chemical potential, are constant, well defined in each point of the system and present no gradient. The occupation probability f_k can thus be decomposed between an equilibrium part, f_{eq}, and an out of balance part that corresponds to the local effect of the electrical field on electronic properties:

$$f_k = f_{eq}(\varepsilon_k) + \delta f_k \tag{4.42}$$

Since $v_k = -v_{-k}$ and $f_{eq}(\varepsilon_k) = f_{eq}(\varepsilon_{-k})$, the equilibrium part of equation (4.41) vanishes and only the out of balance part remains:

$$\vec{J} = e \int_{-\infty}^{+\infty} \delta f_k \vec{v}_k \cdot \frac{d\vec{k}}{4\pi^3} \tag{4.43}$$

In order to get the expression of δf_k, and thus of $\vec{J}$, one has to start from the Boltzmann transport equation in presence of electric field and with no temperature gradient:

$$\left.\frac{\partial f_k}{\partial t}\right|_{\text{field}} + \left.\frac{\partial f_k}{\partial t}\right|_{\text{scatt.}} = 0 \qquad (4.44)$$

Electrons energies will be locally affected by the electric field E, inducing motion, which will be affected by their scattering due to electron–electron, electron–lattice, or electron–impurities interactions. The scattering will be supposed elastic, which implies that the net rate (i.e., the integral over all energies) of the changes at each position of the system is zero.

The field term can be easily determined by first Newton's law of motion:

$$\left.\frac{\partial f_k}{\partial t}\right|_{\text{field}} = \left.\frac{\partial \vec{k}}{\partial t}\right|_{\text{field}} \cdot \frac{\partial \varepsilon_k}{\partial \vec{k}} \frac{\partial f_k}{\partial \varepsilon_k} = -\frac{\partial f_k}{\partial \varepsilon} e\vec{E} \cdot \vec{v_k} \qquad (4.45)$$

where v_k is the speed of an electron of momentum k. Concerning the scattering part, the reasonable phenomenological assumption is made that any out of balance fluctuation of energy states occupation δf_k induced by the electrical field will decay to zero through scattering as

$$\left.\frac{\partial f_k}{\partial t}\right|_{\text{scatt}} = \frac{\delta f_k}{\tau_k}, \qquad (4.46)$$

where τ_k is a relaxation time that may depend on k (see for instance reference [5] for detailed demonstration). Introducing (4.45) and (4.46) into (4.44) leads to the following expression of the out of balance fluctuation of energy states occupation:

$$\delta f_k = -\frac{\partial f_k}{\partial \varepsilon_k} \tau_k e\vec{E} \cdot \vec{v_k} \qquad (4.47)$$

Introducing (4.47) into (4.43) leads to

$$\vec{J} = -e^2 \int_{-\infty}^{+\infty} \frac{\partial f_k}{\partial \varepsilon_k} \tau_k v_k^2 E \frac{d\vec{k}}{4\pi^3}. \qquad (4.48)$$

Comparing (4.48) with (4.40) allows writing the following expression of the conductivity:

$$\sigma_k = -e^2 \tau_k v_k^2 \frac{\partial f_{eq}}{\partial \varepsilon_k} \qquad (4.49)$$

$$\sigma = \left\| \int \sigma_k \frac{d\vec{k}}{4\pi^3} \right\| \qquad (4.50)$$

These equations in k-space are limited to the consideration of a single band in a few $k_B T$ around the chemical potential η. In case of complicated band structure, like overlapping bands or narrow band gaps, there might be more than one band contributing to the conductivity, in which case several energies can correspond to a single k-value. It would thus be needed to rewrite the previous equations as function of energy, leading to:

$$\sigma(\varepsilon) = -e^2 \tau(\varepsilon) v(\varepsilon)^2 g(\varepsilon) \frac{\partial f_{eq}}{\partial \varepsilon} \qquad (4.51)$$

$$\sigma = \int \sigma(\varepsilon) d\varepsilon, \qquad (4.52)$$

where $g(\varepsilon)$ is the density of states at energy ε, which originates from the change of variable from k to ε and directly depends on the band structure; see example of Fig. 4.8.

In the low temperature limit ($k_B T \ll (\eta - E_C)$, $(\eta \approx \varepsilon_F)$, equations (4.51) and (4.52) become:

$$\sigma \approx e^2 \tau(\varepsilon_F) v(\varepsilon_F)^2 g(\varepsilon_F) \qquad (4.53)$$

Putting (4.35), (4.36) and (4.37) into (4.53) an old classical result from the Drude theory of metals is recovered:

$$\sigma \approx \frac{n e^2 \tau}{m^*} = n|e|\mu, \qquad (4.54)$$

where μ is the mobility of the charge carriers, which, according to (4.31) and (4.53), can be defined as follows:

$$\mu = \frac{|e|\tau}{m^*} = \frac{\sigma}{n|e|} = -|e| \frac{\int \tau(\varepsilon) v(\varepsilon)^2 g(\varepsilon) \frac{\partial f_{eq}}{\partial \varepsilon} d\varepsilon}{\int g(\varepsilon) f(\varepsilon) d\varepsilon} \qquad (4.55)$$

As the charge carrier mobility of a material can be experimentally measured (e.g., by Hall measurements), the use of the following alternative expression of the conductivity is quite common:

$$\sigma = |e|\mu \int g(\varepsilon)f(\varepsilon)d\varepsilon \tag{4.56}$$

4.4.4 Thermopower

According to equation (4.5), the thermoelectric power is the coupling term between voltage gradient $\vec{\nabla}V$ and temperature gradient $\vec{\nabla}T$:

$$S = \frac{\vec{\nabla}V}{\vec{\nabla}T} \tag{4.57}$$

The current density will thus depend not only on the electric field E as for electron conductivity but also on the temperature gradient through the thermoelectric power:

$$\vec{J} = \sigma(\vec{E} - S\vec{\nabla}T) \tag{4.58}$$

In order to establish the expression of the thermoelectric power one has to rewrite the Boltzmann transport equation (4.44) including now a diffusion term, since temperature, and consequently chemical potential, is now varying all along the material inducing charge carrier diffusion:

$$\left.\frac{\partial f_k}{\partial t}\right]_{\text{field}} + \left.\frac{\partial f_k}{\partial t}\right]_{\text{scatt.}} + \left.\frac{\partial f_k}{\partial t}\right]_{\text{diff}} = 0 \tag{4.59}$$

$$\left.\frac{\partial f_k}{\partial t}\right]_{\text{diff}} = \frac{\partial r}{\partial t}\frac{\partial f_k}{\partial r} = -\vec{v}_k \cdot \vec{\nabla}f_k, \tag{4.60}$$

where the gradient of electron density takes the following form:

$$\vec{\nabla}f_k = -\frac{\partial f_{eq}}{\partial \varepsilon_k}\left(\vec{\nabla}\eta + \frac{\varepsilon_k - \eta}{T}T\right) \tag{4.61}$$

Following the same path than for the conductivity, one obtains the following expression of the out of balance electron density:

$$\delta f_k = -\frac{\partial f_{eq}}{\partial \varepsilon_k} \tau_k \vec{v}_k \cdot \left(e\left(\vec{E} - \frac{1}{e}\vec{\nabla}\eta\right) - \frac{\varepsilon_k - \eta}{T}\vec{\nabla}T \right) \tag{4.62}$$

Introducing it into expression (4.43) leads to the following expression of the current density:

$$\vec{J} = -e\int \tau_k v_k^2 \frac{\partial f_{eq}}{\partial \varepsilon_k}\left(e\vec{E} - \vec{\nabla}\eta - \frac{\varepsilon_k - \eta}{T}\vec{\nabla}T\right)\frac{d\vec{k}}{4\pi^3} \tag{4.63}$$

From equations (4.58) and (4.63), one can deduce the expression of the thermoelectric power:

$$S = \left\| \int S_k \frac{d\vec{k}}{4\pi^3} \right\| \tag{4.64}$$

$$S_k = \frac{\sigma_k}{\sigma}\frac{\varepsilon_k - \eta}{eT} \tag{4.65}$$

where σ_k and σ are given by (4.49) and (4.50). As function of energy, these equations write as

$$S = \int S(\varepsilon)d\varepsilon \tag{4.66}$$

$$S(\varepsilon) = \frac{\sigma(\varepsilon)}{\sigma}\frac{\varepsilon - \eta}{eT} \tag{4.67}$$

where $\sigma(\varepsilon)$ and σ are given by (4.51) and (4.52). From equation (4.67) one can see that electrons above the chemical potential have a positive contribution to the thermoelectric power while electrons of lower energy have a negative contribution.

4.4.5 The Sommerfeld Expansion:

The previous expressions of the thermoelectric power are exact and allow calculating the electronic properties of a large variety of systems (metals, insulators, semiconductors) from their band structure. However, as for the chemical potential determination, it usually requires numerical calculations and a precisely known band structure, therefore the higher the temperature or the more complex the band structure, the more complicated the calculus.

For low to moderate temperatures (i.e., $k_B T$ much smaller than the energy difference between the chemical potential and the band edges), and in a more general way if no singularity is present in the density of states a few $k_B T$ around the chemical potential (like band overlap, forbidden gap, band edge), one can simplify the expression of the Fermi integral of any energy dependent function $F(\varepsilon)$ using the Sommerfeld expansion:

$$\int_{-\infty}^{+\infty} F(\varepsilon) f(\varepsilon) d\varepsilon = \int_{-\infty}^{\eta} F(\varepsilon) d\varepsilon + \frac{1}{6} \pi^2 (k_B T)^2 \left[\frac{\partial F(\varepsilon)}{\partial \varepsilon} \right]_{\varepsilon = \eta} + \dots \quad (4.68)$$

This approximation is based on a classical Taylor local expansion at the vicinity of the chemical potential and taking advantage of the particular symmetries of the Fermi–Dirac distribution. The detailed demonstration can be found in refs. [5, 6].

Applied to expression (4.31) of the total density of electrons, one obtains

$$n = \int_{-\infty}^{\varepsilon_F} g(\varepsilon) d\varepsilon \approx \int_{-\infty}^{\eta} g(\varepsilon) d\varepsilon + \frac{1}{6} \pi^2 (k_B T)^2 \left[\frac{\partial g(\varepsilon)}{\partial \varepsilon} \right]_{\varepsilon = \eta} \quad (4.69)$$

which leads to

$$0 \approx \int_{-\infty}^{\eta} g(\varepsilon) d\varepsilon - \int_{-\infty}^{\varepsilon_F} g(\varepsilon) d\varepsilon + \frac{1}{6} \pi^2 (k_B T)^2 \left[\frac{\partial g(\varepsilon)}{\partial \varepsilon} \right]_{\varepsilon = \eta} \quad (4.70)$$

$$0 \approx (\eta - \varepsilon_F) g(\mu) + \frac{1}{6} \pi^2 (k_B T)^2 \left[\frac{\partial g(\varepsilon)}{\partial \varepsilon} \right]_{\varepsilon = \eta} \quad (4.71)$$

$$\eta \approx \varepsilon_F - \frac{1}{6} \pi^2 (k_B T)^2 \left[\frac{\partial \ln g(\varepsilon)}{\partial \varepsilon} \right]_{\varepsilon = \eta} \quad (4.72)$$

Supposing that the density of states follows locally a power dependence of the energy ($\alpha = 0.5$ for quasi free electrons):

$$g(\varepsilon) \sim (\varepsilon - E_C)^\alpha, \quad (4.73)$$

where E_C is the band edge position. Equation (4.72) becomes

$$\eta \approx \varepsilon_F - \frac{1}{6} \pi^2 (k_B T)^2 \alpha \frac{1}{\varepsilon_F - E_C}. \quad (4.74)$$

Applying the Sommerfeld expansion to expressions (4.66) and (4.67) of the thermoelectric power the commonly used Mott expression is obtained:

$$S \approx -\frac{\pi^2}{3} \frac{k_B^2 T}{|e|} \left[\frac{\partial \ln\sigma(\varepsilon)}{\partial \varepsilon} \right]_{\varepsilon=\eta} \quad (4.75)$$

This expression has to be used with special care since its validity range suffers several restrictions. First of all, since it is based on the first order term of a Taylor series, it is limited to low temperature, i.e., temperatures low enough so that the linear approximation remains valid in a few $k_B T$ around the chemical potential. Second, no singularity may be present in the relevant energy range, e.g., band edges, sharp bands, impurity levels, etc.; otherwise this expression would not account for the influence of those singularities on the temperature dependence of the chemical potential, which would lead to a bad estimate of the thermoelectric power. It is thus limited to conventional metals and degenerate semiconductors, which typically present broad bands, Fermi levels far from band edges compared to $k_B T$, and single band contribution to electronic properties.

Expression (4.75) can be developed according to (4.51):

$$S \approx -\frac{\pi^2}{3} \frac{k_B^2 T}{|e|} \left[\frac{\partial \ln g(\varepsilon)}{\partial \varepsilon} + \frac{\partial \ln \tau(\varepsilon)}{\partial \varepsilon} + 2 \frac{\partial \ln v(\varepsilon)}{\partial \varepsilon} \right]_{\varepsilon=\eta} \quad (4.76)$$

The first term is similar to expression (4.72), and corresponds thus to the temperature dependence of the chemical potential according to the Femi-Dirac statistics, e.g., for non-interacting hot and cold sides (see Fig. 4.12a). The two other terms are related to the diffusion induced by the thermal gradient (electrons will tend to move from the hot to the cold side) and the associated scattering processes (see Fig. 4.12b).

In the same way, as for the density of states in (4.73), it is quite usual to assume energy power dependences of the relaxation time and the speed as

$$\tau(\varepsilon) \sim (\varepsilon - E_c)^\beta \quad (4.77)$$

$$v(\varepsilon) \sim (\varepsilon - E_c)^{\gamma}. \tag{4.78}$$

Concerning the relaxation time, the value of β depends on the scattering mechanism: $\beta = -0.5$ for scattering by acoustic phonons, $\beta = 0$ for scattering by neutral impurities and $\beta = 1.5$ for scattering by ionized Impurities. For the speed, the most commonly used energy dependence is the effective mass approximation according to equation (4.39), e.g., $\gamma = 1/2$.

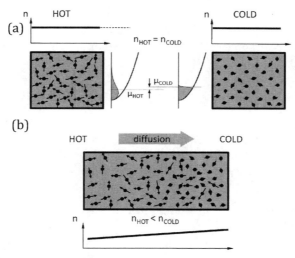

Figure 4.12 Schematic representation of the two main contributions to the thermoelectric power of metals. (a) Contribution of the density of states asymmetry assuming no diffusion. (b) Contribution of the diffusion that leads to a gradient of concentration along the material.

The expression of the thermoelectric power becomes:

$$S \approx -\frac{\pi^2}{3} \frac{k_B^2 T}{|e|} \frac{1}{\eta - E_c} (\alpha + \beta + 2\gamma) \tag{4.79}$$

The thermoelectric power will thus depend on several factors: energy dependence of the density of states and of the speed at the chemical potential through α and 2γ, respectively, and on the scattering mechanism through β. Supposing that the common approximations hold (single band in a few $k_B T$ around E_F, quasi-free electrons ($\alpha = 0.5$), effective mass ($\gamma = 0.5$), scattering by acoustic phonons ($\beta = -0.5$)) the thermoelectric power simplifies as

$$S \approx -\frac{\pi^2}{3} \frac{k_B^2 T}{|e|} \frac{1}{\eta - E_c}. \tag{4.80}$$

4.4.6 Electrons and Holes

The expression (4.41) of the current density in absence of temperature gradient can be decomposed in the following way:

$$\vec{J} = -|e| \int_{-\infty}^{+\infty} \vec{v}_k \cdot \frac{d\vec{k}}{4\pi^3} - \left(-|e| \int_{-\infty}^{+\infty} (1 - f_k) \vec{v}_k \cdot \frac{d\vec{k}}{4\pi^3} \right), \tag{4.81}$$

where the first term on the right side is actually the contribution to the current density of a fully occupied band structure, which therefore vanishes:

$$\vec{J}_{full} = -|e| \int_{-\infty}^{+\infty} \vec{v}_k \cdot \frac{d\vec{k}}{4\pi^3} = -\frac{|e|\hbar}{m^*} \int_{-\infty}^{+\infty} \vec{k} \cdot \frac{d\vec{k}}{4\pi^3} = 0 \tag{4.82}$$

The current density can thus have two symmetrical and equivalent expressions:

$$\vec{J} = -|e| \int_{-\infty}^{+\infty} f_k \vec{v}_k \cdot \frac{d\vec{k}}{4\pi^3} = +|e| \int_{-\infty}^{+\infty} (1 - f_k) \vec{v}_k \cdot \frac{d\vec{k}}{4\pi^3} \tag{4.83}$$

The current density produced by the electrons occupying the band structure actually equals the current density of particles, actually vacancies or *holes* since their occupancy is $(1 - f_k)$, to which the charge $+|e|$ (opposite to the electron charge) would be associated.

If a single energy band is considered in a crystal, as schematically represented in Fig. 4.13, the second derivative of the energy dispersion appears positive at the bottom of the band and negative on the top. According to (4.38) and (4.39), this means that on top of a band charge carriers move in opposite direction. As a result, when the chemical potential is close to the top band edge, the electrons react to a driving field as if they had a negative mass or, differently said, as if they were positively charged particles, e.g., holes. According to equation (4.40) and (4.83) it has no incidence on the sign of the conductivity; however, for the thermoelectric power the situation is different. When the chemical potential is close to the bottom of a band, the density of states increases with energy, leading to a negative thermopower. The material is called n-type, or having an electron character. Conversely, if the chemical potential is close to the band maximum, the density of states decreases with energy leading to a positive thermopower. The material is then called p-type, e.g., having a hole character. This is verified with the Mott expression (4.75) of the thermoelectric power as well as the more general expression given in (4.67). Whether electrons or holes are pertinent to describe the electronic properties of a band depends thus on the position of the chemical potential in respect of the band edges.

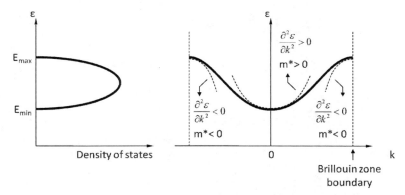

Figure 4.13 Schematic representation of a single band. On the left the density of states, on the right a 1D representation of the energy dispersion within the Brillouin zone. The dashed lines are parabolas representing the effective mass approximation.

4.4.7 Multiband Contribution

Most of the results of the Sommerfeld expansion suppose a unique band contributing to the thermoelectric power that can be described with the quasi-free electron approximation.

When more than one band contributes to the transport properties, the density of states is actually the sum of the contribution of the various bands:

$$g(\varepsilon) = \sum_i g_i(\varepsilon) \qquad (4.84)$$

In consequence, the electronic conductivity will be simply the sum of the individual contribution of each band:

$$\sigma = -e^2 \sum_i \left[\int \tau_i(\varepsilon) v_i(\varepsilon)^2 g_i(\varepsilon) \frac{\partial f_{eq}}{\partial \varepsilon} d\varepsilon \right] = \sum_i \sigma_i \qquad (4.85)$$

Concerning the chemical potential, it cannot be calculated any more from (4.74), and more general expressions need to be used such as (4.29), or (4.72) if the Sommerfeld expansion can be applied on each band:

$$n = \int_{-\infty}^{+\infty} \sum_i g_i(\varepsilon) f(\varepsilon) d\varepsilon \qquad (4.86)$$

$$\eta \approx \varepsilon_F - \frac{1}{6} \pi^2 (k_B T)^2 \left[\frac{\partial \ln \left[\sum_i g_i(\varepsilon) \right]}{\partial \varepsilon} \right]_{\varepsilon = \eta} \qquad (4.87)$$

The overall thermoelectric power will then have the following general expression:

$$S = \frac{1}{\sigma} \int \sum_i \sigma_i(\varepsilon) \frac{\varepsilon - \eta}{eT} d\varepsilon \qquad (4.88)$$

with

$$\sigma_i(\varepsilon) = \tau_i(\varepsilon) v_i(\varepsilon)^2 g_i(\varepsilon) \frac{\partial f_{eq}}{\partial \varepsilon} \qquad (4.89)$$

The contribution of a single band considered alone would then be

$$S_i = \frac{1}{\sigma_i} \int \sigma_i(\varepsilon) \frac{\varepsilon - \eta}{eT} d\varepsilon \approx -\frac{\pi^2}{3} \frac{k_B^2 T}{|e|} \left[\frac{\partial \ln \sigma_i(\varepsilon)}{\partial \varepsilon} \right]_{\varepsilon = \eta} \qquad (4.90)$$

with the overall thermoelectric power given by

$$S = \frac{\sum_i \sigma_i S_i}{\sigma}. \qquad (4.91)$$

The overall thermoelectric power is thus actually the sum of the thermoelectric power of the individual bands considered alone, weighted by their contribution to the overall conductivity. A band might have a high individual thermoelectric power, but it will not contribute notably to the overall thermoelectric power if its contribution to the conductivity is weak.

4.4.8 Thermal Conductivity

The total thermal conductivity κ is the sum of an electronic contribution κ_e, where heat is transmitted by electrons thermally excited above the chemical potential, and a lattice contribution κ_l, where heat is transmitted through lattice vibrations (phonons):

$$\kappa = \kappa_l + \kappa_e \tag{4.92}$$

When a temperature gradient is applied to a thermally conductive material it gives rise to a heat current $\vec{Q}$. According to the two first principles of the thermodynamics, the heat transferred to a system corresponds actually to the total energy transferred minus the work transferred by the system to its surrounding. For electrons of a solid submitted to a temperature gradient, it is translated into the following relation:

$$\vec{Q}_e = \vec{J}_E - \eta \vec{J}_n, \tag{4.93}$$

where $\vec{Q}_e$ is the part of the heat current due to electrons, η their chemical potential, $\vec{J}_E$ the related energy current and $\vec{J}_n$ the electron flux, which is simply the electrical current, defined by expression (4.43), divided by the electron charge:

$$\vec{J}_n = \int_{-\infty}^{+\infty} \delta f v(\varepsilon) g(\varepsilon) d\varepsilon \tag{4.94}$$

By analogy the energy current can be defined as

$$\vec{J}_E = \int_{-\infty}^{+\infty} \delta f \varepsilon v(\varepsilon) g(\varepsilon) d\varepsilon \tag{4.95}$$

Combining the above equations one reaches the expression of the heat current:

$$\vec{Q}_e = \int_{-\infty}^{+\infty} \delta f (\varepsilon - \eta) v(\varepsilon) g(\varepsilon) d\varepsilon \tag{4.96}$$

Replacing the out of balance occupation probability δf_k by its expression (4.62) deduced from Boltzmann transport equations, one reaches the following expression of the heat current:

$$\vec{Q_e} = |e|L_1 \cdot \left(\vec{E} - \frac{1}{|e|}\vec{\nabla}\eta\right) - \frac{1}{T}L_2 \cdot \vec{\nabla}T \qquad (4.97)$$

With the coefficients L_α defined as

$$L_\alpha = -\int_{-\infty}^{+\infty} (\varepsilon - \eta)^\alpha \frac{\partial f_{eq}}{\partial \varepsilon} \tau(\varepsilon) v(\varepsilon)^2 g(\varepsilon) d\varepsilon. \qquad (4.98)$$

Note that according to this formalism, the expressions (4.43), (4.50), and (4.64) of the electrical current, electrical conductivity and thermoelectric power, respectively, would write

$$\vec{J} = e^2 L_0 \cdot \left(\vec{E} - \frac{1}{|e|}\vec{\nabla}\eta\right) - \frac{1}{|e|T}L_1 \cdot \vec{\nabla}T \qquad (4.99)$$

$$\sigma = e^2 L_0 \qquad (4.100)$$

$$S = \frac{1}{|e|T}\frac{L_1}{L_0}. \qquad (4.101)$$

As defined at the beginning of this chapter (expression (4.2)), in absence of electrical current there is a direct relation between the heat current carried by electrons and their thermal conductivity κ_e:

$$\vec{Q_e} = -\kappa_e \vec{\nabla}T \qquad (4.102)$$

Combining expressions (4.97), (4.99) and (4.102) in absence of electrical current, one reaches the following expression of the thermal conductivity:

$$\kappa_e = \frac{1}{T}\left(L_2 - \frac{L_1^2}{L_0}\right) \qquad (4.103)$$

By derivating the Sommerfeld expansion (4.68):

$$-\int_{-\infty}^{+\infty} F(\varepsilon)\frac{\partial f_{eq}}{\partial \varepsilon} d\varepsilon = F(\eta) + \frac{1}{6}\pi^2 (k_B T)^2 \left[\frac{\partial^2 F(\varepsilon)}{\partial \varepsilon^2}\right]_{\varepsilon = \eta} + \dots \qquad (4.104)$$

a simplified expression of the L_2 integral can be obtained:

$$L_2 \approx \frac{\pi^2}{3}(k_B T)^2 \tau(\eta) v(\eta)^2 g(\eta) = \frac{\pi^2}{3}(k_B T)^2 L_0 \qquad (4.105)$$

Combining (4.100), (4.101), (4.103), and (4.105) leads to the following expression of κ_e:

$$\kappa_e = \left(\frac{\pi^2}{3}\frac{k_B^2}{e^2} - S^2\right)\sigma T \qquad (4.106)$$

In the case of metals and degenerate semiconductors, where S is typically negligible in regard of k_B/e, the well-known Wiedemann–Franz empirical law is recovered:

$$\kappa_e = \frac{\pi^2}{3}\frac{k_B^2}{e^2} T\sigma \qquad (4.107)$$

The main consequence of the Wiedemann–Franz law is that metals typically present a large thermal conductivity mainly due to electrons, thanks to their large electrical conductivity and low thermoelectric power. This affects strongly their figure of merit as defined in equation (4.7) by neutralizing the effect of electronic conductivity (σ/κ = constant).

Note that the above expressions of the electronic thermal conductivity apply when the scattering process of electrons is elastic, i.e., electrons do not lose energy when scattered, hypothesis made when the scattering time has been introduced in equation (4.46). In a more practical way, the above expressions of the thermal conductivity are verified as soon as the change of electrons energy after inelastic scattering remains small compared to $k_B T$ [6].

4.4.9 Figure of Merit

4.4.9.1 Optimum chemical potential and quality factor

As defined in Section 4.2, the efficiency of a material for thermoelectric application is characterized by its figure of merit $ZT = TS^2\sigma/\kappa$. The substitution of the semi-classical expressions (4.100), (4.101) and (4.103) for σ, S, and κ, respectively, leads to the following general expression of the figure of merit:

$$ZT = \frac{L_1^2}{L_0 T \kappa_L + L_0 L_2 - L_1^2} \tag{4.108}$$

Figures 4.14a,b present the dependence of σ, S, κ the power factor σS^2 and the figure of merit ZT on the position of the chemical potential with respect to the conduction band edge E_C in the case of a single band contribution, according to calculation made by Pei and collaborators [7].

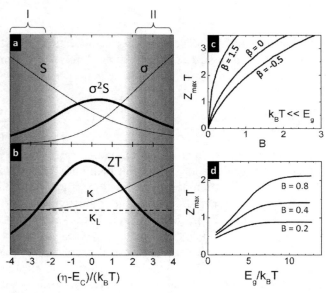

Figure 4.14 (a) Chemical potential dependence of the thermoelectric power, the electrical conductivity, and the power factor, and (b) of the thermal conductivity and the figure of merit (b); both redrawn from ref. [7]. (c) Quality factor dependence of the maximum figure of merit for various charge scattering exponents, redrawn from ref. [8]. (d) Band gap dependence of the maximum figure of merit for various values of the quality factor; redrawn from ref. [9].

As can be seen in Fig. 4.14a, thermoelectric power and electrical conductivity have opposite dependences on the chemical potential position, leading to a maximum of the power factor for a chemical potential at the vicinity of the band edge E_C. This induces a maximum of the figure of merit, although slightly shifted compared to the power factor due to the chemical potential dependence of the electronic part of the thermal conductivity as consequence of the Wiedemann–Franz law (see Section 4.4.8). Figure 4.14c illustrates that the value $Z_m T$ of the maximum of ZT obtained for the optimum chemical potential η_m depends on the exponent β of the energy dependence of the relaxation time as defined in (4.77) and on a factor B usually named "material factor" or "quality factor" [8, 9]:

$$B = N\left(\frac{2\pi}{h^2}\right)^{\frac{3}{2}} \frac{\tau}{\kappa_L} \sqrt{m^*} k_B^{\frac{7}{2}} T^{\frac{5}{2}} \tag{4.109}$$

According to this quality factor, high $Z_m T$ is achieved for materials presenting low lattice thermal conductivity κ_L, high effective mass m^* (which is typically associated with narrow conduction band), large band multiplicity N ($N = 2$ for a single band) and large relaxation time τ. This quality factor has been used since decades as criteria of selection and optimization of thermoelectric materials.

4.4.9.2 Treatment for a single band in the low temperature limit

The optimum chemical potential η_m leading to the maximum figure of merit $Z_m T$ can be determined by the derivation of expression (4.108):

$$\left.\frac{\partial ZT}{\partial \eta}\right]_{\eta=\eta_m} = 0 \tag{4.110}$$

$$ZT_m = ZT(\eta_m) \tag{4.111}$$

The rigorous determination of η_m and ZT_m involves the numerical calculation of the Fermi integrals L_α of expression (4.108). However, in the low temperature limit (i.e., $|E_C - \eta|/k_B T \gg 1$, which corresponds to the gray areas in Fig. 4.14a,b), it is possible to reach analytical expressions for σ, S, κ and ZT.

When the chemical potential is located below the conduction band edge (Fig. 4.15a) and the contribution of other energy bands can be neglected, the low temperature limit condition induces

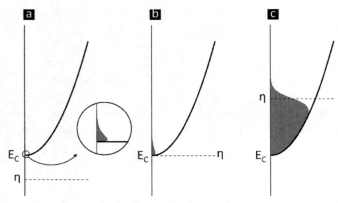

Figure 4.15 Schematic representation of a single quasi-free electron conduction band with the Fermi level is located below the band edge (a), at the band edge (b) and above the band edge (c). The gray areas represent the occupation of the energy states according to the Fermi–Dirac statistics.

$$f(\varepsilon) \xrightarrow[(\varepsilon-\eta) \gg k_B T]{} \exp\left(\frac{\eta-\varepsilon}{k_B T}\right) \tag{4.112}$$

$$\frac{df}{d\varepsilon} \xrightarrow[(\varepsilon-\eta) \gg \kappa_B T]{} -\frac{1}{k_B T}\exp\left(\frac{\eta-\varepsilon}{k_B T}\right). \tag{4.113}$$

This corresponds, for instance, to the case of slightly doped semiconductors with donor levels located below the conduction band edge (see Fig. 4.15a). Expression (4.29) of the density of free charge carriers within the conduction band will then become

$$n \approx n_0 \exp\left(-\frac{E_c - \eta}{k_B T}\right) \tag{4.114}$$

$$n_0 = \int_{E_c}^{+\infty} g(\varepsilon) \exp\left(\frac{E_c - \varepsilon}{k_B T}\right) d\varepsilon. \tag{4.115}$$

In the case of quasi-free electrons (i.e., effective mass approximation), n_0 simplifies as

$$n_0 = 2\left(\frac{2\pi m^* k_B T}{h^2}\right)^{3/2}. \tag{4.116}$$

Expressions (4.52) and (4.66) of the electronic conductivity and thermoelectric power, respectively, become

$$\sigma \approx \mu |e| n_0 \exp\left(-\frac{E_C - \eta}{k_B T}\right) \qquad (4.117)$$

$$S \approx \frac{E_C - \eta + \xi}{eT}, \qquad (4.118)$$

where

$$\xi = \frac{e^2}{\sigma_0 k_B T} \int_{E_C}^{+\infty} (\varepsilon - E_C)\, \tau(\varepsilon)\, v(\varepsilon)^2 g(\varepsilon) \exp\left(\frac{E_C - \varepsilon}{k_B T}\right) d\varepsilon \qquad (4.119)$$

The expression of the figure of merit becomes then

$$ZT \approx \left(\frac{E_C - \eta + \xi}{k_B T}\right)^2 \frac{1}{\Delta + \frac{1}{B} \exp\left(\frac{E_C - \eta}{k_B T}\right)} \qquad (4.120)$$

$$\Delta = \frac{1}{k_B^2 T^2}\left(\frac{L_2}{L_0} - \frac{L_1^2}{L_0^2}\right) \qquad (4.121)$$

$$B = 2\frac{\tau \sqrt{m^*}}{\kappa_L}\left(\frac{2\pi}{h^2}\right)^{\frac{3}{2}} k_B^{\frac{7}{2}} T^{\frac{5}{2}}. \qquad (4.122)$$

Within the low temperature limit, Δ and ξ simplify to [8]:

$$\Delta \approx \xi \approx \beta + \frac{5}{2}. \qquad (4.123)$$

where β is the exponent of the energy dependence of the relaxation time. ξ, Δ and B are thus material's factors that do not depend on η (i.e., on doping level) but on material's properties [8]. Expression (4.120) presents a maximum when the chemical potential is at the vicinity of the conduction band edge, which corresponds to Fig. 4.15b. Although the low temperature limit condition $|E_C - \eta|/k_B T \gg 1$ does not hold, the extrapolation of the above formulas to the vicinity of the band edge allows a fairly good qualitative description of the dependence of ZT on the chemical potential [8]. The approximation get worse for smaller values of B and larger values of β [8]. A more rigorous calculation applied to chalcogenides and based on Kane's band theory can be found in ref. [10].

When the chemical potential is above the conduction band edge, the low temperature limit becomes $(\eta - E_C)/k_B T \gg 1$, see Fig. 4.15c and gray area II of Fig. 4.14a. This corresponds, for example, to classical metals and highly doped semiconductors. Expression (4.53) of the conductivity applies, together with the Sommerfeld expansion of the thermoelectric power (4.79), chemical potential (4.74) and density of charge carriers (4.69), leading to

$$\eta \approx \varepsilon_F \tag{4.124}$$

$$n \sim (\eta - E_c)^{\alpha+1} \tag{4.125}$$

$$\sigma \sim (\eta - E_c)^{\alpha+\beta+2\gamma} \tag{4.126}$$

$$S \sim (\eta - EC)^{-1}. \tag{4.127}$$

This leads to the following expression of the power factor:

$$\sigma S^2 \sim (\eta - E_c)^{\alpha+\beta+2\gamma-2} \tag{4.128}$$

Considering usual values for the energy dependence coefficients (α = 0.5 for quasi-free electrons, γ = 0.5 for the effective mass approximation, and β = −0.5 for acoustic phonons), the power factor decreases when the chemical potential moves away from the conduction band edge, confirming that a maximum of power factor and thus of figure of merit must stand at the vicinity of the conduction band edge.

4.4.9.3 Optimum band gap

According to the previous equations, a temperature raise increases the quality factor B, which in consequence improves the figure of merit ZT. However, when the thermal energy $k_B T$ approaches the band gap E_G, holes are thermally generated in the valence band, which is particularly explicit in the expression of the "np factor" commonly used for semiconductors:

$$np = n_0 p_0 \exp\left(-\frac{E_G}{k_B T}\right) \quad (4.129)$$

As seen in Section 4.4.6, while the presence of holes increases the overall electrical conductivity, the sign of the thermoelectric power of holes is opposite to electrons leading to a saturation and further drop of the figure of merit at high temperature. This is clearly seen for example in Fig. 4.6 where the figure of merit of all materials increases with temperature until reaching a maximum value, which corresponds actually to the material's optimal temperature of use.

This phenomenon is illustrated in Fig. 4.14d, which highlights the energy gap dependence of the maximum figure of merit, as calculated by Mahan et al. [9]. It can be clearly seen that while ZT for a given quality factor is constant for high values of E_G, it drops drastically when the energy gap decreases below 10 $k_B T$. This suggests that E_G should be as high as possible. Unfortunately, scattering time, effective mass, and band gap are correlated. On one side, according to Kane band's theory the larger the energy gap the higher the effective mass, which seems beneficial according to expression (4.122) of the quality factor. However, according to the deformation potential theory of Bardeen–Shockley, the scattering time decreases when the effective mass increases, leading to an actually lower quality factor for most good thermoelectrics [10, 11]. The optimum energy gap for a given working temperature is thus $E_G \sim 10 k_B T$, as determined by Mahan [9].

4.5 Thermoelectric Materials

4.5.1 Historical Overview

As illustrated in Fig. 4.16, the research activity on thermoelectric materials has been subject to a continuous expansion during the first half of the 20th century, reaching its apogee at the end of the 1950s with reported figures of merit approaching the unity for lead and telluride chalcogenides. Although it was still too inefficient for industrial or consumer market applications, it allowed the development of the radio-thermoelectric generators that powered the NASA's first spatial missions.

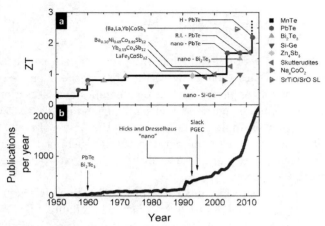

Figure 4.16 (a) Historical evolution of the figure of merit ZT from 1950 to 2014. Symbols represent a selection of major thermoelectric materials as polycrystalline bulks (filled symbols), or multilayer (half-filled symbol). The black line represents the evolution of the state of the art ZT value in polycrystalline bulk materials. Zn_4Sb_3, PbTe and Bi_2Te_3 chalcogenides, and Si–Ge alloys are from ref. [12]. Skutterudites are from refs. [13–16]. Oxides are from ref. [17]. (b) Evolution of the scientific publications per year dedicated to thermoelectrics (*Source*: Web of Knowledge).

Despite of further improvements during the 1960s that allowed to reach $ZT = 1$ for Bi_2Te_3 alloys, two decades followed marked by the absence of major breakthrough allowing to go beyond the $ZT = 1$ limit, inducing a stagnation of the research activity dedicated to the topic. However, a recrudescence of activity occurred during the 1990s associated with the apparition of major new concepts. Until then, the general approach was to look for promising candidates whose figure of merit would be optimized through doping, on the basis of the semiconductor science developed during the first half of the 20th century.

A drastic change of approach occurred after the seminal publications of Hicks and Dresselhaus in 1993, which predicted that a reduction of the dimension to nanoscale should open the door for ZT above unity [18]. Indeed, while the confinement of electrons in nanostructures would lead to improve the power factor by narrowing the energy bands, nanostructured systems should also present lower thermal conductivity by reducing the phonons mean free path. The application of these concepts leaded to impressive values of figure of merit in thin film superlattices reaching 2.4 in both SrTiO/SrO and Bi_2Te_3/Sb_2Te_3 [19, 20], although these examples are hardly upscalable. However, thanks to new bulk synthesis methods inherited from the nanosciences' blast that marked the 2000's decade, nanostructured bulk ceramics have been reported with major improvement of the figure of merit. For example ZT = 1.7 has been reported in 2004 in nanostructured PbTe-based ceramics [21], which was already a huge step forward compared to conventional ceramics of the same material. This has been further improved to ZT = 2.2 very recently in multiscale hierarchically structured PbTe-based ceramics [22].

In 1995 Slack presented the calculus of the maximum figure of merit ZT = 4 of an hypothetical material that would combine the low thermal conductivity of glasses and the electronic properties of the optimum crystal, leading to the novel concept of phonon-glass electron-crystal (PGEC) [23]. Since decades it was known that for approaching ZT = 1, a good thermoelectric material has to contain heavy elements in order to scatter phonons, and gather elements of similar electronegativity so that charge carriers mean free path remain unaffected by phonon scattering. An electronegativity difference would generate charge density oscillation induced by lattice vibration, which would scatter charge carriers and hence limit strongly the possibility to combine low thermal conductivity and high charge carrier mobility. Moreover, mixed covalent crystals, namely crystals presenting elements of different masses occupying the same site, such as substituted chalcogenides or Si–Ge alloys, tend to also present a reduced phonon mean free path compared to their parent compounds [24]. However, the application of these mechanisms did not allow to go beyond the ZT = 1 limit, while in order to approach ZT = 4, Slack calculated that the phonon mean free path would need to be reduced down to the atomic range [23]. In order to avoid such radical phonon scattering to affect the electric conductivity, the electronegativity difference would need to be unrealistically small [23]. In order to induce glass-like thermal conductivity without detrimental decrease of charge carrier scattering, Slack proposed the concept of a ternary compound presenting a binary scaffold with optimum electronic properties which would have voids that could be filled by a third heavy element small enough to "rattle" inducing a strong phonon dampening [23, 25]. Slack's PGEC concept stimulated an intense research in the field and led to increase of ZT for several compounds such as Skutterudites and Clathrates [26]. Interestingly, as the basis of PGEC approach is to reach glass-like thermal conductivity, PGEC related materials such as Skutterudites or Clathrates benefit only marginally from nanostructuring. The reason resides in the fact that such materials present phonon mean free path which are already so small that the scattering by nanoscale grain boundaries would be detrimental to the ratio of electrical over thermal conductivities.

4.5.2 Chalcogenides

In the mid-20th century, concomitantly with the development of the semiconductor science, it appeared clearly that the only way to reach significantly high power factor was combining low charge carrier density with high mobility and low thermal conductivity. As discussed in previous section, this requires small differences of electronegativity of the elements constituting the material in order to ensure a high degree of covalent bonding, and the presence of heavy elements for phonon scattering. These conditions are typically found in chalcogenide compounds since the electronegativity of some of the chalcogen elements (Se or Te) is relatively close to heavy metals such as Bi, Sb, or Pb. Bi_2Te_3 and PbTe have actually been the first materials discovered in the 1950s with sufficient efficiency to foresee real applications, as cooling for Bi_2Te_3 based alloys as they present their best efficiency close to room temperature, and as generation for PbTe-based alloys as their best efficiency is observed at high temperature (see Fig. 4.17). PbTe has been the cornerstone of the first radioisotope thermoelectric generators (RTG) used by the NASA for its spatial missions since the 1960s [27]. Nowadays, both Bi_2Te_3 and PbTe based alloys are still at the state of the art of the thermoelectric technology thanks to recent advances which allowed to boost their efficiency from the initial figure of merit approaching unity, to the present 1.4–2.2 range thanks to doping [28], band structure engineering [7], nanostructuring, and hierarchical structuring [21, 22].

However, in addition to their important weight, these compounds present severe drawbacks: They are constituted of rare, expensive, and toxic elements. If this is not relevant for niche applications such as outer space missions powering, it is a serious issue for industrial use or entering consumer market, which drives the search for alternative materials with similar performance but lower cost and toxicity.

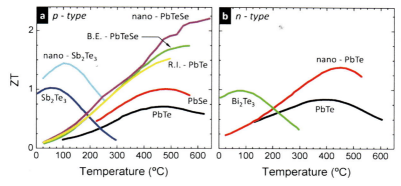

Figure 4.17 Temperature evolution of a selection of n- (a) and p-type (b) chalcogenides. Redrawn from refs. [27] (PbTe, PbSe, B.E.-PbTeSe, Sb_2Te_3, Bi_2Te_3 and nano-PbTe) [29], (nano-Sb_2Te_3), [22] (nano-PbTeSe), [12] (R.I.-PbTe).

4.5.3 Silicon and Si–Ge Alloys

Silicon has been soon identified as a very promising candidate for thermoelectricity as it presents the highest factor $\mu m^{*3/2}$ [9]. However, since it also presents one of the highest thermal conductivity at room temperature (100 W K^{-1}m^{-1}), its figure of merit remained low until the discovery of the unexpectedly low thermal conductivity when alloyed with germanium [9]. The elements Si and Ge being isovalent and presenting relatively small electronegativity difference, the charge carriers are mainly scattered by neutral impurities, their mobility being only slightly affected compared to the single element compounds [30]. On the contrary, the thermal conductivity is greatly affected by the mass fluctuations and strain-field of the randomly mixed Si and Ge atoms [30], leading to an optimum thermoelectric performance for 20% germanium content [31, 32]. According to Mahan's criteria $T_{optimum} \sim E_G/10k_B$, with a band gap around 0.9 eV at RT, Si–Ge alloys are well suited to high temperature application, which is further enhanced by the presence of a minimum of thermal conductivity at 1100 K [9]. Best ZT values reported to date in conventional ceramics of doped Si–Ge alloys reach 0.65 for p-type and 1 for n-type materials (see Fig. 4.18) [31, 32].

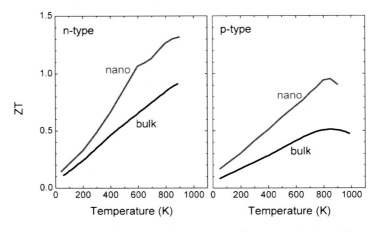

Figure 4.18 Figure of merit of n-type and p-type Si–Ge alloys as standard bulk ceramics (black curves) and nanostructured ceramics (red curves). Redrawn from refs. [31] and [32].

More recently a great advance has been achieved through the engineering of dense ceramics made of nano-sized grains, leading to ZT values of 0.95 for p-type and 1.3 for n-type materials (see Fig. 4.18) [31, 32]. The major effect of the decrease of the grain size of the ceramic is in a considerable increase of the concentration of grain boundaries, which decrease notably the thermal conductivity by providing an important source of phonons scattering. Moreover, the electrical connectivity remains good and the scattering of charge carriers is only slightly affected, leading to a major overall positive effect on the thermoelectric efficiency.

4.5.4 PGEC Related Materials: Skutterudites, Clathrates, and Half-Heusler

Skutterudites are typically binary semiconducting compounds whose crystalline structure has the particularity of presenting large voids that can be filled by heavy metal atoms. Non-filled skutterudites such as $IrSb_3$, $CoSb_3$, or $RhSb_3$ were identified during the first half of the 1990s as promising candidates for thermoelectric application due to the small electronegativity difference of their constituting elements leading to large mobility and the possible thermal conductivity reduction by filling the voids of the structure by phonon dampening "rattling" impurities [25, 23]. Soon after, based on Slack's PGEC approach, Sales et al. reported ZT near unity at 800 K in Fe substituted $CoSb_3$ skutterudites filled with rare earth ions, and predicted ZT as high as 1.4 after optimization [13]. As seen in Fig. 4.16a, high ZT values have been confirmed for skutterudite compounds during the following years, reaching $ZT = 1$ at 600 K for $Yb_{0.19}Co_4Sb_{12}$ [14], $ZT = 1.25$ at 900 K for $Ba_{0.30}Ni_{0.05}Co_{3.95}Sb_{12}$ or $ZT = 1.7$ at 850 K for $(Ba,La,Yb)CoSb_3$ [15, 16]. Although their tendency for degradation in air at high temperature limits their wide application, skutterudites are appealing as they are composed of relatively cheap and non-toxic elements [33].

Clathrates are typically low thermal conductivity compounds thanks to the combination of large unit cells, an open framework, and the presence of cages in their structure that can be filled by "rattling" atoms [26]. Suitable clathrates for thermoelectric application present a covalent semiconducting framework. The majority of the reports are close to $ZT = 1$ [33], best reports reaching 1.35 at 900 K in $Ba_8Ga_{16}Ge_{30}$ or $ZT = 1.25$ for $Ba_{24}Ga_xGe_{100-x}$ [34, 35]. Since the guest atoms influence the electronic properties of the covalent framework, optimization of clathrates is a challenging issue, which remains the subject of an intense research [33, 36].

Half-Heusler compounds are another family of compounds subject to an intense research in the recent years, which tends to be associated with PGEC approach of thermoelectric materials research. The typical Half-Heusler structure consists in the interpenetration of four sub-lattices, one being empty and the three other are filled [26]. Depending on the lattice on which substitutions or doping is made, the electronic and thermal conductivity properties can be tuned [26], although it does not reach values matching the concept of "phonon-glass electron crystal." Even if best reported ZT values hardly reach unity, recent advances in nanostructuring these materials actually hold great promise [37].

4.5.5 Oxides

Oxides constitute a very appealing family of compounds for thermoelectric conversion as they tend to have very good stability at elevated temperature and under oxidant atmosphere, and tend to be relatively low cost and environment friendly. However, the high electronegativity of oxygen is a fundamental handicap for making good thermoelectrics, as it induces ionic bounding as consequence of the inevitably high electronegativity differences between their constituent elements. This results in a strong scattering of charge carriers by acoustic phonons and tends to promote charge localization. For these reasons, oxides where not considered as potential good thermoelectrics until recently.

It is the discovery of a figure of merit approaching unity at 800 K in single crystals of Na_xCoO_2 layered oxides in the beginning of the 2000's decade that really focused the interest of the community into this class of materials [37]. The general structure of layered transition metal oxides of general formula Na_xMO_2 is constituted of MO_2 sheets of edge-sharing MO_6 octahedra in which the transition metal atoms M form a 2D triangular lattice, separated by sodium ions. The amount of Na^+ ions determines the density of charge carrier in the MO_2 layers, which are responsible for the electronic properties. These later can be as varied as superconductivity when water is interleaved in the structure [38], large thermopower around $x = 0.7$ at room temperature attributed to the spin entropy carried by strongly correlated electrons hopping on a triangular lattice [38, 40, 41], or charge ordering associated with electronic insulator character for $x = 0.5$ [42]. In addition to the narrowing of the bands at the vicinity of the chemical potential induced by the combination of the 2D character of the MO_2 layers and the strong electronic correlations, this class of material presents a rather low thermal conductivity due to the strongly anisotropic crystalline structure. It is important to note that the energy bands of the transition metal oxides are typically far from parabolic and the relaxation time is hardly approximated by a power law of the charge carriers' energy, which can lead to rather unexpected results in regard of the classical semiconductor based theory developed in Section 4.4. More complex phenomena involving electron–electron correlation, resonant scattering, and other magnetic interactions have to be taken into account in order to understand and predict the thermoelectric efficiency of these materials. Moreover, contrary to conventional semiconductors in which dopants allow the tuning of the position of the chemical potential with negligible effect on the band structure, in the case of correlated materials any change of the density of electrons may affect non negligibly the density of state.

Many other oxides have recently been tested for thermoelectric performance with relative success, more details can be found for instance in the review of He et al. [17], although for the moment the figure of merit remain modest in bulk materials.

Finally, it is worth noting that the oxide family comprises one of highest figure of merit values reported so far, ZT = 2.4, at the interface of superlattices of SrTiO and SrO [19], see Fig. 4.16. Although such thin film technology is hardly up-scalable at acceptable cost for building thermoelectric generators, these results are quite encouraging in regard of the gain that could be expected from bulk nanostructuring techniques.

4.5.6 Other Materials

Some of the major families of materials studied for thermoelectricity have been listed above in an attempt to cover the major mechanisms leading to good thermoelectric performance, but this is far from being exhaustive. A large variety of materials such as superionic conductors, organic materials or correlated semiconductors are worth mentioning and some of them present fairly good thermoelectric performance. More information about these materials can be found in the reviews of Sootsman et al. [26], Han et al. [33], Zebarjadi et al. or Zhao et al. [2, 43].

4.6 Conclusion

In 2009, more than one century after the discovery of thermoelectric effects, Vining published a quite pessimistic opinion in Nature Materials in regard of the future of thermoelectrics as having a potential impact on climate crisis Vinning et al. [1]. By the time, best reported bulk figure of merit was 1.4, while only hardly up-scalable quantum dots and quantum wells lab-scale samples were able to go beyond $ZT = 2$. Only a few years later, the figure of merit of best bulk thermoelectrics spread in the 1.8–2.3 range thanks to band engineering and hierarchical structuration, respectively. The challenge seems now to combine in a single material the benefits of these individual strategies in an additive way [43]. Such "panoscopic approach" applied to best known thermoelectrics is believed to allow reaching the $ZT = 2.5–3$ range in near future, opening the door of larger scale applications [43]. Even if according to Vining these figure of merits are still not enough to have noticeable impact on global warming, it will surely allow thermoelectrics to play a growing role in our day-to-day life instead of being limited to niche markets. Moreover, the history of thermoelectrics is paved by major discoveries that broke the previously accepted rules that were defining, during a time, the maximum efficiency that could be expected from a thermoelectric material. From this angle of view, one could hope that new mechanisms remain to be discovered to go one step further in combining the seemingly incompatible: the low thermal conductivity of a glass, the high thermoelectric power of a large band gap insulator and the electrical conductivity of a metal.

References

1. C. B. Vining, *Nature Materials*, **8**, 83 (2009).
2. M. Zebarjadi, K. Esfarjani, M. S. Dresselhaus, Z. F. Ren and G. Chen, *Energy & Environmental Science*, **5**, 5147 (2012).
3. G. J. Snyder and E. S. Toberer, *Nature Materials*, **7**, 106 (2008).
4. J. Chelikowsky, D. J. Chadi and M. L. Cohen, *Physical Review B*, **8**, 2786 (1973).
5. J. M. Ziman, "*Principles of the theory of solids*", 2nd edition, Cambridge University Press (1972).
6. N. W. Arshcroft and N. D. Mermin, "*Solid state physics*", Brooks/Cole, Cengage Learning (1976).
7. Y. Pei, H. Wang and G. J. Snyder, *Advanced Materials*, **24**, 6125 (2012).
8. R. P. Chasmar and R. Stratton, *Journal of Electronics and Control*, **7**, 52 (1959).
9. G. D. Mahan, *Solid State Physics*, **51**, 81 (1997).

10. H. Wang, Y. Pei, A. D. LaLonde, and G. J. Snyder, Proceedings of the National Academy of Sciences of the United States of America, **109**, 9705 (2012).
11. Y. Pei, A. D. LaLonde, H. Wang and G. Jeffrey Snyder, Energy and Environmental Science, **5**, 7963 (2012).
12. J. P. Heremans, M. S. Dresselhaus, L. E. Bell and D. T. Morelli, *Nature Nanotechnology*, **8**, 471 (2013).
13. B. C. Sales, D. Mandrus, R. K. Williams, *Science*, **272**, 1325 (1996).
14. G. S. Nolas, M. Kaeser, R. T. Littleton IV, and T. M. Tritt, *Applied Physics Letters*, **77**, 1855 (2000).
15. X. Tang, Q. Zhang, L. Chen, T. Goto and T. Hirai, Journal of Applied Physics, **97**, 093712 (2005).
16. X. Shi, J. Yang, J. R. Salvador, M. Chi, J. Y. Cho, H. Wang, S. Bai, J. Yang, W. Zhang, and L. Chen, *Journal of the American Chemical Society*, **133**, 7837 (2011).
17. J. He, Y. Liu and R. Funahashi, *Journal of Materials Research*, **26**, 1762 (2011).
18. Hicks, L. D. & Dresselhaus, M. S. *Phys. Rev. B,* **47**, 12727 (1993); Hicks, L. D. & Dresselhaus, M. *Phys. Rev. B*, **47**, 16631 (1993).
19. R. Venkatasubramanian E. Siivola, T. Colpitts, B. O'Quinn, *Nature*, **413**, 597 (2001).
20. H. Ohta, S. Kim, Y. Mune, T. Mizoguchi, K. Nomura, S. Ohta, T. Nomura, Y. Nakanishi, Y. Ikuhara, M. Hirano, H. Hosono and K. Koumoto, *Nature Materials*, **6**, 129 (2007).
21. K. F. Hsu, S. Loo, F. Guo, W. Chen, J. S. Dyck, C. Uher, T. Hogan, E. K. Polychroniadis, M. G. Kanatzidis, *Science*, **303**, 818 (2004).
22. B. Kanishka, J. Q. He, D. B. Ivan, *Nature*, **489**, 414 (2012).
23. G. A. Slack, in *CRC Handbook of Thermoelectrics*, (ed. Rowe, D. M.) pp. 407–440 (CRC, 1995).
24. B. Abeles, *Physical Review*, **131**, 1906 (1963).
25. G. A. Slack and Veneta G. Tsoukala, *Journal of Applied Physics*, **76**, 1665 (1994).
26. J. R. Sootsman, D. Y. Chung, and M. G. Kanatzidis, *Angewandte Chemie. Int. Ed.*, **48**, 8616 (2009).
27. A. D. LaLonde, Y. Pei, H. Wang, and G. J. Snyder, *Materials Today*, **14**, 526 (2011).

28. Y. Pei, A. LaLonde, S. Iwanaga and G. J. Snyder, *Energy and Environmental Science*, **4**, 2085 (2011).

29. B. Poudel, Q. Hao, Y. Ma, Y. Lan, A. Minnich, B. Yu, X. Yan, D. Wang, A. Muto, D. Vashaee, X. Chen, J. Liu, M. S. Dresselhaus, G. Chen, and Z. Ren, *Science*, **320**, 634 (2008).

30. G. A. Slack and M. A. Hussain, *J. Appl. Phys.*, **70**, 2694 (1991).

31. G. Joshi, H. Lee, Y. Lan, X. Wang, G. Zhu, D. Wang, R. W. Gould, D. C. Cuff, M. Y. Tang, M. S. Dresselhaus, G. Chen and Z. Ren, *Nano Letters*, **8**, 4670 (2008).

32. X. W. Wang, H. Lee, Y. C. Lan, G. H. Zhu, G. Joshi, D. Z. Wang, J. Yang, A. J. Muto, M. Y. Tang, J. Klatsky, S. Song, M. S. Dresselhaus, G. Chen and Z. F. Ren, *Applied Physics Letters*, **93**, 193121 (2008).

33. C. Han, Z. Li and S X. Dou, *Chinese Science Bulletin*, **59**, 2073 (2014).

34. A. Saramat, G. Svensson, A. E. Palmqvist et al, *Journal of Applied Physics*, **99**, 023708 (2006).

35. J.-H. Kim, N. L. Okamoto, K. Kishida, K. Tanaka, H. Inui, *Acta Materialia*, **54**, 2057 (2006).

36. T. Takabatake and K. Suekuni, *Review of Modern Physics*, **86**, 669 (2014).

37. W. Xie, A. Weidenkaff, X. Tang, Q. Zhang, J. Poon and T. M. Tritt, *Nanomaterials*, **2**, 379 (2012).

38. K. Fujita, T. Mochida and K. Nakamura, *Japanese Journal of Applied Physics*, **40**, 4644 (2001).

39. K. Takada, H. Sakurai, E. Takayama-Muromachi, F. Izumi1, R. A. Dilanian and T. Sasak, *Nature*, **422**, 53 (2003).

40. I. Terasaki, Y. Sasago, and K. Uchinokura, *Physical Review B*, **56**, R12685 (1997).

41. Y. Wang, N. S. Rogado, R. J. Cava and N. P. Ong, *Nature*, **423**, 425 (2003).

42. M. L. Foo, Y. Wang, S. Watauchi, H. W. Zandbergen, T. He, R. J. Cava and N. P. Ong, *Physical Review Letters*, **92**, 247001 (2004).

43. L.-D. Zhao, V. P. Dravid and M. G. Kanatzidis, *Energy and Environmental Science*, **7**, 251 (2014).

Chapter 5

Piezoelectric Conversion

Steven R. Anton

Department of Mechanical Engineering, Tennessee Technological University, Cookeville, Tennessee 38505, USA

santon@tntech.edu

Piezoelectric materials can be used to convert mechanical vibration energy into useful electrical energy through a process known as piezoelectric energy harvesting. The amount of power converted in typical piezoelectric harvesting systems is on the order of microwatts to milliwatts. Conversion of ambient vibration energy to electrical energy through piezoelectric transduction is most often used in the creation of autonomous power supplies for low-power electronic devices.

Several materials exists that exhibit piezoelectric coupling. The most commonly used material is lead zirconate titanate (PZT), a piezoelectric ceramic material with large piezoelectric coupling. Soft, lightweight, and compliant polymer-based piezoelectric materials are also used in energy harvesting systems. The selection of a piezoelectric material for a given vibration energy harvesting application depends on the design constraints of the physical host system upon which the harvester is to be installed, as well as the electrical demands of the device to be powered using the

---

Materials for Sustainable Energy Applications: Conversion, Storage, Transmission, and Consumption
Edited by Xavier Moya and David Muñoz-Rojas
Copyright © 2016 Pan Stanford Publishing Pte. Ltd.
ISBN 978-981-4411-81-3 (Hardcover), 978-981-4411-82-0 (eBook)
www.panstanford.com

harvested energy. In addition to the variety in material selection, the physical attachment of a piezoelectric energy harvester to a host structure can take many forms depending on the application.

Power conditioning circuitry used to convert the electrical output of a piezoelectric material, which is typically high-voltage alternating current, to appropriate regulated direct current that can be used by an electronic load is an important aspect of a piezoelectric energy harvesting system. Energy harvesting circuitry must efficiently extract the generated energy from the piezoelectric material and in most cases deliver it to a storage medium, such as a battery or capacitor. The power output of a piezoelectric harvesting system is often too low to be used directly by an electronic device; therefore, it is common for energy to be intermediately stored and used periodically to power a device operating on a low duty cycle.

Piezoelectric energy harvesting has found a wide range of applications, from providing autonomous power supplies for wireless sensing systems to the development of self-powered biomedical devices. In this chapter, basic concepts in piezoelectric energy harvesting are first reviewed including piezoelectric transduction phenomenon and mathematical modeling of piezoelectric energy harvesters. Energy harvesting circuitry used in piezoelectric systems is discussed. Several applications of piezoelectric energy harvesting are then summarized, and finally, some current research thrusts are reviewed.

5.1 Introduction

5.1.1 Vibration Energy Harvesting Concepts

The conversion of vibratory energy into electrical energy using piezoelectric transduction is typically referred to as *piezoelectric energy harvesting*. Energy harvesting in a general sense can be defined as the conversion of available ambient energy surrounding a system into useful electrical energy through the use of a particular material or transduction mechanism. Several classes of material exist with various conversion mechanisms that can be used to harvest energy. Some of the common materials include those with photovoltaic coupling to convert solar energy to electrical energy, as described in Chapter 2, thermoelectric coupling to convert temperature gradients into electrical energy, which is discussed in Chapter 4, and electromechanical coupling to convert mechanical vibration energy into electrical energy, which is the topic of this chapter.

Persistent vibrations exist in many environments that may benefit from energy harvesting technology. While vibration energy may surround a system as a result of environmental phenomenon (e.g., wind, fluid flow), operational conditions are often the cause of ambient vibrations. This presents an advantage for vibration-based energy harvesting techniques over other methods by decoupling uncontrollable and unpredictable environmental effects from the performance of the harvesting system. Examples of typical environments where vibration energy harvesting may be applied include civil infrastructure such as bridges and buildings [1–3], aerospace structures [4–7], and the human body [8–10].

In order to harness vibratory energy from a structure, a piezoelectric device must mechanically interface with the host structure such that vibration energy is effectively transferred from the structure to the harvester. There are many ways in which this mechanical interface can be established; however, the best method is typically dictated by the characteristics and design constraints of the overall system. Several typical configurations employed in piezoelectric energy harvesting are given in Fig. 5.1. The most commonly used configuration is the piezoelectric *cantilever beam*, shown in Fig. 5.1a in a bimorph configuration (two piezoelectric layers surrounding a central substrate layer) with a tip mass for frequency tuning and increased strain. The cantilever beam can be designed with a specific fundamental frequency such that when attached to a harmonically vibrating structure, the beam will resonate with large amplitude deflections. In systems with a fixed operational frequency, this method of *linear, resonant* harvesting provides strong coupling between the host structure and the piezoelectric harvester. Other configurations include flexible piezoelectric *flags* inserted into flow, surface-mounted piezoelectric *patches* that directly couple to structural flexure, and axial loading of piezoelectric *stacks*, as shown in Fig. 5.1b–d, among others.

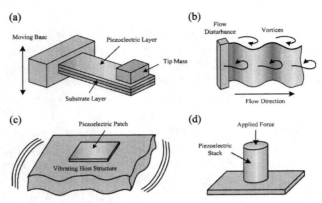

Figure 5.1 Common piezoelectric energy harvesting configurations including (a) piezoelectric *cantilever beam* subjected to base excitations, (b) piezoelectric *flag* subject to flow excitation, (c) surface mounted piezoelectric *patch* on vibrating host, and (d) piezoelectric *stack* under axial force.

Piezoelectric harvesting of vibration energy is typically focused on scavenging low-level energy, on the order of microwatts to milliwatts, in order to provide power to low-power electronic devices. In contrast to many solar and thermal harvesting systems which are capable of generating power on the order of hundreds of watts, piezoelectric materials typical operate at much lower energy levels. The main advantages of piezoelectric transduction over these higher power methods include the fact that, as described above, vibration energy is often an effect of the operational conditions of a system, and is therefore, not reliant on environmental conditions, which may vary significantly over time. Additionally, piezoelectric harvesters can be used in applications where solar and thermal energy is unavailable and are especially useful in embedded systems.

Considering the typical output power of piezoelectric harvesting systems, their most common application is in providing power to low-power electronics including wireless sensor nodes, embedded electronics, implantable biomedical devices, and portable electronic devices. Conventional low-power electronics rely on batteries to provide power to the device. The use of batteries, however, presents several drawbacks including the cost of battery replacement as well as limitations imposed by the need of convenient access to the device for battery replacement purposes. Wireless sensor nodes, for example, are often used in remote locations or embedded into a structure; therefore, access to the device can be difficult or impossible. By scavenging ambient vibration energy surrounding an electronic device, piezoelectric energy harvesting solutions have the ability to provide permanent power sources that do not require periodic replacement. Such systems can operate in an autonomous, self-powered manner, reducing the costs associated with battery replacement, and can easily be placed in remote locations or embedded into host structures.

With recent growth in the development of low-power electronic devices such as microelectronics and wireless sensors, the topic of piezoelectric energy harvesting has attracted much interest in the research community over the past decade. While an exhaustive review of previous literature on piezoelectric energy harvesting is out of the scope of this chapter, the reader is referred to the various review articles that have been published on the topic [11–15], as well the books by Elvin and Erturk [16], Erturk and Inman [17], and Priya and Inman [18].

5.1.2 A Brief History of Piezoelectricity

Piezoelectricity was first discovered in 1880 by Pierre and Jacques Curie when they found that certain naturally occurring crystals, most notably quartz and Rochelle salt, produced a surface charge when subject to a compressive load [19]. This generation of electric charge under mechanical loading is known as the direct piezoelectric effect. One year later, the opposite (or converse) piezoelectric effect, where an induced voltage will cause mechanical deformation of the crystal, was mathematically proven by Gabriel Lippman and soon after experimentally observed by the Curie brothers.

In the following 30 years, researchers worked to fully understand the underlying crystal structures exhibiting piezoelectricity (of which there are 20 natural classes), and all 18 possible macroscopic piezoelectric coefficients were defined, as outlined by Woldemar Voigt in 1910 in his famous publication *"Lehrbuch der Kristallphysik"* [20]. Following this rigorous analysis, the first application of piezoelectricity was during World War I, when in 1917, Paul Langevin and his colleagues developed an underwater sonar system for submarines utilizing quartz actuation. Throughout the next several decades, piezoelectric materials found use in a wide variety of ultrasonic applications.

The natural crystals that were initially discovered exhibit weak coupling between the mechanical and electrical domains. In the 1940s, researchers found that certain synthetic materials, called ferroelectrics, exhibited piezoelectric constants several orders of magnitude larger than the existing natural crystals. Materials such as barium-titanate ($BaTiO_3$) and eventually lead zirconate titanate (PZT) were developed and remain in use today. This finding opened the door to a vast amount of research on new applications for synthetic piezoelectric materials and the continued development of new piezoelectric materials with increased performance.

5.2   Principles of Piezoelectric Transduction

5.2.1   Piezoelectric Transduction Phenomenon

The piezoelectric effect exists mainly in certain crystalline materials due to the polarity of the unit cells within the material. This polarity leads to the production of electric dipoles in the material which, through a poling process (described below), give rise to the piezoelectric properties. The application of mechanical strain in a piezoelectric material causes rotation of the dipoles, leading to the formation of an electric field which manifests as an electric potential across oppositely electroded faces (direct piezoelectric effect). Similarly, the application of a voltage across the material will cause rotation of the dipoles which results in an induced strain in the material (converse piezoelectric effect). Both the direct and converse effects are illustrated in Fig. 5.2.

In order for a crystalline material to exhibit piezoelectricity, not only must it possess the appropriate internal structure in which individual unit cells contain dipole moments, but the material must be poled. Prior to the poling operation, the various regions of the material, called Weiss domains, have aligned dipoles and a net polarity; however, the polarity of neighboring Weiss domains is random, resulting in no overall polarity of the material (Fig. 5.3a). In order to align the dipole moments of all the Weiss domains, a poling treatment is performed by the application of a strong electric field (typically on the order of kV) at a temperature slightly below the Curie temperature of the material (the Curie temperature is the point above which, dipole moments no longer exist, hence all piezoelectric properties are lost). This step of the poling process, which is illustrated in Fig. 5.3b, causes elongation of the sample due to expansion of the domains most closely aligned with the field at the expense of those not well aligned. Once the electric field is removed, the dipoles remain aligned and the material is permanently elongated, shown in Fig. 5.3c, giving the material an overall net polarization that gives rise to piezoelectric properties.

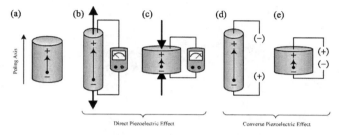

Figure 5.2 Piezoelectric operation; (a) poled, undeformed sample, (b–c) direct piezoelectric effect showing application of force and generation of voltage, (d–e) converse piezoelectric effect showing application of voltage and resulting tensile and compressive stains.

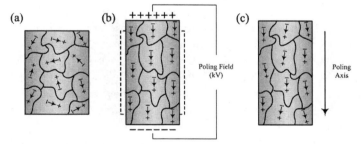

Figure 5.3 Piezoelectric poling treatment; (a) material prior to poling treatment showing randomly oriented Weiss domains, (b) application of poling field near Curie temperature causing alignment of dipoles, (c) dipole alignment after removal of poling voltage and temperature.

Once a piezoelectric material is poled, there are operational limitations that must be considered in order to prevent the loss of piezoelectricity in the material. The application of a large voltage can cause degradation of the piezoelectric properties of the material, and in particular, a large voltage applied in the opposite direction of the poling voltage can cause complete depoling of the material and can even reverse the polarization direction. Additionally, the temperature applied to the material must remain below the Curie temperature to prevent thermal depoling of the material. It is typically suggested that the operating temperature remain less than half of the Curie temperature.

5.2.2 Piezoelectric Material and Transducer Types

A variety of piezoelectric materials exist spanning a range of properties. Natural piezoelectric materials exist, such as quartz, Rochelle salt, and bone; however, these natural materials exhibit very weak piezoelectric coupling. Synthetic piezoelectric materials are, therefore, almost exclusively used today.

The most common piezoelectric material is lead zirconate titanate, a polycrystalline monolithic piezoelectric ceramic known as PZT, shown in Fig. 5.4a. Piezoelectric ceramics, or piezoceramics, are widely used and have the ability to output large voltages on the order of 50 to 100 V (although currents are typically quite small, in the nanoamp to milliamp range). Many compositions of ceramic piezoelectric materials exist, and the development of new compositions is a large ongoing research thrust [21–23]. While piezoelectric ceramics are relatively affordable and provide good coupling, they are extremely brittle and heavy. In order to provide a compliant piezoelectric material, piezoelectric polymers have been developed which include another common piezoelectric material, polyvinylidene fluoride (PVDF), shown in Fig. 5.4b. While piezoelectric polymers are lightweight and flexible, their coupling is considerably lower than their ceramic counterparts.

Improving upon the performance of polycrystalline piezoceramics, piezoelectric single crystals, shown in Fig. 5.4c, have recently been developed and offer superior coupling. The drawback to these materials is their higher cost and their increased brittleness. Finally, a relatively new piezoelectric material is ferroelectret polymer foam, shown in Fig. 5.4d. Ferroelectret foams are extremely lightweight and compliant materials and have been shown to offer around 7× greater coupling than PVDF polymer piezoelectric material.

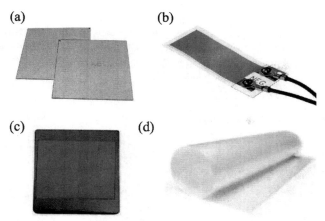

Figure 5.4 Various piezoelectric materials including (a) polycrystalline monolithic lead zirconate titanate (PZT) ceramic, (b) polymer piezoelectric polyvinylidene fluoride (PVDF), (c) single crystal, and (d) piezoelectric foam.

A comparison of typical material properties of the various piezoelectric materials described above is given in Table 5.1.

Piezoelectric polycrystalline ceramics, being the most common piezoelectric material, have drawn much attention in the research community. Several alternative transducer configurations utilizing piezoceramic materials have been developed and are commercially available.

In an effort to improve upon the brittle nature of piezoceramics, companies have developed packaged piezoceramic transducers encapsulated in Kapton with embedded electrodes. The QuickPack transducer, manufactured by Midé Corporation and shown in Fig. 5.5a, is one such transducer consisting of a conventional monolithic piezoceramic with protective Kapton packaging. The Kapton greatly improves the robustness of the transducer by providing electrical isolation as well as isolation from harsh environments. Another transducer utilizing a Kapton coating is the macro fiber composite (MFC) manufactured by Smart Material, Inc., shown in Fig. 5.5b. The MFC greatly improves the flexibility of the transducer while still utilizing piezoceramic material with high coupling by employing a fiber-based design in which many piezoelectric fibers are encapsulated in Kapton. Key to the operation of the MFC is its interdigitated electrode configuration, illustrated in Fig. 5.5b.

Table 5.1 Material properties of various piezoelectric materials (properties taken for Piezo Systems, Inc. PZT-5H [24], Measurement Specialties PVDF film [25], APC International, Ltd. PMN-32%PT single crystal [26], and Emfit HS-01 piezoelectret foam [27–29])

Property	PZT (PZT-5H)	PVDF	Single crystal (PMN-32%PT)	Piezoelectret foam (HS-01)
Density, ρ (kg/m^3)	7800	1780	8200	330
Piezoelectric coefficient, d_{31} (pC/N)	−320	23	−930	2
Piezoelectric coefficient, d_{33} (pC/N)	650	−33	2000	25–250
Dielectric constant, K_3	3800	12	4600	1.2
Elastic modulus Y_1 (GPa)	62	2–4	24.8	0.5–1
Curie/operating temperature (°C)	230	100	150*	50

*Approximate value taken from the literature [18].

Other configurations utilized in piezoelectric energy harvesting exist in which the geometry of the active and/or passive materials are altered to give increased performance. Figures 5.5c,d show unimorph (a single piezoelectric layer and substrate layer) and bimorph (a single substrate layer with two symmetric piezoelectric layers) configurations where a piezoceramic is bonded to a passive substrate to increase the strain in the active material. Pre-curved substrates are also utilized, as shown in Fig. 5.5e, to help apply a pre-strain in the piezoelectric layer. Finally, stack configurations can be used where multiple piezoceramic layers are arranged in series to amplify the performance of the device (Fig. 5.5f).

In addition to the use of various piezoelectric material or transducer configurations, the performance of a piezoelectric device can also be tuned by altering the electrode configuration of the device with respect to the poling axis. It should be noted that the convention in piezoelectric materials is that the material is always poled parallel to the "3" direction, or "z" direction. Figure 5.6 shows the six coordinates assigned when describing piezoelectric materials including the "1," "2," and "3" axes (corresponding to "x," "y," and "z" directions) and "4," "5," and "6," which are shear about the "1," "2," and "3" axes, respectively. The majority of piezoelectric devices utilize the "31" mode (which leads to the common piezoelectric strain coefficient, d_{31}) where the electrodes are perpendicular to the "3" axis and the stress or strain response is in the "1" direction. A popular alternative configuration, however, is the "33" mode. The "33" mode, where the electrodes are perpendicular to the "3" axis and the stress or strain response is also in the "3" direction, is employed, for example, in the MFC device shown in Fig. 5.5b. The advantage of devices utilizing the "33" extensional mode, which is shown schematically along with the "31" extensional mode in Fig. 5.6, is that they exploit the d_{33} piezoelectric coefficient, which is typically larger than the d_{31} coefficient. Several other modes exist depending on transducer geometry and electrode and poling configuration including longitudinal, transverse, and shear modes, and are summarized in a book by Ikeda [30].

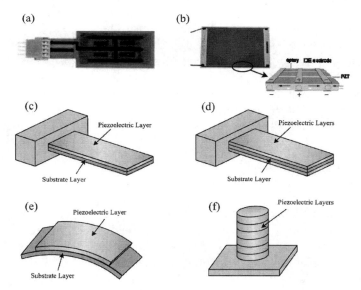

Figure 5.5 Various piezoelectric transducer configurations including (a) QuickPack packaged monolithic ceramic, (b) macro fiber composite (MFC) packaged piezoceramic fiber-based, (c) unimorph, (d) bimorph, (e) curved bender, and (f) stack.

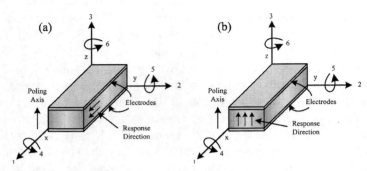

Figure 5.6 Piezoelectric modes of operation including (a) d_{31} mode and (b) d_{33} mode.

5.2.3 Mathematical Modeling of Piezoelectric Energy Harvesters

In order to predict the response of piezoelectric energy harvesting systems, mathematical models can be used as a design tool in the development of optimal harvesting systems. Since the early treatment of piezoelectric energy harvester models which employed simple single degree of freedom approaches [31–33], several advances have been made [34–37]. Some of the original models as well as some modern approaches are outlined in this section.

Both the direct and converse piezoelectric effects can be described mathematically through the piezoelectric constitutive equations, which are well described in the IEEE Standard on Piezoelectricity [38] and given here in their most common form as

$$\begin{Bmatrix} S \\ D \end{Bmatrix} = \begin{bmatrix} s^E & d' \\ d & \varepsilon^T \end{bmatrix} \begin{Bmatrix} T \\ E \end{Bmatrix}, \qquad (5.1)$$

where S is the mechanical strain, D is the electric displacement, T is the mechanical stress, E is the electric field, s^E is the mechanical compliance (reciprocal of the elastic modulus) measured at constant electric field (denoted by the superscript E), d is the piezoelectric strain constant, and ε^T is the dielectric permittivity measured at constant mechanical stress (denoted by the superscript T). While Eq. (5.1) gives the constitutive laws of piezoelectric transduction, electromechanical models including the coupled physics of both the mechanical system and the electric system are needed in order to describe the physical realization of a piezoelectric energy harvester.

Here we will concentrate on efforts used to model the most common configuration in piezoelectric energy harvesting, the cantilever harvester, shown in Fig. 5.7a. Many early works in modeling of piezoelectric energy harvesters considered a lumped parameter single degree of freedom approach, shown in Fig. 5.7b [31–33]. While these models gave insight into the phenomenon of vibration energy harvesting using piezoelectric materials, they produced inaccurate solutions that ignored aspects of the coupled nature of a piezoelectric vibratory system, and furthermore, were only useful in describing a single vibration mode [39].

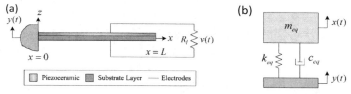

Figure 5.7 (a) Unimorph cantilever piezoelectric energy harvester and (b) lumped parameter single degree of freedom model used in many early works to model piezoelectric energy harvesters.

As an improvement upon lumped-parameter models, distributed parameter models based on the Rayleigh–Ritz discretization have also been investigated in the literature [31, 40, 41]. The Rayleigh–Ritz method gives more accurate approximations compared to the lumped-parameter model. Furthermore, analytical expressions based on Euler–Bernoulli beam theory have been developed in the literature in attempt to derive exact solutions to the cantilever piezoelectric harvester of Fig. 5.7a [42–44]. Several issues are present, however, in the above-mentioned works leading to inaccuracies. These issues are summarized in a work by Erturk and Inman [45].

More recently, an accurate analytical distributed parameter model has been presented by Erturk and Inman for a cantilever unimorph piezoelectric harvester [35], which has become widely accepted in the research community. In the following, an overview of the exact analytical solution for a unimorph piezoelectric cantilever harvester, known as the Erturk–Inman model, will be given along with the expressions for series and parallel connected bimorphs.

5.2.3.1 Unimorph cantilever exact analytical solution

Considering the cantilever unimorph harvester of Fig. 5.7a excited at its base, the mechanical equation of motion based on Euler–Bernoulli beam assumptions is given as

$$\frac{\partial^2 M(x,t)}{\partial x^2} + c_s I \frac{\partial^5 w_{rel}(x,t)}{\partial x^4 \partial t} + c_a \frac{\partial w_{rel}(x,t)}{\partial t} + m \frac{\partial^2 w_{rel}(x,t)}{\partial t^2}$$
$$= -m \frac{\partial^2 w_b(x,t)}{\partial t^2} - c_a \frac{\partial w_b(x,t)}{\partial t}, \quad (5.2)$$

where $M(x, t)$ is the internal bending moment, I is the equivalent area moment of inertia of the composite cross section, $w_b(x, t)$ and $w_{rel}(x, t)$ are the transverse deflection of the base and of the beam relative to the base, respectively, m is the mass per unit length of the composite beam, c_s and c_a are the strain rate damping coefficient and the viscous air damping coefficient, respectively, x is the longitudinal coordinate, and t is time. Considering the first piezoelectric constitutive law given in Eq. (5.1), assuming total electrode coverage and a uniform electric field through the thickness of the piezoceramic layer, and writing the internal moment by integrating the stress distribution over the cross section of the composite beam, one obtains the mechanical equation of motion with electromechanical coupling given by

$$YI \frac{\partial^4 w_{rel}(x,t)}{\partial x^4} + c_s I \frac{\partial^5 w_{rel}(x,t)}{\partial x^4 \partial t} + c_a \frac{\partial w_{rel}(x,t)}{\partial t} + m \frac{\partial^2 w_{rel}(x,t)}{\partial t^2} + \theta v(t)$$
$$\times \left[\frac{d\delta(x)}{dx} - \frac{d\delta(x-L)}{dx} \right] = -m \frac{\partial^2 w_b(x,t)}{\partial t^2} - c_a \frac{\partial w_b(x,t)}{\partial t}, \quad (5.3)$$

where YI is the bending stiffness of the composite cross section, $v(t)$ is the voltage output of the piezoelectric layer, $\delta(x)$ is the Dirac delta function, and θ is the electromechanical coupling term, given by

$$\theta = -\frac{Y_p d_{31} b}{2 h_p} (h_c^2 - h_b^2), \quad (5.4)$$

where Y_p is Young's modulus of the piezoelectric layer, d_{31} is the piezoelectric constant, b is the width of the beam, h_p is the thickness of the piezoelectric layer, h_c is the distance of the top of the piezoelectric layer to the neutral axis of the beam, and h_b is the distance of the bottom of the piezoelectric layer to the neutral axis of the beam.

The electrical circuit equation can be obtained considering the second piezoelectric constitutive law of Eq. (5.1), integrating the electric displacement over the electrode area, and differentiating with respect to time, thus giving

$$\frac{v(t)}{R_1} + C_p \frac{dv(t)}{dt} = -\int_0^L d_{31} Y_p h_{pc} b \frac{\partial^3 w_{rel}(x,t)}{\partial x^2 \partial t} dx, \qquad (5.5)$$

where R_1 is the resistive load, C_p is the capacitance of the piezoelectric layer, h_{pc} is the distance from the center of the piezoelectric layer to the neutral axis, and L is the length of the beam. Equations (5.3) and (5.5) represent the distributed parameter electromechanically coupled equations of the cantilever harvester beam.

The relative motion of the beam, $w_{rel}(x, t)$, can be represented by an absolutely and uniformly convergent series of eigenfunctions given by

$$w_{rel}(x,t) = \sum_{r=1}^{\infty} \phi_r(x) \eta_r(t), \qquad (5.6)$$

where $\phi_r(x)$ and $\eta_r(t)$ are the mass normalized eigenfunctions and the modal coordinates of the r-th mode, respectively. Substituting Eq. (5.6) into Eqs. (5.3) and (5.5), the electromechanically coupled ordinary differential equations for the modal response of the energy harvester beam can be obtained as

$$\frac{d^2 \eta_r(t)}{dt^2} + 2\zeta_r \omega_r \frac{d\eta_r(t)}{dt} + \omega_r^2 \eta_r(t) + \chi_r v(t) = N_r(t)$$
$$\frac{1}{R_1 C_p} v(t) + \frac{dv(t)}{dt} = \sum_{r=1}^{\infty} \varphi_r \frac{d\eta_r(t)}{dt}, \qquad (5.7)$$

where ζ_r and ω_r are the mechanical damping ratio and undamped natural frequency of the r-th mode, respectively. The modal electromechanical coupling coefficient in the mechanical equation is given by χ_r as

$$\chi_r = \theta \left. \frac{d\phi_r(x)}{dx} \right|_{x=L} \tag{5.8}$$

The modal mechanical forcing function, $N_r(t)$, assuming translation and no rotation, is given by

$$N_r(t) = -m\gamma_r^w \frac{d^2 y(t)}{dt^2} - c_a \gamma_r^w \frac{dy(t)}{dt}, \tag{5.9}$$

where

$$\gamma_r^w = \int_0^L \phi_r(x)dx, \tag{5.10}$$

Finally, the modal coupling term in the electrical equation, φ_r, is given by

$$\varphi_r = -\frac{d_{31}Y_p h_{pc} h_p}{\varepsilon_{33}^S L} \left. \frac{d\phi_r(x)}{dx} \right|_{x=L}, \tag{5.11}$$

where ε_{33}^S is the dielectric permittivity measured at constant strain. Assuming harmonic base excitation of the form $y(t) = Y_0 e^{j\omega t}$ is applied to the harvester, where Y_0 is the amplitude of the moving base, j is the unit imaginary number and ω is the excitation frequency, the steady-state voltage response is given by

$$v(t) = \frac{\sum_{r=1}^{\infty} \dfrac{jm\omega^3 \varphi_r \gamma_r^w Y_0}{\omega_r^2 - \omega^2 + j2\zeta_r \omega_r \omega}}{\sum_{r=1}^{\infty} \dfrac{j\omega \chi_r \varphi_r}{\omega_r^2 - \omega^2 + j2\zeta_r \omega_r \omega} + \dfrac{1}{C_p R_l} + j\omega} e^{j\omega t} \tag{5.12}$$

and the steady-state vibration response of the beam relative to the base is given by

$$w_{rel}(x,t) = \sum_{r=1}^{\infty} \left[\gamma_r^w - \chi_r \frac{\sum_{r=1}^{\infty} \dfrac{j\omega \varphi_r \gamma_r^w Y_0}{\omega_r^2 - \omega^2 + j2\zeta_r \omega_r \omega}}{\sum_{r=1}^{\infty} \dfrac{j\omega \chi_r \varphi_r}{\omega_r^2 - \omega^2 + j2\zeta_r \omega_r \omega} + \dfrac{1}{C_p R_l} + j\omega} \right]$$
$$\times \frac{m\omega^2 \phi_r(x)}{\omega_r^2 - \omega^2 + j2\zeta_r \omega_r \omega} e^{j\omega t} \tag{5.13}$$

Equations (5.12) and (5.13) give the exact analytical solution for the cantilever unimorph harvester of Fig. 5.7a subject to harmonic base excitation.

5.2.3.2 Bimorph cantilever exact analytical solution for series connection of electrodes

Exact solutions for symmetric bimorph cantilever piezoelectric harvesters, where a substrate is sandwiched between two piezoelectric layers as shown in Fig. 5.8, have also been given by Erturk and Inman [36]. These solutions consider a tip mass added to the tip of the beam, although solutions for beams with no tip mass can be easily derived from the equations by setting $M_t = 0$. Expressions for the series connection of the electrodes, as shown in Fig. 5.8a are given below.

Following a similar procedure as outlined in the previous section, the steady-state voltage response for a harmonic base excitation of the form $y(t) = Y_0 e^{j\omega t}$ is given by

$$v_s(t) = \frac{\sum_{r=1}^{\infty} \dfrac{j\omega \kappa_r F_r}{\omega_r^2 - \omega^2 + j2\zeta_r \omega_r \omega}}{\sum_{r=1}^{\infty} \dfrac{j\omega \kappa_r \chi_r^s}{\omega_r^2 - \omega^2 + j2\zeta_r \omega_r \omega} + \dfrac{1}{R_1} + \dfrac{j\omega C_p}{2}} e^{j\omega t} \qquad (5.14)$$

where the modal coupling term, κ_r, is given by

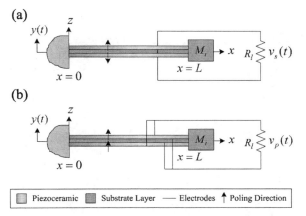

Figure 5.8 Bimorph cantilever piezoelectric energy harvesters with tip mass including (a) series connection and (b) parallel connection of the electrodes.

$$\kappa_r = -d_{31}Y_p h_{pc} b \frac{d\phi_r(x)}{dx}\bigg|_{x=L} \tag{5.15}$$

The forcing term, F_r, is given by

$$F_r = \omega^2 \left(mY_0 \int_0^L \phi_r(x)dx + M_t Y_0 \phi_r(t) \right) \tag{5.16}$$

and the modal electromechanical coupling coefficient for the series connection case is given by χ_r^s as

$$\chi_r^s = \theta_s \frac{d\phi_r(x)}{dx}\bigg|_{x=L}, \tag{5.17}$$

where the piezoelectric coupling term for the series connection, θ_s, is

$$\theta_s = \frac{d_{31}Y_p b}{2h_p}\left[\frac{h_s^2}{4} - \left(h_p + \frac{h_s}{2} \right)^2 \right] \tag{5.18}$$

In Eq. (5.18), h_p is the thickness of a piezoelectric layer and h_s is the thickness of the substrate.

The steady-state vibration response of the bimorph is

$$w_{rel}^s(x,t) = \sum_{r=1}^{\infty} \left[\frac{\left(F_r - \chi_r^s \frac{\sum_{r=1}^{\infty} \frac{j\omega\kappa_r F_r}{\omega_r^2 - \omega^2 + j2\zeta_r\omega_r\omega}}{\sum_{r=1}^{\infty} \frac{j\omega\kappa_r \chi_r^s}{\omega_r^2 - \omega^2 + j2\zeta_r\omega_r\omega} + \frac{1}{R_1} + \frac{j\omega C_p}{2}} \right) \times \frac{\phi_r(x)}{\omega_r^2 - \omega^2 + j2\zeta_r\omega_r\omega} \right] e^{j\omega t} \tag{5.19}$$

5.2.3.3 Bimorph cantilever exact analytical solution for parallel connection of electrodes

Similarly, the steady-state voltage response for a harmonic base excitation of a bimorph with parallel electrode connections, shown in Fig. 5.8b, is

$$v_p(t) = \frac{\sum_{r=1}^{\infty}\dfrac{j\omega\kappa_r F_r}{\omega_r^2 - \omega^2 + j2\zeta_r\omega_r\omega}}{\sum_{r=1}^{\infty}\dfrac{j\omega\kappa_r \chi_r^p}{\omega_r^2 - \omega^2 + j2\zeta_r\omega_r\omega} + \dfrac{1}{2R_l} + j\omega C_p} e^{j\omega t} \qquad (5.20)$$

where the modal electromechanical coupling coefficient for the parallel connection case is given by χ_r^p as

$$\chi_r^p = \theta_p \left.\frac{d\phi_r(x)}{dx}\right|_{x=L}, \qquad (5.21)$$

where the piezoelectric coupling term for the parallel connection, θ_p is

$$\theta_p = 2\vartheta_s = \frac{d_{31}Y_p b}{h_p}\left[\frac{h_s^2}{4} - \left(h_p + \frac{h_s}{2}\right)^2\right] \qquad (5.22)$$

The steady-state vibration response of the bimorph is

$$w_{rel}^p(x,t) = \sum_{r=1}^{\infty}\left[\left(F_r - \chi_r^p \frac{\sum_{r=1}^{\infty}\dfrac{j\omega\kappa_r F_r}{\omega_r^2 - \omega^2 + j2\zeta_r\omega_r\omega}}{\sum_{r=1}^{\infty}\dfrac{j\omega\kappa_r \chi_r^p}{\omega_r^2 - \omega^2 + j2\zeta_r\omega_r\omega} + \dfrac{1}{2R_l} + j\omega C_p}\right)\right.$$
$$\left.\times \frac{\phi_r(x)}{\omega_r^2 - \omega^2 + j2\zeta_r\omega_r\omega} e^{j\omega t}\right] \qquad (5.23)$$

5.2.3.4 Approximate distributed parameter solutions

While the exact analytical solutions presented in the previous sections provide accurate results for modeling simple uniform cantilever unimorph and bimorph harvesters, there are many configurations employed in cantilever piezoelectric harvester design in which an exact solution cannot be found. In order to model thick cantilever beams (not conforming to Euler–Bernoulli assumptions), tapered beams, and asymmetric configurations, for example, an approximate solution is necessary. Such approximate solutions have been presented in the literature. The work presented by duToit et al. [31] gives a Rayleigh–Ritz type approximation for the symmetric bimorph configuration. Elvin and Elvin [34] present an approximate Rayleigh–Ritz formulation for the piezoelectric unimorph. Assumed modes formulations for unimorph, bimorph, and asymmetric cantilever harvesters are given by Erturk and Inman [37].

5.2.3.5 Summary

The exact analytical solution for a unimorph cantilever beam under base excitation has been described in this section. Expressions for bimorph cantilever beams including a tip mass for both series and parallel connection of the electrodes are also given. In addition to exact solutions, approximate solutions based on Rayleigh–Ritz formulation or assumed modes formulation are useful in cases where no exact solution can be derived for a given harvester configuration. A comprehensive summary of piezoelectric cantilever harvester modeling is given in the book by Erturk [17].

5.3 Energy Conditioning Circuitry

The nature of electromechanical coupling in piezoelectric materials results in the production of charge when the material is strained. Most piezoelectric vibration energy harvesting systems place a piezoelectric harvester in a vibration rich environment such that the transducer is subjected to oscillatory strain, resulting in the production of an alternating voltage, or AC power. Electronic devices and energy storage elements require a DC power supply; therefore, the direct output of a piezoelectric energy harvester is not suitable for use in powering electronics and must be conditioned. A typical energy harvesting circuit, shown in Fig. 5.9, consists of AC/DC rectification and DC/DC conversion or regulation. Often times, control electronics are included to *tune* various circuit parameters in order to optimize the energy extraction from the piezoelectric harvester and the energy delivery to a storage device. Advanced circuits aim to maximize the efficiency of the energy conversion while minimizing the quiescent current consumed by the electronics. The most basic harvesting circuits, on the other hand, may simply contain rectification. It should also be noted that in the majority of piezoelectric energy harvesting

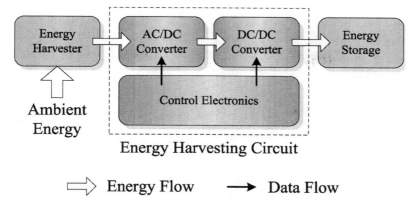

Figure 5.9 Basic piezoelectric energy harvesting circuitry.

systems, the energy generated by the harvester must be stored in a storage element, such as a battery or capacitor, prior to use. The energy level generated by the harvester is often too low to be used directly to power an electronic device, or the system may run on a specified duty cycle. Energy storage is the topic of Part III of this book. The focus of this section is to outline several energy harvesting circuit topologies.

5.3.1 Rectification

Piezoelectric energy harvesters output AC power when subject to vibration energy. This AC output must be rectified before use in electronic devices or to charge energy storage elements, such as batteries or capacitors, which require DC power. The simplest piezoelectric energy harvesting circuit contains a rectifier and a smoothing capacitor to perform AC/DC conversion, and is shown schematically in Fig. 5.10a.

The various stages of voltage output in this simple circuit are shown in Fig. 5.10b. Initially, the piezoelectric device outputs an AC signal, indicated in step 1. A full wave rectifier consisting of a diode bridge made of four diodes is placed after the piezoelectric harvester to convert the AC signal to one with constant polarity, as shown in step 2. The signal, although completely positive, is not yet DC; therefore, a smoothing capacitor is used to create a quasi-DC signal, shown in step 3.

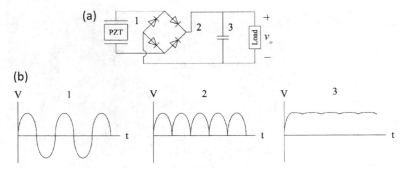

Figure 5.10 Simple energy harvesting circuit containing diode bridge rectifier and smoothing capacitor showing both (a) schematic and (b) voltage signal at various stages.

While this circuit represents a potential implementation in piezoelectric energy harvesting systems, it does not provide an optimized solution for maximum power transfer from the piezoelectric element to the load. Additionally, there is no attempt to regulate the voltage level presented to the load. Most circuits being used in piezoelectric harvesting contain additional components to address these issues. Furthermore, energy harvesting circuitry used in self-powered systems may contain a wide array of additional features from sleep modes and wake-up functionality to battery overvoltage and undervoltage protection. While a large variety of additional features may be introduced to the energy management circuitry depending on the application, in the following, we focus on DC–DC conversion as well as methodologies for improving the energy extraction from the piezoelectric harvester through the concept of impedance matching.

5.3.2 DC–DC Conversion

In a typical piezoelectric energy harvesting system, the output voltage of the harvester is outside of the range required by the load. In this case, the rectified voltage cannot be directly applied to the load and a DC–DC converter is required to further condition the output of the harvester.

There are several DC–DC converter topologies that can be utilized depending on the output voltage of the piezoelectric harvester and the input voltage required by the load electronics. In the case where the voltage generated by the harvester is larger than the voltage required by the load, the simplest DC–DC converter that can be used is the linear voltage regulator. The linear voltage regulator acts as a variable resistor where the resistance is continually adjusted in order to maintain a constant output voltage. The regulator simply dissipates heat in order to reduce the input voltage to the desired output voltage level. While linear regulators have been used in piezoelectric harvesting systems, they have been shown to be inefficient and suboptimal solutions for low-power energy harvesting systems [46, 47].

As an alternative to linear voltage regulators, DC–DC switching converters can be used in energy harvesting applications. In the case where the piezoelectric output voltage is larger than the voltage required by the load, a buck converter, also known as a step-down converter, can be utilized [48]. A buck converter is a switch mode step-down DC–DC converter utilizing an inductor, a switch, and a diode, as shown in Fig. 5.11a. When the switch is closed, energy from the source is stored as a magnetic field in the inductor. When the switch opens, the inductor discharges its stored energy into the load. In order to adjust the voltage output across the load, which is always less than that of the source, the duty cycle of the switch is adjusted.

If the voltage generated by the harvester is lower than the required voltage for the load, a boost converter, or step-up converter, can be used. A boost converter is quite similar to a buck converter in that it utilizes the same components: an inductor, a switch, and a diode; however, the components are arranged in a different configuration as shown in Fig. 5.11b. When the switch is open, energy is transferred from the source to the inductor and stored as a magnetic field. When the switch closes, the inductor and the source are placed in series and discharge together into the load. As with the buck converter, the output voltage is adjusted by varying the duty cycle of the switch; however, the output voltage is always larger than the source voltage. The key operating principle in both the buck and boost converters is the tendency of an inductor to resist changes in current. Details of the operation of these switching DC–DC converters can be found in the text by Erickson [49].

While the buck and boost converters offer ways of handling voltage level mismatch between the source and the load, in practical systems, both the source and load voltage may vary; therefore, a single step-up or step-down converter may not always be appropriate. In this case, a buck-boost converter may be used [50]. A buck-boost converter, shown in Fig. 5.11c, combines the functionality of both buck and boost converter designs and presents an implementation where the output voltage can either be higher or lower than the source voltage [49]. This implementation increases the functionality of the energy harvesting system over the other designs.

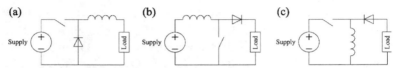

Figure 5.11 Switch-mode DC–DC converter topologies including (a) buck, (b) boost, and (c) buck-boost converters.

In the DC–DC converter circuits described above, the output voltage is controlled based on the duty cycle of the switch. Adjusting the duty cycle also affects the efficiency of the circuit. Several researchers have investigated adaptive circuit designs in which the duty cycle is optimized on the fly in order to improve the efficiency of the circuit [51–53]. Increases on the order of several hundred percent in the amount of power transferred compared to direct charging, which is shown in Fig. 5.10a, have been reported for these adaptive switching DC–DC converters [53].

5.3.3 Synchronous Extraction

An important concept in piezoelectric energy harvesting circuitry was introduced in 2005 by Lefeuvre et al. [54] when the idea of *synchronous electric charge extraction* (SECE) was proposed. The concept of SECE involves synchronizing the extraction of energy from the piezoelectric with the maxima and minima of the displacement (i.e., generated voltage) of the harvester. The circuit proposed in this study is shown in Fig. 5.12, and consists of a diode rectifier combined with a flyback DC–DC converter. A control circuit is used to sense the voltage across the diode rectifier, and when the generated voltage reaches a maximum, the switch is closed and energy transfer through the flyback converter to the load begins. When the charge on the piezoelectric has been completely extracted, the switch is opened and energy transfer stops. Similarly, when the piezoelectric voltage is at a minimum, the process repeats. By extracting energy at the maxima and minima, the circuit was found to provide over 4× increase in energy conversion compared to direct charging [54].

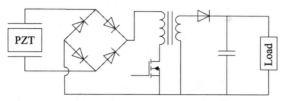

Figure 5.12 Synchronous electric charge extraction (SECE) circuit developed by Lefeuvre et al. [54].

Following the work of Lefeuvre et al. [54], several researchers developed synchronous energy harvesting circuits in which an inductor is placed between the piezoelectric device and the rectifier bridge. Badel et al. and Guyomar et al. in 2005 [55, 56] first introduced the *synchronized switch harvesting on inductor* (SSHI) technique. The works described what would become known as the *parallel-SSHI* circuit topology, shown in Fig. 5.13a, where an inductor and switch are included before the diode rectifier in parallel with the piezoelectric. Like the SECE circuit, energy is extracted from the piezoelectric element when the generated voltage is at a maximum or minimum and the switch opens when all of the charge has been extracted from the generator. In a subsequent study, Lefeuvre et al. in 2006 [57] describe the *series-SSHI* topology in which the switch and inductor are placed in series with the piezoelectric device prior to the rectifier. The study investigates both series-SSHI and parallel-SSHI techniques and notes that they both provide up to 15× improvement in energy transfer over direct charging [57].

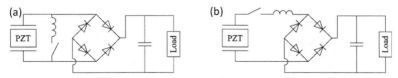

Figure 5.13 Synchronized switch harvesting on inductor (SSHI) circuits including (a) parallel-SSHI [55, 56] and (b) series-SSHI [57].

Several adaptations to the *SSHI* technique have been presented in the literature including the *double synchronized switch harvesting* (DSSH) technique by Lallart et al. in 2008 [58], the *enhanced synchronized switch harvesting* (ESSH) method by Shen et al. in 2010 [59], the *synchronized switch harvesting on inductor using magnetic rectifier* (SSHI-MR) technique developed by Garbuio et al. in 2009 [60], the *hybrid-SSHI* method by Lallart et al. in 2011 [61], and the *self-powered synchronized switch harvesting on inductor* (SP-SSHI) technique developed by Liang and Liao in 2012 [62], among others. Most of these works are summarized in the review article presented by Chao in 2011 [63]. Synchronous energy extraction schemes for piezoelectric harvesting represent a significant portion of the literature on energy harvesting circuitry.

5.3.4 Impedance Matching

In electronic systems, maximum energy is transferred between a source and a load if the impedances of the two are matched [64]. When an impedance mismatch is present, reflections occur between the source and the load. In the case of complex impedances, the following relationship holds for maximum power transfer

$$Z_s = Z_l^*, \quad (5.24)$$

where Z_s and Z_l are the source and load impedances, respectively, and * denotes the complex conjugate.

In order to improve upon basic energy harvesting circuitry utilizing rectification and DC–DC conversion, impedance matching techniques can be employed to increase the efficiency of the harvesting electronics. Impedance matching has been applied to a buck-boost converter topology by Kong et al. in 2010 [65]. In this work, an energy harvesting circuit is developed consisting of a standard diode rectifier followed by a buck-boost converter operating in discontinuous conduction mode, as shown in Fig. 5.14. By adjusting the duty cycle of the buck-boost converter, the effective impedance of the circuit seen by the piezoelectric device can be modified. While complex conjugate matching, shown in Eq. (5.24), provides optimal power transfer, the impedance matching circuit developed by Kong et al. [65] targets resistive impedance matching, or matching of the real part of the impedance. This is mainly due to the large (and impractical) inductance required to match the large capacitance of a piezoelectric harvester. Impedance matching has also been investigated for buck converters by Kim et al. in 2007 [66].

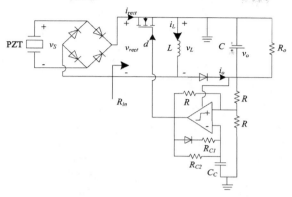

Figure 5.14 Resistive impedance matching circuit consisting of diode rectifier and buck-boost converter operating in discontinuous conduction mode [65].

The concept of utilizing switch mode DC–DC converters for impedance matching purposes has also been applied to the synchronous extraction schemes discussed in the previous section. Lallart et al. in 2008 [58] introduced the *double synchronous switch harvesting* (DSSH) technique which combines a series-SSHI front end with a buck-boost converter back end, as shown in Fig. 5.15. By adjusting the capacitor values in the circuit, the effective impedance of the circuit seen by the piezoelectric element can be modified to provide optimal power transfer. The DSSH technique was shown to provide an increase in power transfer of more than 5× compared to direct charging [58].

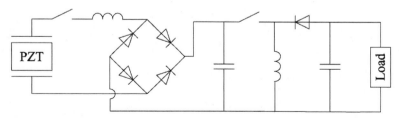

Figure 5.15 Double synchronized switch harvesting (DSSH) circuit composed of series-SSHI front end and buck-boost converter back end [58].

5.3.5 Summary

The electrical circuit components required for the conditioning of piezoelectric output for use in electronic devices or storage elements has been discussed in this section. From the most basic direct charging circuits containing a diode rectifier bridge and smoothing capacitor to advanced switching and impedance matching circuits, a wide variety of circuit topologies have been explored in the literature. In addition to the techniques described here, researchers have investigated many other options for piezoelectric energy harvesting circuitry including the replacement of diodes with switches for rectification, self-powered switching circuitry that does not rely on power-hungry microcontrollers, and *active* energy harvesting where energy is input to the piezoelectric harvester at the beginning of the harvesting cycle to *precharge* the system, to name a few. Thorough reviews of the various energy harvesting circuit topology in piezoelectric harvesting systems have been presented by Chao in 2011 [63] and Szarka in 2012 [67].

5.4 Applications of Piezoelectric Energy Harvesting

Over the past decade, researchers have investigated a variety of applications in which piezoelectric vibration energy harvesting is employed. These applications range from harvesting kinetic energy in the human body to harvesting vibration energy in civil infrastructure. In most cases, the energy scavenged by a piezoelectric harvester is used to provide power to some low-power electronic device. The transfer of kinetic energy in a host structure to a piezoelectric harvester can be achieved using several different physical realizations, many of which will be discussed in this section. In the following, various applications of piezoelectric energy harvesting from the literature are presented.

5.4.1 Self-Powered Sensing Systems

With recent advances in low-power electronics, energy harvesting systems can be considered feasible solutions in providing power to self-powered systems. One of the most common applications of piezoelectric energy harvesting is in providing power for autonomous, self-powered wireless sensing systems. The major components necessary for the realization of self-powered sensing are shown in Fig. 5.16. While conventional wireless sensing systems rely on batteries for power, the use of batteries presents several limitations including the cost of battery replacement as well as limitations imposed by the need for convenient access to the device for battery replacement purposes. In the case where a wireless sensor is embedded in a structure, access to the device may be impossible. The goal of piezoelectric energy harvesting for wireless sensing is to provide maintenance-free, self-powered solutions.

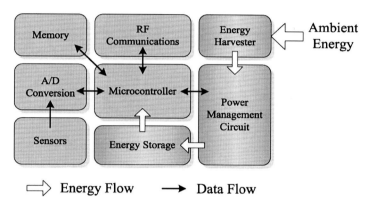

Figure 5.16 Schematic representation of a self-powered sensing system with energy harvesting components highlighted.

One of the pioneering works on the use of piezoelectric energy harvesting to power embedded electronics dates back to 1998 with the work of Paradiso at the MIT Media Laboratory [47, 68]. Paradiso's group investigated piezoelectric energy harvesting in shoes to provide power to an onboard radio frequency identification (RFID) tag. The system developed, shown in Fig. 5.17a, consisted of two piezoelectric generators; a PVDF harvester installed in the toe of the shoe and a PZT harvester installed in the heel, and contained appropriate energy harvesting circuitry to condition the power generated by the piezoelectrics in order to be used by the RFID tag. Experimental testing proved the ability of the shoe harvesters to scavenge an average of around 1–10 mW of power; enough to enable the RFID tag to transmit a 12-bit identification code every 4–5 steps.

Roundy and Wright in 2004 [32] also investigated powering a radio device with energy harvested using piezoelectrics. A small (1 cm^3) piezoelectric cantilever device with a tip mass, shown in Fig. 5.17b, was designed to provide power to a radio transmitter capable of broadcasting a 1.9 GHz signal a distance of 10 m. Experimental testing of the cantilever showed that excitation at 120 Hz with acceleration magnitude of 0.25 g resulted in the generation of around 350 μw; enough power to operate the radio at a 1.6% duty cycle.

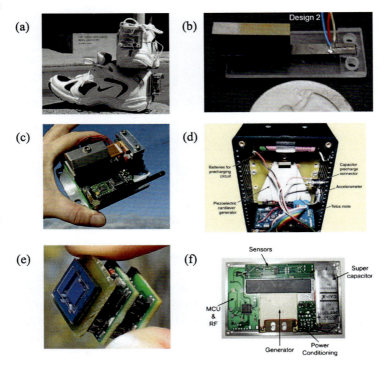

Figure 5.17 Various self-powered sensors utilizing piezoelectric energy harvesting [32, 47, 69–72].

While these previous studies consider piezoelectric energy harvesting to power RF transmission electronics, more recent works have investigated the powering of wireless sensor nodes consisting of sensing, processing, and RF transmission. One of the earlier studies in which a wireless sensor node was powered by piezoelectric vibration energy harvesting was presented by

Arms in 2005 [69]. A complete wireless sensor consisting of a microprocessor, onboard memory, data logging capability, temperature and humidity sensors, a 2.4 GHz radio, and an energy harvesting component was developed, as shown in Fig. 5.17c. A cantilever bimorph harvester utilizing MFC piezoelectric devices was developed and installed on the sensor node. The bimorph harvester produced over 2 mW when excited at 57 Hz with an excitation amplitude of 0.1 g; which enabled perpetual operation of the sensor node with low duty cycle RF transmissions.

Discenzo et al. in 2006 [70] also developed a vibration-powered structural health monitoring node capable of harvesting power from and monitoring the condition of an oil pump on an oil tanker ship. The sensor node, shown in Fig. 5.17d, consisted of a commercially available wireless transceiver, an accelerometer, a tapered piezoelectric cantilever generator tuned to resonate at the operating frequency of the pump (130 Hz), and a storage capacitor bank. During a four month field trail of the system installed on the ship's oil pump, the self-powered node operated continuously; sampling and transmitting acceleration data.

A more recent work presented by Elfrink et al. in 2010 [71] describes a self-powered microelectromechanical (MEMS) sensor node incorporating piezoelectric energy harvesting using aluminum nitride piezoelectric material, wireless transmission, and an on board temperature sensor. The sensor node is shown in Fig. 5.17e. The piezoelectric cantilever has a length of 1.2 mm, a width of 7 mm, a thickness of 42 μm, and contains a large tip mass. When excited at 353 Hz with an amplitude of 0.64 g, the device is able to produce 10 μW of regulated power, which is enough to operate the sensor node to record and transmit temperature data once every 15 s over a distance of 15 m.

Zhu et al. in 2011 [72] describe the development of a credit card sized self-powered sensor node, shown in Fig. 5.17f, that includes a microcontroller, RF communication, an accelerometer, a temperature sensor, a pressure sensor, a piezoelectric cantilever energy harvester, and storage capacitors. The piezoelectric harvester installed in the compact device is designed to resonate at 67 Hz, and was found to generate 240 μW when excited at 0.4 g input, which allowed operation of the sensor node once every 15 min.

The works summarized above represent a sampling of research on the topic of self-powered sensing systems. A more comprehensive review of the topic is given by Chao [63]. The development of autonomous, self-powered electronics is one of the primary research topics in the energy harvesting community and is sure to continue to attract attention in the future.

5.4.2 Biological and Wearable Energy Harvesting

Biological systems present a unique area where energy harvesting can be utilized. The development of more power-efficient personal and portable electronics, along with advances in energy harvesting technology allow piezoelectric vibration energy harvesters to be considered as viable options to provide power to portable and wearable electronics. Additionally, improvements in implantable medical devices also help facilitate the use of energy harvesting inside the body. Implantable devices, such as pacemakers, can benefit greatly from energy harvesting technology as a means of reducing subsequent maintenance surgeries required to replace batteries.

In 1996, Starner [10] presented one of the first surveys of the power available in the human body for harvesting energy to power wearable electronics. Various energy sources on the body were considered including walking, breathing, and finger and upper limb motion, as well as energy available from body heat and blood pressure. Several of the sources were found to produce low amounts of available energy (<1.0 W) including breathing, finger motion, and blood pressure. Of the higher energy sources including body heat, walking, and upper limb motion, walking was determined to hold the most potential for a practical harvesting system. Since this initial work, other studies have also investigated the energy available for harvesting in the human body [73,74] and similar conclusions were made.

One of the original works on wearable energy harvesting was presented by Paradiso's research group at MIT Media Laboratories [47, 68], as described in the previous section, in which piezoelectric devices were integrated into shoes. Similarly, work has been presented by Granstrom et al. in 2007 [75] investigating the use of piezoelectric energy harvesting in backpacks. The work aims to create a backpack capable of generating power during walking as a means of reducing the amount of batteries that must be carried by soldiers, emergency personnel, field workers, etc.

Energy harvesting straps containing PVDF generators were tested in the laboratory and results were used to verify a model of the harvester straps, which predicted that an average of 45.6 mW of power can be generated by the backpack carrying a 444 N load. Extending upon this initial work, Feenstra et al. [76] investigated a backpack with an amplified piezoelectric stack harvester installed in the strap, as shown in Fig. 5.18a. Results of simulations showed that an average power of 0.4 mW could be obtained from the piezoelectric stacks.

In addition to wearable energy harvesting systems, researchers have also investigated implantable piezoelectric energy harvesters aimed at powering electronic devices embedded in the body. One of the original works on in vivo piezoelectric energy harvesting in humans was presented by Platt et al. in 2005 [77] with the development of a total knee replacement unit containing an integrated, self-powered health monitoring system. The knee replacement system, shown in Fig. 5.18b, contains three piezoelectric stacks that are compressed under normal loading of the knee, which can exhibit forces up to three times body weight. The stacks not only harvest energy, but can sense important phenomenon in the knee such as abnormally high forces exerted on the joint, degradation, and misalignment. Testing performed on an experimental model of the knee replacement system showed that when subject to a 900 N standard force profile, 4.8 mW of energy was generated by the stacks and 850 µW of regulated power could be provided for use by low power health monitoring electronics.

Harvesting energy from blood flow using flexible PVDF was presented by Sohn et al. in 2005 [78]. Thin PVDF film was investigated in various configurations and experimental results showed that a circular harvester of radius 5.62 mm and thickness of 28 µm subjected to a sinusoidal pressure of 5333 N/m^2 (similar to that of blood flow) at 1 Hz was able to generate 0.33 µW of power. The experimental setup used in this study is shown in Fig. 5.18c. Calculations were carried out that showed a chip requiring 10 mW of power to transmit data such as DNA information could be operated twice a day when powered by the harvester.

More recently, Karami and Inman in 2012 [8] proposed the use of nonlinear piezoelectric energy harvesting (discussed further in Section 5.5.1) to scavenge energy from heartbeat vibrations to power pacemakers. Simulations performed using measured heartbeat data published in the literature show that a nonlinear cantilever harvester with dimensions of 27 mm × 27 mm with a thickness of 80 µm, shown in Fig. 5.18d, is able to generate over 3 µW of power over a broad range of heart rates from 7 beats per minute to 700 beats per minute.

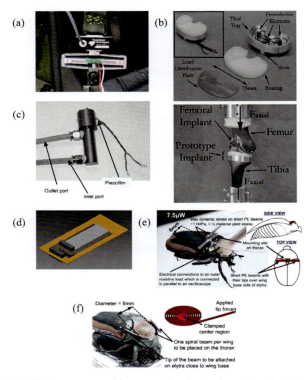

Figure 5.18 Various wearable and implantable piezoelectric energy harvesting devices [8, 75, 77–79].

Research has recently been performed not only on harvesting energy in the human body, but on harvesting energy in insects to create autonomous, living, micro air vehicles (MAVs) that can be controlled. First proposed by Reissman and Garcia in 2008 [80], the possibility of harvesting vibration energy during flight of a moth using piezoelectric materials was investigated. The concept of surgically adding energy harvesting and storage as well as neurological control capability to the insect during the pupa phase of development, prior to becoming an adult moth, is discussed. In a similar study, Aktakka et al. in (2011) [79] have performed experimental studies in which piezoelectric cantilevers as well as spiral harvesters are fixed to a beetle, as shown in Fig. 5.18e,f, respectively. Results of tethered flight testing show that a cantilever piezoelectric harvester with total volume of 11 mm^3 is able to harvest 11.5 µW of power during flight at 85 Hz flapping frequency. Additionally, a spiral harvester concept was investigated through bench top simulations and found to generate around 20 µW of power for typical beetle flight input.

The works presented in this section provide examples of the various research studies performed in biological and wearable piezoelectric energy harvesting systems. A more comprehensive review of energy harvesting systems to power biomedical sensors is given by Romero et al. [81]. Biological systems present exciting environments in which piezoelectric energy harvesting can be used to create autonomous electronics.

5.4.3 Piezoelectric Harvesting in Microelectromechanical Systems

Advances in low-power electronics based on VLSI (very large scale integration) circuitry have led to significant decreases in the power consumption of various electronic components [82]. This progress has allowed chips to draw small amounts of current, down to tens to hundreds of microwatts. With this decrease in power consumption, small-scale microelectronics give rise to the creation of miniature vibration energy harvesters at the microelectromechanical systems (MEMS) scale. Piezoelectric MEMS harvesters can enable self-powered microelectronic sensors and electronics, and have use in biomedical applications.

The first MEMS scale piezoelectric harvester to be fabricated and experimentally tested was presented by Jeon et al. in 2005 [82]. A micro-scale thin-film PZT cantilever harvester was fabricated, as shown in Fig. 5.19a, and included interdigitated electrodes such that the device employs the d_{33} mode of operation. The cantilever has a length of 100 µm, width of 60 µm, thickness of 0.48 µm, and resonance frequency of 13.9 kHz. When excited at resonance with a tip displacement of 2.56 µm, 1.01 µW of power was generated at 2.4 V. The primary drawback to the micro harvester is its extremely high natural frequency. Most mechanical systems exhibit operating frequencies in the tens or perhaps hundreds of Hz; therefore, excitation in the kHz range is impractical.

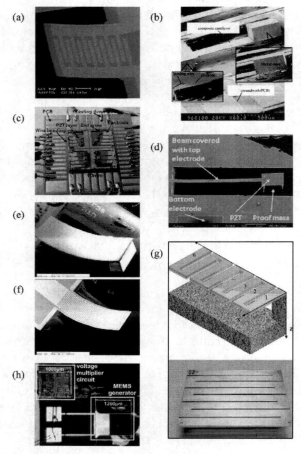

Figure 5.19 An assortment of MEMS piezoelectric energy harvesters from the literature [82–88].

In an effort to address the high resonance frequency of MEMS harvesters, Fang et al. in 2006 [83] developed a cantilever harvester with a very high aspect ratio. The device, shown in Fig. 5.19b, has a length of 2000 μm, width of 600 μm, substrate thickness of 12 μm, piezoelectric thickness of 1.64 μm, and a nickel tip mass, resulting in a natural frequency of 609 Hz. A voltage output of 0.898 V and a power of 2.16 µW were found when excited at resonance under 1 g of excitation. The dramatic decrease in resonance frequency from the work of Jeon et al. [82] marked a significant improvement in the design of practical MEMS vibration harvesters; however, the frequency is still high for most mechanical systems. The work of Fang et al. was continued by Liu et al. in 2008 [84] with the presentation of a MEMS piezoelectric cantilever array, shown in Fig. 5.19c, with a larger frequency bandwidth from 200–400 Hz. Broadband piezoelectric harvesting systems, which are discussed in more detail in Section 5.5.1, are useful in practical applications where the host frequency often drifts over time.

Shen et al. in 2009 [88] pushed the resonance frequency of their silicon oxide wafer-based PZT cantilever harvester below 200 Hz. The cantilever, shown in Fig. 5.19d, has a length of 4800 µm, width of 400 µm, and thickness of 22 µm (of which 1 µm is PZT). Experimental testing showed that the harvester has a resonance frequency of 183.8 Hz and produces 0.32 µW of power when excited at 0.75 g.

Lee et al. in 2009 [86] explored the difference between using the d_{31} and d_{33} piezoelectric operating modes in MEMS harvesters. Through fabrication of both d_{31} and d_{33} PZT cantilevers, shown in Fig. 5.19e,f, respectively, using an aerosol deposition process, direct comparison between the two harvesters was performed. The d_{31} harvester, with dimensions of 3000 × 1500 × 11 µm³ and a proof mass, has a resonance frequency of 255.9 Hz and is able to generate 2.765 µW of power at 2.675 V under 2.5 g excitation. The d_{33} harvester with interdigitated electrodes has the same dimensions, a similar tip mass, and a resonance frequency of 214 Hz, and is able to harvest 1.288 µW of power at 2.292 V under 2.0 g excitation. Results of further comparisons showed that while the d_{33} device gives a larger voltage output, the d_{31} device is capable of harvesting more power.

Recently, work by Karami and Inman in 2011 [85] has investigated the use of novel geometry as a way to decrease the natural frequency of MEMS-based piezoelectric energy harvesters. The proposed zig-zag structure, shown in Fig. 5.19g, is a means of increasing the effective length of the harvester without increasing the overall length dimension of the device. While the geometry is posed for MEMS harvesting, several macro scale harvesters with varying number of zig-zag beam elements (Fig. 5.19g shows an 8 member structure) were experimentally tested and results show that a 17× reduction in natural frequency can be achieved for an 11-member device compared to a cantilever beam with the same thickness and beam length.

The integration of power electronics on a MEMS platform along with a MEMS harvester was investigated by Marzencki et al. in 2008 [87] as a means of miniaturizing the entire harvesting system. A micro cantilever along with a miniature voltage multiplier circuit were fabricated and included in a single System on a Package (SoP), shown in Fig. 5.19h. While the cantilever exhibited a high resonance frequency of 1511 Hz, excitation at resonance with amplitude of 0.4 g resulted in around 30 nW of regulated power at 3.0 V, and proved the ability of the integrated system to harvest and condition energy.

While several MEMS-based piezoelectric harvesters are reviewed in this section with the goal of reducing the natural frequency of MEMS harvesters and integrating MEMS harvesting and power conditioning electronics, several other studies have been performed in the literature. A comprehensive review of MEMS piezoelectric harvesting is presented by Saadon and Sidek [89].

5.4.4 Harvesting Fluid Flow Using Piezoelectric Transduction

Environments containing kinetic energy in the form of fluid flow present opportunities for harvesting energy using piezoelectric materials. Particularly, there are environments rich in fluid flow but lacking in structural vibration where harvesting flow energy can lead to autonomous power sources for electronic devices. Flow of both liquid and gas can be considered as ambient energy sources that can be converted to useable electrical energy via piezoelectric transduction.

5.4.4.1 Harvesting of liquid flow

One of the earliest works on harvesting fluid flow using piezoelectric materials was performed by Taylor et al. in 2001 [90]. In their work, an *energy harvesting eel* composed of a PVDF strip emerged in water flow, was developed. A schematic view of the deployed eel system can be seen in Fig. 5.20a. One of the key operating conditions of the system is that a bluff body must be inserted upstream of the harvester such that vortex shedding causes excitation of the eel. A prototype eel that was 24 cm long, 7.6 cm wide, and 150 μm thick was created and tested in a flow tank. The voltage generated by the eel was measured and recorded for a water velocity of 0.5 m/s, and a peak voltage of around 3.0 V was recorded.

Pobering and Schwesinger in 2004 [91] proposed two types of hydropower piezoelectric devices including a PVDF *fluttering flag*, shown in Fig. 5.20b, and a PZT bimorph, shown in Fig. 5.20c. While the devices were not realized, simulations showed that the PVDF flag could likely generate between 11–32 W/m^2 of power, while a PZT cantilever with 5 mm length, 3 mm width, and 60 μm thickness could generate around 7 μW of power.

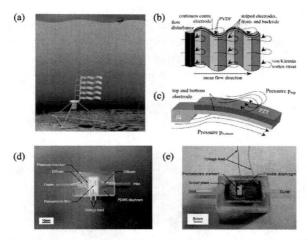

Figure 5.20 Assortment of liquid flow-based energy harvesters [90–93].

In a more recent study, Wang and Ko in 2010 [92] investigated energy harvesting of liquid flow in channels using PVDF piezoelectric film. Energy can be harvested from pressure fluctuations created in a fluid system operated on by a pump. As the fluid pressure changes, energy can be transferred to a piezoelectric diaphragm, which can harvest the energy from pressure change. Experiments performed on the flow-based harvester, shown in Fig. 5.20d and containing a PVDF harvester with a length of 25 mm, width of 13 mm and thickness of 150 μm (of which 28 μm is PVDF, the remainder is a polymer coating), showed that 0.2 μW of power at 2.2 V was harvested for pressure oscillations of 1.196 kPa at 26 Hz. This work was continued by Wang and Liu in 2011 [93] where a PZT film was utilized to harvest energy from fluid fluctuations in a similar fluid flow system. Experiments on the device, shown in Fig. 5.20e and having dimensions of 8 mm × 3 mm × 200 μm, revealed that the device can harvest 0.45 nW of power at 72 mV for a fluctuation pressure of 20.8 kPa at 45 Hz.

5.4.4.2 Harvesting air flow using windmill-style harvesters

In addition to harvesting from liquid flow, research has also been performed on piezoelectric harvesting of wind energy, and typically falls into one of two categories; windmill-style harvesting and flutter-style harvesting.

Priya et al. in 2005 [94, 95] presented the original work on piezoelectric windmill harvesting. A windmill was proposed in which a conventional fan that rotates in the presence of wind is linked to a piezoelectric windmill that contains multiple cantilever piezoelectric bimorphs. As the piezoelectric windmill rotates, the free end of each bimorph, which is near the hub of the windmill, is periodically bent and subsequently released causing free oscillations. Experimental testing of a piezoelectric windmill, shown in Fig. 5.21a, containing 10 PZT bimorphs, each with dimensions of 60 × 20 × 0.6 mm^3, showed that the windmill is capable of harvesting a maximum of 7.5 mW of power for a wind speed of 10 mph (4.47 m/s).

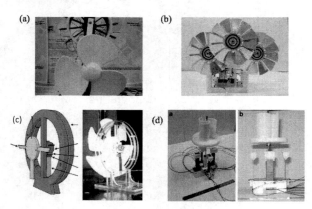

Figure 5.21 Several windmill-style piezoelectric energy harvesting systems from the literature [95–98].

Continuing the work of Priya et al. [94, 95], Myers et al. in 2007 [96] described an optimized piezoelectric windmill consisting of a much simpler design with fewer parts than the previous work. Several designs are considered and the optimal design, shown in Fig. 5.21b, is found to consist of three fans of 12.7 cm diameter that when rotated by wind, cause translation in the harvesting unit that excites two rows of nine bimorphs each. When subject to a wind speed of 10 mph (4.47 m/s), the windmill is able to generate 5 mW of power.

Tien and Goo in 2010 [97] presented a similar windmill harvester in which a single PZT composite cantilever is directly excited by the rotation of a conventional fan through exciter teeth installed on the hub of the turbine, as shown in Fig. 5.21c. Experimental wind tunnel testing of the system showed that the bimorph with dimensions of 72 × 12 × 0.5 mm^3 and a piezoelectric thickness of 0.1 mm was able to harvest a maximum of 8.5 mW of power at 26 V; however, the corresponding wind speed was not specified.

Karami et al. in 2013 [98] developed a novel piezoelectric windmill utilizing nonlinear piezomagnetic harvesting. The noncontact windmill consists of four vertical cantilever PZT bimorphs with a length of 58 mm, width of 12.7 mm, and thickness of 0.38 mm mounted below a miniature Savonius vertical axis wind turbine. Magnets are attached to the tips of the cantilevers and to the bottom of the turbine such that when rotated, the turbine excites each cantilever through magnetic interaction. The magnetic forces cause the beam to operate in a nonlinear fashion. More details on nonlinear energy harvesting are discussed in Section 5.5.1. Two versions of the piezoelectric windmill, both shown in Fig. 5.21d, were tested in a wind tunnel and it was found that for the optimal design, a single bimorph can harvest up to 4 mW of power at a wind speed of 10 mph (4.47 m/s). Additionally, a low startup speed of 2 m/s was reported for the piezomagnetic windmill.

5.4.4.3 Harvesting of air flow using flutter-style harvesters

In addition to the windmill-style harvesters presented in the previous section, various flutter-style piezoelectric wind harvesting systems have been developed in the literature. In an initial investigation into flutter-style wind harvesting, Tan and Panda in 2007 [99] subjected a cantilever PZT beam mounted at an angle to transverse airflow, as shown in Fig. 5.22a. A plastic *flapper* is installed on the end of the beam, presumably because the compliance of the plastic couples well with the excitation from air flow, causing the PZT beam to oscillate. The harvester, which is 76.7 mm long, 12.7 mm wide and 2.2 mm thick, is found to produce around 150 µW of power when subject to an optimal wind speed of 6.7 m/s.

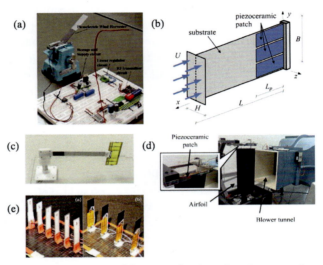

Figure 5.22 An assortment of flutter-style piezoelectric energy harvesting systems from the literature [99–103].

Kwon in 2010 [100] developed a T-shaped cantilever beam, shown schematically in Fig. 5.22b, that facilitates aeroelastic flutter under wind excitation. The T-shaped end of the device causes vortex shedding when placed in flow, which causes excitation of the cantilever leading to flutter. Experimental wind tunnel testing was performed on a T-shaped harvester with overall length of 100 mm, width of 60 mm, and width of the T-shaped section of 30 mm. Six MFC harvesters of 28 mm length, 14 mm width and 300 µm thickness were installed at the root of the harvester. Wind tunnel results showed an initial cut-in wind speed corresponding to flutter of 4 m/s, which provided a maximum power output of 4.0 mW.

The work of Bryant and Garcia in 2011 [101] expands upon the work presented by Tan and Panda [99] through the development of an aeroelastic flutter energy harvester consisting of a rigid flap connected by a ball bearing revolute joint to a cantilever piezoelectric harvester beam, as shown in Fig. 5.22c. When inserted into a flow field, the harvester flap will oscillate causing flutter of the system. The beam has a length of 25.4 cm and a width of 2.54 cm. The flap has a width (span) of 13.6 cm and a length (semichord) of 2.97 cm. Two PZT harvesters with dimensions of 4.6 × 2.06 × 0.254 cm^3 are attached on either side of the root of the beam. Experimental wind tunnel testing showed that the harvester can generate a maximum of 2.2 mW of power at a wind speed of 7.9 m/s, and that the device has a cut-in speed of 2.6 m/s.

Sousa et al. in 2011 [102] present the investigation of a piezoelectric energy harvesting system that couples a conventional airfoil section with piezoelectric energy harvesters. As shown in Fig. 5.22d, piezoelectric devices are installed on the support beams of the airfoil in the *plunge* degree of freedom. Testing in the linear regime near the flutter speed is performed. Additionally, nonlinearities are introduced in the *pitch* degree of freedom to improve the performance of the harvesting system by introducing chaotic vibrations, and testing is repeated. Wind tunnel testing of the system shows that the introduction of nonlinearities results in twice the power output compared to the linear system. Experimental results show that the harvesters are able to generate 27 mW of power for a wind speed of 10 m/s. Although the system tested is quite large and impractical, it is simply meant to demonstrate the benefits of nonlinearity in piezoaeroelastic energy harvesting.

Hobeck and Inman in 2012 [103] introduced an array of piezoelectric cantilever harvesters, dubbed *piezoelectric grass*, that harvest energy from turbulence induced vibration. The concept involves the development of robust cantilever arrays that can be inserted into low-velocity, turbulent flows to harvest energy. Two types of arrays are investigated including PVDF cantilevers and PZT cantilevers, as shown in Fig. 5.22e. Wind tunnel testing results showed that an array of four PZT cantilevers with substrate dimensions of 101.6 mm × 25.4 mm × 101.6 µm and piezoelectric dimensions of 45.97 mm × 20.57 mm × 152.4 µm was able to harvest 1.0 mW per cantilever at a mean wind speed of 11.5 m/s.

Several works focusing on piezoelectric harvesting of fluid flow have been reviewed in this section. Studies have investigated harvesting of liquid flow and pressure fluctuations as well as air flow using both windmill-style harvesters and flutter-style harvesters. These harvesters are useful in environments where structural vibrations may be lacking but kinetic energy in the form of fluid flow is present.

5.4.5 Summary

The works reviewed in this section represent a sampling of the various topics being considered by researchers in the field of piezoelectric vibration-based energy harvesting. From the research summarized here, it is clear that there are many applications in which piezoelectric transducers can be utilized for vibration-to-electrical energy conversion, many of which aim to provide power to low-power electronic devices such as wireless sensor nodes and health monitoring systems. While this section is not intended to be an exhaustive review of piezoelectric harvesting, the reader is referred to various review articles and texts published in the literature for further details and examples [11–15, 18, 63, 81, 89, 104, 105].

5.5 Current Research Thrusts

In the past few years, several research thrusts have emerged in the field of piezoelectric energy harvesting. Many of these areas of research aim to advance the efficiency or increase the functionality of piezoelectric energy harvesting, or to find new materials and methods for conversion of vibration energy to electrical energy through the piezoelectric effect. A summary of several of these thrust areas is included in this section.

5.5.1 Broadband and Nonlinear Harvesting

Typical piezoelectric energy harvesting systems are based on linear resonance phenomenon. The most widely used piezoelectric harvester configuration, the cantilever, relies on ambient excitation at or extremely near the fundamental resonance frequency of the harvester. When the excitation frequency deviates from the ideal resonant frequency, power output can decrease by several orders of magnitude [33]. In order to improve upon the limitations of linear resonant systems, broadband and nonlinear energy harvesting has emerged as a popular topic of research.

5.5.1.1 Broadband piezoelectric energy harvesting

Broadband energy harvesting systems aim to increase the response bandwidth of the harvester to reduce the need of the excitation frequency to be very close to the resonant frequency of the harvester. Broadening of the response can be achieved using passive or active methods. In passive systems, the dynamics of the system are manipulated such that the harvester responds over a range of frequencies. In active systems, energy is input into the system to actively alter the resonant frequency of the harvester to match the current excitation frequency.

Passive broadband systems may be realized through various mechanical configurations. Some of the original works simply attempted to tune the resonance frequency of the harvester using either compressive axial force [106] or attractive/repulsive magnetic forces [107]. In these configurations, the resonance frequency of the harvester could be tuned to match the excitation frequency; however, the active bandwidth of the harvester was not increased. In order to increase the active bandwidth of a piezoelectric harvesting system, a *mechanical bandpass* system was developed by Shahruz in 2006 [108], as shown in Fig. 5.23a, where multiple cantilever beams of difference length and resonant frequency are attached to a common base. The drawback to this configuration is that only a few of the harvesters contribute to the total power output at any given excitation frequency, leaving many unused.

Active tuning methods have also been investigated. In an early attempt by Roundy et al. in 2005 [109, 110], active tuning systems incorporating actuators to actively alter the mechanical properties of the harvester as a means of tuning the resonance frequency were found to consume too much power to be beneficial. In a more recent study by Lallart et al. in 2010 [111], however, an active tuning method utilizing acceleration and displacement feedback of a cantilever harvester along with an actuator controlled via a low power control circuit, as shown in Fig. 5.23b, was shown to effectively increase the bandwidth of a cantilever harvester without significant power draw.

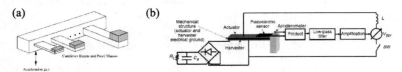

Figure 5.23 Broadband harvesting systems including (a) passive *mechanical bandpass* system [108] and (b) active tuning method [111].

5.5.1.2 Nonlinear piezoelectric energy harvesting

In addition to passive and active broadband energy harvesting systems, nonlinear piezoelectric energy harvesting has also been investigated in recent years by several researchers. The concept of nonlinear harvesting is to add nonlinearities to the system and exploit nonlinear phenomenon to increase the overall response of the system, thus yielding larger power output compared to a similar linear system [112].

Barton et al. in 2010 [113] introduced a type of *mono-stable Duffing oscillator* in the form of a cantilever harvester utilizing magnetic interactions to create a single point of stability in the system, shown in Fig. 5.24a along with its tip velocity response at varying excitation amplitudes. The nonlinear effects are present in the response of the system for increasing excitation amplitude, and it can be seen that under large excitation, a high energy state can be achieved which provides a large increase in tip velocity and, correspondingly, output power.

Erturk et al. in 2009 [114] presented a *bi-stable Duffing oscillator* consisting of a magnetic cantilever piezoelectric harvester and two magnets near the free end of the beam to create two points of stability, shown in Fig. 5.24b along with its voltage output response for varying frequencies of excitation compared to a linear cantilever harvester. From the response of the system, it is clear that the effective bandwidth of the harvester is increased over the linear harvester design. Additionally, the nonlinear effects present in the system allow for high energy states in the form of limit cycle oscillations to be achieved under large amplitude excitation, thus generating significantly more power than a comparable linear harvester. A similar *bi-stable* system was presented by Stanton et al. in 2010 [115].

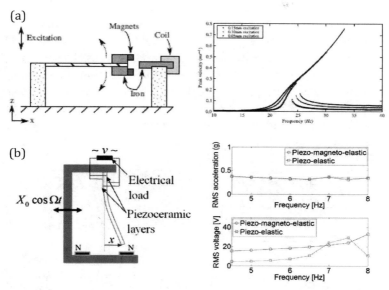

Figure 5.24 Nonlinear piezoelectric energy harvesters including (a) *monostable* [113] and (b) *bi-stable* [114] configurations.

Other researchers have investigated nonlinear piezoelectric energy harvesting systems of varying configurations. Several works in nonlinear harvesting were reviewed by Daqaq [116] in 2014.

5.5.2 Multifunctional Harvesting

Traditional piezoelectric energy harvesting systems consist of an active harvesting element, conditioning circuitry, and a storage medium, where the sole function of the combined system is to convert ambient mechanical energy into usable electrical energy. Furthermore, conventional systems are designed as add-on components to a host structure, often causing undesirable mass loading effects and consuming valuable space. A method of improving the functionality of conventional harvesting designs that has gained interest in the research community involves the use of a multifunctional approach in which the system not only harvests energy, but also performs additional tasks such as storing the scavenged energy or supporting mechanical load in the structure.

A multifunctional approach to vibration energy harvesting using piezoelectric material was presented by Anton et al. in 2010 [117] in which a composite layered device consisting of a substrate, piezoelectric layers for energy generation, and thin-film battery layers for energy storage was developed. The *self-charging structures*, shown in Fig. 5.25a, are capable of simultaneously harvesting vibration energy and storing the harvested energy in the battery layers, thus a single device can both harvest and store energy.

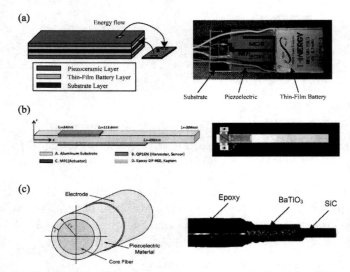

Figure 5.25 Multifunctional energy harvesters including (a) *self-charging structures* [117], (b) energy harvesting and gust alleviation wing spar [118], and (c) structural piezoelectric fibers [122].

The *self-charging structure* concept was applied to multifunctional energy harvesting in unmanned aerial vehicles (UAVs) by Anton et al. in 2012 [4] through the integration into a UAV wing spar, such that during flight, energy is harvested and stored from vibrations in the wings. This concept was expanded upon by Wang and Inman in 2012 [118] when actuation capabilities were added to the multifunctional wing spar, as shown in Fig. 5.25b, in order to provide energy harvesting as well as gust alleviation capabilities. The concept of self-powered gust alleviation involves harvesting vibration energy in the wing of the aircraft and subsequently using the harvested energy to power a piezoelectric actuator that effectively cancels out gust forces in the wing.

Lin and Sodano in 2008 [119] introduced another method of creating multifunctional piezoelectric energy harvesting devices through the development of novel piezoelectric structural fiber. The fiber contains a conventional conductive fiber core which is then coated with piezoelectric material and an outer electrode material, as shown in Fig. 5.25c. The fiber core acts as a structural stiffening layer as well as an inner electrode, and the piezoelectric layer allows the active fiber to possess sensing, actuation, and energy harvesting ability. Fabrication and experimental testing of an active fiber containing a silicon carbide core coated with barium titanate piezoelectric material, an outer electrode composed of silver paint, and an outer polymer coating verified the piezoelectric properties of the fiber [120]. Finally, Lin and Sodano in 2009 [121] investigated the energy storage ability of the fibers. The dielectric properties of the active barium titanate shell were studied in order to create a structural fiber capacitor. Results of experimental testing showed that the fibers are capable of storing harvested energy, thus proving the ability of the multifunctional fiber to both harvest and store energy.

5.5.3 Multi-Source Energy Harvesting

Conventional energy harvesting systems focus on a single mode of energy conversion and are thus highly susceptible to fluctuations in ambient energy levels that greatly affect the performance of the harvesting system. If the ambient energy source fades, the harvester may not produce sufficient energy to power a device, causing it to go "offline." As a means of addressing this concern, multi-source energy harvesting systems capable of simultaneous harvesting from multiple ambient energy sources have been investigated by the research community. Such multi-source systems, shown schematically in Fig. 5.26, are more robust against varying environmental and operational conditions as their performance is not tied to a single ambient source.

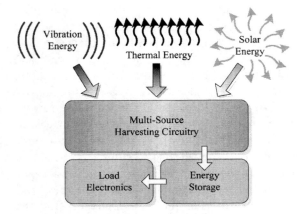

Figure 5.26 Multi-source energy harvesting concept.

Multi-source energy harvesting systems combining vibration harvesting using piezoelectric material with harvesting of other types of ambient energy have been explored recently in the literature. Gambier et al. in 2012 [123] expanded upon the *self-charging structure* concept developed by Anton et al. [117] by including thin-film solar layers in the multi-layered composite device to combine both piezoelectric harvesting and solar harvesting in a single device. Additionally, harvesting of thermal energy using thermoelectric generators was considered; however, the generator was external to the composite device. Experimental testing of the multi-source harvester, shown in Fig. 5.27a, showed that the device was capable of harvesting both solar and vibration energy; however, testing of each harvester type was conducted separately; therefore, simultaneous harvesting of multiple energy sources was not evaluated.

Schlichting et al. in 2012 [124] explored several simple circuit topologies to simultaneously combine the output of solar and vibration energy harvesting devices, including series and parallel electrical configurations. Although no physical system was built or tested, analytical modeling of the multi-source schemes provides insight into issues related to the combination of disparate energy sources into a single storage element.

Experimental field testing of a multi-source energy harvesting system combining piezoelectric vibration harvesting and photovoltaic solar harvesting applied to a structural health monitoring node on a wind turbine was presented by Anton et al. in 2013 [125]. The multi-source system, shown in Fig. 5.27b, consists of several piezoelectric harvesters as well as a thin-film solar panel mounted near the root of a residential-scale wind turbine blade, a hysteretic multi-source energy harvesting circuit with four independent input stages and one common storage capacitor, and a WID 3.0 (wireless impedance device) sensor node [126] mounted to the hub of the turbine, as well as a wireless data acquisition system to monitor the performance of the system. Field testing of the multi-source harvester showed that while the output of the piezoelectric harvester is orders of magnitude lower than the solar harvester, the system is capable of simultaneously harvesting both vibration and solar energy and powering the WID sensor node during operation of the turbine.

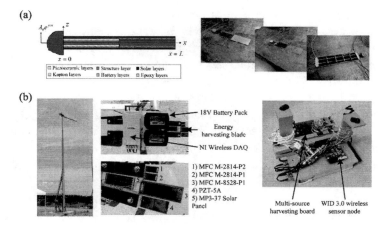

Figure 5.27 Multi-source energy harvesters including (a) multi-source self-charging structures with piezoelectric and solar harvesting combined with energy storage [123] and (b) combined piezoelectric and solar harvesting to power a wind turbine structural health monitoring node [125].

In addition to the development of multi-source energy harvesting systems utilizing piezoelectric vibration harvesting, researchers have also investigated multi-source systems that combine photovoltaic harvesting and wind harvesting [127] as well as photovoltaic, wind, and hydroelectric harvesting [128] to create robust energy systems for wireless sensors.

5.5.4 Novel Piezoelectric Materials

While the majority of published work in piezoelectric energy harvesting deals with the use of conventional piezoelectric materials, such as lead zirconate titanate (PZT) ceramic or PVDF polymer, there is a current thrust in the literature to develop novel piezoelectric energy harvesting materials and composite structures. These novel materials aim to increase the effective piezoelectric coupling, improve the mechanical properties of the material or composite, and increase the safety and biocompatibility of the material. In the following sections, recent research into various novel piezoelectric materials and structures is summarized.

5.5.4.1 Piezoelectric single crystals

While PZT ceramic offers the largest coupling amongst conventional piezoelectric materials, researchers have recently begun to investigate the use of piezoelectric single crystals in vibration energy harvesting systems which possess significantly larger coupling compared to PZT. Single crystal piezoelectric materials have been shown to possess electromechanical coupling coefficients around an order of magnitude larger than PZT [129]. Badel et al. in 2006 [130] compared identical cantilever unimorphs of lead magnesium niobate-lead titanate (PMN-PT) single crystal and PMN-PT ceramic and found that the power output of the single crystal was 20× greater than the ceramic. The drawbacks to single crystals, however, include their fragility and high cost compared to PZT.

Ren et al. in 2006 [131] fabricated a composite energy harvester composed of PMN-PT single crystals placed in an epoxy matrix in order to utilize the longitudinal d_{33} mode of the single crystals. The single crystal material used in this work exhibits a piezoelectric coupling coefficient of d_{33} = 2060 pC/N, compared to the typical coupling coefficient of PZT in the range of d_{33} = 300–600 pC/N [132]. Experimental testing is performed on the PMN-PT composite, which is shown in Fig. 5.28a, along with a PZT-epoxy composite for comparison. When subjected to a sinusoidal compressive load of around 90 MPa at 4 Hz, the PMN-PT single crystal composite was found to generate around 100 mW/cm^3, which is about two times more power across a resistive load than the PZT composite.

In order to utilize the shear mode of a PMN-PT single crystal, which exhibits very large coupling around d_{15} = 3000 pC/N, Ren et al. in 2010 [133] developed the unimorph cantilever shown in Fig. 5.28b. The poling direction of the sample is indicated by the arrow. When subjected to sinusoidal excitation, the unimorph, which contains a brass shim with dimensions of 50.0 × 6.0 × 0.3 mm³, a PMN-PT wafer with dimensions of 13.0 × 6.0 × 1.0 mm³, and a tip mass of 0.5 g, is able to generate 4.16 mW of power for excitation at 60 Hz with a cyclic force of 0.05 N. The peak voltage output of the device was measured as 91.2 V. Comparisons were made to a similar PMN-PT cantilever operating in the transverse d_{31} mode, and the shear mode device was found to generate significantly more power (on the order of 8× more energy).

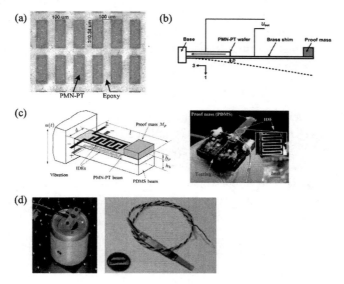

Figure 5.28 Piezoelectric single crystal energy harvesters including (a) PMN-PT/epoxy composite [131], (b) PMN-PT shear-mode harvester [133], (c) PMN-PT harvester with interdigitated electrodes [134], and (d) PMN-PZT unimorph harvester [129].

Mathers et al. in 2009 [134] investigated the use of interdigitated electrodes on a micro-scale PMN-PT cantilever harvester. The device, shown in Fig. 5.28c, consists of a PMN-PT beam with dimensions of 7.4 mm × 2.0 mm × 110 μm, a polydimethylsiloxane

(PDMS) polymer coating layer, a PDMS tip mass, and interdigitated electrodes. When excited at its natural frequency of 1340 Hz with a base displacement of 1 mm, the device generates around 0.3 mW of power at a peak voltage of 10 V.

In addition to PMN-PT, researchers have also investigated vibration energy harvesting using lead magnesium niobate-lead zirconate titanate (PMN-PZT) single crystals. Erturk et al. in 2008 [129] presented results of an investigation using a PMN-PZT cantilever unimorph, shown in Fig. 5.28d, to harvest vibration energy. In their work, a small cantilever with piezoelectric dimensions of 20.0 × 5.0 × 0.5 mm^3 and an aluminum substrate with thickness of 0.79 mm was excited at resonance (1744 Hz) and found to generate a maximum power per base acceleration of 14.7 µW/g^2. Moon et al. in 2009 [135] investigated a similar PMN-PZT cantilever with piezoelectric dimensions of 25.0 × 5.0 × 0.25 mm^3 and a substrate thickness of 0.55 mm, and found the device to generate 0.28 mW of power when excited at resonance (630 Hz) under 1 g acceleration.

5.5.4.2 Piezoelectric nanocomposites

In the past few years, researchers have investigated the development of piezoelectric nanocomposites utilizing piezoelectric nanowires and nanofibers for improved piezoelectric and mechanical performance. One of the original works on piezoelectric nanocomposites was presented by Wang and Song in 2006 [136]. A zinc oxide (ZnO) nanowire array was fabricated and experimentally tested using atomic force microscopy by deflecting a single nanowire, as shown in Fig. 5.29a, and found to possess the electromechanical coupling required to create nanomaterial-based energy harvesting devices.

Feenstra and Sodano in 2008 [137] developed a barium titanate (BaTiO$_3$) piezoelectric nanocomposite paint. Nanocomposite paint is advantageous over conventional piezoelectric materials in that it can be sprayed on to any surface; however, piezoelectric paints often possess weak electromechanical coupling compared to conventional materials. The piezoelectric paint was formed by distributing electrospun piezoelectric wires, as shown in Fig. 5.29b, in an epoxy matrix. The BaTiO$_3$ nanowire paint developed in this work was compared experimentally to paint utilizing piezoelectric nanoparticles and, although it possessed a lower sensitivity than PVDF, it was found to provide as high as 3× increase in electromechanical coupling over the previous nanoparticle composite paint.

PVDF nanofibers have also been investigated in the literature for use in creating nanogenerators for energy harvesting. Fang et al. in 2011 [138] created a PVDF membrane nanogenerator using electrospun PVDF nanofibers sandwiched between two aluminum foil membranes, as shown in Fig. 5.29c. The device, which has a working area of 2.0 cm^2 and a thickness of 140 µm, was able to output a maximum of around 7 V, and was shown to be capable of illuminating a commercial LED, shown in Fig. 5.29c, through the use of a simple energy harvesting circuit and storage capacitor.

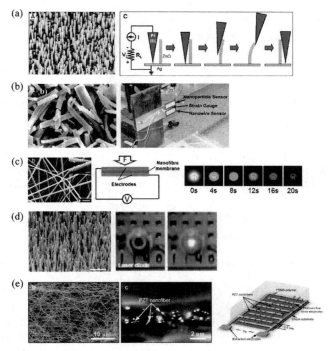

Figure 5.29 Piezoelectric nanofiber-based energy harvesters including (a) ZnO nanowire array [136], (b) BaTiO$_3$ nanocomposite paint [137], (c) PVDF nanofiber membrane [138], (d) PZT nanowire array [139], and (e) aligned PZT nanofiber device with interdigitated electrodes [140].

The development of a nanowire array utilizing the high piezoelectric coupling of lead zirconate titanate was presented by Xu et al. in 2010 [139]. While the fabrication of PZT thin films typically requires high temperature on the order of 650°C, Xu et al. were able to grow vertically aligned single crystal PZT nanowire arrays at 230°C, thus broadening their use by allowing integration with soft materials. A nanowire array, shown in Fig. 5.29d, with an active area of 6 mm^2 was fabricated and experimentally tested and found to generate a peak output voltage of around 0.7 V and an average power density of 2.8 mW/cm^3. Additionally, a seven-layer nanogenerator was found to be capable of powering a commercial laser diode when subject to compressive force, as shown in Fig. 5.29d, by using a rectifying circuit with storage capacitors.

Chen et al. in 2010 [140] also investigated the use of PZT nanofibers to create an energy harvesting device. Electrospun PZT nanofibers were deposited and aligned onto interdigitated electrodes and encased in a PDMS polymer matrix, as shown in Fig. 5.29e. Experimental testing of the nanogenerator showed a maximum voltage generation of 1.63 V when the harvester was pressed with a finger, and a power output of around 0.03 μW when subject to 12% strain at 39.8 Hz. Dimensions of the harvester were not reported.

Many of the works performed on nanofiber piezoelectric energy harvesters are summarized in a review article by Chang et al. in 2012 [141]. The article reviews several types of piezoelectric nanofiber materials, but focuses on PVDF and PZT nanofiber generators. Manufacturing methods and material characterization techniques are also discussed in the review article.

5.5.4.3 Piezoelectret foams

A material that has just recently gained attention for use in vibration energy harvesting systems is piezoelectret foam. Piezoelectret foams, first developed in the 1980s in Finland [142], are a class of electret material; a dielectric material that contains permanent electric charge or polarization analogous to the magnetic fields found in permanent magnets. Piezoelectricity is observed in piezoelectrets due to the deposition of charge on internal voids in the structure. When subject to mechanical or electrical stimuli, the charged voids respond as macroscopic dipoles, thus yielding piezoelectric properties. A schematic of piezoelectret foam showing both the cross section as well as the piezoelectric behavior is given in Fig. 5.30. The major advantage of piezoelectret foam over conventional piezoelectric polymers is its large piezoelectric coupling coefficient, d_{33}, up to 250 pC/N compared to −33 pC/N for PVDF.

Anton et al. in 2014 [143] recently explored the possibility of creating a vibration-based energy harvester using piezoelectret foam. Commercially available piezoelectret foam from Emfit, Corp., with a thickness of 85 µm was used to create a pre-tensioned harvester with dimensions of 15.24 × 15.24 cm², as shown in Fig. 5.31a. When excited longitudinally at 60 Hz with a displacement of ± 73 µm, corresponding to an acceleration of ± 10.38 g, the harvester generates a peak voltage of about 8 V. Experiments were performed in which the harvester was found to be able to charge a 1 mF capacitor up to 4.67 V in 30 min, results of which are shown in Fig. 5.31a (note, these results correspond to the "1-direction"). The harvester delivered an average power of 6.0 µW, which is comparable to the output of many conventional piezoceramic and piezoelectric polymer harvesters.

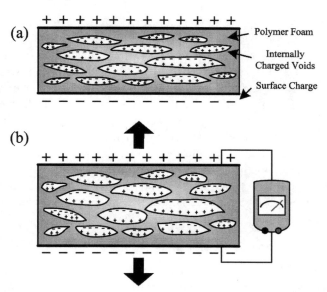

Figure 5.30 Piezoelectret foam schematic showing (a) cross-sectional composition and (b) piezoelectric response.

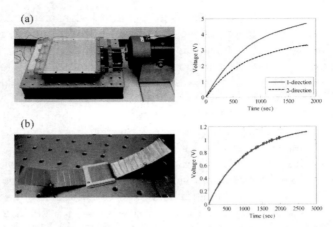

Figure 5.31 Piezoelectret foam energy harvester and capacitor charging results from (a) single layer harvester under transverse excitation [143] and (b) 20-layer stack harvester under compressive excitation [145].

Researchers have also recently investigated the creation of piezoelectret foam stacks that directly utilize the "33" mode of operation for increased energy harvesting performance. Pondrom et al. in 2014 [144] presented results of 9- and 10-layer piezoelectret stack harvesters excited harmonically in compression. The devices were able to generate around 1.3 µW/g^2 across a load resistance of 100 MΩ. Ray et al. in 2015 [145] extended the work of Pondrom et al. [144] by testing a stack with increased layer count (20-layers) as shown in Fig. 5.31b and performing energy harvesting experimentation in which the stack was harmonically excited in compression and used to charge a capacitor. Results of the testing showed that the 20-layer piezoelectret foam stack could generate around 3.8 mW/g^2 across an optimal load resistance of 650 kΩ and was capable of charging a 1 mF capacitor to 1.2 V in about 45 min when excited at resonance (124.4 Hz) at 0.5g, as shown in Fig. 5.31b.

5.5.4.4 Lead-free piezoelectrics

The majority of ceramic-based piezoelectric materials used in energy harvesting and other applications contain lead. In fact, the most commonly used piezoelectric material is PZT; lead zirconate titanate. While PZT offers superior piezoelectric properties to many alternatives, the toxicity of lead leads to inherent health risks in the use of PZT and other lead-based piezoelectrics. This is especially true in biomedical applications. A significant topic of research lies in the development of lead-free piezoelectric materials. Creating a compound with comparable piezoelectric coupling and mechanical durability remain the significant challenges in the development of lead-free piezoelectrics. Much of the recent research on the topic of lead-free piezoelectric material is summarized in a review article by Panda in 2009 [22], as well as a book by Priya and Nahm in 2012 [23].

5.6 Summary and Future Visions

The field of piezoelectric energy harvesting has attracted much attention from the research community throughout the last decade. Piezoelectric energy harvesting is defined as the conversion of ambient mechanical vibration energy to useful electrical energy through the use of a piezoelectric material. Piezoelectric harvesting systems typically generate on the order of microwatts to milliwatts of power and are often used to create autonomous power supplies for self-powered electronics.

A wide variety of applications utilizing piezoelectric energy harvesting have been presented in the literature, ranging from the development of self-powered wireless sensor nodes and biomedical devices, to harvesting energy in microelectromechanical systems and from fluid flow using piezoelectric transduction. In addition, several research thrusts have emerged in the community in the past few years including the development of broadband and nonlinear energy harvesters, multifunctional and multi-source harvesters, and the investigation of novel piezoelectric materials with enhanced mechanical and piezoelectric properties.

The ultimate goal of most piezoelectric energy harvesting systems is to provide adequate electric power to low-power electronics in order to completely eliminate the need for a battery, thus creating truly autonomous devices. Not only are there cost advantages associated with the elimination of batteries as well as gains by reducing environmental waste, but without the restriction of accessibility for battery replacement, new designs for low-power electronics emerge. In particular, piezoelectric energy harvesting solutions will allow electronics to be completely embedded inside vibration-rich environments (i.e., the wing of an aircraft, inside a wind turbine blade, inside the human body, etc.), thus opening the door to many new possibilities.

As piezoelectric energy harvesting technology progresses, new harvester configurations encompassing novel piezoelectric materials, excitation techniques, efficient circuitry, and storage elements will continue to improve the power output and conversion efficiency of piezoelectric harvesting. With this progression, an increasing number of piezoelectric vibration harvesting systems will be capable of reaching usable milliwatt power output levels under reasonable excitation levels (i.e., tens of hertz and tenths of g's). While a small assortment of commercial energy harvesting devices exist today, the continued progression of energy harvesting technology aims to move energy harvesting from the research domain to the commercial domain, where consumer products will eventually contain piezoelectric energy harvesting power sources.

References

1. Ali, S. F., Friswell, M. I., and Adhikari, S. (2011) Analysis of energy harvesters for highway bridges, *J. Intell. Mater. Syst. Struct.*, **22**(16), 1929–1938.
2. Elvin, N. G., Lajnef, N., and Elvin, A. A. (2006) Feasibility of structural monitoring with vibration powered sensors, *Smart Mater. Struct.*, **15**(4), 977–986.
3. Erturk, A. (2011) Piezoelectric energy harvesting for civil infrastructure system applications: Moving loads and surface strain fluctuations, *J. Intell. Mater. Syst. Struct.*, **22**(17), 1959–1973.
4. Anton, S. R., Erturk, A., and Inman, D. J. (2012) Multifunctional unmanned aerial vehicle wing spar for low-power generation and storage, *J. Aircraft*, **49**(1), 292–301.
5. Erturk, A., Renno, J. M., and Inman, D. J. (2009) Modeling of piezoelectric energy harvesting from an L-shaped Beam-mass Structure with an application to UAVs, *J. Intell. Mater. Syst. Struct.*, **20**(5), 529–544.
6. Featherston, C. A., Holford, K. M., and Greaves, B. (2009) Harvesting vibration energy for structural health monitoring in aircraft, *Key Eng. Mater.*, **413–414**, 439–446.

7. Lee, S., and Youn, B. D. (2011) A new piezoelectric energy harvesting design concept: Multimodal energy harvesting skin, *IEEE Trans. Ultrason. Ferroelectr. Freq. Control.*, **58**(3), 629–645.
8. Karami, M. A., and Inman, D. J. (2012) Powering pacemakers from heartbeat vibrations using linear and nonlinear energy harvesters, *Appl. Phys. Lett.*, **100**(4), 042901 (4 pp.).
9. Qi, Y., Jafferis, N. T., Lyons, K., Lee, C. M., Ahmad, H., and McAlpine, M. C. (2010) Piezoelectric ribbons printed onto rubber for flexible energy conversion, *Nano Lett.*, **10**(2), 524–528.
10. Starner, T. (1996) Human-powered wearable computing, *IBM Syst. J.*, **35**(3–4), 618–629.
11. Anton, S. R., and Sodano, H. A. (2007) A review of power harvesting using piezoelectric materials (2003–2006), *Smart Mater. Struct.*, **16**(3), R1–R21.
12. Beeby, S. P., Tudor, M. J., and White, N. M. (2006) Energy harvesting vibration sources for microsystems applications, *Meas. Sci. Technol.*, **17**(12), R175–R195.
13. Cook-Chennault, K. A., Thambi, N., and Sastry, A. M. (2008) Powering MEMS portable devices-A review of non-regenerative and regenerative power supply systems with special emphasis on piezoelectric energy harvesting systems, *Smart Mater. Struct.*, **17**(4), 043001 (33 pp.).
14. Priya, S. (2007) Advances in energy harvesting using low profile piezoelectric transducers, *J. Electroceramics*, **19**(1), 167–184.
15. Sodano, H. A., Inman, D. J., and Park, G. (2004) A review of power harvesting from vibration using piezoelectric materials, *Shock Vibration Dig.*, **36**(3), 197–205.
16. Elvin, N., and Erturk, A. (2013) *Advances in Energy Harvesting Methods*, Springer Science+Business Media, New York.
17. Erturk, A., and Inman, D. J. (2011) *Piezoelectric Energy Harvesting*, John Wiley & Sons, Ltd.
18. Priya, S., and Inman, D. J. (2009) *Energy Harvesting Technologies*, Springer Science+Business Media, LLC, New York.
19. Cady, W. G. (1946) *Piezoelectricity: An Introduction to the Theory and Applications of Electromechancial Phenomena in Crystals*, Dover, New York.
20. Voigt, W. (1910) *Luhrbuch der Kristallphysik*, B.G. Teubner, Leipzig.
21. Oakley, C. G., and Zipparo, M. J. (2000) Single crystal piezoelectrics: A revolutionary development for transducers, in *2000 IEEE Ultrasonics Symposium*. vol. 2, pp. 1157–1167.

22. Panda, P. (2009) Review: Environmental friendly lead-free piezoelectric materials, *J. Mater. Sci.*, **44**(19), 5049–5062.
23. Priya, S., and Nahm, S. (2012) *Lead-Free Piezoelectrics*, Springer Science+Business Media, LLC, New York.
24. Piezo Systems, I., *Piezo Systems, Incorporated*, www.piezo.com.
25. Measurement Specialties, I., *Measurement Specialties, Incorporated*, www.meas-spec.com.
26. Apc International, L., *APC International, Limited*, www.americanpiezo.com.
27. Emfit, L., "Emfit, Ltd." www.emfit.com.
28. Neugschwandtner, G. S., Schwodiauer, R., Bauer-Gogonea, S., Bauer, S., Paajanen, M., and Lekkala, J. (2001) Piezo- and pyroelectricity of a polymer-foam space-charge electret, *J. Appl. Phys.*, **89**(8), 4503–4511.
29. Paajanen, M., Lekkala, J., and Kirjavainen, K. (2000) ElectroMechanical film (EMFi): A new multipurpose electret material, *Sens. Actuators A Phys.*, 40, **84**(1–2), 95–102.
30. Ikeda, T. (1990) *Fundamentals of Piezoelectricity*, Oxford Science Publications, Oxford.
31. duToit, N. E., Wardle, B. L., and Kim, S.-G. (2005) Design considerations for MEMS-scale piezoelectric mechanical vibration energy harvesters, *Integr. Ferroelectr.*, **71**(1), 121–160.
32. Roundy, S. J., and Wright, P. K. (2004) A piezoelectric vibration based generator for wireless electronics, *Smart Mater. Struct.*, **13**(5), 1131–1142.
33. Stephen, N. G. (2006) On energy harvesting from ambient vibration, *J. Sound Vibration*, **293**(1–2), 409–425.
34. Elvin, N. G., and Elvin, A. A. (2009) A general equivalent circuit model for piezoelectric generators, *J. Intell. Mater. Syst. Struct.*, **20**(1), 3–9.
35. Erturk, A., and Inman, D. J. (2008) A distributed parameter electromechanical model for cantilevered piezoelectric energy harvesters, *J. Vibration Acoust.*, **130**(4), 041002 (15 pp.).
36. Erturk, A., and Inman, D. J. (2009) An experimentally validated bimorph cantilever model for piezoelectric energy harvesting from base excitations, *Smart Mater. Struct.*, **18**(2), 025009 (18 pp.).
37. Erturk, A., and Inman, D. J. (2010) Assumed-modes formulation of piezoelectric energy harvesters: Euler-Bernoulli, Rayleigh and Timoshenko Models with axial deformations, in *Proceedings of the*

ASME 2010 ESDA 10th Biennial Conference on Engineering Systems, Design and Analysis. pp. 405–414.
38. *IEEE Standard on Piezoelectricity.* 1987, IEEE.
39. Erturk, A., and Inman, D. J. (2008) On mechanical modeling of cantilevered piezoelectric vibration energy harvesters, *J. Intell. Mater. Syst. Struct.*, **19**(11), 1311–1325.
40. duToit, N. E., and Wardle, B. L. (2007) Experimental verification of models for microfabricated piezoelectric vibration energy harvesters, *AIAA J.*, **45**(5), 1126–1137.
41. Sodano, H. A., Park, G., and Inman, D. J. (2004) Estimation of electric charge output for piezoelectric energy harvesting, *Strain*, **40**(2), 49–58.
42. Ajitsaria, J., Choe, S. Y., Shen, D., and Kim, D. J. (2007) Modeling and analysis of a bimorph piezoelectric cantilever beam for voltage generation, *Smart Mater. Struct.*, **16**(2), 447–454.
43. Chen, S.-N., Wang, G.-J., and Chien, M.-C. (2006) Analytical modeling of piezoelectric vibration-induced micro power generator, *Mechatronics*, **16**(7), 379–387.
44. Lu, F., Lee, H. P., and Lim, S. P. (2004) Modeling and analysis of micro piezoelectric power generators for micro-electromechanical-systems applications, *Smart Mater. Struct.*, **13**(1), 57–63.
45. Erturk, A., and Inman, D. J. (2008) Issues in mathematical modeling of piezoelectric energy harvesters, *Smart Mater. Struct.*, **17**(6), 065016 (14 pp.).
46. Anton, S. R., Erturk, A., and Inman, D. J. (2009) Piezoelectric energy harvesting from multifunctional wing spars for UAVs-Part 2: Experiments and storage applications, in *Proceedings of the 16th SPIE Annual International Symposium on Smart Structures and Materials & Nondestructive Evaluation and Health Monitoring*. San Diego, CA, USA, vol. 7288, pp. 72880D (12 pp.).
47. Shenck, N. S., and Paradiso, J. A. (2001) Energy scavenging with shoe-mounted piezoelectrics, *IEEE Micro*, **21**(3), 30–42.
48. Tabesh, A., and Frechette, L. G. (2010) A low-power stand-alone adaptive circuit for harvesting energy from a piezoelectric micropower generator, *IEEE Trans. Ind. Electron.*, **57**(3), 840–849.
49. Erickson, R. W., and Maksimovic, D. (2001) *Fundamentals of Power Electronics*, 2nd ed, Kluwer Academic Publishers, Norwell, MA.
50. Lefeuvre, E., Audigier, D., Richard, C., and Guyomar, D. (2007) Buck-boost converter for sensorless power optimization of piezoelectric energy harvester, *IEEE Trans. Power Electron.*, **22**(5), 2018–2025.

51. Ammar, Y., Buhrig, A., Marzencki, M., Charlot, B., Basrour, S., Matou, K., and Renaudin, M. (2005) Wireless sensor network node with asynchronous architecture and vibration harvesting micro power generator, in *Proceedings of the 2005 Joint Conference on Smart Objects and Ambient Intelligence: Innovative Context-Aware Services: Usages and Technologies*. Grenoble, France, pp. 287–292.

52. Lesieutre, G. A., Ottman, G. K., and Hofmann, H. F. (2004) Damping as a result of piezoelectric energy harvesting, *J. Sound Vibration*, **269**(3–5), 991–1001.

53. Ottman, G. K., Hofmann, H. F., Bhatt, A. C., and Lesieutre, G. A. (2002) Adaptive piezoelectric energy harvesting circuit for wireless remote power supply, *IEEE Trans. Power Electron.*, **17**(5), 669–676.

54. Lefeuvre, E., Badel, A., Richard, C., and Guyomar, D. (2005) Piezoelectric energy harvesting device optimization by synchronous electric charge extraction, *J. Intell. Mater. Syst. Struct.*, **16**(10), 865–876.

55. Badel, A., Guyomar, D., Lefeuvre, E., and Richard, C. (2005) Efficiency enhancement of a piezoelectric energy harvesting device in pulsed operation by synchronous charge inversion, *J. Intell. Mater. Syst. Struct.*, **16**(10), 889–901.

56. Guyomar, D., Badel, A., Lefeuvre, E., and Richard, C. (2005) Toward energy harvesting using active materials and conversion improvement by nonlinear processing, *IEEE Trans. Ultrason. Ferroelectr. Freq. Control.*, **52**(4), 584–595.

57. Lefeuvre, E., Badel, A., Richard, C., Petit, L., and Guyomar, D. (2006) A comparison between several vibration-powered piezoelectric generators for standalone systems, *Sens. Actuators A Phys.*, **126**(2), 405–416.

58. Lallart, M., Garbuio, L., Petit, L., Richard, C., and Guyomar, D. (2008) Double synchronized switch harvesting (DSSH): A new energy harvesting scheme for efficient energy extraction, *IEEE Trans. Ultrason. Ferroelectr. Freq. Control.*, **55**(10), 2119–2130.

59. Shen, H., Qiu, J., Ji, H., Zhu, K., and Balsi, M. (2010) Enhanced synchronized switch harvesting: A new energy harvesting scheme for efficient energy extraction, *Smart Mater. Struct.*, **19**(11), 115017 (14 pp.).

60. Garbuio, L., Lallart, M., Guyomar, D., Richard, C., and Audigier, D. (2009) Mechanical energy harvester with ultralow threshold rectification based on SSHI nonlinear technique, *IEEE Trans. Ind. Electron.*, **56**(4), 1048–1056.

61. Lallart, M., Richard, C., Garbuio, L., Petit, L., and Guyomar, D. (2011) High efficiency, wide load bandwidth piezoelectric energy scavenging by a hybrid nonlinear approach, *Sens. Actuators A Phys.*, **165**(2), 294–302.
62. Liang, J., and Liao, W.-H. (2012) Improved design and analysis of self-powered synchronized switch interface circuit for piezoelectric energy harvesting systems, *IEEE Trans. Ind. Electron.*, **59**(4), 1950–1960.
63. Chao, P. C. (2011) Energy harvesting electronics for vibratory devices in self-powered sensors, *IEEE Sens. J.*, **11**(12), 3106–3121.
64. Jackson, H. W. (1959) *Introduction to Electric Circuits*, Prentice-Hall, Englewood Cliffs, New Jersey.
65. Kong, N., Ha, D. S., Erturk, A., and Inman, D. J. (2010) Resistive impedance matching circuit for piezoelectric energy harvesting, *J. Intell. Mater. Syst. Struct.*, **21**(13), 1293–1302.
66. Kim, H., Priya, S., Stephanou, H., and Uchino, K. (2007) Consideration of impedance matching techniques for efficient piezoelectric energy harvesting, *IEEE Trans. Ultrason. Ferroelectr. Freq. Control.*, **54**(9), 1851–1859.
67. Szarka, G. D., Stark, B. H., and Burrow, S. G. (2012) Review of power conditioning for kinetic energy harvesting systems, *IEEE Trans. Power Electron.*, **27**(2), 803–815.
68. Kymissis, J., Kendall, C., Paradiso, J., and Gershenfeld, N. (1998) Parasitic power harvesting in shoes, in *Proceedings of the 2nd IEEE International Conference on Wearable Computing*. pp. 132–139.
69. Arms, S. W., Townsend, C. P., Churchill, D. L., Galbreath, J. H., and Mundell, S. W. (2005) Power management for energy harvesting wireless sensors, in *Proceedings of the 12th SPIE Annual International Symposium on Smart Structures and Materials & Nondestructive Evaluation and Health Monitoring*. San Diego, CA, vol. 5763, pp. 267–275.
70. Discenzo, F. M., Chung, D., and Loparo, K. A. (2006) Pump condition monitoring using self-powered wireless sensors, *Sound Vibration*, **40**(5), 12–15.
71. Elfrink, R., Renaud, M., Kamel, T. M., de Nooijer, C., Jambunathan, M., Goedbloed, M., Hohlfeld, D., Matova, S., Pop, V., Caballero, L., and van Schaijk, R. (2010) Vacuum-packaged piezoelectric vibration energy harvesters: Damping contributions and autonomy for a wireless sensor system, *J. Micromech. Microeng.*, **20**(10), 104001 (7 pp.).

72. Zhu, D., Beeby, S. P., Tudor, M. J., and Harris, N. R. (2011) A credit card sized self powered smart sensor node, *Sens. Actuators A Phys.*, **169**(2), 317–325.

73. Gonzalez, J. L., Rubio, A., and Moll, F. (2002) Human powered piezoelectric batteries to supply power to wearable electronic devices, *Int. J. Soc. Mater. Eng. Resources*, **10**(1), 34–40.

74. Niu, P., Chapman, P., Riemer, R., and Zhang, X. (2004) Evaluation of motions and actuation methods for biomechanical energy harvesting, in *IEEE 35th Annual Power Electronics Specialists Conference*. Aachen, Germany, vol. 3, pp. 2100–2106.

75. Granstrom, J., Feenstra, J., Sodano, H. A., and Farinholt, K. (2007) Energy harvesting from a backpack instrumented with piezoelectric shoulder straps, *Smart Mater. Struct.*, **16**(5), 1810–1820.

76. Feenstra, J., Granstrom, J., and Sodano, H. (2008) Energy harvesting through a backpack employing a mechanically amplified piezoelectric stack, *Mechan. Syst. Signal Process.*, **22**(3), 721–734.

77. Platt, S. R., Farritor, S., and Haider, H. (2005) On low-frequency electric power generation with PZT ceramics, *IEEE/ASME Trans. Mechatron.*, **10**(2), 240–252.

78. Sohn, J. W., Choi, S. B., and Lee, D. Y. (2005) An investigation on piezoelectric energy harvesting for MEMS power sources, *Proc. Inst. Mechan. Eng. Part C J. Mechan. Eng. Sci.*, **219**(4), 429–436.

79. Aktakka, E. E., Kim, H., and Najafi, K. (2011) Energy scavenging from insect flight, *J. Micromech. Microeng.*, **21**(9), 095016 (11 pp.).

80. Reissman, T., and Garcia, E. (2008) Cyborg MAVs using power harvesting and behavioral control schemes, *Adv. Sci. Technol.*, **58**, 159–164.

81. Romero, E., Warrington, R. O., and Neuman, M. R. (2009) Energy scavenging sources for biomedical sensors, *Physiol. Meas.*, **30**(9), R35–R62.

82. Jeon, Y. B., Sood, R., Jeong, J. H., and Kim, S. G. (2005) MEMS power generator with transverse mode thin film PZT, *Sens. Actuators A Phys.*, **122**(1), 16–22.

83. Fang, H.-B., Liu, J.-Q., Xu, Z.-Y., Dong, L., Wang, L., Chen, D., Cai, B.-C., and Liu, Y. (2006) Fabrication and performance of MEMS-based piezoelectric power generator for vibration energy harvesting, *Microelectronics Journal*, **37**(11), 1280–1284.

84. Liu, J.-Q., Fang, H.-B., Xu, Z.-Y., Mao, X.-H., Shen, X.-C., Chen, D., Liao, H., and Cai, B.-C. (2008) A MEMS-based piezoelectric power generator

array for vibration energy harvesting, *Microelectron. J.*, **39**(5), 802–806.

85. Karami, M. A., and Inman, D. J. (2011) Analytical modeling and experimental verification of the vibrations of the zigzag microstructure for energy harvesting, *J. Vibration Acoust.*, **133**(1), 011002 (10 pp.).

86. Lee, B. S., Lin, S. C., Wu, W. J., Wang, X. Y., Chang, P. Z., and Lee, C. K. (2009) Piezoelectric MEMS generators fabricated with an aerosol deposition PZT thin film, *J. Micromech. Microeng.*, **19**(6), 065014 (8 pp.).

87. Marzencki, M., Ammar, Y., and Basrour, S. (2008) Integrated power harvesting system including a MEMS generator and a power management circuit, *Sens. Actuators A Phys.*, **145–146**, 363–370.

88. Shen, D., Park, J.-H., Noh, J. H., Choe, S.-Y., Kim, S.-H., Wikle Iii, H. C., and Kim, D.-J. (2009) Micromachined PZT cantilever based on SOI structure for low frequency vibration energy harvesting, *Sens. Actuators A Phys.*, **154**(1), 103–108.

89. Saadon, S., and Sidek, O. (2011) A review of vibration-based MEMS piezoelectric energy harvesters, *Energy Conversion Manag.*, **52**(1), 500–504.

90. Taylor, G. W., Burns, J. R., Kammann, S. A., Powers, W. B., and Welsh, T. R. (2001) The Energy Harvesting Eel: A small subsurface ocean/river power generator, *IEEE J. Oceanic Eng.*, **26**(4), 539–547.

91. Pobering, S., and Schwesinger, N. (2004) A novel hydropower harvesting device, in *Proceedings of the 2004 International Conference on MEMS, NANO and Smart Systems*. pp. 480–485.

92. Wang, D.-A., and Ko, H.-H. (2010) Piezoelectric energy harvesting from flow-induced vibration, *J. Micromech. Microeng.*, **20**(2), 025019 (9 pp.).

93. Wang, D.-A., and Liu, N.-Z. (2011) A shear mode piezoelectric energy harvester based on a pressurized water flow, *Sens. Actuators A Phys.*, **167**(2), 449–458.

94. Priya, S. (2005) Modeling of electric energy harvesting using piezoelectric windmill, *Appl. Phys. Lett.*, **87**(18), 184101 (3 pp.).

95. Priya, S., Chen, C.-T., Fye, D., and Zahnd, J. (2005) Piezoelectric windmill: A novel solution to remote sensing, *Jap. J. Appl. Phys.*, **44**(3), L104–L107.

96. Myers, R., Vickers, M., Kim, H., and Priya, S. (2007) Small scale windmill, *Appl. Phys. Lett.*, **90**(5), 054106 (3 pp.).

97. Tien, C. M. T., and Goo, N. S. (2010) Use of a piezo-composite generating element for harvesting wind energy in an urban region, *Aircraft Eng. Aerosp. Technol.*, **82**(6), 376–381.

98. Karami, M. A., Farmer, J. R., and Inman, D. J. (2013) Parametrically excited nonlinear piezoelectric compact wind turbine, *Renew. Energy*, **50**, 977–987.

99. Tan, Y. K., and Panda, S. K. (2007) A novel piezoelectric based wind energy harvester for low-power autonomous wind speed sensor, in *33rd Annual Conference of the IEEE Industrial Electronics Society, 2007*. pp. 2175–2180.

100. Kwon, S.-D. (2010) A T-shaped piezoelectric cantilever for fluid energy harvesting, *Appl. Phys. Lett.*, **97**(16), 164102 (3 pp.).

101. Bryant, M., and Garcia, E. (2011) Modeling and testing of a novel aeroelastic flutter energy harvester, *J. Vibration Acoust.*, **133**(1), 011010 (11 pp.).

102. Sousa, V. C., Anicézio, M. d. M., De Marqui Jr, C., and Erturk, A. (2011) Enhanced aeroelastic energy harvesting by exploiting combined nonlinearities: Theory and experiment, *Smart Mater. Struct.*, **20**(9), 094007 (8 pp.).

103. Hobeck, J. D., and Inman, D. J. (2012) Artificial piezoelectric grass for energy harvesting from turbulence-induced vibration, *Smart Mater. Struct.*, **21**(10), 105024 (10 pp.).

104. Anderson, I. A., Ieropoulos, I. A., McKay, T., O'Brien, B., and Melhuish, C. (2011) Power for robotic artificial muscles, *IEEE/ASME Trans. Mechatron.*, **16**(1), 107–111.

105. Mitcheson, P. D., Yeatman, E. M., Rao, G. K., Holmes, A. S., and Green, T. C. (2008) Energy harvesting from human and machine motion for wireless electronic devices, *Proc. IEEE*, **96**(9), 1457–1486.

106. Leland, E. S., and Wright, P. K. (2006) Resonance tuning of piezoelectric vibration energy scavenging generators using compressive axial preload, *Smart Mater. Struct.*, **15**(5), 1413–1420.

107. Challa, V. R., Prasad, M. G., Shi, Y., and Fisher, F. T. (2008) A vibration energy harvesting device with bidirectional resonance frequency tunability, *Smart Mater. Struct.*, **17**(1), 015035 (10 pp.).

108. Shahruz, S. M. (2006) Design of mechanical band-pass filters with large frequency bands for energy scavenging, *Mechatronics*, **16**(9), 523–531.

109. Roundy, S., Leland, E. S., Baker, J., Carleton, E., Reilly, E., Lai, E., Otis, B., Rabaey, J. M., Wright, P. K., and Sundararajan, V. (2005) Improving

power output for vibration-based energy scavengers, *Pervasive Comput. IEEE*, **4**(1), 28–36.

110. Roundy, S., and Zhang, Y. (2005) Toward self-tuning adaptive vibration-based microgenerators, in *Proceedings of the 12th SPIE Annual International Symposium on Smart Structures and Materials & Nondestructive Evaluation and Health Monitoring*. vol. 5649, pp. 373–384.

111. Lallart, M., Anton, S. R., and Inman, D. J. (2010) Frequency self-tuning scheme for broadband vibration energy harvesting, *J. Intell. Mater. Syst. Struct.*, **21**(9), 897–906.

112. Quinn, D. D., Triplett, A. L., Bergman, L. A., and Vakakis, A. F. (2011) Comparing linear and essentially nonlinear vibration-based energy harvesting, *J. Vibration Acoust.*, **133**(1), 011001 (8 pp.).

113. Barton, D. A. W., Burrow, S. G., and Clare, L. R. (2010) Energy harvesting from vibrations with a nonlinear oscillator, *J. Vibration Acoust.*, **132**(2), 021009 (7 pp.).

114. Erturk, A., Hoffmann, J., and Inman, D. J. (2009) A piezomagnetoelastic structure for broadband vibration energy harvesting, *Appl. Phys. Lett.*, **94**(25), 254102 (3 pp.).

115. Stanton, S. C., McGehee, C. C., and Mann, B. P. (2010) Nonlinear dynamics for broadband energy harvesting: Investigation of a bistable piezoelectric inertial generator, *Phys. D Nonlinear Phenomena*, **239**(10), 640–653.

116. Daqaq, M. F., Masana, R., Erturk, A., and Dane Quinn, D. (2014) On the role of nonlinearities in vibratory energy harvesting: A critical review and discussion, *Appl. Mech. Rev.*, **66**(4), 040801 (23 pp.).

117. Anton, S. R., Erturk, A., and Inman, D. J. (2010) Multifunctional self-charging structures using piezoceramics and thin-film batteries, *Smart Mater. Struct.*, **19**(11), 115021 (15 pp.).

118. Wang, Y., and Inman, D. J. (2012) Experimental characterization of simultaneous gust alleviation and energy harvesting for multifunctional wing spars, in *Proceedings of the 19th SPIE Annual International Symposium on Smart Structures and Materials & Nondestructive Evaluation and Health Monitoring*. San Diego, CA, vol. 8341, pp. 834114 (11 pp.).

119. Lin, Y., and Sodano, H. A. (2008) Concept and model of a piezoelectric structural fiber for multifunctional composites, *Composites Sci. Technol.*, **68**(7–8), 1911–1918.

120. Lin, Y., and Sodano, H. A. (2009) Fabrication and electromechanical characterization of a piezoelectric structural fiber for multifunctional composites, *Adv. Funct. Mater.*, **19**(4), 592–598.

121. Lin, Y., and Sodano, H. A. (2009) Characterization of multifunctional structural capacitors for embedded energy storage, *J. Appl. Phys.*, **106**(11), 114108 (5 pp.).

122. Lin, Y., and Sodano, H. A. (2009) Electromechanical characterization of a active structural fiber lamina for multifunctional composites, *Composites Sci. Technol.*, **69**(11–12), 1825–1830.

123. Gambier, P., Anton, S. R., Kong, N., Erturk, A., and Inman, D. J. (2012) Piezoelectric, solar and thermal energy harvesting for hybrid low-power generator systems with thin-film batteries, *Meas. Sci. Technol.*, **23**(1), 015101 (11 pp.).

124. Schlichting, A., Tiwari, R., and Garcia, E. (2012) Passive multi-source energy harvesting schemes, *J. Intell. Mater. Syst. Struct.*, **23**(17), 1921–1935.

125. Anton, S. R., Taylor, S. G., Raby, E. Y., and Farinholt, K. M. (2013) Powering embedded electronics for wind turbine monitoring using multi-source energy harvesting techniques, in *Proceedings of the 20th SPIE Annual International Symposium on Smart Structures and Materials & Nondestructive Evaluation and Health Monitoring*. San Diego, CA, USA, vol. 8690, pp. 869007 (9 pp.).

126. Taylor, S. G., Farinholt, K. M., Park, G., Todd, M. D., and Farrar, C. R. (2010) Multi-scale wireless sensor node for health monitoring of civil infrastructure and mechanical systems, *Smart Struct. Syst.*, **6**(5), 661–673.

127. Park, C., and Chou, P. H. (2006) AmbiMax: Autonomous energy harvesting platform for multi-supply wireless sensor nodes, in *3rd Annual IEEE Communications Society on Sensor and Ad Hoc Communications and Networks*. vol. 1, pp. 168–177.

128. Morais, R., Fernandes, M. A., Matos, S. G., Serôdio, C., Ferreira, P. J. S. G., and Reis, M. J. C. S. (2008) A ZigBee multi-powered wireless acquisition device for remote sensing applications in precision viticulture, *Comput. Electron. Agric.*, **62**(2), 94–106.

129. Erturk, A., Bilgen, O., and Inman, D. J. (2008) Power generation and shunt damping performance of a single crystal lead magnesium niobate-lead zirconate titanate unimorph: Analysis and experiment, *Appl. Phys. Lett.*, **93**(22), 224102 (3 pp.).

130. Badel, A., Benayad, A., Lefeuvre, E., Lebrun, L., Richard, C., and Guyomar, D. (2006) Single crystals and nonlinear process for

outstanding vibration-powered electrical generators, *IEEE Trans. Ultrason. Ferroelectr. Freq. Control.*, **53**(4), 673–684.

131. Ren, K., Liu, Y., Geng, X., Hofmann, H. F., and Zhang, Q. M. (2006) Single crystal PMN-PT/Epoxy 1-3 composite for energy-harvesting application, *IEEE Trans. Ultrason. Ferroelectr. Freq. Control.*, **53**(3), 631–638.

132. Hooker, M. W., *Properties of PZT based piezoelectric ceramics between -150 and 250C.* 1998, NASA Report No. NASA/CR-1998-208708, NASA (30 pp.).

133. Ren, B., Or, S. W., Zhang, Y., Zhang, Q., Li, X., Jiao, J., Wang, W., Liu, D., Zhao, X., and Luo, H. (2010) Piezoelectric energy harvesting using shear mode $0.71Pb(Mg_{1/3}Nb_{2/3})O_3$-$0.29PbTiO_3$ single crystal cantilever, *Appl. Phys. Lett.*, **96**(8), 083502 (3 pp.).

134. Mathers, A., Moon, K. S., and Yi, J. (2009) A vibration-based PMN-PT energy harvester, *IEEE Sens. J.*, **9**(7), 731–739.

135. Moon, S. E., Lee, S. Q., Lee, S.-K., Lee, Y.-G., Yang, Y. S., Park, K.-H., and Kim, J. (2009) Sustainable vibration energy harvesting based on Zr-doped PMN-PT piezoelectric single crystal cantilevers, *ETRI J.*, **31**(6), 688–694.

136. Wang, Z. L., and Song, J. (2006) Piezoelectric nanogenerators based on zinc oxide nanowire arrays, *Science*, **312**(5771), 242–246.

137. Feenstra, J., and Sodano, H. A. (2008) Enhanced active piezoelectric 0-3 nanocomposites fabricated through electrospun nanowires, *J. Appl. Phys.*, **103**(12), 124108 (5 pp.).

138. Fang, J., Wang, X., and Lin, T. (2011) Electrical power generator from randomly oriented electrospun poly(vinylidene fluoride) nanofibre membranes, *J. Mater. Chem.*, **21**(30), 11088–11091.

139. Xu, S., Hansen, B. J., and Wang, Z. L. (2010) Piezoelectric-nanowire-enabled power source for driving wireless microelectronics, *Nat. Commun.*, **1**, 93 (5 pp.).

140. Chen, X., Xu, S., Yao, N., and Shi, Y. (2010) 1.6 V nanogenerator for mechanical energy harvesting using PZT nanofibers, *Nano Lett.*, **10**(6), 2133–2137.

141. Chang, J., Dommer, M., Chang, C., and Lin, L. (2012) Piezoelectric nanofibers for energy scavenging applications, *Nano Energy*, **1**(3), 356–371.

142. Savolainen, A., and Kirjavainen, K. (1989) Electrothermomechanical film. Part I. Design and characteristics, *J. Macromol. Sci. Part A Chem.*, **26**(2–3), 583–591.

143. Anton, S. R., Farinholt, K. M., and Erturk, A. (2014) Piezoelectret foam–based vibration energy harvesting, *J. Intell. Mater. Syst. Struct.*, **25**(14), 1681–1692.
144. Pondrom, P., Hillenbrand, J., Sessler, G. M., Bös, J., and Melz, T. (2014) Vibration-based energy harvesting with stacked piezoelectrets, *Appl. Phys. Lett.*, **104**(17), 172901 (5 pp.).
145. Ray, C. A., and Anton, S. R. (2015) Evaluation of piezoelectret foam in a multilayer stack configuration for low-level vibration energy harvesting applications, in *Proceedings of SPIE Smart Structures & Nondestructive Evaluation Conference*. San Diego, CA, vol. 9431, pp. 943111 (11 pp.).

Chapter 6

Fuel cells

Jesús Canales-Vázquez[a] and Juan Carlos Ruiz-Morales[b]

[a]Renewable Energy Research Institute, University of Castilla-La Mancha, 02071 Albacete, Spain
[b]Department of Chemistry, University of La Laguna, 38200 La Laguna, Spain

jesus.canales@uclm.es, jcruiz@ull.es

6.1 Introduction

The outstanding development of renewable energies in the last few decades is the response to the strong driving force for a change in the energy production model currently based upon the combustion of fossil fuels in the search for cleaner and more efficient routes of power generation. Indeed, in the last few years the contribution of renewable resources to the overall power production has dramatically increased up to cover 20–30% of the energy consumption in some western-Europe countries and it should be stressed their potential to cover most of our demand in the mid- and long-term.

Nevertheless, the unpredictable nature of renewables or renewable resources can be certainly considered as a drawback which should be taken into consideration. Indeed, it may happen that there will not be power generation in a moment of peak demand. In addition, most of renewable energies are still growing rather than fully developed and therefore must overcome a consolidation stage prior they cover most of the energy power demand worldwide. As a consequence, it is mandatory to implement technologies to ensure energy storage.

Materials for Sustainable Energy Applications: Conversion, Storage, Transmission, and Consumption
Edited by Xavier Moya and David Muñoz-Rojas
Copyright © 2016 Pan Stanford Publishing Pte. Ltd.
ISBN 978-981-4411-81-3 (Hardcover), 978-981-4411-82-0 (eBook)
www.panstanford.com

In this context, hydrogen and fuel cells may play a highly relevant role to consolidate a new energy scenario dominated by renewable energy sources (RES), as hydrogen is a well-known energy vector and fuel cells are highly efficient devices for power generation. In an ideal situation, the energy produced out of RES during peak periods may be used to cover the power demand and the remnants converted into hydrogen via water electrolysis. The so-produced hydrogen will be used afterwards, i.e., during low production periods, in a fuel cell to generate the electricity to cover the demand. Nowadays it could be stated that such scenario is not realistic and a model based exclusively on hydrogen is not feasible. Nevertheless, both hydrogen and fuel cells can certainly be considered as very important pieces to complete the energy jigsaw.

Fuel cells are electrochemical devices that convert chemical energy into electrical energy. A basic fuel cell consists of two porous electrodes (anode and cathode) separated by a dense electrolyte, a material that ideally is an ionic conductor but electronic insulator. The operating principles are similar to those of batteries. However, while batteries are energy accumulators that require recharging, fuel cells are power generators that use gases as reactants and work as long as fuel and oxidant are supplied to the electrodes.

In the case of a proton conductor working as electrolyte, a fuel containing hydrogen (hydrogen gas, methanol, hydrazine, etc.) flows to the anode, where it is oxidised to protons. The electrons travel through an external circuit while the ions diffuse through the electrolyte. At the cathode, the electrons combine with protons and oxygen to produce water (by-product).

Similarly, if the electrolyte is an oxide-ion conductor, fuel is also oxidised at the anode, but in that case oxide ions coming from the reduction of the oxidant (O_2) move through the electrolyte from the cathode to the anode to also form water as shown in Fig. 6.1.

These are the typical reactions that take place in the electrodes, when the inlet fuel is hydrogen:

$$\text{Anode} \quad 2H_2 \rightarrow 4H^+ + 4e^- \tag{6.1}$$

$$\text{Cathode} \quad 4H^+ + 4e^- + O_2 \rightarrow 2H_2O \tag{6.2}$$

The overall reaction is

$$2H_2 + O_2 \rightarrow 2H_2O \tag{6.3}$$

Introduction

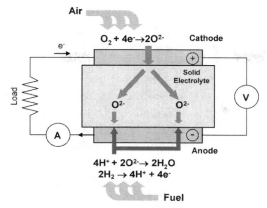

Figure 6.1 Principle of operation of a fuel cell with an oxide-ion conductor as electrolyte.

Therefore, fuel cells directly convert chemical energy into electrical energy avoiding intermediate thermal-to-movement conversions due to combustion, which in turn bypasses the limitations dictated by the Carnot cycle and therefore, the efficiency is higher, particularly at low temperatures (Fig. 6.2).

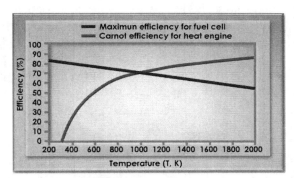

Figure 6.2 Thermodynamic efficiency for a fuel cell (H_2/O_2, $\eta_{therm} = \Delta G°/\Delta H°$) and Carnot efficiency ($\eta_{carnot} = 1 - T_C/T_H$, $T_C = 300$ K) for heat engines [1], where η_{therm} is the thermodynamic efficiency, $\Delta H°$ is the formation enthalpy and $\Delta G°$ is the corresponding Gibbs free energy. η_{carnot} is the Carnot efficiency, whereas T_C and T_H are the temperatures of the cold and hot reservoirs, respectively.

Despite that there exist several types of fuel cells (see Section 6.3), they show several common features as described below:

- **High efficiency.** One of the main advantages of fuel cells is their high efficiency, typically ranging between 40 and 60%, although in the case of co-generation, i.e., the use of residual high quality heat to produce electricity via steam turbines, it would be possible to achieve 80–85%.
- **Low environmental impact.** In addition to hydrogen, fuel cells may use conventional fuels with much lower emissions to those produced by conventional energy generators, e.g., heat engines. Moreover, the production of the harmful NO_x and SO_x is greatly reduced.
- **Modularity.** Fuel cells are very flexible systems regarding size and can be designed/fabricated for any power requirement from MWs down to the µWs.
- **Location.** As a consequence of their flexibility regarding size they can be placed almost anywhere, without restrictions. Moreover, they have no moving parts and are therefore noiseless, which helps in their implementation even in residential urban areas or hospitals (Fig. 6.3).
- **Fuel Choice.** Fuel cells may operate using a wide range of fuels, particularly at high temperatures, as hydrocarbon reforming, i.e., oxidation of hydrocarbons via steam resulting in the generation of hydrogen, may occur in situ rather than using expensive external reforming units. In addition to hydrogen, fuel cells may operate with methane, propane, butane, liquefied petroleum gas (LPG), diesel, alcohols, industrial coal gas (syngas), ammonia, etc.

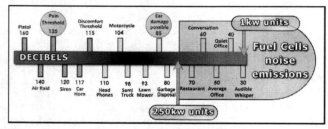

Figure 6.3 Noise emissions chart, showing the level of normal conversation, i.e., 60 dB, which is a common value for all fuel cells applications –1 kW to 250 kW units [2].

In the next pages, fuel cell technologies will be briefly introduced to the reader, starting with history of fuel cells and then a description of the different types of fuel cells and their potential applications. Then, the thermodynamics of fuel cells will be detailed highlighting some important concepts such as efficiency, open circuit voltage, polarisation, etc. Finally, the reader will find a review of projects devoted to fuel cell integration and the state of the art of the technology.

6.2 History

Despite fuel cells are considered as an emergent technology, the principles were discovered by Christian Friedich Schönbein and the first cells were developed by Sir William Grove in 1838 and 1839, respectively [3-4]. Grove cells used Zn and Pt electrodes in contact with two different acid media as depicted in Fig. 6.4. Grove observed that current flowed between the electrodes when they were in contact with oxygen and hydrogen, respectively while the gases were consumed. With such device, Grove was able to produce 12 A at 1.8 V.

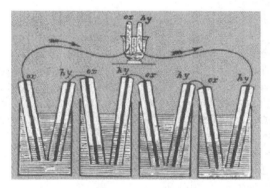

Figure 6.4 Fuel Cells scheme published by Grove [5]. Ox and Hy refer to oxygen and hydrogen in contact with the electrodes, where they are reduced and oxidised, respectively, whilst the arrows indicate the electron flow.

The term "fuel cell" was first coined by L. Mond and C. Langer in 1889 when they designed a device fed with air and syngas. This fuel cell could produce about 0.6 A/cm^2 at 0.73 V and introduced some relevant changes which are still considered in modern fuel cells as is the use of catalysts with a high specific surface (black platinum) and a liquid electrolyte impregnating a porous solid backbone. Several similar prototypes were developed almost simultaneously, but at that time very low performances associated to gas leaking and the use of precious metals at the electrodes appeared as crucial issues that needed to be addressed.

In 1896, W. W. Jacques designed a fuel cell to produce electricity directly from carbon [6]. Air was supplied through the electrolyte (phosphoric acid) to "react" with a carbon electrode. This device delivered 16 A at 90 V, with very high efficiencies, apparently above 80%. Nevertheless, real efficiencies were rather low, i.e., 8%, when considering the energy consumed by furnace, air injector and, especially when discovering that most of the current produced was due to thermoelectric effects.

The development of ceramic fuel cells started with the discovery of solid-state electrolytes by Nernst in 1899. Nernst observed that doped ZrO_2, e.g., 15% Y_2O_3, was an electrical conductor at high temperatures [7] and three decades later, Schottky suggested that such oxides may be used as electrolytes in high-temperature fuel cells [8], although the first patent on fuel cells based on solid electrolytes was filed by Haber in 1905 [9]. It should be highlighted that nowadays the most common electrolyte in solid oxide fuel cells (SOFCs) is still doped ZrO_2. The first operating ceramic fuel cell was built by Swiss engineers Bauer and Preiss in 1937 [10]. They used tubular ZrO_2 electrolytes, Fe/C as anode and Fe_3O_4 as cathode, whereas air and H_2 or CO were used as oxidising agent and fuel, respectively. Using such configuration, they obtained open circuit voltages as high as 1.1 V at 1000 °C, though the current densities were rather modest due to the large internal resistance. O. K. Davtyan tried to overcome such a problem by adding alkaline oxides, W_2O_3 and carbonates to the electrolyte in the search for solid electrolytes exhibiting both high conductivities and also high mechanical stability [11].

Simultaneously and along the 1930s and early 1940s, Bacon developed a simple and economical operating system based on alkaline electrolytes instead of acid media, allowing the use of inexpensive nickel as the catalyst, replacing costly noble metals. The so-called "Bacon cell" operated using pure oxygen and hydrogen at high pressure and temperature and emerged as a benchmark in fuel cell technology, more particularly the 5 kW model, which was subsequently licensed by Pratt and Whitney to power the Apollo space vehicle [12].

The discovery of polymer electrolytes such as sulphonate polystyrene membranes led to the development of polymer electrolyte membrane (PEM) fuel cells in the early 1960s by T. Grubb and L. Niedrach (General Electric) for the Gemini project (NASA) [13, 14]. The first units were fuelled by hydrogen generated by mixing water and metal hydride in a canister. The cell was compact and portable, but they required a considerably large amount of platinum catalysts and membranes degraded rapidly under reducing conditions. In the late 1960s, DuPont developed Nafion™ membranes which led to the first commercial PEM fuel cells.

During the 1970s and 1980s, there were a large number of advances regarding potential applications of fuel cells. For instance, in ceramic fuel cells there was a strong drive towards the use of tubular configurations rather the more conventional planar configuration with thicker electrolytes (Fig. 6.5). Such a change allowed devices with higher efficiencies as a consequence of the smaller internal resistance. Another example is the use of weatherproof materials added to polymer electrolytes to produce more robust PEM cells or the reduction in the Pt-load at the electrodes.

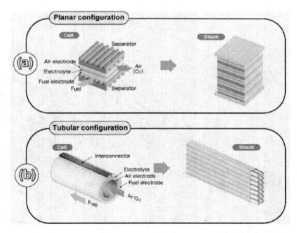

Figure 6.5 (a) Planar and (b) Tubular SOFC configurations [15].

As a consequence of those advances, in the 1990s the first projects to develop vehicles propelled with fuel cells were launched. In 1995, the Canadian company Ballard Systems tested PEM cells in buses in Vancouver and Chicago and later in experimental vehicles made by Daimler-Chrysler. Since then, many companies have announced prototypes and vehicles based on fuel cell technology and also a number of international cooperation projects have been carried out to evaluate the implantation at commercial scale of this technology, such is the case of the Clean Urban Transport for Europe (CUTE), the HyFLEET or the more recent CHIC (Clean Hydrogen in European Cities) [16–18].

Nevertheless, fuel cells have also found application outside the automotive sector as is the case of stationary power units (up to several MWs), hybrid systems coupling renewable energies and fuel cells for a greener power production, remote/off-grid generation (back-up units for telecom towers, isolated houses, etc.) or portable applications (laptops, mobiles, etc.).

6.3 Types of Fuel Cells

There are several types of fuel cell that are usually classified as a function of both the electrolyte and the corresponding operation temperature range. According to this, there are five main fuel cell types: alkaline fuel cells (AFCs), polymer electrolyte membrane fuel cells (PEMFCs), phosphoric acid fuel cells (PAFCs), molten carbonate fuel cells (MCFCs) and SOFCs (shown in Fig. 6.6, along with their range of operation).

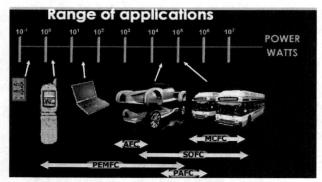

Figure 6.6 Power range and applications of fuel cells [19].

The first three types operate at low temperatures, e.g., below 250°C and the charge carriers are protons or hydroxyl anions. On the other hand, high-temperature fuel cells, MCFCs and SOFCs, present carbonate or oxide ions as charge carriers. The main features of the different fuel cells are summarised in Table 6.1.

Table 6.1 Summary of the different types/features of Fuel Cells [20]

Properties	AFC	PEMFC	PAFC	MCFC	SOFC
Electrolyte	KOH(aq)	Perfluoro sulphonic acid	H_3PO_4	Li, Na and/or K carbonates	Y-stabilised zirconia
Fuel	H_2	$H_2/CO/CH_3OH$	H_2/CO	H_2/CO	$H_2/CO/CH_x$
T (°C)	100–250	50–100	150–200	600–700	600–1000
Efficiency	60%	30–60%	40–80%	50–85%	65–85%
Stack size	10–100 kW	1–250 kW	400 kW	300 kW–3 MW	1 kW–2 MW
Applications	– Military – Space	– Backup power – PP & DG – Transportation	– DG	– Electric utility – DG	– Auxiliary power – Electric utility – DG
Advantages	– Cathode reaction faster in KOH $\Rightarrow$ high performances	– Low temperature – Quick start-up – Low corrosion	– Increased tolerance to fuel impurities – Higher T enables CHP	– High efficiency – Fuel flexibility – Suitable for CHP	– High efficiency – Fuel flexibility – Suitable for CHP – Solid electrolyte
Challenges	– Sensitive to CO_2 in fuel and air – Electrolyte management	– Sensitive to fuel impurities – Expensive catalysts (Pt) – Water management	– Pt catalyst – Long start-up – S Sensitive	– Two gas flow – Long start-up – High-temperature corrosion	– Long start-up – High-temperature corrosion

PP: Portable power; DG: Distributed generation; CHP: Combined heat and power generation.

6.3.1 Alkaline Fuel Cells

Alkaline fuel cells were one of the first types of fuel cells developed by Francis Bacon's pioneering work in the 1950s and their application in NASA projects. AFCs use concentrated KOH (85% w/w) as electrolyte in the 100–250 °C temperature range, though the latest models allow operating temperatures below 70 °C. The electrolyte is contained in an asbestos matrix and may use Ni, Ag, metal oxides and noble metals as catalyst at the electrodes. Due to the low operating temperature, CO may deactivate the catalyst even at very low concentrations and CO_2 reacts with the electrolyte. Therefore AFCs operate with high purity H_2 and O_2.

AFCs have proved reliability operating with high efficiencies (>50%) after 8000 h cycles, though they are rather costly and should operate for over 40000 h to be economically feasible, which is rather difficult to date due to component degradation. Nevertheless, this type of fuel cell has found a niche of application in aerospatial/military applications and the development of zero emission vehicles (ZEVCO).

6.3.2 Polymer Electrolyte Membrane Fuel Cells

Polymer electrolyte membrane fuel cells use an ion-exchange resin as electrolyte (Fig. 6.7). The most common choice is a perfluorinated membrane containing sulphonic acid groups, typically Nafion™, which absorbs water. The electrodes are typically Pt or Pt alloys over C. The operating temperature is around 80 °C with efficiencies of around 40–50%. As the temperature is rather low and water is the only liquid in the system, the problems associated to corrosion phenomena are minimised. On the other hand, the electrolyte requires a certain degree of hydration and therefore care must be taken to prevent the complete evaporation of water from the membrane. Additionally, the use of Pt catalysts at low temperatures imposes strict requirements on fuel purity as CO may also ruin the electrodes, even at very low concentrations (below 100 ppm). Nowadays there is a growing interest in developing electrolyte membranes to operate at higher temperatures. This would imply that CO may be oxidised at the electrodes, which in turn may find alternatives to the costly Pt. In the meantime, the development of fuel cell electrodes with a gradually lower Pt load is also a very relevant issue to reduce fabrication costs as proved by Ballard in their 1 kW stacks, with Pt loads as low as 0.5 mg/cm^2 [21], though the current target is below 0.15 mg/cm^2 as discussed in Section 6.4. PEMFCs can also be used to set up biological fuel cells. In these fuel cells, a microorganism is the catalyst for the reaction avoiding the use of Pt [22].

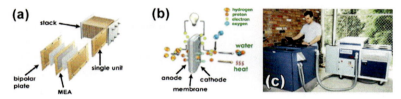

Figure 6.7 (a) Simple scheme of a polymer electrolyte fuel cell stack, (b) each single unit is known as a membrane electrode assembly (MEA). (c) First prototype of a PEM fuel cell power generation, Latham, N.Y., 1998 [23].

Regarding potential applications, the low operating temperatures in PEMFCs facilitate their implementation in portable and transport applications. Indeed, a good deal of the demonstration project throughout the world to power vehicles with fuel cells are based on PEMFCs and companies such as BMW, Toyota, Daimler-Chrysler, Fiat, Ford, Honda, General Motors, Renault, etc., have commercial vehicles based on PEMFC technology.

In PEMFCs, one may use methanol as fuel. These PEMFCs are usually known as direct methanol fuel cells (DMFCs). DMFC technology is relatively new compared with that of fuel cells powered by pure hydrogen and hence their research and development is a few years behind PEMFCs.

DMFCs exhibit several advantages: As they operate with a liquid fuel, they avoid most of the problems related to fuel storage because methanol has a higher energy density than hydrogen. Also methanol is easier to transport and supply to the public using our current infrastructure because it is a liquid, like gasoline. On the other hand, DMFCs usually show problems related to methanol crossover (from the anode to the cathode), which results in lower efficiencies.

6.3.3 Phosphoric Acid

Phosphoric acid fuel cells can be considered as one of the most mature fuel cell technologies with several hundreds of operating devices. In the present case, the electrolyte is concentrated phosphoric acid in a SiC matrix, whereas the electrodes are made of porous carbon impregnated with Pt. The operating temperatures range between 150 and 200 °C. The possibility of operating at around 200 °C helps to minimise the problems due to CO poisoning of the catalysts and also allows co-generation of electricity from steam, which results in efficiencies of up to 80%.

Nevertheless, the use of Pt-based electrodes is still an important disadvantage and the corrosive nature of the acid limits the materials choice and accelerates degradation phenomena. Moreover, PAFCs are rather heavy, bulky and expensive when compared to other fuel cell technologies, with estimated costs of 2500–4000 €/kW (Fig. 6.8).

Figure 6.8 200 kW PAFC unit developed by Toshiba [24].

6.3.4 Molten Carbonate

Molten carbonate fuel cells are high-temperature fuel cells that use molten carbonates within a ceramic matrix (typically $LiAlO_2$) as electrolyte. At relatively high temperatures, i.e., 600 °C, the salts melt and conduct carbonate anions. As they operate at fairly high temperatures, noble metals are not necessary at the electrodes which are usually made of Ni (anode) and NiO (cathode). Considering the possibility of co-generation, the efficiency of MCFCs can achieve 80–85% (Fig. 6.9).

One of the main disadvantages associated to MCFCs is the management of two gas flows at the cathode (CO and O_2) and the formation of H_2O at the anode which may dilute the fuel.

Additionally, the combination of high temperatures and the corrosive nature of the electrolyte leads to rapid degradation of the components. On the other hand, at fairly high temperatures reactions such as steam reforming are favoured, and therefore conventional fuels, e.g., natural gas, may be fed in the fuel cell without bulky and costly external reformers.

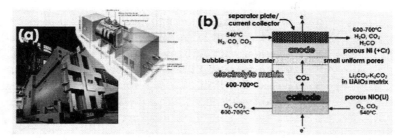

Figure 6.9 (a) 300 kW MCFC unit developed by CFC Solutions Gmbh. (b) Scheme of a MCFC.

MCFCs find application in power plants (static generation) in the range of the MW, as those successfully tested in Italy or Japan [25]. The major MCFC companies in the world such as Fuel Cell Energy and Ansaldo Fuel Cells are able to fabricate prototypes to supply up to 100 MWs for over 25000 h, although the anode must be replaced every 5000 h due to degradation.

6.3.5 Solid Oxide Fuel Cells

Solid oxide fuel cells were first developed by Bauer and Preiss in the 30s [10], although their potential applications in the market are very recent and cannot be considered as a mature technology. SOFCs are based on the possibility of ionic transport in some oxides at high temperatures, typically 600–1000 °C. The most common electrolytes are based on ZrO_2, typically yttria-stabilised zirconia (YSZ), whereas the electrodes are usually a lanthanum strontium manganite (LSM) as cathode and a composite ceramic-metal Ni-YSZ as anode.

As described in the case of MCFCs, the efficiencies in SOFCs may be up to 85% considering co-generation and they can be fed with conventional fuels as they can be oxidised in situ and some common impurities such as CO become a complementary fuel supply [26]. On the other hand, high temperatures limit the materials' choice, i.e., steels cannot be used at temperatures close to 1000 °C, and also accelerate their degradation. Consequently, in recent years the development of SOFCs operating in the intermediate temperature range, i.e., 500–800 °C, has been considered as a priority in this research field.

SOFCs find application mostly in power plants up to MW scale, as those developed by Siemens Westinghouse (Fig. 6.10) and the so-called Bloom boxes developed by Bloomenergy [27] installed in several flagship companies as Ebay, Coca-Cola, Walmart, AT&T, etc. Nevertheless, there are also transport applications in vehicles or auxiliary power units (APU) in trucks or submarines. Moreover, the development of micro-SOFCs has resulted even in portable units as the developed by Prof. Kendall at the University of Birmingham (UK) and Adelan Ltd [28, 29]. Recent work by NECTAR [30] has led to the integration of SOFCs in silicon, which allows covering the 1 W range for portable applications.

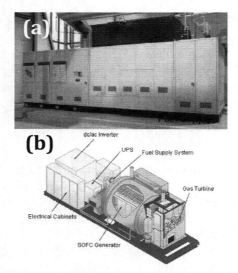

Figure 6.10 (a), (b) 100 kW-class pressurised SOFC-microturbine hybrid power system by Siemens-Westinghouse, 60,000 h operation in various places [27].

It is worth to mention a special type of high-temperature fuel cell, i.e., the **direct carbon fuel cells (DCFC)**. This type of fuel cell directly converts the chemical energy stored in solid carbon directly into electricity through its electrochemical oxidation in the 600–950 °C temperature range. In this type of fuel cell, the same material is used as the anode and the fuel. DCFCs may achieve higher efficiencies (80%) than the molten carbonate and solid oxide fuel cells and low greenhouse gas emissions than conventional coal-burning power plants. More important, solid carbon-rich fuels (e.g., coal, biomass, organic garbage) are readily available and abundant [31]. There are three basic types of direct carbon fuel cells as a function of the electrolyte used as (1) molten salt (KOH, NaOH); operating temperature 500–600 °C; (2) molten carbonate (Li, Na, K); operating temperature 750–800 °C; and (3) oxygen ion conducting ceramic (doped zirconia, ceria); operating temperature 800–950 °C [32].

6.4 Thermodynamics

As previously mentioned, a fuel cell is an electrochemical device that allows the generation of electricity directly from the chemical energy of a fuel and an oxidising agent. Each cell consists of two electrodes, an anode and a cathode, attached to an electrolyte (Fig. 6.11). To promote the electrochemical reactions, oxidation and reduction, respectively, there must be a concentration gradient and the subsequent transport phenomena involving electronic and/or ionic species through both the electrodes and the electrolyte. Therefore, processes involving electrical conduction and the interaction of reacting gas molecules with the electrodes will determine the cell response.

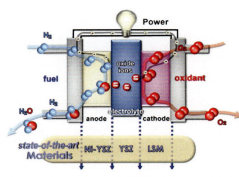

Figure 6.11 Scheme of a traditional SOFC with an oxide-ion conductor as electrolyte [19].

Hydrogen and oxygen are a common choice as fuel and oxidising agent in fuel cells. As mentioned earlier, the overall reaction (reaction 6.3) can be described by the equilibrium constant (K_T) defined as

$$K_T = \frac{\{pH_2O(a)\}^2}{\{pH_2(a)\}^2 \cdot pO_2(a)} \tag{6.4}$$

Nernst established the relation between the standard potential (E^0) of an electrochemical cell and the Gibbs energy ($\Delta G°$):

$$E° = -\frac{\Delta G°}{n \cdot F} = -\frac{\Delta H° - T \cdot \Delta S°}{n \cdot F}, \tag{6.5}$$

where $\Delta H°$ is the standard enthalpy, T is the temperature, $\Delta S°$ is the standard entropy, n is the number of electrons involved in the process and F is the Faraday constant 86484.6 C/mol, being the standard conditions $P = 1$ atm and $T = 298$ K.

All these expressions refer to equilibrium and standard conditions and thus the so-called thermodynamic potential of the system (E_r), also called the open circuit voltage (OCV), can be defined as

$$E_r = \frac{RT}{2F}\ln(K_T) + \frac{RT}{4F}\ln pO_2(c) + \frac{RT}{2F}\ln\frac{pH_2(a)}{pH_2O(a)}, \tag{6.6}$$

considering that

$$\Delta G° = RT \cdot \ln K_T = -nFE° \tag{6.7}$$

$$E_r = E° + \frac{RT}{4F}\ln pO_2(c) + \frac{RT}{2F}\ln\frac{pH_2(a)}{pH_2O(a)}, \tag{6.8}$$

where E_r and $E°$ are different magnitudes and they are equal only in given conditions. On the other hand, the OCV corresponds to E_r and not to $E°$ as sometimes stated. $E°$ is constant for a given reaction and temperature conditions and in theory it should not depend on the gas composition, whereas the OCV varies as a function of the gas composition.

Figure 6.12 shows how the OCV changes as a function of the H_2 and O_2 partial pressures at 1000 °C. On the other hand, $E°$ remains constant at a given temperature.

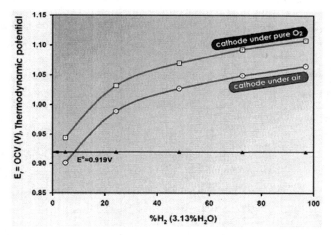

Figure 6.12 Comparison between $E°$- and E_r (OCV)—as a function of pH_2 and pO_2 (in this case pH_2O is constant, 3.13%, at 20 °C) [19].

If CO is the fuel, the number of electrons involved is $n = 2$ and consequently,

$$\left.\begin{array}{l} CO + O^{2-} \to CO^{2-} + 2e^- \\ \frac{1}{2}O_2 + 2e^- \to O^{2-} \end{array}\right\} CO + \frac{1}{2}O_2 \to CO_2 \quad (6.9)$$

The corresponding Nernst equation would be

$$E_r = E° + \frac{RT}{4F}\ln pO_2(c) + \frac{RT}{2F}\ln\frac{pCO(a)}{pCO_2(a)} \quad (6.10)$$

A further example is methane direct oxidation, with $n = 8$:

$$\left.\begin{array}{l} CH_4 + 4O^{2-} \to CO_2 + 2H_2O + 8e^- \\ 2O_2 + 8e^- \to 4O^{2-} \end{array}\right\} CH_4 + 2O_2 \to CO_2 + 2H_2O \quad (6.11)$$

And therefore the Nernst equation is as follows:

$$E_r = E° + \frac{RT}{4F}\ln pO_2(c) + \frac{RT}{8F}\ln\frac{pCH_4(a)}{\{pH_2O(a)\}^2 \, pCO_2(a)} \qquad (6.12)$$

A summary of the most common electrode half-reactions has been presented in Table 6.2.

Table 6.2   The two half-reactions in each type of fuel cell

Fuel cell	Half-reactions	
	Anode	Cathode
PEMFC	$H_2 \rightarrow 2H^+ + 2e^-$	$\frac{1}{2}O_2 + 2H^+ + 2e^- \rightarrow H_2O$
PAFC		
AFC	$H_2 + 2(OH)^- \rightarrow 2H_2O + 2e^-$	$\frac{1}{2}O_2 + H_2O + 2e^- \rightarrow 2(OH)^-$
MCFC	$H_2 + 2CO_3^{2-} \rightarrow H_2O + CO_2 + 2e^-$ $CO + CO_3^{2-} \rightarrow 2CO_2 + 2e^-$	$\frac{1}{2}O_2 + CO_2 + 2e^- \rightarrow CO_3^{2-}$
SOFC	$H_2 + O^{2-} \rightarrow H_2O + 2e^-$ $CO + O^{2-} \rightarrow CO_2 + 2e^-$ $CH_4 + 4O^{2-} \rightarrow 2H_2O + CO_2 + 8e^-$	$\frac{1}{2}O_2 + 2e^- \rightarrow O^{2-}$

6.5   Fuel Cell Efficiency

The global efficiency of a fuel cell can be defined as the product of two contributions, i.e., the electrochemical efficiency (ε_E) and the heating value (ε_C). Nevertheless, the electrochemical efficiency depends on three further contributions, namely the thermodynamic efficiency (ε_T), the voltaic efficiency (ε_V) and the Faradaic efficiency (ε_F). As a consequence, the overall fuel cell efficiency can be described according to

$$\varepsilon_{FC} = \varepsilon_E \cdot \varepsilon_C = \varepsilon_T \cdot \varepsilon_V \cdot \varepsilon_F \cdot \varepsilon_C \qquad (6.13)$$

Obviously, fuel cells operate well below 100% of efficiency as large values of fuel utilisation causes an increase in the diffusion polarisation (voltaic efficiency). The ratio of fuel used (α) is sometimes included within the Faradaic efficiency. In the present case, this will be considered as an independent parameter and therefore the global equation for efficiency is the following:

$$\varepsilon_{FC} = \varepsilon_E \cdot \varepsilon_C = \varepsilon_T \cdot \varepsilon_V \cdot \varepsilon_F \cdot \varepsilon_C \cdot \alpha \qquad (6.14)$$

The different contributions to the efficiency will be described next.

6.5.1 Thermodynamic Efficiency

The variation of Gibbs energy of the chemical reaction occurring in a fuel cell can be transformed into electricity. Therefore, the maximum thermodynamic efficiency can be defined as:

$$\varepsilon_T = \frac{\Delta G}{\Delta H} = 1 - \frac{T\Delta S}{\Delta H} \qquad (6.15)$$

Depending on whether the overall reaction is exothermic or endothermic the apparent thermodynamic efficiency in a fuel cell may be >1, as shown in Table 6.3.

Table 6.3 Thermodynamic parameters for some common reactions in a fuel cell.

Reaction	T (°C)	ΔG, kJ/mol	ΔH kJ/mol	E°, V	ε_T (%)
$H_2 + \frac{1}{2}O_2 \leftrightarrows H_2O$	500	−204.94	−246.21	1.062	83.24
	600	−199.61	−246.95	1.034	80.83
	700	−194.18	−247.63	1.006	78.42
	800	−188.64	−248.25	0.978	75.99
	900	−183.00	−248.80	0.948	73.55
	1000	−177.24	−249.29	0.919	71.10
$CO + \frac{1}{2}O_2 \leftrightarrows CO_2$	500	−215.43	−283.33	1.116	76.04
	600	−206.68	−283.04	1.071	73.02
	700	−197.96	−282.71	1.026	70.02
	800	−189.26	−282.34	0.981	67.03
	900	−180.58	−281.93	0.936	64.05
	1000	−171.92	−281.49	0.891	61.08
$CH_4 + 2O_2 \leftrightarrows CO_2 + 2H_2O$	500	−800.54	−800.22	1.037	100.04
	600	−800.48	−800.49	1.037	100.00
	700	−800.42	−801.00	1.037	99.93
	800	−800.36	−801.70	1.037	99.83
	900	−800.29	−802.55	1.037	99.72
	1000	−800.21	−803.45	1.037	99.59

(Continued)

Table 6.3 (*Continued*)

Reaction	T (°C)	ΔG, kJ/mol	ΔH kJ/mol	E^o, V	ε_T (%)
$C + O_2 \leftrightarrows CO_2$	500	−395.51	−394.08	1.025	100.36
	600	−395.68	−394.28	1.025	100.35
	700	−395.83	−394.50	1.026	100.34
	800	−395.96	−394.74	1.026	100.31
	900	−396.06	−394.98	1.026	100.27
	1000	−396.14	−395.20	1.026	100.24
$C_3H_8 + 5O_2 \leftrightarrows 3CO_2 + 4H_2O$	500	−2125.6	−2041.7	1.102	104.11
	600	−2135.9	−2043.1	1.107	104.54
	700	−2146.2	−2045.0	1.112	104.95
	800	−2156.5	−2047.1	1.118	105.35
	900	−2166.9	−2049.4	1.123	105.73
	1000	−2177.3	−2051.8	1.128	106.12

6.5.2 Voltaic Efficiency

Voltaic efficiency may be defined as the ratio between the voltage under working conditions and the voltage in equilibrium. Under working conditions, the cell voltage is lower than the so-called reversible voltage. As current is drawn from the cell, the voltage drops due to polarisation phenomena, as it happens in any electrochemical cell. These losses depend on a number of parameters such as temperature, pressure, flow, gas composition and cell components.

$$\varepsilon_V = E/E_r \tag{6.16}$$

The difference between those two values are generically called polarisation, η.

$$\eta = E_r - E \tag{6.17}$$

One may consider that E_r is usually identical to the open circuit voltage (OCV) and therefore the voltage under experimental conditions may be described as

$$E = OCV - \eta \tag{6.18}$$

There are four main contributions to the overall polarisation in an electrochemical cell: activation polarisation (η_{act}, charge transfer), diffusion/concentration polarisation (η_{dif}), reaction polarisation (η_{re}) and ohmic/resistive polarisation (η_Ω).

$$\eta = \eta_{act} + \eta_{dif} + \eta_{re} + \eta_\Omega \tag{6.19}$$

Obviously, all these contributions have a negative impact on the fuel cell performance as depicted in Fig. 6.13. Under ideal conditions, the overall polarisation would be null, hence voltage would be constant and equal to the value defined by the Nernst equation (E_r). Nevertheless, such situation never occurs and voltage shows a dependence upon the current density dragged out of the cell, which in turn may be divided in several regions. Indeed, at low current densities, losses are mostly due to activation polarisation. After that, ohmic resistance is the most important contribution as reflected by linear dependence between the voltage and the current density. At large current density values, current density depends strongly on mass transport, which usually results in a marked voltage drop.

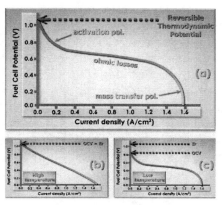

Figure 6.13 (a) Representation of the dependence of the fuel cell voltage with the current density, showing the main losses responsible for the performance of a fuel cell. In this case the reversible thermodynamic potential (E_r) coincides with the open circuit potential (OCV). (b) The activation and mass transfer processes are negligible at high temperature (and with good electrodes); hence only ohmic losses are observed. (c) At low temperature all the processes can be observed and the OCV is lower than E_r [19, 20, 26].

One should note that temperature plays a crucial role in the magnitude of the efficiency losses. Activation and mass transport are thermally activated and therefore their values decrease with increasing the temperature, i.e., their impact on the cell performance is smaller. In other words, high-temperature fuel cells will usually show almost linear E–j curves, which indicates that their performance depends mostly on the ohmic resistance (usually due to the electrolyte) (Fig. 6.13b). In those situations, the OCV and the thermodynamic voltage defined by the Nernst equation, E_r, are almost identical and the only differences are usually related to electronic conductivity of the electrolyte causing short circuit and/or gas leaks.

At lower temperatures, all polarisation phenomena are more important and they may cause the OCV value to be certainly lower compared to the thermodynamic value (Fig. 6.13c).

- **Activation polarisation**: To produce a chemical reaction, an energy barrier must be overcome. Such activation energy may be considered as the extra potential that must be applied to facilitate the process that controls the overall reaction kinetics.

  The activation polarisation is usually related to the slowest reaction/s occurring at the electrode surface, which could be adsorption of the reactants to the electrode surface, electrode ionic transfer, product desorption, etc. The reaction rate depends on several parameters as temperature, pressure or electrode materials. As mentioned above, reaction rates are generally faster with increasing the temperature and consequently the activation polarisation is rather small.

  The activation polarisation is related to the current density through the Butler–Volmer equation:

  $$j = j_o \cdot e^{\beta \cdot \eta_{act} \cdot n \cdot F/R \cdot T} - j_o \cdot e^{(1-\beta) \cdot \eta_{act} \cdot n \cdot F/R \cdot T}, \qquad (6.20)$$

  where β is the charge transfer coefficient, j_o is the exchange current density, n is the number of electrons involved. j_o can be calculated by extrapolation when representing $\log(j)$ vs. η, $\eta = 0$. Large j_o values indicate high electrochemical reaction rates, hence good fuel cell performance.

- **Diffusion/concentration polarisation.** This contribution is important when the electrode reaction rates are dominated by mass transport, e.g., the rate of reagents supply/product withdraw to/from the active sites is lower than the corresponding discharge current. When the electrode process are fully controlled by diffusion, i.e., low concentration of reagents or full conversion to products, a limit current is reached, which implies a marked drop in the cell voltage.

  In the present case, mass transport also depends on temperature, pressure, concentration of reagents/products, physical properties of the system. In fuel cells, reagents and products diffuse through porous electrodes and therefore the microstructure is a crucial parameter to minimise losses associated to mass transport.

- **Reaction polarisation.** Performance losses associated to the reaction appear when the rate of supply/withdraw of reagents/products is low. This is somewhat related to the previous contribution to the overall polarisation and therefore, operating at high temperatures help to minimise this problem.

- **Ohmic polarisation.** Also called ohmic losses, this is caused by the internal resistance of the electrolyte and the electrodes/bipolar plates and also the contact resistance between consecutive cells in stacks. The ohmic polarisation can be described by

$$\eta_\Omega = j \cdot R_{tot}, \tag{6.21}$$

where R_{tot} is the internal cell resistance including the electrolyte and electrodes:

$$\eta_\Omega = j \cdot R_{tot} = j \cdot (R_\Omega^{electrolyte} + R_\Omega^{anode} + R_\Omega^{cathode}) \tag{6.22}$$

An adequate choice of materials for the fuel cell components results on negligible contributions from the anode and cathode compared to the electrolyte and consequently, the ohmic polarisation depends almost exclusively on the electrolyte resistance.

6.5.3 Faradaic Efficiency

The Faradaic efficiency considers the possibility of parallel reactions occurring at the electrodes, resulting in current densities lower than expected. For instance, during the direct oxidation of methanol to CO_2 and water, formic acid or formaldehyde may form and in that situation the number of electrons involved changes from 6 to 2 or 4 in the corresponding parallel reactions, which will result in lower current densities.

Additional effects due to heterogeneous catalysis processes at the electrodes may also appear within this contribution to the overall polarisation.

In the present case, the polarisation may be expressed as the ratio between the produced current density and the current produced for 100% fuel conversion.

$$\varepsilon_F = j/j_F, \tag{6.23}$$

where j_F is the current density for 100% fuel conversion and is given by Faraday's law:

$$j_F = n \cdot F \cdot v_F \tag{6.24}$$

6.5.4 Heat Efficiency

This parameter should be taken into account in the case of commercial fuels containing inert gases, impurities or even other electrochemically active species. It is defined as the ratio between the enthalpy of the active species at the cell (e.g., H_2), ΔH_{react}, and the enthalpy of all the chemicals supplied to the cell (e.g., CO impurities plus H_2), ΔH_{comb}.

$$\varepsilon_c = \frac{\Delta H_{react}}{\Delta H_{comb}} \tag{6.25}$$

In high-temperature fuel cells, this value may be close to 1 as most of the impurities accompanying the fuel can be oxidised, as is the case of CO. In the low temperature range, this value may drop quite markedly.

According to all the contributions to the efficiency losses defined above, we may propose an equation to define the overall fuel cell efficiency as a result of substituting in the previous relation:

$$\varepsilon_{FC} = \frac{\Delta G}{\Delta H} \cdot \frac{E}{E_r} \cdot \frac{j}{j_F} \cdot \frac{\Delta H_{react}}{\Delta H_{comb}} \cdot \alpha \tag{6.26}$$

From Eq. (6.26), we may assume that all the magnitudes are constant, except for E and j, and they can be grouped in a constant K_{FC}:

$$\varepsilon_{FC} = K_{FC} \cdot E \cdot j \tag{6.27}$$

On the other hand, the product of E and j is the power density, P, which in other words indicates that the maximum fuel cell efficiency will occur in the experimental conditions of maximum power density.

$$\varepsilon_{FC} = K_{FC} \cdot P \tag{6.28}$$

Despite all these factors having a negative impact on the performance, fuel cells are still very efficient compared to other more conventional systems for power generation, achieving values in excess of 50%.

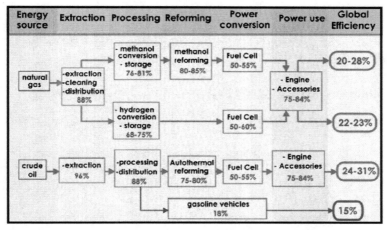

Figure 6.14 Overall and efficiency of each stage in the electrical power of a typical vehicle fuelled by gasoline and compared with a fuel cell–based vehicle.

Nevertheless, there are several other important parameters related to fuel production, transport, storage, etc., that should be considered when defining the efficiency of systems powered by fuel cells. Fuel cells may achieve electrochemical efficiencies above

50%, but the overall efficiency is usually lower as described in Fig. 6.14 for vehicles powered by fuel cells compared to regular gasoline cars. For instance, power plants based on fuel cells operating with pressurised hydrogen show efficiencies of approximately 22%, whilst operating with liquid hydrogen the efficiency drops to 17%. High-temperature fuel cells may increase their efficiency due to the possibility of co-generating electricity via microturbines using the heat exhaust. A clear example of the so-called Hybrid Fuel Cell/Gas Turbine Systems developed by Siemens Westinghouse at the National Fuel Cell Research Center, University of Irvine (US), which is capable of delivering 220 kW with fuel to electricity efficiencies of up to 53%.

6.6 Applications

Fuel cells have been regarded as a key technology in a new energy scenario free of fossil fuels. This has been translated into a relatively large investment to develop different fuel cell technologies and applications; we may find some of them today in the market. Additionally, the search for mass-scale commercialisation and implementation in industry has led to several important demonstration projects. Some relevant examples will be briefly described.

6.6.1 HyFLEET-CUTE [17]

European project (2006–2009) based on previous experience that was used to evaluate the advantages and disadvantages of hydrogen internal combustion engine (ICE) buses with fuel cell buses (Fig. 6.15). CUTE stands for **C**lean **U**rban **T**ransport for **E**urope and during the project hydrogen buses were tested and demonstrated in 10 different cities in Europe, Asia and Australia. Hydrogen has been produced from both renewable and non-renewable sources.

The project has been quite successful with over 2.5 million kilometres travelled, 170,947 h of bus operation and 555 tonnes of hydrogen dispensed—all accomplished safely, using a fleet of 150–200 kW buses. Nevertheless, some issues should be addressed to make this project scalable to a larger vehicle fleet:

- The bus technology must operate with a certain degree of support in a standard public transport bus fleet.

- The purchase price of the buses must be significantly reduced.
- Lifetime operational costs should be considered.
- Hydrogen must be produced from renewables at low cost.
- Hydrogen infrastructure, especially the electrolyser and steam reformer units, which are the key components of on-site H_2 production and also the hydrogen compressors and dispensing equipment, must be as reliable as the buses themselves.

It should be noted that the HYFLEET: CUTE project has been replaced with the CHIC (Clean Hydrogen In European Cities) project, which seeks for mass-scale commercialisation of hydrogen-powered vehicles.

Figure 6.15 Buses of the HyFLEET CUTE project which operated in several European cities.

6.6.2 UTSIRA [34]

In 2004, the Norwegian company Norsk Hydro installed a combined wind–hydrogen system on the island of Utsira (southwest coast of Norway). It can be considered as the first operational system based exclusively on wind power and hydrogen/fuel cells. The target of this project was to demonstrate that efficient and reliable power supply may be achieved in remote areas exclusively via renewable sources.

The two wind turbines installed produce more energy than required by the inhabitants of the island. The surplus energy is transformed to hydrogen via electrolysers and this is stored (pressurised) and then used in fuel cells when the turbines do not work.

6.6.3 Present and Future

Some of the advances made in the past few years can be seen in the marketplace today, for example, in cars, forklifts, etc. Nevertheless, there is a growing interest in further demonstration projects and programs to reduce costs, and thus fuel cells become a competitive technology, e.g., the Strategic European Hydrogen Transport Projects, DOE Hydrogen and Fuel Cells Program, the Hydrogen Energy Test and Research Center in Japan (Hy-TREC) or the U.S. DRIVE (Driving Research and Innovation for Vehicle efficiency and Energy sustainability related to hydrogen/fuel and their implementation at large scale).

As a consequence, hydrogen and fuel cells technology are making outstanding progress. Although this is not a mature technology, the range of markets will be greatly expanded through improvements in performance and durability and obviously through reductions in manufacturing cost and all the processes related to the production, storage and supply of hydrogen. Successful entry into new markets also depends on overcoming some other barriers such as the need for codes and standards, the lack of public awareness and understanding of the technologies.

6.6.4 Last Trends in Fuel Cell Technology

6.6.4.1 Alkaline fuel cells

As mentioned before, AFCs were used decades ago in several NASA projects and have the advantage of faster kinetics for oxygen reduction due to the alkaline media [20, 35]. The possibility of fabricating stacks free of noble metals allowed a dramatic reduction of production costs, through the use of solid electrolytes, preventing leakages, and higher efficiencies of PEMFCs and SOFCs, respectively, made them less attractive for R&D projects. Nevertheless, improved stack designs have revived programs to develop a new generation of AFCs [36, 37]. Indeed, one of the hottest topics in AFC technology nowadays is the development of AFC operating with anion-exchange membranes (AEMs) that replace aqueous KOH electrolytes. Apart from their relatively low working temperature, these novel AFCs, also referred as alkaline polymer electrolyte fuel cell (APEFC) or solid alkaline fuel cell (SAFC), have generated great attention due to their ability to operate directly with alcohols, i.e., ethanol, as fuels [38–40].

6.6.4.2 Phosphoric acid fuel cells

Although PAFCs were considered the first generation of commercial modern fuel cells, interest in them has waned notably with the turn of the twenty-first century. Nevertheless, PAFC stacks installed exceed any other fuel cell technology, e.g., over a hundred plants in the 1995–2005 period in the US or over 40 systems sold by UTC between 2008 and 2012, most of them 400 kW systems. Similar numbers have been achieved in Japan during the same period [41]. The future mass-scale commercialisation of PAFC technology is uncertain as PAFCs' electrical efficiency is inferior to that of other fuel cells. Nevertheless, higher efficiency may be achieved via the development of novel electrocatalysts that may also lower the production costs [42].

6.6.4.3 Polymer electrolyte membrane fuel cells

PEMFCs have undergone phenomenal progress in the last 20 years due as they may be used in portable applications, including vehicles, and small stationary power generators. The use of a solid electrolyte, a polymer, implies several advantages compared to other low-temperature fuel cells that use liquids, including high power density and low corrosion. However, they also show some disadvantages, as is the use of very expensive catalysts based on Pt, which imply around 50% of the total fabrication costs. Therefore, a large number of research laboratories work on the optimisation of the Pt load at the electrodes via the use of dispersed nanoparticles, with targets of <0.15 mg/cm^2 according to the US Department of Energy for 2015. Related to this and due to the low operation temperature, Pt is very sensitive to CO and novel alloys based on Pt and other precious metals such as Rh and Ru apparently lead to higher tolerances [43], although the fabrication costs do not change significantly. An alternative strategy is the use of Pt-free catalysts [44], such as Pd–Ni alloys, Ir–V alloys, carbides and nitrides [45] or transition metal chalcogenides [46, 47].

Alternatively, the performance of PEMFCs may be improved via higher operation temperatures, which requires novel electrolytes to substitute the *state-of-the-art* Nafion. Among others, the performance of polybenzimidazoles (PBI) [48], polyarylene ethers, styrene derivatives, sulphonated polyimides, aromatic polyethers, polyether ketones (PEKs), etc. [41], should be highlighted.

6.6.4.4 Molten carbonate fuel cells

High-temperature fuel cells are devoted mainly to stationary power generation in stacks that range from tens of kilowatts to several megawatts [35]. This imposes very restrictive requirements in the stack design for commercial application as degradation becomes a very relevant issue. For the durability of MCFCs at the commercial level, a lifetime of 40,000 h and a voltage decay rate of less than 10% after this period are required [49]. Therefore one of the key issues in MCFCs is the development of electrodes with improved chemical and physicochemical stability. More particularly, one of the trending topics considers methods to increase the stability of NiO cathodes, particularly in direct internal reforming MCFCs as is the addition of oxides to the carbonate electrolytes to prevent NiO dissolution. Similarly, higher efficiencies may be achieved via the optimisation of the fuel electrode. Thinner anodes will result in enhanced efficiencies as the polarisation is minimised due to the improved mass transfer resistance. On the other hand, the electrodes must exhibit mechanical strength to produce robust devices and an adequate strategy may be the use of thin Ni-based alloys reinforced with metal meshes [49–51].

On the other hand, high temperatures also imply the use of hydrocarbons, including biogas, and therefore the development of MCFC stacks fed with hydrocarbons containing S is a key issue nowadays [52].

6.6.4.5 Solid oxide fuel cells

SOFCs operate at relatively high temperatures (700–900 °C) and therefore the materials' choice is rather restrictive and costly.

Lowering the operation temperature will open up the opportunity to use steels and metal alloys substituting ceramic components, which would help to reduce fabrications costs [26]. Moreover, reducing temperatures would allow the use of SOFCs in small power generation units or even portable applications as mentioned in Section 6.3.5. To achieve this, better electrolyte and electrodes materials are required, which can be achieved via two main routes, *viz.* microstructural improvement and novel materials.

Regarding the electrolyte, the state of the art is yttria-doped zirconia (YSZ), which exhibits high ionic conductivity at 900 °C. To operate at temperatures of 600 °C or below, YSZ electrolytes may be produced using techniques such as PVD and magnetron sputtering, which render fully dense ceramic films down to a few nanometres. Alternatively, YSZ may be replaced by other ion conductors such as those based on cerium oxide, apatites, lanthanum gallates, etc., which exhibit higher ionic conductivity than YSZ in the intermediate temperature range, although they also show some drawbacks related to stability (both chemical and mechanical) [19, 26, 35].

The electrodes also need optimisation to operate at lower temperatures, and the conventional Ni-YSZ anodes have been substituted by all-ceramic anodes based on perovskite materials such as $(La,Sr)(Cr,Mn)O_3$ (LSCM) [53], $Sr_2(Mg,Mo)O_6$ [54] and substituted $(La,Sr)TiO_3$ [55]. Another strategy would be the fabrication of anodes by impregnating with active metal/oxides porous backbones of ionic conductors as is the case of Cu-impregnated fluorites [56].

Nevertheless, one of the major issues concerning intermediate-temperature SOFCs is the cathode. The LSM cathodes are usually responsible for the largest source of potential losses due to their limited ionic conductivity and electrochemical activity at temperatures below 700 °C. Therefore, much effort has been dedicated to the development of mixed-ion conductors such as ferrites and Co-ferrites [26].

References

1. www.fz-juelich.de.
2. www.fuelcells.org.
3. W. R. Grove, *Philos. Mag. Ser.*, **3**(14), 1839, p. 127.

4. C. F. Schönbein, *Philos. Mag. Ser.*, **3**(14), 1839, p. 43.
5. W. R. Grove, *Philos. Mag. Ser.*, **3**(21), 1842, p. 417.
6. W. W. Jacques, *Electrical Rev.*, **38**(970), 1896, p. 826.
7. W. Nernst, *Z. Elektrochem.*, **6**, 1899, p. 41.
8. W. Schottky, Über stromliefernde Prozesse im Konzentrationsgefälle fester Electrolyte, *Wiss Veröff Siemens-Werke* **14**(2), p. 1.
9. Haber F. (1907) Verfahren zur Erzeugung von elektrischer Energie aus Kohle und gasförmigen Brennstoff en. Öster Pat 27 743, filed 5.8.1905.
10. E. Bauer and H. Preis, *Z. Elektrochem.*, **43**, 1937, p. 727.
11. O. K. Davtyan, The problem of the direct conversion of fuel chemical energy to electric energy (in Russian), Izd. AN SSSR, Moscow 1947.
12. F. T. Bacon, *Electrochim. Acta*, **14**, 1969, p. 569.
13. W. T. Grubb, U.S. Patent 2,913,511 (1959).
14. W. T. Grubb and L. W. Niedrach, *J. Electrochem. Soc.*, **107**, 1960, p. 131.
15. http://www.osakagas.co.jp/en/rd/fuelcell/sofc/.
16. www.transport-research.info.
17. www.hyfleetcute.com.
18. www.chic-project.eu.
19. J. C. Ruiz-Morales, J. Canales-Vázquez, D. Marrero-López, J. Peña-Martínez, D. Pérez-Coll, P. Núñez, C. Savaniu, C. Rodríguez-Placeres, V. I. Dorta-Martín and B. Ballesteros, *Pilas de Combustibles de Óxidos Sólidos (SOFC)* (in Spanish), ed. Centro de la Cultura Popular Canaria (CCPC), La Laguna (Spain), 2nd ed. 2008.
20. J. Larminie and A. Dicks, *Fuel Cells Systems Explained*, Ed. Wiley & Sons, Chichester (England), 2nd ed. 2003.
21. T. Ralph, *Platinum Metal Rev.*, **41**, 1997, p. 102.
22. K. Rabaey and W. Verstraete, *Trends Biotechnol.*, **23**(6), 2005, p. 291.
23. www.plugpower.com.
24. R. A. Bajura, *FETC Perspective on the DOE Stationary Power Fuel Cell Program*, 1997.
25. A. Moreno, S. McPhail and R. Bove, *International Status of Molten Carbonate Fuel Cell (MCFC) Technology*, ENEA Report, RSE/2009/181, Rome (Italy).
26. S. C. Singhal and K. Kendall, *High Temperature Solid Oxide Fuel Cells: Fundamentals, Design and Applications*, Elsevier, Oxford (England), 2004.

27. http://www.bloomenergy.com/.
28. K. Kendall, Solid Oxide Fuel Cell Structures, U.S. Patent 5827620, 1998.
29. K. Kendall and I. Kilbride, Fuel Cell Power Generating System, U.S. Patent 20020081472, U.S. Patent, 2002.
30. http://www.nectarpower.com/.
31. D. Cao, Y. Sun and G. Wang, *J. Power Sources*, **167**(2), 2007, p. 250.
32. S. P. S. Badwal and S. Giddey, *Materials Forum*, Ed. J. Feng-Nie, **34**, 2010, p. 181.
33. www.hydro.com.
34. D. Stölten and B. Emonts, *Fuel Cell Science & Engineering: Materials, Processes, Systems and Technology*, Wiley-VCH, Weinheim (Germany), 2012.
35. J. R. Varcoe and R. C. T. Slade, *Fuel Cells*, **5**, 2005, p. 187.
36. O. I. Deavin, S. Murphy, A. L. Ong, S. D. Poynton, R. Zeng, H. Herman, and J. R. Varcoe, *Energy Environ. Sci.*, **9**, 2012, p. 8584.
37. E. Antolini and E. R. Gonzalez, *J. Power Sources*, **195**, 2010, p. 3431.
38. N. Fujirawa, Z. Siroma, S. Yamazaki, T. Ioroi, H. Senoh and K. Yasuda, *J. Power Sources*, **185**, 2008, p. 621.
39. A. D. Modestov, M. R. Tarasevich, A. Y. Leikin and V. Y. Filimonov, *J. Power Sources*, **188**, 2009, p. 502.
40. N. Hikosaka Behling, *Fuel Cells: Current Technology Challenges and Future Research Needs*, Elsevier, 1st ed., 2012.
41. Q. He, S. Mukerjee, R. Zeis, S. Parres-Esclapez, M. J. Illán-Gómez and A. Bueno-López, *Appl. Catal. A: General*, **381** (1–2) 2010, p. 54.
42. *Fuel Cell Handbook*, NETL, the US Department of Energy, 7th ed., 2004.
43. A. Brouzgou, S. Q. Song and P. Tsiakaras, *Appl. Catal. B: Environ.*, **127**, 2012, p. 371.
44. C. W. B. Bezerra, L. Zhang, K. Lee, H. Liu, A. A. L. B. Marques, E. P. Marques, H. Wang and J. Zhang, *Electrochim. Acta*, **53**(15), 2008, p. 4937.
45. N. A. Vante and H. Tributsch, *Nature*, **323** (6087), 1986, p. 431.
46. L. Jiu, J.-W. Lee and B. N. Popov, *J. Power Sources*, **162**(2), 2006, p. 1099.
47. J. A. Asensio, S. Borros and P. Gómez-Romero, *Electrochim. Acta*, 49 2004, p. 4461.
48. E. Antolini, *Appl. Energies*, **88**, 2011, p. 4274.

49. H. V. P. Nguyen, S. A. Song, D.-N. Park, H. C. Ham, J. Han, S. P. Yoon, M. R. Othman and J. Kim, *Int. J. Hydrogen Energy*, **37**, 2012, p. 16161.
50. Y. S. Kim and H. S. Chun, *J. Power Sources*, **84**, 1999, p. 80.
51. F. Zaza, C. Paoletti, R. Lo Presti, E. Simonetti, and M. Pasquali, *J. Power Sources*, **195**(13), 2010, p. 4043.
52. S. Tao and J. T. S. Irvine, *Nat. Mater.*, **2**, 2003, p. 320.
53. Y.-H. Huang, R. I. Dass, Z.-L. Xing and J. B. Goodenough, *Science*, **312**, 2006; p. 254.
54. J. C. Ruiz-Morales, J. Canales-Vázquez, C. D. Savaniu, D. Marrero-López, W. Zhou and J. T. S. Irvine, *Nature*, **439**, 2006, p. 568.
55. S. D. Park, J. M. Vohs and R. J. Gorte, *Nature*, **404**, 2000, p. 265.

Part 3
Energy Storage

Chapter 7

Batteries: Fundamentals and Materials Aspects

Montse Casas-Cabanas[a] and Jordi Cabana[b]

[a]*Structure and Surface Analysis Group, CIC EnergiGUNE, Albert Einstein 48, Miñano, 01510, Spain*
[b]*Environmental Energy Technologies Division, Lawrence Berkeley National Laboratory, 1 Cyclotron Rd. MS62R0203, Berkeley, California 94720-8168, USA*

mcasas@cicenergigune.com, jcabana@lbnl.gov

7.1 Introduction

7.1.1 What Is a Battery?

Batteries are devices that store electrical energy through redox reactions. In order to convert the electron exchange into current, the two half-reactions are separated at opposing electrodes by a dielectric. Spontaneously, the chemical species at the negative electrode are oxidized, releasing electrons that can only flow through an external circuit providing electrical current and are consumed by the chemical species at the positive electrode. For every electron that goes through the external circuit, an equivalent amount of positive charge, carried in the form of ions, is consumed in order to balance charges. This process leads to the discharge of the electrochemical cell. The chemical compounds involved in

---

Materials for Sustainable Energy Applications: Conversion, Storage, Transmission, and Consumption
Edited by Xavier Moya and David Muñoz-Rojas
Copyright © 2016 Pan Stanford Publishing Pte. Ltd.
ISBN 978-981-4411-81-3 (Hardcover), 978-981-4411-82-0 (eBook)
www.panstanford.com

the reaction are called *electroactive materials*. The dielectric that prevents short-circuiting is called electrolyte and must enable the mobility of ions toward and away from the electrodes. When the electroactive compounds are in liquid form, i.e., dissolved in the electrolyte, the electrodes are typically metallic foils. When they are solid, they are assembled in a complex electrode mixture that also contains additional components such as conducting and performance enhancing additives. These complex mixtures are assembled into a porous architecture, most often with the help of polymeric binders, so as to enable electrolyte wetting and homogeneous ion transport. Because ion conduction within the electrolyte is not infinite, the positive and negative electrodes have to be as close as possible in order to minimize the internal resistance of the cell. When liquid phases are involved, a thin, porous, separator made of insulating material impregnated by the electrolyte is typically employed in order to avoid physical contact between the electrodes, which would also short-circuit the cell. An alternative is to host the two electroactive compounds in separate containers, either electrically isolated or ionically connected by a saline bridge, as in the classical Daniell electrochemical cell [1]. If both electroactive compounds are dissolved in the electrolyte, they are referred to as catholyte and anolyte, respectively, and may require physical pumping through the electrode assembly. All the structural components of this set-up (those described and also current collectors, the container and terminals) are what constitute an electrochemical cell. An example of a possible configuration is shown in Fig. 7.1.

The large domain of battery applications has led to a wide variety of configurations and sizes. Figure 7.2 shows different Li-ion cell configurations. The single-cell design would be the simplest battery pack. A module consists of several cells connected in an appropriate arrangement (series or parallel) to provide the required operating voltage and current used in specific applications. A battery pack is assembled by connecting modules together, again either in series or parallel. Series configurations lead to an increase of the voltage, while in parallel configurations each cell adds to the total current while voltage remains the same. Serial/parallel connections are common as allow superior flexibility.

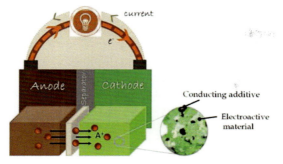

Figure 7.1 Scheme of a battery. Upon battery discharge electrons and cations are released from the anode. Transport of cations occurs through the electrolyte, crossing the separator toward the cathode side, while electrons are driven to an external circuit from which the electrical current can be obtained.

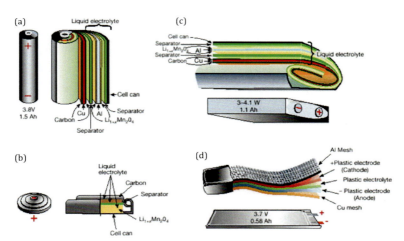

Figure 7.2 Schematic drawing showing the shape of various Li-ion cell configurations: (a) Cylindrical, (b) coin, (c) prismatic, and (d) thin and flat. Reprinted by permission from Macmillan Publishers Ltd: Nature [2], copyright (2001).

When one of the redox reactions is finished the electron flow stops. In some cases the battery can be electrically recharged by applying an external current in the opposite direction to that of the discharge current, reversing the whole process. Rechargeable batteries are also termed secondary batteries or accumulators, while non-rechargeable batteries are called primary batteries. In primary batteries, the positive electrode is usually called cathode and the negative electrode, anode. However, in secondary batteries the designation depends of charge or discharge, therefore this terminology is avoided (or refers to discharge).

The amount of energy that can be stored in a battery is given by

$$\Delta G° = -nFE°, \qquad (7.1)$$

where n is the number of electrons transferred per mole of reactants, F is the Faraday constant being equal to the charge of 1 equivalent of electrons (96497 C/mol) and $E°(V)$ is the standard cell potential or electromotive force of the specific chemical reaction. By convention spontaneous processes have a negative free energy and a positive cell potential. $\Delta G°$ can either be given in J or Wh (1 Wh = 3600 J) and the terms of specific energy Whkg^{-1} or energy density Whl^{-1} are frequently used.

The amount of charge produced, i.e., the capacity ($Q = nF$), is determined by the amount of electroactive material available for reaction and is usually expressed per unit of mass in Ah/kg (1 Ah = 3600C). The theoretical capacity of an electrochemical cell, based only on the active materials participating in the electrochemical reaction, is calculated from the molar mass of the reactants ($Q = nF/M$).

When conditions are other than in the standard state, the voltage E of a cell is given by the Nernst equation:

$$E = E° - RT \ln \frac{A_P}{A_R}, \qquad (7.2)$$

where R is the gas constant, T the absolute temperature, A_P the activity product of the products and A_R the activity product of the reactants. The voltage of the cell is unique for each reaction couple and corresponds to the difference between the chemical potentials (or Fermi levels) of the anode (μ_A) and the cathode (μ_C) divided by the charge of an electron (e):

$$E = \frac{(\mu_A - \mu_C)}{e} \qquad (7.3)$$

These concepts are summarized in Fig. 7.3, where a scheme of the relative electron energies in the electrodes and the electrolyte of an electrochemical cell with an aqueous electrolyte are shown.

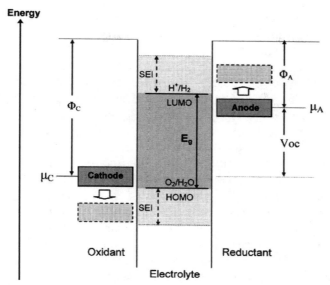

Figure 7.3 Open circuit energy diagram of an aqueous electrolyte. E_g is the window of electrolyte thermodynamic stability, delimited by its HOMO and LUMO orbitals. Φ_A and Φ_C are the anode and the cathode work functions. Chemical potentials η_A and η_C outside this window can only be kinetically stabilized by formation of an SEI layer. Reprinted with permission from [3]. Copyright 2010 American Chemical Society.

Upon battery discharge, the chemical potentials of each electrode will tend to equilibrate by approaching each other, thus reducing the cell voltage. The energy separation of the lowest unoccupied molecular orbital and the highest occupied molecular orbital of the electrolyte (LUMO and HOMO, respectively) determines its electrochemical stability window. Exceeding this window results in electrolyte decomposition: An anode with a η_A above the LUMO will reduce an aqueous electrolyte to H_2 and a cathode with a η_C below the HOMO will oxidize it to O_2 [3]. Thermodynamically, aqueous electrolytes show an electrochemical stability window of 1.23 V although kinetic effects may expand the stability limit to ≈2 V. Organic electrolytes used in Li-based batteries have a voltage stability window > 4.5 V as result in different decomposition products than aqueous electrolytes. A passivating solid layer may block the electron transfer between the electrode and the electrolyte and prevent the decomposition of the latter outside its stability window (see Section 7.3.3.2). However, these layers may also block ion transfer, thereby preventing the redox reaction.

Direct measurement of single (absolute) electrode potentials is considered practically impossible. All standard potentials are given by relative to the standard hydrogen electrode (SHE) whose reaction $H_2/H^+(aq)$ is taken by convention as zero.

The open-circuit voltage (OCV) is the voltage under a no-load condition and is usually a close approximation of the theoretical voltage. In ideal conditions, and if the battery operates through a first-order transition (see Section 7.1.2), the discharge of the battery would proceed at the OCV and would drop to zero once the electroactive materials were consumed. Under actual conditions the voltage of the battery under load (working voltage) is always lower than the open-circuit voltage due to polarization and ohmic losses, which consume part of the energy as waste heat. The resulting working voltage E can be written as

$$E = \text{OCV} - [(\eta_{ct})_A + (\eta_c)_A] - [(\eta_{ct})_C + (\eta_c)_C] - iR_i, \quad (7.4)$$

where $(\eta_{ct})_A$ and $(\eta_{ct})_C$ are charge-transfer or activation polarization at anode and cathode; $(\eta_c)_A$ and $(\eta_c)_A$ are concentration polarization at anode and cathode, i is the operating current of cell on load and R_i is the internal resistance of cell [4].

Activation polarization is caused by the kinetic limitations of the charge transfer process taking place at the electrode/electrolyte interfaces. Concentration polarization refers to mass transport limitations during cell operation, as the concentration at the electrode surface drops faster than the bulk concentration as a consequence of the slow time constant for transport. From the Nernst equation it is clear that as the surface concentration decreases, the potential also decreases. Ohmic polarization (or IR drop) refers to the internal impedance of the cell, which is the sum of the ionic resistance of the electrolyte, the electronic resistances of the active materials and current collectors and the contact resistance between the active mass and the current collector. These resistances follow Ohm's law and are proportional to the current drawn from the system, disappearing instantaneously when current ceases (see Fig. 7.4).

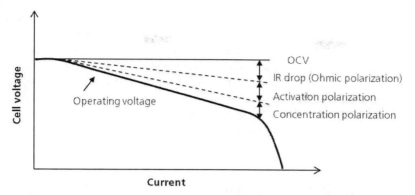

Figure 7.4 Cell polarization as a function of operating current.

7.1.2 Materials Aspects

Since the ultimate goal with batteries is to maximize energy storage, configurations that maximize capacity and cell voltage are continuously sought. Ideally, electrode materials should be capable of sustaining extensive charge/discharge cycling, chemically stable toward the electrolyte, inexpensive, environmentally benign and lightweight. These simple requirements are not easy to accomplish. Proof of it is the fact that despite the huge number of compounds that have been studied as potential electrode materials, only a few have reached the market (and some of the most common battery chemistries were developed more than a century ago).

The electrolyte must have pure ionic conductivity, and any fraction of electronic conductivity would lead to an internal short-circuit. In most cases it is composed of a solvent with dissolved salts (solute) although some systems use polymers or ceramic materials. Ideally the electrolyte should not react with the other cell components and its physical properties (viscosity, etc.) should not change within the battery temperature operation range. Other requirements are thermal and electrochemical stability, low flammability, vapor pressure, toxicity, and cost. Lead-acid batteries and alkaline batteries use aqueous electrolytes, which fulfill in most cases the stability window requirement and whose conductivities are always high. On the other hand, lithium-ion batteries do not operate with solvents containing active protons such as water or ethanol because the reduction of protons and the oxidation of the corresponding anions occur within 2–4 V vs. Li (while many electrode materials react outside this window leading to increased energy densities). Non-protogenic solvents, shortened to "aprotic" electrolytes, are thus the usual choice, though they display lower conductivities than water (see Fig. 7.5).

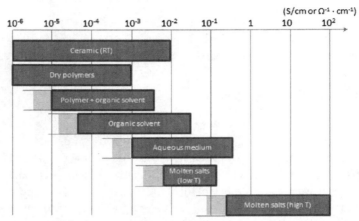

Figure 7.5 Conductivity ranges of different types of electrolyte. Adapted from reference [5].

Electrode materials can react through different reaction mechanisms:

(a) Solution reactions: These reactions involve species that remain in solution during all stages. The most technologically relevant example are redox flow cells.

(b) Intercalation reactions: A guest species is topotactically introduced into a host material without major atomic rearrangement (e.g., H^+ in NiO_2H_x, Li^+ in $Li_4Ti_5O_{12}$). Intercalation reactions can in turn be homogeneous (solid solution) or heterogeneous (two-phase reaction). The former involves the formation of a non-stoichiometric compound whose insertion rate continuously varies throughout the intercalation domain $0 > x > x_{max}$. The host structure does not suffer from major structural changes other than a continuous variation in volume to accommodate the change in composition. For heterogeneous reaction, the closer the two phases are, the better the reversibility.

(c) Displacement reactions: These reactions can be considered as intermediate between conversion and intercalation. While one of the ions in the initial electroactive compound is completely reduced to its elemental state, another ion takes its place in the structure so that the framework is largely preserved. Thus, the reaction is still topotactic. The reaction of Li with Cu_6Sn_5 [6] and the extended family of silver vanadium oxides used in pacemaker batteries [7] are examples of displacement reactions.

(d) Conversion reactions: They occur when the crystalline or molecular structure of the electroactive compound is completely rearranged. The result is a mixture of compounds. A group of reversible conversion reactions corresponds to binary transition metal compounds of the type M_nX_p (X = H, P, N, O, F, S) that react with lithium to yield Li_qX ($0 < q < 3$) and the metal [8], so that reagents and products remain in the solid state. However, these complex reactions can also involve transitions between physical states. These occur at the positive electrode in $Li/SOCl_2$ batteries or during the reaction of an alkali or alkaline earth metal with O_2 to form the corresponding oxidized species [9, 10].

(e) Alloying reactions: These reactions involve the electrochemical formation of an alloy between a metal and another metal or semi-metal. Thus, they could be considered a variation of electroplating where the deposited species is not in pure form, but forming a compound with a different host. Arguably the currently most popular example of these reactions is the alloying of Li with Si [11].

(f) Electroplating and stripping: This mechanism is leveraged in other technological applications, such as in metallurgy. In the context of energy storage, an example of this type of reactions is the lithium metal electrode. When Li is oxidized to Li^+, Li^+ ions are dissolved in the electrolyte. When the reaction is reversed, metallic lithium is redeposited (plated) on the electrode surface. Another example is the lead electrode, but in this case Pb^{+2} cations will react with SO_4^{4-} anions from the electrolyte and precipitate on the surface of the electrode. Since the metal electrode is coupled to the precipitation equilibrium of a metal salt, these are called electrodes of the second kind (as opposed to the former, termed electrodes of the first kind, e.g., lithium metal).

The mechanism of reaction can be initially assessed by a measurement of the change in electrochemical potential. According to the Gibbs phase rule, the number of degrees of freedom F of the system is defined by

$$F = C - P + 2, \tag{7.5}$$

where C is the number of components and P the number of phases. In a homogeneous reaction $F = 3$ ($C = 2$, $P = 1$), while in a heterogeneous reaction that proceeds through the movement of phase boundaries $F = 2$ (e.g., $C = 2$, $P = 2$ in alloying reactions). Considering pressure and temperature as variables, in a heterogeneous reaction there is no degree of freedom left and the electrochemical potential has to be constant. In contrast, the voltage will continuously vary throughout the intercalation of a homogenous reaction, with no phase boundaries formally involved (see Fig. 7.6).

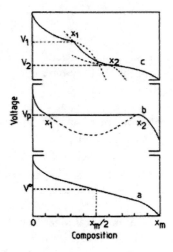

Figure 7.6 Voltage–composition curves for the different types of insertion reactions. (a) Solid solution, with a sloping voltage curve, (b) two-phase reaction, with a flat voltage plateau and (c) compound with different adjacent domains resulting in a multistep reaction. From reference [12].

7.1.3 Methods for Battery Testing

Four basic performance metrics are evaluated during the testing of a battery: capacity, voltage, power, and durability (i.e., cycle life). A variety of electroanalytical tools are also available to correlate these metrics with the mechanism by which the device operates. All these protocols are based on classical electrochemistry techniques [13].

7.1.3.1 Chronopotentiometry

These experiments rely on the application of a constant current to the cell while the evolution of potential is measured. Typically, the current is applied until a cutoff value of potential, followed by cell resting before reversing the polarization. The duration of each constant current step defines the total charge passed in the cell and, thus, its specific capacity. It translates into the composition of the electrodes through Faraday's law (vide supra). Chronopotentiometry is the most widely used tool to evaluate the ability of the materials in the cell to retain their capacity upon extended cycling (oftentimes, thousands of cycles are required for application). Observation of the potential difference between charge and discharge provides an idea not only of the average voltage of the cell, but also of its internal resistance; large differences typically indicate considerable ohmic losses to be minimized. Further, comparison of the capacity obtained upon discharge and charge leads to the coulombic efficiency of the reaction(s) that occur during cycling. This magnitude is extensively used as a first indication of the existence of side reactions in the electrodes, such as those involving electrolyte decomposition. Prolonged coulombic inefficiencies can result in sustained losses of capacity with cycling and, thus, are used to predict cell life.

Chronopotentiometry also affords the opportunity to evaluate the power capability of a cell, and, indirectly, its material constituents. Many protocols are available in the literature to collect this information, and they are commonly referred as rate capability tests. In general, they involve the performance of a finite number of cycles at increasing current densities, and the comparison of the capacity delivered at the highest and lowest rates. It bypasses the need to carry out separate experiments at each rate. An abbreviated version was proposed by Doyle et al. [14]. It involves successive discharges at decreasing rates with a brief relaxation period between each discharge but no charging step. The capacity at each rate is taken as the total amount accumulated in all the preceding discharge steps.

More or less sophisticated variations of these protocols have been put forth by end users of the technologies in order to benchmark new chemistries against internal performance standards [15]. Finally, chronopotentiometry can provide an insight into the mechanism of reaction operating at the electrodes. If a third reference electrode is used or if the activity of one of the two electrodes can be considered largely constant (as is the case of the Li metal electrode), the potential-composition traces, in the limit of negligible ohmic loss, are direct measures of the chemical potential variation in the system of study, so that arguments on the homogeneous character of the reaction can be made. If the ohmic loss is considerable, the current can be applied in constant pulses, typically several minutes long, followed by periods of open circuit. This technique is known as galvanostatic intermittent titration technique (GITT) [16]. The periods of rest are designed to lead to the complete relaxation of the cell, so that a pseudo-OCV value can be acquired. This technique should be applied with caution, as partial relaxation of the electrodes can occur during the rest in certain conditions, resulting in misleading results.

7.1.3.2 Chronoamperometry

These experiments rely on the application of a constant voltage to the cell while the evolution of current is measured. In cases where the redox reaction is highly impeded, holding the voltage of the cell so that there is a significant overpotential is the only means to induce it. In general, chronoamperometry is used in mechanistic studies. The most widely used chronoamperometric technique is potentiostatic intermittent titration technique (PITT) [17]. It involves stepping the potential at very small increments, followed by a holding period while the current reaches very low values, indicating that the reaction is largely complete. As a result, PITT is typically considered to provide close-to-equilibrium values. The resulting profiles, together with the analysis of the current decays, provide powerful hints of the existence of nucleation events or homogeneous transitions. Because they involve potential and current relaxations, both PITT and GITT have also been used to measure diffusion coefficients in materials [18].

7.1.3.3 Cyclic voltammetry

This classical electroanalytical tool relies on the measurement of currents while the potential is swept at a constant rate. Plotting I vs. E reveals the number of redox processes occurring at each sweep (i.e., charge or discharge), which appear as peaks. It is also commonly used to evaluate the ratio of capacitive (electrostatic) vs. faradaic (redox) storage, as the former results in square current–potential curves. Integration of the current with time leads to the total charge passed. In turn, integration of the current at each finite potential value builds a capacity-potential trace that is directly comparable to those obtained in galvanostatic cycling, albeit obtained in very different conditions. Very sharp peaks result in flat traces, thus identifying the existence of first-order (two-phase) phase transitions in the cell. Cyclic voltammetry is also very commonly used to evaluate the kinetics of the system; plotting the potential of a given peak with current serves, for instance, as indication of ohmic vs. mass transport-limited processes.

7.1.3.4 Electrochemical impedance spectroscopy

Electrochemical impedance spectroscopy (EIS) is a tool based on alternate current cycling of an electrochemical cell at very small bias. For extensive details on the technique, the reader is referred to monographs in the literature [19]. Variations in the frequency of the AC probe processes occurring at different timescales, which, in turn, are associated with the magnitude of the electrical resistance to those processes. EIS is a popular tool for battery characterization because it can be used to detect resistance growth in a cell upon extended cycling and directly identify the culprit. It is best used in three-electrode configurations so that the two working electrodes can be monitored separately. Otherwise, the result is a highly convoluted impedance spectrum that can lead to highly misleading conclusions.

7.2 Rechargeable Battery Systems

7.2.1 Lead Acid Batteries

The lead-acid battery was the first practical rechargeable battery to be developed, in 1859, by Gaston Planté. Since then, it has been the most important rechargeable electrochemical storage system for large-scale applications such as automotive, until the advent of lithium-based chemistries. Lead-acid batteries however, still account for more than half of the rechargeable market (in $) [20] and are witnessing increasing demand for stationary applications (grid storage). The long lasting success of this system lies on the good price-performance-life characteristics, with the help of the abundance of lead and its ease of recycling, its excessive weight being its major limitation [21, 22].

Lead-acid batteries are based in the reduction of PbO_2 to $PbSO_4$ and the oxidation of Pb to also $PbSO_4$ in the positive and the negative electrode, respectively, with an output voltage of 2 V. The overall reaction can be written as

$$PbO_2 + Pb + H_2SO_4 \rightarrow 2PbSO_4 + 2H_2O.$$

As described in Section 7.1.2, these processes involve a dissolution-precipitation mechanism. The electrolyte takes part in the reaction, leading to a variation in its properties (density, conductance, and pH) upon battery operation. Indeed, the state of charge is determined by measuring the relative density of the electrolyte, which falls from 1.28 to 1.1 g/cm^3 when discharged [23].

Fundamentally, automotive batteries are very similar to those used 100 years ago, the main differences being related to the materials of construction and its design, leading to improved performances and reduced weight. Most automotive batteries are composed of six series connected cells made of six plates each, to give an output voltage of 12 V.

The lead-acid battery operates outside the stability window of the electrolyte, as shown in Fig. 7.7, which is a scheme of the reactions taking place together with the corresponding equilibrium potentials. These potentials are dependent on the acid concentration. Oxygen is evolved at the positive electrode and hydrogen at the negative, at a rate dependent on temperature, acid concentration and the presence of certain impurities. Conventional lead-acid cells have thus to be vented to release the gases produced from the decomposition of the electrolyte and, as a result, must be used only in upright position to avoid electrolyte leakage. The most successful approach to develop a sealed battery is to recombine oxygen at the negative electrode back to water via the "oxygen cycle" in the so-called valve-regulated-lead-acid battery (VRLA). To do so, the electrolyte is immobilized either by gel formers such as SiO_2 or by using AGM separators (Absorptive Glass Micro-fiber). Therefore, the VRLA battery can be employed in any orientation. In gel batteries oxygen transfer from the positive to the negative occurs through fissures in the gel while in AGM oxygen diffuses through the pores of the separator.

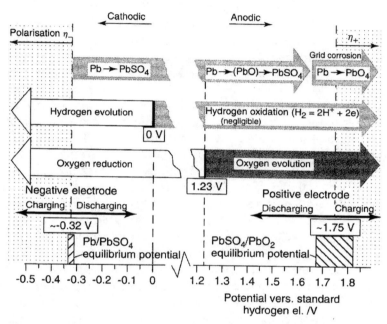

Figure 7.7 Reactions that occur in lead-acid batteries versus electrode potential (thermodynamic situation). Equilibrium potentials of the charge/discharge reactions are represented by columns to indicate their dependence on acid concentration. The inserted values correspond to an acid density of 1.23 g/cm³. Reprinted with permission from [23], John Wiley and Sons.

Most electrodes consist of pasted plates, in which a metallic mesh is coated with active material. The latter is a mixture of PbO and metallic lead, called gray oxide or lead dust. It can be produced by a milling process in which lead pieces are put into a rotary mechanical mill, and the erosion of the pieces causes fine metallic flakes to form. These are oxidized by an airflow, which is also used to remove the formed particles to collection. In another process, the Barton pot process, a fine stream of molten lead is swept around inside a heated vessel, and oxygen from the air reacts with fine droplets to produce an oxide coating around each droplet. In both processes, the gray oxide is further wetted with water and sulfuric acid, resulting in a paste that is pressed into the interstices of the mesh, leading to what is referred to as a battery plate. After a "curing" process during which water is evaporated and lead partially oxidized, the plates are inserted in tanks filled with acid and connected. The active materials are then electrochemically transformed into PbO_2 in the positive electrode and spongy metallic lead in the negative [23]. Typically, a charged positive electrode contains both α-PbO_2 (orthorhombic) and β-PbO_2 (tetragonal), with almost the same equilibrium potential. Neither of the two forms is fully stoichiometric, their composition can be represented by PbO_x, with x varying between 1.85 and 2.05.

The purpose of the mesh is to provide mechanical support for the active material and electronic conductivity between the active material and the cell terminals. Meshes are also made of lead alloys (other more conducting and lighter metals have been considered but corrode in H_2SO_4). They are used instead of foils in order to facilitate impregnation, which requires superior mechanical properties than those of pure lead. Early on Pb–Sb alloys were used as the formation of a eutectic phase at the grain boundaries resulted in improved strength and mechanical resistance of the alloy. However, Sb dissolves (corrodes) during battery charge and deposits on the negative electrode, lowering the overpotential for H_2 evolution. As a consequence, nowadays Pb–Ca alloys are used instead, which show improved conductivities at the expense of worse mechanical properties than Sb counterparts [22].

Acid stratification and sulfation are two common causes of battery failure in lead-acid batteries when they remain partially discharged for long periods. The former occurs as a result of the different densities of water and acid, as the dense concentrated acid accumulates in the base of the cell. This process is restored by diffusion and equilibrium is slowly recovered. Sulfation occurs when $PbSO_4$ crystals form an insulating layer around the battery plates leading to a reduction of the amount of active material. High temperatures or deep discharges can also contribute to sulfation.

7.2.2 Alkaline Rechargeable Batteries

Alkaline rechargeable batteries group all those aqueous batteries that use a basic electrolyte, typically KOH (although NaOH is cheaper, KOH is preferred for its higher conductance). These batteries are more expensive than lead-acid batteries and are typically used in consumer products with low capacity requirements or, more recently, in HEV vehicles and in renewable energy systems. Their success lies in their long cycle life and low maintenance, as well as the fact that they are commercialized in sealed form (although are vented for large capacities) and can operate in a wide temperature range. In many portable applications they have been displaced by Li-ion batteries as the latter exhibit much better energy densities (albeit at a higher price).

The most common positive electrode is the nickel oxyhydroxide electrode (NOE), which can be combined with different negative electrodes such as cadmium, iron, metal hydride (MH) or hydrogen. Nickel-cadmium (NiCad) and nickel-iron batteries have also been in the market for more than a hundred years. Nonetheless, environmental problems associated with Cd have driven manufacturers to replace it by MH since late 1990s, and, as a result, they are now almost obsolete. AgO and MnO_2 are other commercialized positive electrodes although among all alkaline rechargeable batteries, the Ni-MH system is the most widely used. In the following section, the positive and negative electrodes of the Ni/MH battery are described.

The reaction occurring in the positive electrode is based in the transformation of $Ni(OH)_2$ into NiOOH:

$$Ni(OH)_2 \text{ (s)} + OH^- \text{ (aq)} \rightarrow NiOOH + H_2O + e^-$$

This reaction is more complex in practice. Indeed, both $Ni(OH)_2$ and NiOOH exist in two polymorphs and thus four different phases can be involved in the system depending on the reaction conditions. All four phases exhibit lamellar structures and differ in the stacking of the layers, the nickel oxidation state and the amount of intercalated H^+ or other species. The relationship between the different phases was summarized by Bode et al. [24, 25] in the late 1960s and is schematized in Fig. 7.8.

β-$Ni(OH)_2$ is the manufactured product. It is a well-defined compound that exhibits the brucite structure in which slabs of

NiO$_6$ octahedra are stacked in an AB stacking sequence and H$^+$ atoms occupy tetrahedral positions in between the slabs. β-Ni(OH)$_2$ is only electroactive when it is "badly crystallized", that is when it is prepared with a small particle size (in the range of a few tenths of nanometers) and it is believed that structural defects play a beneficial role in its performance [26, 27]. When β-Ni(OH)$_2$ is oxidized, β-NiOOH is obtained by removal of H$^+$ atoms from the structure. The low crystallinity of this material has long hampered its structural characterization and it was believed that the biphasic reaction occurred without structural modifications. However, it has been recently shown that the stacking of the layers is different in β-NiOOH and thus gliding of the layers occurs concomitantly to H$^+$ extraction, leading to an ABCA stacking sequence [28], as shown in Fig. 7.8. This reaction is accompanied by a strong textural transformation in the first cycles that results in the formation of polycrystalline particles by the relaxation of strains [29].

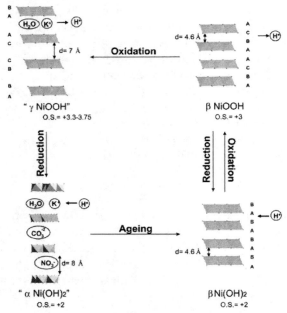

Figure 7.8 Update of the Bode diagram: scheme of the crystal structures of the different phases involved in the nickel electrode and the relationship among them. The nickel oxidation state (O.S.) is indicated for each of them.

β-NiOOH can be further oxidized to form γ-NiOOH. This general stoichiometry designates a variety of compositions in which the nickel oxidation state can oscillate between 3.3 and 3.7 and contains alkaline ions and water molecules in the interlayer spacing [30]. This reaction is accompanied by more severe structural changes than the β/β reaction as not only the stacking sequence changes to ABBCCA (see Fig. 7.8) but also the interlayer spacing increases more than 50%. When reduced, γ-NiOOH transforms into α-Ni(OH)$_2$, a metastable material with a turbostratic structure (layers randomly stacked) and with anionic species, in addition to water and alkaline ions, in the interlayer space [31, 32]. Due to the structural changes involved, the formation of γ-NiOOH and α-Ni(OH)$_2$ quickly degrades the material and this is the reason why, despite providing extra capacity, the α/γ reaction is avoided and commercial batteries operate solely with the β/β reaction, with a theoretical capacity of 289 mAh/g.

Metal hydrides are the materials used in the negative electrode. They lead to a cell voltage of 1.2–1.3 V combined with the nickel oxyhydroxide electrode, with the benefit that a higher energy density is obtained when compared to cadmium. Many metals and alloys are able to reversibly adsorb hydrogen atoms, with different affinity. On one hand, A type materials (rare earths and early period transition metals) form a metal hydride that is stable at ambient conditions and the reaction can only be reverted at high temperature. On the other hand, B elements (high period transition metals) can only form hydrides at high pressure because of their low affinity for hydrogen. Combining both types of element, hydrides of intermediate stability at ambient pressure and temperatures can be obtained. Intermetallic alloys of the type AB_5 are currently used as electrode material, with a capacity of around 300 mAh/g. LaNi$_5$ is the paradigmatic example of AB$_5$ intermetallic alloys, but several drawbacks (cost, corrosion) have led to the empirical development of multielement compositions (LaCePrNdNiCoMnAl) with better thermodynamic stability and corrosion resistance. The partial substitution of Ni by Co, Mn, Al, or other transition metals allows tuning the thermodynamic and electrochemical properties, while the use of Mischmetal (i.e., mixed metal, a combination of rare earth elements) as partial replacement for La has reduced the cost of the material. Other families of intermetallic alloys such as AB_2 or intergrowths of the type $[A_2B_4]\cdot n[AB_5]$ with higher capacities (up to 440 mAh/g) have been developed, but their performance is worse than AB_5. In all these materials hydrogen occupies interstitial positions, therefore in Ni/MH batteries both electrodes operate through the intercalation mechanism [23].

Despite this topotactic mechanism, metal hydrides suffer from strong volumetric changes upon hydrogen uptake (sometimes resulting in amorphization). Therefore they are subject to an activation process prior to their use so as to pulverize them into fine particles by repeated formation and decomposition of the MH, either electrochemically or in gas/solid systems [33].

The operating voltage of these batteries is almost identical to the stability window of the electrolyte. Overcharge of the negative electrode would result in the formation of excess H_2, thereby creating a safety hazard through over-pressurization. Ni-MH batteries are thus limited by the mass of the positive electrode, so that once fully charged, O_2 evolution occurs at the positive electrode, which subsequently diffuses toward the negative and recombines with any excess hydrogen to form water:

$$4\,MH + O_2 \rightarrow 4\,M + 2\,H_2O$$

7.2.3 Lithium Rechargeable Batteries

7.2.3.1 From Li metal to Li-ion

Lithium exhibits two unique characteristics: It has the lowest reduction potential ($E° = -3.04$ V) and is the lightest metal ($M = 46.94$ g mol^{-1}). Therefore, a Li-based battery chemistry has two important benefits: the potential to reach a higher cell voltage (up to about 4 V per cell) with the use of non-aqueous electrolytes and a high energy density (currently up to 210 Wh/kg, 650 Wh/l). In addition to a longer charge retention or shelf life compared to conventional aqueous technologies, these batteries have captured the market despite their higher cost [4, 34].

The first rechargeable Li batteries used a lithium metal anode in combination with intercalation cathodes. However, the success of rechargeable lithium batteries was soon proven to be unviable despite their excellent storage characteristics. The main reason was that lithium has a tendency to grow in the form of dendrites (such as that of Fig. 7.9), which could short the cell and result in serious risk of fire [2, 35], and to react with most electrolyte solvents without passivating, which also leads to highly uneven and inefficient deposition.

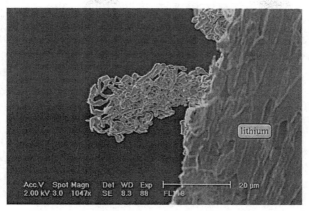

Figure 7.9 Lithium dendrites being formed on electrodeposition of lithium. Reprinted from [36] with permission from Elsevier.

Substitution of lithium metal by another intercalation material like carbon was the most successful approach to circumvent these safety issues. Carbons are lightweight, conducting, and inexpensive. More importantly they are able to accommodate lithium at an anode potential close to that of lithium, thus the energy penalty of replacing lithium by a carbonaceous material is minimized. Because both electrodes operate through an intercalation mechanism, these batteries were referred as rocking chair batteries and are now commonly called Li-ion batteries. As carbon still operates below the HOMO of the electrolyte, another material with better stability, $Li_4Ti_5O_{12}$, is also commercialized.

While there are few choices for negative electrodes, several competing chemistries are currently commercialized as positive electrodes, intended for different applications. A description of the most common chemistries for the different cell components (positive and negative electrodes and electrolyte) is reviewed in the following paragraphs.

7.2.3.2 Negative electrodes

Graphite is still the most commonly used anode in commercial Li-ion batteries. Lithium intercalation in the interlayer space leads to the final nominal composition LiC$_6$ (corresponding to a capacity of 372 mAh/g). The maximum value of 1 Li$^+$ per 6 C atoms results from avoiding occupation in the nearest neighbor positions as shown in Fig. 7.10a, leading to a hexagonal ring. The voltage–composition curve exhibits different potential steps that are related to a staging phenomenon: In order to avoid repulsion between adjacent Li layers, insertion occurs in an ordered manner in which the more distant graphene layers are preferentially occupied before reaching the final composition LiC$_6$ as shown in Fig. 7.10b,c. The reaction proceeds through several two-phase reactions from stage s = IV until stage s = I [37].

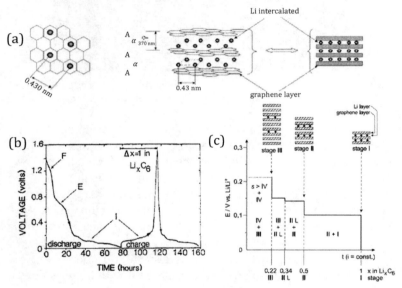

Figure 7.10 (a) Structure of LiC$_6$. Left: view perpendicular to the basal plane of LiC$_6$, Right: schematic drawing showing the AA layer stacking sequence and the $\alpha\alpha$ interlayer ordering of the intercalated lithium. (b) Initial $1\frac{1}{2}$ cycles of a Li/graphite cell. The cell was cycled at a rate of 40 h for $\Delta x = 0.5$ in Li$_x$C$_6$ in 1.0 M LiAsF$_6$ in EC/PC. F denotes the irreversible capacity associated with SEI formation, E the irreversible capacity due to exfoliation, and I the reversible capacity due to lithium intercalation into carbon. (c) Schematic galvanostatic curve of stage formation during electrochemical intercalation of lithium into graphite. Adapted with permission from references [40] and [38].

As shown in Fig. 7.10b, the charge consumed in the first cycle significantly exceeds the theoretical capacity. This phenomenon is ascribed to the formation of a decomposition layer as a result of the reduction of the electrolyte (denoted as F in Fig. 7.10b). Because charges are consumed during this irreversible decomposition, part of the capacity is not recovered in the following discharge. [38]. The extent of the irreversible capacity will depend on the accessible surface area by the electrolyte. After the first cycle, and thanks to the small volume changes undergone by the graphite electrode, the decomposition layer remains largely intact leading to passivation of the surface and coulombic efficiencies (defined as the ratio between discharge and charge capacity) approach 100% in the following cycles [38]. While this surface layer is electronically insulating it conducts ions, thereby allowing them to reach the bulk of the graphite electrode. As a consequence, it is commonly known as solid electrolyte interphase (SEI). In practice the SEI acts as a protective layer and its stability will have a strong impact in the shelf and cycling life of the battery [39]. These unique properties lie at the origin of the commercial success of graphite electrodes, despite the fact that they operate outside the window of stability of the organic electrolytes. Indeed, it was not until electrolytes including ethylene carbonate in their composition were developed that graphite could be used as a practical negative electrode, as only EC provides a stable SEI and avoids exfoliation of graphite (shown as E in Fig. 7.10b) [38].

The composition of the SEI is complex; it contains solvent and solute decomposition products such as Li_2O, $LiOH$, Li_2CO_3, and Li_3N or $(CH_2OCO_2Li)_2$, among others, and depending on the solvent and the lithium salt. The distribution of these components is known to vary greatly in space, especially in depth. It is known to be highly dependent on cycling conditions, but complete control has been elusive and typically involves empirical approaches. As a result, Li-ion cells in the consumer market typically undergo a so-called "formation process" at the manufacture sites, which involves a carefully designed protocol of few cycles and diagnostics [35].

A commercial alternative to carbon anodes is $Li_4Ti_5O_{12}$. This compound has a defect spinel structure that can be written as $Li[Li_{1/3}Ti_{5/3}]O_4$ and can reversibly intercalate lithium to yield the rock-salt type phase $Li_2[Li_{1/3}Ti_{5/3}]O_4$, yielding a capacity of 175 mAh/g. As the reaction takes place above the HOMO of the electrolyte (1.55 vs. Li/Li$^+$), the SEI layer does not form [41]. Another particularity of this material is the fact that the cell volume remains unchaged during the intercalation of lithium, making it a zero strain material. Both caracteristics are strongly beneficial, leading to improved thermal stability and to a competitive performance for high rate designs. On the other hand, because the potential for intercalation is approximately 1.3 V higher than that of graphite, the operating voltage of the cell is compromised and the cell's energy is reduced accordingly [42].

Alloying reactions represent another alternative to graphite. Alloying reactions have been known for about 40 years [43] and some materials are able to react with several moles of lithium per metallic atom leading to binaries of the form Li$_x$M. These reactions deliver huge theoretical capacities but the drastic volumetric changes associated to the reaction result in the pulverization of the electrode material, loss of electrical contact between particles and rapid capacity loss. Several strategies can be adopted to minimize pulverization such as downsizing particle sizes, moving to nanotextured/nanostructured composites, or designing new electrode concepts with matrices and new binders [44]. The most appealing electrode materials are Si, with the highest gravimetric and volumetric capacities (around 3500 mAh g^{-1} and 8500 Ah l^{-1}), Sn and Pb, both with comparable volumetric densities (around 7000 Ah l^{-1} and 6000 Ah l^{-1}, respectively). The first commercial application was in SONY's Nexelion camcorder battery pack in which the anode is a composite of a carbon matrix and an amorphous Sn–Co alloy containing some titanium, in which the coarsening is prevented and is therefore able to cycle well with a 50% increase in volumetric capacity. Silicon seems to be part of the next step as numerous announcements have been made about commercializing silicon anodes [45].

7.2.3.3 Positive electrodes

In batteries using carbon as the negative electrode, the positive side must act as a source of Li, thus requiring use of air-stable Li-based intercalation compounds to facilitate the cell assembly [2]. Therefore, current Li-ion batteries are assembled in the discharged state and must be charged before use.

The first Li-ion commercial battery (Sony 1991) used LiCoO$_2$ [46] as positive electrode. This material exhibits a layered structure in which Co^{+3} and Li^+ occupy alternate layers of octahedral voids in a cubic close packed array of oxide ions. It operates within the range 4.5 to 3.0 V vs. Li with a maximum practical specific capacity of about 180 mAh/g. This value results from the reversible extraction of 0.5 Li^+ per formula unit through a series of structural transformations. Higher capacities could be achieved by further Li^+ extraction but has detrimental effects upon cycle life as a result of structural collapse [47]. Moreover, overcharge leads to electrolyte and cathode decomposition leading to the exothermic formation of acidic species and oxygen, causing thermal runaway of the cells [48].

Isostructural less costly materials such as $LiNiO_2$ were considered as an alternative [49]. This material yields higher capacities (220 mAh/g) albeit the operating voltage of the Ni^{+3}/Ni^{+4} couple is 0.2 V lower than that of the Co^{+3}/Co^{+4} couple in $LiCoO_2$. $LiNiO_2$ usually crystallizes as $Li_{1-y}Ni_{1+y}O_2$ because of the instability of Ni^{+3}. Due to the similar size between Ni^{+2} and Li^+, these atoms often interchange positions in the structure affecting lithium diffusion [50]. Moreover, similarly to $LiCoO_2$, because of the chemical instability of Ni^{+4}, lattice oxide ions can be preferentially oxidized resulting in the formation of oxygen and NiO through an exothermic reaction that seriously compromises cell safety. Replacement of Ni^{+3} by other metals results in more stable materials and several compositions such as $Li[Ni_{0.8}Co_{0.15}Al_{0.05}]O_2$ (NCA) [51] and $Li[Ni_{1/3}Mn_{1/3}Co_{1/3}]O_2$ (NMC) [52] are now a commercial success. The former exhibits a capacity of 180–200 mAh/g at a voltage range of 4.25–3.5 V, but giving off heat as low as 200°C when in contact with the electrolyte so thermal runaway is a real concern with it [45]. The second, less expensive and more stable (does not release heat until around 350°C), has a capacity of 170 mAh/g in the voltage range 3.6–4.7 V that results from oxidation of Ni^{+2} to Ni^{+4} and Co^{+3} to Co^{+4}, while Mn ions remain unchanged [53].

$LiMn_2O_4$ (LMO) is the thermodynamically stable form of Mn oxide and crystallizes in the spinel structure. This material (and its doped variants) is also a well-established commercial lithium-ion battery cathode material. Removal of lithium to form λ-MnO_2 occurs at 4 V vs. Li/Li^+, with a capacity of about 120 mAh/g, which corresponds to the extraction/insertion of 0.8 Li^+ per formula unit. Below 3 V vs. Li/Li^+ additional lithium can be inserted to form the Jahn–Teller distorted tetragonal phase Li_2MnO_4, but the strong distortion this material exhibits strongly affects stability and the reaction is therefore avoided [54]. While the $LiMn_2O_4$ spinel has several advantages (such as an excellent power capability thanks to the 3D diffusion of lithium ions within the structure, the fact that Mn is inexpensive and environmentally benign and the better thermal stability), its weakness is capacity fading. Its origin is Mn dissolution in the electrolyte through the disproportionation reaction $2Mn^{+3}$ -> $Mn^{+2} + Mn^{+4}$ causing a progressive capacity fade as Mn^{+2} leaches out into the electrolyte in the presence of trace amounts of HF (formed by decomposition of $LiPF_6$ in the electrolyte—see below) [55]. Diffusion of Mn^{+2} toward the negative electrode might also result in safety hazards as the resistance of the SEI increases and causes a local increase of temperature that can turn into a thermal runaway. It has been shown that substitution of Mn by Li [56] or Al [57] in order to reduce the Mn^{+3}: Mn^{+4} ratio or substitution of O by F greatly improves the stability of the material. Another strategy is to protect the material with a surface coating of inert oxides such as Al_2O_3 in order to avoid direct contact with the electrolyte [23].

$LiFePO_4$ is now considered one of the best available positive electrode because of its cycling performance, safe operation, low cost and non-toxicity. This material crystallizes in the olivine structure from which Li ions can be reversibly extracted yielding a capacity of 170 mAh/g [58]. The reaction occurs through a biphasic mechanism between $LiFePO_4$ and $FePO_4$, resulting in a flat voltage at 3.5 V vs. Li/Li^+. Although the Fe^{+2}/Fe^{+2} couple in a simple oxide would normally operate at a voltage < 2.5 V vs. Li/Li^+, polyanion hosts of the XO_4^{-n} type (X = S, P, Mo, W) induce a lowering of the Fe^{+3}/Fe^{+2} redox energy. This phenomenon, called the inductive effect, is attributed to the influence of the strength of the X–O bonding on the covalency of the Fe–O bond (and thereby on the position of the Fermi level). When the X–O bond is the strongest, the cell voltage is higher [58]. In order to overcome the capacity limitations that result from the poor transport of electrons and ions in $LiFePO_4$, two strategies are generally used in combination. The first one consists in the coating the particles of $LiFePO_4$ with carbon in order to eliminate the obstacle of electronic conduction. The second one refers to the reduction of the particle size which results in shorter diffusion lengths for Li ions, an approach that has a strong impact in this material since LiFePO$_4$ is a one dimensional Li-ion conductor [59].

7.2.3.4 Electrolytes

In order to dissolve sufficient amounts of salt, only solvents with polar groups and high dielectric constants can be considered and they must be able to solvate the cation (Li$^+$). Organic liquid electrolytes based on carbonate solvents are the usual choice in Li-ion batteries (although polymer electrolytes working at > 50°C are also commercial and can be used with metallic lithium anodes). Most of those that have been employed are cyclic esters (ethylene carbonate, propylene carbonate), linear esters (methyl formate, methyl acetate, dimethyl carbonate), cyclic ethers (dioxolane) and linear ethers (dimethoxyethane). Besides viscosity and conductivity, the choice of the solute-solvent combination will be highly determined by the particular chemistry of the interface reaction and the passivating layer stability. Moreover, a mixed solvent is often preferred to fulfill diverse and sometimes contradicting requirements. As an example ethylene carbonate, which is the most popular solvent, can be combined with a linear carbonate like dimethyl carbonate to widen the electrochemical stability up to 5.0 V, or with low viscosity solvents (more fluid) such as dimethoxyethane to ensure adequate ion transport [35, 47]. Neither co-solvents alone have these advantages. A typically used salt is LiPF$_6$, which offers well-balanced properties such as high conductivity with little ion-pairing, solubility and formation of an effective SEI. However, thermal instability and decomposition to the highly resistive LiF and PF$_5$ (which in the presence of traces of moisture produce a series of corrosive products such as HF) are important drawbacks [60]. Despite a rather limited choice due to the low solubility/dissociation of lithium salts, many solutes have been investigated as replacement such as LiClO$_4$, although hazardous; LiAsF$_6$, but toxic; LiBF$_4$, with moderate ion conductivity and thermally unstable; LiCF$_3$SO$_3$ (LiTf), also with poor ion conductivity and intensive aluminum current collectors corrosion, LiN(SO$_2$CF$_3$)$_2$ (LiTFSI) with the same handicap despite its conversely high conductivity or lithium bis(oxalato)borate (LiBOB) which protects graphite from exfoliation even in PC, among others [35]. Although none has been able to meet all the requirements simultaneously like does $LiPF_6$, some of them show great promise, are still being intensively investigated and are being used increasingly as additives. Intensive research is also currently being devoted to other forms of electrolyte such as ceramic, polymeric or room temperature molten salts (ionic liquids).

7.2.3.5 ... and back to Li metal

The technological push toward the design of battery-based vehicles has brought an intense interest in leaps in the energy storage capability of current technologies. While Li-ion continues to be the battery of choice in commercial applications, [61] scientists decided to revisit lithium metal-based chemistries in hopes of enabling the desired increases in energy density. The use of Li^0 as the anode creates the opportunity of employing positive electrodes that are Li-free, unlike existing Li-ion couples. Among them, the choices attracting the highest scrutiny are O_2 and S_8 due to their very high theoretical capacities. The mechanisms operating these electrodes resemble in a general sense, as they both involve reactions through states of matter, as will be discussed below. They both also result in species that can severely poison the Li metal electrode, potentially leading to extremely hazardous reactions. Together with the instability of the metal toward the majority of electrolyte solvents, these issues pose a tremendous challenge. The most successful strategy to circumvent them was found when the use of protected lithium electrodes (PLE) was first proposed [62]. This concept is based on use of multi-phase electrolytes, one of them being solid and forming a stable interface with Li metal. This strategy is being very actively pursued by many research groups and has already led to the first prototypes of rechargeable cells.

In non-aqueous solvents, O_2 gas reacts with Li to form solid Li_2O_2, through the formation of LiO_2 as intermediate. The electrically insulating properties of the peroxide have completely precluded the demonstration of a full four-electron electrochemical cell through the formation of Li_2O. While the reaction is theoretically simple, in practice, it is plagued with humongous challenges, larger even than those of the oxygen reduction and evolution processes (OER, ORR) in aqueous solvents, which have been a classical object of study in the field of electrocatalysis [63].

These challenges have recently been extensively reviewed [9, 64]. To start up with, in order to prevent any contact between O_2 and Li, PLEs are required. While formation of Li_2O_2 has been reported in several cases, side reactions also occur involving the active species, the electrolyte and/or the electrode support that lead to by-products such as Li_2CO_3. The formation of insulating products also limits the total capacity of the cell, as the reaction is self-limiting and stops when the active sites are covered. Solvent instabilities also come into play during the OER, so that high electrochemical efficiency remains to be demonstrated. Finally, huge potential differences between ORR and OER (hysteresis) also produce a significant penalty in roundtrip efficiency. Lower hysteresis has been reported in systems based on other alkali metals, such as Na and K [65, 66] because the reaction stops with the formation of NaO_2 and KO_2, respectively. However, these superoxide species are still very reactive and do not bypass the problem of electrolyte instabilities. Some groups are pursuing devices that are based on aqueous electrolytes by taking advantage of PLEs. The main advantage is that they involve the formation of LiOH, which is soluble in water so that the storage capacity is limited by the saturation point of the solution, and, thus, potentially higher than in non-aqueous cells. The trade-off is a lower cell voltage, leading to lower theoretical energy storage capability. Also, the need to recharge the cell creates issues with the poor efficiency of the OER. Finally, engineering considerations also make the road to practical O_2-based batteries in electric vehicles long and winding; they relate to the need to install gas diffusion layers in a complicated system to purify the gas from air or the alternative installation of potentially explosive gas tanks on board.

The reduction of solid S_8 by Li also results in solid (and also insulating) Li_2S, but intermediate polysulfides are formed that are highly soluble in a large variety of solvents. Because these compounds would also be insulating in solid form, their soluble nature is, in practice, the key to achieving the full capacity of the sulfur electrode. However, they still complicate cell design. These dissolution-precipitation reactions induce shape changes at the positive electrode that result in isolated domains of insulating sulfur species. The soluble polysulfides act as shuttles in the cell and can diffuse to the counter electrode, where they react to form insulating layers. The use of PLEs bypasses this latter problem at the expense of a higher internal resistance of the cell as a result of multiple electrolyte phases, but is justified by the need to control the dissolution and precipitation close to the positive electrode structure.

7.3 Beyond Li-Ion: From Single to Multivalent Ion Chemistries

The limitations encountered with Li-ion batteries have also pushed research interests in a different direction. By leveraging the knowledge created during 30 years of development, scientists are turning their interest toward systems based on other ions, the most prominent being Na^+ and Mg^{2+}. The reader is referred to specialized reviews available in the literature for extensive discussions of these technologies [67, 68], and only a general overview will be presented here. As for Li, in principle, both species can be incorporated in systems based on intercalation chemistries as well as in metal/air (O_2) or sulfur.

The case for Na-based chemistries is driven by the much higher abundance and lower price of the ores containing this alkali metal compared to Li. It also opens the possibility of finding couples that offer higher storage capacity because of differences in solid-state chemistry between compounds with these two ions. Finally, minor advantages such as the ability to use less expensive current collectors such as Al in both electrodes are also under consideration. The main disadvantage is that there is no clear theoretical path toward devices that can really surpass the energy density of Li ion, especially when considering that a small penalty in cell voltage typically occurs when the equivalent couples are used, due to the slightly higher potential of reduction of Na than Li.

Devices based on Mg present the attractive that the ionic carrier, Mg^{2+}, has twice the charge density of Li^+. Thus, twice as many electrons can be moved in the cell while keeping the mass transport constant. Since electrode instabilities in Li-ion chemistries are driven by a "hollowing out" of the crystal structure upon removal of large amounts of Li to achieve high capacity, it is possible that the same structures could be cycled in a reversible manner should Mg^{2+} be the shuttle species instead. The high charge density of the ion is also the main obstacle toward practical application. It results in large barriers to ion mobility in the host structure as well as in the electrolyte phases, so that the redox reaction is highly impeded. Very few materials have been shown to reversibly accommodate Mg [69]. In order to produce a clear advantage over Li-ion batteries, the use of Mg metal as the anode is proposed. Mg^0 is more stable toward certain solvents and plating is more homogeneous than Li^0, so that, in principle, the barriers to practicality are lower. However, Mg^0 also forms an extremely stable MgO passivation layer on its surface. Its low conductivity requires etching to expose fresh metal surfaces. Highly acidic electrolyte compounds, such as Grignard reagents, are typically used to remove this layer, but these species are also very reactive toward many possible positive electrode materials, such as oxides. In the end, all these limitations have reduced the choice of couples to those that involve sulfur-containing positive electrodes [69, 70].

7.4 Redox Flow Batteries

The need for storage devices that can be integrated in an electrical grid powered by renewable energy sources has created a technological market with specific demands [71]. In this case, energy density is less critical than in automotive or portable electronics applications, and, instead, the design rules are dominated by the cost of storing the energy. The obvious consequence is the choice of aqueous-based cells, which is why legacy lead-acid and Ni-based batteries are finding a niche. In applications where the load varies considerably depending on the time of the day, the ability to deliver energy fast (i.e., high power technologies) is another constraint. Redox flow cells offer an elegant way of addressing these power requirements. They are based on flowing solutions containing electroactive solutes through a small electrode stack that is composed of porous metal electrodes (Fig. 7.11) [71]. The main consequence of this design is that the energy storage capability of the battery is defined by the size of the tanks that hold catholyte and anolyte solutions, whereas the power density depends on the electrode stack size (more specifically, its surface areas). Typically, both redox forms of the electroactive species are soluble in the electrode. As a result, no mass transport occurs between the solid metal electrode and the catholyte/anolyte, eliminating one of the most common sources of power limitations in other devices. The fact that the electrochemical reactions always occur in the liquid state also bypasses a common mechanism of fatigue in batteries based on solid-state processes: that of mechanical cycling and degradation. However, the active solutions in the tanks cannot communicate, so they are flown through separate systems, and membranes are needed to prevent crossover while allowing inactive ionic charge-carriers to transport through to balance the movement of electrons between electrodes. Finding appropriate materials and configurations for the latter is probably the single most pressing technical need to fully enable the penetration of redox flow batteries in the market, as crossover is the most prominent mechanism of failure and the most effective membranes also significantly raise the cost of the device.

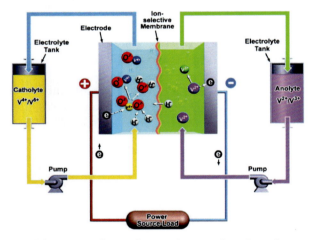

Figure 7.11 Schematic of an all vanadium redox flow battery as an example of redox flow batteries. Reprinted with permission from [71]. Copyright (2011) American Chemical Society.

In principle, any redox couple that involves soluble species could be used in a redox flow cell. In practice, couples are sought that combine cheap and abundant elements with high charge-to-mass ratios (i.e., energy density). These requirements inevitably lead to couples where ions in high oxidation states are involved. These ions tend to be highly unstable in aqueous solutions unless very low pH are used [72] because they undergo Lewis acid-base reactions with H_2O and OH^- to form oxidic species that are much less soluble. As a result, the solutions that are typically used contain highly corrosive species, which pose design constraints to ensure the sustained integrity of the containers, pumps and flow systems. Redox flow batteries are increasingly the focus of research efforts, so that innovative new concepts are continuously emerging. These concepts go from variations in the redox couples, including metal plating [73] to flowable solid dispersions of classical intercalation electrodes in very high concentrations [74]. As interest grows, it is expected that a parallel increase in the fundamental understanding of all the complex processes involved in this technology will come about.

Acknowledgments

MCC wishes to acknowledge financial support by the Departamento de Desarrollo Económico y Competitividad of the Gobierno Vasco as well as Dr. M. Armand for critical reviewing part of this chapter. JC wishes to acknowledge financial support by the Assistant Secretary for Energy Efficiency and Renewable Energy, Office of Vehicle Technologies of the U.S. Department of Energy, under Contract DE-AC02-05CH11231, as part of the Batteries for Advanced Transportation Technologies (BATT) Program. The program is managed by LBNL for the Department of Energy.

References

1. J. Frederic Daniell. On voltaic combinations. *Philos. Trans. R. Soc. Lond.*, **126**, 107–124, 1836.
2. J. M. Tarascon and M. Armand. Issues and challenges facing rechargeable lithium batteries. *Nature*, **414**(6861), 359–367, 2001.
3. J. B. Goodenough and Y. Kim. Challenges for rechargeable Li batteries. *Chem. Mater.*, **22**(3), 587–603, 2010.
4. D. Linden and T. B. Reddy, eds. *Handbook of Batteries*, 3rd ed., McGraw-Hill, 2002) and therefore the number of each reference in the text does not need to be changed (but needs to be changed in the list of references).
5. Techniques de l'ingénieur, editor. *Accumulateurs d'Énergie Portable*, volume base documentaire: 42243210. Editions T.I., 2013. fre.

6. K. D. Kepler, J. T. Vaughey, and M. M. Thackeray. $Li_xCu_6Sn_5$ ($0 < x < 13$): An intermetallic insertion electrode for rechargeable lithium batteries. *Electrochem. Solid-State Lett.*, **2**(7), 307–309, 1999.
7. K. J. Takeuchi, A. C. Marschilok, S. M. Davis, R. A. Leising, and E. S. Takeuchi. Silver vanadium oxides and related battery applications. *Coordination Chem. Rev.*, **219–221**(0), 283–310, 2001.
8. P. Poizot, S. Laruelle, S. Grugeon, L. Dupont, and J. M. Tarascon. Nano-sized transition-metaloxides as negative-electrode materials for lithium-ion batteries. *Nature*, **407**(6803), 496–499, 2000.
9. G. Girishkumar, B. McCloskey, A. C. Luntz, S. Swanson, and W. Wilcke. Lithium-air battery: Promise and challenges. *J. Phys. Chem. Lett.*, **1**(14), 2193–2203, 2010.
10. N. Marincic. Lithium batteries with liquid depolarizers. In (R. E. White, J. O'M. Bockris, and B. E. Conway, eds.), *Modern Aspects of Electrochemistry*, Springer US, pp. 167–233, 1983.
11. C. M. Park, J. H. Kim, H. Kim, and H. J. Sohn. Li-alloy based anode materials for Li secondary batteries. *Chem. Soc. Rev.*, **39**(8), 3115–3141, 2010.
12. G.-A. Nazri and C. Julien eds. *Solid State Batteries: Materials Design and Optimization*. Kluwer, 1997.
13. A. J. Bard and L. R. Faulkner. *Electrochemical Methods: Fundamentals and Applications*. Wiley, 1980.
14. M. Doyle, J. Newman, and J. Reimers. A quick method of measuring the capacity versus discharge rate for a dual lithium-ion insertion cell undergoing cycling. *J. Power Sources*, **52**(2), 211–216, 1994.
15. http://www.uscar.org/guest/article_view.php?articles_id=86.
16. W. Weppner and R. A. Huggins. Determination of the kinetic parameters of mixed-conducting electrodes and application to the system Li_3Sb. *J. Electrochem. Soc.*, **124**(10), 1569–1578, 1977.
17. C. John Wen, B. A. Boukamp, R. A. Huggins, and W. Weppner. Thermodynamic and mass transport properties of "lial". *J. Electrochem. Soc.*, **126**(12), 2258–2266, 1979.
18. C. Delacourt, M. Ati, and J. M. Tarascon. Measurement of lithium diffusion coefficient in Li_yFeSO_4F. *J. Electrochem. Soc.*, **158**(6), A741–A749, 2011.
19. M. E. Orazem and B. Tribollet. *Electrochemical Impedance Spectroscopy*. The ECS Series of Texts and Monographs. Wiley, 2011.
20. http://www.pikeresearch.com/wordpress/wp-content/uploads/2012/12/ALAB 12-Executive-Summary.pdf.

21. P. Ruetschi. Review on lead-acid-battery science and technology. *J. Power Sources*, **2**(1), 3–24, 1977.
22. R. M. Dell, D. Antony, and J. Rand. *Understanding Batteries*. RSC Paperbacks. The Royal Society of Chemistry, 2001.
23. C. Daniel and J. O. Besenhard, eds. *Handbook of Battery Materials*. Wiley-VCH Verlag GmbH & Co. KGaA, 2011.
24. H. Bode, K. Dehmelt, and J. Witte. Nickel hydroxide electrodes 2. Oxidation products of nickel (2) hydroxides. *Zeitschrift Fur Anorganische Und Allgemeine Chemie*, **366**(1–2), 1–21, 1969.
25. H. Bode, K. Dehmelt, and J. Witte. Zur kenntnis der nickelhydroxidelektrode. i. Über das nickel (ii)-hydroxidhydrat. *Electrochim. Acta*, **11**(8), 1079–1087, 1966.
26. S. U. Falk and A. J. Salkind. *Alkaline Storage Batteries*. Electrochemical Society series. Wiley, 1969.
27. M. Casas-Cabanas, J. Rodriguez-Carvajal, J. Canales-Vazquez, and M. R. Palacin. New insights on the microstructural characterisation of nickel hydroxides and correlation with electrochemical properties. *J. Mater. Chem.*, **16**(28), 2925–2939, 2006.
28. M. Casas-Cabanas, J. Canales-Vazquez, J. Rodriguez-Carvajal, and M. R. Palacin. Deciphering the structural transformations during nickel oxyhydroxide electrode operation. *J. Am. Chem. Soc.*, **129**(18), 5840–5842, 2007.
29. A. Delahaye Vidal, B. Beaudoin, and M. Figlarz. Textural and structural studies on nickel-hydroxide electrodes: 1. crystallized nickel-hydroxide materials submitted to chemical and electrochemical redox cycling. *Reactivity Solids*, **2**(3), 223–233, 1986.
30. H. Bartl, H. Bode, G. Sterr, and J. Witte. Nickel hydroxide electrode: 4. Crystal structure study of highly oxidized gamma1 nickel hydroxide. *Electrochim. Acta*, **16**(5), 615–621, 1971.
31. S. Le Bihan, J. Guenot, and M. Figlarz. Crystallogenesis of nickel hydroxide $Ni(OH)_2$. *Comptes Rendus Hebdomadaires Des Seances De L Academie Des Sciences Serie C*, **270**(26), 2131–3133, 1970.
32. S. Le Bihan and M. Figlarz. Croissance de l'hydroxyde de nickel $Ni(OH)_2$ à partir d'un hydroxyde de nickel turbostratique. *J. Cryst. Growth*, **13/14**(0), 458–461, 1972.
33. R. A. Huggins. *Advanced Batteries. Materials Science Aspects*, Springer, 2009.
34. J. M. Tarascon. Key challenges in future Li-battery research. *Philos. Trans. R. Soc. A Math. Phys. Eng. Sci.*, **368**(1923), 3227–3241, 2010.

35. K. Xu. Nonaqueous liquid electrolytes for lithium-based rechargeable batteries. *Chem. Rev.*, **104**(10), 4303–4417, 2004.
36. F. Orsini, A. Du Pasquier, B. Beaudoin, J. M. Tarascon, M. Trentin, N. Langenhuizen, E. De Beer, and P. Notten. In situ scanning electron microscopy (SEM) observation of interfaces within plastic lithium batteries. *J. Power Sources*, **76**(1), 19–29, 1998.
37. M. Winter and R. J. Brodd. What are batteries, fuel cells, and supercapacitors? *Chem. Rev.*, **104**(10), 4245–4269, 2004.
38. R. Fong, U. Vonsacken, and J. R. Dahn. Studies of lithium intercalation into carbons using nonaqueous electrochemical-cells. *J. Electrochem. Soc.*, **137**(7), 2009–2013, 1990.
39. P. Verma, P. Maire, and P. Novak. A review of the features and analyses of the solid electrolyte interphase in Li-ion batteries. *Electrochim. Acta*, **55**(22), 6332–6341, 2010.
40. M. Winter, J. O. Besenhard, M. E. Spahr, and P. Novak. Insertion electrode materials for rechargeable lithium batteries. *Adv. Mater.*, **10**(10), 725–763, 1998.
41. M. Hirayama, K. Kim, T. Toujigamori, W. Cho, and R. Kanno. Epitaxial growth and electrochemical properties of $Li_4Ti_5O_{12}$ thin-film lithium battery anodes. *Dalton Trans.*, **40**(12), 2882–2887, 2011.
42. J. Christensen, V. Srinivasan, and J. Newman. Optimization of lithium titanate electrodes for high-power cells. *J. Electrochem. Soc.*, **153**(3), A560–A565, 2006.
43. A. N. Dey. Electrochemical alloying of lithium in organic electrolytes. *J. Electrochem. Soc.*, **118**(10), 1547–1549, 1971.
44. D. Larcher, S. Beattie, M. Morcrette, K. Edstroem, J. C. Jumas, and J. M. Tarascon. Recent findings and prospects in the field of pure metals as negative electrodes for Li-ion batteries. *J. Mater. Chem.*, **17**(36), 3759–3772, 2007.
45. M. S. Whittingham. History, evolution, and future status of energy storage. *Proc. IEEE*, **100**, 1518–1534, 2012.
46. K. Mizushima, P. C. Jones, P. J. Wiseman, and J. B. Goodenough. Li_xCoO_2 ($0 < x < 1$): A new cathode material for batteries of high energy density. *Mater. Res. Bull.*, **15**(6), 783–789, 1980.
47. C. Vincent and B. Scrosati, eds. *Modern Batteries. An Introduction to Electrochemical Power Sources*, 2nd ed., Butterworth-Heinemann, 1997.
48. D. D. MacNeil and J. R. Dahn. The reaction of charged cathodes with nonaqueous solvents and electrolytes: I. $Li_{0.5}CoO_2$. *J. Electrochem. Soc.*, **148**(11), A1205–A1210, 2001.

49. J. R. Dahn, U. von Sacken, and C. A. Michal. Structure and electrochemistry of $Li_{1\pm y}NiO_2$ and a new Li_2NiO_2 phase with the $Ni(OH)_2$ structure. *Solid State Ionics*, **44**(1-2), 87–97, 1990.

50. A. Rougier, P. Gravereau, and C. Delmas. Optimization of the composition of the $Li_{1-z}Ni_{1+z}O_2$ electrode materials: Structural, magnetic, and electrochemical studies. *J. Electrochem. Soc.*, **143**(4), 1168–1175, 1996.

51. Tsutomu Ohzuku, Atsushi Ueda, and Masaru Kouguchi. Synthesis and characterization of $LiAl_{1/4}Ni_{3/4}O_2$ (R3m) for lithium-ion (shuttlecock) batteries. *J. Electrochem. Soc.*, **142**(12), 4033–4039, 1995.

52. T. Ohzuku and Y. Makimura. Layered lithium insertion material of $LiCo_{1/3}Ni_{1/3}Mn_{1/3}O_2$ for lithium-ion batteries. *Chem. Lett.*, (7), 642–643, 2001.

53. B. L. Ellis, K. T. Lee, and L. F. Nazar. Positive electrode materials for Li-ion and Li-batteries. *Chem. Mater.*, **22**(3), 691–714, 2010.

54. M. M. Thackeray, W. I. F. David, P. G. Bruce, and J. B. Goodenough. Lithium insertion into manganese spinels. *Mater. Res. Bull.*, **18**(4), 461–472, 1983.

55. D. H. Jang, Y. J. Shin, and S. M. Oh. Dissolution of spinel oxides and capacity losses in 4 V $Li/Li_xMN_2O_4$ coils. *J. Electrochem. Soc.*, **143**(7), 2204–2211, 1996.

56. R. J. Gummow, A. de Kock, and M. M. Thackeray. Improved capacity retention in rechargeable 4 V lithium/lithium-manganese oxide (spinel) cells. *Solid State Ionics*, **69**(1), 59–67, 1994.

57. Y. S. Lee and M. Yoshio. Effect of Mn source and peculiar cycle characterization for $LiAl_{0.1}Mn_{1.9}O_4$ material in the 3 V region. *Electrochem. Solid-State Lett.*, **4**(10), A155–A158, 2001.

58. A. K. Padhi, K. S. Nanjundaswamy, and J. B. Goodenough. Phospho-olivines as positive-electrode materials for rechargeable lithium batteries. *J. Electrochem. Soc.*, **144**(4), 1188–1194, 1997.

59. J. M. Tarascon. Key challenges in future Li-battery research. *Phil. Trans. R. Soc. A*, **368**(3227), DOI: 10.1098/rsta.2010.0112, 2010.

60. Y. Sasaki. Organic electrolytes of secondary lithium batteries. *Electrochemistry*, **76**(1), 2–15, 2008.

61. F. T. Wagner, B. Lakshmanan, and M. F. Mathias. Electrochemistry and the future of the automobile. *J. Phys. Chem. Lett.*, **1**(14), 2204–2219, 2010.

62. S. J. Visco and Y. S. Nimon, and B. D. Katz. US Patent 2008/0057386.

63. J. Lee, B. Jeong, and J. D. Ocon. Oxygen electrocatalysis in chemical energy conversion and storage technologies. *Curr. Appl. Phys.*, **13**(2), 309–321, 2013.

64. Alexander Kraytsberg and Yair Ein-Eli. Review on Li-air batteries: Opportunities, limitations and perspective. *J. Power Sources*, **196**(3), 886–893, 2011.

65. P. Hartmann, C. L. Bender, M. Vracar, A. K. Durr, A. Garsuch, J. Janek, and P. Adelhelm. A rechargeable room-temperature sodium superoxide (NaO_2) battery. *Nat. Mater.*, **12**(3), 228–232, 2013.

66. X. Ren and Y. Wu. A low-overpotential potassium oxygen battery based on potassium superoxide. *J. Am. Chem. Soc.*, **135**(8), 2923–2926, 2013.

67. S.-W. Kim, D.-H. Seo, X. Ma, G. Ceder, and K. Kang. Electrode materials for rechargeable sodium-ion batteries: Potential alternatives to current lithium-ion batteries. *Adv. Energy Mater.*, **2**(7), 710–721, 2012.

68. N. Amir, Y. Vestfrid, O. Chusid, Y. Gofer, and D. Aurbach. Progress in nonaqueous magnesium electrochemistry. *J. Power Sources*, **174**(2), 1234–1240, 2007. 13th International Meeting on Lithium Batteries.

69. E. Levi, Y. Gofer, and D. Aurbach. On the way to rechargeable mg batteries: The challenge of new cathode materials. *Chem. Mater.*, **22**(3), 860–868, 2010.

70. H. S. Kim, T. S. Arthur, G. D. Allred, J. Zajicek, J. G. Newman, A. E. Rodnyansky, A. G. Oliver, W. C. Boggess, and J. Muldoon. Structure and compatibility of a magnesium electrolyte with a sulphur cathode. *Nat. Commun.*, **2**, 427, 2011.

71. Z. Yang, J. Zhang, M. C. W. Kintner-Meyer, X. Lu, D. Choi, J. P. Lemmon, and J. Liu. Electrochemical energy storage for green grid. *Chem. Rev.*, **111**(5), 3577–3613, 2011.

72. M. Pourbaix. *Atlas of Electrochemical Equilibria in Aqueous Solutions*. National Association of Corrosion Engineers, 1974.

73. Y. Wang, P. He, and H. Zhou. Li-redox flow batteries based on hybrid electrolytes: At the cross road between Li-ion and redox flow batteries. *Adv. Energy Mater.*, **2**(7), 770–779, 2012.

74. M. Duduta, B. Ho, V. C. Wood, P. Limthongkul, V. E. Brunini, W. C. Carter, and Y.-M. Chiang. Semi-solid lithium rechargeable flow battery. *Adv. Energy Mater.*, **1**(4), 511–516, 2011.

Chapter 8

Environmentally Friendly Supercapacitors

Ana Karina Cuentas-Gallegos,[a] Daniella Pacheco-Catalán,[b] and Margarita Miranda-Hernández[a]

[a]*Instituto de Energías Renovables Universidad Nacional Autónoma de México, Privada Xochicalco S/N Col. Centro, AP 34, Temixco, Morelos, 62580, México*
[b]*Unidad de Energía Renovable, Centro de Investigación Científica de Yucatán A.C., Calle 43 No. 130 Chuburná de Hidalgo, Mérida, Yucatán, C.P. 97200, México*

akcg@ier.unam.mx

8.1 Introduction

Supercapacitors (SC), also known as electrochemical capacitors or ultracapacitors, are energy storage devices that have the same basic components than a battery: two electrodes, an electrolyte, and a separator. These devices are thought to cover the gap between traditional electrostatic capacitors with high specific power (up to 10,000 kW/kg) and low specific energy (up to 0.08 Wh/kg) [1], and batteries with higher specific energy (20–200 Wh/kg) [2]. Most of commercial available SC are constructed with a symmetric configuration using high-surface-area carbons as the active electrode material, and an organic or non-aqueous electrolyte. The main disadvantages of these SC are the electrolyte

---

Materials for Sustainable Energy Applications: Conversion, Storage, Transmission, and Consumption
Edited by Xavier Moya and David Muñoz-Rojas
Copyright © 2016 Pan Stanford Publishing Pte. Ltd.
ISBN 978-981-4411-81-3 (Hardcover), 978-981-4411-82-0 (eBook)
www.panstanford.com

toxicity and their low specific energy (5–10 Wh/Kg) [2, 3]. More recently, aqueous electrolytes and hybrid or asymmetric SC are being introduced slowly to the market in order to overcome these disadvantages, and move to a more environmentally friendly device (Cellergy Israel, Elton Super Capacitors, Inmatech).

This chapter is focused on promoting these environmentally friendly devices based on aqueous electrolytes, using an asymmetric configuration, and promote the use of less toxic components during assembly (current collectors, separators, and binders) [4]; in order to move forward to a greener device with high specific energy, without compromising the characteristic high specific power of >10 kW/kg [2]. Basic concepts of energy storage devices, a background of SC, and a detail description on how energy is stored are explained. In addition, an inclusive classification is proposed where electrode materials as well as the electrolyte type, storage mechanism, and cell configuration is presented and discussed. Finally, the construction and characterization of electrode materials and lab-scale cell devices are presented, followed by a discussion on how to improve this type of device based on an environmentally friendly supercapacitor (EFSC) concept.

8.2 Energy Storage Devices

Nowadays, global warming and the continuous decrease of fossil fuels are forcing societies to move toward the development and use of renewal and sustainable energy sources. As a result the increase in renewal energy from the sun and the wind has been observed, as well as the development of electric vehicles (EV) and hybrid electric vehicles (HEV) with lower CO_2 emissions. Nevertheless, since the sun does not shine at night and the wind does not blow when we want it to, and we expect to drive our EV or HEV at least for many hours, energy storage devices are of great importance for the development of these technologies.

Energy storage devices can be described in terms of specific energy (Wh/kg) and power (W/kg) as in the well-known Ragone plot (Fig. 8.1). Even though the cycling performance is not revealed in this Ragone plot, it serves the purpose of comparing the performance of a range of energy storage devices, such as conventional capacitors, supercapacitors (SC), batteries (B), and fuel cells (FC). Conventional capacitors have the ability to give energy in the smallest time range giving way to the highest specific power (up to 10,000 kW/kg) of all energy storage devices, but their specific energy (up to 0.08 Wh/kg) is the lowest [1]. On the other hand, B and FC have the ability to give the maximum amount of charge resulting in devices with the highest specific energy, but the time required to obtain this energy is too long, resulting in the lowest specific power (down to 100 W/kg). For example, Li ion B shows a greater specific energy value (120–170 Wh/kg), followed by Ni metal hydroxide (40–100 Wh/kg), and Pb-acid B (20–25 Wh/kg) [5]. It must be clarified that FC are not energy storage devices per se, since they do not store any charge. Instead, they are more like an energy convertor, since fuels like hydrogen or alcohol injected in the cathode and oxygen in the anode, results in a chemical reaction that provides energy. Nevertheless, FC are included in Fig. 8.1 in terms of their capability in delivering energy. Finally, SC fill in the gap between the high specific energy devices (B and FC) and the high specific power of conventional capacitors.

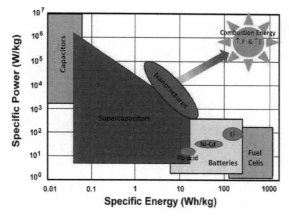

Figure 8.1 Ragone plot of different energy storage devices, where devices with nanotectures are included as well as different rechargeable battery technologies (Pb-Acid, Ni-Cd, and Li) and combustion energy engines [6, 7].

SC have some advantages and disadvantages when compared with rechargeable B. Figure 8.2 shows a charge-discharge cycle using galvanostatic measurements for a conventional capacitor, an ideal rechargeable B, and a SC. In a conventional capacitor the voltage linearly drops independently of charge due to the absence of a redox process dependent of a current, only the migration of ions occur to form the double layer. In the case of an ideal B, the voltage is maintained constant, known as a "plateau," while two phases of the electrode active material are maintained and in equilibrium. Here the redox process involves phase transformations that are dependent on the faradaic processes involved. On the other hand, in a SC the voltage might not drop linearly during discharge as a conventional capacitor, but it follows this linear drop with some slope changes related with pseudocapacitive contributions that do not involve phase transformations as in batteries, but that are dependent on current due to ion diffusion.

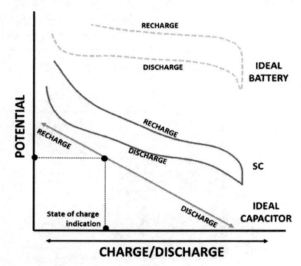

Figure 8.2 Charge/discharge profiles of an ideal capacitor, an ideal battery, and an example of a SC with pseudocapacitance. Adapted from reference [1].

The main advantages of SC when compared with B are the fast charge (Q) propagation related with their characteristic high power density, long cycle life due to the absence of phase transformation in the bulk electrode materials, and that they can be completely discharged. On the other hand, the main disadvantages of SC are their low energy density and in second degree the absence of a plateau during discharge, which makes the self-discharge effect more important. Therefore, the energy that is stored has to be used as soon as possible in order to avoid its waste, or use a B combined with a SC in a hybrid system to store the energy that is lost during this self-discharge.

Rechargeable B have been the energy storage device most commonly used until now. They have been the technological choice for a major number of applications, mainly due to their ability to provide suitable power. Cycle life is their main problem and technological limitation, aside from the high toxicity of their components. Nevertheless, society has learned to tolerate these deficiencies due to the lack of alternatives. More recently, power demands in many applications have increased, exceeding the performance of B, making them an insufficient alternative. SC can provide 3–30 times more power, an extended cycle life, more environmentally friendly materials and components [4], and lower weight due to the lack of heavy metals as used in batteries, making them an attractive technology. In Table 8.1 a comparison between the properties of rechargeable B and SCs is shown.

Table 8.1 Property comparison of rechargeable batteries and supercapacitors

	Devices	
Property	**Battery**	**Supercapacitor**
Storage mechanism	Chemical (REDOX)	Physical (double layer)
Power restriction	Reaction kinetics and mass transport	Chemical (pseudocapacitance)
Energy storage	High (in volume)	Electrolyte conductivity
Charging rate	Kinetically limited	Limited (surface area and pore size)
Limiting cycling life	Mechanical stability, chemical reversibility	High, equal to discharging rate

Based on the advantages and disadvantages of SC, they have been used in different technological applications for different sectors. A brief overview is described in the following sentences. At a consumer level, NessCap has developed car audio, remote controls, and a SolarCap tile, among others. In the industrial sector, SC have been applied in combination with renewal energy generation as wind turbines, since these devices are not voltage dependent as batteries and can support (absorb, accommodate) the changes in wind velocity. Also, in all repetitive applications like service lift in maritime harbors from NIPON Company to unload boats, where the SC is discharged during the lift and charged during the drop (energy recovery). In the automobile sector, SC have been used as starters, in motorcycles, assembled in modules for cars, busses, and trucks. In cars, SC are adjusted in the braking system in order to recharge during braking, and provide energy when acceleration is required (regenerative braking).

Combustion energy used in vehicles provides high specific energy and power values (Fig. 8.1), making its replacement with energy storage devices a difficult issue when developing EVs and/or HEVs, for cities with all types of topography. A single energy storage device will make this task very difficult, but instead a combination of such devices could make the trick. For example, SC have 3–6 times higher power capabilities than Li B [6], can provide the power peaks required for acceleration, and can be recharged during braking. There have been studies where either a FC or a rechargeable B has been arranged with a SC to obtain a hybrid system. A FC–SC hybrid system is the least desirable choice due to the low specific energy of the SC compared to recent advancements in specific power, efficiency, and cost of Li-ion B [7]. Nevertheless, a FC used as the main power source and the SC as auxiliary power source has been studied with good results [8, 9]. In addition, FC-B and FC-B-SC hybrid systems are a better choice because they provide higher fuel economy, and the B lifetime can be extended due to less battery strain [7]. Another hybrid combination based on a SC-B, where a Pb-Acid battery is used, makes use of the ability of the SC to be charged to a higher voltage than the B, shielding the B from most of the current pulses increasing its cycle life [10]. Therefore, a hybrid combination using different energy storage devices is a promising line of research that requires attention. In addition, a hybrid device approach combining a B and a SC electrode material in a single device assembly, as well as, the use of nanostructures can be explored aside from this hybrid system combination. These hybrid devices and nanostructured materials will be the focus in this chapter, aside from the environmentally friendly approach.

8.3 Supercapacitors Background

SC have their origin on conventional capacitors, which go back to 1745, when the concept of electricity storage in a charged capacitor was understood [1]. A typical capacitor, conventional capacitor, or a dielectric capacitor consists of two electric conductors that are usually metallic plates, separated and insulated by a dielectric material. These devices are presently applied in electric circuits within the field of refrigeration and air conditioning in the compressor motor [12, 13]; some of them are shown in Fig. 8.3.

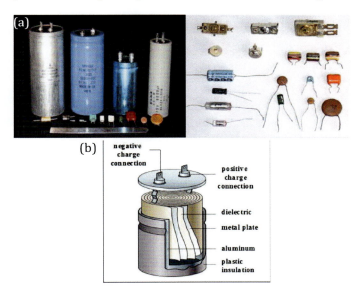

Figure 8.3 Different types of capacitors for electric applications and their essential parts where (a) shows commercial capacitors and (b) is an example of the assembly and components of a typical capacitor. Images taken from different web pages © 2001--2014 Softpedia; © Copyright 2014 Splung.com; www.mds975.co.uk © 2005-2014 [14-16].

Energy is stored in capacitors because the dielectric material is polarized during charging, holding the charges on the electrodes. That is, when a potential difference across the conductors (metal plates) is presented a static electric field develops across the dielectric material, causing a positive charge on one plate and a negative charge on the other plate, and energy is stored in the electrostatic field. This energy is described in terms of a constant capacitance and is measured in farad, which is the ratio between the electric charge on each conductor and the potential difference between them (see Fig. 8.4).

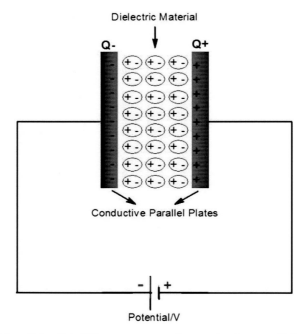

Figure 8.4 Potential difference across the conducting parallel plates in conventional capacitors.

The total capacitance (C) is given by the following equation:

$$C = \varepsilon_r A/d, \tag{8.1}$$

where C is the capacitance, A is the total area of the metallic plates, d is the separation between these plates, and ε_r is the permittivity or dielectric constant (different capacity for establishing an electric field) of the dielectric material used as separator. The unit of measurement is the farad and in practical applications submultiples like millifarad (mF), microfarad (µF), nanofarad (nF), and picofarad (pF) are generally used. From this equation, it can be deduced that the capacitance depends on the physical characteristics of the capacitor. If the area of the parallel plates increases, the distance between them decreases, and the applied potential is increased, the charge that can be stored increases giving greater capacitance values. In addition, the type of dielectric material used between the plates affects the charge storage, as the permittivity value gets larger the capacitance value gets improved. The main limitation of this type of device is the narrow variety of dielectric materials (e.g., air, ceramics, polymeric films, organic electrolytes), and that in order to achieve high charge storage, the area of the metal plates needs to be increased resulting in a higher cost and a larger device, which is in conflict with the current device development trend to reduce size.

The work of Helmholtz in 1853 revealed that the electric charge could be stored not only on the surface of a conductor, but also at the double layer interface developed between the electrode and the electrolyte [17, 18]. Then, almost one century later the storage of electric energy for practical applications in the form of cells or stacks was claimed in 1957 in a patent granted to Becker (General Electric Corp.) [19]. This patent described the storage of electric energy by the charge retained in the double layer of a porous carbon using aqueous electrolyte, and capacitance was greatly improved (6 orders of magnitude compared to conventional capacitors) due to the high surface carbon. After Becker's patent, another patent by Sohio Corporation in 1969 elucidated the use of high-surface-area carbon electrodes with organic electrolytes to obtain higher operational voltage (3.4–4.0 V). These organic electrolytes show higher decomposition voltage compared to aqueous electrolytes, improving energy density of these double layer capacitors. This technology was license to Nippon Electric Company, who finally introduced the term Supercapacitor. A new concept, introduced in 1975–1981 by Conway, based on a different storage mechanism known as pseudocapacitance, describes electrochemical adsorption of species (H, Pb, Bi, Cu) and/or solid oxide redox systems (RuO_2 electrodes in aqueous H_2SO_4) with higher specific capacitance values. Pinnacle Research developed in 1984 an electrochemical capacitor with electrodes made of RuO_2 for military applications, but the high Ru cost limited the development of large-scale capacitors.

In SC, the electrolyte and separators replaces the dielectric material of conventional capacitors, and the polarization takes place at the interface between the porous electrode material and the electrolyte. The ions from the electrolyte move to the external surface of the pores holding the charge electrostatically at the inner wall of the electrode surface forming what is known as the double layer (dl) storage mechanism. If the ions go beyond this dl through a pseudocapacive contribution, the capacitance value increases as mentioned before.

8.4 Charge Storage Mechanisms

In this section, a detailed description of the two charge storage mechanisms involved in SC will be presented: double layer (dl) and pseudocapacitance (psc). Different double layer models (Helmholtz, Gouy–Chapman, Stern, Grahame), and different phenomena related with pseudocapacitance (e.g., redox reactions, electrosorption of ions, and intercalation process) will be described.

8.4.1 The Electric Double Layer

Figure 8.5 shows a representation of a double layer capacitor, where a relative accumulation of cations and anions takes place at the solution side in response to a negative and positive electric polarization of the electrodes, respectively. An electrostatic equilibrium is established, giving rise to a "double layer." To understand this equilibrium, it is necessary to describe the electrode–electrolyte interface.

All electrochemical processes take place at the electrode/solution region so-called interface, which generates a charge distribution; this dl is formed even when no potential is applied to the electrode. The concept and model of the double layer emerged from Von Helmholtz (1853) who reported a work related with the interfaces of colloidal suspensions. This was extended to the surface of metal electrodes by Gouy, Chapman, and Stern, and later in the notable work of Grahame around 1947 [20]. The concepts of double layer are very important to understand the charge storage mechanism in electrochemical capacitors. A brief description of these different dl models, which describe the characteristics and properties of electrochemical interfaces are explained below.

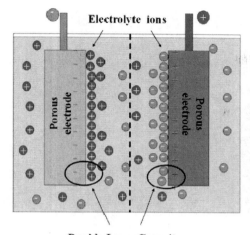

Figure 8.5 Representation of charge accumulation through the *dl* formation in an electrochemical capacitor.

8.4.1.1 Helmholtz model

The Helmholtz model was the first to describe the double layer formation [21, 22], and it is known as the compact layer model. In this model, the interface is considered similar to a conventional capacitor. That is, two charged parallel plates separated by a distance d. According to this analogy, we can define the capacity per unit area of the electrical double layer as

$$C = \frac{\varepsilon}{4\pi d}, \tag{8.2}$$

where C = capacitance, ε = dielectric constant, and d = separation between the plates. It is important to note that C is independent of the electric potential and of the electrolyte composition. In Eq. 8.2, the composition is not explicitly considered; therefore there is no influence on the dielectric constant. This model assumes a linear profile of the potential within the double layer as shown in Fig. 8.6, where positively charged electrode material is on one side and negative ions are on the other side of the double layer. However, this model does not consider the adsorption of water molecules and counter ions.

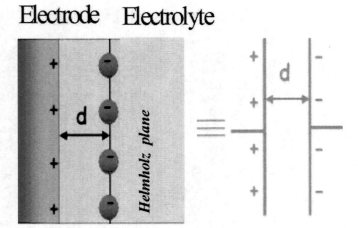

Figure 8.6 Double layer Helmholtz model.

8.4.1.2 Gouy–Chapman model

This model defines a diffuse layer of ions inside the Helmholtz plane. The potential drop at the interface is not a linear function with distance. In a metal the charges with the same sign forms a plane, while the opposite sign are distributed across the electrolyte solution along a certain distance (Helmholtz plane). This model uses two equations:

(i) Electric potential (Poisson law):

$$\Delta\phi = \frac{4\pi\rho}{\varepsilon}, \qquad (8.3)$$

where $\Delta\phi$ = potential difference, ε = dielectric constant and ρ = charge density.

(ii) Ionic concentration (Maxwell-Boltzmann equation):

$$C_i = C_{i0}\exp(-ze\varphi/kT), \qquad (8.4)$$

where C_i = ion concentration, C_{i0} = ion concentration at the interface, z = ion charge, e = electron charge, k = Boltzmann constant, T = Temperature, and φ = potential. Considering that both parallel surfaces are equipotential, the equation is defined as

$$\Delta\varphi = \partial\varphi^2/\partial x^2 = (-4\pi/\varepsilon)\sum z_i e C_{i0}\exp(-z_i e\varphi/kT), \qquad (8.5)$$

where x = distance from the charged interface to the bulk of the electrolyte solution (diffuse layer, Fig. 8.7); for small potential variations, Eq. (8.5) is reduced to

$$-z_i e\varphi/kT \ll 1 \tag{8.6}$$

$$\varphi = \varphi_0 \exp(-kx) \tag{8.7}$$

At 25°C the next approximation represents the potential associated with $x = 0$, that is to the interface: the different phenomenon as the Van der Waals interactions, hydrogen bonds, attraction-repulsion forces between ions, ion size, and charge occurring in the electrolyte solution due to the presence of the solid electrode.

$$\varphi_0 \leq 50/z \text{ mV} \tag{8.8}$$

On the other hand, the limitation of this model is related with low ionic concentration (C_{i0}) due to the presence of less interactions between the ions, making the dielectric constant change throughout the electrolyte medium. The ions are considered as point particles of the same size, which have the possibility of randomly approaching the electrode surface. This description does not resemble reality, because the ions have different ionic radius and solvation spheres, and cannot approach the electrode surface as near as their own size. In addition, this model assumes that charges decay rapidly and continuously from the conductive material to the electrolyte, without a distinctive layer separation as in the Helmholtz model. Nevertheless, this model successfully predicts the temperature and potential effects on the capacitance, even though it does not explain the effect of low electrolyte concentrations (low surface charge density), and the distorted structure by steric effect and hydration forces occurring in the nanopores.

8.4.1.3 Stern and modern models

Stern proposed an electric double layer model in which the two structures mentioned previously are merged. Thus, the total potential difference across the interface takes into consideration a constant potential near the interface (Helmholtz compact layer), and an exponential potential decay as the ions get far away from the interface (diffuse layer). In the Helmholtz or Stern compact layer the ions defines a plane, known as the Helmholtz inner plane, where the ions are adsorbed on the surface of the electrode. Similar to the Helmholtz model, this model does not take into consideration the adsorption of water molecules and other adsorbed ions. In more recent models, these two effects are considered and accounted for the capacitance calculation. Figure 8.7 shows a comparison among all models described in this section. It is important to consider that all models describe the different surface interactions that occur in the electrode due to the presence of ions in solution, whereas the more recent models that are still being under study, describe the different phenomenon as the Van der Waals interactions, hydrogen bonds, attraction-repulsion forces between ions, ion size and charge occurring in the electrolyte solution due to the presence of the solid electrode.

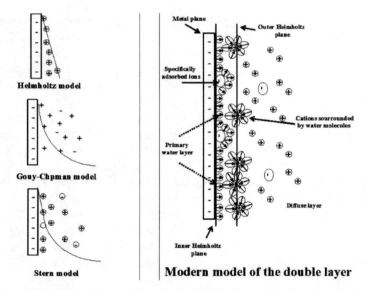

Figure 8.7 Different models of the double layer, the line in the first three models resemble the charge distribution, and in the modern model the inner Helmholtz plane. Figure adapted from different figures of Conway's Book, Chapter 6 [1].

The double layer structure can affect the charge transfer rate processes that occur on the electrode. For example, electroactive species not specifically adsorbed on the electrode can approach it only to the outer Helmholtz plane and therefore "feel" a different potential, which is established between the electrode surface and the solution. This difference may be important or not, depending on the process under study.

The models described are useful to give criteria of the charge accumulation process in electrochemical capacitors. However is important to consider that the modern models still cannot be used to elucidate the situation inside the nanopores, and to predict the complications associated with ion pairing and limited mobility. Only the use of in situ characterization techniques and computational modeling will help to advance in the understanding of the double layer at carbon nanopores [2].

8.4.2 Pseudocapacitance Mechanism

Pseudocapacitance is a common concept within the field of SC, which involves a faradaic process that takes place on the electrode surface. The difference between pseudocapacitance and double layer capacitance was not clear in the field of applied electrochemistry. However, from a fundamental electrochemistry point of view this concept was understood in the work of Trasatti and Buzzanca [23]. Their work showed that RuO_2 DSA electrodes resulted in a cyclic voltammetry profile characteristic of a capacitor response. Pseudocapacitance associated with the underpotential (potential shift to more positive values, a process with less energy in relation to the thermodynamic conditions) related with deposition of H adatoms (and later of metal atoms) was treated in detail by Conway and Gileadi in 1962 [24]. As noted earlier, the faradaic charge (Q) obtained for example in an electrosorption process, in a quasi-two-dimensional intercalation process, or in a surface redox process as in RuO_2, is a function of electrode potential and its derivative dq/dV, giving way to a capacitance value. This kind of capacitance has a faradaic origin, rather than being associated with accumulation of electrostatic charge as in a double layer capacitor. Therefore, this capacitance is referred to as pseudocapacitance, where the term "pseudo" has been used only as formalism to distinguish it from electrostatic capacitance.

It is clear that an electron transfer at the interface (electrode–electrolyte) causes changes in the oxidation states of electrolyte electroactive species, as well as in the electrode material. Pseudocapacitace increases when mechanisms like ion electroadsorption, redox reactions, and intercalation-deintercalation on the surface of the metallic electrodes or oxides (RuO_2, IrO_2) are presented. Charge transfer in these different mechanisms depends on the potential, such as in the capacitive phenomena, and is described below [24–28]. Figure 8.8 shows in a general way these three different processes: adsorption, redox, and three-dimensional intercalation.

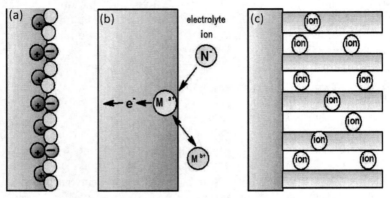

Figure 8.8 Representation of three different processes of pseudocapacitance: (a) Ion electrosorption or adsorption, (b) surface redox process, and (c) intercalation.

8.4.2.1 Redox reactions

In a reduction/oxidation reaction (redox reaction) two species are considered: an oxidant (Ox) and a reductor (Red) that participates in the charge transfer process (Eq. 8.9).

$$Ox + ne^- \leftrightarrow Red \tag{8.9}$$

The potential (E) is given by Nernst equation:

$$E = E^0 + \frac{RT}{zF} \ln \frac{\mathfrak{R}}{1-\mathfrak{R}}, \tag{8.10}$$

where E_0 is the standard potential, z is the number of electrons involved in the redox reaction, R the gas constant, T is absolute temperature, F Faraday constant, $\mathfrak{R}$ is the ratio $\frac{[Ox]}{[Ox]+[Red]}$, and the charge Q (given by zF) is a function of potential (E) in Eq. (8.10). Figure 8.9 shows two different examples of a typical surface redox reaction. Figure 8.9a shows a RuO_2 grown on carbon paste electrode in 1M $HClO_4$ electrolyte [29], where the two shoulders (I, II) are associated to charge transfer of RuO_2 [26], and are reversible responses (oxidation/reduction) according to following equation:

$$RuO_x(H_2O) \rightleftarrows RuO_{(x+\delta)}H_2O_{(y-\delta)} + 2\delta H^+ + 2\delta e \qquad (8.11)$$

In Fig. 8.9b cyclic voltammetry corresponding to a carbon paste electrode with quinone groups in 1 M $HClO_4$ is shown, where peak I and II correspond to a charge transfer processes associated to quinone (C–Q)/hydroquinone (C–QH$_2$) redox couple, according to Eq. 8.12:

$$C-Q + 2H_2O + 2e^- \rightleftarrows C-QH_2 + 2OH^- \qquad (8.12)$$

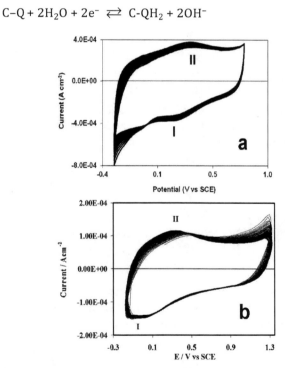

Figure 8.9 Typical response of a surface redox reaction, obtained by cyclic voltammetry for the system 1 M $HClO_4$ at 10 mV/s. (a) RuO_2 and (b) carbon paste electrode. Equations 8.11 and 8.12 show the chemical species associated in each redox pair for each case.

Redox reactions occurring in pseudocapacitive systems should be generated only at the electrode–electrolyte interface. This leads to a difficulty in controlling the redox reaction, since otherwise the system would suffer a complete chemical transformation. That is, the Gibbs free energy of the pseudocapacitive process must be less and faster than any other process occurring in the system.

8.4.2.2 Ion electrosorption

Absorption of ions forming a monolayer on the electrode is a reversible process that results in a faradaic charge transfer, giving rise to a pseudocapacitance similarly as in redox reactions [24, 26, 27]. The adsorption/desorption process can be describe as follows:

$$A_c^{\pm} + S_{1-\theta} \pm E \rightleftarrows S_{\theta_A} A_{abs}, \tag{8.13}$$

where A is the ionic species (the subheading c is the available ion concentration, and ± is the ions charge), S the substrate, $(1 - \theta_A)$ is the free available area for absorption, and θ_A the covered surface for ions), E the electrode potential, and A_{abs} absorbed ions. Considering that the sites are occupied in a random way, the equation of Langmuir is used to determine the surface coverage:

$$\theta_A / 1 - \theta_A = K_c \, (-EF/RT), \tag{8.14}$$

where K_c is the electrochemical equilibrium constant and the coverage change $(d\theta)$ is proportional to the charge dQ, expressed by

$$dQ = Q_1 d\theta, \tag{8.15}$$

where Q_1 is the charge required to form a monolayer, θ is a function of potential (E), and deriving Eq. 8.14 the capacitance with pseudocapacitive origin can be obtained as in Eq. 8.16.

$$C_\varphi = Q_1 F / RT \; K_c \pm \exp\left\{\left(\frac{-EF}{RT}\right) \middle/ \left(1 + K_c \pm \exp\left(\frac{-EF}{RT}\right)\right)^2\right\} \tag{8.16}$$

In order for this absorption process to occur, the Gibbs free energy has to be lower than any other free energy corresponding to another process (e.g., formation of hydrogen when H^+ is the cation), the hydration energy or the energy of any redox process, and so on. The system must be in thermodynamic equilibrium in order to obtain an ideal pseudocapacitive behavior [23, 25, 30, 31]. This implies that the adsorption rate and the desorption rate must be equal and greater than the velocity of any other process.

8.4.2.3 Intercalation

Finally, intercalation/deintercalation processes may also give pseudocapacitance. Figure 8.10 shows the intercalation of Li^+ ion, obtained in 1 M $LiClO_4$ at 10 mV/s for a carbon paste electrode (CPE). A1 and C1 are associated with intercalation/deintercalation processes.

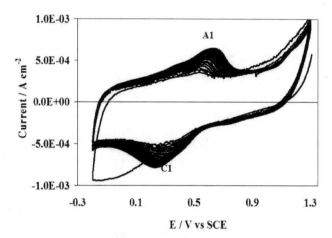

Figure 8.10 Cyclic voltammetry of a Li^+ intercalation process into carbon paste electrode (CPE), 25 successive cycles obtained at 10 mV/s and 1M $LiClO_4$ [32].

Unlike in the adsorption phenomena, in the intercalation process the cations are inserted between the structural layers of the material held together by van der Waals forces. The diffusion of cations, to and from the layers decreases the speed of charge and discharge loading and unloading of the SC.

The three mechanisms described above for the pseudocapacitive phenomena involve changes in the oxidation states of the species held on the interface. This pseudocapacitance can store from two to three electrons per atom, or even more, depending on the oxidation states of the atoms available on the electrode/electrolyte interface. This increases the capacitance from 10 to 100 times the value obtained with the electric double layer capacitors [28, 33], since in these SC one electron per atom accessible on the surface of the electrode can be stored. However, the charge and discharge rate is lower in a pseudocapacitor due to the nature of its mechanism. Table 8.2 summarizes the thermodynamic relation for each type of process involved in pseudocapacitance.

Table 8.2 Correlation for three pseudocapacitance processes [1]

System type	Thermodynamical relation
(a) Redox system $Ox + ze^{-1} \rightleftharpoons Red$	$E = E_0 + (RT/zF) \ln \Re/(1-R)$ $\Re = \dfrac{[Ox]}{([Ox]+[Red])}; \dfrac{\Re}{(1-\Re)} = \dfrac{[Ox]}{[Red]}$
(b) Underpotenctial deposition $M^{z+} + S + ze^- \rightleftharpoons S \times M$ ($S \equiv$ surface lattice sites)	$E = E_0 + (RT/zF) \ln \theta/(1-\theta)$ θ = two-dimensional site occupancy fraction
(c) Intercalation system Li^+ into MA_2^+	$E = E_0 + (RT/zF) \ln X/(1-X)$ X = occupancy fraction of layer lattice sites (e.g., for Li^+ in TiS_2)

8.5 Classification

SC are traditionally classified based on the type of electroactive material used in the electrodes: Carbons, metal oxides, conducting polymers, and nanostructured composites or hybrid materials. Also, they have been classified based on the cell assembly: symmetric when using the same electroactive material in both electrodes, or asymmetric when different materials are used in each electrode. Other classifications are based on the type of electrolyte used (aqueous or organic, solid or liquid) or on the type of storage mechanism ruling the performance of the device: double layer capacitors or pseudocapacitors. Halper et al. [34] have proposed a more complete classification where they take into consideration mainly the type of electrode materials, the storage mechanisms, the electrolyte type and cell assembly, in good agreement with a recent book on supercapacitors [35]. Based on this more complete classification, some modifications are proposed in order to have an actualized version and are shown in Fig. 8.11. Aside from this proposed general classification, for a better understanding of all concepts, this section has been organized as follows: (1) SC based on their mechanism of charge storage, followed by (2) electrolyte type. Before describing in detail all of these concepts, there are some general aspects related with electrode materials for SC cells that need to be mentioned.

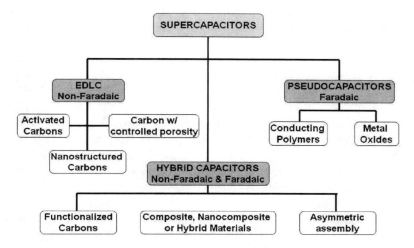

Figure 8.11 Schematic classification of supercapacitors, where the storage mechanism, cell configuration, and electrode materials are considered.

In general, materials used as electrodes in supercapacitor cells must fulfill certain fundamental properties, such as great surface area and adequate porosity distribution. The pores need to be accessible to the electrolyte in order to increase the electrode–electrolyte interface, and achieve greater capacitance values, by either double layer formation or pseudocapacitance effects. If pseudocapacitance is presented, their processes must be reversible and fast in order to capitalize this contribution to the whole capacitance value. Finally, the electroactive materials must have the ability to perform at least 10^5 charge-discharge cycles with good coulombic efficiency.

Nanotechnology has provided materials with all these properties. As shown in Fig. 8.1, nanotectures used as electrodes in supercapacitor cells have slightly moved the area of SC in the

Ragone plot toward the area of gasoline. There are four research lines in which these nanotectures or nanomaterials have been studied: porous carbon materials that store energy by the formation of the double layer; metallic oxides and conducting polymers where mainly their contribution is through a faradaic contribution (pseudocapacitance); and more recently nanocomposite and/or hybrid materials where both contributions, non-faradaic and faradaic are tuned and are participating in the total capacitance value. All these concepts will be explained in detail in the following sections.

8.5.1 Charge Storage Mechanism

As explained in previous sections, there are two main charge storage mechanisms involved in SC (double layer and pseudocapacitance) that have resulted in the development of two different devices: EDLC (electric double layer capacitors) where high-specific-area carbon materials are used as electrodes, and pseudocapacitors using electrode materials as transition metal oxides or conducting polymers. It is important to mention that these devices do not work on a total 100% of a certain storage mechanism, but instead on a predominant mechanism. For example, in EDLC the charge is stored mainly in the *dl* and shows perhaps 1% to 5% of pseudocapacitance phenomena. *dl* is always present in any interphase between a conductor and electrolyte. Therefore, in pseudocapacitors the *dl* is always implicit, and is about 5% to 10% of the total capacitance value, while the rest of the charge storage comes from pseudocapacitance [1]. It is important to point out that these two devices are constructed mainly using a symmetric assembly, that is, using the same materials in both electrodes.

There is a third type of SC, called hybrid capacitors, where both charge storage mechanisms are tuned and involved in a more equilibrated way. There are different approaches to build this type of capacitors. The first one is by introducing pseudocapacitance to carbon materials through electroactive functional groups, transition metal oxides or conducting polymers to obtain nanocomposite electrode materials. A second approach is using an asymmetric assembly (different materials in each electrode), with one electrode based on *dl* charge storage (carbon) and the other based on pseudocapacitance or a battery-like electrode material. In the following sections, these three types of SC based on the different storage mechanisms and assemblies will be discussed, making emphasis on the materials involved in each one of them.

8.5.1.1 Electric double layer capacitors

An electrochemical double-layer capacitor (EDLC) is a capacitor with relatively high specific energy, assembled with high-surface-area porous electrodes. Their specific energy is typically hundreds of times greater than conventional electrolytic capacitors (<0.5 Wh/kg), the charge and discharge is carried out in interfaces of high specific area. Materials such as porous carbons are used in this type of capacitors. The electric charge of these EDLC can be stored in a reversible way at high current densities, and can be operated at higher specific power (W/kg) than batteries [1, 36, 37]. The energy storage mechanism involved in these devices is non-faradaic and is carried out by the charge of the double layer, as implied in their name. By applying a potential difference to EDLC, one of the electrodes is positively charged and the other negatively, this provokes the ions diffusion from the bulk of the solution to the electrode surface. On the electrode/electrolyte interface, a double layer is formed (Fig. 8.12), where the ions from the electrolyte are attracted by the opposite charges of the electrodes. In this process, electron transfers do not occur on the interface and the charging process is purely a non-faradaic process, and therefore no chemical transformations are presented. The conduction band electrons of the electrode materials are the only ones involved in the electric double layer formation.

In this case, this electrostatic charge storage occurs by reversible adsorption of electrolyte ions onto active and electrochemically stable electrode materials. This interfacial process is very fast (10^{-8} s) because involves only a rearrangement of ions from the electrolyte (Fig. 8.13) to the electrode surface, without heterogeneous charge-transfer and chemical phase changes as presented in batteries. This confers to the EDLC, properties such as a fast response to potential changes, which results in a higher speed of charge and discharge. In this process, being free of chemical reactions, ideally there is no degradation of the materials (although minimal contribution of redox reactions are always involved), thereby EDLC shows a great ability to carry out successive charge-discharge cycles in a reversible way (500,000 cycles or more) showing a higher useful life time than batteries.

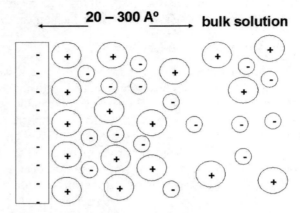

Figure 8.12 Electrode/electrolyte interface (double layer): The ions from electrolyte are attracted to the opposite charges of electrodes.

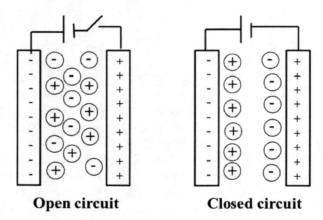

Figure 8.13 Double layer charge storage Mechanism in EDLC s.

In EDLC with two parallel electrodes, the charge stored is called capacitance and is defined by the following equation:

$$C = Q/E = \varepsilon_1 A/d \qquad (8.17)$$

and the energy (E_c) is calculated from

$$E_c = 1/2QE = 1/2(\varepsilon_r AE^2/d) \tag{8.18}$$

For both equations, Q is the charge, E the voltage window, A the surface area of electrode, and d the thickness of the electric double layer.

Generally, the distance of the compact double layer or interface is ~3 Å of thickness or up to about 1000 Å, including the diffuse part [38]; because this distance is in the order of angstroms (Å), and capacitance is inversely proportional to the distance of charge separation, the capacitance increases to about 10,000 times comparing with a conventional capacitor. Other characteristics of these systems are that their electrode materials show high conductivity, high surface area (up to 3000 m^2/g), good corrosion resistance, high-temperature stability, controlled pore structure that is adapted to the solvated ions size for easy access to the electrode surface, processability and compatibility in nanocomposite materials, absence of secondary reactions, and relatively low cost [1, 35, 39].

Most of the research in EDLC has been focused on the use of different carbon materials with high surface areas, with different forms and shapes (particles, fibers, cloth, sponges, aerogels, nanocarbons, etc.) due to their low cost, availability, established electrode production technologies [40], and environmental friendliness [41]. The most common carbons are carbon blacks [42], activated carbons [39], carbons with controlled porosity including carbon aerogels/xerogels [3, 43, 44], single-walled carbon nanotubes (SWCNTs) and multiwalled carbon nanotubes (MWCNT) [45–47], and more recently graphene [48–50].

Commercial SC are generally EDLC assembled in a symmetric configuration with various types of carbons, separated by a polymeric or cellulose membrane impregnated with organic electrolytes. These EDLC have good stability during charge-discharge cycles and high specific power as mentioned above, but low specific energy are obtained as observed in Table 8.3. The main drawbacks of these commercial devices are the utilization of organic solvent that makes them non-environmentally friendly devices, and their low specific energy compared to batteries, which is limiting their introduction to the market in a massive way.

Table 8.3 Commercial supercapacitors based on carbon electrodes and organic electrolytes

Company	Voltage (V)	Capacitance (F/kg)	Energy density (Wh/Kg)	Power density (W/Kg)
Panasonic	2.5	6329	3.7	1035
Maxwell	2.7	5895	4.45	900
Ness	2.7	5713	4.3	958
Asahi Glass	2.7	6548	4.9	390
EPCOS	2.7	5667	4.3	760
LS Cable	2.8	5079	3.7	1400
Power Sys	3.3	8571	8	825

Note: The best devices based on specific energy were selected for each company [51].

8.5.1.1.1 Activated carbons

Activated carbons (ACs) are well-known as electrode materials for storing electrical charge by the double layer mechanism. They provide high surface area due to their microporous structrure, and are low-cost materials [52–59]. The ratio between pore size distribution and its electrochemical behavior has been deeply studied [60]. The synthesis process used for their preparation involves two steps: (1) pyrolysis of organic precursors, (2) followed by a process known as activation [61]. Several precursors have been used from natural sources known also as biomass, to produce biochar and obtain ACs in a completely eco-friendly process [62], such as coconut shells [63, 64], sunflower seed shells [65], banana fibers [66], sugar cane [67, 68], coffee endocarp [68, 69], seaweed [70], rice husk, beet sugar and straw [41, 71], peach stones [72], lapsi seed [61], bamboo [61], cherry stones [73], cotton stalks [74–77], date palm seeds [78], pine-cone [4, 41], wood [77], among others. Most of these biomasses are agricultural waste, and are considered as a very important feedstock to obtain environmentally friendly activated carbons, as they are renewable as well as low-cost materials [79]. Nevertheless, ACs have been also obtained from coal-based carbon, resin-based carbon, phenolic resins, coal, wood [79, 80], pitch, coke, from synthetic precursors such as selected polymers [81], or from industrial residues [69].

The conversion of biomass into carbon by pyrolysis involves a thermal controlled decomposition, where most of the non-carbon elements are eliminated. The residual carbon atoms group themselves into sheets of condensed aromatic ring systems, often bent, where the arrangement is irregular, leaving free interstices that may be filled or blocked by disorganized carbon from deposition and decomposition of tars [71]. The systematic order of breaking down the main components of biomass is as follows: 200–800°C hemicellulose, 260–350°C cellulose, and 280–500°C lignin [64]. The porosity of the obtained carbons is not completely accessible, and the objective of the activation process is to enhance this pore structure. This activation process can be carried out with a physical or chemical activation, where in both methods a reaction of the precursor with the activating agent to develop porosity takes place, but the practical procedure and the mechanism by which the activating agent develops porosity differs [3, 74, 82].

Physical or thermal activation is a two-stage process, where pyrolysis of the precursor at 400–900°C is performed before activation, followed by a thermal treatment between 700 and 1100°C in a controlled atmosphere (steam, CO_2, air, oxygen, or mixtures). During this activation, elimination by partial gasification of a large amount of internal carbon mass is carried out to develop the carbon structure with enhance porosity, where the activating agent has an influence [35, 78, 83, 84]. For example, carbon activated using steam produces carbon oxides, and these oxides diffuse out of the carbon resulting in a partial gasification, which opens pores that were previous closed, and further develops the carbon internal structure [85]. In the case of using air and/or oxygen, an exothermic reaction with the carbon is observed making the reaction difficult to control, resulting in a decreased activated carbon yield, rapid enlargement of micropores into mesopores, but the activation is a low energy/cost process compared to steam and CO_2 [86, 87]. The chemical and physical nature of these activated carbons is dependent on the precursor, the activating agent, the temperature of activation, and degree of activation. In general, the higher the activation temperature/activation time, higher degree of porosity development. Nevertheless, higher porosity developments involve the broadening of the pore size distribution [86], and ACs with microporosity cannot be developed [85].

On the other hand, in the chemical activation the pyrolysis and activation step is carried out in a one-step process, but can occur separately [88]. Normally, biomass precursors (eco-friendly) are exposed to a chemical agent (H_3PO_4, $ZnCl_2$, KOH, NaOH, among others) by impregnation or physical mixing, and the pyrolysis is carried out using lower temperatures ($\approx$400–900°C) [61, 82, 83]. These chemical agents influence the pyrolytic decomposition, inhibiting the formation of tar, enhancing the yield of carbon with higher porosity [82, 84]. It is important to point out that the washing procedure after activation is a very important step to accomplish the characteristic high surface area of ACs, of up to 3000 m^2/g with KOH [39, 59]. However, it is well known that only a surface area between 1000 and 2000 m^2/g can be used in practical applications [81]. Regarding the eco-friendliness of this activation, the use of $ZnCl_2$ has been recently avoided [85], and the activation with H_3PO_4, KOH, and NaOH has been preferred. In addition, comparing with the physical activation, chemical activation is a lower energy consumption process that involves lower pyrolysis temperatures, higher carbon yield is obtained with higher surface area, and microporosity [35, 83, 89]. These last two characteristics of the chemical activation process are the key parameters for applications in SC [87].

The electrochemical behavior of ACs revealed from the specific capacitance values is highly influenced by the total specific surface area, porosity nature, pore size distribution, and electric conductivity [90]. Control of pore size in ACs is a challenging quest, and as a result ACs show a very complex pore structure composed of a mixture of different pore sizes: micro or nano pores (<2 nm), mesopores (2–50 nm), and macropores (>50 nm). It has been observed that increasing the surface area by traditional activating processes, where no control of porosity is achieved, the capacitance can increase linearly up to a certain value, and then a plateau is reached [91–94]. Therefore, it is believed that the disordered pore structure in ACs is responsible for the limited capacitance values obtained with very high surface areas [95, 96]. In addition, regarding ion diffusion into the carbon porosity, it is well known that mesopores are good at high charge-discharge rates [96–100], while micropores contribute to higher capacitances at lower cycling rates. Thus, having only micropores is not desirable because ion diffusion into the deeper region of the electrode or bulk of the material gets rougher, and higher cycling rates cannot be achieved due to a faster ion movement requirement. Therefore, for an improved capacitance value a desirable concentration of mesopores is necessary to provide a good entry for ions from the electrolyte into the bulk, to then access the micropores [91, 101]. This ideal AC material with high surface area and suitable micro-mesopore size distribution is very tricky to obtain [3, 96], making the synthesis procedure of ACs the limiting factor to increase capacitance [1, 39, 60, 92, 102–111].

8.5.1.1.2 Carbons with controlled porosity

Carbons with complex porosity as ACs, limits their performance as electrodes in SC cells. As previously mentioned, capacitance values of ACs can increase up to a certain point by increasing the surface area, after this critical point a plateau is reached, and a further increase of the surface area will not result in an enhanced capacitance. Therefore, a pore design approach for carbon materials is needed to further improve the capacitance values, as well as the cycling rate. It is very important to take into consideration that when trying to control the porosity properties in a carbon material, if the porosity is highly increased the electric conductivity will decay, resulting in low volumetric capacitances (F/cm^3) [39, 112]. Carbons obtained from templates, carbide, or aerogels from carbon gel precursors are some examples where the porosity can be highly tuned.

Ordered mesoporous carbons (OMCs) are synthesized by using different types of templates, such as nanostructured or mesostructured silica materials, obtaining nano-meso porous carbon materials with high surface area, large pore volumes, and uniform mesoporosity. The main advantage of OMCs compared to ACs is their rapid ion diffusion at high rates, resulting in higher capacitance values. Thus, at similar conditions capacitances exceeding 200 F/g in sulfuric acid (1 mA/cm^2) [97–100], and 190 F/g for KOH (20 mV/s) [96, 97] aqueous electrolytes have been obtain for OMCs. Clearly, W. Xing et al. have shown that at lower scan rates (5–10 mV/s) in aqueous KOH electrolyte, better capacitance values are obtained for the AC known as Maxsorb (333–289 F/g) compared to an OMC (211–207 F/g), but as the scan rate increased up to 50 mV/s the OMC showed a greater capacitance (180 F/g), outperforming the AC (73 F/g) [96–99, 113]. This rapid ion diffusion due to their particular pore structure, and the lower content of oxygen-based functionalities [99] are the main factors contributing to a better cycling rate performance for high power output, outperforming microporous structures with better performance at lower rates [96, 100].

On the other hand, when the main goal is to obtain higher specific capacitances values, microporous carbons are preferred (less than 2 nm) [91, 101, 114]. Initially, it was assumed that the capacitance would decrease with decreasing the pore size, but it has been proven with carbide-derived carbons that this is true for pore sizes above 1 nm, below this critical point the capacitance surprisingly increases [92]. It is important to point out that the porosity in these materials can be tunable with an accuracy of 0.05 nm, making these carbide-derived carbons model compounds to study charge storage in nanopores between 0.6 and 1.1 nm. Therefore, the enhanced capacitance due to microporosity has been explained based on the distortion of the ion solvation shell, which results in a closer interaction between the ion and the carbon surface [91, 95, 115, 116].

Carbon aerogels (CA) are highly porous materials with good electric conductivity, where their porosity properties are controlled by modifying the synthesis parameters. They are usually synthesized by using a system based on resorcinol-formaldehyde (RF). A polycondensation of this system takes place via a sol-gel process, where a posterior pyrolysis of the resulting organic gels (RF gel) is needed to obtain the characteristic highly porous and high-surface-area (400–1000 m^2/g) carbon aerogels [117, 118]. The density, pore size, and particle size are controlled during the sol-gel process, and are influenced by the reagents and catalyst concentration. The radii and morphologies of the primary particles formed on the RF gel determine the pore characteristics. The CA matrix is constructed with interconnected colloidal-like carbon particles or polymeric chains that after pyrolysis result in a higher electric conductivity compared to activated carbons [119], where normally binders are needed for the electrode elaboration introducing resistivity to the material. Electric conductivities of up to 100 S/cm have been obtained for these carbon aerogel low-density materials. CA can be produced as monoliths, composites, thin films, powders, or micro-spheres [52, 120–123].

The electrochemical performance of carbon aerogels, as in other carbonaceous materials, greatly depend on factors such as specific surface area, pore size distribution, nature and concentration of surface functional groups [124]. These parameters can be tuned during the aerogel synthesis, or with the posterior activation step that is used also in activated carbons [125]. CA with pore sizes between 2–50 nm have been obtained with high electrical conductivity, but lower specific surface areas (400–900 m^2/g) compared to ACs. Therefore, it has been found that the capacitance values of CA are more correlated with the interconnected ordered pore structure than with the total specific surface area [39]. Specific capacitances calculated from cyclic voltammetry or galvanostatic charge-dischage cycles using half cells (three-electrode arrangement) and aqueous electrolytes, results in values between 100 and 200 F/g (1–5 mV/s, 0.5–20 A/g) [81, 118, 126–129]. The electrolyte can access without any difficulty to the interconnected pore structure of these materials, promoting a high ionic conductivity that is reflected in high power capabilities [130].

CA have been assembled in SC stacks of up to eight cells, labeled as Aerocapacitor providing good cell capacitances of up to 38 F/g [131]. Nevertheless, their low-density results in low volumetric capacitance, making these materials more suitable for their use as a matrix to disperse, immobilize, or anchor pseudocapacitive materials [2].

Another type of carbon with controlled porosity are the hierarchical porous carbons (HPCs), which have a 3D texture with combined macroporous cores, mesoporous walls, and micropores. In this monolithic structure, macropores are used as temporary buffering storage of ions for the micropores, reducing the diffusion distances. Mesoporous walls provide low-resistance pathways for ions, and the micropores increases the double layer. As in CA, HPCs can also be obtained from resorcinol-formaldehyde systems, resulting in carbon materials with comparable performance to ACs and CA in terms of power and energy, and capacitances values using aqueous electrolyte of 250–100 F/g (two-electrode cells, galvanostatic cycling) [132, 133].

8.5.1.1.3 Nanostructured carbons

Within this group of carbons, we can find nanofibers (NFs), SWCNTs, MWCNTs, and graphene just to mention some of them.

Carbon nanotubes (CNTs) show high conductivity, unique mechanical properties, high surface areas and large accessible mesopores due to their entangled structure and nanoscale tubular morphology, and are synthesized by catalytic decomposition of hydrocarbons. Therefore, there is a considerable interest in their application as electrode material for SC. CNTs show lower surface areas than ACs (~400 and >3000 m^2/g, respectively), but in some cases, the capacitance values for CNTs reach values as ACs (4 to 180 F/g in aqueous electrolytes). This is possible because both, the surface area, and the pore size, affect directly the capacitance values. In ACs a portion of their larger surface area does not contribute to the capacitance value due to a mismatch between the pore size and pore distribution with the ion size of the electrolyte [3]. In the case of CNTs most of the surface area is available for its contribution to the capacitance [45].

CNTs have been studied in organic and aqueous electrolytes, showing a dependency of the capacitance with their morphology and purity [134]. The specific capacitance is not impressive for purified nanotubes (without amorphous carbon and catalyst residues), varying from 15 to 135 F/g with surface areas between 120 and 400 m^2/g [134, 135]. This has been attributed mainly to the hydrophobic property of CNT surface [81]. For this reason, the introduction of functional groups as a strategy to develop hydrophilicity increases the surface area up to one order of magnitude, and introduction of pseudocapacitance by this functional groups enhance the specific capacitance [39]. Nevertheless, functionalization of CNTs introduces defects as cracks, and surface irregularities by partial erosion of the outer carbon layer. These defects can affect the capacitance values, obtaining similar values than without any modification. For example, as for the other carbons, chemical activation with KOH has been used to increase the surface area of these nanotubes up to 1035 m^2/g, resulting in a low capacitance (90 F/g) in alkaline media [39]. In addition, CNTs functionalized with different groups, such as carboxyl and alkyl groups of different chain lengths to change hydrophobicity degree (butyl, oxtyl, dodecyl), showed higher capacitances for CNTs with carboxylic groups due to hydrophilicity [9]. Therefore, suitable conditions for the introduction of functional groups are requiered to obtain good capacitance values.

Different forms of carbon nanotubes have been characterized as electrode materials: directly grown on current collectors, as porous tablets, aligned, entangled, and assembled structures. The main advantage of directly growing carbon nanotubes on current collectors is that the contact resistance is minimized [136]. In addition, aligned nanotubes have shown lower ESR values, enhanced cycling rate due to larger pore sizes, more regular pore structures, and ion conductive pathways compared to entangled nanotubes [137].

SWCNTs with surface areas considerably higher than for MWCNTs of 240–1250 m^2/g have shown capacitances of 180 F/g, high specific power (20 kW/kg) at a specific energy range of 7–605 Wh/kg [138, 139]. Nevertheless, the capacitance values are low because the nanotubes tend to aggregate in bundles, decreasing the surface area. In addition, SWCNTs usually contain amorphous carbon, which contributes to the surface area decrease [140, 141]. Physical activation of these SWCNTs has shown higher capacitance values (180 F/g) than without any activation (113 F/g), due to a greater surface area and more mesoporosity (3–5 nm) [142]. On the other hand, MWCNTs catalytically grown with a surface area of 430 m^2/g have shown a lower capacitance value of 113 F/g and a lower specific power of 8 kW/kg at an energy density of 0.56 Wh/kg compared to SWCNTs [143]. The limitations in their effective surface-area have driven their use into the preparation of nanocomposites based on the incorporation of pseudocapacitive materials, as in the case of CA [39]. CNTs are high-cost materials that have limited their application in SC technology, but they might become commercially viable when the prices fall, the purity is improved, and the non-environmental procedures for their synthesis are addressed [144].

Graphene is defined as a two-dimensional material with tree-coordinated sp^2 carbons formed by six-carbon rings joined in a honeycomb form, and in some cases taking part of other carbon-based structures such as nanotubes, fullerenes and graphite (see Fig. 8.14). In general, graphene consists of one exfoliated sheet from graphite structure, but also 2 to 10 sheets of graphite with an interlayer spacing of 0.34 nm is considered as graphene.

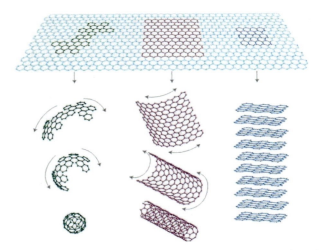

Figure 8.14 Graphene matrix with different structures of carbon as fullerenes, carbon nanotube, and graphite, respectively. Copyright [145].

Graphene was discovered in 2004 by Andre Geim and Konstantin Novoselov [146, 147]. In order to understand its electric behavior, as well as its mechanical properties, a great effort is being carried out to obtain isolated graphene. Amongst the most significant properties, graphene has a theoretical specific surface area up to 2600 m^2/g [148], thermal conductivity of 3000 W/mK dominated by photons with high-speed electron mobility and mechanical stress up to 1060 GPa [149–151], a flexible electric conductivity of 0.96 × 10^6 S/cm [147, 152], and a density of 2.2 g/cm^3 [152].

The electronic structure of graphene is rather different from usual three-dimensional materials. Its Fermi surface is characterized by six double cones (Fig. 8.15). Intrinsic graphene (undoped) is situated at the connection points of these cones (Fig. 8.15a–b). However, the Fermi level can be changed by an electric field so that the material becomes either n-doped (with electrons) or p-doped (with holes), depending on the polarity of the applied field. Graphene can also be doped by adsorbing, for example, water or ammonia [147, 153]. The electric conductivity for doped graphene is potentially quite high, at room temperature and it may even be higher than for copper. Therefore, graphene is making possible to go beyond the silicon age [154].

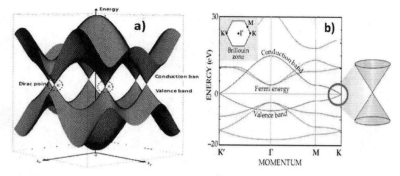

Figure 8.15 Bloch band description of graphene's electronic structure. The orbital energies depend on the momentum of carriers in the crystal Brillouin zone, and the formation of conical valleys between the conduction and valence band of the K at K' points. Copyright [155] and [156], respectively.

Graphene can be obtained using four principal methods [157]: (1) Epitaxial growth on a SiC single crystal [158–160] or chemical vapor deposition on metal substrates [161–163] to obtain high-quality large-scale graphene sheets or films; (2) the oxidation, exfoliation, and reduction of graphite to obtain graphene powder through thermal and chemical oxidation procedures, this method is known as the modified Hummer method [164, 165]; (3) Direct exfoliation of graphite with different procedures as sonication with different solvents [166], intercalation [167], electrochemically [168, 169] or by a quenching method [170], and finally (4) a methodology for obtaining "nanosize" graphene by organic coupling reactions [171, 172].

The most relevant properties of graphene for its use as electrode material for SC are the high specific surface area, electric conductivity, and its atomic thickness. These properties are important because they provide (1) higher specific capacitance, (2) low resistance, and (3) ultrathin thickness that obviously shorten the diffusion distance of ions improving ionic conductivity. Aside from these properties, the thermal and chemical stability, structural flexibility for constructing flexible electrodes, the possibility of abundant surface functional groups that make it hydrophilic in aqueous electrolytes, and finally a wide electrochemical window that is critical for increasing energy density, are key properties of graphene that makes it a promising electrode material for SC.

Different theoretical specific surface area values for graphene have been reported by different authors, between 2620 and 2965 m^2/g. These values depend on the calculation method or some considerations; for example, for 2965 m^2/g it was considered both sides of the graphene sheet [148, 173, 174]. For the case of graphene electric conductivity, this can be increased by thermal treatments at elevated temperatures, similarly to the reported for SWCNT's, and as a consequence the specific surface increases, decreasing the internal resistance, and therefore favoring the energy storage [175].

The first works based on the use of graphene as electrode material for supercapacitors cells were reported in 2007 [176]. These began with the use of carbon allotropes, for example: "carbon onion" or "fullerenes onion" in an organic electrolyte, showing specific capacitances of 22–40 F/g, while the MWCNTs provided 12–22 F/g. However, when the current density applied increased up to 50 mA/cm^2, the capacitance of graphene decreased a 10–35%, while in MWCNT only 10% [177]. For the case of SWCNT's, with structure more similar to graphene and prepared at 1000°C, different values of specific capacitance were obtained. For example, values up to 180 F/g at 1 mA/cm^2 and a specific power of 20 kW/kg was obtained using a symmetric configuration with a thin polymer (Celgard) separator in 7.5 N KOH aqueous solution as the electrolyte [175]. Moreover, a specific capacitance of 30 F/g at 1 mA cm^{-2} was obtained using a solid electrolyte for printable and flexible thin film SC assembled with SWCNTs [178], and a specific capacitance value of 160 F/g for electrodes based on packed sheets of aligned and pure SWCNT's obtained by chemical vapor deposition (CVD) [179]. More recently, the performance of SC based on graphene has been highly improved due to three different preparation methods: thermal exfoliation of graphitic oxide (EG), heating of nanodiamonds in helium atmosphere (DG), and decomposition of camphor on nickel nanoparticles (CG). The capacitance of these SC cells showed specific capacitance values up to 117 F/g for EG in H_2SO_4, followed by DG (35 F/g), and CG (6 F/g). In addition, the reported values for the maximum specific energy were 31.9 and 17 Wh·Kg^{-1} using ionic liquid $PYR_{14}TFSI$ by EG and DG, respectively [180]. Chemically modified graphene SC cells gave specific capacitances between 135 and 99 F/g in an aqueous and organic electrolyte, respectively at 10 mA [181]. In general, SC assembled with graphene electrodes have shown specific capacitance of 100–205 F/g, similar and in some cases higher than for SWCNTs [180–182].

As mentioned above, the capacitance of graphene, as in the other carbon materials, is affected by several aspects as pore size and pore distribution, available surface area, conductivity that can be affected by defects in its structure, and the restacking of their sheets that considerably decreases its theoretical capacitance value of 550 to ~100 F/g [5, 183, 184]. In order to avoid this restacking, it is necessary a functionalization, oxidation, doping, or thermal treatment to obtain high capacitance and energy density. Nevertheless, it has been observed that some functional groups decrease their power density and cyclability [185, 186]. The specific capacitance values reported for graphene are between 100 and 348 F/g [180, 181, 187], and are dependent mainly on the preparation method, structure modification or doping level, used electrolyte, and system employed for the characterization: (three-electrode) half-cell or complete cell [181, 182, 188–192]. In addition, higher capacitance from graphene compared to MWCNT's has been obtained. For example, MWCNTs treated with H_2SO_4 and/or HNO_3, resulted in a specific capacitance around 40 F/g (three-electrode cell) obtained by EIS data fitting [193]. Furthermore, analysis of parameters affecting the specific capacitance for pristine MWCNTs resulted in values no more than 80 F/g [194], lower than the lowest value obtained for graphene (100 F/g) [181, 184]. Comparing graphene with ACs, the capacitance values in the majority of cases are similar to the typical values of ACs (100–120 F/g) in organic electrolytes [97, 181].

8.5.1.2 Pseudocapacitors

Redox SC, also known as pseudocapacitors, show an increase in their specific energy compared to EDLC due to an additional storage process known as "pseudocapacitance." The formation of the non-faradic double layer storage process is always implicit, but when pseudocapacitance is presented in a material the capacitance value increases up to 10–100 times, doubling the energy density compared to that of EDLCs [36]. The prefix "pseudo" is used to distinguish it from the pure electrostatic process. This behavior is associated with three different processes, as mentioned in Section 8.4.2: (1) ion electrosorption on the surface from the electrolyte, as H atoms; (2) redox reactions involving ions from the electrolyte, for example metal oxides or the doping-undoping process of conducting polymer materials in the electrode, and (3) three-dimensional intercalation, as for the Li-ion batteries and more recently in high ordered nanometal oxides [38, 195–198]. Nevertheless, the redox process is the pseudocapacitive process most commonly known and the reason why this type of devices have been named generally as redox SC. The main difference with hybrid SC, where double layer and pseudocapacitance is also present, is that in pseudocapacitors the major contribution to the whole capacitance value is due to a pseudocapacitance single material, as in conducting polymers and metal oxides.

Generally, the pseudocapacitive processes are primarily surface mechanisms of electrochemical charge–transfer processes (ionic or electronic) between electrolyte and the electrode. Therefore, the specific capacitance is highly dependent on the electrode material surface area [39, 195]. The main drawback of pseudocapacitors is their limited cycle life compared to EDLC, but their higher specific energy is their main attractiveness.

Materials that may exhibit these types of processes are: transition metal oxides (as RuO_2, MnO_2, NiO, Co_3O_4, Fe_3O_4, V_2O_5, IrO_2, PtO_2, PbO_2, $NiCo_2O_4$) [38, 199–205]; metal hydroxides (as $Ni(OH)_2$, $Co(OH)_2$) [206, 207], and intrinsically conducting polymers. The most common polymers are poly(aniline) [208, 209], poly(pyrrole) [210–213] or derivates of poly(thiophene) as poly(3-methylthiophene) [214, 215] or poly(3,4-ethylenedioxythiophene) [216, 217].

8.5.1.3 Conducting organic polymers

Intrinsically conducting polymers, also known generally as conducting polymers (CPs), are materials that by their chemical structure have the capacity of conducting electrons. These materials have been under development with more intensity since 1971, when Shirakawa and Hikeda, prepared conductive films from polymerization of acetylene [218]. Later, MacDiarmid in 1977 observed the effect of using a treatment with I_2 as dopant agent, where the polymer experienced an increase of the electric conductivity values up to ten orders of magnitude [219]. All these works received the Nobel prize in 2000 (A. J. Hegger, A. G. McDiarmid, and H. Shirakawa), and since then many polymers have been synthesized with different methodologies, and their properties are still being elucidated as well as their possible applications.

CPs show high electric conductivities, with values corresponding from semiconductors (10^{-11} to 10^{-3} S·cm^{-1}) to metals (10^{-1} to 10^{6} S·cm^{-1}). These values depend on the polymer chemical nature, synthetic method, concentration and chemical nature of the dopant agent, temperature, pH, etc. In general, the structure of these types of polymers are linear with conjugated double bonds, aromatic rings, and/or heterocyclic compounds, such as poly(acetylene), poly(anile), and poly(pyrrole) (Table 8.4).

Table 8.4 Common intrinsically conducting polymers

Polymer	Structure
Poly(acetylene)	
Poly(pyrrole)	
Poly(thiophene)	
Poly(p-phenylene)	
Poly(p- phenylene vinylene)	
Poly(aniline)	

In order for polymers to conduct electricity, they must present conjugated $\pi-\pi^*$ bonds in their backbone structure, which favors the electric charge delocalization in presence of a dopant agent. The dopants can be cations or anions that are added during the synthesis of CPs. These ions are derived from salts or by dissociation of strong acids. The term doping is used to describe the result of reversible electrochemical oxidation or reduction, and in analogy with semiconductors they are named as p-doping and n-doping, respectively (Fig. 8.16). When the polymer is doped, it promotes the formation of new energy levels, diminishing the band gap between the conduction and valence bands, and the electron flux from the valence to the conduction bands is permitted. Nevertheless, the dopant agent is only at the polymer surface and does not alter its structure as it does in typical inorganic semiconductors.

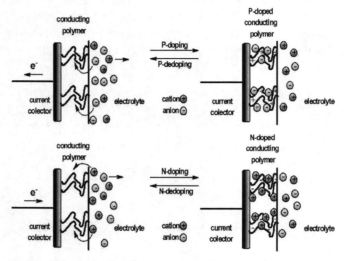

Figure 8.16 Charge/discharge process in two different types of doped polymeric electrodes in a supercapacitor cell [1].

These materials can be obtained by three main synthetic procedures: direct synthesis, chemical oxidation, and electrochemical oxidation synthesis. The direct synthesis was developed by Shirakawa in 1971 and consisted of covering the beaker inner wall with a Ziegler–Natta catalyst followed by the addition of acetylene monomer, forming a brilliant poly(acetylene) film [220]. On the other hand, the chemical oxidation methology is carried out with the monomer in solution, where a strong oxidant agent is added. This reaction results in a precipitate, which is the polymer. This chemical method has the advantage of obtaining large amounts of polymer powder or colloidal dispersions with moderate conductivity (10^{-5}–10^{-7} S cm^{-1}), but the doping level is difficult to control [221]. Finally, the electrochemical oxidation is similar to the chemical oxidation. This heterogeneous process is carried out in an anode surface arranged in a typical electrochemical cell in presence of an electrolyte (strong acid or dissolved salt) and the monomer. The electric current is applied, favoring the polymer oxidation that is dependent on the applied potential or current [221, 222]. This last method can provide high conducting films, controlled polymerization rate with the applied current or potential, but is not possible to obtain great amounts of polymer, making this process very expensive. Nevertheless, there are other less popular methods to obtain CPs, as partial pyrolysis of non-conducting polymers, photoinduced polymerization [223], or by condensation [224].

The possibility of using CPs as electrodes for redox SC was first suggested by Gottesfeld and Conway [36, 225]. CPs are very attractive materials as electrode due to its high electric and ionic conductivity as mentioned before, and low cost. In addition, these materials have been proposed for the fabrication of lightweight, compact and flexible devices that can be environmentally friendly due to the lack of heavy metals. Other attractive properties of these polymers in advantage with metal oxides are

- Low cost
- High conductivity in doped state
- High charge/discharge cycling rate, due to the reversible doped/dedoped process
- Easy to obtain a great amount of material
- Possibility of obtaining high conducting films with great surface and less thickness
- High specific capacitance with theoretical values of $\approx$ 700 F/g, where experimental values up to 500 F/g have been obtained [226, 227].
- Good reversibility of the systems
- Infrastructure already exists for battery manufacturing that can be used.

Capacitance for these polymers can be increased by controlling the synthetic parameters, and by designing different morphologies. These materials have shown problems when used as electrode in SC due to detrimental effect during cycling, which is related to a capacitance loss after some charge/discharge cycles (from 500 to 20,000 cycles depending on the conducting polymer nature and electrolyte used). This decrease through cycling is due mainly to swelling of the polymers leading to their degradation [226].

These materials can be synthesized electrochemically or chemically. For the first case, the conducting polymers can be electrochemically synthesized directly on the current collector surface as a thin layer, with high electric conductivity and controlled doping level. Also, with this synthetic approach, they can be obtained in their reduced state for their later use in asymmetric SC [228], or with different morphologies, such as nanofibers [229], highly porous structures [230], microtubes, nanowires, or nanotubes [231]. In the second case, a greater amount of material can be obtained. In most cases, the chemically obtained polymer results insoluble in typical solvents, and shows a moderate conductivity due to less doping level control. Nevertheless, the electrode fabrication process (Section 8.7.1) to assemble a SC cell involves a homogeneous mixture of the CP and a conducting additive (as graphite) to decrease the ohmic resistance of the electrode.

SC assembled with CPs as electrodes can be classified in three different types according to their oxidation state and cell configuration, showing particular properties [226, 232]:

- *Type I (symmetric)*: Both electrodes used to assemble the SC cell are elaborated with the same polymer in the same oxidation state (p-doped state), in a cell configuration known as symmetric assembly.
- *Type II (asymmetric)*: The electrodes used in the SC cell have different chemical polymeric structure, but with the same oxidation state (p-doped)
- *Type III (symmetric)*: Both electrodes assembled in the SC cell are made with the same polymer, but one is in its oxidized state (p-doped) and the other in its reduced state (n-doped).
- *Type IV (asymmetric)*: One electrode is a p-doped polymer and the other is n-doped.

Each type of SC offers different operational voltage window due to the nature of the electrode materials used in their assembly. For example, type I SC cell shows their best performance when a voltage window of 1 V is used, and 1.5 V is observed for type II. Nevertheless, these two types of cells can also work with organic, as well as with aqueous electrolytes, reaching up to 2.4 and 1.25 V, respectively [209, 213, 226, 232–235]. On the other hand, type III SC cells can work up to 3 V and type IV from 1.3 to 3.5 V, due to this increase in the operational voltage window, only organic electrolytes can be used [21, 233, 236, 237]. Therefore, this last cell is the most promising in terms of energy and power density, but the main drawback is the difficulty to obtain n-doped polymers in an efficient way [209, 232]. In addition, all cells have shown good capacitance retention near 100%, for up to 5000 charge/discharge cycles in asymmetric and symmetric SC based on Poly(3-methylthiophene), (pMeT), poly[3-(4-fluorophenyl) thiophene] (pFPT) and poly(thiophene) (pTh) [236, 237]. Table 8.5 shows conductivity, intrinsic theoretic specific capacitance, and operational voltage window values reported for the most common CPs. The highest theoretical intrinsic specific capacitance is presented by poly(aniline), followed by poly(pyrrole). This theoretical capacitance considers the polymer doping level ($\alpha < 1$), the total electrons withdrawn per monomer unit (β, between 2.0 and 2.7 depending on the nature of the polymer), and has been calculated as follows:

$$\beta - 2 = \alpha \tag{8.19}$$

$$C_{th} = \frac{\alpha \cdot F}{\Delta E \cdot M}, \tag{8.20}$$

where F is the Faraday constant (96,485 C/mol), M the molecular mass of the monomer unit in the polymer, and ΔE the potential range [238].

Table 8.5 Conductivity, voltage range and theoretical specific intrinsic capacitance values reported for the most common polymers used as electrodes in supercapacitors

Polymer	Conductivity (S cm^{-1})	Voltage vs. SCE (V)	Theoretical specifc intrinsic capacitance (F g^{-1})	Ref.
Poly(pyrrole)	10–50	−0.1–0.8	620	[221, 226]
Poly(thiophene)	300–400	0.1–0.8	485	[226, 239]
Poly(3-methylthiophene)	400–600	−0.2–1.15	220	[226, 240]
Poly(aniline)	0.1–5	0–0.7	750	[226, 239]

8.5.1.4 Transition metal oxides

Pseudocapacitors based on metal oxides have received a lot of attention due to the higher capacitances and energy densities compared to carbon materials in EDLCs, and better electrochemical stability than conducting polymers. These oxides store energy through charge transfer reactions at the interphase between the electrode surface, and ions from the electrolyte in a specific voltage window [203]. For this reason, transition metal oxides have been considered as good candidates for the development of pseudocapacitors [201, 241, 242]. Nevertheless, their improvement is still needed for their practical use in SC, and the trend nowadays is their design at the nanoscale. This approach involves morphology and chemical composition design in transition metal oxides, in order to improve the electrolyte penetration to the entire electrode material promoting the electric double layer capacitance, as well as increasing the surface electroactive sites participating in the pseudocapacitive reactions [242].

In general, metal oxides need to fulfill certain requirements for their application in SC, such as electric conductivity, show two or more oxidation states coexisting in a continuous range with no phase changes involving irreversible structural modifications, and protons that can freely intercalate into the oxide lattice on reduction; allowing an easy $O^{2-} \leftrightarrow OH^-$ reversible reaction [1]. The pseudocapacitive properties of metal oxides are related with their multiple oxidation states, and are typically classified as noble transition metal oxides and base metal oxides. Noble transition metal oxides as RuO_2 and IrO_2 show good conductivities, the best capacitive properties (higher than CPs and ten times higher than carbon) [39], excellent power densities, but they result very expensive and harmful to the environment [243]. On the other hand, base metal oxides as MnO_2, NiO, Fe_3O_4, etc., are environmentally friendly and cheaper materials with good capacitive properties comparable to noble transition metal oxides [201, 242, 244].

In the last few decades, RuO_2 has been the most promising electrode material due to its highly reversible redox reactions, high proton conductivity, high specific capacitance, good thermal stability, high cycle life, high electric conductivity, and high rate capability [243, 245–247]. Nevertheless, its low abundance has resulted in a high-cost production, which is a disadvantage for commercial purposes. Other noble transition metal oxide as IrO_2 has not drawn much attention in comparison with RuO_2 due its heavy atomic mass causing a low specific capacitance [248].

The electrochemical characterization of RuO_2 results in cyclic voltammograms with a rectangular shape as shown in Fig. 8.17, due to multiple redox states with good electric conductivity, resembling the carbon-based electric double layer capacitors [23]. These multiple redox states are related to three different oxidation states accessible in 1.35 V for amorphous ruthenium oxides [249], where electro-adsorption of protons participate reversibly in the redox reaction involving Ru(II) to Ru(IV) with a theoretical specific capacitance up to 2200 F/g (Table 8.6), and the reaction is as follows [23, 195, 250–256]:

$$RuO_2 + xH^+ + xe^+ \leftrightarrow RuO_{2-x}(OH)_x, 0 \le x \le 2 \qquad (8.21)$$

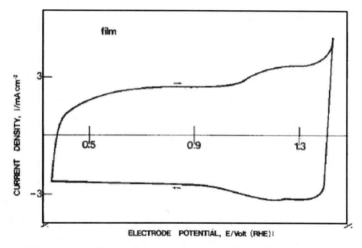

Figure 8.17 Square-like cyclic voltammogram of pseudocapacitive behavior of RuO_2 film in $HClO_4$ at 40 mV/s. Copyright [23].

In order to improve the electrochemical behavior of Ru oxides regardless of the electrolyte used, the following factors play a key role: the specific surface area, the combined water to improve protonic conduction, the crystallinity degree, and the particle size [257]. On the other hand, in order to decrease the fabrication cost of RuO_2, different synthetic approaches (Table 8.6) have been proposed focusing mainly on the factors mentioned above, and on the addition of other cheaper metal oxides (SnO_2, MnO_2, NiO, VO_x, TiO_2, MoO_3, WO_3, CaO, etc.) to improve their specific capacitance [243, 258–261]. In this last case, an important material is $Ru_{1-y}Cr_yO_2/TiO_2$ that has allowed easier electrochemical reactions due to a more expose active sites of the active material resulting in capacitances as high as 1272 F/g when using TiO_2 nanotubes as the matrix [262].

Nowadays, manganese oxides have attracted much attention due to their low cost, eco-friendly properties compared to the rest of metal oxides, easy fabrication, and high theoretical specific capacitance (up to 1370 F/g, Table 8.7) [282]. The pseudocapacitance contribution of MnO_2 has been attributed to the redox oxidation of surface oxy-cation species of its various oxidation states, and currently there are two different theories related with its charge storage mechanism. The first theory is similar to the observed in RuO_2 and is related to the intercalation of protons or/and alkali metal cations (C^+ as Li^+, Na^+, K^+) to the bulk of the material upon reduction and deintercalation upon oxidation, as follows [269, 273, 283]:

$$MnO_2 + \delta C^+ + 5e^- \leftrightarrow MnOOC^+ \tag{8.22}$$

On the other hand, the alternative theory states that the redox process occurs only on the surface of MnO_2 as an adsorption-desorption process, as the following reaction [282, 284].

$$(MnO_2)_{surface} + C^+ + e^- \leftrightarrow (MnO_2^-C^+)_{surface} \tag{8.23}$$

The proposed mechanisms are based mainly in a redox reaction between the oxidation states III and IV of manganese [242], but the oxidation states II/III, IV/VI cannot be ruled out [285, 286]. In addition, as observed for RuO_2 cyclic voltammogram (Fig. 8.18), MnO_x-based electrodes show also a rectangular-shape profile analogous to non-faradaic energy storage mechanism. Moreover, the cycle stability of Mn oxides relies on the control of the microstructure, and the specific capacitance from pseudocapacitance are affected by several factors, such as crystallinity, crystal structure, morphology, thickness of the electrode layer, specific surface area and pore structure, and chemical factors. Mn oxides with high crystallinity show higher conductivity, but the lower surface area makes the storage mechanism more limited. Therefore, the higher electric conductivity observed in the crystalline solid phase and the higher ionic transport observed in the higher porosity of the amorphous phase should be controlled, by controlling the annealing temperature during the synthesis [257]. The synthesis of Mn oxides is complicated due to a great diversity of allotropic forms (α, β, γ, λ). Mixtures of all allotropic forms during the synthesis are obtained, complicating the charge capacity evaluation for each phase, and increasing controversy in the scientific community. Nevertheless, the specific capacitance evaluated for the different crystalline forms of MnO_2 prepared by an aqueous-phase method, resulted as follows: α (240 F/g) $\approx$ δ (236 F/g) > γ (107 F/g) > λ (21 F/g) > β (9 F/g) [287].

Most of research in the area is moving toward different preparation methods resulting in nano-micro materials with controlled sizes and ordered morphologies due to the relation with specific surface area, and therefore with capacitance performance [270]. It is well known that higher surface area and good pore size distribution will result in higher capacitance values due to an enhanced ionic conductivity. Therefore, surface area and pore distribution in MnO_2 has to be controlled, and an approach used has been based on the adjustment of the reaction time and the surfactant concentration when using an aqueous system [271]. On the other hand, physically and chemically bounded water molecules seem to play an important role on implementing the ion transport, as well as the chemical state of Mn in the oxide. Finally, the poor conductivity of these oxides requires the fabrication of thin electrodes in order to have a higher effective utilization of the electrode material, resulting in higher capacitance values [249, 282]. Different synthesis techniques can be applied to obtain different manganese oxides, as shown in Table 8.7, where the sol-gel route seems to result in the highest capacitance value.

Nickel oxide is another environmentally friendly potential candidate for its use as electrode material in pseudocapacitors [201, 277], is cheap and abundant, has chemical and thermal stability, and has shown good electrochemical activity based on the transformation of $Ni(OH)_2$/NiOOH in alkaline media [266, 288]. The storage mechanism involved results in a theoretical capacitance of up to 3750 F/g (Table 8.7) that depends strongly on crystallinity [289], and the reactions are as follows [242]:

$$Ni(OH)_2 + OH^- \leftrightarrow NiOOH + H_2O + e^- \qquad (8.24)$$

$$NiO + OH^- \leftrightarrow NiOOH + e^- \qquad (8.25)$$

Nevertheless, its instability at high current densities (formation of Ni(OH) in the surface) results in a drastic decrease of the specific capacitance. Although, it shows a high specific capacitance, its cycling life is limited due to microstructure deterioration (4.5% capacitance loss after 400 charge/discharge cycles) [203]. Thus, the utilization of this oxide is limited to alkaline aqueous electrolytes; dissolution of Ni^{+2} is carried out below a pH of 9. It is important to point out that a clear redox couple is observed, compared with the rest of the oxides with a square-like voltammogram. In Table 8.7, capacitance values obtained using different syntheses are shown, where the sol-gel route resulted in an electrode material with enhance performance similar as in the case of MnO_2.

Iron oxides as Fe_2O_3 and Fe_3O_4 are other attractive electrode materials for pseudocapacitors due to their low cost and minimal environmental impact. Fe_2O_3 has poor electronic conductivity, restricting its application in high power storage devices. On the other hand, Fe_3O_4 in alkaline sulfite and sulfate aqueous media exhibits pseudocapacitance, which is very sensible to the anion species but not to either alkaline cations or electrolyte with high pH < 11 [249, 290]. This behavior suggests that a different mechanism is taking place during the charge storage, compared to the one presented in RuO_2 and MnO_2. Charge storage mechanism of Fe_3O_4 in Na_2SO_3 (170 F/g), Na_2SO_4 (25 F/g), and KOH (3 F/g) aqueous solutions in function of different Fe_3O_4 stoichiometries have been studied [291], and compared to other oxides (Mn, Fe, Co, Ni) with similar crystalline structure, showing an influence of the crystalline structure in the obtained capacitance value [292]. Nevertheless, none of the iron oxides seems realistic materials for SC devices due to their low capacitance values and cycle stability [257].

V_2O_5 has been also studied as electrode material for SC due to their variable oxidation states, but with less emphasis. It has been synthesized by a co-precipitation method and a posterior heat treatment above 300°C [293], resulting in high surface area with a maximum capacitance value of 262 F/g in KCl aqueous electrolyte. Moreover, higher capacitance values have been obtained (350 F/g) for amorphous powders obtained by quenching V_2O_5

heated at 950°C in the same electrolyte [294]. This capacitance changes by using different electrolytes as 1 M NaCl and 2 M LiCl. Nevertheless, the challenge for this vanadium oxide is still its low conductivity.

Table 8.6 Comparison of different transition metal oxides, where their theoretical specific capacitance are compared to different experimental values obtained by different synthetic routes

Transition metal oxide	Theoretical specific capacitance (F/g)	Synthesis method and experimental specific capacitance
RuO_2	1300–2200 [263]	chemical deposition 325 F/g [264]
		Sol-gel (38%w loading) 570 F/g [264]
		Electrochemical deposition 1300 F/g [263]
MnO_2	1100–1370 [266–270]	Hydrothermal 168 F/g [271]
		Sol-gel 698 F/g [267]
		Electrochemical deposition 310 F/g [266]
		Sonochemistry 350 F/g [272, 273]
NiO	2573–3750 [274–279]	Chemical Deposition 390 F/g [280]
		Sol-gel 698 F/g [281]
		Electrochemical Deposition 277 F/g [275] 100–200 F/g [201, 277]

Summarizing, the research focus of transition metal oxides can be as follows: (1) Noble metal oxides (RuO_2, IrO_2) with great specific surface area prepared by different methodologies, (2) Noble metal oxides mixed with other metal compounds to increase the specific capacitance, and to reduce cost, (3) the use of other metal oxides to reduce cost, as manganese oxides, iron oxides, nickel oxides, and vanadium pentoxide among others [242, 249, 257].

Figure 8.18 shows the specific intrinsic capacitance of different types of materials used as electrodes in SC, where pseudocapacitive materials (CP and Metal oxides) result in higher specific capacitance values compared to carbon. Based on this data, RuO_2 and PAni are the best pseudocapacitive materials.

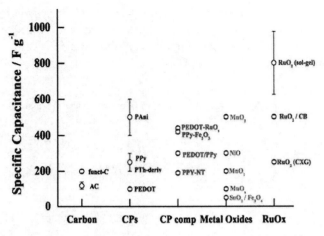

Figure 8.18 Specific intrinsic capacitance values reported for different electrode materials. Copyright [226].

8.5.1.5 Hybrid supercapacitors

A hybrid supercapacitor is a concept where both energy storage mechanisms, double layer and pseudocapcitance, are in combination and tuned in a device [285]. There are different approaches to tuning or combining these mechanisms in a device: (1) through material design or engineering by introduction of pseudocapacitive contributions to carbon matrices (e.g., functionalization with electroactive functional groups, introduction of metal oxides or CPs) to obtain in some cases nanocomposite or hybrid materials, and (2) through the design of an asymmetric assembly or hybrid device, where one electrode is based on a double layer material and the other electrode based on a pseudocapacitive (e.g., metal oxides or conducting organic polymers) or battery type (e.g., PbO_2 material). In the following sections, a detailed description of each type of hybrid supercapacitors used to advance in this line of research will be presented.

8.5.1.6 Functionalized carbons

The performance of carbon as electrode material for SC can be further improved by introducing faradaic pseudocapacitance, as it has been explained before. The increase of capacitance of any carbon material is carried out by chemical modification to create surface groups that change the interfacial state of carbon and its double layer properties, such as wettability, point of zero charge, electrical contact resistance, adsorption of ions, and self-discharge characteristics [1, 54, 56–58]. In addition, the introduction of surface groups can induce redox reactions (pseudocapacitance) with electroactive centers, such as functional groups containing nitrogen, oxygen, phosphorous, or sulfur [295–299]. This functionalization process is sometimes more useful than provoking high porosity in carbon structures due to a detrimental effect on electric conductivity values. The methodologies commonly used for the introduction of electroactive functional groups onto a carbon surface can take place during the synthesis, posterior oxidation treatments, and grafting [300], among others [288, 289].

Carbons used for EDLC are synthesized by pyrolysis of different precursors, as previously described (Section 8.5.1.1). The composition of these precursors sometimes can present certain concentration of heteroatoms (O, N, S, halogens), which can become part of the chemical structure as a result of partial pyrolysis. Most of the obtained carbons doped with these heteroatoms can be included within the activated carbon group classification (Section 8.5.1.1.1), if the functionalities introduced are electrochemically inert in the potential range of operation. Thus, improvement on specific capacitances is due to improved pore access (greater surface utilization) as a consequence of enhanced wettability [301]. Nevertheless, some of these doping heteroatoms introduce pseudocapacitance, additional to the double layer storage mechanism. For example, research related with nitrogen enrichment has resulted in an interesting route to improve capacitance [302]. In this sense, nitrogen atoms act as electron donors that interact with protons from the electrolyte, giving rise to a pseudocapacitance [303]. For example, N-doped mesoporous carbon by ammoxidation has resulted in an enhance capacitance from 119 F/g up to 182 F/g [304]. Moreover, carbons doped with boron and nitrogen have shown a stable pseudocapacitive behavior capable of increasing hydrogen overpotential evolution to almost –1.4 V vs. Ag/AgCl [290], making this type of carbon suitable for their use as negative electrode in an asymmetric configuration with aqueous electrolyte.

Functional groups can also be included to carbon after their synthesis. As in nitrogen doped carbons, oxygen enriched carbons also show higher capacitances due to a pseudocapacitive contribution. Carbon–oxygen complexes are by far the most important surface group among carbons, showing ketone, lactol, ether, carboxylic acid, carboxylic anhydride, lactonic, and phenolic groups as shown in Fig. 8.19. These groups are formed when carbons are exposed to an oxidizing gas between 200 and 700°C or by a controlled oxidation using treatments in solution at room temperature up to 100°C. The degree of oxidation or functionalization depends on the strength of the oxidants that include oxidizing gases (air, oxygen, nitrous oxide, nitric oxide) and oxidizing solutions (HNO_3, H_2O_2, persulfates, hypochlorite, permanganates, dichromates, and chlorates). Functionalization conditions must be carefully chosen in order to prevent excessive corrosion and structural collapse of the carbon skeleton. Moreover, the type of oxidant determines the types of surface groups, which can be characterized by XPS, FTIR, electrochemical techniques, and by thermogravimetric analysis (TGA) giving information on CO and CO_2 desorption [292]. For example, the commonly used oxidant HNO_3 is a highly efficient and controllable agent mainly creating carboxylic groups [294, 305].

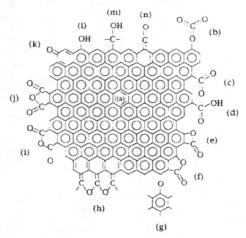

Figure 8.19 Schematic diagram of oxygen-based functional groups on oxidized carbon: (a) Aromatic C=C stretching; (b) and (c) carboxylic-carbonates; (d) carboxylic acid, (e) lactone (four-membered ring), (f) lactone (five-membered ring); (g) ether bridge; (h) cyclic ethers; (i) cyclic anhydride (six-membered ring); (j) cyclic anhydride (five-membered ring), and (k) quinone. Copyright [291].

This introduction of oxygen-based functional groups turns the surface of carbons from hydrophobic to hydrophilic in nature, resulting in larger capacitances [3]. For example, carbon fibers treated under oxygen at 250°C showed a 25% capacitance increase (from 120 to 150 F/g, at 0.5 mA/cm^2) in H_2SO_4 aqueous electrolyte [295], and a 58% increase (from 163.5 to 280.7 F/g at 50 mA/g) when using KOH electrolyte [62]. Also, OMCs have shown an 11.2% increase (from 222 to 250 F/g, at 1 mV/s) in KOH aqueous electrolyte when treated with HNO_3 [305]. For the case of nanocarbons, the introduction of these types of functional groups improves their dispersion aside from the hydrophilic nature, resulting also in increased capacitance values. For example, graphene treated with HNO_3 showed 30 times higher capacitance (65 F/g at 0.5 A/g in KOH) compared with untreated graphene [306], and for MWCNTs treated in H_2SO_4 + HNO_3 solution almost three times higher capacitance was obtained (33.5 F/g at 25 mV/s) [72]. Although capacitance values are not impressive for bulk nanocarbons, it clearly shows the effect of these functional groups for all carbon materials.

Nevertheless, some of these oxygen-based groups are chemically unstable and may serve as active sites for the redox catalysis in carbon, or for the decomposition of the electrolyte components resulting in a high self-discharge rate when used in a SC cell [39, 295, 307]. Therefore, a proper degree of carbon oxidation will result in an enhanced capacitance, greater operational voltage window increasing the specific energy (explained in Section 8.6) [308], and improving power capability due to an enhance ion diffusion coefficient within the pore structure [309].

Conventional oxidative treatments traditionally used in carbon matrices do not precisely control the creation of oxygen-based functional groups, obtaining a cocktail of such groups. In addition, this methodology decreases the electric conductivity due to a significant breakage of the carbon structure causing mechanical deterioration and discontinuity of the conjugated sp^2 bonds [300]. On the other hand, chemical modification of carbon following the diazonium salt route represents an innovative procedure that allows the incorporation of functional groups in a selective way and with controlled coverage, without significant surface destruction [310, 311]. This route, known as "grafting," commonly takes place by the in situ diazonium cation generation from the reduction of an amine salt, which allows the modification of carbon surface by a substituted aryl group, as shown in Fig. 8.20 [311–319].

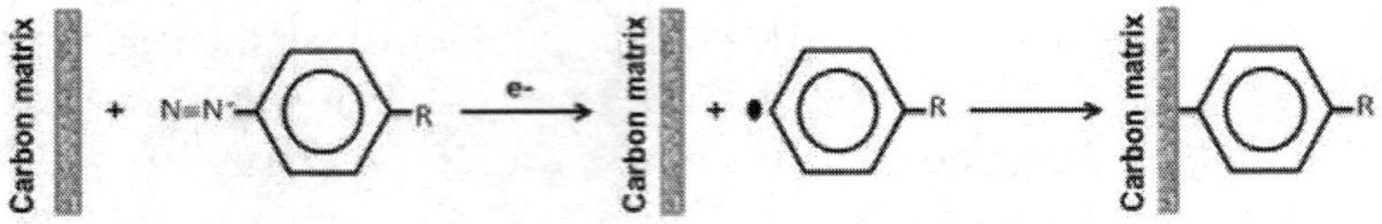

Figure 8.20 Schematic representation of grafting of carbon through the incorporation of functional groups by in situ diazonium cation generation, where R is the specific functional group to incorporate.

In Fig. 8.20, R is the functional group of interest, such as anthraquinone [315, 319], cyano groups [320], amino groups [317, 320, 321], nitro groups [314, 316, 319, 321], halogens [313, 316, 319, 321], carboxylic groups [321, 322], sulfur-based groups [312, 321], etc. It is important to mention that in this methodology the functional group is "added" to the carbon surface, affecting more the surface area than the double layer capacitance of the carbon. This is because the grafting procedure takes place preferable at the entrance of the ultramicropores and micropores when using anthraquinone as the functional group [315, 323]. Nevertheless, if this grafted functional group exceeds a certain concentration limit, the double layer capacitance can decrease, but the total capacitance can be compensated or improved by the pseudocapacitive contribution of such grafted group [315, 323]. The grafting of carbon with a sulfur-based group has improved the ionic conductivity, resulting in electrodes with higher charge-discharge rates [312].

8.5.1.7 Nanocomposites and/or hybrid materials

From a materials science point of view, nanocomposites and/or hybrid materials have been the latest approach to enhance the performance of electrode materials for SC cells. The addition of a faradaic contribution (pseudocapacitance) to the double layer capacitance is an attractive approach for performance improvement. This can be carried out by highly dispersing electroactive molecules with fast electron transfer, at a molecular level in a carbon matrix. In this sense, it has been observed that the addition of electroactive species as metal oxides or CPs on to carbon matrices to obtain nanocomposites and/or hybrid materials, overcomes some of the pseudocapacitive disadvantages as cycle life, without compromising the increased specific capacitance [144].

Different carbons have been used for the immobilization and/or dispersion of pseudocapacitive materials, where the most popular matrices are based on nanocarbons, and some work has been carried out in low-cost carbons with controlled porosity as carbon aerogels [144, 324]. For instance, CNTs are excellent 3D host for CPs and metal oxides for the design of these materials. They have played an electric conductive role for these pseudocapacitive materials, which has helped to improve the rate capability and by consequence the specific power of SC. However, there is an optimum concentration of CNTs that needs to be evaluated in order to prevent a detrimental effect on capacitance values due to their well-known agglomeration [3].

Different metal oxides have been incorporated to diverse carbon matrices, where their dispersion degree and concentration of each component have been the key parameters to improve SC performance. The most commonly metal oxides chosen for this hybrid design have been RuO_2 [200, 263, 325–327], SnO_2 [258, 328], MnO_2 [329–333], NiO[280] [334–336], MoO_3 [198, 337], cobalt oxide [333, 338, 339], polyoxometalates [340–343], VO_5, V_2O_5 [344, 345] and ZnO [346, 347], among others with good results.

Carbon nanotubes can play different and important role in metal oxide based hybrid nanocomposite materials, such as (1) electric percolating network with mechanical robustness that could avoid the use of conductive additives, and even be used as conducting enhancer (2) self-standing electrodes eliminating the use of mechanical binders and current collectors, which has a direct impact on lowering de ESR values, and (3) 3D matrix with tunable macro/mesoporosity for accommodating or supporting metal oxide particles [144, 178, 348–350]. The use of CNTs as a matrix for metal oxides improves the charge/electron transport, as well as inter-particle contact and mechanical integrity during extended cycling, improving the specific capacitance in these nanocomposite materials compared to the pure oxide electrodes [351]. For instance, MnO_2 shows a high resistivity and large electron transfer resistance in the electrolyte when the redox reactions occur, and the capacitive performance cannot meet the expected performance for practical applications. Therefore, the synergic effects of MnO_2-based nanocomposites and carbon materials have resolved the problematic of each component, making this nanocomposite route an efficient way for SC improvement [282]. In addition, it has been observed that when using aligned CNTs networks to directly deposit metal oxides in order to promote charge transfer, a decrease in ESR values are observed due to a better contact between the current collector and the electrode material. A high capacitance of up to 784 F/g has been obtained for γ-MnO_2 particles dispersed on to this type of arrangement [352]. For the case of the expensive RuO_2, nanocrystallites supported on CNTs sheets has resulted in specific capacitances of 1715 F/g [351], and the decreased amount used of RuO_2 in the nanocomposite has driven to a cheaper material. On the other hand, NiO is known for its instability in neutral and lower pH values, and has been supported in carbon materials not only to improve their stability but to obtain a higher capacitance value of 1329 F/g in alkaline electrolytes [353]. Moreover, NiO deposited on CNTs resulted in an ultrahigh specific capacitance of 1701 F/g normalized by the oxide content on an aqueous electrolyte [279].

Summarizing, metal oxide nanoparticles dispersed and/or immobilized in the surface of CNTs is an excellent approach to improve performance, because not only metal oxides are dispersed but also CNTs, which have a tendency to agglomerate. Thus, the surface area of CNTs can increase contributing more to the total capacitance of the hybrid material, as shown in Fig. 8.21, for some nanostructures.

Another novel carbon matrix that can be used to prepare hybrid materials is graphene, which is generally considered as an ideal building block in nanocomposite materials that combined with a variety of metal oxides show an exceptional performance. The main advantages of these nanocomposites are the agglomeration suppression of metal oxides and the re-stacking of graphene as shown in Fig. 8.22. The use of graphene as a matrix promotes a uniform dispersion of metal oxide particles, a high conductive and flexible network, a high capacitance, good rate capability, an improved cycle life, and an improved specific energy and power [354].

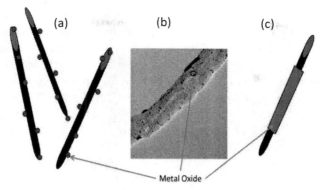

Figure 8.21 Schematic structures for some Hybrid nanocomposite materials based on metal oxide nanoparticles dispersed on CNTs (a) or wrapping the nanotube (c). (b) TEM image of an example of a hybrid material based on polyoxometalate nanoparticles anchored to the surface of a multiwall carbon nanotube.

Nanocomposites based on graphene/metal oxides, such as Co_3O_4 [355], RuO_2 [356], MnO_2, [357], Mn_3O_4 [358], MoO_3 [359], and ZnO [360], obtained with different synthesis procedures, have resulted in diverse capacitance values for materials and cells, from 11.3 to 570 F/g [354]. For example, PtRu/SnO–graphene nanocomposites prepared with a microwave-assisted one-pot reaction process provided a value of specific capacitance of 140 F/g at 40 mV/s [361], and a maximum specific capacitance of 243.2 F/g at 10 mV/s in 6 M KOH aqueous solution for graphene nanosheet (GNS)/Co_3O_4 nanocomposite [362]. In addition, nanocomposites based on hydrous ruthenium oxide/graphene sheets resulted in higher specific capacitance (570 F/g at 1 mV/s) in acidic aqueous electrolyte with a 38 wt% Ru loading, and an enhance cyclic stability was observed compared to pure graphene and pristine RuO_2 [264].

Nanocarbons matrices are ideal materials to incorporate metal oxides, and can be prepared by environmentally friendly processes [19, 20, 363]. Nevertheless, these matrices are expensive and the use of low-cost carbons obtained from biomass, represents an attractive approach for designing hybrid materials. MnO_2 and Fe_2O_3 are the most environmentally friendly of all oxides and they have been studied in these types of matrices. For instance,

MnO$_2$ with AC [364] and carbon aerogels [365–367] have been synthesized, resulting in high capacitance values (374 F/g at 3 mA/cm^2 and 226.4 F/g, respectively) with good charging-discharging characteristics, and at least 91% capacitance retention. For the case of Fe$_2$O$_3$, a green method has been used to obtain hybrids from mushroom-derived porous carbon [368], showing a maximum specific capacitance of 367 F/g (0.5 A/g) and at higher rates of 214 F/g (5 A/g) with good capacitance stability over 1500 cycles (82.7% capacitance retention). The capacitance for this nanocomposite was much higher than when using nanocarbons [369, 370]. Moreover, a facile template-free greener route has been used to fabricate natural polysaccharide based biocompatible mesh-like Fe$_2$O$_3$/C nanocomposite for SC [371], obtaining 295 F/g (0.5 A/g) and 100 F/g (5 A/g) with capacity retention of 88.9% after 1500 cycles. This same oxide with AC showed 177 F/g (5 A/g) with an 82.3% capacitance retention [372], and with well-ordered mesoporous carbon, 235 F/g at 0.5 A/g and 150 F/g at 5 A/g [372]. It is important to point that all these materials were studied using aqueous electrolytes.

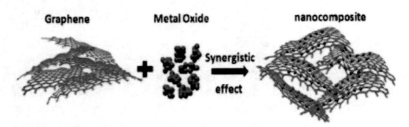

Figure 8.22 Schematic representation of graphene/metal oxide nanocomposites. Copyright [354].

In order to increase the performance of CP, research in the last 10 years has focused on the development of CP based nanocomposites [373, 374], or hybrid materials based on nanostructures of carbon (MWCNT, SWCNT) with CPs [216, 375, 376]. Table 8.7 shows some examples of composites based on carbon structures (carbon nanotubes, activated carbon, and/or mesoporous carbon) with conducting polymers, such as polypyrrole (PPy), poly-3-methylthiophene (P3MTP), and polyaninline (PAni).

One of the main drawbacks of using pure conducting polymers (CPs) is its well-known detrimental cycle life, which has been attributed mainly to swelling effects in the polymer during charge/discharge cycling. The use of carbon matrices to obtain nanocomposite materials with CPs has been a good approach to overcome in some extent the poor cycle life of these polymers. For example, CNTs have offered the mechanical properties necessary to adapt to any volume change during electrochemical performance resulting in an extended cycle life [381].

Table 8.7 Specific capacitance values reported for some nanocomposites or hybrid materials based on carbon structures and intrinsically conducting polymers

Electrode	Electrolyte	Specific capacitance ($F\ g^{-1}$)	Ref.
MWCNTs/PPy Nanocomposite	1 M LiClO$_4$	87*	[377]
MWCNTs/P3MeT Nanocomposite	1 M LiClO$_4$	45*	[377]
AC/PPy Nanocomposite	6 M KOH	300**	[378]
MWCNTs/PPy Nanocomposite	1 M H$_2$SO$_4$	165*	[379]
MesoC/PPy hybrid	2 M H$_2$SO$_4$	83.8*	[380]
MesoC/P3MeT Hybrid	2 M H$_2$SO$_4$	52.3*	[380]
MesoC/PAni hybrid	2 M H$_2$SO$_4$	57.6*	[380]

*Specific capacitance calculated from two-electrode cells.
**Intrinsic specific capacitance of the material calculated from three-electrode cells.

SC electrodes based on graphene/CP nanocomposite materials have been synthesized with different polymers obtaining different specific capacitance values. For example, hybrids using polypyrrole (G/PPy) with GO (graphene oxide) resulted in a higher capacitance of 267 F/g (at 100 mV/s) in aqueous acidic electrolyte, and good cyclability (90% retention after 500 charge-discharge cycles) compared with pristine PPy (137 mV/s and 47.4% retention) [382]. For the case of hybrids with aniline [383] or polyaniline (PAni) [384], and graphene (G/Ani and G/PAni), capacitance values higher than for PPy hybrids were obtained using the same acidic aqueous electrolyte (531 F/g at 200 mA/g for G/Ani, and 480 F/g at 0.1 A/g for G/PAni) [384]. Also, doubled intrinsic capacitance values were found just by re-doping PAni in the nanocomposite (1126 F/g) with 84% cycling retention after 1000 cycles [385].

Other type of nanocomposite materials have been the ones based on CPs used as binding agents for metal oxides, which enhanced the conductivity of metal oxides such as MnO_2 and NiO, and the capacitance values considerably increased making these combinations suitable electrode materials for SC. For example, higher capacitance values for MnO_2/PPy nanocomposite of 328 and 600 F/g were found compared to their pure components [386, 387]. Moreover, the latest trend in these nanocomposite materials approach is based on a triple hybrid nanocomposite concept, which is based on the use of a carbon matrix such as CNTs, CP, and/or metal oxides, resulting in an excellent capacitive behavior, cycle life, mechanical flexibility, and durability [370, 388]. MWCNT/PAni/MnO_2, MWCNT-PSS/Ppy-MnO_2, and SnO_2-V_2O_5-CNT are some examples of these types of materials, where the proper combination and concentration of each component needs to be optimized for an improved performance in SC cells.

8.5.1.8 Asymmetric assembly

The term asymmetric is derived from the US Patent 6,222,723 titled "Asymmetric Electrochemical Capacitor and Method of Making" by Razoumov et al. [389]. The main idea when developing asymmetric capacitors is to take advantage of the best properties of each type of electrode, [390] trying to decrease or eliminate their drawbacks to get in an ideal way a synergic effect, which can result in a novel device development with an enhanced performance. Asymmetric cells are usually assembled with one double layer type carbon material (dl), and a pseudocapacitative material (Pc) or battery-type (Bt) electrode (also known as hybrid device) with high rate charge-discharge characteristics. That is, one electrode stores charge through a reversible non faradaic process of ion movement to the surface of carbon maintaining fast charge-discharge capacitive response that is reflected in high specific power values, while the other electrode uses reversible faradaic reactions of fast redox reactions to increase the capacitance and specific energy of the device as in metal oxides, intercalation compounds, or conducting polymers [391–393]. This asymmetric assembly concept extends the effective operational voltage window, resulting in significantly higher specific energy than for EDLC and higher specific power than conventional battery systems, with low internal resistance and good cyclability [394–396]. Although, the typical assembly is with a *dl* electrode in one side and a Ps or Bt material on the other side, there are also all-carbon based asymmetric capacitors using the same or different carbon material in the positive and negative electrodes. This type of capacitor arises from the asymmetry observed on EDLC [33, 392], which can be optimized by balancing the mass in each electrode. Normally, this mass balance is carried out due to the different charge contributions of each electrode (positive and negative) with complementary voltage windows. In this way, this asymmetric device is optimized [397, 398]. In Table 8.8, the first three asymmetric assemblies are based on the same carbon material, but different masses ($R_{+/-}$) are used to enhance the performance. On the other hand, different carbons optimized for the positive and negative electrodes through different specific area and pore size (Table 8.8, row 7, 8, and 12) [102, 399, 400], or functionalization to introduce specific pseudocapacitive processes (Table 8.8, row 4 to 6, and 9 to 11) as redox reactions or H electrosorption [396] can be used for an asymmetric configuration in order to improve the energy densities [395, 398, 401–405]. Therefore, the design of carbon materials with controlled porosity properties and/or functionalities, together with the correct cell design with mass balance of both electrodes is needed to obtain optimized carbon asymmetric capacitors with increased capacitance, energy and power density values comparable or even better than for the well-known EDLCs.

Asymmetric configurations based on carbon materials in the negative electrode, and pseudocapacitive or battery-like materials in the positive electrode have been under study for many years, but more recently have shown promising results. Most of these asymmetric assemblies have used activated carbon as negative electrode due to its high hydrogen evolution overpotential, which extends the negative potential limit. On the other hand, carbon xerogels, nanotubes, or more recently graphene have been used with no significant advantages in their performance compared to conventional activated carbons. Nevertheless, the introduction of pseudocapacitance through electroactive functional groups to different carbon electrode materials has made a positive effect on the performance of this asymmetric configuration.

Table 8.8 Carbon-based asymmetric capacitors based on mass optimization for each electrode (negative and positive, $R_{+/-}$ is the mass ratio), or using different carbons with different porosity properties, or introduction of different functionalities

Asymmetric assembly -/+ experimental conditions	Electrolyte type	Voltage window ΔE	Specific capacitance F/g	Specific energy Wh/kg	Specific power kW/Kg	Ref.
1. C_{DTC}/C_{DTC}, $R_{+/-} = 1.3$ GC (20 mA/cm^2)	IL	3.9	30	47	13	[397]
2. C_X/C_X, $R_{+/-} = 1.4$ GC (20 mA/cm^2)	IL	3.4	26	31	9.5	[44]
3. C_{AOx}/C_{AOx}, $R_{+/-} = 2$CV (2 mV/s)	H$^+$/aqueous	1.5	136	10.6	—	[398]
4. C_{AOx}/C_{BOx} $R_{+/-} = 1.4$ CV (2 mV/s)	H$^+$/aqueous	1.6	321	28.6	37.5	[398]
5. G/AC, $R_{+/-} = 5$ GC (650 mA/g)	Organic	3	95	103.8	10	[401]
6. AC/G, $R_{+/-} = 1$ GC (0.5 mA/cm^2)	Organic	3.5	34.3	—	—	[402]
7. AC/C_{meso} GC (2.22 mA/cm^2 = 100 mA/g)	Organic	2.6	30	—	—	[399]
8. AC2/AC1 GC (2.22 mA/cm^2 = 100 mA/g)	Organic	2.6	36	—	—	[400]
9. G_{OH}/AC	Organic	2.3	40	145.1	15.09	[406]
10. AC/AC$_{1000}$ GC (500 mA/g)	H$^+$/aqueous	1	210*	7	0.09	[403]
11. C_{micro}/C_{meso} $R_{+/-} = 1$ CV (5 mV/s)	aqueous	1.2	22	4.5	0.35	[102]

Note: C_{DTC} = disordered carbon, C_x = carbon xerogels, C_{AOX} = Super 50 Norit carbon oxidized, C_{BOX} = Maxsorb carbon oxidized, G = graphene, G_{OH} = graphitizable carbon activated with KOH, AC = activated carbon, C_{micro} = microporous carbon, C_{meso} = mesoporous carbon, IL = ionic liquid.

*No specification related with the use of the masses of both electrodes to calculate capacitance, energy and power values

The first asymmetric assemblies studied were using battery-type materials as PbO_2, $Ni(OH)_2$, or lithium intercalated compounds coupled with activated carbon. The first two asymmetric capacitors have been tested by esma-cap company and Axion Power in particular applications (i.e., railroad applications, hybrid vehicles) [407]. showing specific energies from 8–10 Wh/Kg for carbon/NiOOH to 25 Wh/Kg for carbon/PbO_2. Nevertheless, the bulkier individual electrodes that need to be fabricated in these asymmetric configurations ensures a long-term cycling stability, but longer cycling time than for typical EDLC [396].

Energy and power capabilities of these types C/Battery-type asymmetric devices can be superior to symmetric assemblies, as long as the operating voltage is raised through a proper selection and adjustment of the battery-type electrode [392]. Table 8.9 shows some examples of research carried out using asymmetric configurations, where nanocomposite materials or introduction of metals to the nickel hydroxide have been also included. It is important to point out the difficulty to normalize the capacitance, energy, and power values found in the literature due to variations in electrode fabrication, active material loading, or electrochemical testing procedures. Therefore, the following tables are only for data compilation purposes of the latest research carried out, where the capacitance, specific energy, and power values were calculated based on galvanostatic cycling measurements.

The cost and toxicity of battery-like electrodes make the use of pseudocapacitive materials, as metal oxides or CPs, attractive materials for asymmetric capacitors. Metal oxides based on ruthenium are the best candidates based on pseudocapacitive performance. Nevertheless, the higher cost makes them not suitable for commercial applications. Manganese oxides are a low-cost alternative to ruthenium oxide and environmentally friendly, but their numerous polymorphs due to their varying crystalline-amorphous structure, water content, and the presence of intercalated cations complicates the pseudocapacitance contribution interpretation. Therefore, a deeper understanding of this storage mechanism needs to be elucidated. In Table 8.10, data collected from asymmetric assemblies based on carbon and metal oxides, such as AC/MnO_2 [414, 415], AC/$NaMnO_2$ [416], AC/RuO_2 [417], and AC/CoAl double hydroxide [418] are the most studied. Although it is difficult to obtain conclusive information from this table, aqueous electrolyte (aq) asymmetric assemblies show an extended effective operational voltage window beyond the thermodynamic limit (1.23 V) up to 2 V, which results in higher energy values compared to EDLC. In addition, manganese oxide based asymmetric assemblies show higher specific capacitance values with alkaline or Li-based electrolyte, or with graphene as negative electrode or as conducting agent for manganese oxide through nanocomposite materials. The addition of metals to MnO_2 ($NaMnO_2$, $K_{0.27}MnO_2$) does not result in a further improved capacitance. Comparing capacitance values for discharge times between 200 and 400 s, RuO_2 and $Co(OH)$ asymmetric-based devices show the best performance as well as with MnO_2. In contrast, Fe_3O_4 and V_2O_5 asymmetric devices show low capacitance values. Up to this point, MnO_2 has shown to be the environmentally friendly option for SC using aqueous electrolytes and asymmetric assemblies. It is important to point out that more work on MnO_2 has been carried out, and other oxides will require more research before they can be considered for this type of device.

Table 8.9 Asymmetric capacitors based on activated carbon as negative electrode and battery-type materials as the positive electrode

Asymmetric assembly -/+ (experimental conditions)	Electrolyte type	Voltage window ΔE	Specific capacitance F/g discharge time	Specific energy Wh/kg	Specific power kW/Kg	Ref.
AC/α-Ni(OH)$_2$ (5 mA/cm^2)	KOH(aq)	1.2	127≈1200 s	42	—	[408]
AC/CNT-Ni(OH)$_2$ (0.2 A)	KOH(aq)	1.6	60≈150 s	25.8	2.8	[409]
AC/C$_{Aerogel}$-Ni(OH)$_2$ (30 mA)	KOH(aq)	1.6	66.5≈120 s	22.3	0.35	[410]
AC/Zn-Co-Ni(OH)$_2$ (20 mA/cm^2)	KOH(aq)	0.8	150	35.7	0.9	[411]
AC/PbO$_2$ (10–0.75 mA/cm^2)	H$_2$SO$_4$(aq)	1	32.2, 34.7 —, ≈3000 s	7.8, 11.7	0.25, 0.02	[412]
Li$_4$Ti$_5$O$_{12}$–C$_{NF}$/AC 26.8 A/g = 30 mA/cm^2	Organic	1.5	13 F/L	15 Wh/L	6 kW/L	[413]

All values were calculated from galvanostatic cycling measurements.

Table 8.10 Asymmetric capacitor-based on pseudocapacitive metal oxides

Asymmetric assembly −/+ (experimental conditions)	Electrolyte type	Voltage window ΔE	Capacitance F/g discharge time	Specific energy Wh/kg	Specific power kW/Kg	Ref.
AC/MnO$_2$ (2.5 mA/cm^2)	K$_2$SO$_4$(aq)	2	21≈450 s	—	—	[415]
AC/MnO$_{2-nanorods}$ 2C	K$_2$SO$_4$(aq)	1.8	53.7≈1800 s	17	2	[419]
AC/MnO$_2$ (200 mA/g)	Na$_2$SO$_4$(aq)	2	≈25 ≈2600 s	—	—	[420]
AC/AC-MnO$_2$	Na$_2$SO$_4$(aq)	2	33.2≈120 s	18.2	18.5	[421]
C$_{xerogel}$/MnO$_2$ (0.8 A/g)	Na$_2$SO$_4$(aq)	1.6	106.5	—	—	[422]
AC/λ-MnO$_2$ (10 mA/cm^2)	Li$_2$SO$_4$(aq)	2.2	53≈450 s	36	0.31	[414]
AC/MnO$_2$	(NH$_4$)$_2$SO$_4$(aq)	1	40≈5 s	—	—	[423]
AC/MnO$_2$ (0.25 A/g)	KCl(aq)	2	52≈400 s	28.8	9	[424]
AC/MnO$_2$ (100 mA/g)	KNO$_3$(aq)	2	≈30 ≈600 s	1.9	3.8	[425]
AC/MnO$_2$ (0.1 A/g)	KOH(aq)	1.8	70≈480 s	31.3	0.09	[426]
AC/MnO$_2$	LiNO$_3$(ac)	2	33≈60 s	—	—	[427]
AC/MnO$_2$ (1.5 A/g)	MgCl$_2$ CaCl$_2$(ac)	2	≈40 ≈40 s	22	11	[428]
AC/MnO$_2$ (0.3 A/g)	Ca(NO$_3$)$_2$(aq)	2	33.75≈225 s	21	—	[429]

(Continued)

Table 8.10 (Continued)

Asymmetric assembly −/+ (experimental conditions)	Electrolyte type	Voltage window ΔE	Capacitance F/g discharge time	Specific energy Wh/kg	Specific power kW/Kg	Ref.
CNT-MnO$_2$/MnO$_2$ [10 mA/cm^2]	Organic	2.6	33	32.91	—	[330]
AC/MnO$_{2\text{-Graphene}}$ CV (5 mV/s)	Na$_2$SO$_4$(aq)	1.8	79	51.1	—	[430]
G$_{\text{raphene}}$/MnO$_{2\text{-Graphene}}$	Na$_2$SO$_4$(aq)	2	28	30.4	5–0.007	[431]
G$_{\text{raphene}}$/MnO$_{2\text{-Graphene}}$ [223 mA/g]	Na$_2$SO$_4$(aq)	1.7	≈30 ≈300 s	20	0.5	[432]
CNT/MnO$_{2\text{-Graphene}}$ (2.23 A/g)	Na$_2$SO$_4$(aq)	1.5	132≈90 s	10	—	[433]
AC/NaMnO$_2$ (10C)	Na$_2$SO$_4$(aq)	1.9	38.9	19.5	0.13	[416]
AC/K$_{0.27}$MnO$_2$0.6H$_2$O(2C)	K$_2$SO$_4$(aq)	1.8	57.7≈1800 s	26	0.05	[434]
C-LiTi$_2$(PO$_4$)$_3$/MnO$_2$ (2 mA/cm^2)	Li$_2$SO$_4$(aq)	1.3	99.7≈1000 s	47	0.2	[435]
AC/Li$_4$Mn$_5$O$_{12}$ (100 mA/g)	Li$_2$SO$_4$(aq)	1.4	43≈750 s	—	—	[436]
C$_{\text{NT}}$MnO$_2$/LiMn$_2$O$_4$ (1.95 A/g)	Organic	2.5	30≈75 s	26	2.4	[437]
AC/Ni (1 mA/cm^2)	KOH(aq)	1	30≈100 s	28.8	0.33	[438]
AC/NiO$_2$ (1 mA/cm^2)	KOH(aq)	1	40	35	0.33	[438]
AC/h-NiO$_2$ (1 A/g)	KOH(aq)	1.3	25 F/g≈75 s	—	—	[439]

Classification | 417

Asymmetric assembly -/+ (experimental conditions)	Electrolyte type	Voltage window ΔE	Capacitance F/g discharge time	Specific energy Wh/kg	Specific power kW/Kg	Ref.
AC/NiO$_2$ (0.1 A/g)	Gel(aq)	1.6	73.4≈2000 s	26.1	—	[336]
AC/Co$_{0.56}$Ni$_{0.44}$O (166 mA/g)	KOH(aq)	1.6	97≈1000 s	34.5	≈1	[440]
AC/RuO$_2$ (2.5 & 10 mA/cm^2)	Solid (aq)	1	41.5–35.8	—	—	[441]
C$_{anthr}$/RuO$_2$ (10 mA)	H$_2$SO$_4$(aq)	1.3	109≈300 s	26.7–12.7	17.3	[404]
AC/RuO$_2$-TiO$_2$ (120 mA/cm^2)	KOH(aq)	1.4	21≈20 s	5.7	1.2	[442]
AC/Co(OH)$_2$ (5 mA/cm^2)	KOH(aq)	1.6	72.4≈200 s	92.7	≈2	[443]
AC/Co(OH)$_2$-USY (5 mA/cm^2)	KOH(aq)	1.5	105≈500 s	30.62	0.52	[444]
AC/CoAl$_{hydroxide}$ (25 mA/cm^2)	KOH(aq)	0.4	77≈130 s	15.5	0.25	[418]
C/NiZnCoO-OH (10 A/g)	KOH(aq)	1.5	90≈120 s	16.62	2.9	[445]
AC/LiO$_2$, (750 A/g)	Organic	2	60	—	—	[446]
AC/Ni$_{1/3}$Co$_{1/3}$Mn$_{1/3}$(OH)$_2$, (100 mA/g)	LiOH(aq)	1	35.9≈330 s	—	—	[447]
AC/Fe$_3$O$_4$ (2 mA/cm^2)	KOH(aq)	1.2	36≈425 s	—	—	[284]
AC/V$_2$O$_5$ 2C	K$_2$SO$_4$(aq)	1.8	64.4≈1800 s	29–20	0.1–2	[448]
C-LiTi$_2$(PO$_4$)$_3$/AC (2 mA/cm^2)	Li$_2$SO$_4$(aq)	1.2	84≈1000 s	24	0.2	[449]
Li$_2$FeSiO$_4$/AC (1 mA/cm^2)	Organic	3	49≈1200 s	43	0.2	[71]

Note: All values were obtained with galvanostatic cycling measurements.

CPs have also been used as pseudocapacitative based electrode materials for asymmetric configurations. The main draback of devices assembled with this type of polymers is their detrimental effect on cyclability due to a swelling effect on the polymer caused by the electrolyte as previously mentioned in Section 8.5.1.7. There have been different aproaches to solving this problem. Nanocomposite materiales based on these polymers and different carbons have helped to improve this property as previously mentioned due to the mechanical support that carbon gives to the polymer absorbing their swelling effect [450]. Also, the use of an assymetric asembly using a carbon material in the negative electrode and a nanocomposite material in the positive electrode has resulted in longer cycle life. Table 8.11 shows a compilation of asymmetric cells assembled with a carbon material or carbon-based nanocomposite material for the negative electrode with different CPs. These cells have been studied with organic, solid, or aqueous electrolytes, showing similar results for all electrolytes. Therefore, it is recommended to use aqueous or solid electrolytes for a more environmentally friendly device.

In all asymmetric assemblies described in this section, the use of organic electrolyte does not enhance their performance compared to conventional EDLC, and more importantly compared with aqueous asymmetric capacitors (Tables 8.9, 8.10, and 8.11).

In Fig. 8.23, a schematic representation of cyclic voltammograms of positive and negative electrodes for the three different asymmetric configurations (aqueous-based electrolyte) described until now are shown. The symmetric cell based on manganese oxide (Fig. 8.23a) can be used as an analogy for the symmetric EDLC, where the same material is used as the positive and negative electrode, and therefore each electrode contributes with half of the charge observed from the cyclic voltammogram of the material (divided in two complementary cell operational voltage window). On the other hand, when a carbon material is used instead of MnO_2 (Fig. 8.23a) as negative electrode, an asymmetric device arises (Fig. 8.23b) increasing the contribution of the positive MnO_2 electrode (from 0.5 V, Fig. 8.23a, to 1 V, Fig. 8.23b increased voltage window) as well as for the cell (from 1 to 2 V). This asymmetric configuration can be also assembled with a battery-like electrode (Fig. 8.23c) instead of a pseudocapacitive material. There is a clear difference on the cyclic voltammogram behavior but always a mass balance based on total charge from cyclic voltammetry for each electrode will be required (Section 8.7).

Table 8.11 Asymmetric cells using conducting polymers as pseudocapacitive electrode material

Asymmetric assembly −/+ experimental conditions	Electrolyte type	Potential window ΔE	Capacitance F/g discharge time (s)	Specific energy Wh/kg	Specific power kW/Kg	Ref.
AC/pMeT (5 mA/cm^2)	Organic	2	35 ≈210 s	30	0.5	[451]
C/p-Th-CNV-EDOT2 (2 mA)	Organic	3	16.6 ≈50 s	42	3	[452]
TiO$_2$-C$_{NT}$/P$_{Ani}$-C$_{NT}$ (10 mA)	Solid	1	345 ≈100 s	—	—	[453]
C/P$_{Ani}$-C (5 mA/cm^2)	H$_2$SO$_4$(aq)	1.4	87.4 ≈390 s	23.8	0.2	[454]
P$_{Py}$-C$_{NT}$/P$_{Ani}$-C$_{NT}$ (1 mA)	H$_2$SO$_4$(aq)	0.6	76 ≈650 s	—	—	[455]
G$_{raphene}$/P$_{Ani\ nanofibers}$ CV(500 mV/s)	HCl (Aq)			4.86	8.75	[456]
AC/P$_{Ani}$-LiPF$_6$ (0.25 mA/cm^2)	Organic	3	58 ≈750 s	—	—	[457]
AC/P$_{F-C}$ (2 mA/cm^2)	Organic	3.2	34 ≈30 s	47	4–6	[458]
G$_{raphene}$/P$_{Ani\ nanofibers}$ CV(500 mV/s)	H$_2$SO$_4$(aq)	1		4.86	8.75	[456]
AC/P$_{Ani}$-MnO$_2$-AC (0.17 A/g)	Organic	2	≈115 ≈1350 s	61	0.17	[459]

Note: All values have been normalized by the mass of both electrodes and using galvanostatic discharge curve.

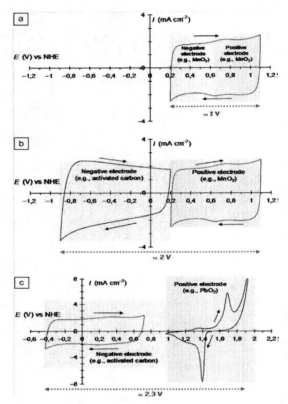

Figure 8.23 Schematic representation of cyclic voltammograms for three different configurations of asymmetric supercapacitors (aqueous-based electrolyte) in which areas shaded in red and blue represent the potential window of the positive and negative electrodes, respectively for the assembly of: (a) symmetric $MnO_2//MnO_2$ cell in 0.5 M K_2SO_4, (b) asymmetric $AC//MnO_2$ cell in 0.5N K_2SO_4, and (c) $AC//PbO_2$ cell in 1M H_2SO_4. NHE, normal hydrogen electrode; I, measured current; and E, electrode potential. Copyright [396].

More electrode combinations have been explored, e.g., the use of iron oxides as negative electrode instead of carbon (Table 8.12), and this has provided some advantages in terms of volumetric capacitance and energy density. However, in terms of specific energy (Wh/kg), the value is penalized by the higher weight mass of the oxide. In Table 8.12, different electrode combinations used for alternative asymmetric capacitors are summarized.

Table 8.12 Asymmetric cells with pseudocapacitive materials in both electrodes

Asymmetric assembly -/+ experimental conditions	Electrolyte type	Potential window ΔE	Capacitance F/g discharge time (s)	Specific energy Wh/kg	Specific power kW/Kg	Ref.
Fe_3O_4/MnO_2 (360 mA/g)	K_2SO_4(aq)	1.8	20 ≈80 s	7	0.82	[460]
$LiFeO_2/MnO_2$ (5 mA/cm²)	Li_2SO_4(aq)	1.5	10 ≈35 s	2	0.2	[461]
$MnFe_2O_4/LiMn_2O_4$ (1 A/g)	$LiNO_3$(aq)	1.3	50 ≈55 s	5.5–10	0.3–1.8	[462]
MnO_2/PEDOT (250 mA/g)	KNO_3(aq)	1.8	60	13.5	120	[463]
WO_3-PAni/PAni	H_2SO_4(aq)	1.2	48.6	9.72	0.053	[464]
$Ru_{0.35}V_{0.65}O_2$/NiO (7.5 mA/cm²)	KOH(aq)	1.7	102.6 ≈320 s	23–41.2	1.41	[465]
C_{NT}-SnO_2/ C_{NT}-MnO_2	KCl(aq)	1.7	49	20.3	143.7	[328]

All capacitance values are calculated based on galvanostatic cycling using the formula $C = It/Vm$, where I is the current applied, t is the discharge time, V is the potential window, and m is the mass of both electrodes.

The most important consideration when designing an asymmetric assembly is finding the proper electrode materials with complementary potential windows, using the effective mass in each electrode for optimum performance in order to accomplish a higher operational voltage window of the device. The specific capacitance and specific energy of an asymmetric capacitor is improved by increasing the operational voltage window and/or the capacitance value through pseudocapacitive contributions, making this asymmetric approach a viable pathway for a superior performance in supercapacitors.

8.5.2 Electrolyte

The selected electrolyte to assemble a SC cell is of great importance because it determines the operational voltage window that together with the resistance, determines the power and energy density. Electrolytes can be classified based on the physical state as solid, liquid or gel. Although this classification is useful, a more common classification is based on the chemical nature of the electrolyte, organic or aqueous electrolytes. Nevertheless, other types of electrolytes as solid electrolytes as mentioned before, and ionic liquids have been investigated more recently because they represent an emerging field within electrolytes for SC cells. All electrolytes mentioned will be described in the following sections in more detail.

8.5.2.1 Organic electrolytes

Organic electrolytes are composed mainly of inorganic salts dissolved in organic solvents. Nevertheless, organic salts such as quaternary ammonium salts are preferred due to their better solubility and conductivity [466]. The main advantage of these electrolytes is the high energy densities, which is due to a major voltage window stability (up to ΔE = 2.5–4.0 V). Thus, six to nine times the energy density can be accomplished compared to aqueous electrolyte SC. However, these types of electrolyte show lower conductivity values (10^{-3} S/cm, higher resistance) at ambient temperature, big ionic radius, higher cost, and higher toxicity compared to aqueous electrolytes which makes them hazardous to the environment. The larger solvated ion size found in these electrolytes needs the porosity tuning in the carbon electrodes that are used to assemble the SC cell [3]. In addition, their poor stability in air makes the use of a dry box necessary to diminish water content, making the procedure more complex. These electrolytes are the most common in commercial devices, and their composition is a BF_4 in acetonitrile or propylene carbonate [2]. In the case of acetonitrile, it shows a low flash point making this solvent unsafe. Therefore, in countries as Japan where this solvent is prohibited, its substitution with propylene carbonate has been proposed at the expense of the power density due its lower conductivity value [467].

8.5.2.2 Ionic liquids

Electrolytes based on ionic liquids (ILs) are solvent-free electrolytes that are liquid at room temperature and are composed solely by anions and/or cations. These electrolytes are eco-friendly, and are attractive for their use in SC because the ions are not solvated (solvent-free), and they can interact directly with the surface of the electrode. Typical cations used in these ILs are tetraalkylammonium, imidazolium, pyrrolidinum, piperidinum, or pyridinium, while anions can be based on cyano groups, halides, trifluoroacetic, trifluoromethanesulfonic acid (triflate), amides, methide, tetrafluoroborate, etc. [468, 469]. Generally, ILs are imidazole salts with ionic conductivity values of 0.1–10^{-4} S/cm, lower than in aqueous electrolytes and comparable with organic electrolytes [470]. Nevertheless, the operational voltage window of such ILs are as high as 4 V resulting in high energy densities (90 Wh/kg), and show low vapor pressures making them useful in a wide range of temperatures [2, 44, 471]. They are not corrosive, and are thermally and hydrolytically stable [468]. In general, the physical and chemical properties of these electrolytes can be tuned by choosing different combinations of cations, anion, and substituents [469]. Regarding their eco-friendliness, it is important to point out that these electrolytes have been considered as green electrolytes mainly due to their negligible vapor pressure. Nevertheless, their high solubility in water and toxicity to aquatic organisms are a great concern [73]. Therefore, ILs biodegradation has been the main focus nowadays within the scientific community [74, 75]. If we add this unresolved toxicity to their high cost, ILs are not attractive yet for commercial SC.

8.5.2.3 Polymeric electrolytes

Cheap polymer-based electrolytes are highly desirable for practical SC assembly. Comparing with all liquid electrolytes, polymer electrolytes have the possibility of forming thin layers with great surfaces, higher corrosion resistance in metal current collectors due to the lack of dissolvent, and less moisture-sensitive compared to organic electrolytes [472]. Poly(benzimidazole) PBI [473, 474], and sulfonated tetrafluoroethylene also denominated as Nafion® [475] are some examples of these type of electrolytes. They show simultaneously high backbone flexibility and low cohesive energy, making this membrane a mechanically flexible electrolyte, easy to use. These types of polymers contain a heteroatom in the main backbone, which can be solvated by small cations or anions. The salts completely dissociate and dissolve in the polymeric matrix. Nevertheless, they show low ionic conductivity (10^{-3}–10^{-4} S/cm), low charge transfer rate [472–474], they are not cheap (PBI, NAFION), and not that eco-friendly as aqueous electrolytes [4].

In order to use solid electrolytes, the hydrophobicity at the electrode–electrolyte interface needs to be decreased. Thus, alternative materials with high electrochemical stability in a wide potential window (4 V) are the gel polymeric electrolytes [476]. These materials are systems of organic liquid electrolyte inclusions based on alkali salts in a polymeric matrix, such as poly(acrilonitrile) (PAN) [477], poly(vinylpyrrolidone) PVP [478], poly(tetraethylene glycol dimethacrylate) (PEGDA), poly(methyl methacrylate) (PMMA), poly(tetrafluoroethylene) (PFTE), poly(vinyl alcohol) (PVA) and poly(ethylene oxide) (PEO) [479], Nylon [480], and poly(vinylidene fluoride-hexafluoropropylene) (PVdF-HFP) [481, 482]. These materials have an amorphous character, an ionic conductivity at ambient temperature lower than in aqueous electrolyte (between 10^{-4} and 10^{-3} S/cm), and good swelling stability [476]. When water is used as plasticizer, these electrolytes are denominated as hydrogel polymer electrolytes. Examples of these electrolytes are based on silica gels doped with acids and show lower resistance values [483, 484]. These materials are more environmentally friendly due to their water content. However, they are limited to a low operating potential window of 1.23 V.

8.5.2.4 Aqueous electrolytes

Aqueous electrolytes have small solvated ions providing access to different pores, have high dielectric constants promoting storage, and both effects contribute to capacitance. These types of electrolytes are composed mainly of acidic (H_2SO_4) or alkaline (KOH, NaOH) solutions, but also neutral salts (Na_2SO_4, K_2SO_4) have been used. Chloride salts are not commonly used due to their adsorption and reduction nature of chloride ions [485]. These aqueous electrolytes show high ionic conductivity (0.5–1 S/cm) due to their small ionic size as mention before (10–20 Å, hydrated ions), making the micropores accessible to these ions ensuring a high ionic diffusion coefficient (10^5 cm^2/s). In addition, their lower electric resistance and higher conductivity is revealed in higher capacitance and power values compared to other electrolyte types [485]. On the other hand, they show low performance at low operating temperature, corrosion of current collectors and electrode materials may occur through cycling, and energy density is decreased due to a smaller voltage window limited by water decomposition (1.23 V). Nevertheless, these types of electrolytes are environmentally friendly compared to the organic type, making them an emerging research field in combination with asymmetric SC. In this sense, there has been active research related to the increase of the voltage window using aqueous electrolyte in order to improve the energy density, obtaining up to 2 V with an asymmetric assembly [226, 232, 414, 415, 420, 421, 424, 425, 427, 431].

Regarding the corrosion nature of aqueous electrolytes, it is known that the use of KOH or H_2SO_4 is responsible, making neutral electrolytes an option to solve this problem. The use of sodium acetate (CH_3COONa) as a neutral and eco-friendly aqueous electrolyte has been proposed with better performance than for sodium sulfate ($NaSO_4$) [4]. Its smaller desolvated acetate anion (CH_3COO^-) has a diameter of 0.15 nm (0.71 nm solvated) [486], compared to SO_4^{2-} anion (0.28 nm desolvated and 0.73 nm solvated) [78], and has shown its best performance using a 1M solution, even at high rates [4]. Therefore, the operational voltage window increase, together with the use of neutral aqueous electrolytes as eco-friendly alternatives opens the possibility of designing EFSC with truly commercial opportunities.

8.6 Designing High-Performance Environmentally Friendly Supercapacitors

Nowadays, worldwide research in SC has been focused on increasing power and energy density, using low-cost eco-friendly materials for the electrodes, electrolytes, and cell assembly. Specific power (W/kg) of a device can be optimized based on cell design. In order to obtain high power devices, several parameters need to be tuned: the ESR (equivalent series resistance) needs to be decreased as much as possible, and the voltage window increased as shown in the following equation:

$$P = \Delta E^2/4R, \qquad (8.26)$$

where ΔE is the operating voltage window, and R is the equivalent series resistance (ESR). Several contributions are included in the ESR, as the electrolyte resistance (including the separator), the electrode material resistance (ionic and electronic contributions), the current collector-electrode material interface resistance, and physical parameters as the pressure applied during the cell assembly.

As described in the electrolyte section, environmentally friendly aqueous electrolytes (neutral, i.e., CH_3COONa) show lower electric resistance and higher ionic conductivity, and therefore higher power values can be obtained. This kind of electrolytes combined with the use of eco-friendly separators, that facilitates ion mobility and double layer formation is highly desirable [79] for high SC performance. B. Dyatkin et al. studied different porous membranes available for chemical filtration and used them as separator in SC assemblies. The results showed that cellulose esters membranes terminated with acetate groups, gave the best combined performance of environmental capability and electrochemical behavior when used as separator [4].

Other important factors where optimization is needed to decrease as much as possible the internal resistance contributing to the whole ESR are the electrode material porosity, the use of binder to elaborate the electrodes, and the thickness of such electrodes. Porosity is an important quality of electrode materials due to the high surface area available for charge storage. Thus, the pore size distribution is very important when using organic or ionic liquids as electrolytes, and the porosity needs to be tuned for high ionic conductivity. For the case of eco-friendly aqueous electrolytes, the pore size is not that relevant due to their smaller solvated ions, making most of the nano-meso pores accessible. Nevertheless, porosity means also a discontinuity or the introduction of defects in the structure of such electrode materials resulting in a higher electric resistance. Therefore, the porosity will need to be adjusted in order to obtain the highest area for charge storage without compromising the electric conductivity of the material, or otherwise the introduction of an electric conductive additive will be necessary.

The use of binders to make electrodes is required to give mechanical properties to the electroactive materials, which are normally in powder form. Different binders have been used for the elaboration of electrodes, such as PVDF or Kynar Flex® (polyvinylidene difluoride), PTFE or Teflon® (Polytetrafluoroethylene), and Nafion®. Most of these binders are insulators and their addition implicates a contribution to the ESR. Only Nafion® can contribute with its well-known ionic conductivity, which can improve in some cases the storage mechanisms [487]. Nevertheless, none of these binders are environmentally friendly due to their F content and are high cost. Some eco-friendly alternatives have been proposed, such as natural cellulose [488, 489], sodium-carboxymethyl cellulose (Na-CMC) [76, 490, 491], Na-alignate [492], egg white [4], chitosan [493], and PVA and PVAc [4]. Cellulose-based electrodes show thermal stabilities comparable to those of conventional electrodes (PTFE, PVdF), are compatible with all types of electrolytes, and most importantly they show high performance in terms of specific capacitance and cycling stability compared to PVdF binders. In addition, cellulose is a cheap and safe material and is soluble only in certain ILs, making the replacement of traditional binders in the electrode manufacturing process a difficult task [488, 489]. Na-CMC is also a cheap binder, and its performance in SC is comparable with electrodes based on traditional binders in several types of electrolytes. Nevertheless, their solubility in water makes them not compatible with aqueous electrolytes [76, 490, 491]. Another eco-friendly binder incompatible with aqueous electrolytes is Na-alignate, which has been used only with organic electrolytes [492]. All these binders (cellulose, Na-CMC, Na-alignate), including egg white, have been cast directly on current collectors. This casting approach has shown good electrochemical performance, but fabrication of freestanding films offer the possibility of controlling the thickness and mass loading in the electrode, shows better flexibility, and makes it easy for a large-scale production. For these freestanding eco-friendly electrodes, cross-linking of chitosan, PVA and PVAc binders can be use with neutral aqueous electrolytes (eco-friendly). It is important that these binders retain the electrode surface area when forming the elastic structural matrix.

Chitosan binder is obtained from chitin present in arthropods skeleton, is the second most abundant natural polymer after cellulose [494], and is cheap, biodegradable, and non-toxic [495]. In addition, the high electrochemical performance of chitosan comes from its high water retention capability [493]. PVAc and PVA binders can produce freestanding films with 20 and 12 wt%, respectively. Nevertheless, high resistivity and pore clogging affecting the surface area of the electrode has been observed, resulting in poor electrochemical performance in SC compared to PTFE binder. This has been explained based on the fiber-like structure in PTFE, which is not present in PVA and PVAc. Until now, all these eco-friendly binders to produce free-standing electrodes have not match the performance of commonly used binders. Therefore, more research on these binders, as well as on the development of new eco-friendly polymers is needed [4].

On the other hand, binder-free or binder-less electrodes have been elaborated with carbon monoliths [496, 497] or carbon nanotube networks [498–500], where the electric conductivity of the electrode has been enhanced but the ohmic resistance related with the current collector needs to be decreased. The thickness of the electroactive material deposited on to the surface of a current collector is a parameter that has to be adjusted based on the properties needed on the SC device. It is well known that thicker electrodes results in higher specific energy values but with higher resistance due to longer ion diffusion pathways, giving lower power values. On the other hand, thinner electrodes show less resistance and therefore are most commonly used for high power applications.

There have been many works related to the improvement of surface contact between a metallic current collector and the electroactive material to obtain an upgraded electric contact with less ohmic drop. In this sense, electrochemical or chemical etching treatments in an aluminum current collector to introduce roughness has been carried out, followed by a carbonaceous conducting thin film coating via a sol-gel procedure to use activated carbon with the addition of carbon nanotubes [501–503], or to grow carbon nanofibers as electroactive materials [501] resulting in a lower electric resistance interface. Metallic current collectors are not desirable if an environmentally friendly device wants to be developed. Therefore, a good eco-friendly alternative is based on carbon materials with high conductivity and good interaction with the electrode material. In this sense, aligned carbon nanotubes [504], carbon nanofiber mat, and graphite foil sheets have been studied [4]. Results showed that current collectors made of aligned carbon nanotubes or carbon nanofiber did not reveal good bulk electron conductivity, and are high cost. On the other hand, graphite foil with a mechanical roughening treatment, showed better performance than stainless steel current collector due to an improved interaction with a carbon electrode and high conductivity. Therefore, graphite foil has been preferred as the environmentally friendly option to carbon-treated metal current collectors [4].

Finally, the ESR can be decreased by external physical parameters as pressure during cell assembly, because the physical contact between the electrodes and the porous separator can be modulated. If an exceeded pressure is applied, a displacement of the electrolyte from the electroactive electrode material and/or a decrease or collapse of the accessible porosity can be observed, and the performance of the SC is affected [505]. On the other hand, if no pressure is applied an increase on the ESR is observed due to low contact between the current collectors, electroactive material electrodes, separator, and electrolyte. Therefore, a proper pressure needs to be applied in the cell assembly to decrease ESR and obtain a higher power in the device.

Specific energy (Wh/kg) is the main parameter to improve in SC in order to capture or gain the battery market. This parameter is basically influenced by two factors, as shown in the following equation:

$$E = 1/2 C \Delta E^2, \tag{8.27}$$

where C is the capacitance (F/g or F/cm^2) and ΔE the operational voltage window. The first approach used to increase the specific energy was the improvement of the capacitance value, which was focused on improving the electrode materials and/or by developing novel materials. Research around the world introduced pseudocapacitive contributions to improve the capacitance values, introducing CPs, oxides, or functional groups to different carbons. More recently, research has focused on increasing the voltage window in which the device operates, and this has even more effect on the energy value since in Eq. 8.27, ΔE is elevated to the square. The ΔE not only has a straight influence on the energy density value but also affects the power density. A wider voltage window can be obtained by designing environmentally friendly devices, using aqueous neutral electrolytes (eco-friendly) and an asymmetric assembly (Fig. 8.11) to increase capacitance, energy, and power values.

8.7 Characterization

Electrode materials are characterized using different techniques commonly used in the field of material science, such as XRD (X ray diffraction), FTIR (infrared spectroscopy), nitrogen and/or CO_2 adsorption isotherms to obtain surface area and porosity properties, SEM and/or TEM (scanning or transmission electron microscopy, respectively), etc. In this section, only the electrochemical characterization needed to understand the charge storage properties of electrode materials and their behavior in supercapacitor cells will be defined. Conventional three-electrode cells are suitable for fundamental studies on the energy storage properties of a given electroactive material, whereas a two-electrode cell is recommended for the characterization of cell performance because simulates the setup of actual SC. In order to characterize the electrochemical performance of SC, cyclic voltammetry (CV), galvanostatic cycling (GC), and electrochemical impedance spectroscopy (EIS) have been the most popular techniques [506]. Nevertheless, in this section only CV and GC will be described, as well as, a detail description of some techniques to fabricate electrodes, on how to assemble and characterize three-electrode and two-electrode cell to obtain capacitance, ESR, specific energy and power, cycle life, and self-discharge rate.

8.7.1 Electrode Fabrication

Electrode fabrication involves the use of current collectors (CCs), conductive additives, binders, and physical techniques to obtain the electroactive film electrodes with thickness control. Different CCs have been used, such as metallic (Al, Stainless steel, Ni, Cu, Ag, Ti) in different forms and shapes (plates, mesh, foil, sponges), and carbon paper and cloth that are eco-friendly options, just to mention some of them. The most important factor to take into consideration when selecting a CC is their chemical and electrochemical stability in the chosen electrolyte.

Cyclic voltammetry helps to determine the operational voltage window in which the current collector is stable, using a three-electrode arrangement with a reference and counter electrode, and the CC as the working electrode in the selected electrolyte. Metallic CCs are usually used with organic electrolytes (commercial devices), and some of them in specific aqueous electrolytes depending on stability. In acidic aqueous electrolytes, metallic CCs are not recommended due to dissolution and/or corrosion during long-term cycling, and carbon CCs are more recommended in addition to its eco-friendliness. As explained in Section 8.6, the electric contact between the CC and the electroactive film electrode needs to be very efficient in order to not affect the specific power value of the device, and some approaches have been mentioned. On the other hand, conductive additives based on graphitic carbon have been used to enhance the electric conductivity of some electroactive materials showing a more efficient charge storage mechanism, which is reflected on the capacitance values. Carbon nanotubes have been added to different electroactive carbon electrodes to improve conductivity, by lowering the contribution to the ESR [502, 507]. In Fig. 8.24, cyclic voltammograms are shown for different carbon nanotube loadings in carbon aerogels, where the voltammetric profile of pure carbon aerogel (CA) turns more rectangular with carbon nanotubes addition due to an enhanced electric conductivity. Moreover, a 3% loading of carbon nanotubes (CA-3wt%MWNT) shows the best performance with a rectangular voltammogram profile typical of a capacitive behavior with a higher current range (related with higher capacitance value), and a lower ESR as observed in the voltage drop during a galvanostatic discharge (Fig. 8.24). Therefore, the use of these types of conducting additives needs to be optimized adjusting the concentration.

Normally, electroactive film electrodes are made from the electroactive material in powder form with binders to obtain a self-standing film or deposit on to the current collector with good mechanical properties. By using a binder, it is ensured that the electroactive power does not detach. Section 8.6 explains the different binders commonly used to make these film electrodes and their environmentally friendly options, how they introduce resistance, and why they need to be optimized based on concentration. Here, a description of different techniques on how to make these electroactive films will be described.

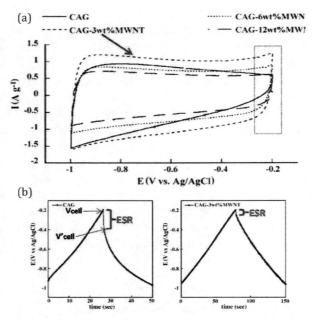

Figure 8.24 (a) Cyclic voltammograms of bare carbon aerogel electrode (CA) and carbon nanotube modified CA (CA-MWNT), in 5M KOH electrolyte using a scan rate of 5 mV/s. (b) Typical galvanostatic charge/discharge curves at 1 mA for CA (left) and CA with the addition of 3 wt% of MWNT, where the voltage drop during discharge is indicated as the ESR. Copyright [507].

Usually, the first step on fabricating these film electrodes is to make a slurry or paste from the electroactive material, a minimum amount of binder to ensure mechanical stability, a conductive additive, if needed, and a solvent. A green solvent as ethanol [508] or 1,3-Dioxolane [509] is preferred, or the recycling of the solvent [488]. Different deposition techniques have been used to make a film from this type of slurry, where the most common are tape-casting, spin-coating [510], screen-printing (Fig. 8.25), airbrush, and cold-rolled (Fig. 8.26). As mentioned before, the thickness of these films need to be controlled considering that thin films will result in higher power and thick electrodes in higher capacitance and energy devices. Thinner electrodes can be obtained with low viscosity slurry and deposition techniques like tape-casting, spin-coating, and airbrush. On the other hand, thicker electrodes are obtained with paste-consistency slurries, using a screen-printing or a cold-rolled technique.

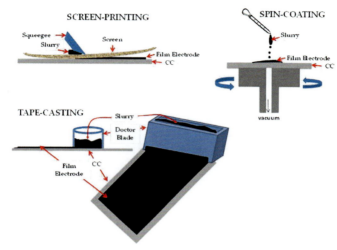

Figure 8.25 Schematic representation of some deposition techniques are illustrated, such as screen-printing, spin-coating, and tape casting.

Figure 8.26 Images of the airbrush and cold-rolled technique are illustrated.

8.7.2 Electrode Material Characterization

Materials designed for their application as electrodes in SC cells, need to be first electrochemically characterized in a typical three-electrode cell arrangement (reference electrode, counter electrode, and the electroactive material as working electrode) before they are assembled and characterized in a two-electrode SC cell. The main objective of this characterization is to understand how they can store charge in a specific voltage window. That is, the charge storage mechanism needs to be first determined: double layer, pseudocapacitance, or battery-like behavior. The electrochemical technique commonly used in a first approach is cyclic voltammetry. The obtained voltammograms can result in a rectangular shape, indicating a capacitive behavior from the double layer formed at the electrode/electrolyte interphase, whereas pseudocapacitive contributions are more commonly revealed from the appearance of wide redox peaks with a deviation from the rectangular shape [506]. Battery-like materials show intense redox peaks, also related with faradaic phenomena.

In general, CV technique in a three-electrode configuration helps to determine if the electrode material can store any charge, how the charge is stored, in what voltage window, and with these results capacitance of the material can be calculated. This value will be called the intrinsic capacitance of the material, which is different from the capacitance of a two-electrode SC cell that will be explained in Section 8.7.3.

Different cyclic voltammetry profiles (Fig. 8.23) can be obtained depending on the charge storage mechanism, and different approaches can be applied to calculate the intrinsic capacitance values from this technique. For a typical double layer mechanism or capacitive profile, different scan rates (v = mV/s) are used to obtain their correspondent cyclic voltammograms, as shown in Fig. 8.27. In this case, a potential or voltage range is selected in a place where the current is constant in the voltammogram (red line in Fig. 8.27), and then the capacitance is evaluated by Eq. 8.28.

$$\Delta_i = I_a - I_c \text{ vs } v \text{ (mV/s)}, \tag{8.28}$$

where Δ_i is the difference between the anodic (I_a) and cathodic (I_c) current for a given voltammogram using a specific scan rate, v (mV/s). The intrinsic capacitance (F/g) is evaluated from the slope of the linear relation obtained from Δ_i vs. v (mV/s) and divided by the electroactive mass of the electrode.

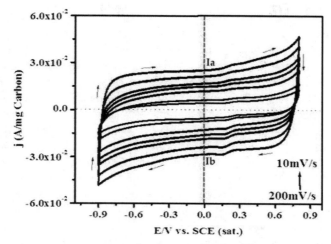

Figure 8.27 Cyclic voltammetry profiles for a typical capacitive response obtained at different scan rates v(mV/s), for a Norit activated carbon electrode prepared with Nafion® as binder in a 0.01 M $K_3Fe(CN)_6$, 1M KCl electrolyte.

Other approaches to calculate the intrinsic capacitance value from cyclic voltammetry can be applied to either charge storage mechanism profile at a given scan rate. One of them is using the following equation:

$$C = \frac{Q}{\Delta E \times m}, \tag{8.29}$$

where ΔE is the voltage window, m is the mass of the total electrode, and Q is the voltammetric charge determined by integrating the cyclic voltammogram curve as follows:

$$Q = \int_{t0}^{t1} It, \tag{8.30}$$

where I is the voltammetric current and t the time in seconds. Another approach is to divide the voltammetric current by the scan rate and by the total mass of the electrode. There are some cases where a double layer charge storage mechanism is mixed and differentiated from the pseudocapacitive contribution, as in Fig. 8.28. If the pseudocapacitive or faradaic contribution wants to be separated from the double layer or capacitive contribution, the total charge (Q, Eq. 8.30) is needed to calculate the total intrinsic capacitance (C, Eq. 8.29), and then two different pathways can be followed. The first approach is to obtain different cyclic voltammograms from varying the scan rate (in a similar way as for Fig. 8.27) and selecting a voltage where the current is constant. Then, the double layer capacitance (C_{dl}) is calculated from the slope of the linear relation obtained from Δ_i vs. v(mV/s) and divided by the total mass of the electrode as previously mention. This C_{dl} capacitance is subtracted from the total intrinsic capacitance (C) to obtain the pseudocapacitive contribution C_p. The second approach is to subtract the charge related with the pseudocapacitive contribution (Q_{AQ}) hatched area on Fig. 8.28 from the total voltammetric charge (Q) to obtain the charge related with the capacitive behavior (rectangular shape voltammetric profile), related with the double layer contribution (Q_{dl}); and with Q_{AQ} and Q_{dl} the different capacitance contributions (C_{dl} and C_p) can be obtained with Eq. 8.27. Thus, the sum of these capacitances will result in the total intrinsic capacitance value (C). It is important to mention that sometimes both mechanisms are hard to separate, and in this case only the total intrinsic capacitance can be calculated from the total voltammetric charge.

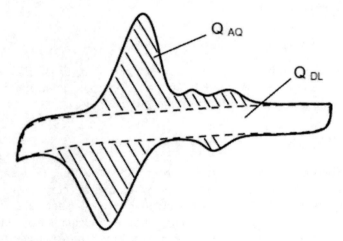

Figure 8.28 Representative cyclic voltammogram for an antraquinone (AQ) modified carbon giving the double layer voltammetric charge (Q_{dl}) as well as the faradaic charge associated to AQ-grafted groups Q_{AQ}). Copyright [315].

8.7.3 Cell Characterization

In this section, cell assembly and characterization will be focused on asymmetric cell configurations with aqueous electrolytes. In general, a two-electrode cell is assembled with the two electrodes (current collector and electrode material) separated by a membrane impregnated by the electrolyte, and pressed for good contact. In Fig. 8.29 different types of lab-scale SC cells are illustrated, where a planar cell made of two plates as support (acrylic, Teflon) using four screws to make the pressure to the assembly, a Teflon Swagelok cell, or a button cell can be used among others. The first one (planar cell) is not a sealed cell and requires to be immersed in the electrolyte, whereas the last two (Swagelok and button cells) are sealed. As previously mentioned, when using an aqueous electrolyte, corrosion of metal (contacts, current collectors, and package) can be presented, and this issue needs to be evaluated for a specific cell arrangement. For clarity, the characterization of lab-scale cell assemblies will only be described, and for prototype testing the work of Andrew Burke should be reviewed [51, 511, 512].

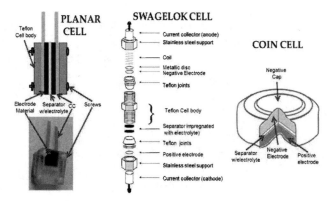

Figure 8.29 Supercapacitor lab-scale cells, where each component has been marked for clarity (cell body, electrode materials, separator, electrolyte, current collectors (CC), and other cell specific pieces).

8.7.3.1 Cyclic voltammetry

The characterization of asymmetric assemblies involves first the evaluation of each electrode by cyclic voltammetry in order to determine the voltage range in which every electrode will act. Once each voltage range is fixed so that they have a common limit (the upper limit of the negative is the lower limit of the positive) resulting in a complementary voltage window, the specific or coulombic charge (F/g) of each electrode is calculated (Fig. 8.23). This characterization is carried out as explained in the previous Section 8.7.2 and once the coulombic charge for each electrode is calculated Eq. 8.28, a charge balance is required before assembling the asymmetric cell. This charge balance is carried out by adjusting the mass ratio ($R = m^+/m^-$) between the negative (m^-) and positive (m^+) electrodes, to get exactly the same coulombic charge on each electrode upon charging and discharging. It is important to mention that the mass of each electrode is the total mass including binder and conducting additives.

The theoretical capacitance (C_a) of an asymmetric cell can be calculated as explained in the work of Thierry Brousse et al. [415]:

$$C_a = \frac{C_+ C_-}{C_+ + C_-} \qquad (8.31)$$

where C_+ and C_- are the capacitance of the positive and negative electrodes in F (farad). Taking into consideration Eq. (8.29) and the weight ratio of the electrode materials, this equation transforms into

$$C_a = \frac{C_+ C_-}{C_+ + RC_-} \times \frac{R}{1+R}, \qquad (8.32)$$

where C_+, C_- and C_a are expressed in F/g and $R = m^+/m^-$ is the weight ratio.

The experimental value of capacitance (C_{cell}) can be obtained from cyclic voltammetry, taking into consideration that the cell voltage is the sum of the potential or voltage interval of each electrode, when the intervals are totally complementary. As an example, C/MnO$_2$ asymmetric cell (Fig. 8.23b) with complementary potential intervals for the negative and positive electrodes, results in an operational cell voltage window of 2 V (Fig. 8.30a). The theoretical (C_a) and the experimental capacitance (C_{cell}) should result almost the same [415].

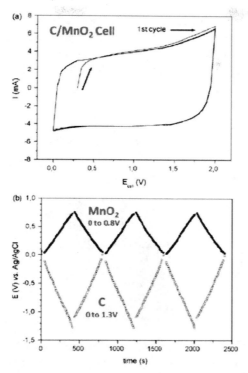

Figure 8.30 (a) Cyclic voltammogram of the asymmetric AC/MnO$_2$ cell (scan rate = 2 mV/s) and (b) corresponding plots for each electrode. Copyright [415].

An interesting and very useful parameter that can monitored during cyclic voltammetry of the SC cell, is to know the real voltage interval or window in which each electrode is really cycling or operating, similar to the results shown in Fig. 8.30b. This evaluation can be carried out using two different channels in the potentiostat: one to carry out the cyclic voltammetry of the two-electrode asymmetric SC cell, and the other channel connected in a three-electrode configuration to monitor the open circuit potential of only one electrode (for example the positive electrode), and by subtraction obtaining the voltage window of the other electrode (negative electrode). Some potentiostats can carry out this characterization by just using one channel, and obtain the voltage window automatically for each electrode. The three-electrode configuration is carried out by using a reference electrode placed as near as possible to the two-electrode cell as shown in Fig. 8.31, as the working electrode one of the electrodes of the SC cell (for example, the positive electrode), and a counter electrode that can be a Pt mesh or the other electrode of the SC cell.

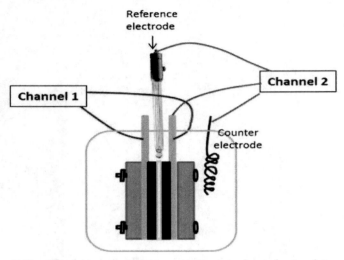

Figure 8.31 Arrangement of two potentiostat channels to determine the operational voltage window of each electrode during cyclic voltammetry of a supercapacitor cell.

Normally, the voltage windows obtained from cyclic voltammetry for each individual electrode can slightly change when assembled in the asymmetric SC cell, probably due to the individual material electrode OCP values (open circuit potential), or other factors that will need more detailed study [415].

8.7.3.2 Galvanostatic measurements

For more practical characterization of these SC, a complementary technique known as galvanostatic cycling is usually employed to record charge/discharge cycles at constant currents. Galvanostatic studies are carried out by applying a constant current in their initial cell state (ocv) to the upper limit cell voltage for the charge (i.e., 2 V for C/MnO_2), and then the same current with negative signal is applied to discharge the cell.

Different currents can be applied in galvanostatic cycling measurements to determine capacitance, and to make the Ragone plot from specific power and energy values. It is important to select currents that will result in charge-discharge cycles in the range of seconds to several minutes for SC applications [51]. Cyclic voltammetry of the complete cell will provide the charge (Q) in the device, and by multiplying by time in seconds the current can be obtained. Therefore, aside from the capacitance, specific power and energy values to obtain the Ragone plot, this galvanostatic technique is used also to obtain the ESR value, and self-discharge rate of the SC.

(a) **ESR**

The ESR is measured in ohm/cm^2 and represents the ohmic resistance of the SC from contributions associated with electric properties of the electrodes and electrolyte, mass transfer resistance of electrolyte ions, and the contact resistance between the CC and electrode material. It is common to find in the literature the ESR value measured by using the AC impedance technique. Nevertheless, the obtained ESR value is always significantly lower than the DC value obtained from galvanostatic measurements, by about a factor of two [511]. Therefore, for a proper ESR measurement the galvanostatic methodology using the current interrupt technique at each end of charge is highly recommended.

The cell is first charged to the fully charge stage (V_{cell}) and then the current is switched (I^-), observing a voltage drop (V'_{cell}) (similar as in Fig. 8.24), and the ESR can be calculated by the following equation:

$$\text{ESR} = \frac{V_{cell} - V'_{cell}}{I} \tag{8.33}$$

It is important to mention that the cell must be maintained for 5 s at least at the fully charge stage (V_{cell}), and then reverse the current to discharge. This procedure does not take into account the leakage current but is suitable when using potentiostats that cannot shift from positive to negative current in 1 ms. SC Cells resulting in significant ESR and/or potential dependent capacitance as presented with pseudocapacitive contributions, the charge/discharge curves will deviate from linearity [513] as expected in some asymmetric assemblies.

(b) **Specific capacitance**

Specific capacitance (C) value for SC cells using the galvanostatic technique is calculated from the discharge curve as presented in Fig. 8.32, and using the following equation:

$$C = \frac{It}{\Delta E}, \tag{8.34}$$

where I is the applied constant current (A), t is the time interval measured from the initial stage of discharge (t_0) after the ESR drop to the totally discharge stage (t_f), and ΔE is the cell operational voltage window. Capacitance obtained from very low current densities (≈ 10 mA/cm^2) usually overestimates the performance on carbon materials and can lead to optimistic predictions of SC performance [51]. Therefore, higher currents are recommended for practical purposes.

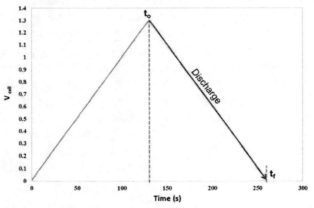

Figure 8.32 Schematic representation of a charge-discharge curve from galvanostatic cycling of a cell.

It is known that two-electrode SC cells are considered as two capacitors connected in series, where the total specific capacitance (C_T) can be calculated according to Eq. 8.35, where C_1 and C_2 are the intrinsic capacitances of the two electrodes [506].

$$\frac{1}{C_T} = \frac{1}{C_1} + \frac{1}{C_2} \tag{8.35}$$

In some papers, a calculation from this C_T value is used to calculate the intrinsic capacitance of the electrode materials, since $4C_T = C_{electrode}$. This calculation can work for symmetric devices, where only takes into consideration the mass of only one electrode [514]. Nevertheless, it is not recommended to use this approach for asymmetric devices.

(c) **Ragone plots (specific energy and power)**

Ragone plots are useful and essential to compare different energy storage devices, in terms of specific power (kW/kg) and specific energy (Wh/kg). High power is always achieved at the expense of lower energy and vice versa [513]. The specific energy in asymmetric SC shows a dependence on the charge/discharge rate [392], and the normal approach to obtain the specific energy and power of the SC cell is based on the capacitance values obtained at different current densities. The specific energy (E) which is the capacity to perform work, and the specific power (P) that shows how fast the energy is delivered, have been calculated using the following equations:

$$E = \frac{\left(\frac{1}{2}\right)CV^2}{3600}\left(\frac{Wh}{kg}\right), \quad P = \frac{V^2}{4(ESR)m} = \frac{3600\,E}{t}(W/kg), \quad (8.36)$$

where V is the cell operational voltage window, C is the specific capacitance value (F/kg) at each galvanostatic current applied, ESR is the resistance of the cell, and m the total weight of both electrodes (electroactive materials, conductive additives, and binders). It has been reported that the maximum energy (E_{max}) can be calculated from the theoretical capacitance value obtained from CV (Ca, Eqs. 8.31 and 8.32), and with this the maximum power (P_{max}) is obtained [415].

(d) **Long-term cycling or cyclability measurements**

SC are known for their good cyclability during a long period of time and numerous charge/discharge cycles ($\approx 10^5$) can be achieved. Nevertheless, for asymmetric cells the cycle life is affected by the pseudocapacitance contribution. The deterioration during cycling involves a variety of processes, which depend on design and operation conditions for a specific application. A way to evaluate the cycle life of a cell is through their capacitance and/or ESR values, which changes during cycling and can help to determine the end of the SC life. The approach used to determine cycle life is normally based on galvanostatic conditions, and the selected current for this characterization should be in the range of SC applications (discharge between 60 s and several minutes as already mentioned). These cycles are carried out considering the voltage limits of V_{cell}, and during a defined number of complete charge-discharge cycles, capacitance or ESR values are calculated. Cyclability or cycle life is terminated when the capacitance decreases at least 20–30% in comparison with the initial capacitance [390].

(e) **Self-discharge**

SC as well as batteries in their charged stage show the highest Gibbs (G) energy compared to their discharge stage, in which there is a thermodynamic driving force for their self-discharge on open-circuit. That is, a spontaneous decline of G with time. Although, self-discharge is of fundamental interest, its relevance depends on the specific application. When the recharge is carried out after a long period of rest-time, as in standby applications, self-discharge is of great importance for device performance specifications. Self-discharge rates can greatly vary depending on the mechanisms of the processes taking place at the electrode interfaces, as well as in the kinetics of the electrochemical processes involved. Self-discharge has a relation with the leakage current in a way that if the self-discharge increases the leakage current results higher [513]. The simplest methodology to characterize self-discharge is by measuring the open-circuit decline of electrode potential or state-of-charge with time. That is, the two-electrode cell is charge to the highest potential (V_{cell}), maintaining for a certain time, and then the potential is monitored at open-circuit. This methodology does not give information of the processes involved in each electrode during self-discharge, and sometimes this information is important taking into consideration that electrodes operating mainly with a capacitive behavior suffer faster self-discharge than in electrodes with pseudocapacitive contributions. Therefore, for an asymmetric configuration is important to obtain information on self-discharge individually for each electrode, by using a three-electrode configuration where a reference electrode is included [515]. Nevertheless, the net result of self-discharge for an asymmetric configuration using battery-like electrodes or electrodes with pseudocapacitance is usually an improvement compared to EDLC [392].

All characterization techniques and evaluation methodologies exposed here for SC, are focused on materials characterization. More detail analysis within the field of SC are described elsewhere [51, 506, 511, 512].

8.8 Future Perspectives

SC still face several technical challenges, as well as high cost issues at the current stage of technology, which has enabled their massive commercialization. Nevertheless, they still offer many advantages over batteries. Among the technical aspects that need to be addressed are their low specific energy, and their high self-discharge rate. The low specific energy is the major challenge for this energy storage device in the immediate and medium time range. Most of the commercially available SC can provide only up to 3–5 Wh/kg (Table 8.3) and if a greater energy is required, a larger SC must be constructed increasing the cost [257]. As for the high self-discharge rate, the problem arises from the electrostatic nature of double layer storage mechanism in which commercial SC are based, and the introduction of pseudocapacitive contributions has helped to minimize this effect. Different approaches have been used or are under development to resolve these problems without compromising the specific power, such as (1) making a hybrid system with a SC combined to a battery, where the SC contributes with their characteristic high peak power and the battery with its higher specific energy and low self-discharge rate; (2) using a nanoscale approach to design carbon functionalities or elegant nanocomposite materials, where a balance on energy and power values, and decreasing the self-discharge rate can be carried out through synergic effects from double layer and pseudocapacitive contributions; and (3) using an asymmetric assembly with neutral aqueous electrolytes and eco-friendly components, as suggested in Section 8.6, to what it has been called "environmentally friendly supercapacitors".

Suitable performance of SC is not sufficient for its effective marketing due to their high cost, which has been a critical issue for market widespread diffusion. Comparing the performance of SC with other technologies, it is well known that the power and cycle life capabilities are superior to those of high power batteries. Only lithium-ion batteries have demonstrated a close power capability to that of SC. The cost of an energy storage device has been defined in terms of $/Wh (energy cost), $/kW (power cost), or $/kg (material cost). Therefore, batteries have shown lower cost in terms of energy ($/Wh) compared to SC due to the larger specific energy of batteries. However, the increased power capability of SC has decreased their power cost ($/kW) compared to batteries [51].

The main cost of a SC comes from its electrode materials, separators, the use of organic electrolytes, and more importantly the small-scale production. On the other hand, the price of battery-like or pseudocapacitive materials has shown more relation with the cost of batteries that use similar materials and electrolytes and has driven the cost down compared to EDLC as shown in Table 8.13 [51].

Table 8.13 Cost in terms of $/Wh(Energy cost), $/W (power cost), and $/kg (material cost) extracted from the work of A. Burke [51]

Device	$/Wh	$/kW	$/kg
C/Organic Electrolyte/C	2.5	10	10
C/Aqueous Electrolyte/PbO_2	0.34	1.3	4.6

The future improvement in specific energy and the cost reduction should open the market for SC. The improvement based on activated carbon electrodes provides a rigid alternative for any emerging device in terms of cost and simpler manufacture procedures. Therefore, the key factors for improvement are the use of lower-cost electrode materials based on carbon and metal oxides compatible with the environment and the use of environmentally friendly aqueous electrolytes that are cheaper [467]. Moreover, the use of nanostructured hybrid electrode materials and/or involving asymmetric configurations opens the opportunity to develop devices that can compete with commercial SC and batteries for particular specific energy and power combinations [144]. These asymmetric configurations that can be EFSC is a more flexible alternative for SC improvement, where simpler fabrication and packaging procedures are involved due to the use of an environmentally friendly and cheaper electrolyte, higher degree of safety during cycling [516], and most importantly a specific energy greater that for organic based electrolyte EDLCs. Therefore, EFSC are the following emerging generation of SC technology, but will need the continuing research on the development of new high performance electrode materials, cell additives and components, and a more detail understanding of the electrochemical processes involved in high charge storage capacity and cycle life, in order to bring together the energy/power performance gap of conventional batteries and EDLCs [396].

Acknowledgments

We acknowledge the financial support granted from CONACYT Basic Science Project 154259, PAPIIT project IN112414, the University of Nantes for providing invited professorship to AKCG, and the Universidad Nacional Autónoma de Mexico for the financial support granted for DPC visit through the DGCI program.

References

1. Conway, B. E. (1999). *Electrochemical Supercapacitors*. (Kluwer Academic/Plenum, New York).
2. Simon, P., and Y. Gogotsi (2013). Capacitive energy storage in nanostructured carbon–electrolyte Systems. *Accounts of Chemical Research*, **46**(5), pp. 1094–1103.
3. Ghosh, A., and Y. H. Lee (2012). Carbon-based electrochemical capacitors. *ChemSusChem*, **5**(3), pp. 480–499.
4. Li, J., and H. Xie (2012). Synthesis of graphene oxide/polypyrrole nanowire composites for supercapacitors. *Materials Letters*, **78**(0), pp. 106–109.
5. Liu, C., Z. Yu, D. Neff, A. Zhamu, and B. Z. Jang (2010). Graphene-based supercapacitor with an ultrahigh energy density. *Nano Letters*, **10**(12), pp. 4863–4868.
6. Burke, A., and M. Miller (2011). The power capability of ultracapacitors and lithium batteries for electric and hybrid vehicle applications. *Journal of Power Sources*, **196**(1), pp. 514–522.
7. Bauman, J., and M. Kazerani (2008). A comparative study of fuel-cell–battery, fuel-cell–ultracapacitor, and fuel-cell–battery-ultracapacitor vehicles. *Vehicular Technology, IEEE Transactions on*, **57**(2), pp. 760–769.

8. Thounthong, P., S. Raël, and B. Davat (2006). Control strategy of fuel cell/supercapacitors hybrid power sources for electric vehicle. *Journal of Power Sources*, **158**, pp. 806–814.

9. Kim, Y.-T., Y. Ito, K. Tadai, T. Mitani, U.-S. Kim, H.-S. Kim, and B.-W. Cho (2005). Drastic change of electric double layer capacitance by surface functionalization of carbon nanotubes. *Applied Physics Letters*, **87**(23), pp. http://dx.doi.org/10.1063/1.2139839.

10. Dumitrescu, I., P. R. Unwin, and J. V. Macpherson (2009). Electrochemistry at carbon nanotubes: Perspective and issues. *Chemical Communications*, **45**, pp. 6886–6901.

11. Stienecker, A. (2009). *Hybrid Energy Storage Systems: An Ultracapacitor-Battery Energy Storage System for Hybrid Electric Vehicles* (VDM Verlag USA).

12. Dorf, R. C., and J. A. Svoboda (2001). *Introduction to Electric Circuits*, 5th ed., Wiley and Sons, Inc., New York.

13. Huelsman, L. P. (1972). *Basic Circuit Theory with Digital Computations. Series in Computer Applications in Electrical Engineering* (Prentice-Hall).

14. Splung.com. *Capacitors*. [cited 2014 22-10-2014]; Available from: http://www.splung.com/content/sid/3/page/capacitors.

15. Softpedia. *How Capacitors Work*. 2008 [cited 2014 22-10-2014]; Available from: http://news.softpedia.com/news/How-Capacitors-Work-82563.shtml#sgal_2.

16. Smith, M. *Resistor and capacitor tables*. [cited 2014 22-oct-2014]; Available from: http://www.mds975.co.uk/Content/components01.html.

17. Helmholtz, H. (1853). Ueber einige Gesetze der Vertheilung elektrischer Ströme in körperlichen Leitern mit Anwendung auf die thierisch-elektrischen Versuche. *Annalen der Physik*, **165**(6), pp. 211–233.

18. Helmholtz, H. (1879). Studien über electrische Grenzschichten. *Annalen der Physik*, **243**(7), pp. 337–382.

19. Fan, Z., K. Wang, T. Wei, J. Yan, L. Song, and B. Shao (2010). An environmentally friendly and efficient route for the reduction of graphene oxide by aluminum powder. *Carbon*, **48**(5), pp. 1686–1689.

20. Paredes, J. I., S. Villar-Rodil, M. J. Fernandez-Merino, L. Guardia, A. Martinez-Alonso, and J. M. D. Tascon (2011). Environmentally friendly approaches toward the mass production of processable graphene from graphite oxide. *Journal of Materials Chemistry*, **21**(2), pp. 298–306.

21. Jennifer, A. I., J. I. David, and D. S.-S. John (2006). *Conjugated Polymers Electroactive Polymers for Batteries and Supercapacitors,* CRC Press, pp. 1–29.
22. Deshmukh, A. B., and M. V. Shelke (2013). Synthesis and electrochemical performance of a single walled carbon nanohorn-Fe_3O_4 nanocomposite supercapacitor electrode. *RSC Advances*, **3**(44), pp. 21390–21393.
23. Trasatti, S., and G. Buzzanca (1971). Ruthenium dioxide: A new interesting electrode material. Solid state structure and electrochemical behaviour. *Journal of Electroanalytical Chemistry and Interfacial Electrochemistry*, **29**(2), pp. A1–A5.
24. Conway, B. E., and E. Gileadi (1962). Kinetic theory of pseudocapacitance and electrode reactions at appreciable surface coverage. *Transactions of the Faraday Society*, **58**(0), pp. 2493–2509.
25. Winter, M., and R. J. Brodd (2004). What are batteries, fuel cells, and supercapacitors? *Chemical Reviews*, **104**(10), pp. 4245–4270.
26. Gileadi, E., and B. E. Conway (1963). Kinetic theory of adsorption of intermediates in electrochemical catalysis. *The Journal of Chemical Physics*, **39**(12), pp. 3420–3430.
27. Srinivasan, S., and E. Gileadi (1966). The potential-sweep method: A theoretical analysis. *Electrochimica Acta*, **11**(3), pp. 321–335.
28. Angerstein-Kozlowska, H., J. Klinger, and B. E. Conway (1977). Computer simulation of the kinetic behaviour of surface reactions driven by a linear potential sweep: Part I. Model 1-electron reaction with a single adsorbed species. *Journal of Electroanalytical Chemistry and Interfacial Electrochemistry*, **75**(1), pp. 45–60.
29. Martínez Álvarez, O., *Development and design of modified carbon composites for application as electrochemical capacitors*, in *Centro de Investigación en Energía—UNAM* 2010, Universidad Nacional Autonoma de México: Temixco, Morelos.
30. Wen, J., and Z. Zhou (2006). Pseudocapacitance characterization of hydrous ruthenium oxide prepared via cyclic voltammetric deposition. *Materials Chemistry and Physics*, **98**(2–3), pp. 442–446.
31. Hu, C.-C., M.-J. Liu, and K.-H. Chang (2007). Anodic deposition of hydrous ruthenium oxide for supercapacitors. *Journal of Power Sources*, **163**(2), pp. 1126–1131.
32. Martinez-Alvarez, O., and M. Miranda-Hernandez (2008). Characterization of carbon pastes as matrices in composite electrodes for use in electrochemical capacitors. *Carbon: Science and Technology*, **1**(1), pp. 30–38.

33. Conway, B. E., and W. G. Pell (2003). Double-layer and pseudocapacitance types of electrochemical capacitors and their applications to the development of hybrid devices. *Journal of Solid State Electrochemistry*, **7**(9), pp. 637–644.

34. Halper, M. S., and J. C. Ellenbogen (2006). *Supercapacitors: A Brief Overview* (Group, M. N., ed.), The MITRE Corporation McLean, Virginia, USA. p. 34.

35. Caturla, F., M. Molina-Sabio, and F. Rodríguez-Reinoso (1991). Preparation of activated carbon by chemical activation with $ZnCl2$. *Carbon*, **29**(7), pp. 999–1007.

36. Conway, B. E. (1991). Transition from "Supercapacitor" to "battery" behavior in electrochemical energy storage. *Journal of The Electrochemical Society*, **138**(6), pp. 1539–1548.

37. Pletcher, D., R. Greff, R. Peat, L. M. Peter, and J. Robinson (2010). Instrumental Methods in Electrochemistry "5–The electrical double layer", (Woodhead Publishing), pp. 149–177.

38. Conway, B. E., V. Birss, and J. Wojtowicz (1997). The role and utilization of pseudocapacitance for energy storage by supercapacitors. *Journal of Power Sources*, **66**(1–2), pp. 1–14.

39. Pandolfo, A. G., and A. F. Hollenkamp (2006). Carbon properties and their role in supercapacitors. *Journal of Power Sources*, **157**(1), pp. 11–27.

40. Kötz, R., and M. Carlen (2000). Principles and applications of electrochemical capacitors. *Electrochimica Acta*, **45**(15–16), pp. 2483–2498.

41. Dyatkin, B., V. Presser, M. Heon, M. R. Lukatskaya, M. Beidaghi, and Y. Gogotsi (2013). Development of a green supercapacitor composed entirely of environmentally Friendly materials. *ChemSusChem*, **6**(12), pp. 2269–2280.

42. Kossyrev, P. (2012). Carbon black supercapacitors employing thin electrodes. *Journal of Power Sources*, **201**(1), pp. 347–352.

43. Pekala, R. W., J. C. Farmer, C. T. Alviso, T. D. Tran, S. T. Mayer, J. M. Miller, and B. Dunn (1998). Carbon aerogels for electrochemical applications. *Journal of Non-Crystalline Solids*, **225**(1), pp. 74–80.

44. Lazzari, M., F. Soavi, and M. Mastragostino (2008). High voltage, asymmetric EDLCs based on xerogel carbon and hydrophobic IL electrolytes. *Journal of Power Sources*, **178**(1), pp. 490–496.

45. Candelaria, S. L., Y. Shao, W. Zhou, X. Li, J. Xiao, J.-G. Zhang, Y. Wang, J. Liu, J. Li, and G. Cao (2012). Nanostructured carbon for energy storage and conversion. *Nano Energy*, **1**(2), pp. 195–220.

46. Frackowiak, E., J. Machnikowski, and F. Béguin (2006) *New Carbon Based Materials for Electrochemical Energy Storage Systems: Batteries, Supercapacitors and Fuel Cells Novel Carbonaceous Materials for Application in the Electrochemical Supercapacitors*, Springer Netherlands, pp. 5–20.

47. Inoue, T., S. Mori, and S. Kawasaki (2011). Electric double layer capacitance of Graphene-Like materials derived from single-walled carbon nanotubes. *Japanese Journal of Applied Physics*, **50**(1), pp. 01AF07-01AF07-04.

48. Yang, X., F. Zhang, L. Zhang, T. Zhang, Y. Huang, and Y. Chen (2013). A high-performance Graphene oxide-doped ion gel as gel polymer electrolyte for all-solid-state supercapacitor applications. *Advanced Functional Materials*, **23**(26), pp. 3353–3360.

49. Tamilarasan, P., and S. Ramaprabhu (2013). Graphene based all-solid-state supercapacitors with ionic liquid incorporated polyacrylonitrile electrolyte. *Energy*, **51**(0), pp. 374–381.

50. Lee, J. S., S. I. Kim, J. C. Yoon, and J. H. Jang (2013). Chemical Vapor Deposition of Mesoporous Graphene Nanoballs for Supercapacitor. *ACS Nano*, **7**(7), pp. 6047–6055.

51. Burke, A. (2007). R&D considerations for the performance and application of electrochemical capacitors. *Electrochimica Acta*, **53**(3), pp. 1083–1091.

52. Wei, Y. Z., B. Fang, S. Iwasa, and M. Kumagai (2005). A novel electrode material for electric double-layer capacitors. *Journal of Power Sources*, **141**(2), pp. 386–391.

53. Chmiola, J., C. Largeot, P.-L. Taberna, P. Simon, and Y. Gogotsi (2008). Desolvation of ions in subnanometer pores and its effect on capacitance and double-layer theory. *Angewandte Chemie*, **120**(18), pp. 3440–3443.

54. Boehm, H. P. (1994). Some aspects of the surface chemistry of carbon blacks and other carbons. *Carbon*, **32**(5), pp. 759–769.

55. Nakajima, T., Y. Matsuo, S. Kasamatsu, and K. Nakanishi (1994). Carbon-fluorine bonding of fullerene C60 fluorinated by elemental fluorine, with HF gas, under UV irradiation and in chlorofluoro-carbon solvent. *Carbon*, **32**(6), pp. 1177–1180.

56. Momma, T., X. Liu, T. Osaka, Y. Ushio, and Y. Sawada (1996). Electrochemical modification of active carbon fiber electrode and its application to double-layer capacitor. *Journal of Power Sources*, **60**(2), pp. 249–253.

57. Ma, R. Z., J. Liang, B. Q. Wei, B. Zhang, C. L. Xu, and D. H. Wu (1999). Study of electrochemical capacitors utilizing carbon nanotube electrodes. *Journal of Power Sources*, **84**(1), pp. 126–129.

58. Fang, B., Y. Z. Wei, and M. Kumagai (2006). Modified carbon materials for high-rate EDLCs application. *Journal of Power Sources*, **155**(2), pp. 487–491.

59. Lozano-Castelló, D., D. Cazorla-Amorós, A. Linares-Solano, S. Shiraishi, H. Kurihara, and A. Oya (2003). Influence of pore structure and surface chemistry on electric double layer capacitance in non-aqueous electrolyte. *Carbon*, **41**(9), pp. 1765–1775.

60. Marañón Di Leo, J., and J. Marañón (2005). Hydration and diffusion of cations in nanopores. *Journal of Molecular Structure: THEOCHEM*, **729**(1–2), pp. 53–57.

61. Rajbhandari, R., L. K. Shrestha, B. P. Pokharel, and R. R. Pradhananga (2013). Development of nanoporous structure in carbons by chemical activation with zinc chloride. *J Nanosci Nanotechno*, **13**(4), pp. 2613–2623.

62. Liu, C.-L., W.-S. Dong, G.-P. Cao, J.-R. Song, L. Liu, and Y.-S. Yang (2007). Influence of KOH followed by oxidation pretreatment on the electrochemical performance of phenolic based activated carbon fibers. *Journal of Electroanalytical Chemistry*, **611**(1–2), pp. 225–231.

63. Zhou, P. W., B. H. Li, F. Y. Kang, and Y. Q. Zeng (2006). The development of supercapacitors from coconut-shell activated carbon. *New Carbon Materials*, **21**(2), pp. 125–131.

64. Jain, A., and S. K. Tripathi (2014). Fabrication and characterization of energy storing supercapacitor devices using coconut shell based activated charcoal electrode. *Materials Science and Engineering B*, **183**(0), pp. 54–60.

65. Li, X., W. Xing, S. Zhuo, J. Zhou, F. Li, S.-Z. Qiao, and G.-Q. Lu (2011). Preparation of capacitor's electrode from sunflower seed shell. *Bioresource Technology*, **102**(2), pp. 1118–1123.

66. Subramanian, V., C. Luo, A. M. Stephan, K. S. Nahm, S. Thomas, and B. Wei (2007). Supercapacitors from activated carbon derived from banana fibers. *The Journal of Physical Chemistry C*, **111**(20), pp. 7527–7531.

67. Rufford, T. E., D. Hulicova-Jurcakova, K. Khosla, Z. Zhu, and G. Q. Lu (2010). Microstructure and electrochemical double-layer capacitance of carbon electrodes prepared by zinc chloride activation of sugar cane bagasse. *Journal of Power Sources*, **195**(3), pp. 912–918.

68. Thambidurai, A., J. Lourdusamy, J. John, and S. Ganesan (2014). Preparation and electrochemical behaviour of biomass based porous carbons as electrodes for supercapacitors—a comparative investigation. *Korean Journal of Chemical Engineering*, **31**(2), pp. 268–275.

69. Valente Nabais, J. M., J. G. Teixeira, and I. Almeida (2011). Development of easy made low cost bindless monolithic electrodes from biomass with controlled properties to be used as electrochemical capacitors. *Bioresource Technology*, **102**(3), pp. 2781–2787.

70. Bichat, M. P., E. Raymundo-Piñero, and F. Béguin (2010). High voltage supercapacitor built with seaweed carbons in neutral aqueous electrolyte. *Carbon*, **48**(15), pp. 4351–4361.

71. Karthikeyan, K., V. Aravindan, S. B. Lee, I. C. Jang, H. H. Lim, G. J. Park, M. Yoshio, and Y. S. Lee (2010). A novel asymmetric hybrid supercapacitor based on Li_2FeSiO_4 and activated carbon electrodes. *Journal of Alloys and Compounds*, **504**(1), pp. 224–227.

72. Hu, C.-C., J.-H. Su, and T.-C. Wen (2007). Modification of multi-walled carbon nanotubes for electric double-layer capacitors: Tube opening and surface functionalization. *Journal of Physics and Chemistry of Solids*, **68**(12), pp. 2353–2362.

73. Thuy Pham, T. P., C.-W. Cho, and Y.-S. Yun (2010). Environmental fate and toxicity of ionic liquids: A review. *Water Research*, **44**(2), pp. 352–372.

74. Stolte, S., S. Steudte, A. Igartua, and P. Stepnowski (2011). The biodegradation of ionic liquids—the view from a chemical structure perspective. *Current Organic Chemistry*, **15**(12), pp. 1946–1973.

75. Stolte, S., J. Arning, and J. Thoming (2011). Biodegradability of ionic liquids—test procedures and structural design. *Chemie Ingenieur Technik*, **83**(9), pp. 1454–1467.

76. Krause, A., and A. Balducci (2011). High voltage electrochemical double layer capacitor containing mixtures of ionic liquids and organic carbonate as electrolytes. *Electrochemistry Communications*, **13**(8), pp. 814–817.

77. Nahil, M. A., and P. T. Williams (2012). Pore characteristics of activated carbons from the phosphoric acid chemical activation of cotton stalks. *Biomass and Bioenergy*, **37**(0), pp. 142–149.

78. Fedotova, M. V., and S. E. Kruchinin (2011). Hydration of acetic acid and acetate ion in water studied by 1D-RISM theory. *Journal of Molecular Liquids*, **164**(3), pp. 201–206.

79. Tõnurist, K., T. Thomberg, A. Jänes, T. Romann, V. Sammelselg, and E. Lust (2013). Influence of separator properties on electrochemical performance of electrical double-layer capacitors. *Journal of Electroanalytical Chemistry*, **689**(0), pp. 8–20.

80. Wu, F.-C., R.-L. Tseng, C.-C. Hu, and C.-C. Wang (2004). Physical and electrochemical characterization of activated carbons prepared from firwoods for supercapacitors. *Journal of Power Sources*, **138**(1–2), pp. 351–359.

81. Simon, P., and A. Burke (2008). Nanostructured carbons: Double-layer capacitance and more. *Interface*, spring, pp. 38–43.

82. Girgis, B. S., and M. F. Ishak (1999). Activated carbon from cotton stalks by impregnation with phosphoric acid. *Materials Letters*, **39**(2), pp. 107–114.

83. Kuppireddy, S. K. R., K. Rashid, A. Al Shoaibi, and C. Srinivasakannan (2013). Production and characterization of porous carbon from date palm seeds by chemical activation with H_3PO_4: Process optimization for maximizing adsorption of methylene blue. *Chemical Engineering Communications*, **201**(8), pp. 1021–1040.

84. Kandiyoti, R., J. I. Lazaridis, B. Dyrvold, and C. R. Weerasinghe (1984). Pyrolysis of a $ZnCl_2$-impregnated coal in an inert atmosphere. *Fuel*, **63**(11), pp. 1583–1587.

85. El-Hendawy, A.-N. A., A. J. Alexander, R. J. Andrews, and G. Forrest (2008). Effects of activation schemes on porous, surface and thermal properties of activated carbons prepared from cotton stalks. *Journal of Analytical and Applied Pyrolysis*, **82**(2), pp. 272–278.

86. Jiménez-Cordero, D., F. Heras, M. A. Gilarranz, and E. Raymundo-Piñero (2014). Grape seed carbons for studying the influence of texture on supercapacitor behaviour in aqueous electrolytes. *Carbon*, **71**(0), pp. 127–138.

87. Sevilla, M., and R. Mokaya (2014). Energy storage applications of activated carbons: Supercapacitors and hydrogen storage. *Energy & Environmental Science*, **7**(4), pp. 1250–1280.

88. Jankowska, H., A. Swiatkowski, and J. Choma (1991). *Active Carbon* (Ellis Horwood Ltd, NJ).

89. Ahmadpour, A., and D. D. Do (1996). The preparation of active carbons from coal by chemical and physical activation. *Carbon*, **34**(4), pp. 471–479.

90. Toupin, M., D. Bélanger, I. R. Hill, and D. Quinn (2005). Performance of experimental carbon blacks in aqueous supercapacitors. *Journal of Power Sources*, **140**(1), pp. 203–210.

91. Salitra, G., A. Soffer, L. Eliad, Y. Cohen, and D. Aurbach (2000). Carbon electrodes for double-layer capacitors I. Relations between ion and pore dimensions. *Journal of The Electrochemical Society*, **147**(7), pp. 2486–2493.
92. Chmiola, J., G. Yushin, Y. Gogotsi, C. Portet, P. Simon, and P. L. Taberna (2006). Anomalous increase in carbon capacitance at pore sizes less than 1 nanometer. *Science*, **313**(5794), pp. 1760–1763.
93. Barbieri, O., M. Hahn, A. Herzog, and R. Kötz (2005). Capacitance limits of high surface area activated carbons for double layer capacitors. *Carbon*, **43**(6), pp. 1303–1310.
94. Endo, M., T. Maeda, T. Takeda, Y. J. Kim, K. Koshiba, H. Hara, and M. S. Dresselhaus (2001). Capacitance and pore-size distribution in aqueous and nonaqueous electrolytes using various activated carbon electrodes. *Journal of the Electrochemical Society*, **148**(8), pp. A910–A914.
95. Simon, P., and A. Burke (2008). Nanostructured carbons: Double-layer capacitance and more. *Interface*, **17**(1), pp. 38–43.
96. Xing, W., S. Z. Qiao, R. G. Ding, F. Li, G. Q. Lu, Z. F. Yan, and H. M. Cheng (2006). Superior electric double layer capacitors using ordered mesoporous carbons. *Carbon*, **44**(2), pp. 216–224.
97. Fernández, J. A., T. Morishita, M. Toyoda, M. Inagaki, F. Stoeckli, and T. A. Centeno (2008). Performance of mesoporous carbons derived from poly(vinyl alcohol) in electrochemical capacitors. *Journal of Power Sources*, **175**(1), pp. 675–679.
98. Álvarez, S., M. C. Blanco-López, A. J. Miranda-Ordieres, A. B. Fuertes, and T. A. Centeno (2005). Electrochemical capacitor performance of mesoporous carbons obtained by templating technique. *Carbon*, **43**(4), pp. 866–870.
99. Sevilla, M., S. Alvarez, T. A. Centeno, A. B. Fuertes, and F. Stoeckli (2007). Performance of templated mesoporous carbons in supercapacitors. *Electrochimica Acta*, **52**, pp. 3207–3215.
100. Fuertes, A. B., G. Lota, T. A. Centeno, and E. Frackowiak (2005). Templated mesoporous carbons for supercapacitor application. *Electrochimica Acta*, **50**(14), pp. 2799–2805.
101. Vix-Guterl, C., E. Frackowiak, K. Jurewicz, M. Friebe, J. Parmentier, and F. Béguin (2005). Electrochemical energy storage in ordered porous carbon materials. *Carbon*, **43**(6), pp. 1293–1302.
102. Zheng, C., L. Qi, M. Yoshio, and H. Wang (2010). Cooperation of micro- and meso-porous carbon electrode materials in electric double-layer capacitors. *Journal of Power Sources*, **195**(13), pp. 4406–4409.

103. Shi, H. (1996). Activated carbons and double layer capacitance. *Electrochimica Acta*, **41**(10), pp. 1633–1639.

104. Chmiola, J., G. Yushin, R. Dash, and Y. Gogotsi (2006). Effect of pore size and surface area of carbide derived carbons on specific capacitance. *Journal of Power Sources*, **158**(1), pp. 765–772.

105. Eliad, L., G. Salitra, A. Soffer, and D. Aurbach (2001). Ion Sieving effects in the electrical double layer of porous carbon electrodes: Estimating effective ion size in electrolytic solutions. *The Journal of Physical Chemistry B*, **105**(29), pp. 6880–6887.

106. Largeot, C., C. Portet, J. Chmiola, P.-L. Taberna, Y. Gogotsi, and P. Simon (2008). Relation between the ion size and pore size for an electric double-layer capacitor. *Journal of the American Chemical Society*, **130**(9), pp. 2730–2731.

107. Gryglewicz, G., J. Machnikowski, E. Lorenc-Grabowska, G. Lota, and E. Frackowiak (2005). Effect of pore size distribution of coal-based activated carbons on double layer capacitance. *Electrochimica Acta*, **50**(5), pp. 1197–1206.

108. Lee, J., S. Han, and T. Hyeon (2004). Synthesis of new nanoporous carbon materials using nanostructured silica materials as templates. *Journal of Materials Chemistry*, **14**(4), pp. 478–486.

109. Vix-Guterl, C., S. Saadallah, K. Jurewicz, E. Frackowiak, M. Reda, J. Parmentier, J. Patarin, and F. Beguin (2004). Supercapacitor electrodes from new ordered porous carbon materials obtained by a templating procedure. *Materials Science and Engineering B*, **108**(1–2), pp. 148–155.

110. de Levie, R. (1963). On porous electrodes in electrolyte solutions: I. Capacitance effects. *Electrochimica Acta*, **8**(10), pp. 751–780.

111. Conway, B. E. (2002). Reflections on directions of electrochemical surface science as a leading edge of surface chemistry. *Journal of Electroanalytical Chemistry*, **524–525**(0), pp. 4–19.

112. Donnet, J.-B., R. C. Bansal, and M. J. Wang, eds. *Carbon Black: Science and Technology*, 2nd ed., 1993, CRC Press: NY. 461.

113. Liang, Y., F. Liang, Z. Li, D. Wu, F. Yan, S. Li, and R. Fu (2010). The role of mass transport pathway in wormholelike mesoporous carbon for supercapacitors. *Physical Chemistry Chemical Physics*, **12**(36), pp. 10842–10845.

114. Jänes, A., L. Permann, M. Arulepp, and E. Lust (2004). Electrochemical characteristics of nanoporous carbide-derived carbon materials in

non-aqueous electrolyte solutions. *Electrochemistry Communications*, **6**(3), pp. 313–318.

115. Ruiz, V., C. Blanco, R. Santamaría, J. M. Juárez-Galán, A. Sepúlveda-Escribano, and F. Rodríguez-Reinoso (2008). Carbon molecular sieves as model active electrode materials in supercapacitors. *Microporous and Mesoporous Materials*, **110**(2–3), pp. 431–435.

116. Koresh, J., and A. Soffer (1983). Stereoselectivity in ion electro-adsorption and in double-layer charging of molecular sieve carbon electrodes. *Journal of Electroanalytical Chemistry and Interfacial Electrochemistry*, **147**(1–2), pp. 223–234.

117. Pekala, R. W. (1989). Organic aerogels from the polycondensation of resorcinol with formaldehyde. *Journal of Materials Science*, **24**(9), pp. 3221–3227.

118. Pröbstle, H., M. Wiener, and J. Fricke (2003). Carbon aerogels for electrochemical double layer capacitors. *Journal of Porous Materials*, **10**(4), pp. 213–222.

119. Fischer, U., R. Saliger, V. Bock, R. Petricevic, and J. Fricke (1997). Carbon aerogels as electrode material in supercapacitors. *Journal of Porous Materials*, **4**(4), pp. 281–285.

120. Kirkbir, F., H. Murata, D. Meyers, and S. R. Chaudhuri (1998). Drying of aerogels in different solvents between atmospheric and supercritical pressures. *Journal of Non-Crystalline Solids*, **225**(0), pp. 14–18.

121. Gavalda, S., K. E. Gubbins, Y. Hanzawa, K. Kaneko, and K. T. Thomson (2002). Nitrogen adsorption in carbon aerogels: A molecular simulation study. *Langmuir*, **18**(6), pp. 2141–2151.

122. Dresselhaus, M. S. (1997). Future directions in carbon science. *Annual Review of Materials Science*, **27**(1), pp. 1–34.

123. Baumann, T. F., and J. H. Satcher Jr (2004). Template-directed synthesis of periodic macroporous organic and carbon aerogels. *Journal of Non-Crystalline Solids*, **350**(0), pp. 120–125.

124. Wang, J., X. Yang, D. Wu, R. Fu, M. S. Dresselhaus, and G. Dresselhaus (2008). The porous structures of activated carbon aerogels and their effects on electrochemical performance. *Journal of Power Sources*, **185**(1), pp. 589–594.

125. Saliger, R., U. Fischer, C. Herta, and J. Fricke (1998). High surface area carbon aerogels for supercapacitors. *Journal of Non-Crystalline Solids*, **225**(0), pp. 81–85.

126. Li, W., G. Reichenauer, and J. Fricke (2002). Carbon aerogels derived from cresol–resorcinol–formaldehyde for supercapacitors. *Carbon*, **40**(15), pp. 2955–2959.

127. Lee, Y. J., J. C. Jung, J. Yi, S.-H. Baeck, J. R. Yoon, and I. K. Song (2010). Preparation of carbon aerogel in ambient conditions for electrical double-layer capacitor. *Current Applied Physics*, **10**(2), pp. 682–686.
128. Li, J., X. Wang, Y. Wang, Q. Huang, C. Dai, S. Gamboa, and P. J. Sebastian (2008). Structure and electrochemical properties of carbon aerogels synthesized at ambient temperatures as supercapacitors. *Journal of Non-Crystalline Solids*, **354**(1), pp. 19–24.
129. Li, J., X. Wang, Q. Huang, S. Gamboa, and P. J. Sebastian (2006). Studies on preparation and performances of carbon aerogel electrodes for the application of supercapacitor. *Journal of Power Sources*, **158**(1), pp. 784–788.
130. Pröbstle, H., R. Saliger, and J. Fricke (2000) *Studies in Surface Science and Catalysis Electrochemical Investigation of Carbon Aerogels and Their Activated Derivatives,* Elsevier, pp. 371–379.
131. Mayer, S. T., R. W. Pekala, and J. L. Kaschmitter (1993). The aerocapacitor: An electrochemical double-layer energy-storage device. *Journal of The Electrochemical Society*, **140**(2), pp. 446–451.
132. Carriazo, D., F. Pico, M. C. Gutierrez, F. Rubio, J. M. Rojo, and F. del Monte (2010). Block-Copolymer assisted synthesis of hierarchical carbon monoliths suitable as supercapacitor electrodes. *Journal of Materials Chemistry*, **20**(4), pp. 773–780.
133. Liu, D., J. Shen, N. Liu, H. Yang, and A. Du (2013). Preparation of activated carbon aerogels with hierarchically porous structures for electrical double layer capacitors. *Electrochimica Acta*, **89**(0), pp. 571–576.
134. Frackowiak, E., K. Jurewicz, K. Szostak, S. Delpeux, and F. Béguin (2002). Nanotubular materials as electrodes for supercapacitors. *Fuel Processing Technology*, **77–78**(0), pp. 213–219.
135. Frackowiak, E., S. Delpeux, K. Jurewicz, K. Szostak, D. Cazorla-Amoros, and F. Béguin (2002). Enhanced capacitance of carbon nanotubes through chemical activation. *Chemical Physics Letters*, **361**(1–2), pp. 35–41.
136. Chen, J. H., W. Z. Li, D. Z. Wang, S. X. Yang, J. G. Wen, and Z. F. Ren (2002). Electrochemical characterization of carbon nanotubes as electrode in electrochemical double-layer capacitors. *Carbon*, **40**(8), pp. 1193–1197.
137. Zhang, H., G. Cao, Y. Yang, and Z. Gu (2008). Comparison between electrochemical properties of aligned carbon nanotube array and entangled carbon nanotube electrodes. *Journal of The Electrochemical Society*, **155**(2), pp. K19–K22.

138. Fujiwara, A., K. Ishii, H. Suematsu, H. Kataura, Y. Maniwa, S. Suzuki, and Y. Achiba (2001). Gas adsorption in the inside and outside of single-walled carbon nanotubes. *Chemical Physics Letters*, **336**(3–4), pp. 205–211.

139. Byl, O., J. Liu, and J. T. Yates (2005). Etching of carbon nanotubes by ozone a surface area study. *Langmuir*, **21**(9), pp. 4200–4204.

140. Pico, F., C. Pecharroman, A. Ansón, M. T. Martinez, and J. M. Rojo (2007). Understanding carbon–carbon composites as electrodes of supercapacitors. *Journal of The Electrochemical Society*, **154**(6), pp. A579–A586.

141. Picó, F., J. M. Rojo, M. L. Sanjuán, A. Ansón, A. M. Benito, M. A. Callejas, W. K. Maser, and M. T. Martínez (2004). Single-walled carbon nanotubes as electrodes in supercapacitors. *Journal of the Electrochemical Society*, **151**(6), pp. A831–A837.

142. An, K. H., W. S. Kim, Y. S. Park, J. M. Moon, D. J. Bae, S. C. Lim, Y. S. Lee, and Y. H. Lee (2001). Electrochemical properties of high-power supercapacitors using single-walled carbon nanotube electrodes. *Advanced Functional Materials*, **11**(5), pp. 387–392.

143. Niu, C., E. K. Sichel, R. Hoch, D. Moy, and H. Tennent (1997). High power electrochemical capacitors based on carbon nanotube electrodes. *Applied Physics Letters*, **70**(11), pp. 1480–1482.

144. Zhao, X., B. M. Sanchez, P. J. Dobson, and P. S. Grant (2011). The role of nanomaterials in redox-based supercapacitors for next generation energy storage devices. *Nanoscale*, **3**(3), pp. 839–855.

145. Geim, A. K., and K. S. Novoselov (2007). The rise of graphene. *Nature Materials*, **6**(3), pp. 183–191.

146. Geim, A. K., S. V. Dubonos, I. V. Grigorieva, K. S. Novoselov, A. A. Zhukov, and S. Y. Shapoval (2003). Microfabricated adhesive mimicking gecko foot-hair. *Nature Materials*, **2**(7), pp. 461–463.

147. Novoselov, K. S., A. K. Geim, S. V. Morozov, D. Jiang, Y. Zhang, S. V. Dubonos, I. V. Grigorieva, and A. A. Firsov (2004). Electric field effect in atomically thin carbon films. *Science*, **306**(5696), pp. 666–669.

148. Chae, H. K., D. Y. Siberio-Pérez, J. Kim, Y. Go, M. Eddaoudi, A. J. Matzger, M. O'Keeffe, and O. M. Yaghi (2004). A route to high surface area, porosity and inclusion of large molecules in crystals. *Nature*, **427**, pp. 523–527.

149. Frank, I. W., D. M. Tanenbaum, A. M. van der Zande, and P. L. McEuen (2007). Mechanical properties of suspended graphene sheets. *Journal of Vacuum Science & Technology B*, **25**(6), pp. 2558–2561.

149. Frank, I. W., D. M. Tanenbaum, A. M. V. D. Zande, and P. L. McEuen (2007). Mechanical properties of suspended graphene sheets AVS.

150. Salvetat, J. P., J. M. Bonard, N. H. Thomson, A. J. Kulik, L. Forró, W. Benoit, and L. Zuppiroli (1999). Mechanical properties of carbon nanotubes. *Applied Physics A: Materials Science & Processing*, **69**(3), pp. 255–260.

151. Hirata, M., T. Gotou, S. Horiuchi, M. Fujiwara, and M. Ohba (2004). Thin-film particles of graphite oxide 1: High-yield synthesis and flexibility of the particles. *Carbon*, **42**(14), pp. 2929–2937.

152. Novoselov, K. S., A. K. Geim, S. V. Morozov, D. Jiang, M. I. Katsnelson, I. V. Grigorieva, S. V. Dubonos, and A. A. Firsov (2005). Two-dimensional gas of massless Dirac fermions in graphene. *Nature*, **438**(7065), pp. 197–200.

153. Novoselov, K. S., D. Jiang, F. Schedin, T. J. Booth, V. V. Khotkevich, S. V. Morozov, and A. K. Geim (2005). Two-dimensional atomic crystals. *Proceedings of the National Academy of Sciences of the United States of America*, **102**(30), pp. 10451–10453.

154. Dong, L.-X., and Q. Chen (2010). Properties, synthesis, and characterization of graphene. *Frontiers of Materials Science in China*, **4**(1), pp. 45–51.

155. Ando, T. (2009). The electronic properties of graphene and carbon nanotubes. *NPG Asia Materials*, **1**(1), pp. 17–21.

156. Geim, A. K., and A. H. MacDonald (2007). Graphene: Exploring carbon flatland. *Physics Today*, **60**(8), pp. 35–41.

157. Sun, Y., Q. Wu, and G. Shi (2011). Graphene based new energy materials. *Energy & Environmental Science*, **4**(4), pp. 1113–1132.

158. Jernigan, G. G., B. L. VanMil, J. L. Tedesco, J. G. Tischler, E. R. Glaser, A. Davidson, P. M. Campbell, and D. K. Gaskill (2009). Comparison of epitaxial graphene on Si-face and C-face 4H SiC formed by ultrahigh vacuum and RF furnace production. *Nano Letters*, **9**(7), pp. 2605–2609.

159. Huang, H., W. Chen, S. Chen, and A. T. S. Wee (2008). Bottom-up growth of epitaxial graphene on 6H-SiC(0001). *ACS Nano*, **2**(12), pp. 2513–2518.

160. Berger, C., Z. Song, X. Li, X. Wu, N. Brown, C. Naud, D. Mayou, T. Li, J. Hass, A. N. Marchenkov, E. H. Conrad, P. N. First, and W. A. de Heer (2006). Electronic confinement and coherence in patterned epitaxial graphene. *Science*, **312**(5777), pp. 1191–1196.

161. Somani, P. R., S. P. Somani, and M. Umeno (2006). Planer nanographenes from camphor by CVD. *Chemical Physics Letters*, **430**(1–3), pp. 56–59.

162. Coraux, J., A. T. N'Diaye, M. Engler, C. Busse, D. Wall, N. Buckanie, F.-J. Meyer zu Heringdorf, R. van Gastel, B. Poelsema, and T. Michely (2009). Growth of graphene on Ir(111). *New Journal of Physics*, **11**(2), pp. 023006.

163. Yuan, G. D., W. J. Zhang, Y. Yang, Y. B. Tang, Y. Q. Li, J. X. Wang, X. M. Meng, Z. B. He, C. M. L. Wu, I. Bello, C. S. Lee, and S. T. Lee (2009). Graphene sheets via microwave chemical vapor deposition. *Chemical Physics Letters*, **467**(4–6), pp. 361–364.

164. Hummers, W. S., and R. E. Offeman (1958). Preparation of graphitic oxide. *Journal of the American Chemical Society*, **80**(6), pp. 1339–1339.

165. Dreyer, D. R., S. Park, C. W. Bielawski, and R. S. Ruoff (2010). The chemistry of graphene oxide. *Chemical Society Reviews*, **39**(1), pp. 228–240.

166. Hernandez, Y., V. Nicolosi, M. Lotya, F. M. Blighe, Z. Sun, S. De, I. T. McGovern, B. Holland, M. Byrne, Y. K. Gun'Ko, J. J. Boland, P. Niraj, G. Duesberg, S. Krishnamurthy, R. Goodhue, J. Hutchison, V. Scardaci, A. C. Ferrari, and J. N. Coleman (2008). High-yield production of graphene by liquid-phase exfoliation of graphite. *Nature Nano*, **3**(9), pp. 563–568.

167. Li, X., G. Zhang, X. Bai, X. Sun, X. Wang, E. Wang, and H. Dai (2008). Highly conducting graphene sheets and Langmuir–Blodgett films. *Nature Nanotechnology*, **3**(9), pp. 538–542.

168. Su, C.-Y., A.-Y. Lu, Y. Xu, F.-R. Chen, A. N. Khlobystov, and L.-J. Li (2011). High-quality thin graphene films from fast electrochemical exfoliation. *ACS Nano*, **5**(3), pp. 2332–2339.

169. Wei, D., L. Grande, V. Chundi, R. White, C. Bower, P. Andrew, and T. Ryhanen (2012). Graphene from electrochemical exfoliation and its direct applications in enhanced energy storage devices. *Chemical Communications*, **48**(9), pp. 1239–1241.

170. Tang, Y. B., C. S. Lee, Z. H. Chen, G. D. Yuan, Z. H. Kang, L. B. Luo, H. S. Song, Y. Liu, Z. B. He, W. J. Zhang, I. Bello, and S. T. Lee (2009). High-quality graphenes via a facile quenching method for field-effect transistors. *Nano Letters*, **9**(4), pp. 1374–1377.

171. Feng, X., M. Liu, W. Pisula, M. Takase, J. Li, and K. Müllen (2008). Supramolecular organization and photovoltaics of triangle-shaped discotic graphenes with swallow-tailed alkyl substituents. *Advanced Materials*, **20**(14), pp. 2684–2689.

172. Yan, X., X. Cui, B. Li, and L.-S. Li (2010). Large, solution-processable graphene quantum dots as light absorbers for photovoltaics. *Nano Letters*, **10**(5), pp. 1869–1873.

173. Matranga, K. R., A. L. Myers, and E. D. Glandt (1992). Storage of natural gas by adsorption on activated carbon. *Chemical Engineering Science*, **47**(7), pp. 1569–1579.

174. Connolly, M. L. (1983). Solvent-accessible surfaces of proteins and nucleic acids. *Science*, **221**(4612), pp. 709–713.

175. An, K. H., W. S. Kim, Y. S. Park, Y. C. Choi, S. M. Lee, D. C. Chung, D. J. Bae, S. C. Lim, and Y. H. Lee (2001). Supercapacitors using single-walled carbon nanotube electrodes. *Advanced Materials*, **13**(7), pp. 497–500.

176. Pletcher, D., R. Greff, R. Peat, L. M. Peter, and J. Robinson (2010). Instrumental Methods in Electrochemistry "9–Electrocrystallisation", (Woodhead Publishing), pp. 283–316.

177. Portet, C., G. Yushin, and Y. Gogotsi (2007). Electrochemical performance of carbon onions, nanodiamonds, carbon black and multiwalled nanotubes in electrical double layer capacitors. *Carbon*, **45**(13), pp. 2511–2518.

178. Kaempgen, M., C. K. Chan, J. Ma, Y. Cui, and G. Gruner (2009). Printable thin film supercapacitors using single-walled carbon nanotubes. *Nano Letters*, **9**(5), pp. 1872–1876.

179. Izadi-Najafabadi, A., S. Yasuda, K. Kobashi, T. Yamada, D. N. Futaba, H. Hatori, M. Yumura, S. Iijima, and K. Hata (2010). Extracting the full potential of single-walled carbon nanotubes as durable supercapacitor electrodes operable at 4 V with high power and energy density. *Advanced Materials*, **22**(35), pp. E235–E241.

180. Vivekchand, S., C. Rout, K. Subrahmanyam, A. Govindaraj, and C. Rao (2008). Graphene-based electrochemical supercapacitors. *Journal of Chemical Sciences*, **120**(1), pp. 9–13.

181. Stoller, M. D., S. Park, Y. Zhu, J. An, and R. S. Ruoff (2008). Graphene-based ultracapacitors. *Nano Letters*, **8**(10), pp. 3498–3502.

182. Wang, Y., Z. Shi, Y. Huang, Y. Ma, C. Wang, M. Chen, and Y. Chen (2009). Supercapacitor devices based on graphene materials. *The Journal of Physical Chemistry C*, **113**(30), pp. 13103–13107.

183. Yang, X., J. Zhu, L. Qiu, and D. Li (2011). Graphene assembly: Bioinspired effective prevention of restacking in multilayered graphene films: Towards the next generation of high-performance supercapacitors (*Adv. Mater.* 25/2011). *Advanced Materials*, **23**(25), pp. 2771–2771.

184. Zhao, B., P. Liu, Y. Jiang, D. Pan, H. Tao, J. Song, T. Fang, and W. Xu (2012). Supercapacitor performances of thermally reduced graphene oxide. *Journal of Power Sources*, **198**(0), pp. 423–427.

185. Mishra, A. K., and S. Ramaprabhu (2011). Functionalized graphene-based nanocomposites for supercapacitor application. *The Journal of Physical Chemistry C*, **115**(29), pp. 14006–14013.

186. Miller, J. R., R. A. Outlaw, and B. C. Holloway (2010). Graphene double-layer capacitor with ac line-filtering performance. *Science*, **329**(5999), pp. 1637–1639.

187. Yu, A., I. Roes, A. Davies, and Z. Chen (2010). Ultrathin, transparent, and flexible graphene films for supercapacitor application. *Applied Physics Letters*, **96**(25), pp. 253105–253103.

188. Xue, L. P., M. B. Zheng, C. F. Shen, H. L. Lu, N. W. Li, L. J. Pan, and J. M. Cao (2010). Preparation of functionalized graphene sheets via microwave-assisted solid-state process and their electrochemical capacitive behaviors. *Chinese Journal of Inorganic Chemistry*, **26**(8), pp. 1375–1381.

189. Zhao, B., P. Liu, Y. Jiang, D. Pan, H. Tao, J. Song, T. Fang, and W. Xu (2012). Supercapacitor performances of thermally reduced graphene oxide. *Journal of Power Sources*, **198**(5), pp. 423–427.

190. Chen, Y., X. Zhang, D. Zhang, P. Yu, and Y. Ma (2011). High performance supercapacitors based on reduced graphene oxide in aqueous and ionic liquid electrolytes. *Carbon*, **49**(2), pp. 573–580.

191. Du, X. A., P. Guo, H. H. Song, and X. H. Chen (2010). Graphene nanosheets as electrode material for electric double-layer capacitors. *Electrochimica Acta*, **55**(16), pp. 4812–4819.

192. Hantel, M. M., T. Kaspar, R. Nesper, A. Wokaun, and R. Kötz (2011). Partially reduced graphite oxide for supercapacitor electrodes: Effect of graphene layer spacing and huge specific capacitance. *Electrochemistry Communications*, **13**(1), pp. 90–92.

193. Obradovic, M. D., G. D. Vukovic, S. I. Stevanovic, V. V. Panic, P. S. Uskokovic, A. Kowal, and S. L. Gojkovic (2009). A comparative study of the electrochemical properties of carbon nanotubes and carbon black. *Journal of Electroanalytical Chemistry*, **634**(1), pp. 22–30.

194. Pan, H., J. Li, and Y. Feng (2010). Carbon Nanotubes for Supercapacitor. *Nanoscale Research Letters*, **5**(3), pp. 654–668.

195. Simon, P., and Y. Gogotsi (2008). Materials for electrochemical capacitors. *Nature Materials*, **7**(11), pp. 845–854.

196. Rudge, A., I. Raistrick, S. Gottesfeld, and J. P. Ferraris (1994). A study of the electrochemical properties of conducting polymers for application in electrochemical capacitors. *Electrochimica Acta*, **39**(2), pp. 273–287.

197. Lu, P., D. Xue, H. Yang, and Y. Liu (2012). Supercapacitor and nanoscale research towards electrochemical energy storage. *International Journal of Smart and Nano Materials*, pp. 1–25.

198. Brezesinski, T., J. Wang, S. H. Tolbert, and B. Dunn (2010). Ordered mesoporous [alpha]-MoO_3 with iso-oriented nanocrystalline walls for thin-film pseudocapacitors. *Nature Materials*, **9**(2), pp. 146–151.

199. Hu, G., C. Tang, C. Li, H. Li, Y. Wang, and H. Gong (2011). The sol-gel-derived nickel-cobalt oxides with high supercapacitor performances. *Journal of The Electrochemical Society*, **158**(6), pp. A695–A699.

200. Kurzweil, P. (2009). Precious metal oxides for electrochemical energy converters: Pseudocapacitance and pH dependence of redox processes. *Journal of Power Sources*, **190**(1), pp. 189–200.

201. Wu, M.-S., Y.-A. Huang, C.-H. Yang, and J.-J. Jow (2007). Electrodeposition of nanoporous nickel oxide film for electrochemical capacitors. *International Journal of Hydrogen Energy*, **32**(17), pp. 4153–4159.

202. Z. J. Lao, K. Konstantinov, Y. Tournaire, S. H. Ng, G. X. Wang, and H. K. Liu (2006). Synthesis of vanadium pentoxide powders with enhanced surface-area for electrochemical capacitors. *Journal of Power Sources*, **162**(2), pp. 1451–1454.

203. Zhao, D.-D., S.-J. Bao, W.-J. Zhou, and H.-L. Li (2007). Preparation of hexagonal nanoporous nickel hydroxide film and its application for electrochemical capacitor. *Electrochemistry Communications*, **9**(5), pp. 869–874.

204. Dong, W., D. R. Rolison, and B. Dunna (2000). Electrochemical properties of high surface area vanadium oxide aerogels. *Electrochemical and Solid-State Letters*, **3**(10), pp. 457–459.

205. Naoi, K., and P. Simon (2008). New materials and new configurations for advanced electrochemical capacitors. *Interface*, **17**(1), pp. 34–37.

206. Jiang, H., C. Li, T. Sun, and J. Ma (2012). High-performance supercapacitor material based on $Ni(OH)_2$ nanowire-MnO_2 nanoflakes core-shell nanostructures. *Chemical Communications*, **48**(20), pp. 2606–2608.

207. Chang, J.-K., C.-M. Wu, and I. W. Sun (2010). Nano-architectured $Co(OH)_2$ electrodes constructed using an easily-manipulated electrochemical protocol for high-performance energy storage applications. *Journal of Materials Chemistry*, **20**(18), pp. 3729–3735.

208. Prasad, K. R., and N. Munichandraiah (2002). Electrochemical studies of polyaniline in a Gel polymer electrolyte. *Electrochemical and Solid-State Letters*, **5**(12), pp. A271–A274.

209. Zhou, H., H. Chen, S. Luo, G. Lu, W. Wei, and Y. Kuang (2005). The effect of the polyaniline morphology on the performance of polyaniline supercapacitors. *Journal of Solid State Electrochemistry*, **9**(8), pp. 574–580.

210. Yuan, L., B. Yao, B. Hu, K. Huo, W. Chen, and J. Zhou (2013). Polypyrrole-coated paper for flexible solid-state energy storage. *Energy & Environmental Science*, **6**(2), pp. 470–476.

211. Zhang, J., L.-B. Kong, H. Li, Y.-C. Luo, and L. Kang (2010). Synthesis of polypyrrole film by pulse galvanostatic method and its application as supercapacitor electrode materials. *Journal of Materials Science*, **45**(7), pp. 1947–1954.

212. Muthulakshmi, B., D. Kalpana, S. Pitchumani, and N. G. Renganathan (2006). Electrochemical deposition of polypyrrole for symmetric supercapacitors. *Journal of Power Sources*, **158**(2), pp. 1533–1537.

213. Li-Zhen, F., and M. Joachim (2006). High-performance polypyrrole electrode materials for redox supercapacitors. *Electrochemistry Communications*, **8**(6), pp. 937–940.

214. Hashmi, S. A., and H. M. Upadhyaya (2002). Polypyrrole and poly(3-methyl thiophene)-based solid state redox supercapacitors using ion conducting polymer electrolyte. *Solid State Ionics*, 152–153, pp. 883–889.

215. Hashmi, S. A., A. Kumar, and S. K. Tripathi (2005). Investigations on electrochemical supercapacitors using polypyrrole redox electrodes and PMMA based gel electrolytes. *European Polymer Journal*, **41**(6), pp. 1373–1379.

216. Wang, J., Y. Xu, X. Chen, and X. Du (2007). Electrochemical supercapacitor electrode material based on poly(3,4-ethylenedioxythiophene)/polypyrrole composite. *Journal of Power Sources*, **163**, pp. 1120–1125.

217. Ryu, K. S., Y.-G. Lee, Y.-S. Hong, Y. J. Park, X. Wu, K. M. Kim, M. G. Kang, N.-G. Park, and S. H. Chang (2004). Poly(ethylenedioxythiophene) (PEDOT) as polymer electrode in redox supercapacitor. *Electrochimica Acta*, **50**(2-3), pp. 843–847.

218. Shirakawa, H., and S. Ikeda (1970). Infrared spectra of poly(acetylene). *Polymer Journal*, **2**(2), pp. 231–244.

219. Shirakawa, H., E. J. Louis, A. G. MacDiarmid, C. K. Chiang, and A. J. Heeger (1977). Synthesis of electrically conducting organic polymers:

halogen derivatives of polyacetylene (CH). *Journal of the Chemical Society, Chemical Communications*, **16**, pp. 578–580.

220. Shirakawa, H. (2001). The discovery of polyacetylene film: The dawning of an era of conducting polymers. *Synthetic Metals*, **125**(1), pp. 3–10.

221. Novák, P., K. Müller, K. S. V. Santhanam, and O. Haas (1997). Electrochemically active polymers for rechargeable batteries. *Chemical Reviews*, **97**(1), pp. 207–282.

222. Gurunathan, K., A. V. Murugan, R. Marimuthu, U. P. Mulik, and D. P. Amalnerkar (1999). Electrochemically synthesised conducting polymeric materials for applications towards technology in electronics, optoelectronics and energy storage devices. *Materials Chemistry and Physics*, **61**(3), pp. 173–191.

223. Potje-Kamloth, K., B. J. Polk, M. Josowicz, and J. Janata (2002). Doping of polyaniline in the solid state with photogenerated triflic acid. *Chemistry of Materials*, **14**(6), pp. 2782–2787.

224. Skotheim, T. A. (1998). Handbook of Conducting Polymers, Second Edition. Schlüter, A.-D. "Syntehsis of poly(para-phenylene)s" (Taylor & Francis), pp. 209–224.

225. Gottesfeld, S., A. Redondo, and S. W. Feldberg (1987). On the mechanism of electrochemical switching in films of polyaniline. *Journal of The Electrochemical Society*, **134**(1), pp. 271–272.

226. Snook, G. A., P. Kao, and A. S. Best (2011). Conducting-polymer-based supercapacitor devices and electrodes. *Journal of Power Sources*, **196**(1), pp. 1–12.

227. Snook, G. A., and G. Z. Chen (2008). The measurement of specific capacitances of conducting polymers using the quartz crystal microbalance. *Journal of Electroanalytical Chemistry*, **612**(1), pp. 140–146.

228. Chae, J. H., K. C. Ng, and G. Z. Chen (2010). Nanostructured materials for the construction of asymmetrical supercapacitors. *Proceedings of the Institution of Mechanical Engineers Part a-Journal of Power and Energy*, **224**(A4), pp. 479–503.

229. Ghenaatian, H. R., M. F. Mousavi, and M. S. Rahmanifar (2012). High performance hybrid supercapacitor based on two nanostructured conducting polymers: Self-doped polyaniline and polypyrrole nanofibers. *Electrochimica Acta*, **78**(0), pp. 212–222.

230. Liu, J., M. Zhou, L.-Z. Fan, P. Li, and X. Qu (2010). Porous polyaniline exhibits highly enhanced electrochemical capacitance performance. *Electrochimica Acta*, **55**(20), pp. 5819–5822.

231. Pan, L., H. Qiu, C. Dou, Y. Li, L. Pu, J. Xu, and Y. Shi (2010). Conducting polymer nanostructures: Template synthesis and applications in energy storage. *International Journal of Molecular Sciences*, **11**(7), pp. 2636–2657.

232. Arbizzani, C., M. Mastragostino, and L. Meneghello (1996). Polymer-based redox supercapacitors: A comparative study. *Electrochimica Acta*, **41**(1), pp. 21–26.

233. Mastragostino, M., C. Arbizzani, and F. Soavi (2002). Conducting polymers as electrode materials in supercapacitors. *Solid State Ionics*, **148**(3–4), pp. 493–498.

234. Sivaraman, P., V. R. Hande, V. S. Mishra, C. S. Rao, and A. B. Samui (2003). All-solid supercapacitor based on polyaniline and sulfonated poly(ether ether ketone). *Journal of Power Sources*, **124**(1), pp. 351–354.

235. Hussain, A. M. P., and A. Kumar (2006). Enhanced electrochemical stability of all-polymer redox supercapacitors with modified polypyrrole electrodes. *Journal of Power Sources*, **161**(2), pp. 1486–1492.

236. Bhat, D., and M. Selva Kumar (2007). N and p doped poly(3,4-ethylenedioxythiophene) electrode materials for symmetric redox supercapacitors. *Journal of Materials Science*, **42**(19), pp. 8158–8162.

237. Ferraris, J. P., M. M. Eissa, I. D. Brotherston, and D. C. Loveday (1998). Performance evaluation of poly 3-(Phenylthiophene) derivatives as active materials for electrochemical capacitor applications. *Chemistry of Materials*, **10**(11), pp. 3528–3535.

238. Peng, C., D. Hu, and G. Z. Chen (2011). Theoretical specific capacitance based on charge storage mechanisms of conducting polymers: Comment on Vertically oriented arrays of polyaniline nanorods and their super electrochemical properties. *Chemical Communications*, **47**(14), pp. 4105–4107.

239. MacDiarmid, A. G. (2001). Synthetic metals: A novel role for organic polymers. *Synthetic Metals*, **125**(1), pp. 11–22.

240. Bongini, A., G. Barbarella, L. Favaretto, G. Sotgiu, M. Zambianchi, M. Mastragostino, C. Arbizzani, and F. Soavi (1999). New n-dopable thiophene based polymers. *Synthetic Metals*, **101**(1–3), pp. 13–14.

241. Wahdame, B., D. Candusso, X. François, F. Harel, J.-M. Kauffmann, and G. Coquery (2009). Design of experiment techniques for fuel cell characterisation and development. *International Journal of Hydrogen Energy*, **34**(2), pp. 967–980.

242. Deng, W., X. Ji, Q. Chen, and C. E. Banks (2011). Electrochemical capacitors utilising transition metal oxides: An update of recent developments. *RSC Advances*, **1**(7), pp. 1171–1178.

243. Kim, I.-H., and K.-B. Kim (2006). Electrochemical characterization of hydrous ruthenium oxide thin-film electrodes for electrochemical capacitor applications. *Journal of The Electrochemical Society*, **153**(2), pp. A383–A389.

244. Yang, X.-H., Y.-G. Wang, H.-M. Xiong, and Y.-Y. Xia (2007). Interfacial synthesis of porous MnO_2 and its application in electrochemical capacitor. *Electrochimica Acta*, **53**(2), pp. 752–757.

245. Lee, H., M. S. Cho, I. H. Kim, J. D. Nam, and Y. Lee (2010). RuO_x/polypyrrole nanocomposite electrode for electrochemical capacitors. *Synthetic Metals*, **160**(9–10), pp. 1055–1059.

246. Jia, Q. X., S. G. Song, X. D. Wu, J. H. Cho, S. R. Foltyn, A. T. Findikoglu, and J. L. Smith (1996). Epitaxial growth of highly conductive RuO[sub 2] thin films on (100) Si. *Applied Physics Letters*, **68**(8), pp. 1069–1071.

247. Sakiyama, K., S. Onishi, K. Ishihara, K. Orita, T. Kajiyama, N. Hosoda, and T. Hara (1993). Deposition and properties of reactively sputtered ruthenium dioxide films. *Journal of The Electrochemical Society*, **140**(3), pp. 834–839.

248. Chen, Y.-M., J.-H. Cai, Y.-S. Huang, K.-Y. Lee, D.-S. Tsai, and K.-K. Tiong (2011). A nanostructured electrode of IrO × foil on the carbon nanotubes for supercapacitors. *Nanotechnology*, **22**(35), pp. 355708.

249. Lokhande, C. D., D. P. Dubal, and O.-S. Joo (2011). Metal oxide thin film based supercapacitors. *Current Applied Physics*, **11**(3), pp. 255–270.

250. Zheng, J. P., P. J. Cygan, and T. R. Jow (1995). Hydrous ruthenium oxide as an electrode material for electrochemical capacitors. *Journal of The Electrochemical Society*, **142**(8), pp. 2699–2703.

251. Trasatti, S. (1991). Physical electrochemistry of ceramic oxides. *Electrochimica Acta*, **36**(2), pp. 225–241.

252. Galizzioli, D., F. Tantardini, and S. Trasatti (1974). Ruthenium dioxide: A new electrode material. I. Behaviour in acid solutions of inert electrolytes. *Journal of Applied Electrochemistry*, **4**(1), pp. 57–67.

253. Hu, C. C., C. H. Lee, and T. C. Wen (1996). Oxygen evolution and hypochlorite production on Ru-Pt binary oxides. *Journal of Applied Electrochemistry*, **26**(1), pp. 72–82.

254. Ramani, M., B. S. Haran, R. E. White, and B. N. Popov (2001). Synthesis and characterization of hydrous ruthenium oxide-carbon supercapacitors. *Journal of The Electrochemical Society*, **148**(4), pp. A374–A380.

255. Ferro, S., and A. De Battisti (2002). Electrocatalysis and chlorine evolution reaction at ruthenium dioxide deposited on conductive diamond. *The Journal of Physical Chemistry B*, **106**(9), pp. 2249–2254.
256. Wen, T. C., and C. C. Hu (1992). Hydrogen and oxygen evolutions on Ru-Ir binary oxides. *Journal of The Electrochemical Society*, **139**(8), pp. 2158–2163.
257. Wang, G., L. Zhang, and J. Zhang (2012). A review of electrode materials for electrochemical supercapacitors. *Chemical Society Reviews*, **41**(2), pp. 797–828.
258. Wu, N.-L., S.-L. Kuo, and M.-H. Lee (2002). Preparation and optimization of RuO_2-impregnated SnO_2 xerogel supercapacitor. *Journal of Power Sources*, **104**(1), pp. 62–65.
259. Sugimoto, W., H. Iwata, Y. Murakami, and Y. Takasu (2004). Electrochemical capacitor behavior of layered ruthenic acid hydrate. *Journal of The Electrochemical Society*, **151**(8), pp. A1181–A1187.
260. Sugimoto, W., T. Shibutani, Y. Murakami, and Y. Takasu (2002). Charge storage capabilities of rutile-type RuO_2 VO_2 solid solution for electrochemical supercapacitors. *Electrochemical and Solid-State Letters*, **5**(7), pp. A170–A172.
261. Takasu, Y., and Y. Murakami (2000). Design of oxide electrodes with large surface area. *Electrochimica Acta*, **45**(25–26), pp. 4135–4141.
262. Bo, G., Z. Xiaogang, Y. Changzhou, L. Juan, and Y. Long (2006). Amorphous $Ru_{1-y}CryO_2$ loaded on TiO_2 nanotubes for electrochemical capacitors. *Electrochimica Acta*, **52**(3), pp. 1028–1032.
263. Hu, C.-C., K.-H. Chang, M.-C. Lin, and Y.-T. Wu (2006). Design and tailoring of the nanotubular arrayed architecture of hydrous RuO_2 for next generation supercapacitors. *Nano Letters*, **6**(12), pp. 2690–2695.
264. Lee, W., Mane, R. S., Todkar, V. V., Lee, S., Egorov, O., Chae, W. S., Han, S. H. (2007). Implication of liquid-phase deposited amorphous RuO_2 electrode for electrochemical supercapacitor. *Electrochemical and Solid-State Letters*, **10**, pp. A225–A227.
265. Wu, Z.-S., D.-W. Wang, W. Ren, J. Zhao, G. Zhou, F. Li, and H.-M. Cheng (2010). Anchoring hydrous RuO_2 on graphene sheets for high-performance electrochemical capacitors. *Advanced Functional Materials*, **20**(20), pp. 3595–3602.
266. Ataherian, F., K.-T. Lee, and N.-L. Wu (2010). Long-term electrochemical behaviors of manganese oxide aqueous electrochemical capacitor under reducing potentials. *Electrochimica Acta*, **55**(25), pp. 7429–7435.

267. Pang, S. C., M. A. Anderson, and T. W. Chapman (2000). Novel electrode materials for thin-film ultracapacitors: Comparison of electrochemical properties of sol-gel-derived and electrodeposited manganese dioxide. *Journal of The Electrochemical Society*, **147**(2), pp. 444–450.

268. Devaraj, S., and N. Munichandraiah (2005). High capacitance of electrodeposited MnO_2 by the effect of a surface-active agent. *Electrochemical and Solid-State Letters*, **8**(7), pp. A373–A377.

269. Toupin, M., T. Brousse, and D. Bélanger (2004). Charge storage mechanism of MnO_2 electrode used in aqueous electrochemical capacitor. *Chemistry of Materials*, **16**(16), pp. 3184–3190.

270. Chang, J.-K., M.-T. Lee, and W.-T. Tsai (2007). In situ Mn K-edge X-ray absorption spectroscopic studies of anodically deposited manganese oxide with relevance to supercapacitor applications. *Journal of Power Sources*, **166**(2), pp. 590–594.

271. Chen, P.-C., G. Shen, Y. Shi, H. Chen, and C. Zhou (2010). Preparation and characterization of flexible asymmetric supercapacitors based on transition-metal-oxide nanowire/single-walled carbon nanotube hybrid thin-film electrodes. *ACS Nano*, **4**(8), pp. 4403–4411.

272. Zolfaghari, A., F. Ataherian, M. Ghaemi, and A. Gholami (2007). Capacitive behavior of nanostructured MnO2 prepared by sonochemistry method. *Electrochimica Acta*, **52**(8), pp. 2806–2814.

273. Ghaemi, M., F. Ataherian, A. Zolfaghari, and S. M. Jafari (2008). Charge storage mechanism of sonochemically prepared MnO_2 as supercapacitor electrode: Effects of physisorbed water and proton conduction. *Electrochimica Acta*, **53**(14), pp. 4607–4614.

274. Liu, K. C., and M. A. Anderson (1996). Porous nickel oxide/nickel films for electrochemical capacitors. *Journal of The Electrochemical Society*, **143**(1), pp. 124–130.

275. Nam, K.-W., and K.-B. Kim (2002). A study of the preparation of NiO_x electrode via electrochemical route for supercapacitor applications and their charge storage mechanism. *Journal of The Electrochemical Society*, **149**(3), pp. A346–A354.

276. Castro, E. B., S. G. Real, and L. F. Pinheiro Dick (2004). Electrochemical characterization of porous nickel–cobalt oxide electrodes. *International Journal of Hydrogen Energy*, **29**(3), pp. 255–261.

277. Wu, M.-S., Y.-A. Huang, J.-J. Jow, W.-D. Yang, C.-Y. Hsieh, and H.-M. Tsai (2008). Anodically potentiostatic deposition of flaky nickel oxide nanostructures and their electrochemical performances. *International Journal of Hydrogen Energy*, **33**(12), pp. 2921–2926.

278. Jayashree, R. S., and P. V. Kamath (2001). Suppression of the $\alpha \to \beta$-nickel hydroxide transformation in concentrated alkali: Role of dissolved cations. *Journal of Applied Electrochemistry*, **31**(12), pp. 1315–1320.

279. Nam, K.-W., K.-H. Kim, E.-S. Lee, W.-S. Yoon, X.-Q. Yang, and K.-B. Kim (2008). Pseudocapacitive properties of electrochemically prepared nickel oxides on 3-dimensional carbon nanotube film substrates. *Journal of Power Sources*, **182**(2), pp. 642–652.

280. Zhang, X., W. Shi, J. Zhu, W. Zhao, J. Ma, S. Mhaisalkar, T. Maria, Y. Yang, H. Zhang, H. Hng, and Q. Yan (2010). Synthesis of porous NiO nanocrystals with controllable surface area and their application as supercapacitor electrodes. *Nano Research*, **3**(9), pp. 643–652.

281. Cheng, J., G.-P. Cao, and Y.-S. Yang (2006). Characterization of sol-gel-derived NiO$_x$ xerogels as supercapacitors. *Journal of Power Sources*, **159**(1), pp. 734–741.

282. Li, S.-H., Q.-H. Liu, L. Qi, L.-H. Lu, and H.-Y. Wang (2012). Progress in research on manganese dioxide electrode materials for electrochemical capacitors. *Chinese Journal of Analytical Chemistry*, **40**(3), pp. 339–346.

283. Toupin, M., T. Brousse, and D. Belanger (2002). Influence of microstucture on the charge storage properties of chemically synthesized manganese dioxide. *Chemistry of Materials*, **14**(9), pp. 3946–3952.

284. Du, X., C. Wang, M. Chen, Y. Jiao, and J. Wang (2009). Electrochemical performances of nanoparticle Fe$_3$O$_4$/activated carbon supercapacitor using KOH Electrolyte Solution. *The Journal of Physical Chemistry C*, **113**(6), pp. 2643–2646.

285. Hu, C.-C., and T.-W. Tsou (2002). Ideal capacitive behavior of hydrous manganese oxide prepared by anodic deposition. *Electrochemistry Communications*, **4**(2), pp. 105–109.

286. Messaoudi, B., S. Joiret, M. Keddam, and H. Takenouti (2001). Anodic behaviour of manganese in alkaline medium. *Electrochimica Acta*, **46**(16), pp. 2487–2498.

287. Devaraj, S., and N. Munichandraiah (2008). Effect of crystallographic structure of MnO$_2$ on its electrochemical capacitance properties. *The Journal of Physical Chemistry C*, **112**(11), pp. 4406–4417.

288. Isikli, S., and R. Díaz (2012). Substrate-dependent performance of supercapacitors based on an organic redox couple impregnated on carbon. *Journal of Power Sources*, **206**(0), pp. 53–58.

289. Leitner, K. W., B. Gollas, M. Winter, and J. O. Besenhard (2004). Combination of redox capacity and double layer capacitance in composite electrodes through immobilization of an organic redox couple on carbon black. *Electrochimica Acta*, **50**(1), pp. 199–204.

290. Tomko, T., R. Rajagopalan, P. Aksoy, and H. C. Foley (2011). Synthesis of boron/nitrogen substituted carbons for aqueous asymmetric capacitors. *Electrochimica Acta*, **56**(15), pp. 5369–5375.

291. Fanning, P. E., and M. A. Vannice (1993). A DRIFTS study of the formation of surface groups on carbon by oxidation. *Carbon*, **31**(5), pp. 721–730.

292. Wildgoose, G. G., P. Abiman, and R. G. Compton (2009). Characterising chemical functionality on carbon surfaces. *Journal of Materials Chemistry*, **19**(28), pp. 4875–4886.

293. Lao, Z. J., K. Konstantinov, Y. Tournaire, S. H. Ng, G. X. Wang, and H. K. Liu (2006). Synthesis of vanadium pentoxide powders with enhanced surface-area for electrochemical capacitors. *Journal of Power Sources*, **162**(2), pp. 1451–1454.

294. Bazuła, P. A., A.-H. Lu, J.-J. Nitz, and F. Schüth (2008). Surface and pore structure modification of ordered mesoporous carbons via a chemical oxidation approach. *Microporous and Mesoporous Materials*, **108**(1–3), pp. 266–275.

295. Hsieh, C.-T., and H. Teng (2002). Influence of oxygen treatment on electric double-layer capacitance of activated carbon fabrics. *Carbon*, **40**(5), pp. 667–674.

296. Jurewicz, K., K. Babel, A. Ziólkowski, and H. Wachowska (2003). Ammoxidation of active carbons for improvement of supercapacitor characteristics. *Electrochimica Acta*, **48**(11), pp. 1491–1498.

297. Bleda-Martínez, M. J., D. Lozano-Castelló, E. Morallón, D. Cazorla-Amorós, and A. Linares-Solano (2006). Chemical and electrochemical characterization of porous carbon materials. *Carbon*, **44**(13), pp. 2642–2651.

298. Hulicova-Jurcakova, D., A. M. Puziy, O. I. Poddubnaya, F. Suárez-García, J. M. D. Tascón, and G. Q. Lu (2009). Highly stable performance of supercapacitors from phosphorus-enriched carbons. *Journal of the American Chemical Society*, **131**(14), pp. 5026–5027.

299. Lota, G., B. Grzyb, H. Machnikowska, J. Machnikowski, and E. Frackowiak (2005). Effect of nitrogen in carbon electrode on the supercapacitor performance. *Chemical Physics Letters*, **404**(1–3), pp. 53–58.

300. Stein, A., Z. Wang, and M. A. Fierke (2009). Functionalization of porous carbon materials with designed Pore Architecture. *Advanced Materials*, **21**(3), pp. 265–293.

301. Qu, D. (2002). Studies of the activated carbons used in double-layer supercapacitors. *Journal of Power Sources*, **109**(2), pp. 403–411.

302. Frackowiak, E. (2007). Carbon materials for supercapacitor application. *Physical Chemistry Chemical Physics*, **9**, pp. 1774–1785.

303. Hulicova, D., M. Kodama, and H. Hatori (2006). Electrochemical performance of nitrogen-enriched carbons in aqueous and non-aqueous supercapacitors. *Chemistry of Materials*, **18**(9), pp. 2318–2326.

304. Kim, N. D., W. Kim, J. B. Joo, S. Oh, P. Kim, Y. Kim, and J. Yi (2008). Electrochemical capacitor performance of N-doped mesoporous carbons prepared by ammoxidation. *Journal of Power Sources*, **180**(1), pp. 671–675.

305. Li, H., H. A. Xi, S. Zhu, Z. Wen, and R. Wang (2006). Preparation, structural characterization, and electrochemical properties of chemically modified mesoporous carbon. *Microporous and Mesoporous Materials*, **96**(1–3), pp. 357–362.

306. Shen, B. S. W. J. Feng, J. W. Lang, R. T. Wang, Z. X. Tai, and X. B. Yan (2012). Nitric acid modification of graphene nanosheets prepared by arc-discharge method and their enhanced electrochemical properties. *Acta Physico Chimica Sinica.*, **28**(7), pp. 1726–1732.

307. Morimoto, T., K. Hiratsuka, Y. Sanada, and K. Kurihara (1996). Electric double-layer capacitor using organic electrolyte. *Journal of Power Sources*, **60**(2), pp. 239–247.

308. Raymundo-Piñero, E., F. Leroux, and F. Béguin (2006). A high-performance carbon for supercapacitors obtained by carbonization of a seaweed biopolymer. *Advanced Materials*, **18**(14), pp. 1877–1882.

309. Liu, X., Y. Wang, L. Zhan, W. Qiao, X. Liang, and L. Ling (2011). Effect of oxygen-containing functional groups on the impedance behavior of activated carbon-based electric double-layer capacitors. *Journal of Solid State Electrochemistry*, **15**(2), pp. 413–419.

310. Delamar, M., R. Hitmi, J. Pinson, and J. M. Saveant (1992). Covalent modification of carbon surfaces by grafting of functionalized aryl radicals produced from electrochemical reduction of diazonium salts. *Journal of the American Chemical Society*, **114**(14), pp. 5883–5884.

311. Downard, A. J. (2000). Electrochemically assisted covalent modification of carbon electrodes. *Electroanalysis*, **12**(14), pp. 1085–1096.

312. Pech, D., D. Guay, T. Brousse, and D. Belanger (2008). Concept for charge storage in electrochemical capacitors with functionalized carbon electrodes. *Electrochemical and Solid-State Letters*, **11**(11), pp. A202–A205.

313. Ghodbane, O., G. Chamoulaud, and D. Bélanger (2004). Chemical reactivity of 4-bromophenyl modified glassy carbon electrode. *Electrochemistry Communications*, **6**(3), pp. 254–258.

314. Toupin, M., and D. Belanger (2008). Spontaneous functionalization of carbon black by reaction with 4-nitrophenyldiazonium cations. *Langmuir*, **24**(5), pp. 1910–1917.

315. Pognon, G., T. Brousse, and D. Bélanger (2011). Effect of molecular grafting on the pore size distribution and the double layer capacitance of activated carbon for electrochemical double layer capacitors. *Carbon*, **49**(4), pp. 1340–1348.

316. Louault, C., M. D'Amours, and D. Bélanger (2008). The electrochemical grafting of a mixture of substituted phenyl groups at a glassy carbon electrode surface. *ChemPhysChem*, **9**(8), pp. 1164–1170.

317. Breton, T., and D. Bélanger (2008). Modification of carbon electrode with aryl groups having an aliphatic amine by electrochemical reduction of in situ generated diazonium cations. *Langmuir*, **24**(16), pp. 8711–8718.

318. Chamoulaud, G., and D. Bélanger (2007). Spontaneous derivatization of a copper electrode with in situ generated diazonium cations in aprotic and aqueous media. *The Journal of Physical Chemistry C*, **111**(20), pp. 7501–7507.

319. Baranton, S., and D. Bélanger (2008). In situ generation of diazonium cations in organic electrolyte for electrochemical modification of electrode surface. *Electrochimica Acta*, **53**(23), pp. 6961–6967.

320. Lyskawa, J., A. Grondein, and D. Bélanger (2010). Chemical modifications of carbon powders with aminophenyl and cyanophenyl groups and a study of their reactivity. *Carbon*, **48**(4), pp. 1271–1278.

321. Baranton, S., and D. Bélanger (2005). Electrochemical derivatization of carbon surface by reduction of in situ generated diazonium cations. *The Journal of Physical Chemistry B*, **109**(51), pp. 24401–24410.

322. Cheng, L., J. Liu, and S. Dong (2000). Layer-by-layer assembly of multilayer films consisting of silicotungstate and a cationic redox polymer on 4-aminobenzoic acid modified glassy carbon electrode and their electrocatalytic effects. *Analytica Chimica Acta*, **417**(2), pp. 133–142.

323. Pognon, G., T. Brousse, L. Demarconnay, and D. Bélanger (2011). Performance and stability of electrochemical capacitor based on anthraquinone modified activated carbon. *Journal of Power Sources*, **196**(8), pp. 4117–4122.

324. Baeza-Rostro, D. A., and A. K. Cuentas-Gallegos (2013). Capacitance improvement of carbon aerogels by the immobilization of polyoxometalates nanoparticles. *Journal of New Materials for Electrochemical Systems*, **16**(3), pp. 203–207.

325. Qin, X., S. Durbach, and G. T. Wu (2004). Electrochemical characterization on $RuO_2 \cdot xH_2O$/carbon nanotubes composite electrodes for high energy density supercapacitors. *Carbon*, **42**(2), pp. 451–453.

326. Chen, P., H. Chen, J. Qiu, and C. Zhou (2010). Inkjet printing of single-walled carbon nanotube/RuO_2 nanowire supercapacitors on cloth fabrics and flexible substrates. *Nano Research*, **3**(8), pp. 594–603.

327. Gujar, T. P., W.-Y. Kim, I. Puspitasari, K.-D. Jung, and O.-S. Joo (2007). Electrochemically deposited nanograin ruthenium oxide as a pseudocapacitive electrode *International Journal of Electrochemical Science*, **2**, pp. 666–673.

328. Ng, K. C., S. Zhang, C. Peng, and G. Z. Chen (2009). Individual and bipolarly stacked asymmetrical aqueous supercapacitors of CNTs/SnO_2 and CNTs/MnO_2 nanocomposites. *Journal of The Electrochemical Society*, **156**(11), pp. A846–A853.

329. Ma, S.-B., K.-Y. Ahn, E.-S. Lee, K.-H. Oh, and K.-B. Kim (2007). Synthesis and characterization of manganese dioxide spontaneously coated on carbon nanotubes. *Carbon*, **45**(2), pp. 375–382.

330. Wang, G.-X., B.-L. Zhang, Z.-L. Yu, and M.-Z. Qu (2005). Manganese oxide/MWNTs composite electrodes for supercapacitors. *Solid State Ionics*, **176**(11–12), pp. 1169–1174.

331. Chang, J.-K., C.-T. Lin, and W.-T. Tsai (2004). Manganese oxide/carbon composite electrodes for electrochemical capacitors. *Electrochemistry Communications*, **6**(7), pp. 666–671.

332. Fan, Z., J. Chen, M. Wang, K. Cui, H. Zhou, and Y. Kuang (2006). Preparation and characterization of manganese oxide/CNT composites as supercapacitive materials. *Diamond and Related Materials*, **15**(9), pp. 1478–1483.

333. Wang, G. P., L. X. Liu, L. Zhang, and J. J. Zhang (2013). Nickel, cobalt, and manganese oxide composite as an electrode material for electrochemical supercapacitors. *Ionics*, **19**(4), pp. 689–695.

334. Lee, J. Y., K. Liang, K. H. An, and Y. H. Lee (2005). Nickel oxide/carbon nanotubes nanocomposite for electrochemical capacitance. *Synthetic Metals*, **150**(2), pp. 153–157.

335. Liang, K., N. Wang, M. Zhou, Z. Cao, T. Gu, Q. Zhang, X. Tang, W. Hu, and B. Wei (2013). Mesoporous LaNiO$_3$/NiO nanostructured thin films for high-performance supercapacitors. *Journal of Materials Chemistry A*, **1**(34), pp. 9730–9736.

336. Yuan, C., X. Zhang, Q. Wu, and B. Gao (2006). Effect of temperature on the hybrid supercapacitor based on NiO and activated carbon with alkaline polymer gel electrolyte. *Solid State Ionics*, **177**(13–14), pp. 1237–1242.

337. Shakir, I., M. Shahid, S. Cherevko, C.-H. Chung, and D. J. Kang (2011). Ultrahigh-energy and stable supercapacitors based on intertwined porous MoO$_3$–MWCNT nanocomposites. *Electrochimica Acta*, **58**(0), pp. 76–80.

338. Kim, D.-H., and K. Waki (2010). Crystal defects on multi-walled carbon nanotubes by cobalt oxide. *Journal of Nanoscience and Nanotechnology*, **10**(4), pp. 2375–2380.

339. Xu, J., L. Gao, J. Cao, W. Wang, and Z. Chen (2010). Preparation and electrochemical capacitance of cobalt oxide (Co$_3$O$_4$) nanotubes as supercapacitor material. *Electrochimica Acta*, **56**(2), pp. 732–736.

340. Cuentas-Gallegos, A. K., S. Jimenez-Penaloza, D. A. Baeza-Rostro, and A. German-Garcia (2010). Influence of the functionalization degree of multiwalled carbon nanotubes on the immobilization of polyoxometalates and its effect on their electrochemical behavior. *Journal of New Materials for Electrochemical Systems*, **13**(4), pp. 369–376.

341. Cuentas-Gallegos, A. K., R. Martínez-Rosales, M. Baibarac, P. Gómez-Romero, and M. E. Rincón (2007). Electrochemical supercapacitors based on novel hybrid materials made of carbon nanotubes and polyoxometalates. *Electrochemistry Communications*, **9**(8), pp. 2088–2092.

342. Cuentas-Gallegos, A. K., R. Martínez-Rosales, M. E. Rincón, G. A. Hirata, and G. Orozco (2006). Design of hybrid materials based on carbon nanotubes and polyoxometalates. *Optical Materials*, **29**(1), pp. 126–133.

343. Ma, H., J. Peng, Y. Chen, Y. Feng, and E. Wang (2004). Photoluminescent multilayer film based on polyoxometalate and tris(2,2-bipyridine) ruthenium. *Journal of Solid State Chemistry*, **177**(10), pp. 3333–3338.

344. Yu, L., C. Zhao, X. Long, and W. Chen (2009). Ultrasonic synthesis and electrochemical characterization of V_2O_5/mesoporous carbon composites. *Microporous and Mesoporous Materials*, **126**(1-2), pp. 58–64.

345. Kudo, T., Y. Ikeda, T. Watanabe, M. Hibino, M. Miyayama, H. Abe, and K. Kajita (2002). Amorphous V_2O_5/carbon composites as electrochemical supercapacitor electrodes. *Solid State Ionics*, **152–153**(0), pp. 833–841.

346. Zhang, Y., X. Sun, L. Pan, H. Li, Z. Sun, C. Sun, and B. K. Tay (2009). Carbon nanotube-ZnO nanocomposite electrodes for supercapacitors. *Solid State Ionics*, **180**(32-35), pp. 1525–1528.

347. Kalpana, D., K. S. Omkumar, S. S. Kumar, and N. G. Renganathan (2006). A novel high power symmetric ZnO/carbon aerogel composite electrode for electrochemical supercapacitor. *Electrochimica Acta*, **52**, pp. 1309–1315.

348. Du, C., J. Yeh, and N. Pan (2005). Carbon nanotube thin films with ordered structures. *Journal of Materials Chemistry*, **15**(5), pp. 548–550.

349. Du, C., and N. Pan (2006). Supercapacitors using carbon nanotubes films by electrophoretic deposition. *Journal of Power Sources*, **160**(2), pp. 1487–1494.

350. Chen, J., A. I. Minett, Y. Liu, C. Lynam, P. Sherrell, C. Wang, and G. G. Wallace (2008). Direct growth of flexible carbon nanotube electrodes. *Advanced Materials*, **20**(3), pp. 566–570.

351. Reddy, A. L. M., and S. Ramaprabhu (2007). Nanocrystalline metal oxides dispersed multiwalled carbon nanotubes as supercapacitor electrodes. *The Journal of Physical Chemistry C*, **111**(21), pp. 7727–7734.

352. Fan, Z., J. Chen, B. Zhang, B. Liu, X. Zhong, and Y. Kuang (2008). High dispersion of γ-MnO_2 on well-aligned carbon nanotube arrays and its application in supercapacitors. *Diamond and Related Materials*, **17**(11), pp. 1943–1948.

353. Lin, P., Q. She, B. Hong, X. Liu, Y. Shi, Z. Shi, M. Zheng, and Q. Dong (2010). The nickel oxide/CNT composites with high capacitance for supercapacitor. *Journal of The Electrochemical Society*, **157**(7), pp. A818–A823.

354. Wu, Z.-S., G. Zhou, L.-C. Yin, W. Ren, F. Li, and H.-M. Cheng (2012). Graphene/metal oxide composite electrode materials for energy storage. *Nano Energy*, **1**(1), pp. 107–131.

355. Yan, J., T. Wei, W. M. Qiao, B. Shao, Q. K. Zhao, L. J. Zhang, and Z. J. Fan (2010). Rapid microwave-assisted synthesis of graphene

nanosheet/Co$_3$O$_4$ composite for supercapacitors. *Electrochimica Acta*, **55**(23), pp. 6973–6978.

356. Wu, Z. S., D. W. Wang, W. Ren, J. Zhao, G. Zhou, F. Li, and H. M. Cheng (2010). Anchoring hydrous RuO$_2$ on graphene sheets for high-performance electrochemical capacitors. *Advanced Functional Materials*, **20** (20), pp. 3595–3602.

357. Qian, Y., S. B. Lu, and F. L. Gao (2011). Preparation of MnO$_2$/graphene composite as electrode material for supercapacitors. *Journal of Materials Science*, **46**(10), pp. 3517–3522.

358. Wang, B., J. Park, C. Y. Wang, H. Ahn, and G. X. Wang (2010). Mn$_3$O$_4$ nanoparticles embedded into graphene nanosheets: Preparation, characterization, and electrochemical properties for supercapacitors. *Electrochimica Acta*, **55**(22), pp. 6812–6817.

359. Hu, J., A. Ramadan, F. Luo, B. Qi, X. Deng, and J. Chen (2011). One-step molybdate ion assisted electrochemical synthesis of [small alpha]-MoO$_3$-decorated graphene sheets and its potential applications. *Journal of Materials Chemistry*, **21**(38), pp. 15009–15014.

360. Wang, J., Z. Gao, Z. Li, B. Wang, Y. Yan, Q. Liu, T. Mann, M. Zhang, and Z. Jiang (2011). Green synthesis of graphene nanosheets/ZnO composites and electrochemical properties. *Journal of Solid State Chemistry*, **184**(6), pp. 1421–1427.

361. Wang, S., S. P. Jiang, and X. Wang (2011). Microwave-assisted one-pot synthesis of metal/metal oxide nanoparticles on graphene and their electrochemical applications. *Electrochimica Acta*, **56**(9), pp. 3338–3344.

362. Yan, J., T. Wei, W. Qiao, B. Shao, Q. Zhao, L. Zhang, and Z. Fan (2010). Rapid microwave-assisted synthesis of graphene nanosheet/Co$_3$O$_4$ composite for supercapacitors. *Electrochimica Acta*, **55**(23), pp. 6973–6978.

363. Kumar, M., and Y. Ando (2007). Carbon nanotubes from camphor: An environment-friendly nanotechnology. *Journal of Physics: Conference Series*, **61**(1), p. 643.

364. Wen, Z., Y. Liu, Z.-F. Hu, A. Liu, C. Wang, and X. Xu (2010). Preparation and electrochemical performance of thin layer crystal α-MnO$_2$/AC composite electrode materials. *Acta Chimica Sinica*, **68**(15), pp. 1473–1480.

365. Li, G.-R., Z.-P. Feng, Y.-N. Ou, D. Wu, R. Fu, and Y.-X. Tong (2010). Mesoporous MnO$_2$/carbon aerogel composites as promising electrode materials for high-performance supercapacitors. *Langmuir*, **26**(4), pp. 2209–2213.

366. Lin, Y.-H., T.-Y. Wei, H.-C. Chien, and S.-Y. Lu (2011). Manganese oxide/carbon aerogel composite: An outstanding supercapacitor electrode material. *Advanced Energy Materials*, **1**(5), pp. 901–907.

367. Liu, T.-T., G.-J. Shao, M.-T. Ji, and A. Z.-P. Ma (2013). Research progress in nano-structured MnO_2 as electrode materials for supercapacitors: A review. *Asian Journal of Chemistry*, **25**(13), pp. 7065–7070.

368. Li, J., Q. Wu, and G. Zan (2014). Facile synthesis and high electrochemical performance of porous carbon composites for supercapacitors. *RSC Advances*, **4**(66), pp. 35186–35192.

369. Xia, X., Q. Hao, W. Lei, W. Wang, D. Sun, and X. Wang (2012). Nanostructured ternary composites of graphene/Fe_2O_3/polyaniline for high-performance supercapacitors. *Journal of Materials Chemistry*, **22**(33), pp. 16844–16850.

370. Zhao, X., C. Johnston, and P. S. Grant (2009). A novel hybrid supercapacitor with a carbon nanotube cathode and an iron oxide/carbon nanotube composite anode. *Journal of Materials Chemistry*, **19**(46), pp. 8755–8760.

371. Sethuraman, B., K. K. Purushothaman, and G. Muralidharan (2014). Synthesis of mesh-like Fe_2O_3/C nanocomposite via greener route for high performance supercapacitors. *RSC Advances*, **4**(9), pp. 4631–4637.

372. Li, Y., L. Kang, G. Bai, P. Li, J. Deng, X. Liu, Y. Yang, F. Gao, and W. Liang (2014). Solvothermal synthesis of Fe_2O_3 loaded activated carbon as electrode materials for high-performance electrochemical capacitors. *Electrochimica Acta*, **134**(0), pp. 67–75.

373. Zhang, H., H. Li, F. Zhang, J. Wang, Z. Wang, and S. Wang (2008). Polyaniline nanofibers prepared by a facile electrochemical approach and their supercapacitor performance. *Journal of Materials Research*, **23**(9), pp. 2326–2332.

374. Gupta, V., and N. Miura (2006). High performance electrochemical supercapacitor from electrochemically synthesized nanostructured polyaniline. *Materials Letters*, **60**(12), pp. 1466–1469.

375. Mi, H., X. Zhang, X. Ye, and S. Yang (2008). Preparation and enhanced capacitance of core-shell polypyrrole/polyaniline composite electrode for supercapacitors. *Journal of Power Sources*, **176**(1), pp. 403–409.

376. Xu, Y., J. Wang, W. Sun, and S. Wang (2006). Capacitance properties of poly(3,4-ethylenedioxythiophene)/polypyrrole composites. *Journal of Power Sources*, **159**(1), pp. 370–373.

377. Xiao, Q. and X. Zhou (2003). The study of multiwalled carbon nanotube deposited with conducting polymer for supercapacitor. *Electrochimica Acta*, **48**(5), pp. 575–580.
378. Jong-Huy, K., L. Yong-Sung, A. K. Sharma, and C. G. Liu (2006). Polypyrrole/carbon composite electrode for high-power electrochemical capacitors. *Electrochimica Acta*, **52**, pp. 1727–1732.
379. Frackowiak, E., K. Jurewicz, S. Delpeux, V. Bertagna, S. Bonnamy, and F. Béguin (2002). Synergy of components in supercapacitors based on nanotube/polypyrrole composites. *Molecular Crystals and Liquid Crystals*, **387**(1), pp. 73–78.
380. Pacheco-Catalán, D. E., Mascha A. Smit, and E. Morales (2011). Characterization of composite mesoporous carbon/conducting polymer electrodes prepared by chemical oxidation of gas-phase absorbed monomer for electrochemical capacitors. *International Journal of Electrochemistry Science*, **6**(1), pp. 78–90.
381. Lota, K., V. Khomenko, and E. Frackowiak (2004). Capacitance properties of poly(3,4-ethylenedioxythiophene)/carbon nanotubes composites. *Journal of Physics and Chemistry of Solids*, **65**(2–3), pp. 295–301.
382. Bose, S., N. H. Kim, T. Kuila, K.-T. Lau, and J. H. Lee (2011). Electrochemical performance of a graphene–polypyrrole nanocomposite as a supercapacitor electrode. *Nanotechnology*, **22**(29), pp. 295202.
383. Wang, H., Q. Hao, X. Yang, L. Lu, and X. Wang (2009). Graphene oxide doped polyaniline for supercapacitors. *Electrochemistry Communications*, **11**(6), pp. 1158–1161.
384. Zhang, K., L. L. Zhang, X. S. Zhao, and J. Wu (2010). Graphene/polyaniline nanofiber composites as supercapacitor electrodes. *Chemistry of Materials*, **22**(4), pp. 1392–1401.
385. Wang, H., Q. Hao, X. Yang, L. Lu, and X. Wang (2010). A nanostructured graphene/polyaniline hybrid material for supercapacitors. *Nanoscale*, **2**(10), pp. 2164–2170.
386. Li, J., L. Cui, and X. Zhang (2010). Preparation and electrochemistry of one-dimensional nanostructured MnO_2/PPy composite for electrochemical capacitor. *Applied Surface Science*, **256**(13), pp. 4339–4343.
387. Sharma, R. K., A. C. Rastogi, and S. B. Desu (2008). Manganese oxide embedded polypyrrole nanocomposites for electrochemical supercapacitor. *Electrochimica Acta*, **53**(26), pp. 7690–7695.
388. Zhao, X., C. Johnston, A. Crossley, and P. S. Grant (2010). Printable magnetite and pyrrole treated magnetite based electrodes for supercapacitors. *Journal of Materials Chemistry*, **20**(36), pp. 7637–7644.

389. Razoumov, S., A. Klementov, S. Litvinenko, and A. Beliakov, *Asymmetric Electrochemical Capacitor and Method of Making*, in *Google Patents*, "Elton", J. S. C., 2001: E.E.U.U. p.9.
390. Conte, M. (2010). Supercapacitors technical requirements for new applications. *Fuel Cells*, **10**(5), pp. 806–818.
391. Zheng, J. P. (2003). The limitations of energy density of battery/double-layer capacitor asymmetric cells. *Journal of The Electrochemical Society*, **150**(4), pp. A484–A492.
392. Pell, W. G., and B. E. Conway (2004). Peculiarities and requirements of asymmetric capacitor devices based on combination of capacitor and battery-type electrodes. *Journal of Power Sources*, **136**(2), pp. 334–345.
393. Fabio, A. D., A. Giorgi, M. Mastragostino, and F. Soavi (2001). Carbon-poly(3-methylthiophene) hybrid supercapacitors. *Journal of The Electrochemical Society*, **148**(8), pp. A845–A850.
394. Arbizzani, C., M. Mastragostino, and F. Soavi (2001). New trends in electrochemical supercapacitors. *Journal of Power Sources*, **100**(1–2), pp. 164–170.
395. Aida, T., I. Murayama, K. Yamada, and M. Morita (2007). Analyses of capacity loss and improvement of cycle performance for a high-voltage hybrid electrochemical capacitor. *Journal of The Electrochemical Society*, **154**(8), pp. A798–A804.
396. Long, J. W., D. Bélanger, T. Brousse, W. Sugimoto, M. B. Sassin, and O. Crosnier (2011). Asymmetric electrochemical capacitors—Stretching the limits of aqueous electrolytes. *MRS Bulletin*, **36**(07), pp. 513–522.
397. Lazzari, M., F. Soavi, and M. Mastragostino (2010). Mesoporous carbon design for ionic liquid-based, double-layer supercapacitors. *Fuel Cells*, **10**(5), pp. 840–847.
398. Khomenko, V., E. Raymundo-Piñero, and F. Béguin (2010). A new type of high energy asymmetric capacitor with nanoporous carbon electrodes in aqueous electrolyte. *Journal of Power Sources*, **195**(13), pp. 4234–4241.
399. Wang, L., T. Morishita, M. Toyoda, and M. Inagaki (2007). Asymmetric electric double layer capacitors using carbon electrodes with different pore size distributions. *Electrochimica Acta*, **53**(2), pp. 882–886.
400. Wang, L., M. Toyoda, and M. Inagaki (2008). Performance of asymmetric electric double layer capacitors—predominant contribution of the negative electrode. *Adsorption Science & Technology*, **26**(7), pp. 491–500.

401. Khomenko, V., E. Raymundo-Piñero, and F. Béguin (2008). High-energy density graphite/AC capacitor in organic electrolyte. *Journal of Power Sources*, **177**(2), pp. 643–651.

402. Wang, H., M. Yoshio, A. K. Thapa, and H. Nakamura (2007). From symmetric AC/AC to asymmetric AC/graphite, a progress in electrochemical capacitors. *Journal of Power Sources*, **169**(2), pp. 375–380.

403. Ruiz, V., C. Blanco, M. Granda, and R. Santamaría (2008). Enhanced life-cycle supercapacitors by thermal treatment of mesophase-derived activated carbons. *Electrochimica Acta*, **54**(2), pp. 305–310.

404. Algharaibeh, Z., and P. G. Pickup (2011). An asymmetric supercapacitor with anthraquinone and dihydroxybenzene modified carbon fabric electrodes. *Electrochemistry Communications*, **13**(2), pp. 147–149.

405. Sun G.-W., W.-H. Song, X.-J., Liu, W.-M., Qiao, and L.-C. Ling (2011). Asymmetric capacitance behavior based on the relationship between ion dimension and pore size. *Acta Physico-Chimica Sinica.*, **27**(02), pp. 449–454.

406. Aida, T., I. Murayama, K. Yamada, and M. Morita (2007). High-energy-density hybrid electrochemical capacitor using graphitizable carbon activated with KOH for positive electrode. *Journal of Power Sources*, **166**(2), pp. 462–470.

407. Power, A. *Lead-Carbon Energy Storage Device Overview*. [cited 2012 26/12/2012]; Available from: http://www.axionpower.com/Technology.

408. Lang, J.-W., L.-B. Kong, M. Liu, Y.-C. Luo, and L. Kang (2010). Asymmetric supercapacitors based on stabilized α-Ni(OH)$_2$ and activated carbon. *Journal of Solid State Electrochemistry*, **14**(8), pp. 1533–1539.

409. Wang, X.-F., D.-B. Ruan, and Z. You (2006). Application of spherical Ni(OH)$_2$/CNTs composite electrode in asymmetric supercapacitor. *Transactions of Nonferrous Metals Society of China*, **16**(5), pp. 1129–1134.

410. Qin, C., X. Bai, G. Yin, Y. Liu, Z. Jin, and H. Niu (2009). Electrochemical supercapacitors based on carbon aerogels/Ni(OH)$_2$ composites and activated carbon. *Pigment & Resin Technology*, **38**(4), pp. 230–235.

411. Park, J. H., S. Kim, O. O. Park, and J. M. Ko (2006). Improved asymmetric electrochemical capacitor using Zn-Co co-doped Ni(OH)$_2$ positive electrode material. *Applied Physics A*, **82**(4), pp. 593–597.

412. Yu, N., L. Gao, S. Zhao, and Z. Wang (2009). Electrodeposited PbO$_2$ thin film as positive electrode in PbO$_2$/AC hybrid capacitor. *Electrochimica Acta*, **54**(14), pp. 3835–3841.

413. Naoi, K. (2010). Nanohybrid capacitor: The next generation electrochemical capacitors. *Fuel Cells*, **10**(5), pp. 825–833.
414. Xue, Y., Y. Chen, M.-L. Zhang, and Y.-D. Yan (2008). A new asymmetric supercapacitor based on λ-MnO_2 and activated carbon electrodes. *Materials Letters*, **62**(23), pp. 3884–3886.
415. Brousse, T., P.-L. Taberna, O. Crosnier, R. Dugas, P. Guillemet, Y. Scudeller, Y. Zhou, F. Favier, D. Bélanger, and P. Simon (2007). Long-term cycling behavior of asymmetric activated carbon/MnO_2 aqueous electrochemical supercapacitor. *Journal of Power Sources*, **173**(1), pp. 633–641.
416. Qu, Q. T., Y. Shi, S. Tian, Y. H. Chen, Y. P. Wu, and R. Holze (2009). A new cheap asymmetric aqueous supercapacitor: Activated carbon//$NaMnO_2$. *Journal of Power Sources*, **194**(2), pp. 1222–1225.
417. Algharaibeh, Z., X. Liu, and P. G. Pickup (2009). An asymmetric anthraquinone-modified carbon/ruthenium oxide supercapacitor. *Journal of Power Sources*, **187**(2), pp. 640–643.
418. Wang, Y.-G., L. Cheng, and Y.-Y. Xia (2006). Electrochemical profile of nano-particle CoAl double hydroxide/active carbon supercapacitor using KOH electrolyte solution. *Journal of Power Sources*, **153**(1), pp. 191–196.
419. Qu, Q., P. Zhang, B. Wang, Y. Chen, S. Tian, Y. Wu, and R. Holze (2009). Electrochemical performance of MnO_2 nanorods in neutral aqueous electrolytes as a cathode for asymmetric supercapacitors. *The Journal of Physical Chemistry C*, **113**(31), pp. 14020–14027.
420. Demarconnay, L., E. Raymundo-Piñero, and F. Béguin (2011). Adjustment of electrodes potential window in an asymmetric carbon/MnO_2 supercapacitor. *Journal of Power Sources*, **196**(1), pp. 580–586.
421. Gao, P.-C., A.-H. Lu, and W.-C. Li (2011). Dual functions of activated carbon in a positive electrode for MnO_2-based hybrid supercapacitor. *Journal of Power Sources*, **196**(8), pp. 4095–4101.
422. Lufrano, F., P. Staiti, E. G. Calvo, E. J. Juárez-Pérez, J. A. Menéndez, and A. Arenillas (2011). Carbon xerogel and manganese oxide capacitive materials for advanced supercapacitors. *Int. J. Electrochem. Sci.*, **6**, pp. 596–612.
423. Roberts, A. J., and R. C. T. Slade (2010). Effect of specific surface area on capacitance in asymmetric carbon/α-MnO_2 supercapacitors. *Electrochimica Acta*, **55**(25), pp. 7460–7469.
424. Hong, M. S., S. H. Lee, and S. W. Kim (2002). Use of KCl aqueous electrolyte for 2 V manganese oxide/activated carbon hybrid capacitor. *Electrochemical and Solid-State Letters*, **5**(10), pp. A227–A230.

425. Khomenko, V., E. Raymundo-Piñero, and F. Béguin (2006). Optimisation of an asymmetric manganese oxide/activated carbon capacitor working at 2 V in aqueous medium. *Journal of Power Sources*, **153**(1), pp. 183–190.
426. Sun Z., K.-Y. Liu, H.-F., Zhang, A.-S. Li, and X.-C. Xu (2009). Study on meso-C/MnO_2 asymmetric supercapacitors. *Acta Physico-Chimica Sinica*, **25**(10), pp. 1991–1997.
427. Mosqueda, H. A., O. Crosnier, L. Athouël, Y. Dandeville, Y. Scudeller, P. Guillemet, D. M. Schleich, and T. Brousse (2010). Electrolytes for hybrid carbon–MnO_2 electrochemical capacitors. *Electrochimica Acta*, **55**(25), pp. 7479–7483.
428. Tomko, T., R. Rajagopalan, M. Lanagan, and H. C. Foley (2011). High energy density capacitor using coal tar pitch derived nanoporous carbon/MnO_2 electrodes in aqueous electrolytes. *Journal of Power Sources*, **196**(4), pp. 2380–2386.
429. Xu, C., H. Du, B. Li, F. Kang, and Y. Zeng (2009). Asymmetric activated carbon-manganese dioxide capacitors in mild aqueous electrolytes containing alkaline-earth cations. *Journal of The Electrochemical Society*, **156**(6), pp. A435–A441.
430. Fan, Z., J. Yan, T. Wei, L. Zhi, G. Ning, T. Li, and F. Wei (2011). Asymmetric supercapacitors based on graphene/MnO_2 and activated carbon nanofiber electrodes with high power and energy density. *Advanced Functional Materials*, **21**(12), pp. 2366–2375.
431. Wu, Z.-S., W. Ren, D.-W. Wang, F. Li, B. Liu, and H.-M. Cheng (2010). High-energy MnO_2 nanowire/graphene and graphene asymmetric electrochemical capacitors. *ACS Nano*, **4**(10), pp. 5835–5842.
432. Deng, L., G. Zhu, J. Wang, L. Kang, Z.-H. Liu, Z. Yang, and Z. Wang (2011). Graphene–MnO_2 and graphene asymmetrical electrochemical capacitor with a high energy density in aqueous electrolyte. *Journal of Power Sources*, **196**(24), pp. 10782–10787.
433. Yu, G., L. Hu, M. Vosgueritchian, H. Wang, X. Xie, J. R. McDonough, X. Cui, Y. Cui, and Z. Bao (2011). Solution-processed graphene/MnO_2 nanostructured textiles for high-performance electrochemical capacitors. *Nano Letters*, **11**(7), pp. 2905–2911.
434. Qu, Q., L. Li, S. Tian, W. Guo, Y. Wu, and R. Holze (2010). A cheap asymmetric supercapacitor with high energy at high power: Activated carbon//$K_{0.27}MnO_2 \cdot 0.6H_2O$. *Journal of Power Sources*, **195**(9), pp. 2789–2794.
435. Luo, J.-Y., J.-L. Liu, P. He, and Y.-Y. Xia (2008). A novel $LiTi_2(PO4)_3$/MnO_2 hybrid supercapacitor in lithium sulfate aqueous electrolyte. *Electrochimica Acta*, **53**(28), pp. 8128–8133.

436. Hao, Y.-J., Y.-Y. Wang, Q.-Y. Lai, Y. Zhao, L.-M. Chen, and X.-Y. Ji (2009). Study of capacitive properties for LT-$Li_4Mn_5O_{12}$ in hybrid supercapacitor. *Journal of Solid State Electrochemistry*, **13**(6), pp. 905–912.

437. Ma, S.-B., K.-W. Nam, W.-S. Yoon, X.-Q. Yang, K.-Y. Ahn, K.-H. Oh, and K.-B. Kim (2007). A novel concept of hybrid capacitor based on manganese oxide materials. *Electrochemistry Communications*, **9**(12), pp. 2807–2811.

438. Ganesh, V., S. Pitchumani, and V. Lakshminarayanan (2006). New symmetric and asymmetric supercapacitors based on high surface area porous nickel and activated carbon. *Journal of Power Sources*, **158**(2), pp. 1523–1532.

439. Wang, D.-W., F. Li, and H.-M. Cheng (2008). Hierarchical porous nickel oxide and carbon as electrode materials for asymmetric supercapacitor. *Journal of Power Sources*, **185**(2), pp. 1563–1568.

440. Lang, J.-W., L.-B. Kong, M. Liu, Y.-C. Luo, and L. Kang (2010). $Co_{0.56}Ni_{0.44}$ oxide nanoflake materials and activated carbon for asymmetric supercapacitor. *Journal of The Electrochemical Society*, **157**(12), pp. A1341–A1346.

441. Staiti, P., and F. Lufrano (2007). A study of the electrochemical behaviour of electrodes in operating solid-state supercapacitors. *Electrochimica Acta*, **53**(2), pp. 710–719.

442. Wang, Y.-G., Z.-D. Wang, and Y.-Y. Xia (2005). An asymmetric supercapacitor using RuO_2/TiO_2 nanotube composite and activated carbon electrodes. *Electrochimica Acta*, **50**(28), pp. 5641–5646.

443. Kong, L.-B., M. Liu, J.-W. Lang, Y.-C. Luo, and L. Kang (2009). Asymmetric supercapacitor based on loose-packed cobalt hydroxide nanoflake materials and activated carbon. *Journal of The Electrochemical Society*, **156**(12), pp. A1000–A1004.

444. Liang, Y.-Y., H.-L. Li, and X.-G. Zhang (2008). A novel asymmetric capacitor based on $Co(OH)_2$/USY composite and activated carbon electrodes. *Materials Science and Engineering A*, **473**(1–2), pp. 317–322.

445. Wang, H., Q. Gao, and J. Hu (2010). Asymmetric capacitor based on superior porous Ni–Zn–Co oxide/hydroxide and carbon electrodes. *Journal of Power Sources*, **195**(9), pp. 3017–3024.

446. Yoon, J. H., H. J. Bang, J. Prakash, and Y. K. Sun (2008). Comparative study of $Li[Ni_{1/3}Co_{1/3}Mn_{1/3}]O_2$ cathode material synthesized via different synthetic routes for asymmetric electrochemical capacitor applications. *Materials Chemistry and Physics*, **110**(2–3), pp. 222–227.

447. Zhao, Y., Q. Y. Lai, Y. J. Hao, and X. Y. Ji (2009). Study of electrochemical performance for AC/(Ni$_1$/3Co$_{1/3}$Mn$_{1/3}$)(OH)$_2$. *Journal of Alloys and Compounds*, **471**(1–2), pp. 466–469.

448. Qu, Q. T., Y. Shi, L. L. Li, W. L. Guo, Y. P. Wu, H. P. Zhang, S. Y. Guan, and R. Holze (2009). V$_2$O$_5 \cdot$0.6H$_2$O nanoribbons as cathode material for asymmetric supercapacitor in K$_2$SO$_4$ solution. *Electrochemistry Communications*, **11**(6), pp. 1325–1328.

449. Luo, J.-Y., and Y.-Y. Xia (2009). Electrochemical profile of an asymmetric supercapacitor using carbon-coated LiTi$_2$(PO$_4$)$_3$ and active carbon electrodes. *Journal of Power Sources*, **186**(1), pp. 224–227.

450. Cuentas Gallegos, A. K., and M. E. Rincón (2006). Carbon nanofiber and PEDOT-PSS bilayer systems as electrodes for symmetric and asymmetric electrochemical capacitor cells. *Journal of Power Sources*, **162**, pp. 743–747.

451. Laforgue, A., P. Simon, J. F. Fauvarque, M. Mastragostino, F. Soavi, J. F. Sarrau, P. Lailler, M. Conte, E. Rossi, and S. Saguatti (2003). Activated carbon/conducting polymer hybrid supercapacitors. *Journal of The Electrochemical Society*, **150**(5), pp. A645–A651.

452. Villers, D., D. Jobin, C. Soucy, D. Cossement, R. Chahine, L. Breau, and D. Bélanger (2003). The influence of the range of electroactivity and capacitance of conducting polymers on the performance of carbon conducting polymer hybrid supercapacitor. *Journal of The Electrochemical Society*, **150**(6), pp. A747–A752.

453. Estaline Amitha, F., A. Leela Mohana Reddy, and S. Ramaprabhu (2009). A non-aqueous electrolyte-based asymmetric supercapacitor with polymer and metal oxide/multiwalled carbon nanotube electrodes. *Journal of Nanoparticle Research*, **11**(3), pp. 725–729.

454. Cai, J. J., L. B. Kong, J. Zhang, Y. C. Luo, and L. Kang (2010). A novel polyaniline/mesoporous carbon nano-composite electrode for asymmetric supercapacitor. *Chinese Chemical Letters*, **21**(12), pp. 1509–1512.

455. Khomenko, V., E. Frackowiak, and F. Béguin (2005). Determination of the specific capacitance of conducting polymer/nanotubes composite electrodes using different cell configurations. *Electrochimica Acta*, **50**(12), pp. 2499–2506.

456. Hung, P.-J., K.-H. Chang, Y.-F. Lee, C.-C. Hu, and K.-M. Lin (2010). Ideal asymmetric supercapacitors consisting of polyaniline nanofibers and graphene nanosheets with proper complementary potential windows. *Electrochimica Acta*, **55**(20), pp. 6015–6021.

457. Ryu, K. S., Y. Lee, K.-S. Han, Y. J. Park, M. G. Kang, N.-G. Park, and S. H. Chang (2004). Electrochemical supercapacitor based on polyaniline doped with lithium salt and active carbon electrodes. *Solid State Ionics*, **175**(1–4), pp. 765–768.

458. Machida, K., S. Suematsu, S. Ishimoto, and K. Tamamitsu (2008). High-voltage asymmetric electrochemical capacitor based on polyfluorene nanocomposite and activated carbon. *Journal of The Electrochemical Society*, **155**(12), pp. A970–A974.

459. Zou, W.-Y., W. Wang, B.-L. He, M.-L. Sun, and Y.-S. Yin (2010). Supercapacitive properties of hybrid films of manganese dioxide and polyaniline based on active carbon in organic electrolyte. *Journal of Power Sources*, **195**(21), pp. 7489–7493.

460. Brousse, T., and D. Bélanger (2003). A hybrid Fe_3O_4–MnO_2 capacitor in mild aqueous electrolyte. *Electrochemical and Solid-State Letters*, **6**(11), pp. A244–A248.

461. Santos-Peña, J., O. Crosnier, and T. Brousse (2010). Nanosized α-$LiFeO_2$ as electrochemical supercapacitor electrode in neutral sulfate electrolytes. *Electrochimica Acta*, **55**(25), pp. 7511–7515.

462. Lin, Y.-P., and N.-L. Wu (2011). Characterization of $MnFe_2O_4$/$LiMn_2O_4$ aqueous asymmetric supercapacitor. *Journal of Power Sources*, **196**(2), pp. 851–854.

463. Khomenko, V., E. Raymundo-Piñero, E. Frackowiak, and F. Béguin (2006). High-voltage asymmetric supercapacitors operating in aqueous electrolyte. *Applied Physics A*, **82**(4), pp. 567–573.

464. Zou, B.-X., Y. Liang, X.-X. Liu, D. Diamond, and K.-T. Lau (2011). Electrodeposition and pseudocapacitive properties of tungsten oxide/polyaniline composite. *Journal of Power Sources*, **196**(10), pp. 4842–4848.

465. Yuan, C.-Z., B. Gao, and X.-G. Zhang (2007). Electrochemical capacitance of NiO/$Ru_{0.35}V_{0.65}O_2$ asymmetric electrochemical capacitor. *Journal of Power Sources*, **173**(1), pp. 606–612.

466. Ue, M., K. Ida, and S. Mori (1994). Electrochemical properties of organic liquid electrolytes based on quaternary onium salts for electrical double-layer capacitors. *Journal of The Electrochemical Society*, **141**(11), pp. 2989–2996.

467. Burke, A. (2000). Ultracapacitors: why, how, and where is the technology. *Journal of Power Sources*, **91**(1), pp. 37–50.

468. Galiński, M., A. Lewandowski, and I. Stępniak (2006). Ionic liquids as electrolytes. *Electrochimica Acta*, **51**(26), pp. 5567–5580.

469. Denshchikov, K. K., M. Y. Izmaylova, A. Z. Zhuk, Y. S. Vygodskii, V. T. Novikov, and A. F. Gerasimov (2010). 1-Methyl-3-butylimidazolium tetraflouroborate with activated carbon for electrochemical double layer supercapacitors. *Electrochimica Acta*, **55**(25), pp. 7506–7510.
470. Liu, H., Y. Liu, and J. Li (2010). Ionic liquids in surface electrochemistry. *Physical Chemistry Chemical Physics*, **12**(8), pp. 1685–1697.
471. Arbizzani, C., S. Beninati, M. Lazzari, F. Soavi, and M. Mastragostino (2007). Electrode materials for ionic liquid-based supercapacitors. *Journal of Power Sources*, **174**(2), pp. 648–652.
472. Lufrano, F., P. Staiti, and M. Minutoli (2003). Evaluation of nafion based double layer capacitors by electrochemical impedance spectroscopy. *Journal of Power Sources*, **124**(1), pp. 314–320.
473. Rathod, D., M. Vijay, N. Islam, R. Kannan, U. Kharul, S. Kurungot, and V. Pillai (2009). Design of an "all solid-state" supercapacitor based on phosphoric acid doped polybenzimidazole (PBI) electrolyte. *Journal of Applied Electrochemistry*, **39**(7), pp. 1097–1103.
474. Gómez-Romero, P., M. Chojak, K. Cuentas-Gallegos, J. A. Asensio, P. J. Kulesza, N. Casañ-Pastor, and M. Lira-Cantú (2003). Hybrid organic-inorganic nanocomposite materials for application in solid state electrochemical supercapacitors. *Electrochemistry Communications*, **5**(2), pp. 149–153.
475. Staiti, P., M. Minutoli, and F. Lufrano (2002). All solid electric double layer capacitors based on Nafion ionomer. *Electrochimica Acta*, **47**(17), pp. 2795–2800.
476. Choudhury, N. A., S. Sampath, and A. K. Shukla (2009). Hydrogel-polymer electrolytes for electrochemical capacitors: An overview. *Energy & Environmental Science*, **2**(1), pp. 55–67.
477. Groce, F., F. Gerace, G. Dautzemberg, S. Passerini, G. B. Appetecchi, and B. Scrosati (1994). Synthesis and characterization of highly conducting gel electrolytes. *Electrochimica Acta*, **39**(14), pp. 2187–2194.
478. Alamgir, M., and K. M. Abraham (1993). Li ion conductive electrolytes based on poly(vinyl chloride). *Journal of The Electrochemical Society*, **140**(6), pp. L96–L97.
479. Hashmi, S. A., R. J. Latham, R. G. Linford, and W. S. Schlindwein (1997). Studies on all solid state electric double layer capacitors using proton and lithium ion conducting polymer electrolytes. *Journal of the Chemical Society, Faraday Transactions*, **93**(23), pp. 4177–4182.
480. Lassègues, J. C., J. Grondin, T. Becker, L. Servant, and M. Hernandez (1995). Supercapacitor using a proton conducting polymer electrolyte. *Solid State Ionics*, **77**(4), pp. 311–317.

481. Osaka, T., X. Liu, M. Nojima, and T. Momma (1999). An electrochemical double layer capacitor using an activated carbon electrode with gel electrolyte binder. *Journal of The Electrochemical Society*, **146**(5), pp. 1724–1729.
482. Chojnacka, J., J. L. Acosta, and E. Morales (2001). New gel electrolytes for batteries and supercapacitor applications. *Journal of Power Sources*, **97–98**(0), pp. 819–821.
483. Matsuda, A., H. Honjo, M. Tatsumisago, and T. Minami (1998). Electric double-layer capacitors using $HClO_4$-doped silica gels as a solid electrolyte. *Solid State Ionics*, **113–115**(0), pp. 97–102.
484. Matsuda, A., H. Honjo, K. Hirata, M. Tatsumisago, and T. Minami (1999). Electric double-layer capacitor using composites composed of phosphoric acid-doped silica gel and styrene–ethylene–butylene–styrene elastomer as a solid electrolyte. *Journal of Power Sources*, **77**(1), pp. 12–16.
485. Inagaki, M., H. Konno, and O. Tanaike (2010). Carbon materials for electrochemical capacitors. *Journal of Power Sources*, **195**(24), pp. 7880–7903.
486. Cannon, W. R., B. M. Pettitt, and J. A. McCammon (1994). Sulfate anion in water: Model structural, thermodynamic, and dynamic properties. *The Journal of Physical Chemistry*, **98**(24), pp. 6225–6230.
487. López-Chavéz, R., and A. K. Cuentas-Gallegos (2013). The effect of binder in electrode materials for Capacitance improvement and EDLC binder-free cell design. *Journal of New Materials for Electrochemical Systems*, pp.
488. Varzi, A., A. Balducci, and S. Passerini (2014). Natural cellulose: A green alternative binder for high voltage electrochemical double layer capacitors containing ionic liquid-based electrolytes. *Journal of the Electrochemical Society*, **161**(3), pp. A368–A375.
489. Böckenfeld, N., S. S. Jeong, M. Winter, S. Passerini, and A. Balducci (2013). Natural, cheap and environmentally friendly binder for supercapacitors. *Journal of Power Sources*, **221**(0), pp. 14–20.
490. Krause, A., P. Kossyrev, M. Oljaca, S. Passerini, M. Winter, and A. Balducci (2011). Electrochemical double layer capacitor and lithium-ion capacitor based on carbon black. *Journal of Power Sources*, **196**(20), pp. 8836–8842.
491. Brandt, A., P. Isken, A. Lex-Balducci, and A. Balducci (2012). Adiponitrile-based electrochemical double layer capacitor. *Journal of Power Sources*, **204**(0), pp. 213–219.

492. Yamagata, M., S. Ikebe, K. Soeda, and M. Ishikawa (2013). Ultrahigh-performance nonaqueous electric double-layer capacitors using an activated carbon composite electrode with alginate. *RSC Advances*, **3**(4), pp. 1037–1040.

493. Choudhury, N. A., P. W. C. Northrop, A. C. Crothers, S. Jain, and V. R. Subramanian (2012). Chitosan hydrogel-based electrode binder and electrolyte membrane for EDLCs: Experimental studies and model validation. *Journal of Applied Electrochemistry*, **42**(11), pp. 935–943.

494. Wan, Y., K. A. M. Creber, B. Peppley, and V. T. Bui (2003). Ionic conductivity of chitosan membranes. *Polymer*, **44**(4), pp. 1057–1065.

495. Qin, C., H. Li, Q. Xiao, Y. Liu, J. Zhu, and Y. Du (2006). Water-solubility of chitosan and its antimicrobial activity. *Carbohydrate Polymers*, **63**(3), pp. 367–374.

496. Chmiola, J., C. Largeot, P.-L. Taberna, P. Simon, and Y. Gogotsi (2010). Monolithic carbide-derived carbon films for micro-supercapacitors. *Science*, **328**(5977), pp. 480–483.

497. Pröbstle, H., C. Schmitt, and J. Fricke (2002). Button cell supercapacitors with monolithic carbon aerogels. *Journal of Power Sources*, **105**(2), pp. 189–194.

498. Kaempgen, M., J. Ma, G. Gruner, G. Wee, and S. G. Mhaisalkar (2007). Bifunctional carbon nanotube networks for supercapacitors. *Applied Physics Letters*, **90**(26), pp. 264104–264102.

499. Pushparaj, V. L., M. M. Shaijumon, A. Kumar, S. Murugesan, L. Ci, R. Vajtai, R. J. Linhardt, O. Nalamasu, and P. M. Ajayan (2007). Flexible energy storage devices based on nanocomposite paper. *Proceedings of the National Academy of Sciences of the United States of America*, **104**(34), pp. 13574–13577.

500. Hu, L., J. W. Choi, Y. Yang, S. Jeong, F. La Mantia, L.-F. Cui, and Y. Cui (2009). Highly conductive paper for energy-storage devices. *Proceedings of the National Academy of Sciences*, **106**(51), pp. 21490–21494.

501. Portet, C., P. L. Taberna, P. Simon, and C. Laberty-Robert (2004). Modification of Al current collector surface by sol–gel deposit for carbon–carbon supercapacitor applications. *Electrochimica Acta*, **49**(6), pp. 905–912.

502. Portet, C., P. L. Taberna, P. Simon, E. Flahaut, and C. Laberty-Robert (2005). High power density electrodes for carbon supercapacitor applications. *Electrochimica Acta*, **50**(20), pp. 4174–4181.

503. Taberna, P. L., C. Portet, and P. Simon (2006). Electrode surface treatment and electrochemical impedance spectroscopy study on carbon/carbon supercapacitors. *Applied Physics A*, **82**(4), pp. 639–646.

504. Zhou, R., C. Meng, F. Zhu, Q. Li, C. Liu, S. Fan, and K. Jiang (2010). High-performance supercapacitors using a nanoporous current collector made from super-aligned carbon nanotubes. *Nanotechnology*, **21**(34), p. 345701.

505. Gourdin, G., A. Meehan, T. Jiang, P. Smith, and D. Qu (2011). Investigation of the impact of stacking pressure on a double-layer supercapacitor. *Journal of Power Sources*, **196**(1), pp. 523–529.

506. Zhang, J., and X. S. Zhao (2012). On the configuration of supercapacitors for maximizing electrochemical performance. *ChemSusChem*, **5**(5), pp. 818–841.

507. Bordjiba, T., M. Mohamedi, and L. H. Dao (2007). Synthesis and electrochemical capacitance of binderless nanocomposite electrodes formed by dispersion of carbon nanotubes and carbon aerogels. *Journal of Power Sources*, **172**(2), pp. 991–998.

508. *American Chemical Society Pharmaceutical Round Table* (2014). American Chemical Society.

509. Moscoso, R., J. Carbajo, and J. A. Squella (2014). 1,3-Dioxolane: A green solvent for the preparation of carbon nanotube-modified electrodes. *Electrochemistry Communications*, **48**(0), pp. 69–72.

510. Leela Mohana Reddy, A., F. Estaline Amitha, I. Jafri, and S. Ramaprabhu (2008). Asymmetric flexible supercapacitor stack. *Nanoscale Research Letters*, **3**(4), pp. 145–151.

511. Burke, A., and M. Miller (2010). Testing of electrochemical capacitors: Capacitance, resistance, energy density, and power capability. *Electrochimica Acta*, **55**(25), pp. 7538–7548.

512. Zhao, S., F. Wu, L. Yang, L. Gao, and A. F. Burke (2010). A measurement method for determination of dc internal resistance of batteries and supercapacitors. *Electrochemistry Communications*, **12**(2), pp. 242–245.

513. Niu, J., W. G. Pell, and B. E. Conway (2006). Requirements for performance characterization of C double-layer supercapacitors: Applications to a high specific-area C-cloth material. *Journal of Power Sources*, **156**(2), pp. 725–740.

514. Staiti, P., and F. Lufrano (2005). Design, fabrication, and evaluation of a 1.5 F and 5 V prototype of solid-state electrochemical supercapacitor. *Journal of The Electrochemical Society*, **152**(3), pp. A617–A621.

515. Niu, J., B. E. Conway, and W. G. Pell (2004). Comparative studies of self-discharge by potential decay and float-current measurements at C double-layer capacitor and battery electrodes. *Journal of Power Sources*, **135**(1–2), pp. 332–343.

516. Guillemet, P., Y. Scudeller, and T. Brousse (2006). Multi-level reduced-order thermal modeling of electrochemical capacitors. *Journal of Power Sources*, **157**(1), pp. 630–640.

Chapter 9

Power-to-Fuel and Artificial Photosynthesis for Chemical Energy Storage

Albert Tarancón, Cristian Fábrega, Alex Morata, Marc Torrell, and Teresa Andreu

Catalonia Institute for Energy Research (IREC), Jardins de les Dones de Negre, 1, Sant Adrià del Besòs (Barcelona), ES-08930, Spain

atarancon@irec.cat

The transformation of green energy into fuels easy to store, handle and transport is probably the most straightforward strategy for a future global energy scenario based on renewable resources and, therefore, has recently attracted a lot of attention from researchers, industrialists and politicians. While the generation of electricity from fuels has been developed for decades and presents a mature status, the production of synthetic gas/fuel is still a matter of study. This chemical storage strategy leaves room for different technological approaches. In this chapter, special attention will be devoted to the conversion of green electricity by following power-to-fuel schemes and the accumulation of solar energy in fuel form by artificial photosynthesis.

Materials for Sustainable Energy Applications: Conversion, Storage, Transmission, and Consumption
Edited by Xavier Moya and David Muñoz-Rojas
Copyright © 2016 Pan Stanford Publishing Pte. Ltd.
ISBN 978-981-4411-81-3 (Hardcover), 978-981-4411-82-0 (eBook)
www.panstanford.com

9.1 Energy Storage in Current and Future Energy Scenarios

The current energy crisis is motivating the exploration of clean and renewable energy sources to mitigate the heavy dependence on finite fossil fuels (oil, coal and natural gas). However, if a substantial amount of the global energy supply is to come from intrinsically intermittent renewable energy sources, it is necessary to develop cost-effective and efficient technologies to store and transport large quantities of energy through existing energy infrastructures, i.e., gas/fuel and electrical grids. Moreover, the massive implementation of renewable energies will likely involve decentralized power production. In this distributed generation model, the complexity of energy management greatly increases and energy storage becomes a crucial component of a reliable and efficient energy supply.

Currently, the main advances in energy storage are primarily focused on small accumulators for the transportation sector and portable applications, particularly, through the recent commercialization of pure electrical and hybrid vehicles as well as the boom of consumer electronics. However, due to the expected major role of large-scale energy storage in the near future, ambitious research programs are being implemented in many countries to consolidate current knowledge and explore new technologies. According to the "Energy Storage Tracker" report by Navigant Research [1], more than 600 storage projects are currently operating (or in development) worldwide for a total deployed capacity of around 130 GW (+40 GW announced) and a total annual production of 600–700 TWh. Since this value is still far from the total annual world electricity consumption of 20,000 TWh per year [2], a pronounced market growth is anticipated in the next years.

In this first section, a brief introduction on the different energy storage systems with special emphasis in chemical energy storage systems will be presented. The production of synthetic fuels from syngas or CO_2 and H_2 and the achievable efficiency of the conversion from chemical energy to electricity will also be covered. Finally, a possible sustainable scenario for energy generation, storage and consumption will be proposed.

9.1.1 Energy Storage Systems

Energy storage systems can be classified according to the type of energy accumulated, namely chemical, electrical, thermal or mechanical:

(i) Chemical energy storage consists of the production of fuels, which can play the role of energy carrier, by using other sources, e.g., hydrogen production via electrolysis or photocatalysis using excess electricity or solar energy, respectively.
(ii) Electrochemical energy storage includes the accumulation of energy from chemicals that react in electrochemical devices such as batteries.
(iii) Electricity can be directly accumulated as electrostatic charges in high energy/power capacitors or as a magnetic field in superconducting magnetic energy storage (SMES) systems.
(iv) Thermal energy storage allows heat to be collected for later use in simple systems like hot water tanks or more complex phase-change materials or storage heaters.
(v) Mechanical energy can be accumulated as rotational energy in flywheels, as potential energy in pumped-storage hydroelectricity (PSH) or as pressure in compressed air energy storage (CAES).

The size and timescale of the application, i.e., the energy capacity and power supply (discharge time), are usually employed to determine the most suitable technology for each application (Fig. 9.1). Low cost and high energy systems such as pumped hydroelectric storage, compressed air energy storage or flow batteries are preferred for energy management purposes where the timing of generation and consumption is decoupled while high power technologies like batteries, capacitors, superconducting magnetic energy storage or flywheels are the favourite for ensuring power quality and reliability, i.e., uninterruptible power supply (UPSs).

Figure 9.2 shows the worldwide installed storage capacity for electrical energy [3]. With almost 99% of the total (~127,000 MW), pumped hydro is the technology with the largest share of the global energy storage capacity. The PSH has a high efficiency (70–85%), huge capacity (see Box 9.1) and long durability but requires large areal land use and is limited by topography. The same drawback is presented by CAES since the compressed air is stored underground in caverns, aquifers or abandoned mines. Although, CAES shows low round-trip efficiencies below 50%, this technology is second in terms of installed capacity with ca. 450 MW (<0.5% of the total). Most of the remaining storage capacity is in the form of batteries, in particular, those based on sodium-sulphur (NaS) with more than 300 MW already installed. NaS battery technology presents very fast response times (on the order of milliseconds) and high efficiency (AC-based round-trip efficiency of 75%) that makes it suitable for grid stabilization purposes. While flow batteries do not represent a significant portion of the installed electrical energy storage capacity, they do have big advantages over conventional secondary batteries. This technology allows decoupling power and capacity since the electrolyte (that contains the electroactive species) is a liquid stored in tanks. This accumulated electrolyte is pumped to the electrochemical cell that converts chemical energy into electricity. Therefore, the power depends on the size of the cell while the capacity is directly related to the volume of the electrolyte tanks. This flexibility could promote large-scale implementation of the technology once it has reached its maturity.

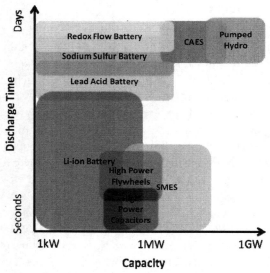

Figure 9.1 Discharge time and energy capacity for different storage technologies.

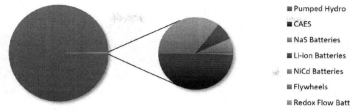

Figure 9.2 Percentage of the worldwide installed storage capacity for electrical energy by technology. Data from ref. [4].

Box 9.1

Calculate the amount of energy (in GWh) stored in a hydroelectric reservoir based on a cylindrical tank (height H = 40 m, diameter d = 2120 m and area A) elevated at h = 20 m from the power turbines. How much power is generated if it provides the energy in one week (the efficiency of conversion in the turbine can be fixed at η = 80%)? How many tonnes of water are necessary to pump?

Solution: The potential energy (E) of a mass (m) elevated at a certain height (x) is defined as $E = mgx$, where g is the standard gravity. Since the tank is full of water (density, ρ) and has a certain height (H) the amount of potential energy accumulated arises from a simple integration:

$$\Delta E = \int_h^H mg\,dx = \rho A g \int_h^H x\,dx = \frac{\rho A g}{2}(H^2 - h^2) = 2.07 \times 10^{13}\,\text{J} = 5.7\,\text{GWh}$$

If this energy is provided in one week, the power generated is

$$P = \frac{\Delta E}{\Delta t}\eta = 27.1\,\text{MW}$$

The amount of water required can be directly calculated from the volume of the tank:

$$m = \rho V = \rho A (H - h) = 7.1 \times 10^5\,\text{tn}$$

9.1.2 Chemical Energy Storage

As previously mentioned, chemical energy storage is based on the artificial production of a chemical that can play the role of secondary energy carrier. Therefore, a highly portable and usable chemical storage media (CSM) together with an efficient energy converter (from the primary energy source to the CSM and from the CSM to electricity) are fundamental for defining a competitive storage system, e.g., hydrogen and water electrolysis/fuel cells. Stable CSMs allow for the accumulation of large amounts of energy (~TWh, see Box 9.2) for long periods of time (even years), contrary to other storage technologies. This makes this technology particularly convenient for grid applications. Moreover, the existence of a flexible CSM is clearly the most straightforward model to fulfil the requirements of fundamental sectors like transport or portable applications. In both sectors, current lithium-ion or NaS batteries seem to be not enough for covering future energy demands.

> **Box 9.2**
>
> How many kilograms and cubic meters of H_2 and methane are necessary for accumulating the amount of chemical energy of the reservoir defined in Box 9.1 and a power backup of 25 kWh? Estimate the tank volume if the gases are liquefied.
>
> *Solution*: Assuming the chemical energy accumulated in hydrogen as the formation enthalpy of liquid water (ΔH_f° = −286 kJ/mol), a simple calculation yields the equivalent litres of gas of H_2:
>
> $$n = \frac{\Delta E}{\Delta H_f^\circ}$$
>
> $$V = n \frac{RT}{P} = \frac{\Delta E}{\Delta H_f^\circ} \frac{RT}{P}$$
>
> | | Reservoir (ΔE = 5.7 GWh) | | Backup (ΔE = 25 kWh) | |
> | --- | --- | --- | --- | --- |
> | STP | 1.6×10^6 m³ | — | 7.0 m³ | — |
> | P = 300 bar | 5.3×10^3 m³ | 1.4×10^5 kg | 0.024 m³ | 1.6 kg |
>
> The same calculation for CH_4 yields (ΔH_c° = −889 kJ/mol):
>
> | | Reservoir (ΔE = 5.7 GWh) | | Backup (ΔE = 25 kWh) | |
> | --- | --- | --- | --- | --- |
> | STP | 5.2×10^5 m³ | — | 2.2 m³ | — |
> | P = 300 bar | 1.8×10^3 m³ | 3.7×10^5 kg | 0.0075 m³ | 0.63 kg |

Assuming the densities of liquid hydrogen, δ_{LH2} = 71 g/l at 20 K, and LNG, δ_{LNG} = 422 g/l at 288 K, it is possible to calculate the corresponding tank volume by using the simple expression:

$$V = m/\delta$$

$V_{LH2} = 2.0 \times 10^3$ m^3 (spherical tank of 8 m of radius)
$V_{LNG} = 8.1 \times 10^2$ m^3 (spherical tank of 6 m of radius)

Another major challenge in chemical energy storage is to have a CSM compatible with current infrastructures, vehicles and devices. Although pure hydrogen could be considered as a convenient CSM since it can be directly injected in the natural gas network (at concentrations below 10%), synthetic gas and liquid hydrocarbons are clearly the most practical short-term energy carriers due to their ability to directly substitute fossil fuels in the current system while keeping a similar high energy density.

A sustainable and clean production of these synthetic hydrocarbon fuels should be based on hydrogen and carbon compounds coming from an abundant and renewable feedstock. Figure 9.3 shows a scheme of the different routes that will be presented in this chapter. Hydrogen production from water and carbon obtained from reduced CO_2 will be considered in Sections 9.2 and 9.3, respectively. Starting from these products, highly efficient catalytic routes will be presented in Section 9.1.4 for producing liquid fuels.

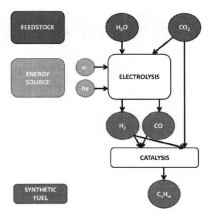

Figure 9.3 Graphical outline of the chapter, from water and carbon dioxide feedstock to synthetic fuel via electrolysis.

9.1.3 Synthetic Fuels Production

As previously mentioned, not only a suitable CSM but an efficient conversion from renewable to chemical energy is crucial to define a feasible chemical energy storage scheme. In this context, one promising CSM for short-term applications is synthetic fuels. While different synthetic fuel production routes can be envisaged, water and carbon dioxide are the preferred feedstock for carbon and hydrogen due to their availability and, in the case of CO_2, poor industrial utilization and deleterious climate change effects [5]. However, the direct use of H_2O or CO_2 as reactants requires large amounts of energy due to the high stability of these two molecules and the endothermic reactions generally associated. To overcome this issue, reduced species like H_2 or CO are preferred to define catalytic routes implementable in the chemical industry. The generation of the necessary CO and H_2 will be the object of the next sections, in particular by using electrolysis or photosynthesis. This section will describe the main two routes for the synthesis of fuels based on (i) CO and H_2, syngas (Fischer–Tropsch synthesis, FTS) and (ii) CO_2 and H_2 (Sabatier reaction).

9.1.3.1 Fischer–Tropsch synthesis

Syngas, or synthetic gas, is a fuel gas mainly consisting of carbon monoxide and hydrogen with a certain amount of carbon dioxide. It is usually employed as intermediate for the generation of fuels such as H_2 (via water-gas-shift reaction, WGS, Eq. 9.1), methanol or dimethyl ether (via methanol synthesis, Eq. 9.2) and different synthetic fuels (via Fischer–Tropsch synthesis, FTS, as described in Eqs. 9.3–9.5 for alkanes, alkenes and alcohols, respectively) [6]. All these fuels can be used as feedstock for chemical-to-energy converters like fuel cells (see Section 9.1.4) thereby providing an efficient way to store energy in fuel form. Due to this versatility, syngas and its production routes are at the centre of most fuel-based energy schemes.

$$CO + H_2O \Leftrightarrow CO_2 + H_2 - 41 \text{ kJ/mol} \tag{9.1}$$

$$CO + 2H_2 \Leftrightarrow CH_3OH - 91 \text{ kJ/mol} \tag{9.2}$$

$$nCO + (2n+1)H_2 \Rightarrow C_nH_{2n+2} + nH_2O \tag{9.3}$$

$$nCO + 2n\,H_2 \Rightarrow C_nH_{2n} + nH_2O \tag{9.4}$$

$$nCO + 2n\,H_2 \Rightarrow C_nH_{2n+1}OH + (n-1)H_2O \tag{9.5}$$

In general, syngas can be obtained from virtually any hydrocarbon feedstock by gasification with an oxidant or steam. Natural gas is the most commonly employed feedstock. Different production processes are typically used including steam methane reforming (SMR), auto-thermal reforming (ATR) and catalytic/partial oxidation (CPO/POX). The catalysts employed in this reaction are varied, the most common being nickel, cobalt, iron and ruthenium with promoters like potassium or copper supported on high surface area silica, alumina or zeolites [7–9].

However, for *sustainable* energy storage purposes pursued in this chapter, it is more interesting to consider producing syngas by reducing CO_2 with H_2 (as previously mentioned) by following the endothermic reverse water-gas-shift (RWGS) reaction (reverse of Eq. 9.1) [10]. The massive implementation of this route would represent a great impact on the future of sustainable energy since (i) it represents a chemical recycling opportunity for carbon dioxide coming from current and future carbon sequestration and storage (CSS) plants, available in large quantities and at a very low-cost and (ii) promotes the use of hydrogen potentially obtained from efficient conversion of renewable energy via electrolysis or photocatalytic water splitting (as explained in the next sections).

Even direct production of syngas from CO_2 and H_2O is being evaluated in highly efficient devices like high-temperature solid oxide electrolysis cells (SOECs, see Section 9.2.3) [11]. This type of devices allow working in thermoneutral conditions, i.e., without any external heat source, achieving conversion efficiencies over 85%.

9.1.3.2 Sabatier reaction

The reaction of CO_2 methanation, i.e., the hydrogenation of CO_2 to methane (Eq. 9.6), is known as the Sabatier reaction. The Sabatier reaction is an exothermic process that requires a $H_2:CO_2$ ratio of 4:1.

$$CO_2 + 4H_2 \rightarrow CH_4 + 2H_2O$$
$$\Delta H_r = -164.9\ \text{kJ/mol};\ \Delta G_{r,298\,K} = -130.8\ \text{kJ/mol} \tag{9.6}$$

This reaction is a combination of the endothermic reverse water gas shift (RWGS) reaction (Eq. 9.7) and the exothermic carbon monoxide hydrogenation (Eq. 9.8), the whole process being exothermic, with around 17% of the energy released as heat. Therefore, unless coupled to an endothermic process the overall energy efficiency of the system is low.

$$CO_2 + H_2 \rightarrow CO + H_2O \quad \Delta H_r = +41.5 \text{ kJ/mol} \tag{9.7}$$

$$CO + 3H_2 \rightarrow CH_4 + H_2O \quad \Delta H_r = -206.4 \text{ kJ/mol} \tag{9.8}$$

Although this reaction has been known for over a century, it is still under development at catalyst and reactor levels. The catalysts used either in the CO_2 or CO hydrogenation is quite similar, mostly based on highly loaded nickel (10–30%) on silica [12] or ceria [13] supports. Noble metals can also catalyse the reaction, and Ru, Rh and Pd are the most suitable when used on silica or alumina with CO_2 promoters like MgO [14] to enhance the adsorption and reactivity.

9.1.4 Efficiency of Converting Chemical Energy into Electricity

Two main schemes for converting chemical energy into electricity can be defined. The most common one is based on a two-step approach that combines a heat engine (chemical-to-mechanical energy conversion) with an electric generator (mechanical-to-electrical energy conversion). An alternative approach that has recently received increasing interest is the direct conversion of chemical energy into electricity by employing electrochemical devices such as fuel cells. These two schemes and their respective conversion efficiencies will be briefly outlined in this section.

9.1.4.1 Chemical-to-mechanical-to-electrical conversion: Heat engines

From the 17th century, when the steam engine was the workhorse of the Industrial Revolution, and through the commercial drilling and petroleum production in the 1850s, the capability of converting chemical energy into mechanical work using heat engines has been a cornerstone for the evolution of humankind to modern day society.

The theoretical efficiency limit for converting heat to mechanical energy using a heat engines is determined by the corresponding idealized thermodynamic cycle. Considering the Carnot cycle, this efficiency maximum is determined as a function of the upper (T_1) and lower operating temperatures (T_2) of the engine (see Eq. 9.9 and Fig. 9.4). Therefore, excluding expensive engines capable of operating at high temperatures and pressures with very high efficiencies, e.g., 70% in rocket engines, most engines achieve efficiencies below 20% [15]. Gas turbines working in thermal power plants typically achieve efficiencies of around 35% when combined-cycle schemes. Furthermore, by combining two or more heat engines in such a way that one heat engine is used as the heat source for another, the efficiency can be increased up to 60% [16].

$$\eta = \frac{T_1 - T_2}{T_1} \quad (9.9)$$

Once the chemical energy is converted into mechanical energy, an electrical generator is employed for the final conversion from mechanical energy into electricity. These electric generators have extremely high efficiencies, typically over 95% [15]. This combination of rotating machinery coupled to electric generators is the typical approach employed in fossil fuel power stations yielding a total efficiency from 33% (ordinary cycle) to 57% (combined-cycle).

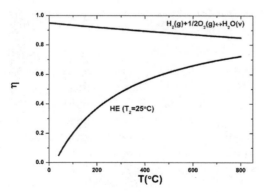

Figure 9.4 Maximum efficiency of an ideal fuel cell and a heat engine (HE) as a function of the operation temperature. The lower heating value is employed for the fuel cell and a Carnot cycle with a thermal sink at room T is supposed for the heat engine.

9.1.4.2 Direct chemical-to-electrical conversion: Fuel cells

Fuel Cells are electrochemical devices that convert energy released from a chemical reaction directly into electrical energy. The ideal efficiency of a fuel cell is achieved when the change in Gibbs free energy (ΔG) of the reaction is available as useful electric energy at the temperature of conversion,

$$\eta = \frac{\Delta G}{\Delta H} \tag{9.10}$$

Considering a cell that uses hydrogen as a fuel, it is possible to calculate both the free energy and the total change in energy between the products and the reactants, i.e., the enthalpy increment (ΔH). Two different reactions should be considered according to the nature of the water produced: liquid (l) or vapour (v). For standard conditions, the relevant thermodynamic magnitudes are listed in Table 9.1, yielding theoretical energy conversion efficiencies of around 90%.

Table 9.1 Enthalpy, entropy, Gibbs free energy and efficiency for the water (liquid or vapour) formation reaction under standard conditions (T = 298 K and P = 1 atm)

Reaction	$-\Delta H°$ (kJ/mol)	$-\Delta S°$ (J/mol·K)	$-\Delta G°$ (kJ/mol)	η_{max}
$H_2(g) + 1/2\ O_2(g) \rightarrow H_2O\ (l)$	285.8	163.2	237.2	0.83
$H_2(g) + 1/2\ O_2(g) \rightarrow H_2O\ (v)$	241.8	44.4	228.6	0.95

Accounting for intrinsic losses that are always present in fuel cells such as ohmic or polarization resistances, and also considering design and stacking issues like fuel utilization or gas diffusion, the efficiency in commercial systems usually ranges between 30% and 60% (see Chapter 6 in this book). Solid oxide fuel cells (SOFCs) are in the upper part of this range with reported efficiencies approaching 60% or even 80–85% when coupled to a gas turbine in cogeneration mode (heat and power) [17].

Although nowadays chemical fuels are converted into electricity solely by internal heat engines, the introduction and expansion of more efficient and environmentally friendly electrochemical devices is expected in the near future. This could promote a new energy model based on fuel cells bridging the present economy based on fossil fuels and a future carbon-free hydrogen economy [11, 18, 19].

9.1.5 One Possible Sustainable Generation/Storage/Consumption Cycle

Based on the previous sections, it is possible to envisage a fascinating *sustainable* energy scenario based on hydrogen and/or hydrocarbons as CSM combined with fuel cells and electrolysers/artificial photosynthesis reactors. A complete generation/storage/consumption cycle based on these elements is sketched in Fig. 9.5. Excess electric–ity produced at the renewable energy generator is used to produce hydrogen or hydrocarbons (after complementary catalytic conversion). After storage or transport of the gas/fuel through the current infrastructures, this fuel could be used in fuel cells (see Box 9.3). Exhaust gases would be steam and carbon dioxide. It is clear that the cycle is closed if some CO_2 concentration can take place since the water will directly enter into its natural cycle when reaching the atmosphere.

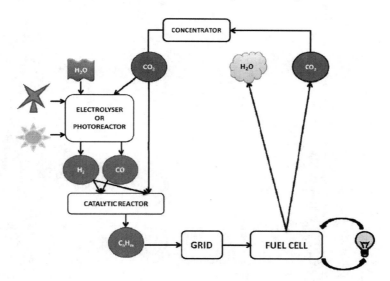

Figure 9.5 Sketch of a sustainable generation/storage/consumption cycle based on renewable energies.

Box 9.3

Considering a country with a natural gas trunkline of 10,000 km (compressed at 72 bar in 28 inch diameter pipelines) (a) calculate the energy autonomy with the gas accumulated in the network if the total electricity consumption of the country is 250 TWh per year (assume a gas-to-electricity conversion efficiency of 60%); calculate the same including an available gas reservoir of 1.5×10^9 m³; (c) calculate how many hours of operation are necessary to generate the total amount of gas accumulated by using a power-to-gas scheme based on wind energy (assume 2.5 GW of wind power installed in the country and a conversion efficiency of 68%).

Solution: (a) The total amount of CH_4 (in moles) accumulated in the proposed gas network can be calculated by employing the next equation

$$n_{network} = \frac{PV}{RT} = 11.6 \times 10^6 \text{ mol of } CH_4,$$

where V is the volume of the 10,000 km of 28 inch diameter pipeline and $P = 72$ bar.

Assuming the chemical energy accumulated in methane as its formation enthalpy ($\Delta H_f^\circ = -889$ kJ/mol), a simple calculation yields the equivalent energy:

$$\Delta E_{network} = n_{network} \Delta H_f^\circ = 1.03 \times 10^{13} \text{ J} = 2.85 \text{ GWh}$$

Therefore, the autonomy of energy accumulated assuming a conversion efficiency of $\eta = 60\%$ is

$$t = \frac{\Delta E_{network}}{\Delta E_{consumption}/\text{year}} \eta = 6.8 \times 10^{-6} \text{ years} = 3.6 \text{ min}$$

(b) The same calculation considering gas reservoirs:

$$n_{reservoir} = \frac{PV}{RT} = 61.3 \times 10^6 \text{ mol of } CH_4$$

$$\Delta E_{total} = (n_{network} + n_{reservoir})\Delta H_f^\circ = 6.48 \times 10^{13} \text{ J} = 18.0 \text{ GWh}$$

$t = 22.7$ min

(c) To calculate the time required to generate this amount of gas by using a power-to-gas scheme with 68% of efficiency based on 2.5 GW of installed wind energy:

$$t = \frac{\Delta E_{total}}{P\eta} = 106 \text{ h}$$

It is clear from this simple calculation that this country will need higher gas storage capability and an increase of the wind power capacity installed (or an equivalent green electricity production system) to cover the needs of the population (without reducing the demand).

Box 9.4

Calculate the efficiency of the transport of electricity through an electric network over 1000 km compared to covering the same distance using a power-to-gas conversion/transport scheme. Use the efficiencies for every step as indicated in the following sketch.

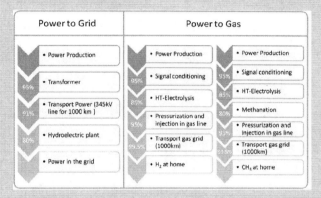

Solution: The efficiency of the different processes can be easily calculated by multiplying the different efficiencies. Therefore,

$$\eta_{\text{Power-to-Power}} = 0.95 \times 0.91 \times 0.80 = 0.69 \ (69\%)$$

$$\eta_{\text{Power-to-H}_2} = 0.95 \times 0.85 \times 0.95 \times 0.995 = 0.76 \ (76\%)$$

$$\eta_{\text{Power-to-CH}_4} = 0.95 \times 0.85 \times 0.80 \times 0.95 \times 0.995 = 0.61 \ (61\%)$$

The lowest efficiency is obtained for the power-to-methane scheme mainly due to the methanation step. A high efficiency for the high-temperature electrolysis has been assumed although this is not a commercial technology and presented values are only based on laboratory-scale studies. It would be necessary to consider the final step in all the cases, i.e., the signal conditioning and transport of the power from the hydroelectric plant to the user and the efficiency of the use of the gas by the customer.

From a technological and environmental point of view, the presented cycle is clearly in the way of an ideal model of sustainability. Focusing on the storage stage, the simple calculations carried out in Box 9.4 show a similar efficiency between the power-to-gas conversion/transport scheme and the classical transport of electricity through an electric network (and storage using a hydroelectric plant). Therefore, this novel approach can be considered a promising candidate to overcome current issues of energy accumulation, grid stability and decentralized distribution strategies. Obviously, a comprehensive techno-economical analysis is always necessary to complete the picture of the feasibility of any new model and this, although out of the scope of this chapter, could be carried out following examples provided in the literature [20].

9.2 Power to Fuel

Power-to-fuel refers to the group of technologies capable of converting electricity into fuel. The generation of fuels that can be transported in the existing natural gas network and infrastructure is of particular interest. In this context, two major alternative energy fuels currently under consideration are hydrogen and, obviously, methane. Since the main routes for generating synthetic methane are also based on hydrogen (see Section 9.1.3), H_2 production represents the core of the present power-to-gas technology. In order to directly produce high value-added carbon-based fuels, it is required to introduce a carbon source, together with water, in the inlet mixture. This more complex approach represents a major challenge to be covered by the research community in the next years.

This section is mainly focused on describing the different systems for hydrogen production based on water electrolysis. After some general aspects on electrolysis and electrolytic cells (Section 9.2.1), the section covers the most promising low and high-temperature water electrolysis technologies (Section 9.2.2). The last part is focused on briefly describing recent approaches for direct carbon dioxide electrolysis and H_2O-CO_2 co-electrolysis (Section 9.2.3).

9.2.1 General Aspects of Electrolytic Cells

9.2.1.1 Fundamentals of electrolysis

An efficient and clean way to reduce water and CO_2 is to use electrolysis. Electrolysis is a chemical decomposition reaction activated by an electrical current. Electrolytic cells are devices that are capable of forcing this non-spontaneous chemical reaction by injection of a DC current. The main parts of an electrolytic cell are the cathode, the anode and the ion-conducting electrolyte. An electrical current is driven between the anode and the cathode forcing the oxidation of species at the anode and reduction at the cathode. The electrons circulate through an external circuit and are used in the reactions taking place at the electrodes while ions flow through the electrolyte. Figure 9.6 shows a scheme of a typical electrolytic cell performing water electrolysis, i.e., water is reduced to H_2 at the cathode while O_2 is produced at the anode.

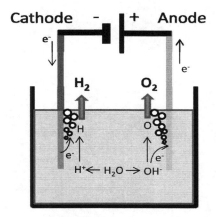

Figure 9.6 Scheme of the electrolysis of water by applying an electric potential.

According to Faraday's first law, the mass of a substance decomposed by passing an electric current through an electrode is directly proportional to the quantity of electricity transferred at that electrode (Eq. 9.11):

$$m = kq = \frac{M}{zF}q, \qquad (9.11)$$

where m is the mass of the substance liberated (in grams), $k = M/(zF)$ is a constant, q is the total electric charge circulating through the circuit, F is the Faraday constant ($F = 96485$ C mol^{-1}, i.e., the electrical charge per mole of electrons), M is the molar mass of the substance and z is the number of electrons transferred per ion. For the particular case of constant current electrolysis, $q = It$, where I is the electrical current that is applied for a duration of time t. See an example of application of Faraday's law in Box 9.5.

Box 9.5

By assuming an ideal electrolyser, calculate the days of operation of a wind farm with a power capacity of 100 MW and an average capacity factor (cf) of 20% for generating the amount of hydrogen accumulated in the reservoirs proposed in Box 9.2.

Solution: Using Eq. 9.11 and assuming a constant current (I), it is possible to obtain an expression for the time to produce a certain amount of electrolysed mass (m):

$$t = \frac{mzF}{MI}$$

By calculating the effective current, i.e., considering a capacity factor of 20%, the times to convert a mass of 1.4×10^5 kg and 0.63 kg (Box 9.2) are

$$I_{eff} = \frac{P}{V} cf = \frac{100 \cdot 10^6}{1.23} 0.2 = 16.3 \text{ MA}$$

$$t = \frac{mzF}{MI_{eff}} = \frac{1.4 \cdot 10^5 \cdot 2 \cdot 96485}{0.002 \cdot 16300000} \frac{1}{3600 \cdot 24} = 9.6 \text{ days}$$

$$t = \frac{mzF}{MI_{eff}} = \frac{0.63 \cdot 2 \cdot 96485}{0.002 \cdot 16300000} = 2 \text{ s}$$

In order to force electrolysis to occur, electrical energy equivalent to the change in the Gibb's free energy (ΔG) of the reaction is required. Since the work done by an electric field is defined as $W = q\Delta E$, the maximum work accomplished by applying a certain electric potential difference between two electrodes ($\Delta E = E_{anode} - E_{cathode}$) is related to the Faraday constant according to

$W_{max} = zN_A F\Delta E$ (where N_A is Avogrado's number). Therefore, the minimum voltage required to drive electrolysis, corresponding to the work associated with the reversible reaction, can be defined as

$$\Delta E_{rev} \equiv E_{rev} = \frac{\Delta G}{zF}. \tag{9.12}$$

This reversible potential (E_{rev}) corresponds to the voltage appearing in an electrochemical cell involving the *direct reaction*, i.e., it corresponds to the so-called open circuit voltage (OCV) defined by the Nernst equation at thermodynamic equilibrium:

$$E_{rev} = E_{rev}^0 - \frac{RT}{zF}\ln Q, \tag{9.13}$$

where R is the universal gas constant (8.314 J/K^{-1}mol^{-1}), T is the temperature in Kelvin and Q is the reaction coefficient of the direct reaction. For the particular case of water, this expression becomes

$$E_{rev} = E_{rev}^0 - \frac{RT}{2F}\ln\frac{pH_2O}{pH_2 pO_2^{1/2}}. \tag{9.14}$$

9.2.1.2 Temperature and pressure effects on electrolysis

The process of water dissociation increases the entropy (S) of the system. This non-reversible contribution is clearly not considered for the calculation of the reversible potential. Therefore, for an applied ΔE_{rev} the electrolysis reaction is endothermic. The extra energy ($T\Delta S$) required to balance this increase can be provided by the environment at temperature T. Of course, when operating at room temperature this is not a problem since the environment acts as a thermal reservoir. However, for high-temperature applications, this forces the necessity of an external heat source coupled to the system.

Alternatively, the entropy change in the reaction can be balanced with the heat generated by the Joule effect in non-ideal cells, i.e., with a certain overpotential (E_{ov}):

$$Q = E_{ov} zF = (E_{cell} - E_{rev})zF \tag{9.15}$$

The voltage that allows the compensation of the entropy change by Joule heating is called thermoneutral potential ($E_{cell} \equiv E_{th}$ for $Q = T\Delta S$):

$$Q = T\Delta S$$
$$(E_{th} - E_{rev})zF = T\Delta S$$
$$E_{tn} = \frac{\Delta H}{zF} \tag{9.16}$$

When the cell is working below E_{tn}, it operates in "*endothermal mode*" and, as previously mentioned, heat must be supplied from an external source. On the other hand, when it operates above E_{tn}, in the "*exothermal mode*", Joule heating arises due to the internal resistance of the cell.

From a thermodynamic point of view, it is convenient to operate in the endothermal mode using heat coming from an inexpensive source (e.g., solar concentrators, nuclear waste heat, etc.). However, the thermal management of an electrolyser running the exothermal mode is much simpler.

Figure 9.7 shows the variation of ΔH, ΔG and $T\Delta S$ as a function of temperature for the particular case of water electrolysis. ΔH increases slightly with temperature, nevertheless, an important variation in ΔG, i.e., electrical demand, is driven by the increase of the entropy term $T\Delta S$, i.e., heat demand. Thus, by operating at elevated temperatures, more thermal energy is supplied and the consumption of electrical energy decreases, thereby reducing costs. In devices operating at temperatures above 100°C i.e., with water in gas phase, the potential is reduced due to the higher entropy of water vapour compared to liquid water. See Box 9.6 for the calculation of the thermoneutral potential associated with the particular case of water electrolysis at low and high temperatures.

Increasing the pressure of an electrolysis cell can also be beneficial for several reasons. For one, the product gas is naturally compressed, reducing the energy cost of pressurization for further transport or storage by about 5% in comparison to high efficiency compressors [21]. Furthermore, the reduction in volume of the gases at the surface of the electrodes improves diffusion, leading to better evacuation. In the case of liquid electrolytes, higher pressure also allows the operating temperature to increase above 100°C, substantially reducing voltage losses in the cells [22].

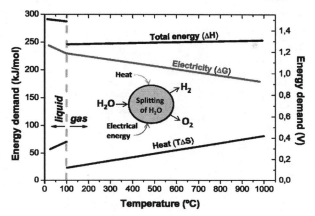

Figure 9.7 Energy required for the electrolysis of water as a function of temperature.

Nevertheless, electrolysis at higher pressures also has some drawbacks. The influence of the pressure on the electrolysers can easily be derived for the reversible potential. Assuming an overall pressure equivalent in both electrodes ($p = P/P^0$), Eq. 9.16 becomes

$$E_{rev} = E^0_{rev} - \frac{RT}{2F} \ln \frac{pH_2O}{pH_2 pO_2^{1/2}} = E^0_{rev} - \frac{RT}{2F} \ln \frac{1}{p^{1/2}}. \qquad (9.17)$$

Therefore, the voltage increases with increasing pressure according to the following equation:

$$\Delta E = (E_{rev} - E^0_{rev}) = -\frac{RT}{2F} \ln \frac{1}{p^{1/2}} \qquad (9.18)$$

According to Eq. 9.7, only small changes in the voltage are expected associated to an increase in the total pressure (see Box 9.6). Therefore, this small inconvenience would be a low price to pay compared to the aforementioned compensations. The most important disadvantages will arise from the difficulties associated with designing the electrolyzing unit to work at higher pressures with severe requirements in terms of strength and air tightness [23].

Box 9.6

Calculate the reversible and thermoneutral potential for the water splitting reaction at standard conditions (25°C, 1 atm) and at 800°C using the enthalpy and entropy values of the following tables:

T (°C)	ΔH (kJ/mol)	ΔS (kJ/mol · K)
25°C, H_2O (l)	285.8	0.165
800°C, H_2O (v)	248.2	0.055

Calculate the change in the open circuit voltage by increasing the operation pressure up to 20 atm.

Solution: The Gibb's free energy can be obtained by operating the changes in enthalpy and entropy:

$$\Delta G = \Delta H - T\Delta S$$

And, therefore, the reversible and thermoneutral potentials

$$E_{rev} = \frac{\Delta G}{zF}$$

$$E_{tn} = \frac{\Delta H}{F}$$

The following table summarizes the results obtained:

9.4T (°C)	E_{rev} (V)	E_{tn} (V)	$E_{tn} - E_{rev}$ (V)
25°C	1.23	1.48	0.25
800°C	0.98	1.29	0.31

Therefore, the minimum voltage to start the electrolysis is lower at higher temperatures ($E_{cell} > E_{rev}$). However, overpotential ($E_{ov} = E_{cell} - E_{rev}$) required to reach the thermoneutral at temperatures. The production rate depends on the current density available at the operating voltage, i.e., on the internal resistance of the cell.

Increasing the pressure increases the open circuit voltage of the cells, according to the following expression:

$$\Delta E = \frac{RT}{2F} \ln \frac{1}{p^{1/2}}$$

The corresponding increments are

$$\Delta E \,(25°C, 20\,atm) = 19\,mV$$
$$\Delta E \,(800°C, 20\,atm) = 69\,mV$$

The beneficial effects of increasing the pressure have to be considered since elevated pressures do not significantly affect the electrolyser performance.

9.2.1.3 Types of electrolysers according to the electrolyte

Electrolysers can be divided into different groups according to the nature of the cell electrolyte. Three systems will be considered in this section, namely, Alkaline Electrolysis Cells (AECs) based on alkaline electrolyte solutions, Proton Exchange Membrane Electrolysis Cells (PEMECs) based on polymeric membranes and SOECs employing oxide electrolytes (see Fig. 9.8 and Table 9.2).

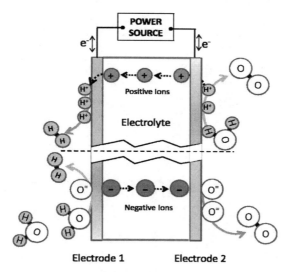

Figure 9.8 Schematic of an individual electrolyser. Two sketches considering positive or negative ions flowing across the electrolyte are presented.

According to Table 9.2, AEC, PEMEC, and SOEC cells operate at different temperatures. While AEC and PEMEC electrolysers typically work at low temperature ($T < 200°C$), SOEC operate at higher temperature ($T > 600°C$). From Fig. 9.9, which shows the ionic conductivity of different electrolytes employed in AECs, PEMECs and SOECs, it is clear that the working temperature is primarily determined by the ionic conductor. This temperature can be established for a reasonable electrolyte conductivity target of 0.1 S/cm (dashed line). Specifically, as shown in Fig. 9.9, the ionic conductivity of Nafion® (used in PEMEC) and PBI-high (used in AEC) reaches 0.1 S/cm at temperatures less than 100 and 200°C, respectively.

Table 9.2 Summary of the different electrolyser types and their particular features

Electrolysers	AEC	PEMEC	SOEC
Electrolyte	Solution of NaOH or KOH	Hydrated polymeric membranes	Ceramic
Charge ion conductor	OH^-	H^+, H_3O^+	O^{2-}
Cathode Reaction (HER)	$2H_2O + 2e^- \rightarrow H_2 + 2OH^-$	$2H^+ + 2e^- \rightarrow H_2$	$H_2O + 2e^- \rightarrow H_2 + O^{2-}$
Anode Reaction (OER)	$2OH^- \rightarrow H_2O + \frac{1}{2}O_2 + 2e^-$	$H_2O \rightarrow \frac{1}{2}O_2 + H_2 + 2e^-$	$O^{2-} \rightarrow \frac{1}{2}O_2 + 2e^-$
Electrodes	Ni, C	C	Ceramic/Cermet
Catalyst	Ni, Fe, Pt	Pt	Electrode material
Interconnector	Metal	Carbon metal	Stainless steel, ceramic
Operating temperatures	40–90°C	20–150°C	600–1000°C

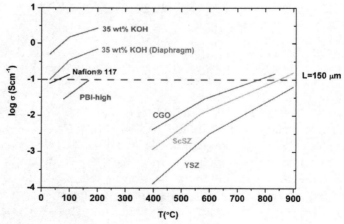

Figure 9.9 Ionic conductivity as a function of temperature for typical electrolytes in AEC (brown), PEMEC (blue) and SOEC (green). The dashed line indicates the target value of area specific resistance for self-supported electrolytes (thickness of 150 μm).

AEC and PEMEC occur at low temperatures for differing reasons. For the case of the PEMEC the ion transported through the electrolyte of a polymeric membrane cell is the proton, which is smaller and more mobile than OH^- or O^{2-} ions. On the other hand, transport of bigger OH^- ions in alkaline electrolysers takes place in the liquid phase, inherently increasing the mobility. Finally, for the case of SOECs, the larger size and double charge of the ion limits its mobility. Only a limited number of materials from certain crystallographic families (fluorites, oxygen-deficient perovskites, aurivillius, pyrochlores and apatites [24]) exhibit reasonable conductivity at elevated temperatures to thermally activate the oxide-ion transport mechanism.

The behaviour of all these electrolytes as a function of temperature is usually described by an Arrhenius law[1]:

$$\sigma = \sigma_0 \exp(-E_a/k_B T), \tag{9.19}$$

where E_a is the activation energy of the transport mechanism, k_B the Boltzmann constant and σ_0 a pre-exponential factor.

9.2.1.4 Non-ideal electrolysers

Thermodynamics gives us information about theoretical limits that can be achieved by electrolyser devices. Nevertheless, deviations from ideal behaviour observed in real devices reduce their performance. Several losses and the corresponding potential change have to be considered since they increase the operating voltage (compared to the reversible potential, see Eq. 9.21). These voltage increments are called "overpotentials" (E_{ov}).

---

[1]In some cases, the Vogel–Tamman–Fulcher (VTF) model is employed at high temperatures. The VTF model involves carrier charge migration assisted by a semi-rigid motion of the counterionic matrix, being useful for polymeric and liquid electrolytes. The conductivity associated to the VTF model is described by the following expression:

$$\sigma = \frac{B}{T^{1/2}} \exp\left(-\frac{E_a}{k_B(T-T_0)}\right), \tag{9.20}$$

where B is a pre-exponential factor, E_a a pseudo-activation energy for the ionic migration and T_0 represents a characteristic temperature.

$$E_{cell} = E_{rev} - E_{ov} \tag{9.21}$$

Figure 9.10 shows a schematic diagram of the most relevant overpotentials affecting V–I curves in electrolysers. The overpotentials arise from several irreversible processes taking place at the different components of the cell. First, ohmic losses appear at the electrolyte, the electrodes, current collectors or any of the interfaces mainly due to charge transport phenomena (ionic/electronic conduction). Non-ohmic contributions are usually associated with the electrodes since (i) a potential is needed to activate the electrode reactions (activation overpotential) and (ii) the gas/liquid diffusion phenomena can control the electrode reaction (concentration overpotentials)[2] [25]. Special attention must be paid to overpotentials introduced by the activation energies of reactions at the electrodes, as this is one of the aspects where the optimization of materials has a critical impact. The oxygen evolution reaction (OER) takes place at the anode while the hydrogen evolution reaction (HER) occurs at the cathode. At a fixed temperature and atmosphere, the rate of the reactions mainly depends on the imposed current per unit area and the microstructure and catalytic properties of the electrode. Research on improving catalytic and microstructural properties of anodes and, especially, cathodes is of fundamental importance for improving performance towards the ideal case.

For some purposes, it is useful to think of an electrolysis cell as a set of elements connected in series. Every one of these constituents introduces, to some extent, an increase in the potential required to drive electrolysis. Usually, at a particular set of operating conditions, one of the components arises as the major restriction to current flow, forcing the introduction of higher voltages to the system. In this situation, the performance of the cell will be limited by this component, no matter how efficient the other elements are. Classical DC measurements are usually not adequate to identify the limiting element, and electrochemical impedance spectroscopy (EIS) is the preferred characterization technique [26]. The analysis of the impedance as a function of the excitation frequency allows one to identify individual contributions

---

[2]Other sources of overpotential can be described for particular types of electrolysers like formation of bubbles in liquid electrolytes (blocking part of the active area) or fuel crossover through the electrolyte in polymeric membranes.

to the total impedance from different elements in an electrolysis cell. Figure 9.11, shows a typical Nyquist plot of an electrolysis cell. A proper deconvolution of the involved phenomena (even those artificially introduced like the inductance) is sketched to show the potential of EIS as an optimization tool for electrolysis cells.

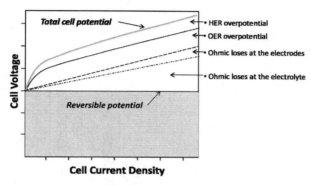

Figure 9.10 Scheme of the typical evolution of the cell potential of an electrolyser as a function of the current density. Overcoming the activation energy of electrodes produces an important increase at low cell currents. Ohmic losses due to the resistance of flow of ions and electrons through all the components are added linearly.

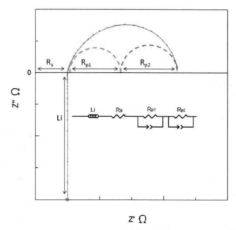

Figure 9.11 Nyquist plot of an EIS measurement of a simple electrolysis cell. Deconvolution of the different phenomena is presented in dashed lines together with the corresponding equivalent circuit.

9.2.1.5 Cell efficiency

The maximum efficiency of an electrolysis process is defined by the maximum electricity-to-chemical conversion, i.e., the enthalpy change of the reaction divided by the work introduced in the system (Gibb's free energy). Using the definition of the thermoneutral and reversible potentials, it is possible to define

$$\eta_{max} = \frac{\Delta H}{\Delta G} = \frac{E_{tn}}{E_{rev}} \quad (9.22)$$

In addition, any overvoltage above the reversible potential will be converted into heat. In this context, it is possible to define a voltaic efficiency:

$$\eta_V = \frac{E_{rev}}{E_{cell}} \quad (9.23)$$

Finally, the typical faradaic efficiency can be defined as the actual amount of gas generated with respect to the theoretical amount of gas expected to be generated according to Faraday's law. It can be expressed as a function of the current density:

$$\eta_F = \frac{j_{cell}}{j_F} \quad (9.24)$$

Therefore, the total efficiency of an electrolyser can be defined as

$$\eta_T = \eta_{max}\eta_V\eta_F = \frac{E_{tn}}{E_{cell}} \frac{j_{cell}}{j_F} \quad (9.25)$$

Assuming 100% faradaic efficiency, the highest efficiencies are obtained by operating close to the reversible potential. Furthermore, the overall efficiency of the electrolysis cell can even exceed unity when operating in the endothermal mode ($E_{rev} < E_{cell} < E_{tn}$). However, it is noted that the endothermal mode is only feasible assuming an external heat source is introduced into the system. Otherwise, the cell would cool below optimal operating temperatures due to the endothermic reactions taking place.

9.2.2 Electrolysis of Water

Nowadays, the most mature technologies for efficient water electrolysis are those based on devices operating at low temperatures such as alkaline and polymer electrolyte membrane electrolysers. Nevertheless, the potentiality of generating H_2 at high production rates via high-temperature electrolysis, principally due to its low overpotentials, is motivating intense research efforts to drive solid oxide cell technology into the market [27] (see Box 9.7). In the following sections, the basics, advantages and main drawbacks of alkaline, polymer electrolyte membrane, and solid oxide electrolysers will be presented in detail.

Box 9.7

Calculate the specific hydrogen production rate (mol H_2/m^2h) and hydrogen production per consumed energy (mol H_2/kWh) for alkaline, polymer membrane and solid oxide electrolysers. The table below lists typical internal resistances at the operating voltage (E_{cell}) and OCVs for the different technologies. The SOEC is coupled to a waste heat source. Calculate the overall efficiencies of the three technologies. Finally, with these overall efficiencies recalculate the days of operation of the wind farm in Box 9.5.

Note: Assume 100% of efficiency in the electricity-to-gas conversion.

	ASR ($\Omega\,cm^2$)	E_{cell} (V)	OCV (V)	T (°C)
AEC	2.5	1.9	1.23	80
PEMEC	0.5	1.7	1.23	80
SOEC	0.15	1.15	0.98	800

Solution: The hydrogen production rate of an electrolyser is not dependent on its efficiency but mainly on its internal resistance. This determines the maximum current to be injected in the cells for electrolysis conversion:

$$j = \frac{(E_{cell} - OCV)}{ASR}$$

By applying Faraday's law (Eq. 9.11), the specific hydrogen production (r) can be obtained in mols unit time and area as follows:

$$r = \frac{j}{2F} = \frac{(E_{cell} - OCV)}{2F \cdot ASR}$$

The production of hydrogen per unit of consumed energy (r') can be obtained dividing r by the power density injected ($p = j(E_{cell} - OCV)$)

On the other hand, assuming the faradaic efficiency $\eta_F = 1$, the overall efficiency of the different technologies can be obtained by employing Eq. 9.25 after calculating the thermoneutral voltage:

$$\eta_T = \eta_{max}\eta_V\eta_F = \frac{E_{tn}}{E_{cell}}$$

Substituting the values in the corresponding equations and taking E_{th} from Box 9.6, it is possible to fill in the following table:

	9.5 J (A/cm^2)	9.6 r (mol H$_2$/m^{2}h)	r' (mol H$_2$/kWh)	E_{tn} (V)	η_T (%)
AEC	0.27	50	27	1.48	78
PEMEC	0.94	175	40	1.48	87
SOEC	1.13	211	110	1.29	112

Due to the lower internal resistance and voltage, cells operating at high temperature (SOEC) are the most efficient and productive technology.

A simple recalculation of the operation days of the wind farm referred to in Box 9.5 yields the following values:

	Days of operation of the wind farm
Ideal electrolyser (Box 9.5)	9.6
AEC	12.3
PEMEC	11.0
SOEC	8.6

9.2.2.1 Low-temperature electrolysers

9.2.2.1.1 Alkaline electrolysis cells

Alkaline water electrolysis is a well-established commercial technology for hydrogen production. High power plants of up to some megawatts (160 MW in Aswan, Egypt [28]) have been successfully running for decades around the world [29–32]. The majority of existing power plants are located in Europe and North America. Germany is one of the most active countries in implementing this technology [33, 34]. Its capacity and reliability is clearly above all other electrolyser technologies.

The alkaline electrolysers are formed by a liquid electrolyte, in which two metal electrodes are immersed (Fig. 9.12). A solution of KOH (20–30%) is the usual choice for the electrolyte. In this configuration, hydroxide ions move from the cathode towards the anode, promoting oxygen evolution. Cost effective electrode materials can be employed, nickel and nickel compounds being the state-of-the-art materials [35–37]. Other non-precious metals like cobalt or iron can be employed for the anode while carbon-supported catalysts are also used in the cathode [38, 39]. Finally, a diaphragm is introduced between the electrodes to separate the generated gas species. The materials employed for the fabrication of the central diaphragm are often nickel oxide or asbestos [40, 41]. The typical operation conditions for these type of electrolysers are summarized in the first column of Table 9.3 [19, 42, 43].

Although alkaline electrolysers are commercially available for hydrogen production, they suffer from several inconveniences and limitations. The main restriction of conventional alkaline electrolysers is their relatively low operating current density. This is mainly due to hydrogen crossover through the liquid electrolyte and the corresponding bubble formation that, apart from producing hydrogen of lower purity, increases the electrolyte resistance and reduces the active area of the electrodes. This fact also prevents the application of high pressures and temperatures to avoid dissolved gas at the electrolyte [25]. However, operating at high temperature and generating pressurized products represent a big advantage since it increases the device efficiency and simplifies the distribution of the fuel, respectively.

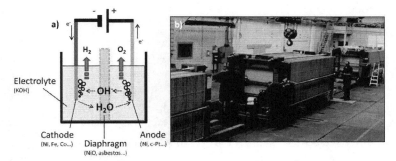

Figure 9.12 (a) Scheme of an alkaline electrolyser and (b) example of a commercial stack [44].

The development in alkaline electrolysis technology is currently focused on system engineering. Some intrinsic difficulties related to the highly corrosive KOH electrolyte or safety issues derived from high oxygen pressure increase the complexity of the balance of plant [45]. Regarding materials optimization, there is still room for improvement. Typical sources of overpotential include overcoming the activation energy of reactions occurring at the electrodes, resistance to ion transport through the electrolyte, diffusion of ions through the gas separation membrane and, probably the most critical one, the formation of gas bubbles on the surface of the electrodes [46].

Table 9.3 Typical operation conditions for different types of electrolysers

	AEC	PEMEC	SOEC
Temperature (°C)	50–80	50–80	750–850
Pressure (bar)	1–30	1–30	1–?
Cell voltage (V)	1.8–2.4	1.7–2.1	1.1–1.5
Current density (A/cm^2)	0.2–0.6	0.5–2.0	0.5–2.0
Power density (W/cm^2)	0.3–1.5	0.8–4.0	0.5–3.0
Efficiency (%)	60–80	70–80	80–90
H$_2$ production rate/commercial stack (Nm^{3}h^{-1})	1–750	0.01–30	—
Commercial stack size (kW)	5–3500	0.5–160	—
Lifetime evaluated (h)	90000	35000	9000 (cell)
Degradation rate (μVh^{-1})	<3	<6	<5

Materials research challenges

One major issue concerning AECs is the degradation and deactivation of the nickel electrodes. The introduction of electrocatalysts like iron has been proposed to enhance and stabilize electrode activity [47]. Another major challenge is to reduce ohmic resistances associated with the electrolyte. In this regard, electrolyte additives have been employed to increase the ion transfer while reducing gas bubble formation at the electrodes (surfactants). Among other strategies for minimizing the major problem of gas bubble formation, circulating the electrolyte or forcing microgravity conditions have demonstrated significant performance improvements [25].

The long life of alkaline electrolysers has been largely demonstrated but only under continuous operation. On the other hand, matching the operation of AECs with fluctuating power sources is inefficient because the liquid electrolyte responds very slowly to load changes and the optimization of electrodes is current dependent. Unfortunately, power-to-gas schemes are based on renewable energies, such as wind turbines, which provide a varying power load. This issue has been partially solved by using DC-to-DC converters, which protect the cells from load oscillations by imposing the optimum current-voltage operation point. Nevertheless, DC-to-DC converters impose an energy loss of around 20%. Therefore, future work on fast-response electrolytes and flexible electrodes represents a big challenge for efficiently coupling alkaline electrolysis to intermittent sources in power-to-fuel systems.

9.2.2.1.2 Proton exchange membrane electrolysis cells

Proton exchange membrane electrolysers are based on PEM fuel cells (Chapter 6 in this book) operating in reverse mode. As shown in Table 9.3, PEMEC operate at the same or higher temperatures than AECs, presenting more efficient and higher hydrogen production rates. This is mainly due to their lower internal resistance (higher current density) associated with the use of a proton conducting polymer membrane instead of a liquid electrolyte and a diaphragm. The PEMEC technology is in a pre-commercial stage, although long term tests have already been carried out (>30,000 h) and small stacks are available under demand.

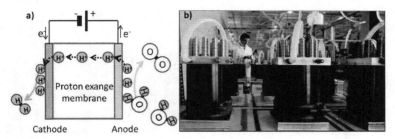

Figure 9.13 (a) Scheme of a polymer water electrolyser and the involved reactions and (b) image of a commercial device.

PEM electrolysers are based on a proton conducting solid polymer membrane working as an electrolyte with two electrodes, an anode and a cathode, attached to both sides (see Fig. 9.13). The protons generated in the anode side reach the cathode through the polymeric membrane to generate hydrogen. Platinum group metals (PGM) are required for catalyzing the reactions at the electrodes thereby increasing the cost of the device.

As previously outlined, these devices present several advantages compared to AEC technology. The most obvious is a non-corrosive solid electrolyte that provides lower resistances and withstands higher operating pressures. The most common membrane material used in these cells is Nafion®, which was developed in the 1960s by DuPont. The membrane thickness ranges from a few tens of microns to several microns, and allows the total resistance of the cell to be reduced to values below 0.5 Ωcm^2. These low internal resistances allow for high current densities and high hydrogen production rates, with low energy consumption. The solid nature of the electrolyte also functions to reduce hydrogen permeation, which leads to high gas purity at atmospheric or moderate pressures. However, increasing the pressure to the upper operating range (>20 Bar) causes degradation of the membrane and has a negative impact on the purity of products due to crossover of the produced gases. The enhanced oxygen production rate leads to a reduction of the anodic hydrogen content, which means that cross permeation is much more important at lower current densities [48]. Therefore, an excessive increase of pressure leads to unacceptable levels of H_2 in O_2, even in balanced pressure configurations. This problem can be partially overcome by an increase of the electrolyte thickness or a reduction of the permeability [49], but, of course, these solutions raise the overpotential of the cell (reduction of affordable current densitie).

It is important to mention here that PEMEC technology offers the possibility of working in a wide range of partial loads as well as quick response times to fluctuating power inputs. This makes this technology a better option than the AEC-based technology for applications in power-to-gas schemes based on intermittent renewable power sources (see Box 9.8 for a simple calculation of the scaling of this technology).

> **Box 9.8**
>
> Given a wind farm overproduction of 20 MW coupled to a PEMEC based on 500 cm² cells operating at a voltage of 1.7 V and a current density of 1 A/cm², (a) how many stacks of 200 cells are needed for converting the electricity surplus, and (b) which is the flow of H_2 that will be produced in total and per stack unit?
>
> *Solution*: The power provided by the wind farm can be easily converted in current by assuming the cell voltage of 1.7 V (all the cells connected in parallel):
>
> $$I_T = P_T/E_{cell} = 20 \text{ MW}/1.7 \text{ V} = 11.76 \text{ MA}$$
>
> $$11760000 \text{ A} \times \frac{1 \text{ cm}^2}{1 \text{ A}} \times \frac{1 \text{ cell}}{500 \text{ cm}^2} \times \frac{1 \text{ stack}}{200 \text{ cells}} = 118 \text{ stacks}$$
>
> Now, it is possible to calculate the total H_2 flow generated by simply applying the Faraday law (assuming a 100% of faradaic efficiency):
>
> $$I_T \times \frac{1 \text{ C/s}}{1 \text{ A}} \times \frac{1 \text{ mol e}^-}{96487 \text{ C}} \times \frac{1 \text{ mol } H_2}{2 \text{ mol e}^-} \times \frac{22.4 \text{ l}}{1 \text{ mol } H_2} = 1365 \text{ l } H_2/s$$
>
> The H_2 flow generated per stack unit is therefore 11.6 lH_2/s (42 Nm³h⁻¹).

Materials research challenges

(1) Polymer Electrolyte Membrane

Commercial devices are assembled using membranes made of a perfluoronitrate sulphonic acid polymer. The state of the art material, due to its exceptional ion conductivity and chemical and thermal stabilities, is Nafion® (Fig. 9.14). Although Nafion® is by far the usual choice, other similar perfluorinated polymers are commercially available on the market: Fumapem® and Fumion® (Fumatech), FlemionR® (Asahi Glass Engineering), XUS® (Dow Chemicals), Aciplex-S® (Asahi Kasei), Aquivion® (Solvay), etc. [50]. The high cost of these fluoronitrated polymers represents a serious restriction to the large-scale commercialization of PEMEC. For this reason, many studies have been carried out to find alternative membranes, mostly driven by the PEMFC field

[51–55]. Also, important efforts are being made to reduce unwanted cross-over through the membranes, mostly at higher pressures.

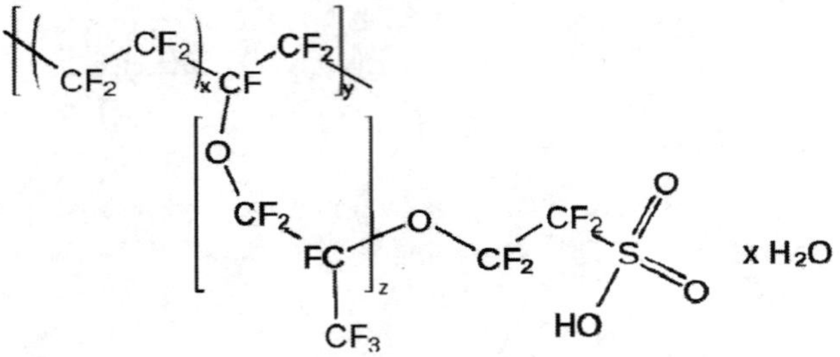

Figure 9.14 Lewis structure of Nafion®.

Increasing temperature in order to improve efficiencies is also a challenge for most electrolyte membranes, as they need to be hydrated. In contrast, Nafion® can maintain a good ionic conductivity if water is replaced by some liquids with a higher boiling point and capable of dissolving ions. Phosphoric acid or ionic liquids can play this role. As a consequence, the temperature can be raised above 100°C with a corresponding increase in efficiency and hydrogen production rate. However, increasing the reliability of high-temperature units is still a challenge.

(2) PEMEC electrodes

The electrodes, particularly the anodes, represent the main source of overpotential and are the most crucial component for PEMEC performance. Up to present, electrode catalysts for oxygen and hydrogen evolution reactions in PEM electrolysers are based on scarce and expensive noble metals of the platinum group. Non-noble metals like Ni or Co cannot be used because they suffer from corrosion in acidic operation environments [56]. Since the first publications in the 1970s, extensive efforts have been devoted to find alternative materials. The aim is to find a good trade-off between electrode activity, stability and cost.

The oxygen evolution reaction (OER) creates a bottleneck for the flow of current in PEMECs due to the irreversibility and the sluggishness of this reaction at the anode. The most active anode materials for this reaction are noble metal oxides. State-of-the-art catalysts are a mixture of IrO_2 and RuO_2. Although pure RuO_2

exhibits the best catalytic performance, it suffers from corrosion at an unacceptable rate and must be mixed with the less efficient but more stable catalyst IrO_2. The first success stabilizing Ir and Ru oxides was achieved using Ti as a support [57]. Nevertheless, in applications where high conductivities are required, Ti suffers from oxidation due to its low electrical conductivity [58]. For this reason, in PEM electrolysis cells where high current densities are demanded, these materials are applied as an ink onto the electrolyte membrane.

As both RuO_2 and IrO_2 are unaffordable for most commercial applications, many studies have been directed towards combining these active catalysts with less expensive oxides. Different fabrication techniques and materials like TiO_2, SnO_2, Nb_2O_5, Sb_2O_5 and Ta_2O_5 have been tested [59]. Although the addition of these materials reduces the catalyst load, very high noble metal proportions are still required to preserve percolation and electron conductivity [60, 61]. Ohmic overpotential usually produced by low electrode conductivity becomes especially important at higher current density operation [62].

The hydrogen evolution reaction (HER), at the cathode side, is less technologically challenging than the OER. Nevertheless, at present, the material choice is still restricted to noble metals and Pt is the standard catalyst for the HER. In order to reduce costs, the same strategy as for the anode has been carried out. That is, Pt loading on the cathode has been reduced to about 0.5 mg/cm^2. This has been achieved with the utilization of carbon scaffolds, where Pt has been introduced in the form of nanoparticles or nanostructured thin films.

Although great reductions of the cost of the components involved in PEMEC technology is in progress, a main drawback compared to AEC technology is the price. The membrane, catalysts, and anode and cathode supports represent almost 80% of the total cost of a PEMEC. Therefore, a clear research goal is to find low-cost alternatives to substitute the state-of-the-art materials.

9.2.2.2 High-temperature electrolysers

9.2.2.2.1 Solid oxide electrolysis cells

Solid oxide electrolysers are the least commercially developed technology at the moment. SOEC is based on the reverse mode of

SOFCs presented in Chapter 6. The main difference compared to AEC or PEMEC is the much higher operating temperature that brings superior energy conversion efficiency and less electricity consumption [63]. The possibility of coupling SOFCs to waste heat sources (see Box 9.9) and their amenability to high pressure operation (above that of AEC or PEMEC due to the ceramic nature of the electrolyte) ensures a cost reduction on hydrogen production and a significant increase of the overall efficiency rendering them very attractive from an economical point of view.

Box 9.9

Given a high-temperature electrolyser with a total power of 100 kW operating in self-sustained conditions at 800°C: (a) Calculate the time to produce 1000 l of H_2; (b) Repeat the calculation if the electrolyser is using Q = 60 KJ/mol of heat produced by a nuclear plant; (d) Determine the overall efficiency of the process.

Solution: To ensure an isothermal mode the system has to operate at the thermoneutral voltage:

$$V_{tn} = \frac{\Delta H}{nF} = 1.48 \text{ V}$$

Assuming cells in parallel,

$$I = \frac{100{,}000 \text{ W}}{1.48 \text{ V}} = 67480 \text{ A}$$

To calculate the process time, Faraday's law can be applied:

$$m = \frac{Vol_{H2}}{22.4 \text{ l}} M$$

$$t = \frac{mnF}{IM} = 128 \text{ s}$$

If the electrolyser is using Q = 60 KJ/mol of heat produced by a nuclear plant, the time to produce the same amount of hydrogen can be calculated as

$$V_w = \frac{\Delta H}{nF} - \frac{Q}{nF} = 1.48 - 0.31 = 1.17 \text{ V}$$

$$I' = \frac{100{,}000 \text{ W}}{1.17 \text{ V}} = 85535 \text{ A}$$

$$t' = \frac{mnF}{I'M} = 101 \text{ s,}$$

which is the total efficiency of the process:

$$\eta_T = \frac{V_{tn}}{V_w} = \frac{t}{t'} = 1.26$$

Efficiencies higher than 100% are possible due to part of the required energy is supplied by the heat source and not supplied by the current density.

A solid oxide cell consists of two electrodes on each side of a dense oxide ion conducting electrolyte (see Fig. 9.15). The anode electrode, typically a mixed ionic conductor like lanthanum strontium cobalt ferrite (LSCF), is where the oxidation reaction takes place producing pure oxygen. At the cathode, which is classically comprised of a Ni-based cermet, the catalytic reduction of the fuel is performed. A ceramic ionic conductor layer is used as the electrolyte. The most common electrolyte being yttria stabilized zirconia. High-temperature sealing and interconnectors become necessary for stacking different cells in high capacity SOEC electrolysers. The typical operation conditions for this type of electrolysers are summarized in the third column of the Table 9.3.

SOECs exhibit low internal resistances in comparison to AECs or PEMECs, which permits a higher current density, i.e., hydrogen production rate, at low operating voltages (see Fig. 9.16). The low values of internal resistance are associated with the solid nature of the electrolyte and electrodes and the high operating temperature, which greatly improves the catalytic activity. However, high operating temperatures usually result in materials degradation issues such as secondary phase formation by reactions occurring at the interfaces or microstructure evolution with time, e.g., agglomeration of metals. Although many advances towards preventing long-term degradation in SOFCs were achieved during the last decades, particularities of the SOEC require further research. In this direction, promising low degradation rates have recently been presented from a 9000 h long test carried out in a SOEC of 45 cm^2 operating at −1 A/cm^2 with an efficiency higher than one [43]. Moreover, Graves et al. [64] claim for a complete elimination of the electrolysis-induced degradation by reversibly

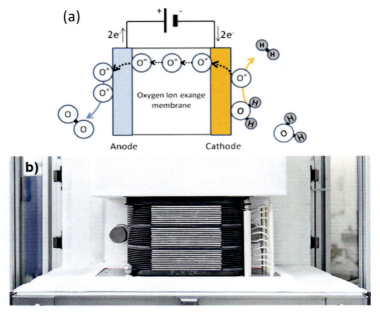

Figure 9.15 (a) Scheme of processes taking place in a solid oxide electrolysis cell. (b) Commercial electrolyser stack from Sunfire.

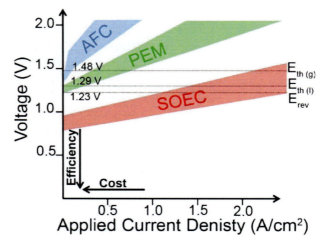

Figure 9.16 Operation range of different types of electrolyser. Adapted from reference 19.

cycling between electrolysis and fuel-cell modes. Reinforcing and extending these recent results to the stack level and for longer times is crucial to improve the reliability of SOEC technology for future implementation in power-to-fuel systems. Its unique capability of electrolyzing CO_2 (or even co-electrolyzing CO_2 and steam together) makes SOEC the only option for a single step generation of syngas or metgas (CO-$2H_2$) for synthetic fuel production (see the next section) [19].

Materials research challenges

Nowadays the state-of-the-art materials for solid electrolyser cells are mainly based on the ones previously developed for SOFCs. However SOEC technologies demand some specific requirements due to the out-of-equilibrium nature of the device, i.e., the operation under forced high current densities, the pure oxygen atmosphere in the anode side and the current-dependent reducing conditions on the cathode.

Oxide-ion conductor electrolyte

Only a few materials fulfil the severe conditions required for use as an electrolyte in SOECs. In particular, it has been especially difficult to identify non-reactive electrode materials that are stable at high operating temperatures in highly reducing and oxidizing atmospheres. In fact, there is still no reliable alternative to the original yttria stabilized zirconia (YSZ) electrolytes employed in SOFCs since the 1930s [65]. Although, more recently other oxide ionic conductors based on scandia/ytterbia doped ZrO_2, doped Ce_2O_3 or doped lanthanum gallates have been proposed as electrolytes for SOEC [63]. Among them, the use of scandia doped zirconia (ScSZ) is being considered in real applications due to its higher ionic conductivity which allows for thicker self-supported electrolytes operating at lower temperatures (T < 800°C). However, ScSZ shows low stability at high operating voltages and low oxygen partial pressures, partially avoidable only with co-doping [66]. Opposite to SOFCs, long-term tests on SOECs show that the degradation of the electrolyte represents an important contribution to the total resistance, being even the most important effect in anode-supported cells [43].

As an alternative to the O^{2-} conducting electrolytes, proton conducting electrolytes have been traditionally studied in SOFCs, e.g., barium zirconates [67]. The main advantage of these materials for SOEC applications is that pure H_2 is produced and the fuel is not diluted with steam. However, although great improvements have taken place in the recent years, they are still problems associated with chemical stability and compatibility that require further research [68].

SOEC electrodes

Typical cathodes for SOEC are cermets based on the YSZ electrolyte for the ceramic part and Ni for the metallic counterpart [63]. The ceramic scaffold inhibits sintering of metal particles at the operation temperature and adjusts the thermal expansion coefficient to avoid mismatch between the electrode and electrolyte layers. Well defined microstructures are critical for avoiding nickel agglomeration, particularly in the high current and high-temperature SOEC application [69]. Alternatives cermets based on nickel and gadolinia doped ceria (CGO) or even Ni-YSZ impregnated with CGO/Rh have also been evaluated and have been shown to exhibit lower polarization resistances due to better catalytic properties [70, 71].

Since the reducing conditions of the cathode side in a SOEC are current dependent, i.e., the amount of hydrogen present depends on the amount of water electrolysed; alternatives to pure metals are required to avoid problems associated with their oxidation/reduction cycles, especially for the injection of low current densities. Few complex metal oxides showing mixed ionic-electronic conduction have been evaluated in SOECs with results comparable to cermets, e.g., $La_{0.35}Sr_{0.65}TiO_3$ [72], $(La_{0.75}Sr_{0.25})_{0.95}Cr_{0.5}Mn_{0.5}O_{3-\delta}$ [73] or $Sr_2Fe_{1.5}Mo_{0.5}O_{6-\delta}$ [74]. However, long-term measurements are still required to present these materials as real alternatives to classical cermets.

The anode has often been described as the limiting component of SOECs due to the accumulation of oxide-ions at the electrolyte-anode interface at high current densities [75–77]. This leads to delamination of the electrode, particularly for those electrodes with insufficient catalytic activity. This seems to be the case for most common materials used as anodes in SOFCs, i.e., $(La_{0.8}Sr_{0.2})_{0.9}MnO_3$ (LSM), also employed in SOEC. In order to achieve high electrode performance LSM-YSZ composite anodes in SOECs are subjected to a current polarization treatment. This treatment improves catalytic activity by promoting the partial reduction of the electrode and altering its defect structure to improve oxygen ion diffusion [78]. Alternatively, oxygen electrodes based on Sr-doped $LaCoO_3$ (LSC), $LaFeO_3$ (LSF) or $LaCoFeO_3$ (LSCF) exhibit good performance without undergoing polarization treatments [79].

Other anode compounds showing excellent catalytic activity at low temperatures have recently been tested in SOECs. Among them, ordered double perovskites such as $LnBaCo_2O_5$ or $Ln_2NiO_{4+\delta}$, which yield excellent results in SOFCs [80, 81], have been tested in SOEC mode showing promising results. Long-term stability under real operation is required before implementation in real devices.

9.2.2.2.2 Molten carbonate electrolyser cells

Molten carbonate electrolysis cells (MCEC) are essentially molten carbonate fuel cells operating in reverse mode, which is a mature and commercialized technology, discussed in Chapter 6. The electrolyte is a molten mixture of alkali metal carbonates comprising a binary mixture of lithium and potassium, or lithium and sodium carbonates. This mixture is retained in a ceramic matrix of $LiAlO_2$. Molten carbonates cells work at temperatures between 600 and 700°C. In this temperature range alkali carbonates form a highly conductive molten salt, with carbonate CO_3^{2-} ions providing ionic conduction. At these high working temperatures, Ni and NiO are adequate to promote the reactions in the cathode and anode, respectively, and noble metals are not required.

One major disadvantage of performing electrolysis with MCECs is that CO_3^{2-} ions transported across the electrolyte reduce H_2O but also produce CO_2, which is mixed with the O_2 at the anode side. It is necessary to separate and recirculate this CO_2 prior to extracting the exhaust gas. Only few studies have been published on the field of the MCEC electrolysis [82]. Kaplan et al. reported the conversion of CO_2 to CO using a cell with a lithium carbonate and lithium oxide molten electrolyte mixture at temperatures of 850°C [83, 84]. Recently, new molten carbonate cells are being developed where the carbonate is also a reactant reducing the amount of CO_2 in the anode exhaust gas [19].

9.2.3 Coelectrolysis of Water and Carbon Dioxide

9.2.3.1 Low-temperature carbon dioxide electrolysis

Although extremely interesting for producing active compounds containing carbon, low-temperature electrolysis of CO_2 has not been studied to the same level of detail as the electrolysis of water. In order to carry out the co-electrolysis of CO_2 in an alkaline electrolyser, CO_2 is dissolved in a liquid electrolyte. In some cases water is substituted by other liquids, such as methanol, that have a higher capacity of dissolving CO_2 [85–87]. A number of products are generated including CO, CH_4, C_2H_4, CO_2H_2 as well as longer chain hydrocarbons and alcohols. The selectivity of cathode materials and conditions for the formation of the desired product is the main concern for current research on the topic [18]. Once again, in this case, the functionalization of materials by means of nanotechnology is playing an important role [88, 89].

Also important efforts are also being carried out to avoid deactivation of cathodes due to carbon deposition [90–92]. Pure metal electrodes suffer from an enhanced accumulation of carbon during operation. Several strategies are used in order to avoid this effect. Using alloys demonstrates a higher durability [19, 93] while a reverse current applied to re-oxidize the deposited carbon partially recovers the original catalytic activity [94, 95].

At present, cell efficiencies and durability must be increased in order to make low-temperature carbon dioxide electrolysis technologies competitive. Although efficiencies up to 80% have been recently reported [96], low catalytic activity, poor stability, reduced product selectivity and limited CO_2 solubility in the electrolyte are still major drawbacks to be solved in the near future [96].

9.2.3.2 High-temperature co-electrolysis of steam and carbon dioxide in SOECs

Besides electrolyzing pure steam at high temperatures, it is also interesting to consider operating SOEC technology under different atmospheres containing CO_2, namely, (i) pure carbon dioxide (CO_2 electrolysis) and (ii) steam and carbon dioxide (CO_2 and H_2O co-electrolysis) (see Fig. 9.17). The main product of co-electrolysis is syngas, which is of great interest in power-to-fuel schemes because it can be used as a precursor to synthesize hydrocarbons (see Section 9.1.3). The overall reaction also includes the production of pure oxygen:

$$H_2O + CO_2 \Leftrightarrow H_2 + CO + O_2 \tag{9.26}$$

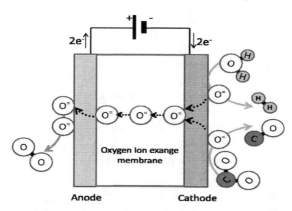

Figure 9.17 Scheme of the processes taking place in a solid oxide electrolyser working in coelectrolysis mode.

Any process involving the electrolysis of carbon dioxide requires greater energy inputs. A simple analysis of the thermodynamics of the system clearly shows the origin of this difference. A higher value of the change of enthalpy of the CO_2 conversion reaction involves a higher thermoneutral potential (see Fig. 9.16). As an example, the thermoneutral voltage at 800°C is 1.29 V for the H_2O conversion reaction while it is 1.47 V for the CO_2 conversion reaction. However, the reversible potential is almost the same for H_2O and CO_2 electrolysis ($\approx$0.9–1 V from 750–850°C) since the Gibbs free energy is similar in the typical range of operation of SOECs (see Fig. 9.18). Therefore, both reactions are compatible and co-electrolysis should be possible above the same minimum potential.

There is still a third reaction competing with the electrolysis of CO_2 and H_2O. This is the reverse water gas shift reaction (RWGSR) that consumes hydrogen and CO_2 to produce water and CO (Eq. 9.27). According to the thermodynamics of the water gas shift reaction (see Fig. 9.18), the equilibrium reaction moves to RWSR above 810–820° and CO is produced [19, 97].

$$H_2 + CO_2 \Leftrightarrow H_2O + CO \tag{9.27}$$

Therefore, co-electrolysis in SOECs is a more complicated process than pure steam or carbon dioxide electrolysis, involving different reactions competing under typical operation conditions. In fact, the co-electrolysis process is not completely understood yet since the controlling mechanism is depending on the kinetics of the multiple reactions, i.e., it is dependent on the catalysts involved. In this sense, it has not yet been shown if the syngas generation by co-electrolysis of steam and CO_2 actually proceeds by simultaneous electrolysis of both CO_2 and H_2O or by predominant steam electrolysis combined to RWSR (hydrogen production by electrolysis that immediately reacts with CO_2 to produce CO) [98]. According to the polarization values obtained in different experiments, the most likely hypothesis is a direct conversion of water into hydrogen and CO production arising from the RWSR.

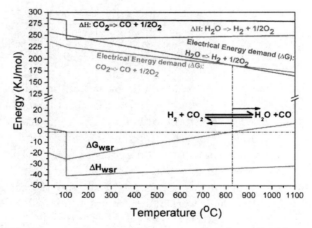

Figure 9.18 Thermodynamics of water and carbon dioxide reduction reactions.

The main advantage of simultaneously electrolysing CO_2 and H_2O is the low polarization resistance (similarly to the case of pure steam electrolysis, which also dominates this co-electrolysis process) and the negligible reduction of CO to carbon. This coke deposition takes place following the Boudouard reaction that is favoured at low temperatures and high concentrations of CO [99], i.e., opposite to high temperature and low CO concentrations typical for solid oxide electrolysis.

Since CO and CO_2 are similarly present in SOFCs, and with carbon monoxide being a typical fuel, the same materials employed in SOFCs and SOECs have been used for the first prototypes of co-electrolysers of solid oxide cells [98]. Long-term tests are still pending to confirm the suitability of this technology for real implementation in power-to-fuel schemes, although first results are promising with <10% of degradation after 1000 h of operation [100].

9.2.3.3 Polygeneration in solid oxide fuel cells

Syngas and electricity coproduction from high-temperature fuel cells is a novel concept that aims to maximize the efficiency of these devices by polygeneration [101, 102]. This polygeneration is based on designing the SOFC for generating high internal heat and high electrical power, i.e., able to operate at low voltages (high currents). At this operation point the electrical efficiency is reduced but the heat generated can be employed for internal reforming of the fuel, for instance, natural gas (see Fig. 9.19). In other words, the solid oxide fuel cell behaves as a fuel reformer able to convert the excess heat into chemical energy, for instance, CO and H_2 from CH_4. This type of devices can be called solid oxide fuel cell reformers (SOFCR).

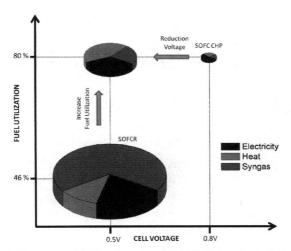

Figure 9.19 Energy polygenerated in SOFC/SOFCR as a function of the fuel utilization and operation voltage conditions. Modified from reference [101].

The net electrical efficiency of a SOFCR coupled, for instance, to a PEMFC can increase above 70% [101, 103]. In addition, the flexibility of this type of devices allows its incorporation in multiple scenarios where chemical energy storage is a priority.

9.3 Artificial Photosynthesis

Artificial photosynthesis is a research field devoted to mimicking the natural process of photosynthesis. In general, this term commonly refers to any scheme for capturing and storing energy from sunlight in the chemical bonds of a fuel. This activity is mainly clustered into two groups: water splitting and carbon dioxide reduction. Both lines are intimately related but differ in the final targeted product: (i) water splitting aims to produce hydrogen for direct use or subsequent combination with CO/CO_2 in Fischer–Tropsch/Sabatier reactions to generate synthetic fuels (see Section 9.1.3), while (ii) carbon dioxide reduction pursues direct generation of carbon-based fuels.

In this part of the chapter, an overview of the different technologies developed during the last decades in the field of artificial photosynthesis is presented. After some general concepts and definitions, the section includes two parts devoted to water splitting systems for hydrogen generation and novel carbon dioxide reduction approaches.

9.3.1 General Aspects of Artificial Photosynthesis

Artificial photosynthesis can be considered as any reaction in which water or carbon dioxide is transformed into a fuel (hydrogen or carbon based) with solar light as the only energy source. The typical schemes adopted for artificial photosynthesis are based on

(i) homogeneous systems in which the photocatalyst and the reactants are present in the same phase, usually liquid.

(ii) heterogeneous systems in which the components are in two separate phases, being usually the photocatalyst in solid phase and the reactants in either liquid or gas phase.

Both schemes can work using direct photocatalysis and photoelectrochemical cells (PEC) as shown in Fig. 9.20. While direct photocatalysis simply requires an illuminated photocatalyst in solution, in a PEC cell both the oxidation and reduction semireactions are decoupled using two different electrodes. For PEC cells, an external bias can be applied (if needed) to optimize the solar-to-fuel conversion efficiency of the materials involved.

 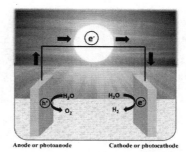

Figure 9.20   Scheme of the two different approaches for artificial photosynthesis: direct photocatalysis (left) and a photoelectrochemical cell (right).

The critical component in any of these photocatalyst systems is the light absorbing material, which plays an analogous role to leaves in plants. This material is, in general, a semiconductor with a band gap suitable to absorb light in the visible. The catalysts supported on the absorbing material are typically metal and metal oxides such as PGMs, Ni or cobalt oxides.

Figure 9.21 shows the standard AM1.5[3] solar spectral irradiance distribution with different reference semiconductors (TiO_2, CdS, Fe_2O_3, GaAs and Si) represented by coloured vertical lines positioned at the wavelength corresponding to their energy band gap. The maximum irradiance they can absorb and convert into current density and the corresponding hydrogen production rate is represented by the intersection of these vertical lines with the solid black line and is referred to on the right axes. This representation allows us to know the ideal performance expected for a given semiconductor. The efficiency in the solar-to-electricity conversion as well as the incident solar irradiance determines the

---

[3]AM1.5 is a standard definition of the solar spectral irradiance distribution. It integrates the component of solar irradiance and hemispherical solar irradiance, consisting of both the diffuse and direct components, that is incident on a sun-facing, 37-tilted surface, considering an air mass of 1.5 and a turbidity of 0.27. These parameters were chosen for this standard because they are representative of average conditions in the 48 contiguous states of the United States.

illuminated surface area required to electrolyse a certain amount of water. Box 9.10 includes a simple calculation showing typical expected land occupation for this technology (compared to a power-to-gas scheme).

Box 9.10

Given a photoelectrochemical module to convert solar light to hydrogen with a maximum photocurrent density of $j = 8$ mA·cm^{-2} and considering a solar irradiance of 2000 h per year and a packing factor[4] of 80%, determine the number of hectares to produce 1.43×10^5 kg of hydrogen in one week (equivalent to the reservoir in Boxes 9.1 and 9.2). Compare this value with the required area to install the wind farm of the Box 9.5 (considering a typical footprint of 0.1 ha for a wind mill of 1 MW).

Solution: The first step is converting the production in mA·cm^{-2} to kg of hydrogen per hectare (ha) and year. To do this, the Faraday constant, which accounts for the electric charge in Coulombs per mole of electrons involved in the reaction, will be employed:

$$\frac{8 \text{ mA}}{\text{cm}^2} \cdot \frac{1 \text{ A}}{10^3 \text{ mA}} \cdot \frac{C}{1 \text{ A} \cdot \text{s}} \cdot \frac{1 \text{ mol e}^-}{96485} \cdot \frac{1 \text{ mol H}_2}{2 \text{ mol e}^-} \cdot \frac{2 \text{g H}^2}{1 \text{ mol H}^2} \cdot \frac{1 \text{ kg}}{10^3 \text{ g}} \cdot$$

$$\frac{3600 \text{ s}}{1 \text{ h}} \cdot \frac{10^8 \text{ cm}^2}{1 \text{ ha}} = 29.84 \frac{\text{kg H}_2}{\text{h} \cdot \text{ha}}$$

By considering an average solar irradiance of 2000 h per year, hydrogen production rate would be

$$29.84 \frac{\text{kg H}_2}{\text{h} \cdot \text{ha}} \cdot \frac{2000 \text{ h}}{\text{year}} = 59690 \frac{\text{kg H}_2}{\text{year} \cdot \text{ha}} = 1147 \frac{\text{kg Hl}}{\text{week} \cdot \text{ha}}.$$

Finally, for producing 1.43×10^5 kg of hydrogen in one week with a packing factor of 80%, the surface required would be

$$S_{solar} = \frac{1.43 \cdot 10^{5 \text{ kg H}_2}/\text{week}}{1147^{\text{kg H}_2}/\text{week} \cdot \text{ha}} \cdot 0.8 = 100 \text{ ha}.$$

On the other hand, the wind farm proposed in the Box 9.5 requires

$$S_{wind} = 300 \text{ MW} \cdot \frac{0.1 \text{ ha}}{1 \text{ MW}} = 30 \text{ ha}.$$

---

[4] The packing factor or density of solar cells in a PV module refers to the area of the module that is covered with solar cells compared to that which is blank

Only considering the 60 MW employed in the electrolyser this values becomes

$$S_{wind} = 60 \text{ MW} \cdot \frac{0.1 \text{ ha}}{1 \text{ MW}} = 6 \text{ ha}$$

Just for comparison, the reservoir defined in Box 9.1 takes up 352 ha.

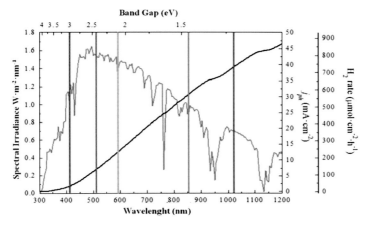

Figure 9.21 AM1.5 Solar Spectrum with different reference semiconductors plotted as vertical lines (TiO$_2$-blue, CdS-olive, Fe$_2$O$_3$-green, GaAs-orange and Si-red). The black solid line corresponds to the theoretical photocurrent density and the associated hydrogen production rate (right axes).

9.3.2 Water Splitting

9.3.2.1 Photolysis of water

Water photolysis systems includes those systems in which water redox reactions take place on a single particle of the catalytic material suspended in an aqueous solution. Under irradiation of photons with energy equivalent to or greater than the band gap of the semiconductor photocatalyst, electrons in the valence band are excited into the conduction band, leaving holes in the valence band. These photogenerated electrons and holes cause reduction and oxidation reactions, respectively (Fig. 9.22).

Photocatalytic semiconductor materials must fulfil certain requirements in order to drive water splitting in a photolysis system: (i) A semiconductor band gap greater than 1.8 eV is needed (thermodynamic potential of 1.23 V plus Ohmic components and overpotentials required); (ii) conduction band and valence band edges must bracket the two redox potentials levels corresponding to the Hydrogen Evolution Reaction (HER) and the Oxygen Evolution Reaction (OER), respectively. This is equivalent to stating that photogenerated electrons have sufficient energy to reduce protons and photogenerated holes have sufficient energy to oxidize water (Fig. 9.23).

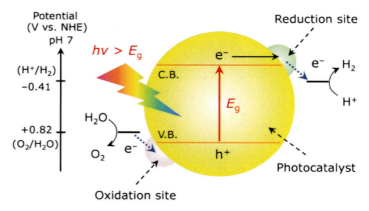

Figure 9.22 Single band gap photolysis system based on a photoabsorber and a catalyst. Extracted from ref. [104].

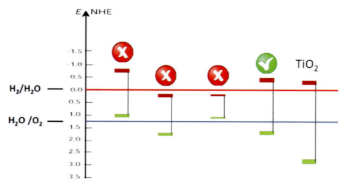

Figure 9.23 Band energy diagram with the position of the conduction and valence band versus water redox pair. The hypothetic materials represented by a cross would not be able to produce one or both half reactions (OER or HER), while the one with a tick has the appropriate band positions for both reactions.

Apart from band gap energy considerations, the ability of the semiconductor material to exhibit high stability against photocorrosion in water is obviously necessary. This requirement limits the number of suitable materials for photocatalysis to a collection of few oxide semiconductors. Among any other choice, titanium dioxide has been traditionally the preferred option, despite its large band gap (3.2 eV) that limits the maximum possible H_2 evolution rate (see Fig. 9.21). Other materials, such as $NaTaO_3$, $SrTiO_3$, TaON, WO_3, Bi_2WO_6 or $BiVO_4$, have also been explored but usually exhibit several disadvantages [105].

Since the number of semiconductor materials that fulfil the above-mentioned material requirements is limited, other system configurations have been proposed to avoid or circumvent some of these constraints. For example, the so-called "Z-Scheme" relies on the photolysis of water in a two-step photooxidation process using two different semiconductor powders and a reversible donor/acceptor pair, called a shuttle redox mediator, which is usually based on organic or inorganic molecules.

A schematic illustration of this Z-scheme water splitting is shown in Fig. 9.24. The reduction of water promoted by conduction band electrons and the oxidation of an electron donor (D) by valence band holes yield the corresponding electron acceptor (A). This can be expressed as follows:

$$2H^+ + 2e^- \xrightarrow{yields} H_2 \text{ (photoreduction)}$$

$$D + nh^+ \xrightarrow{yields} A \text{ (photooxidation)}$$

On the other hand, the reaction of O_2 evolution is

$$H_2O + 4h^+ \xrightarrow{yields} O_2 + 4H^+ \text{ (photooxidation)}$$

$$A + ne^- \xrightarrow{yields} D \text{ (photoreduction)},$$

where the electron acceptor generated by the paired H_2 evolution photocatalyst is converted to its reduced form (D), and water oxidation occurs with the valence band holes. Thus, a cycle of redox pairs (D and A) occurs, and the water-splitting reaction is achieved. The most employed redox couples are Fe^{3+}/Fe^{2+} and

IO_3^-/I^-, both of which have unique features affecting the efficiency of Z-scheme water splitting [104].

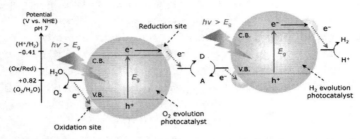

Figure 9.24 "Z-Scheme" configuration of dual band gap photolysis system from ref. 115.

Therefore, in this system, two different photocatalysts are combined using an appropriate redox shuttle mediator, allowing visible light to be utilized more efficiently than in the conventional one-step water splitting system. This is because the energy required to drive each photocatalyst is reduced. It is also possible to apply a photocatalyst that has either water reduction or oxidation potential to one side of the system. For example, WO_3, that does not have the ability to reduce H^+, is capable of producing O_2 from an aqueous solution containing appropriate electron acceptors under visible light and acts as an effective building block for O_2 evolution in Z-scheme water splitting.

9.3.2.1.1 Materials research challenges

Concerning the energetic requirements of the photocatalyst, photo-exited electrons can be utilized efficiently if the redox reduction potential of the reaction is lower than the conductance band of the semiconductor. Additionally, the energy level of the valence band must allow the transference of photogenerated holes to split water into oxygen. Semiconductors like SiC, GaP, CdS, $BiVO_4$, Cu_2O, ZnO and TiO_2 accomplish this requirement. Among them, TiO_2 is one of the most extensively investigated due to its chemical stability, low cost and relatively high lifetime and diffusion length of photogenerated carriers, although its optical absorption is limited to the UV part of the solar spectrum. A number of strategies have been devoted to increase the photoactivity of titania by reducing its recombination rate as to extend its light absorption beyond the UV region. Strategies employed to modify titania can be summarized as follows:

(i) Doping TiO_2 with metals/metal ions (Pt, Pd, Rh, Cu, Ag, Fe, etc.). Metal co-catalyst can act as electron traps and facilitate charge carrier separation, and can induce visible light absorption by introducing impurity states within the band gap of the semiconductor or via surface plasmon resonances.

(ii) Doping TiO_2 with anions such as N, S or F, narrows its band gap by combining p states of these dopant atoms with O 2p states in the valence band of TiO_2 to create impurity states above the valence band.

(iii) Coupling TiO_2 with a narrow band semiconductor, such as Cu_2O, to promote the double effects of extending light absorption to the visible spectral region and enhancing charge carrier separation.

(iv) Sensitization of TiO_2 with light harvesting dyes (phtalocyanines, porphyrins) that can transfer photogenerated electrons directly to the conduction band of titania. However, major concerns associated with this approach are the chemical stability and regenerability of the dyes.

Furthermore, some materials usually employed as photoelectrodes in PECs can be used in powder form for photocatalysis, i.e., $SrTiO_3$ and $KTaO_3$ [106, 107]. Photoelectrodes with perovskite structure can split water even without an external bias because of their high conduction band levels. Domen and co-workers have reported that NiO-loaded $SrTiO_3$ powder can decompose pure water into H_2 and O_2 [108]. The NiO co-catalyst for H_2 evolution is usually activated by H_2 reduction and subsequent O_2 oxidation to form a NiO/Ni double layer structure that is convenient for electron migration from a photocatalyst substrate to a co-catalyst [109]. Rh is also a suitable co-catalyst for the $SrTiO_3$ [110] photocatalyst. TiO_2 and $SrTiO_3$ photocatalysts are also active for reduction of NO_3 using water as an electron donor [111].

The layered perovskite with formula $K_2La_2Ti_3O_{10}$ represents a unique photocatalyst system. H_2 evolution takes place on a pretreated NiOx co-catalyst while O_2 evolves at their hydrated interlayer. Many titanate, niobate and tantalate photocatalysts with layered perovskite structure have been reported after the $K_2La_2Ti_3O_{10}$ parent composition was first discovered. Similar structures like $Sr_3Ti_2O_7$ and $Sr_4Ti_3O_{10}$ with perovskite slabs of $SrTiO_3$ or $Na_2Ti_6O_{13}$ and $BaTi_4O_9$ with tunnel structure are also unique titanate photocatalysts. $Gd_2Ti_2O_7$ and $Y_2Ti_2O_7$ with pyrochlore structure are also active. ZrO_2 is active even without co-catalyst because of its high conduction band level. This photocatalyst is also active for CO_2 reduction to CO accompanied with O_2 evolution by oxidation of water without any sacrificial reagents [112].

9.3.2.2 Photoelectrochemical water spltting

Unlike photolysis of water, photoelectrochemical water splitting systems are those devices in which the two half reactions, oxidation and reduction of water, take place separately at two different electrodes, namely the anode and cathode, respectively. Both electrodes are electrically connected and the circuit is closed through an electrolyte.

In this case, one or both electrodes may comprise a photoactive material. These materials must also fulfil the requirements mentioned in the previous section, with some exceptions regarding the band gap and band edge positions, which will be discussed later.

For simplicity, it is possible to first consider a photoelectrochemical system with only one photoactive electrode (anode or cathode). When a semiconductor is immersed in a redox electrolyte, the electrochemical potential (Fermi level for a semiconductor) is dissimilar across the interface between the semiconductor and the electrolyte. In order to reach equilibrium, the semiconductor Fermi level (E_F) and the redox Fermi level (E_{Fredox}) should be balanced. This equilibrium generates a flow of charge from one phase to the other. After contacting, the net result is a "built-in" voltage (V_{OC}) within the semiconductor phase and a "band bending" close to the surface. A sketch of the band diagrams before and after the contact of the two phases is illustrated in Fig. 9.25. Figure 9.25a shows an n-type (photoanode) while Fig. 9.25b represents a p-type (photocathode) semiconductor.

Figure 9.26 shows the scheme of a photoelectrochemical cell with an n-type photoanode and a metal counter electrode in a water-based electrolyte. When a photoanode or photocathode is illuminated with the appropriate light, an electron-hole pair is created. For a photoanode, electrons are drifted towards the back contact by the "built-in" voltage, while holes are driven to the semiconductor surface. These holes are now transferred to the electrolyte and oxidize water. On the other hand, electrons flow across the circuit to the cathode (typically a metal catalyst) and reduce water. The complementary mechanism applies for photocathodes.

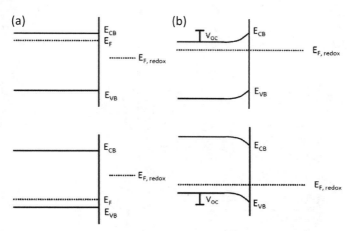

Figure 9.25 Band bending diagram of an n-type semiconductor (top) and a p-type semiconductor (bottom) before (a) and after (b) equilibrium with the electrolyte.

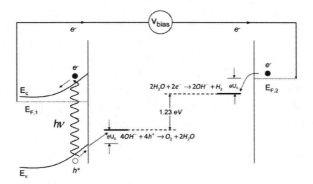

Figure 9.26 Scheme of a photoelectrochemical cell with an n-type photoanode and a metal counter electrode.

The region in which an electric field or a band bending is developed is called the "depletion region". This region is particularly important in a photoelectrochemical system because only those electron-hole pairs generated within this region are going to be efficiently separated. Otherwise, bulk recombination would be dominant.

Similarly to the photolysis system, photoelectrochemical devices can be classified into two groups: (i) single band gap photoelectrodes (previously described) and (ii) dual band gap p-n configuration with an n-type and a p-type photoelectrode connected in series as can be seen in Fig. 9.27.

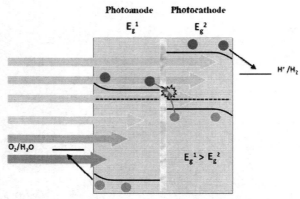

Figure 9.27 Scheme of a dual tandem cell consisting of a photoanode (n-type) and a photocathode (n-type) with different band gaps.

A dual band gap configuration is particularly convenient for obtaining efficient water splitting devices using currently available semiconductor materials, due to the ability to explore various combinations of smaller band gap semiconductor materials that have complementary absorption and stability characteristics. A dual band gap photoanode/cathode configuration also allows for a higher obtainable photovoltage, while partitioning the water splitting half-reactions between two semiconductor/liquid interfaces.

9.3.2.2.1 Materials research challenges

As for the case of photolysis, titanium dioxide was the first material demonstrating water splitting in a photoelectrochemical system [113]. However, its practical application has been limited by its low efficiency and wide band gap, which requires ultraviolet

(UV) radiation as the excitation source. To overcome these limitations many authors have been exploring different approaches from doping with metals [114] and non-metals [115] to loading with noble metals [116]. Other similar materials such as ZnO [117] have been reported to perform photolysis at similar rates as TiO_2 but with poor stability.

Metal oxides with suitable and narrower band gaps have also been studied extensively. Among them, Fe_2O_3 [118], WO_3 [119] and $BiVO_4$ [120] demonstrate the best photoelectrochemical features despite requiring an external bias to assist the water splitting process due to the low reduction power of their conduction band.

Recently, p-type materials have attracted the attention of many researchers regardless their poor photostability. Passivation strategies have been developed to overcome stability issues. Thin layers of stable metal oxides like TiO_2 [121], SiO_2 [122] and Al_2O_3 were used to extend the lifetime of p-Si and CuO [123], among others.

9.3.3 Photoreduction of Carbon Dioxide

As previously mentioned, photocatalytic processes can be utilized to convert solar energy into chemical energy, not only in the form of hydrogen production via water splitting, but also to generate hydrocarbons. That is hydrocarbon fuels can be produced through the photoreduction of carbon dioxide with water, like plants do, in a process which is generally known as "artificial photosynthesis".

Similarly to water splitting, the yield rates of the products depend on the lifetime of photogenerated carriers and on the positions of the band edges (conduction and valence bands) with respect to the redox potentials of the chemical pair of reactants and products. Additionally, once photoinduced electrons move towards the outer surface of the semiconductor, they can be trapped by the adsorbed species. This electron transfer will be more effective if pre-adsorbed species (e.g., CO_2 and H_2O) already exist at the semiconductor surface.

Carbon dioxide is a linear and stable molecule difficult to reduce to other chemicals at low operating temperatures. With a redox potential much higher (Eq. 9.28) than that required for reducing water to H_2 (Eq. 9.29), it is supposed to be more favourable to reduce H_2O than CO_2. Some of the reactions that occur on the surface in aqueous solution at pH 7.0 at 25°C, versus normal hydrogen electrode (NHE) are given below:

$$CO_2 + 1e^- \rightarrow CO_2^- \quad E = -1.90 \text{ V vs. NHE} \tag{9.28}$$

$$2H^+ + 2e^- \rightarrow H_2 \quad E = -0.41 \text{ V vs. NHE} \tag{9.29}$$

$$CO_2 + 2H^+ + 2e^- \rightarrow CO + 2H_2O \quad E = -0.52 \text{ V vs. NHE} \tag{9.30}$$

$$CO_2 + H^+ + 2e^- \rightarrow HCOO^- \quad E = -0.43 \text{ V vs. NHE} \tag{9.31}$$

$$CO_2 + 6H^+ + 6e^- \rightarrow CH_3OH + H_2O \quad E = -0.38 \text{ V vs. NHE} \tag{9.32}$$

$$CO_2 + 8H^+ + 8e^- \rightarrow CH_4 + 2H_2O \quad E = -0.24 \text{ V vs. NHE} \tag{9.33}$$

From the above scheme it is clear that CO_2 photoreduction is not a single step reaction and involves different multi-electronic processes. Pre-adsorbed species (e.g., CO_2 and H_2O) are playing an important role in the competition between water splitting and CO_2 reduction because once the reduction of carbon dioxide is initiated by the electron transfer from the valence band of the semiconductor to the adsorbate, CO_2 is reduced irreversibly. From the above reactions, it can also be inferred that the most probable products from the photoreduction of carbon dioxide are methanol (Eq. 9.32) and methane (Eq. 9.33).

Since the photocatalytic processes highly depend on (i) light absorption by the photocatalyst, (ii) charge separation, (iii) photocarrier trapping at the surface and (iv) surface reaction kinetics; photoreactor design is a key component determining the efficiency of the photocatalytic system, especially for carbon dioxide photoreduction when the CO_2 has to be externally introduced.

The most common types of photoreactors under investigation for carbon dioxide photoreduction are slurry reactors, fixed bed reactors, annular/bubble flow reactors and surface coated reactors.

In 1979, Inoue et al. [124] reported for the first time the photoreduction of carbon dioxide using a slurry reactor, and subsequent works also used this type of reactor. A batch annular reactor is shown in Fig. 9.28a. During the reaction, the vessel is irradiated by an inner lamp and the catalyst is maintained in suspension in water by a magnetic stirrer. CO_2 is bubbled through the water to assure a strong interaction between reactants. However, the main drawbacks are the limited solubility of CO_2 in water (0.033 M) and the need for separation processes (filtration, flocculation) to recover the catalyst. To overcome these limitations, the fixed bed reactor is preferred. In a fixed bed reactor (Fig. 9.28b–d), the catalyst is coated or anchored in the reactor walls, optical fibres or monoliths, for efficient light absorption and the reactants are usually CO_2 and H_2O vapour. Fixed-bed configurations also give lower pressure drops which enable the system to be operated under reduced cost and catalyst stability.

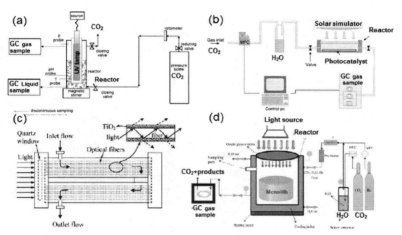

Figure 9.28 Reactor types: (a) batch annular reactor with a suspended catalyst [125], (b) fixed bed reactor [126], (c) optical fibre reactor [127], and (d) monolithic reactor [128].

The direct photoreduction of CO_2 using solar light as energy source has a very low solar-to-chemical conversion efficiency (~0.01%) even considering significant improvements recently achieved. This low efficiency in CO_2 reduction could be attributed to the fact that the hydrocarbons and alcohols formed during the photoreduction can be reoxidized acting as a hole scavenger blocking the evolution of oxygen which requires a higher overpotential, being the whole reaction self-limited. The physical separation of the water photodissociation from the CO_2 reduction stages, in a similar way to the photoelectrochemical (PEC) approach, could instead overcome some of these limitations, as presented by G. Centi and co-workers [10] (Fig. 9.29).

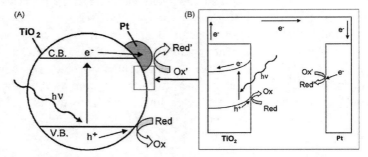

Figure 9.29 Schematic representations of (a) the Pt-loaded TiO_2 (Pt/TiO_2) particle system and (b) the "short-circuited" photoelectrochemical cell (PEC). Source: ref. [10].

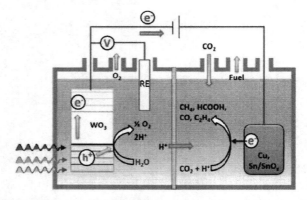

Figure 9.30 Schematic representation of a photoelectrochemical cell using a WO_3 photoanode and Cu or Sn/SnO_x cathode for the CO_2 reduction to fuels using solar light. Extracted from ref. [129].

Using this approach, it is possible to enhance the overall efficiency by using an external bias to favour the charge carrier separation as well as to incorporate a membrane to separate both half-reactions (oxidation and reduction). The challenge here is to optimize both components of the PEC cells in order to achieve efficiencies that are competitive with a photovoltaic cell coupled to an electrolyser (see Box 9.11). On one hand, the optimization of the solar-to-electricity conversion of the photoanode to provide enough energy for the electroreduction in the cathode side. On the other hand, the optimization of the electrocatalytic carbon dioxide reduction process by reducing the overpotential and increasing the faradaic efficiency. Besides this, the introduction of an external bias extends the range of semiconducting materials to be explored, like WO_3 [129] (Fig. 9.30) which can harvest more photons from the solar spectra than classical TiO_2 photocatalysts.

Box 9.11

Consider a power generation system consisting of a photovoltaic panel (η_{PV} = 0.2) coupled to a commercial electrolyser ($\eta_{Electrolyser}$ = 0.7) and a DC-DC converter (η_{DC-DC} = 0.9) to regulate the final signal. Try to design an ideal PEC with the same overall efficiency ($\eta_{PEC} = E_{rev} \cdot j_{ph}/I_{Irradiance}$), i.e., determine the band gap of the semiconductor material to implement in the PEC (use Fig. 9.21). Consider a typical irradiance of 100 mW·cm^{-2}.

Solution: The first step is to determine the overall efficiency of the photovoltaic system:

$$\eta_{total} = \eta_{PV} \cdot \eta_{Electrolyser} \cdot \eta_{DC-DC} = 0.126$$

The theoretical efficiency of a photoelectrochemical cell is naturally given by

$$\eta_{PEC} = \frac{E_{rev}^{H_2O}(V) \cdot j_{ph}(mA \cdot cm^{-2})}{I_{Irradiance}(mW \cdot cm^{-2})}$$

$$j_{ph} = \frac{0.126 \cdot 100 \,(mW \cdot cm^{-2})}{1.23 \,V} = 10.2 \,mA \cdot cm^{-2}$$

According to Fig. 9.21, a photocurrent of 10 mA·cm^{-2} can be achieved by choosing a semiconductor material with a band gap lower than 2.3 eV.

9.4 Concluding Remarks

The only large-scale storage technology commercially available is the pumped-storage hydroelectricity. However, the increasing demand of additional storage capacity associated to intermittent renewable energy schemes is promoting the development of unconventional storage technologies. Among other alternatives, we analysed in this chapter different chemical energy storage approaches based on efficiently converting energy from renewable sources into chemicals.

The increasing interest in chemical energy storage schemes like power-to-gas or power-to-liquid is mainly due to their capability of storing and transporting large amounts of electricity, obtained from renewable energies, over long distances. However, probably the major advantage for their mid-term implementation is the suitability of these technologies to smartly couple the two major energy infrastructures of our modern society, i.e., gas and electricity networks, while covering the necessities of the mobile sector.

The possible near-future implementation of these schemes would bring together a revival of efficient gas-to-power converters such as fuel cells. In particular, high-temperature fuel cell systems able to operate using a variety of fuels, including hydrocarbons, should be favoured in the short-term. Moreover, special efforts should be devoted to improve the performance and durability of reversible cells (able to operate as power generators and electrolysers) since it would represent a major reduction of initial investment costs of any power-to-fuel-to-power system.

In parallel to the commercialization and implementation of these novel storage schemes, it would be necessary to facilitate the deployment of new emerging technologies with excellent long-term prospects such as artificial photosynthesis or direct conversion of CO_2 and H_2O by high-temperature electrolysis. This, bearing in mind that support to CO_2 capture and storage is required to reach the desired totally sustainable energy production/consumption cycle.

References

1. Navigant Research, Energy Storage Tracker 3Q13, http://www.navigantresearch.com.
2. International Energy Agency, International Energy Outlook 2013, DOE, August 2013, www.eia.gov/ieo.
3. Dötsch, C. (2007). Electrical energy storage from 100 kW: State of the art technologies, fields of use. *2nd International Renewable Energy Storage Conference*, Bonn/Germany, 22 November 2007, Electric Power Research Institute: Electric Energy Storage Technology Options White Paper, 2010.

4. Electric Energy Storage Technology Options: A White Paper Primer on Applications, Costs, and Benefits. EPRI, Palo Alto, CA, 2010. 1020676.
5. Arce, G. L., A. F., Carvalho, Jr, J. A., Nascimento, L. F. C. (2014). A time series sequestration and storage model of atmospheric carbon dioxide, *Ecolog. Model.*, **272**, 59–67.
6. The Fischer Tropsch Archive, http://www.fischer-tropsch.org/.
7. Fechete, I., Y. Wang, J. C. Védrine (2012). The past, present and future of heterogeneous catalysis, *Catal. Today*, **189**(1), 2–27.
8. Mosayebi, A., R. Abedini (2013). Partial oxidation of butane to syngas using nano-structure Ni/zeolite catalysts, *J. Ind. Eng. Chem.*, doi:10.1016/j.jiec.2013.07.044.
9. Gallucci, F., E. Fernandez, P. Corengia, M. van Sint Annaland (2013). Recent advances on membranes and membrane reactors for hydrogen production, *Chem. Eng. Sci.*, **92**, 40–66.
10. Centi, G., S. Perathoner (2009). Opportunities and prospects in the chemical recycling of carbon dioxide to fuels, *Catal Today*, **148**(3–4), 191–205.
11. Zhan, Z., W. Kobsiriphat, J. R. Wilson, M. Pillai, I. Kim, S. A. Barnett (2009). Syngas production by coelectrolysis of CO_2/H_2O: The basis for a renewable energy cycle, *Energ. Fuel*, **23**, 3089.
12. Abelló, S., C. Berrueco, D. Montané (2013). High-loaded nickel–alumina catalyst for direct CO_2 hydrogenation into synthetic natural gas (SNG), *Fuel*, **113**, 598–609.
13. Tada, S., T. Shimizu, H. Kameyama, T. Haneda, R. Kikuchi (2012). Ni/CeO_2 catalysts with high CO_2 methanation activity and high CH_4 selectivity at low temperatures, *Int. J. Hydrogen Energy*, **37**, 5527–5531.
14. Kim, H. Y., H. M. Lee, J.-N. Park (2010). Bifunctional mechanism of CO_2 methanation on $Pd-MgO/SiO_2$ catalyst: Independent roles of MgO and Pd on CO_2 methanation, *J. Phys. Chem. C*, **114**, 7128–7131.
15. Eg&G Technical Services (2004). *Fuel Cell Handbook*. 7th ed.
16. Larmine, J., A. Dicks (2003). *Fuel Cell System Explained*. 2nd ed., Wiley.
17. Journal News, CFCL generator achieves 60% efficiency, *Fuel Cells Bulletin*, 2009 (2009) 6.
18. Gattrell, M., N. Gupta, A. Co (2007). Electrochemical reduction of CO_2 to hydrocarbons to store renewable electrical energy and upgrade biogas, *Energy. Convers. Manag.*, **48**, 1255.

19. Graves, C., S. D. Ebbesen, M. Mogensen, K. S. Lackner (2011). Sustainable hydrocarbon fuels by recycling CO_2 and H_2O with renewable or nuclear energy, *Renew. Sust. Energy. Rev.*, **15**, 1–23.
20. Lackner, K., S., C. J. Meinrenken, E. Dahlgren, C. Graves, T. Socci, Closing the carbon cycle: Liquid fuels from air, water and sunshine, White Paper, Lenfest Center for Sustainable Energy, Columbia University, NY, SA, 2010, http://wordpress.ei.columbia.edu/lenfest/files/2012/11/SunlightToFuels_WhitePaper.pdf.
21. Onda, K., T. Kyakuno, K. Hattori, K. Ito (2004). Prediction of production power for high-pressure hydrogen by high-pressure water electrolysis, *J. Power Sources*, **132**, 64–70.
22. Allebrod, F., C. Chatzichristodoulou, M. B. Mogensen (2013). Alkaline electrolysis cell at high temperature and pressure of 250°C and 42 bar, *J. Power Sources*, **229**, 22–31.
23. Grigor'ev, S. A., M. M. Khaliullin, N. V. Kuleshov, V. N. Fateev (2001). Electrolysis of water in a system with a solid polymer electrolyte at elevated pressure, *Russ. J. Electrochem.*, **37**, 819–822.
24. Goodenough, J. B. (2003). Oxide-ion electrolytes, *Annu. Rev. Mater. Res.*, **33**, 91–128.
25. Zeng, K., D. Zhang (2010). Recent progress in alkaline water electrolysis for hydrogen production and applications, *Prog. Energy Combustion Sci.*, **36**, 307–326.
26. Barsoukov, E., R. MacDonald (2005). *Impedance Spectroscopy: Theory, Experiment, and Applications*, Wiley.
27. Graves, C., S. D. Ebbesen, M. Mogensen, K. (2011). Co-electrolysis of CO_2 and H_2O in solid oxide cells: Performance and durability, *Solid State Ionics*, **192**, 398–403.
28. Mosalam Shaltout, M. A. (1998). Solar hydrogen from Lake Nasser for 21st century in Egypt, *Int. J. Hydr. Energy*, **23**, 233–238.
29. Gahleitner, G. (2013). Hydrogen from renewable electricity: An international review of power-to-gas pilot plants for stationary applications, *Int. J. Hydrogen Energy*, **38**, 2039–2061.
30. Szyszka, A. (1998). Ten years of solar hydrogen demonstration project at Neunburg vorm Wald, Germany. *Int J Hydrogen Energy*, **23**, 849–860, Schatz Energy Research Center (SERC).
31. Schatz solar hydrogen project. http://www.schatzlab.org/docs/v1n4_dig_sm.pdf (accessed 03/01/2014).
32. Sotavento project: system to produce hydrogen http://www.sotaventogalicia.com/es/area-tecnica/instalaciones-renovables/planta-de-hidrogeno (accessed 04/01/2014).

33. E.ON 2012 Sustainability Report http://www.eon.com/content/dam/eon-com/Nachhaltigkeit/CS%20Report%202012/Download-Dokumente/E.ON_CSReport2012.pdf (accessed 03/24/2014).

34. Press Release, "World's largest Power-to-Gas plant for generating methane enters operation Stage prior to industrial application achieved", Stuttgart, 30 October 2012, http://www.zsw-bw.de/en/support/press-releases/press-detail/weltweit-groesste-power-to-gas-anlage-zur-methan-erzeugung-geht-in-betrieb.html. (accessed. 3/18/2014).

35. Mazloomi, S., K. N. Sulaiman (2012). Influencing factors of water electrolysis electrical efficiency, *Renew. Sustain Energy Rev.*, **16**, 4257–4263.

36. Salvi, P., P. Nelli, P. Villa, Y. Kiros, G. Zangari, G. Bruni et al. (2011). Hydrogen evolution reaction in PTFE bonded Raney-Ni electrodes, *Int. J. Hydrogen Energy*, **36**, 7816–7821.

37. Hashimoto, K., T. Sasaki, S. Meguro, K. Asami (2004). Nanocrystalline electrodeposited Ni–Mo–C cathodes for hydrogen production, *Mater. Sci. Eng. A*, **375–377**, 942–945.

38. Li, X., B. N. Popov, T. Kawahara, H. Yanagi (2011). Non-precious metal catalysts synthesized from precursors of carbon, nitrogen, and transition metal for oxygen reduction in alkaline fuel cells, *J. Power Sources*, **196**, 1717–1722.

39. Sanetuntikul, J., S. Shanmugam (2014). Prussian blue-carbon hybrid as a non-precious electrocatalyst for the oxygen reduction reaction in alkaline, *Electrochim. Acta*, **119**, 92–98.

40. Montoneri, E., L. Giuffré, G. Modica, E. Tempesti (1986). Reinforced asbestos separators for water electrolysis, *Int. J. Hydrogen Energy*, **11**, 233–240.

41. Manabe, A., M. Kashiwase, T. Hashimoto, T. Hayashida, A. Kato, K. Hirao, I. Shimomura, I. Nagashima (2013). Basic study of alkaline water electrolysis, *Electrochim. Acta*, **100**, 249–256.

42. Bhandari, R., C. A. Trudewind, P. Zapp (2013). Life cycle assessment of hydrogen production via electrolysis: A review, *Int. J. Hydrogen Energy*, **38**, 4901.

43. Schefold, J., A. Brisse, F. Tietz (2012). Nine thousand hours of operation of a solid oxide cell in steam electrolysis mode, *J. Electrochem. Soc.*, **159**, A137.

44. http://www.fuelcelltoday.com/news-events/news-archive/2012/november/enertrag-delivers-three-2-mw-alkaline-electrolysers.

45. Ayers, K., E., E. B. Anderson, C. B. Capuano, B. D. Carter, L. T. Dalton, G. Hanlon, J. Manco, M. Niedzwiecki (2010). Research advances towards low cost, high efficiency PEM electrolysis, *ECS Trans.*, **33**, 3–15.

46. Riegel, H., J. Mitrovic, K. Stephan (1998). Role of mass transfer on hydrogen evolution in aqueous media, *J. Appl. Electrochem.*, **28**, 10–17.

47. Stojic, D. L., T. D. Grozdic, M. P. Marceta Kaninski, A. D. Maksic, N. D. Simic (2006). Intermetallics as advanced cathode materials in hydrogen production via electrolysis, *Int. J. Hydrogen Energy*, **31**, 841–846.

48. Schalenbach, M., M. Carmo, D. L. Fritz, J. Mergel, D. Stolten (2013). Pressurized PEM water electrolysis: Efficiency and gas crossover. *Int. J. Hydrogen Energy*, **38**, 14921–14933.

49. Ito, H., T. Maeda, A. Nakano, H. Takenaka (2011). Properties of Nafion membranes under PEM water electrolysis conditions. *Int. J. Hydrogen Energy*, **36**, 10527–10540.

50. Iulianelli, A., I. Gatto, F. Trotta, M. Biasizzo, E. Passalacqua, A. Carbone, A. Bevilacqua, G. Clarizia, A. Gugliuzza, A. Basile (2013). Electrochemical characterization of sulfonated PEEK-WC membranes for PEM fuel cells, *Int. J. Hydrogen Energy*, **38**, 551–557.

51. Alberti, G., M. Casciola, L. Massinelli, B. Bauer (2001). Polymeric proton conducting membranes for medium temperature fuel cells (110–160°C), *J. Membr. Sci.*, **185**, 73–81.

52. Staiti, P., F. Lufrano, A. S. Arico, E. Passalacqua, V. Antonucci (2001). Sulfonated polybenzimidazole membranes–physico-chemical characterization, *J. Membr. Sci.*, **188**, 71–78.

53. Kobayashi, T., M. Rikukawa, K. Sanui, N. Ogata (1998). Proton-conducting polymers derived from poly(ether-etherketone) and poly(4-phenoxybenzoyl-1,4-phenylene), *Solid State Ionics*, **106**, 219–225.

54. Devanathan, R. (2008). Recent developments in proton exchange membranes for fuel cells, *Energy Environ. Sci.*, **1**, 101–119.

55. Di Vona, M. L., E. Sgreccia, M. Tamilvanan, M. Khadhraoui, C. Chassigneux, P. Knauth (2010). High ionic exchange capacity polyphenylsulfone (SPPSU) and polyethersulfone (SPES) cross-linked by annealing treatment: Thermal stability, hydration level and mechanical properties, *J. Membr. Sci.*, **354**, 134–141.

56. Ma, X., Z. Shao, R. T. Baker, B. Yi (2008). Electrochemical investigation of electrocatalysts for the oxygen evolution reaction in PEM water electrolysers, *Int. J. Hydrogen Energy*, **33**, 4955–4961.
57. Beer, H. B. (1980). The invention and industrial development of metal anodes, *J. Electrochem. Soc.*, **127**, 303C–307C.
58. de Oliveira-Sousa, A., M. A. S. da Silva, S. A. S. Machado, L. A. Avaca, P. de Lima-Neto (2000). Influence of the preparation method on the morphological and electrochemical properties of Ti/IrO_2-coated electrodes, *Electrochim. Acta*, **45**(27), 4467–4473.
59. Kadakia, K., M. K. Datta, O. I. Velikokhatnyi, P. Jampani, S. K. Park, P. Saha, J. A. Poston, A. Manivannan, P. N. Kumta (2012). Novel (Ir, Sn, Nb)O_2 anode electrocatalysts with reduced noble metal content for PEM based water electrolysis, *Int. J. Hydrogen Energy*, **37**, 3001–3013.
60. Yu, R., E. Galyamov, B. Sh. Belova, I. D. Shifrina, R. R. Kozhevnikov, V. B. Bystrov, V. I. Elektrokhimiya (1982). **18**, 1327.
61. Chen, G., X. Chen, P. L. Yue (2002). Electrochemical behavior of novel Ti/IrO_x–Sb_2O_5–SnO_2 anodes, *J. Phys. Chem. B*, **106**(17), 4364–4369.
62. Ma L., Sui S., Zhai Y. (2009). Investigations on high performance proton exchange membrane water electrolyzer. *Int. J. Hydrogen Energy*, **34**, 678–684.
63. Laguna-Bercero, M. A. (2012). Recent advances in high temperature electrolysis using solid oxide fuel cells: A review, *J. Power Sources*, **203**, 4–16.
64. Graves, C., S. D. Ebbesen, S. H. Jensen, S. B. Simonsen, M. B. Mogensen, Eliminating degradation in solid oxide electrochemical cells by reversible operation, *Nat. Mater.*, DOI:10.1038/NMAT4165.
65. Baur, E., H. Preis (1937). Über Brennsto-Ketten mit Festleitern, *Z. Elktrochem*, **43**, 727–732.
66. Laguna-Bercero, M. A., V. M. Orera (2011). Micro-spectroscopic study of the degradation of Scandia and ceria stabilized zirconia electrolytes in solid oxide electrolysis cells, *Int. J. Hydrogen*, 13051.
67. Kreuer, K. D. (2003). Proton-conducting oxides, *Annu. Rev. Mater. Res.*, **33**, 333–359.
68. Fabbri, E., D. Pergolesi, E. Traversa (2010). Materials challenges toward proton conducting oxide fuel cells: A critical review, *Chem. Soc. Rev.*, doi:10.1039/b902343g.
69. Hauch, A., S. B. Ebbesen, S. H. Jensen, M. Mogensen (2008). Solid oxide electrolysis cells: Microstructure and degradation of the Ni/yttria-stabilized zirconia electrode, *J. Electrochem. Soc.*, **155**, B1184.

70. Kim-Lohsoontorn, P., J. Bae (2011). Electrochemical performance of solid oxide electrolysis cell electrodes under high-temperature coelectrolysis of steam and carbon dioxide, *J. Power Sources*, **196**, 7161.
71. Kim-Lohsoontorn, P., Y.-M. Kim, N. Laosiripojan, J. Bae (2011). Gadolinium doped ceria-impregnated nickel-yttria stabilized zirconia cathode for solid oxide electrolysis cell, *Int. J. Hydrogen Energy*, **36**, 9420.
72. Marina, O. A., L. R. Pederson, M. C. Williams, G. W. Coffey, K. D. Meinhardt, C. D. Nguyen, E. C. Thomsen (2007). Electrode performance in reversible solid oxide fuel cells, *J. Electrochem. Soc.*, **154**, B452.
73. Yang, X., J. T. S. Irvine (2008). $(La_{0.75}Sr_{0.25})_{0.95}Mn_{0.5}Cr_{0.5}O_3$ as the cathode of solid oxide electrolysis cells for high temperature hydrogen production from steam, *J. Mater. Chem.*, **18**, 2349.
74. Liu, Q., C. Yang, X. Dong, F. Chen (2010). Perovskite $Sr_2Fe_{1.5}Mo_{0.5}O_{6-\delta}$ as electrode materials for symmetrical solid oxide electrolysis cells, *Int. J. Hydrogen Energy*, **35**, 10039.
75. Momma, A., T. Kato, Y. H. Kaga, S. Nagata (1997). Polarization behavior of high temperature solid oxide electrolysis cells (SOEC), *J. Ceram. Soc. Jpn*, **105**, 369.
76. Sohal, M. S., J. E. O'Brien, C. M. Stoots, V. I. Sharma, B. Yildiz, A. Virkar (2012). Degradation issues in solid oxide cells during high temperature electrolysis, *J. Fuel Cell Sci. Technol.*, **9**, 011017.
77. Mawdsley, J. R., J. D. Carter, A. J. Kropf, B. Yildiz, V. A. Maroni (2009). Post-test evaluation of oxygen electrodes from solid oxide electrolysis stacks, *Int J Hydrogen Energy*, **34**, 198.
78. Wang, W., S. P. Jiang (2006). A mechanistic study on the activation process of (La, Sr)MnO_3 electrodes of solid oxide fuel cells, *Solid State Ionics*, **177**, 1361.
79. Wang, W., Y. Huang, S. Jung, J. M. Vohs, R. J. Gorte (2006). A comparison of LSM, LSF, and LSCo for solid oxide electrolyser anodes, *J. Electrochem. Soc.*, **153**, A2066.
80. Tarancón, A. (2009). Strategies for lowering solid oxide fuel cells operating temperature, *Energies*, **2**, 1130.
81. Tarancón, A., M. Burriel, J. Santiso, S. J. Skinner, J. A. Kilner (2010). Advances in layered oxide cathodes for intermediate temperature solid oxide fuel cells, *J. Mater. Chem.*, **20**, 3799.
82. Hu L, et al. (2014). Electrochemical performance of reversible molten carbonate fuel cells, *Int. J. Hydrogen Energy*, http://dx.doi.org/10.1016/j.ijhydene.2014.02.144.

83. Feldman, Y., I. Lubomirsky (2010). Conversion of CO_2 to CO by electrolysis of molten lithium carbonate, *J. Electrochem. Soc.*, **157**, B552–B556.
84. Kaplan, V., E. Wachtel, I. Lubomirsky (2012). Titanium carbide coating of titanium by cathodic deposition from a carbonate melt, *J. Electrochem. Soc.*, **159**, E159–E161.
85. Iiba, K., H. Katsumata, T. Suzuki, K. Ohta (2006). Electrochemical reduction of high pressure CO_2 at a Cu electrode in cold methanol, **51**, 4880–4885.
86. Ito, K., S. Ikeda, N. Yamauchi, T. Iida, T. Takagi (1985). Electrochemical reduction products of carbon-dioxide at some metallic electrodes in nonaqueous electrolytes, *Bull. Chem. Soc. Jpn.*, **58**, 3027–3028.
87. Rosen, B. A., A. Salehi-Khojin, M. R. Thorson, W. Zhu, D. T. Whipple, P. J. A. Kenis, R. I. Masel (2011). Ionic liquid-mediated selective conversion of CO_2 to CO at low overpotentials, *Science*, **334**, 643–644.
88. Perathoner, S., M. P. Gangeria, G. Lanzafamea (2007). Centi' Nanostructured Electrocatalytic Pt–carbon materials for fuel cells and CO_2 conversion, *Kinet. Catal.*, **48**, 877–883.
89. Centi, G., S. Perathoner, G. Winèa, M. Gangeri (2007). Electrocatalytic conversion of CO_2 to long carbon-chain hydrocarbons, *Green Chem.*, **9**, 671–678.
90. Hori, Y., H. Konishi, T. Futamura, A. Murata, O. Koga, H. Sakurai, et al. (2005). Deactivation of copper electrode" in electrochemical reduction of CO_2, *Electrochim. Acta*, **50**(27), 5354–5369.
91. Wasmus, S., E. Cattaneo, W. Vielstich (1990). Reduction of carbon-dioxide to methane and ethane. An online MS study with rotating electrodes, *Electrochim. Acta*, **35**, 771–775.
92. DeWulf, D. W., T. Jin, A. J. Bard (1989). Electrochemical and surface studies of carbon dioxide reduction to methane and ethylene at copper electrodes in aqueous solutions, *J. Electrochem. Soc.*, **136**, 1686–1691.
93. Kyriacou, G., A. Anagnostopoulos (1992). Electrochemical reduction of carbon dioxide at copper + gold electrodes, *J. Electroanal. Chem.*, **328**(1–2), 233–243.
94. Ishimaru, S., R. Shiratsuchi, G. Nogami (2000). Pulsed electroreduction of CO_2 on Cu–Ag alloy electrodes, *J. Electrochem. Soc.*, **147**, 1864–1867.
95. Shiratsuchi, R., Y. Aikoh, G. Nogami (1993). Pulsed electroreduction of CO_2 on copper electrodes, *J. Electrochem. Soc.*, **140**, 3479–3482.
96. Lee, J., Y. Kwon, R. L. Machunda, H. J. Lee (2009). Electrocatalytic recycling of CO_2 and small organic molecules, *Chem. Asian J.*, **4**, 1516–1523.

97. Li, W., H., Wang, Y. Shi, N. Cai, Performance and methane production characteristics of H_2O-CO_2 co-electrolysis in solid oxide electrolysis cells, *Int. J. Hydrogen Energy*, doi:10.1016/j.ijhydene.2013.01.008.

98. Stoots, C., J. O'Brien, J. Hartvigsen (2009). Results of recent high temperature coelectrolysis studies at the Idaho National Laboratory, *Int. J. Hydrogen Energy*, **34**, 4208–4215.

99. Ebbesen, S. D., M. Mogensen (2009). Electrolysis of carbon dioxide in solid oxide electrolysis cells, *J. Power Sources*, **193**, 349.

100. Ebbesen, S. D., C. Graves, A. Hauch, S. H. Jensen, M. Mogensen (2010). Poisoning of solid oxide electrolysis cells by impurities, *J. Electrochem. Soc.*, **157**, B1419.

101. Vollmar H. E., et al. (2000). Innovative concepts for the coproduction of electricity and syngas with solid oxide fuel cells, *J. Power Sour*, **86**, 90–97.

102. Margalef, P., T. Brown, J. Brouwer, S. Samuelsen (2011). Efficiency of poly-generating high temperature fuel cells, *J. Power Sources*, **196**, 2055.

103. Brouwer J., E. Leal (2006). A thermodynamic analysis of electricity and hydrogen polygeneration using a solid oxide fuel cell, *J. Fuel Cell Sci. Technol.*, **3**, 137–143.

104. Maeda K. (2013). Z-scheme water splitting using two different semiconductor photocatalysts, *ACS Catal.*, **3**, 1486–1503.

105. Kudo A. Y. Miseki (2009). Heterogeneous photocatalyst materials for water splitting, *Chem. Soc. Rev.*, **38**, 253–278.

106. Nozik, A. J. (1978). Photoelectrochemistry: Applications to solar energy conversion, *Annu. Rev. Phys. Chem.*, **29**(1), 189–222.

107. Mavroides, J. G., J. A. Kafalas, D. F. Kolesar (1976). Photoelectrolysis of water in cells with $SrTiO_3$ anodes, *Appl. Phys. Lett.*, **28**(5), 241–243.

108. Domen, K., S. Naito, T. Onishi, K. Tamaru, M. Soma (1982). Study of the photocatalytic decomposition of water vapor over a nickel(II) oxide-strontium titanate ($SrTiO_3$) catalyst, *J. Phys. Chem.*, **86**(18), 3657–3661.

109. Domen, K., A. Kudo, T. Onishi, N. Kosugi, H. Kuroda (1986). Photocatalytic decomposition of water into H_2 and O_2 over NiO-$SrTiO_3$ powder. 1. Structure of the catalyst, *J. Phys. Chem.*, **986**, 90(2), 292–295.

110. Lehn, J. M., J.-P. Sauvage, R. Ziessel (1980). Photochemical generation of carbon monoxide and hydrogen by reduction of carbon dioxide and water under visible light irradiation, *Nouv. J. Chim.*, **4**, 623.

111. Kudo, A., K. Domen, K. Maruya, T. Onishi (1987). Photocatalytic Reduction of NO_3 to form NH_3 over Pt-TiO_2. *Chem. Lett.*, **6**, 1019–1022.
112. Sayama K., H. Arakawa (1996). Effect of carbonate addition on the photocatalytic decomposition of liquid water over a ZrO_2 catalyst, *J. Photochem. Photobiol. A*, **94**, 67–76.
113. Fujishima, A., K. Honda (1972). Electrochemical photolysis of water at a semiconductor electrode, *Nature*, **238**(5358), 37–38.
114. Fàbrega, C., T. Andreu, A. Cabot, J. R. Morante (2010). Location and catalytic role of iron species in TiO_2:Fe photocatalysts: An EPR study, *J. Photochem. Photobiol. A*, **211**, 170–175.
115. Fàbrega, C., T. Andreu, F. Güell, D. Prades, S. Estradé, J. M. Rebled, F. Peiró, J. R. Morante (2011). Effectiveness of nitrogen incorporation to enhance the photoelectrochemical activity of nanostructured TiO_2:NH_3 versus H_2/N_2 annealing, *Nanotechnology*, **22**, 235403.
116. Diebold, U. (2003). The surface science of titanium dioxide, *Surf. Sci. Rep.*, **48**, 53–229.
117. Fan, J., F. Guell, C. Fabrega, A. Shavel, A. Carrete, T. Andreu, J. R. Morante, A. Cabot (2011). Enhancement of the photoelectrochemical properties of Cl-oped ZnO nanowires by tuning their coaxial doping profile, *Appl. Phys. Lett.*, **99**(26), 262102.
118. Park, Y., K. J. McDonald, K.-S. Choi (2013). Progress in bismuth vanadate photoanodes for use in solar water oxidation, *Chem. Soc. Rev.*, **42**(6), 2321–2337.
119. Wang, H., T. Lindgren, J. He, A. Hagfeldt, S.-E. Lindquist (2000). Hotolelectrochemistry of nanostructured WO_3 thin film electrodes for water oxidation: Mechanism of electron transport, *J. Phys. Chem. B*, **104**(24), 5686–5696.
120. Kudo, A., K. Ueda, H. Kato, I. Mikami (1998). Photocatalytic O_2 evolution under visible light irradiation on $BiVO_4$ in aqueous $AgNO_3$ solution, *Catal. Lett.*, **53**, 229–230.
121. Seger, B., T. Pedersen, A. B. Laursen, P. C. K. Vesborg, O. Hansen, I. Chorkendorff. Using TiO_2 as a conductive protective layer for photocathodic H_2 evolution, *J. Am. Chem. Soc.*, **135**(3), 1057–1064.
122. Esposito, D. V., I. Levin, T. P. Moffat, A. A. Talin (2013). H_2 evolution at Si-based metal-insulator semiconductor photoelectrodes enhanced by inversion channel charge collection and H spillover, *Nat. Mater.*, **12**(6), 562–568.
123. Paracchino, A., V. Laporte, K. Sivula, M. Gratzel, E. Thimsen (2011). Highly active oxide photocathode for photoelectrochemical water reduction, *Nat. Mater.*, **10**(6), 456–461.

124. Inoue, T., A. Fujishima, S. Konishi, K. Honda (1979). Photoelectrocatalytic reduction of carbon dioxide in aqueous suspensions of semiconductor powders, *Nature*, **277**, 637–638.

125. Kocí, K., L. Obalová, L. Matejová, D. Plachá, Z. Lacný, J. Jirkovský, O. Solcová (2009). Effect of TiO_2 particle size on the photocatalytic reduction of CO_2, *Appl. Catal. B Environ.*, **89**, 494–502.

126. Manzanares, M., C. Fàbrega, J. O. Ossò, L. F. Vega, T. Andreu, J. R. Morante. (2014). Engineering the TiO_2 outermost layers using magnesium for carbon dioxide photoreduction, *Appl. Catal. B Environ.*, DOI:10.1016/j.apcatb.2013.11.036.

127. Nguyen, T.-V., J. C. S. Wu (2008). Photoreduction of CO_2 in an optical-fiber photoreactor: Effects of metals addition and catalyst carrier, *Appl. Catal. A General*, **335**, 112–120.

128. Tahir, M., N. S. Amin (2013). Photocatalytic CO_2 reduction and kinetic study over In/TiO_2 nanoparticles supported microchannel monolith photoreactor, *Appl. Catal. A General*, **467**, 483–496.

129. Magesh, G., E. S. Kim, H. J. Kang, M. Banu, J. Y. Kim, J. H. Kim, J. S. Lee (2013). A versatile photoanode-driven photoelectrochemical system for conversion of CO_2 to fuels with high faradaic efficiencies at low bias potentials, *J. Mater. Chem. A*, DOI:10.1039/c3ta14408a.

Chapter 10

Hydrogen Storage

Raphaël Janot

Laboratory of Reactivity and Chemistry of Solids (LRCS), UMR 7314 CNRS, University of Picardie Jules Verne, 33 rue st Leu, 80039 Amiens, France
raphael.janot@u-picardie.fr

Generalities on hydrogen

Hydrogen is the first element in the periodic table of the elements. Henry Cavendish was the first in 1766 to recognize hydrogen gas as a discrete substance, by naming the gas from a metal-acid reaction "flammable air." Further findings in 1781 show that the gas produces water when burned. In 1783, Antoine Lavoisier gave the element the name hydrogen (from the Greek *hydro* meaning water and *genes* meaning creator). Nicholson and Carlisle split water into oxygen and hydrogen in 1800 by passing an electric current through it. This was the first electrolysis. Hydrogen was liquefied for the first time by James Dewar in 1898. Using liquid nitrogen he pre-cooled gaseous hydrogen under 180 atmospheres and then expanded it through a valve in an insulated vessel. He produced solid hydrogen the next year.

Hydrogen exists as three different isotopes: protium, deuterium, and tritium. The ordinary isotope of hydrogen H (1 proton and 1 electron) with a natural abundance of 99.985% is named protium.

Materials for Sustainable Energy Applications: Conversion, Storage, Transmission, and Consumption
Edited by Xavier Moya and David Muñoz-Rojas
Copyright © 2016 Pan Stanford Publishing Pte. Ltd.
ISBN 978-981-4411-81-3 (Hardcover), 978-981-4411-82-0 (eBook)
www.panstanford.com

Deuterium (D) is a stable isotope (1 proton, 1 neutron, and 1 electron) with a natural abundance of 0.015%. In 1934, an unstable isotope of hydrogen (1 proton, 2 neutrons, and 1 electron) was discovered. This radioactive element named tritium has a half-life of 12.5 years. Tritium is used as a neutrons source for boosting fission in thermonuclear weapons. This has led to the so-called "hydrogen bombs."

Hydrogen is the most abundant element in the universe: It has been estimated that hydrogen represents more than 90% of all the atoms of the universe. It is found in the sun and most stars. Hydrogen is the third most abundant element on the Earth's surface (behind silicon and oxygen), mostly in the form of chemical compounds such as hydrocarbons and water. Under ordinary conditions, hydrogen gas is very rare in the Earth's atmosphere (1 ppm by volume) because of its light weight, which enables it to escape from Earth's gravity more easily than heavier gases.

As hydrogen gas is not present in the Earth's atmosphere, it has to be produced. Hydrogen is today mainly produced by steam reforming of natural gas (containing mainly methane). In a first step, syngas (CO/H_2 mixture) is formed by the endothermic reaction of CH_4 with H_2O. Then CO is exothermically converted to CO_2 by water gas shift. The global reaction can be summarized as follows:

$$CH_4 + 2\,H_2O \rightarrow CO_2 + 4\,H_2 \tag{10.1}$$

The separation of H_2 from CO_2 is first made by decarbonatation using concentrated alkaline solutions (NaOH) leading to hydrogen with 95–98% purity. Then high-purity hydrogen (>99.999%) is obtained by pressure swing adsorption (PSA) that is the selective adsorption of the gases on zeolithes or activated carbons near ambient temperature. Large reforming plants produce up to 1000 kg H_2 (11000 Nm^3) per hour. In order to transport this hydrogen gas close to the users, pipelines networks can be used. The first hydrogen pipeline was opened in the Ruhr valley in Germany in 1938. Today, a network of about 1500 km is operated in Europe and one of 900 km in the USA. The tubes are built using conventional pipe steels and are filled at pressures up to 20 bars. An alternative way of producing high-purity hydrogen, especially for on-site production, is the water electrolysis. Low-temperature electrolyzers (alkaline or more recently with polymer electrolyte membranes) can produce up to 100 Nm^3/h.

Generalities on hydrogen storage

Hydrogen is stable as a diatomic molecule H_2. The interaction between two hydrogen atoms goes through a minimal energy at an inter-atomic distance of 0.746 Å, when two electrons with opposite spin form a singlet state (total electron spin equal to zero). There are two kinds of hydrogen molecules in the singlet state: ortho-H_2 with parallel nuclear spins and para-H_2 with anti-parallel nuclear spins. At room temperature, hydrogen is a mixture of 75% ortho-H_2 and 25% para-H_2. The conversion of ortho-H_2 to para-H_2 is exothermic. When hydrogen is cooled from room temperature to the boiling point (20.2 K), ortho-H_2 is converted to para-H_2 (only 0.2% of ortho-H_2 at 20 K).

The phase diagram for hydrogen is shown in Fig. 10.1. In hydrogen, the interaction between molecules is weak as compared to other molecules, therefore the critical temperature is low (T_c = 33 K, P_c = 1310 Pa). At 1 bar, the melting point of hydrogen is 14 K and its boiling point is 20.2 K. Liquid hydrogen has a density of 70.8 kg·m^{-3} at the boiling point temperature. At higher temperatures (T > 20.2 K), hydrogen is a gas that can be described by the van der Waals equation:

$$P = nRT/(V - nb) - a \cdot n^2/V^2, \qquad (10.2)$$

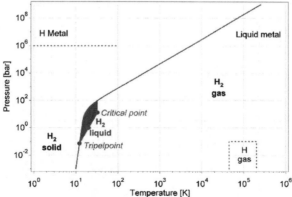

Figure 10.1 Phase diagram for hydrogen showing that liquid hydrogen has a relatively small domain of stability (in blue).

where P is the gas pressure, V the volume, T the temperature, n the number of moles, R the gas constant (R = 8.314 J·K^{-1}·mol^{-1}), a is the dipole interaction or repulsion constant ($a = 2.476 \times 10^{-2}$ $m^6 \cdot Pa \cdot mol^{-2}$), and b is the volume occupied by the H_2 molecules ($b = 2.661 \times 10^{-5}$ $m^3 \cdot mol^{-1}$). The ideal gas law ($PV = nRT$) can only be used at ambient temperature and pressures below 200 bars. Hydrogen gas has a low density of 90 $g \cdot m^{-3}$ at 273 K and a pressure of 1 bar. In order to reduce the enormous volume of hydrogen gas, either we have to compress the hydrogen, or the temperature has to be decreased in order to liquefy hydrogen, or finally, the repulsion between the H_2 molecules has to be reduced by the interaction (reaction) with a material.

10.1 Conventional Hydrogen Storages

10.1.1 Compressed Gas

One kilogram of hydrogen at ambient temperature and atmospheric pressure takes a volume of 11 m^3. The easiest way to decrease this volume is to compress the gas. Figure 10.2 shows the volume occupied by 5 kg hydrogen at 298 K (the weight typically required for car applications) calculated using the ideal gas law (in blue) and the van der Waals equation (in red). Clearly, the ideal gas law cannot be used at pressures above 200 bars as the volume is largely underestimated (by a factor of about two at 1000 bars).

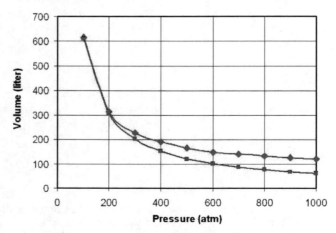

Figure 10.2 Volume occupied by 5 kg of hydrogen at 298 K as a function of the pressure. Calculations done using the ideal gas law (in blue) and the van der Waals equation (in red).

The most common high-pressure gas cylinders are made of steel (called type-I tanks) with a maximum working pressure of 200 bars and a volumetric density of about 16 kg·m^{-3}. In order to further reduce the volume, new lightweight composite cylinders have been developed, which are able to withstand pressures up to 800 bars (80 MPa). At this pressure, the volumetric density of hydrogen is 36 kg·m^{-3}, approximately half the density of liquid hydrogen at 20 K. In theory, the best shape to withstand high pressures is a sphere because stresses and strains are uniformly distributed. However, large-size spherical containers are expensive due to their complex manufacturing process and they fit with difficulty the space available in many applications such as vehicles. Therefore, most of the high-pressure hydrogen tanks are cylindrical with two hemispherical extremities.

The high-pressure composite vessels are usually made of three layers (cf. Fig. 10.3): an inner liner with low hydrogen permeability, overwrapped with a carbon-fiber composite (which is the stress-bearing component), and an outer layer of a cheap material (such as aramide) capable of withstanding mechanical and corrosion damage. Most of the high-pressure hydrogen cylinders use today austenitic stainless steels or aluminum-based alloys for the inner liner. These materials are largely immune to hydrogen effects at ambient conditions. Many other materials with higher tensile strengths, such as martensitic steels or titanium-based alloys, cannot be used due to embrittlement effects (the so-called HIC process: hydrogen-induced cracking, due to the diffusion of H atoms through the metal surface and recombination to form H$_2$ molecules creating bubbles up to cracking).

For lightweight composite hydrogen tanks, Al-based alloys are now often used for the inner liner (density of 2.7 instead of 8 for the steels) and new welding techniques have been developed. Actually, common welding techniques (arc- or gas-welding) involve the melting of the metal and the formation of heat-affected zones (HAZ), where residual stresses can finally lead to cracking with hydrogen. In order to weld Al-based alloys without melting, friction-stir welding (a derivative of forge welding) is used. This technique consists of heating below the melting point of aluminum ($T < 660°C$) and application of a very high download force on the two metallic pieces to weld together. A simultaneous rotation and translatory motion is then exerted.

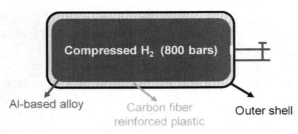

Figure 10.3 Schematic representation of a Type-III lightweight high-pressure hydrogen composite tank.

To withstand the high pressure of the cylinder, the metallic liner is wrapped with carbon fiber reinforced plastics. The tensile strengths can reach up to 5000 MPa, much higher than the values obtained with stainless steels (460 MPa) or even high quality steels (1100 MPa). The composite tanks are called type-II when they are only hoop-wrapped or type-III if fully wrapped (i.e., including the hemispherical extremities). The high cost of carbon fibers is today considered as the main drawback for the large diffusion of lightweight composite hydrogen tanks. Many efforts are devoted to the development of new manufacturing processes for carbon fibers and their composites in order to reduce the cost of high-pressure hydrogen tanks; 70 MPa hydrogen composite tanks have been used on the road with prototype fuel-cell vehicles such as GM HydroGen4 or Toyota FCHV.

The replacement of the inner metallic liner by a polymer liner is also investigated leading to the development of the Type-IV hydrogen tanks. Polymers, such as high-density poly-ethylene (HDPE), have very low hydrogen permeability and can be used for hydrogen storage. These all-polymer tanks are remarkably lightweight and gravimetric storage capacities as high as 14 wt.% have been reported. However, the strong loss of strength of the polymers above 120°C is still today a big issue for the commercialization of type-IV tanks.

Hydrogen can be compressed using standard piston-type mechanical compressors. Slight modifications are, however, sometimes necessary to compensate for the higher diffusivity of hydrogen. The theoretical work necessary for the isothermal compression of hydrogen is given by the following equation:

$$W = P_0 \cdot V_0 \ln (P_1/P_0), \qquad (10.3)$$

where P_1 and P_0 stand for the final and starting pressures, respectively. In a real process, the work needed for compression is much higher because the compression cannot be kept isothermal (cf. Fig. 10.4). In practice, multistage compressions are performed for allowing heat transfers between each pressure step. Figure 10.4 illustrates the energy needed to compress hydrogen up to 80 MPa. A typical multistage process requires 18 MJ·kg^{-1} (5 kWh·kg^{-1}) for compressing hydrogen from 0.1 to 80 MPa, which corresponds to about 12% of HHV (HHV being the high heating value of hydrogen: 143 MJ·kg^{-1}). The energetic cost of hydrogen compression is therefore significant and cannot be neglected when evaluating the possible implementation of the hydrogen economy.

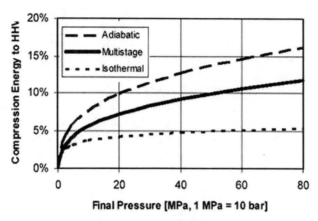

Figure 10.4 Energy needed for the compression of hydrogen up to 800 bars (80 MPa) expressed as the percentage of the HHV of hydrogen.

10.1.2 Liquid Hydrogen

Liquid hydrogen is largely used today for spacecraft applications (US space shuttle, Ariane rockets, ...). Liquid hydrogen is stored in cryogenic tanks at 20 K and ambient pressure. Due to the low critical temperature of hydrogen (33 K), liquid hydrogen can only be stored in open systems. The volumetric density of liquid hydrogen is 70.8 kg·m^{-3} at 20 K. The challenges of liquid hydrogen storage are the thermal insulation of the cryogenic storage vessel in order to reduce the boil-off of hydrogen and the development of an energy efficient liquefaction process.

Three basic mechanisms are responsible for heat transfers: conduction, convection, and radiation. In order to reduce the thermal losses of a liquid hydrogen tank, each of this mechanism has to be fought and this has led to the design of double-walled vacuum-insulated tanks (as a Dewar). Briefly, in order to reduce conduction, materials with low thermal conductivities (for example, expanded polystyrene) are used for the outer layer of the tank. To fight radiation, multi-layers of metallic foils with glass wools are typically used and this constitutes the inner layer of the tank. Between the inner and the outer layers, vacuum is made in order to reduce the convection as much as possible. For this type of like-Dewar cryogenic tanks, boil-off losses are typically of the order of 0.5% per day for a volume of 50 L, but the losses can be much lower for larger tanks (less than 0.1% per day for very large tanks as those used for space applications). Today, due to manufacturing difficulties, the cost of well-insulated liquid hydrogen tank is still high.

The first hydrogen liquefaction was made by Dewar in 1898 and the process was then optimized by Linde and Claude (the latter being the founder of the Air Liquide French company). When hydrogen is cooled from room temperature to the boiling point (20 K), ortho-hydrogen converts to para-hydrogen as shown on Fig. 10.5 (75% O-H_2 at 298 K, 50% at 77 K and finally 0.2% at 20 K). The self-conversion rate is an activated process and is very slow, since the half-life time of the conversion is greater than one year at 77 K. The conversion from ortho to para-hydrogen is exothermic and the heat of conversion is temperature dependent. At 300 K, the heat is 270 $J \cdot g^{-1}$ and increases as the temperature decreases (519 $J \cdot g^{-1}$ at 77 K). At temperatures below 77 K, the heat of conversion is almost constant (about 523 $J \cdot g^{-1}$). As the heat of conversion is greater than the latent heat of vaporization of hydrogen at 20 K (452 $J \cdot g^{-1}$), if unconverted hydrogen is placed in a storage vessel, the heat of conversion will be released in the vessel, which leads to the evaporation of the liquid hydrogen. In order to favor the formation of para-hydrogen, paramagnetic adsorbers can be used as catalysts. The conversion may take only a few minutes if a highly active form of charcoal is used for instance. Other suitable ortho- to para-catalysts are metals such as nickel or any paramagnetic oxides such as chromium oxides. The nuclear spin is reversed without breaking the H–H bond.

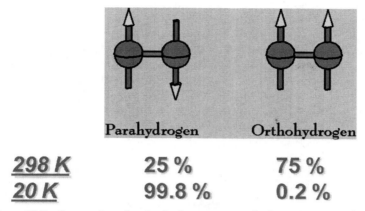

Figure 10.5 Conversion of ortho-hydrogen to para-hydrogen upon cooling.

The most used liquefaction process is the Joule–Thompson cycle. The gas is first compressed and then cooled in a heat exchanger, before it passes through a throttle valve where it undergoes an isenthalpic Joule–Thompson expansion, producing some liquid hydrogen. Very importantly, hydrogen must be first cooled below its inversion temperature (202 K) in order to cool upon expansion. Indeed, contrarily to many gases (as nitrogen) which have inversion temperatures above room temperature, hydrogen warms upon expansion at room temperature. Therefore, hydrogen is usually pre-cooled below its inversion temperature using liquid nitrogen (77 K) before the first expansion step occurs.

The free enthalpy change between hydrogen gas at 300 K and liquid hydrogen at 20 K is 11.6 MJ·kg^{-1}. However, in practice, the work needed to liquefy hydrogen is much higher and is largely dependent of the size of the liquefactor (cf. Fig. 10.6). For small devices (less than 10 kg of liquid H$_2$ per hour), the energy needed exceeds 100 MJ·kg^{-1} and can even exceed the heating values of hydrogen (121 MJ·kg^{-1} for LHV and 143 MJ·kg^{-1} for HHV). For very large liquefactors (more than 1000 kg of liquid hydrogen per hour), the energy needed for liquefaction can decrease to about 40 MJ/kg, which is still about 33% of the LHV (low heating value) of hydrogen. The large amount of energy needed for the liquefaction makes liquid hydrogen not an energy efficient storage medium. In addition, the continuous boil-off of hydrogen limits the possible applications to cases where hydrogen is consumed in a short time, such as space applications.

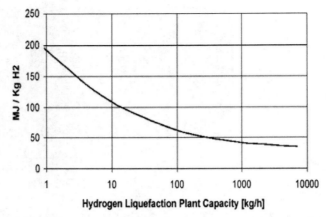

Figure 10.6 Energy needed for hydrogen liquefaction as a function of the plant capacity.

10.2 Hydrogen Physisorption

Physisorption, or physical adsorption, is the mechanism by which hydrogen is stored in the molecular form (without dissociating the H_2 molecules) on the surface of a solid material. The molecular adsorption occurs by van der Waals forces, resulting from the interaction between temporary dipoles, which are formed by the fluctuations in the molecules charge distribution. The combination of attractive van der Waals forces and short-range repulsive interactions between a gas molecule and an atom on the surface can be described by the Lennard–Jones potential (cf. Fig. 10.7). The minimum of potential occurs at a distance from the surface that is approximately the sum of the radii of the adsorbent atom and the adsorbate molecule. Owing to the low polarizability of the H_2 molecule, the interaction with the surface is very weak and leads to values of heat of adsorption in the range of 5–10 kJ·mol^{-1} H_2. Due to this weak interaction, the amount of hydrogen stored is only significant at low temperatures; typically hydrogen physisorption is performed at the temperature of liquid nitrogen (77 K).

Since physisorption occurs at the surface of a solid, only porous materials with very high specific surface areas (i.e., large micro/mesoporous volumes) are regarded as suitable materials for hydrogen storage. Compared to metal hydrides, hydrogen bulk diffusion and structural changes do not take place for the physisorption process. As a result, porous materials used for hydrogen physisorption gave a long lifetime upon cycling: Physisorption is a completely reversible process, which means that the hydrogen can be easily adsorbed and desorbed during several cycles without any capacity losses. Furthermore, no activation energy is involved in the storage process, which means that the adsorption/desorption kinetics are very fast.

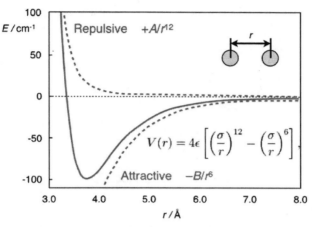

Figure 10.7 Interaction of hydrogen with a solid surface described by the Lennard–Jones potential.

Three main families of porous materials are investigated for hydrogen storage: the carbons, the zeolites, and the metal-organic frameworks (MOFs).

10.2.1 Carbon Materials

For porous carbon materials, the hydrogen storage capacity is usually proportional to the specific surface area at both 77 K and room temperature [1–2]. Considering the BET model for specific surface area, the capacity increment is of the order of 2×10^{-3} wt.%·m^{-2}·g at 77 K and 0.2×10^{-3} wt.%·m^{-2}·g at room temperature. Among all carbon materials, activated carbons appear still as the best adsorbents, because they possess the highest specific surface areas (up to 3000 m^2·g^{-1}). Capacities around 5 wt.% at 77 K/20 bars H$_2$ were reported as early as 1994 for high grade activated carbons [3]. The hydrogen storage capacity was also shown to correlate well with the microporous volume [4–5] (cf. Fig. 10.8). The capacity increment is of the order of 7 wt.%·cm^{-3}·g and therefore a capacity of 7 wt.% could be achieved for highly microporous carbons with porous volume of 1 cm^3.g^{-1}. However, activated carbons are generally porous material with a broad distribution of pores size.

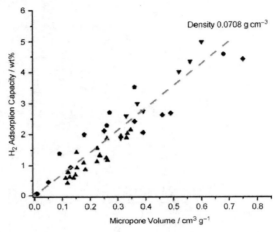

Figure 10.8 Hydrogen adsorption capacity at 77 K of different porous carbons as a function of the microporous volume (pores size less than 2 nm). Reprinted from Ref. [5] with permission from Elsevier.

In contrast, zeolites-templated carbons possess pores in a very narrow range of diameter. A microporous carbons prepared using β–zeolite as hard-template and acetonitrile as the carbon precursor shows a BET specific surface area of 3200 m^2.g^{-1} and very small pores size (0.5–0.9 nm) [6]. This material is able to physisorb 6.9 wt.% of hydrogen at 77 K/20 bars and leads to heat of adsorption up to 7 kJ·mol^{-1} [6], slightly higher than that measured for activated carbons (5 kJ·mol^{-1}), due to the presence of optimum pores size. The influence of the pores size on the carbon/H$_2$ interaction was further investigated on CDC materials (carbide-derived-carbons, i.e., metallic carbides leached with HCl at high temperatures for removing the metallic lattice) [7]. The hydrogen uptake at 77 K is proportional to the surface area of pores smaller than 1 nm, while no correlation is found for larger pores. This clearly shows that ultra micropores ($\varnothing$ < 1 nm) are needed to enhance the hydrogen storage capacity of porous carbons.

Both carbon nanotubes and nanofibers attracted a great interest in the 1990s for hydrogen storage with several reports [8, 9] of exceptionally high storage capacities (up to 67 wt.% by Chambers et al. in 1998 [9]), even at room temperature, but these high values were never reproduced. Actually, it was later assumed that experimental errors (issued mainly from the small amounts of pure sample available) were at the origin of these wrong data [10, 11]. Reproducible results obtained on purified single-wall carbon nanotubes (SWCNTs) do not exceed 0.6 wt.% at room temperature [12–14]. Even at 77 K, their capacity is limited to about 2.0 wt.% under 50 bars H_2, which is a lower value than those obtained with activated carbons or zeolite-templated microporous carbons.

Even though carbon nanotubes do not appear anymore as promising hydrogen storage materials, they are—due to their long-range ordering—still studied for a better understanding of the interaction of hydrogen with carbon surfaces. A bunch of SWCNTs possess different adsorption sites: namely, the outer surface, the channel between the tubes, and the pore inside the tube in the case of opened nanotubes. At least, two different adsorption sites have been identified by Raman spectroscopy [15], by Inelastic Neutron Scattering (INS) [16] and by thermal desorption spectroscopy (TDS) [17]. The stronger adsorption sites (enthalpy up to 8 kJ·mol^{-1}) are assigned to the channels, whereas the lower corresponds to the outer surface, where the hydrogen molecules are more weakly bonded (5 kJ·mol^{-1}).

10.2.2 Zeolites

Zeolites are crystalline alumino-silicates microporous materials, built of AlO_4 and SiO_4 tetrahedrons sharing all four corners. The general formula is $M_{x/n}[(AlO_2)_x(SiO_2)_y]\cdot zH_2O$ where M is an exchangeable non-framework cation (generally Na$^+$, K$^+$, Ca^{2+}) maintaining the electro-neutrality for every Si^{4+} substituted by Al^{3+}. Smaller cations (like Li$^+$) are usually not observed in zeolites because the cavities in the lattice are too large. The replacement of one quarter or one half of the Si atoms is quite common and gives $MAlSi_3O_8$ and $M_2Al_2Si_2O_8$ compositions, respectively.

Zeolites have a very open microporosity (cf. Fig. 10.9) with different types of frameworks depending on the assembly of the building units. Today, 206 unique zeolites frameworks have been identified and over 40 naturally occurring zeolites are known. Zeolites have been investigated for several years for hydrogen storage applications. At 77 K, the hydrogen uptake reaches 2.2 wt.% under 15 bar for CaX zeolithe [18], but the capacity is less than 0.1 wt.% at room temperature [19]. These storage capacities are too low for any possible hydrogen storage application.

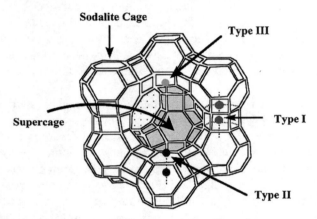

Figure 10.9 Microporosity of X-zeolite with the indication of the different types of cationic sites.

Similarly as for carbon materials, the maximum hydrogen uptake shows a good correlation with the BET specific surface area measured by nitrogen physisorption [18]. Beyond the specific surface area, the interaction of hydrogen with zeolites is also influenced by the type of cations present in the framework channels. For X-zeolites, possessing the same structure but different exchanged cations (Na^+, Ca^{2+}, Zn^{2+}) (cf. Fig. 10.9), a direct correlation between the polarizing effect, as expressed by the ratio between the charge and the radius of the cation (Z/r), and the vibrational frequency of the H_2 molecule (as determined by Inelastic Neutron Scattering) was observed [20]. The interaction increases with the polarizing effect. For example, a Mg^{2+}-exchanged zeolite has an extremely high adsorption enthalpy of 17 kJ·mol^{-1} [21] (much higher than the 5–7 kJ·mol^{-1} values observed for carbons) due the strong polarizing effect of Mg^{2+} cations. However, these strong sites saturate at low hydrogen concentration and no influence is actually observed on the maximum storage capacity at high pressures [20].

10.2.3 Metal-Organic Frameworks

Metal-organic framework materials consist of metal oxide clusters that are connected by organic molecules, which are often carboxylic acids, in a three-dimensional lattice. Since both the metallic center and the organic ligand can be changed, a huge variety of MOFs with different composition and different structure can be obtained. MOFs can possess very high specific surface areas and therefore be used for gas storage. Surface areas exceeding 5000 m$^2 \cdot$g^{-1} (Langmuir or BET specific surface area) have been reported but the exact meaning of these values remains questionable with very small micropores ($\varnothing$ < 2 nm) as those found in MOFs.

Yaghi et al. first reported in 2003 [22] the hydrogen adsorption properties of a Zn-based MOF, called MOF-5, which has a cubic structure formed by $Zn_4O(CO_2)_6$ clusters connected by terephtalic acid ligands (cf. Fig. 10.10). The first investigation of the hydrogen storage properties of this material showed excellent capacities of 1 wt.% at room temperature under 20 bars and 4.5 wt.% at 77 K under only 0.8 bar [22]. These high values were, however, not fully reproduced. Maximum hydrogen storage capacities of about 5 wt.% at 77 K/50 bars are now reproducibly reported for MOF-5 by different groups [23, 24]. The hydrogen storage properties of Cr- and Al-based MOFs called MIL-53 were also reported in 2003 by Férey et al. with maximum storage capacity of 3.8 wt.% at 77 K/16 bars in the case of the aluminum-based compound [25].

Many other MOFs were studied for hydrogen storage with the best materials giving storage capacities higher than 7 wt.% at 77 K. For instance, MOF-177, which consists of $Zn_4O(CO_2)_6$ clusters coordinated by benzene tribenzoate linker, gives a hydrogen gravimetric capacity of 7.5 wt.% [26, 27], a higher value than for any other hydrogen-adsorbing material. Indeed, it was shown that the hydrogen storage capacity is linearly proportional to the specific surface area [28] (cf. Fig. 10.11). This linear correlation was found considering either the BET or the Langmuir surface model for the nitrogen adsorption isotherms at 77 K, even if both models do not appropriately describe the specific surface area of microporous materials: The Langmuir model does not consider multilayer adsorption of N_2, which could occur in larger pores, whereas the BET-specific surface area is obtained from relative pressures P/P_0 in the 0.05–0.3 range, values higher than the saturation pressure of micropores. Nevertheless, a large specific surface area is undoubtedly a necessity for reaching high hydrogen storage capacities.

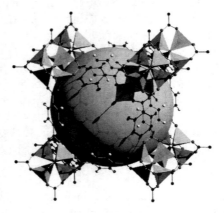

Figure 10.10 Cubic structure of MOF-5 formed by terephtalic acid ligands bridging $Zn_4O(CO_2)_6$ clusters. The specific surface area and the porous volume are 2300 $m^2 \cdot g^{-1}$ and 1.2 $cm^3 \cdot g^{-1}$, respectively.

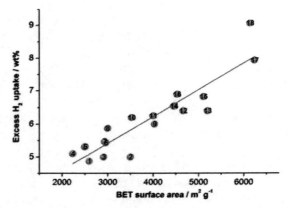

Figure 10.11 Hydrogen-storage capacity at 77 K as a function of the BET specific surface area for different crystalline microporous MOFs. Reproduced from Ref. [28], Energy and Environmental Science (2011), with permission of the Royal Society of Chemistry.

Different studies have been devoted to the understanding of the interaction between MOFs and hydrogen molecules. Using inelastic neutron scattering, it has been shown that the strongest binding sites are present on the inorganic clusters and the weaker adsorption sites are in the vicinity of the organic ligands [29]. X-ray diffraction has also shown that the inorganic clusters are the first adsorption sites for H_2 molecules in MOF-5 [30]. The finding that the metallic clusters may influence the hydrogen uptake led to the development of MOFs with unsaturated metal centers or the so-called MOFs with open metal sites [31]. This can be obtained by the removal of a terminal ligand from the metal site without collapse of the framework. However, as the number of adsorption sites close to these centers is limited, they are saturated at very low hydrogen concentrations [32].

Actually, thermal desorption studies indicate that, at high hydrogen concentration (i.e., close to the maximum capacity), the physisorption of hydrogen into MOFs is more influenced by the pores size than the building units: MOFs with smaller cavities show a higher heat of adsorption [33]. The combined effect of open metal enters and small pores size is considered as a promising route for enhancing the hydrogen storage capacity of MOFs. One well-reported method to decrease the pores size is to form catenated framework [31]. Catenation occurs when the pores in the network accommodate the inorganic cluster of a second network. Two types of catenation can take place: interweaving and interpenetration. In the case of interpenetration, a maximal displacement between the two networks is obtained and thus a significant pores size reduction is possible. Grand Canonical Monte Carlo simulations show that the interpenetration should lead to the formation of a strong adsorption site due to the overlapping of the van der Waals potentials between the two close linkers [32]. However, experimentally, catenation effect has not yet given significant increase of the hydrogen storage capacity of MOFs. On the contrary, in most cases, the catenated frameworks have lower hydrogen uptakes than the non-catenated ones, due to the strong decrease of porous volume [34]. The increase of the affinity to hydrogen by decreasing the pores size, while keeping at the same time a large porous volume, is still today a great challenge for the design of suitable MOFs for hydrogen storage.

To conclude on the possible use of MOFs for hydrogen storage, one of the main drawback, in addition to the low temperature required for physisorption (typically 77 K), is that MOFs are crystalline materials with very low densities, e.g., MOF-5 has a crystallographic density of only 0.61 g·cm^{-3}. Therefore, the volumetric hydrogen storage capacity of MOF-5 is limited to 30 g H$_2$·L^{-1}, much lower than the density of pure liquid hydrogen (70 g H$_2$·L^{-1}).

10.3 Metal Hydrides

Other than adsorption, hydrogen can be stored by absorption into various compounds under different pressure-temperature conditions. Solid-state storage of hydrogen into metals/intermetallics gives the highest volumetric densities as compared to the other systems (up to 150 g H$_2$·L^{-1}, e.g., more than two times the density of liquid hydrogen). Hydrogen atoms fill the interstitial sites within the host lattice (α-phase) leading to a close-packed hydride structure (β-phase) with a volume expansion of 10–20% of the metal lattice, as seen in Fig. 10.12. This causes stresses at the phase boundaries, which in turn results in the decrepitation of the brittle host metals, giving typical particle sizes of 10–40 μm. The amount of hydrogen in the hydride phase depends on the number of interstitial sites available in the host. For that, two empirical conditions have to be satisfied: (1) The distance between two

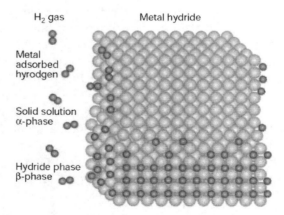

Figure 10.12 Schematic representation of the formation of a metal hydride phase.

interstitial sites has to be at least 2.1 Å (Switendick criterion) [35, 36], and (2) the radius of the largest sphere in the interstitial site touching all neighboring metallic atoms is at least 0.37 Å (Westlake criterion) [37].

The thermodynamics of the metal-hydride reaction is described by the van't Hoff equation, a relation between P_{H2} the hydrogen equilibrium pressure and T the temperature (cf. Eq. 10.4). This is done by first plotting the equilibrium pressure-composition-isotherms (PCI curves) at various temperatures where the hydrogen capacity is recorded with pressure changes at a particular temperature, as shown in Fig. 10.13a. For example, starting on desorption: At the beginning, (1) decrements in pressure cause slight decrease in the hydrogen concentration within the β-phase, before (2) reaching the plateau between the metal and hydride phases (2-phases domain), until the end (3) when only the α-phase is left, decrements in pressure do not cause significant decrease in hydrogen content.

$$\ln(P_{H2}/P_0) = -\Delta H_{des}/RT + \Delta S_{des}/R \qquad (10.4)$$

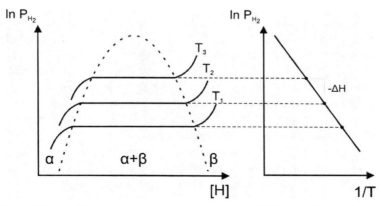

Figure 10.13 (a) Pressure-composition-isotherms (PCI curves) at different temperatures and (b) the corresponding van't Hoff plot allowing the calculation of the enthalpy (gradient) and entropy (intercept) of the reaction.

The plateau pressures from each isotherms are usually placed into a linear plot as presented in Fig. 10.13b: the natural logarithm of the equilibrium pressure against the inverse of the temperature, which allows the change in the enthalpy and entropy values to be obtained through the gradient $-\Delta H_{des}/R$ and intercept $\Delta S_{des}/R$, respectively. With these thermodynamic values, the temperature of desorption (T_{des}) at 1 bar can be calculated by using the van't Hoff equation. Similar experiments and calculations can also be made on absorption. Most of the time, a significant hysteresis is observed with the absorption plateau at a higher pressure than the desorption plateau. Hysteresis is a complex phenomenon that is not fully explained yet. The more common explanation relies on the mechanical strain that has to be overcome upon hydrogenation due to the large lattice expansion (around 20%) [38, 39]. Therefore, the "true" equilibrium pressure is usually considered as the one measured upon desorption. For practical applications, hysteresis is a very important issue, because it can have a large impact on the service pressure of a hydrogen tank. For applications, hysteresis should be as small as possible.

Since the entropy change term corresponds mainly to the change from molecular hydrogen to interstitial hydrogen in the solid phase, it is approximately the value of the standard entropy change of hydrogen, and a typical value of S_0 = 130 J·K^{-1}·mol^{-1} H$_2$ is approximated. However, this value is seldom reached in many known metal hydride systems, e.g., for LaNi$_5$-LaNi$_5$H$_6$ an entropy value of 108 J·K^{-1}·mol^{-1} H$_2$ is found in the literature [40]. The enthalpy change term determines the stability of the metal-hydrogen bond, i.e., the stability of the hydride. For a hydride to be a suitable material for reversible hydrogen storage, the hydrogen desorption is an endothermic process and hydrogen absorption is an exothermic one. If the hydrogen desorption is exothermic, this means that the re-absorption reaction is not feasible; and hence it will be an irreversible process. Larger endothermic ΔH_{des} values represent more stable hydrides and smaller endothermic ΔH_{des} values less stable ones. For the ambient working conditions aimed at applications toward on-board storage of hydrogen, with equilibrium pressures 1–10 bar and temperatures 0–100°C, a typical ΔH_{des} value of +39 kJ·mol^{-1} H$_2$ is required.

Metal hydrides can be classified into two groups following the electronegativity of the element: ionic and metallic hydrides. Group I and II alkali and alkaline earth, together with rare earth elements, form ionic hydrides due to a high electron localization on the hydride anion (H$^-$). These hydrides are stable and easily formed under an atmosphere of hydrogen gas. For example, lithium hydride with a high hydrogen content of 12.5 wt.% is so stable that the ΔH_{des} is +190 kJ·mol^{-1} H$_2$, which corresponds to a T_{des} of 910°C at 1 bar! Binary hydrides of transition elements are predominantly metallic in character. FeH$_{0.5}$ and NiH$_{0.5}$ are two examples, both having negative ΔH_{des} of −20 kJ·mol^{-1}. Therefore, such metallic hydrides are not stable and are only formed under extremely high hydrogen pressures.

Binary ionic hydrides are too stable (high hydrogen affinity) whereas binary metallic hydrides are too unstable (low hydrogen affinity). Therefore to tune the thermodynamics, both hydrides can combine to form a ternary hydride system AB$_x$H$_n$ (with A: ionic and B: metallic). For example, Mg$_2$NiH$_4$ contains 3.6 wt.% H$_2$, requiring a T_{des} of 553 K at 1 bar and LaNi$_5$H$_6$ has a capacity of 1.4 wt.% with a T_{des} of 298 K at 1.8 bar. Although these intermetallic hydrides have relatively low gravimetric capacities, they often have good volumetric density of hydrogen atoms within the host lattice, reaching up to 115 kg·m^{-3} for LaNi$_5$H$_6$ (much higher than the density of liquid hydrogen of 70 kg·m^{-3}, due to the absence of repulsions between hydrogen molecules and thus a closer packing). By comparing LiH and LaNi$_5$H$_6$, there is a compromise to be found between high gravimetric capacity and good working conditions. In the following, the different families of intermetallics studied so far for hydrogen storage, as classified by Sandrock [41], are briefly described.

10.3.1 Elements

Most of the elements can be hydrogenated under appropriate pressure-temperature conditions. Unfortunately, only vanadium is in the range for practical applications (1–10 bar, 0–100°C) through the reversible formation of the VH$_2$ hydride. There is still today some interest in solid solutions of V and other metals (see Section 10.3.6). Palladium has been studied for more than 100 years for its hydrogen storage properties, but this element is very expensive, stores only 0.6 wt.% of hydrogen (PdH$_{0.6}$), and requires heating well above 100°C for releasing hydrogen with a fast rate.

Most of the hydrides of alkali and alkaline-earth metals are very stable and therefore unable to desorb hydrogen at temperatures below 400°C (even under dynamic vacuum). A notable exception is magnesium. Because of its abundance, low cost and very high gravimetric storage capacity, magnesium is still today largely studied for hydrogen storage applications. Even though MgH_2 (7.6 wt.% hydrogen content) was observed at the end of the 19th century, its thermodynamic data were reported only in 1955 [42]. The hydrogenation of Mg is very exothermic (enthalpy of -75 kJ·mol^{-1} H_2) leading to a 1 bar equilibrium pressure at about 330°C. In addition, Mg can be easily oxidized and the hydrogen absorption/desorption kinetics are slow.

The first method for enhancing the hydrogenation kinetic of Mg is ball-milling (mechanical milling), which brings nanocrystallinity [43, 44]. The kinetic improvement is explained by the creation of defects that act as nucleation sites for the hydride phase and grain boundaries that facilitate hydrogen diffusion into the material. In order to further increase the kinetics, a wide range of catalysts has been added to magnesium, such as Pd and other transition metals, oxides or even hydrides. In fact, the best improvement has been obtained by ball-milling Mg with oxides such as Nb_2O_5 [45, 46]. Full hydrogenation of Mg in a few seconds has been reported at 200°C or even lower temperatures for such ball-milled Mg-Nb_2O_5 materials.

The literature on this subject is abundant, and it is now quite well admitted that milling with an oxide gives a smaller particles size than when the milling is performed with a metal [47]. This could be explained by the role of the oxide as a process control agent, which acts as a lubricant and cracking phase upon milling. True catalytic effects could be also responsible for the kinetic enhancement. A possible interpretation relies on the formation of ternary Mg–Nb oxides on the magnesium surface, as evidenced by transmission electron microscopy [48] and neutron diffraction [49]. However, more research is still needed to definitively conclude about the exact catalytic mechanism.

To conclude about the possible use of magnesium for reversible hydrogen storage, the kinetics limitations can be overcome through a careful optimization of the ball-milling conditions and the addition of appropriate catalytic species: Hydrogenation of Mg nanoparticles can be even possible at room temperature. However, the thermodynamic stability of MgH_2 is still a big issue as a temperature of 330°C is required to get 1 bar hydrogen equilibrium pressure, this temperature being too high for mobile applications.

Destabilization of MgH_2 through nano-confinement approach is currently pursued by different groups, but if some slight enthalpy modifications have been noticed, further work is still needed to confirm these results.

10.3.2 AB_5 Intermetallic Compounds

The near ambient hydrogen storage properties of the AB_5 intermetallics were found around 1970 at Philips Eindhoven when studying the $SmCo_5$ magnet alloy [50–52]. These intermetallics generally have a hexagonal crystal structure (Hauke phase, prototype $CaCu_5$, space group P6/mmm). A-element is usually one of the lanthanides, whereas B is Ni with many possible substitutions such as Co, Al, Mn, Fe, Cu, Sn, Ti, ... Most of the commercial AB_5 alloys are used as negative electrode in Ni-MH batteries and are based on the lanthanide mischmetal mixture for the A-side (for saving cost of lanthanide separation and purification) and Ni substituted by a significant amount of Co for the B-side (for improving the resistance to corrosion in the very alkaline KOH liquid electrolyte used in the Ni-MH batteries).

The van't Hoff plots of various representative AB_5 alloys are shown in Fig. 10.14 [41]. The tunability of the thermodynamic data is evident since the 25°C plateau pressure can be varied over three orders of magnitude depending on the composition. $LaNi_5$ has an equilibrium pressure of 1.8 bar at 25°C and the hysteresis (difference between absorption and desorption plateau pressures) is generally quite low for all the AB_5 alloys. In addition, these AB_5 alloys are easy to activate and their hydrogen absorption/desorption kinetics are fast at room temperature. Due to the heavy weight of the lanthanides, the hydrogen storage capacity is, however, limited to about 1.3 wt.%, which hinders the use of these alloys for most of mobile applications.

The AB_5 alloys decrepitate on the first cycle, due to the large volume change (about 20%) occurring upon the hydrogenation/dehydrogenation process, leading to fine powders, which are mildly pyrophoric if suddenly exposed to air. AB_5 metallurgy is rather well controlled and single-phase alloy can be easily prepared by vacuum induction melting techniques. Some AB_5 alloys as $CaNi_5$ are subject to disproportionation (formation of CaH_2 and Ni instead of the ternary $CaNi_5H_6$ hydride). To overcome this issue, it has been shown that the partial substitution of Ni by Sn or Al reduces significantly the risk of disproportionation.

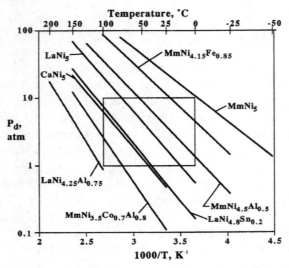

Figure 10.14 Van't Hoff plots for various AB_5 alloys. The near ambient pressure-temperature conditions are highlighted by a rectangle. Reprinted from Ref. [41] with permission from Elsevier.

10.3.3 AB_2 Intermetallic Compounds

The AB_2 alloys have most of the time a Laves phase crystal structure. Laves phase exists as three different structures types, cubic $MgCu_2$ (C15), hexagonal $MgZn_2$ (C14) and hexagonal $MgNi_2$ (C36), and they form a large group of intermetallics with a wide range of properties. The stability depends on various factors such as geometry, packing density, valence electron concentration or the difference in electronegativity. The A-elements are often from the IVA group (Ti, Zr, Hf), whereas the B-elements can be a variety of transition or non-transition metals with a preference for V, Cr, Mn, or Fe. As for the AB_5 alloys, a wide range of substitutions are possible for both A- and B-elements providing a large degree of tuning the hydrogen equilibrium pressure (cf. Fig. 10.15).

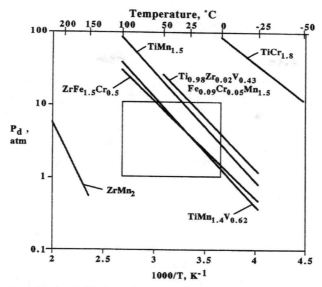

Figure 10.15 Van't Hoff plots for various AB_2 Laves phases. The near ambient pressure-temperature conditions are highlighted by a rectangle. Reprinted from Ref. [41] with permission from Elsevier.

The reversible hydrogen storage capacities of the AB_2 alloys are similar to those of the AB_5 ones (around 1.3 wt.%) even if their hydrogen contents can be higher, due to some non reversible hydrogen uptake [53, 54]. For instance, $ZrMn_2$ can absorb up to 1.2 H atom per metal ($ZrMn_2H_{3.6}$) corresponding to a hydrogen content of 1.77 wt.%, but the reversible capacity is limited to 0.9 wt.% as the hydride is rather stable (1 bar equilibrium pressure at 167°C). The $TiCr_{1.8}$ alloy is also able to absorb a large amount of hydrogen (2.43 wt.%), but only 0.85 wt.% is reversible under moderate pressure-temperature conditions.

AB_2 alloys are generally difficult to activate, more than AB_5, and are quite sensitive to impurities present in hydrogen: The alloys are easily passivated by an oxide layer. For the same reason, the activated powders are highly pyrophoric when exposed to air and therefore the handling of fine powders of AB_2 Laves phases requires a great care. In addition, the large-scale production of AB_2 alloys is more difficult than that of AB_5 alloys. Because of the high melting points of the principal elements (Ti: 1670°C, Zr: 1855°C, Cr: 1863°C), it is very difficult to use vacuum induction melting. More expensive cold crucible vacuum arc melting techniques are usually required.

The AB_2 alloys offer some advantages over the AB_5 in term of cost (as they avoid the use of lanthanides which are becoming nowadays subject to severe politico-economical issues), at least if the A-element is mostly Ti and not Zr. A strong effort is also devoted by many groups to replace vanadium, which is very expensive, by ferrovanadium, a low cost product used by the steel industry.

10.3.4   AB Intermetallic Compounds

The first AB intermetallic hydride was the ZrNi compound reported in 1958. Unfortunately, $ZrNiH_3$ has a 1 bar equilibrium pressure at about 300°C, too high for many applications. The first practical AB hydride was TiFe as demonstrated in 1974 by Reilly et al. [55]. TiFe and its substitutional modifications remain today the best of the AB alloys. TiFe-based AB alloys have a body-centered-cubic structure (bcc) (prototype CsCl, space group Pm-3m). Generally, these intermetallics have two plateaus (two distinct hydrides) with reasonable pressures at room temperature.

The thermodynamics of TiFe can be tuned by partial substitution of Fe by Mn or Ni as shown in Fig. 10.16 [41]. TiFe and $TiFe_{0.85}Mn_{0.15}$ show good volumetric and gravimetric capacities, competitive with the best AB_5 and AB_2 alloys. Namely, reversible hydrogen storage of about 1.5 wt.% is obtained at ambient conditions. $TiFe_{0.8}Ni_{0.2}$ is not so useful as it has a lower reversible capacity (0.8 wt.%), due to the high plateau pressure for the formation of the fully hydrogenated compound ($TiFe_{0.8}Ni_{0.2}H_{1.4}$). Only the lower plateau (0.4 H per metal) is accessible at moderated temperatures.

The activation of the AB alloys is quite difficult and can require high hydrogen pressure for several days in order to break the oxides covering layer. On the positive side, these materials are little pyrophoric. One of the main advantages of the TiFe-based compounds is their low cost, significantly lower than those of the AB_5 (no lanthanide) or AB_2 alloys (no vanadium).

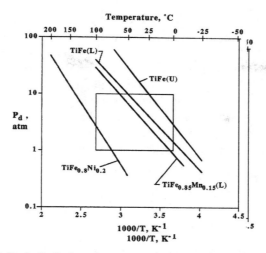

Figure 10.16 Van't Hoff plots for various *bcc* AB alloys. The near ambient pressure-temperature conditions are highlighted by a rectangle. L indicates lower plateau and U upper plateaus. Reprinted from Ref. [41] with permission from Elsevier.

10.3.5 A$_2$B Intermetallic Compounds

Mg$_2$Ni is a well-known example of A$_2$B intermetallic compound able to form a hydride. There has been an extensive study on Mg$_2$Ni since the 1960s [56], both from fundamental and applications point of view. This compound can form Mg$_2$NiH$_4$, which is actually a complex hydride with a well-defined (NiH$_4$)$^{4-}$ entity, similarly to the (CoH$_5$)$^{4-}$ entity found in Mg$_2$CoH$_5$ and (FeH$_6$)$^{4-}$ entity found in Mg$_2$FeH$_6$. The Mg$_2$NiH$_4$ hydride is quite stable with an enthalpy of formation of −65 kJ·mol^{-1} H$_2$, a value more exothermic than the targeted one of −35 kJ·mol^{-1} required for reversible hydrogen storage near ambient conditions. As a consequence, the 1 bar equilibrium pressure is obtained at the high temperature of 255°C. Nevertheless, this hydride is interesting due to its rather high gravimetric capacity of 3.6 wt.%. Numerous attempts have been made to tune the thermodynamics of this system with very little success due to the lack of possible substitution in the Mg$_2$Ni compound.

The hydrogen absorption-desorption kinetics of Mg$_2$Ni can be strongly improved by nano-structuration using ball-milling process

[57, 58] or addition of catalysts such as Pd nanoparticles [59]. Figure 10.17 shows the enhancement of the hydrogenation kinetics of Mg_2Ni at 150°C under 10 bars of hydrogen. Whereas Mg_2Ni is not fully hydrogenated after 240 min of exposure to hydrogen, the sample ball-milled with carbon and coated with 5 wt.% of Pd nanoparticles shows a much faster rate of hydrogen uptake [60]. The ball-milling step with carbon additives [61, 62] has been shown to favor the breaking and even the reduction of the oxides passivating layer present at the surface of the Mg_2Ni particles.

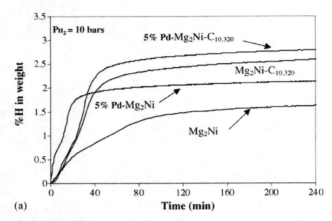

Figure 10.17 Enhancement of the hydrogenation kinetic of Mg_2Ni at 150°C under 10 bars of hydrogen by ball-milling with carbon and deposition of Pd nanoparticles (5 wt.%).

10.3.6 Solid Solutions

Several solid solution alloys form reversible hydrides in particular those based on the elements Pd, Ti, Zr, Nb, and V. Although the thermodynamic properties of Pd-based solid solutions are favorable, they have low gravimetric capacities, never exceeding 1.0 wt.% H_2, and of course they are prohibitively expensive. Ti- and Zr-based solid solutions form too stable hydrides, even when highly alloyed.

Vanadium can form VH_2 with useful reversible hydrogen storage, so it is logical that V-based solid solutions offer further opportunities. The first extensive work on V-based solid solutions was done by Reilly et al. in the early 1970s and a strong regain of interest on these materials is observed since 2000, especially in

Japan [63]. The V-based alloys crystallize in the body-centered-cubic (*bcc*) structure and their di-hydrides generally form a face-centered-cubic (*fcc*) structure. It has been shown that the lattice parameter plays an important role on the absorption properties. For example, in V–Ti–Cr–Fe alloys, the maximum reversible capacity is reached for a parameter $a = 3.036$ Å [64].

The V–Ti–Fe *bcc* alloys appear among the most promising solid solution materials for hydrogen storage. By varying x from 0 to 0.075 in $(V_{0.9}Ti_{0.1})_{1-x}Fe_x$, the equilibrium pressure can be affected by more than one order of magnitude, without affecting the capacity. These alloys are attractive, since they absorb large amounts of hydrogen (up to 3.7 wt.%) with more than 2 wt.% reversible at room temperature [65, 66]. Because, they contain a large proportion of vanadium, these alloys are, however, very costly. In order to reduce the price, the replacement of V by ferrovanadium (FeV) is pursued. The $TiCr_{1.2}(FeV)_{0.6}$ composition can reach 3.2 wt.% of hydrogen content, of which 2.0 wt.% are fully reversible in ambient conditions [67].

10.4 Complex Hydrides

All the systems described in the previous sections suffer from severe drawbacks: low volumetric capacity for high-pressure gas cylinders, low temperature usage for liquid hydrogen in cryogenic tanks, low temperature and low volumetric capacity for the physisorption materials, and low gravimetric capacity in metal hydrides. Therefore, a reversible and high-capacity system close to ambient working conditions is still required. Whereas in metal hydrides the hydrogen atoms are located in the interstitial positions within the crystal structure of the metal host, in complex hydrides, covalent bonds are found between the hydrogen atoms and an element in the host structure, forming anionic molecular clusters that are counter-balanced by alkali/alkaline earth cations, e.g., $LiBH_4$: $Li^+[BH_4]^-$. Complex hydrides are usually formed from light elements, which ensures a high hydrogen storage capacity due to (1) the lightweight of the other elements, and (2) the fact that more than one hydrogen atom can be bonded per metal atom. Some examples of complex hydrides include borohydrides (BH_4^- also called tetrahydroborates), alanates (AlH_4^-), silanides (SiH_3^-) and these compounds are described in the following.

10.4.1 Borohydrides

Schlesinger at al. prepared the first borohydrides (e.g., Al(BH$_4$)$_3$, NaBH$_4$) by reaction between diborane B$_2$H$_6$ and different metal alkyls in the 1940s, but the results were not published before 1953 [68]. Today, the borohydrides (NaBH$_4$) are widely used in organic chemistry for the reduction of aldehydes or ketones to alcohols. About hydrogen storage, the important property is the high gravimetric capacity of the borohydrides reaching 18.4 wt.% for LiBH$_4$ and even 20.7 wt.% for Be(BH$_4$)$_2$, but the toxicity of the latter excludes its use as hydrogen storage material. The stability of the borohydrides is governed by the charge transfer from the metal cation to the BH$_4^-$ anion. Thus, higher is the difference of electronegativity between the metallic cation and boron (χ = 2), higher is the stability of the borohydride. A linear relation has even been found between the heat of formation of the borohydrides and the Pauling electronegativity of the metal cation [69]. For instance, KBH$_4$ (χ_K = 0.82), is very stable and releases hydrogen at 585°C upon melting, whereas Al(BH$_4$)$_3$ (χ_{Al} = 1.47), is a liquid at room temperature and releases hydrogen at only 40°C. Generally, NaBH$_4$ and KBH$_4$ are considered as too stable compounds and are almost not investigated for hydrogen storage applications.

10.4.1.1 Lithium borohydride: LiBH$_4$

At the industrial scale, LiBH$_4$ is produced from NaBH$_4$, so it is of importance to briefly present first how is synthesized NaBH$_4$. Usually, boron is provided as a borate and the hydrogenation is performed through a transfer of hydrogen from an alkali metal hydride. The most common method for the commercial production of NaBH$_4$ is based on the use of a mineral oil suspension of NaH as hydriding agent. Sodium dispersion may be hydrogenated directly in the mineral oil and this oil provides also heat sink, which facilitates the temperature control of the exothermic reaction with borates [68]. The reaction is usually carried out at 250–270°C using trimethyl borate as the initial reactant and the synthesis can be resumed according to the following equation:

$$4\,NaH + B(OCH_3)_3 \rightarrow NaBH_4 + 3\,NaOCH_3 \qquad (10.5)$$

Concerning the synthesis of $LiBH_4$ itself, a wet chemical reaction between $NaBH_4$ and $LiCl$ (metathesis reaction) is made in ethers [70], usually in the presence of isopropyl amine, according to the following reaction:

$$NaBH_4 + LiCl \rightarrow LiBH_4 + NaCl \quad (10.6)$$

This metathesis reaction leads to the formation of sodium chloride salt and this by-product must be separated from $LiBH_4$. This requires a specific solvent for either $LiBH_4$ or the by-product, but most solvents have a finite solubility in the $LiBH_4$ product and are persistent. The solvent-free synthesis of $LiBH_4$ is therefore of a high interest and several recent studies have been devoted to the development of such dry synthesis methods. The synthesis by direct hydrogenation of the elements is very difficult due to the boron chemical inertness, but it has been shown that this issue must be partially overcome when starting from lithium-boron alloys [71]. Another promising route for the $LiBH_4$ synthesis consists of the highly energetic ball-milling of LiH with boron under hydrogen atmosphere [72]. Although the yield of $LiBH_4$ formation reported in this first study is low (only 27%), the optimizations of the milling conditions should largely enhance this value in a near future.

$LiBH_4$ crystallizes in an orthorhombic unit cell (Pnma) at room temperature and undergoes a structural transition at about 116°C leading to a hexagonal phase (P6$_3$mc) [73, 74]. This structural transition is accompanied by a jump of the electrical conductivity by three orders of magnitude [75]: The conductivity reaches 10^{-3} S·cm^{-1} immediately above the transition temperature (cf. Fig. 10.18). It has been shown, especially by ^{7}Li NMR, that this conductivity is purely ionic [76, 77], which makes pristine $LiBH_4$ a good lithium ionic conductor above 116°C. Interestingly, the high-temperature hexagonal phase of $LiBH_4$ has been recently stabilized at room temperature by synthesizing $LiBH_4$-LiI solid solutions [78, 79]. These materials have ionic conductivities of the order of 10^{-4} S·cm^{-1} at 25°C and could be used as solid electrolytes in lithium-ion batteries.

Although $LiBH_4$ is a well-known compound, its thermal decomposition follows a complex process, which is still today the subject of investigations. Briefly, after the $LiBH_4$ melting at 286°C, the dehydriding reaction proceeds slowly from the molten salt and leads in at least two steps to a mixture of amorphous boron and lithium hydride (cf. DSC curve in blue on Fig. 10.19). LiH being a very stable hydride, its decomposition does occur only above 600°C under vacuum and therefore this step is usually not considered in the hydrogen release process, thereby limiting the hydrogen release capacity of $LiBH_4$ to 13.8 wt.% (versus a total hydrogen content of 18.4 wt.%). One of the intermediate compounds formed upon the decomposition process of $LiBH_4$ has been recently identified as lithium dodecaborane ($Li_2B_{12}H_{12}$) by both computational approaches [80] and by experimental results like Raman spectroscopy [81] and X-ray diffraction [82]. The overall desorption process ($LiBH_4$ giving LiH + B) is highly endothermic (70 kJ·mol^{-1} H$_2$) and the rehydrogenation of the LiH + B mixture has been only reported at 350 bars H$_2$ and 600°C [83]. According to these high pressure-temperature conditions, it is evident that bulk $LiBH_4$ cannot be used for reversible hydrogen storage applications.

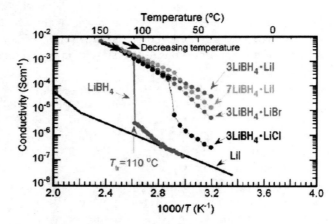

Figure 10.18 Electrical conductivities obtained by AC impedance spectroscopy measurements as a function of the temperature for $LiBH_4$, LiI, and $LiBH_4$-LiX solid solutions (with X = Cl, Br, I). Reprinted with permission from Ref. [79]. Copyright (2009) American Chemical Society.

The hydrogen release kinetic of $LiBH_4$ can be strongly enhanced by ball-milling with metal oxides (SiO_2) or chlorides ($TiCl_3$)

[84, 85]. In the first case, the oxides act as catalysts (no reaction with $LiBH_4$ upon milling) and can lower the onset dehydrogenation temperature of $LiBH_4$ to less than 200°C. The exact mechanism of the catalytic effect with oxides is, however, unidentified. About the chlorides-modified $LiBH_4$, the process is very different since a chemical reaction occurs between $LiBH_4$ and the chlorides ($TiCl_3$) upon milling. The resulting borohydrides, like $Ti(BH_4)_3$ for instance, can be decomposed at remarkably low temperatures (60°C), but these reactions are fully irreversible.

Another strategy for the enhancement of the desorption kinetic is the synthesis of $LiBH_4$ nanoparticles into nanoporous hosts [86–89]. This approach has been achieved using various mesoporous carbons (aerogels, silica replica) and various methods for the carbon impregnation (molten $LiBH_4$ at 300°C, $LiBH_4$ dissolved in ethers). In all cases, the rate of hydrogen release of $LiBH_4$ has been strongly enhanced together with a decrease of the onset dehydrogenation temperature to about 200°C (so below the melting point of $LiBH_4$). In addition to the kinetic effect, a modification of the desorption mechanism is noticed: It seems that the formation of intermediate compounds like $Li_2B_{12}H_{12}$ does not occur upon the hydrogen release for the carbon-encapsulated $LiBH_4$ nanoparticles [88, 89].

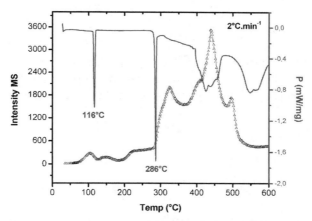

Figure 10.19 DSC profile (in blue) and TPD-MS curve of hydrogen emission (in black), for $LiBH_4$ heated at $2°C \cdot min^{-1}$. The two sharp endothermic peaks correspond to the structural transition of $LiBH_4$ at 116°C and the melting at 286°C.

10.4.1.2 Magnesium borohydride

With a hydrogen content of 14.8 wt.% and a more suitable enthalpy of formation than $LiBH_4$, $Mg(BH_4)_2$ seems promising for hydrogen storage. $Mg(BH_4)_2$ can be easily prepared from $MgCl_2$ and $LiBH_4$ in ether solutions. After drying at 145°C, a solvent-free crystalline α-$Mg(BH_4)_2$ material can be obtained [90, 91]. Upon heating, α-$Mg(BH_4)_2$ undergoes a phase transformation at about 190°C, before it releases hydrogen in several steps to form finally MgB_2 [92] (cf. Eq. 10.7). The weight loss between 290 and 500°C is 14 wt.% H_2 with a maximum release between 300 and 400°C. Small amounts of B_2H_6 can be detected by mass spectroscopy in the hydrogen desorbing flow, which is a strong drawback for applications such as PEM fuel cells. The decomposition mechanism can be described as follows:

$$Mg(BH_4)_2 \rightarrow MgH_2 + 2\,B + 3\,H_2 \rightarrow MgB_2 + 4\,H_2 \quad (10.7)$$

For the first decompositions step, the formation of $(B_{12}H_{12})^{2-}$ anion in an intermediate solid phase is possible [93] (as similarly demonstrated in the case of $LiBH_4$). Only the second step between MgH_2 and MgB_2 shows some reversibility with experimental capacities of about 4 wt.% at 350°C [94].

10.4.2 Alanates

The most intensively studied complex aluminum hydride is the sodium compound $NaAlH_4$, with a hydrogen content of 7.5 wt.% H_2. It decomposes in three steps according to the following equations:

$$3\,NaAlH_4 \rightarrow Na_3AlH_6 + 2\,Al + 3\,H_2 \quad (10.8)$$

$$Na_3AlH_6 \rightarrow 3\,NaH + Al + 3/2\,H_2 \quad (10.9)$$

$$3\,NaH \rightarrow 3\,Na + 3/2\,H_2 \quad (10.10)$$

The first decomposition step with a capacity of 3.7 wt.% has an equilibrium pressure of 1 bar at around 30°C. The second decomposition step releases additionally 1.9 wt.% of hydrogen with an equilibrium pressure of 1 bar at about 130°C. As NaH is a very stable hydride, the decomposition temperature for the third step is too high for technical applications and is therefore not considered as reversible. $NaAlH_4$ shows the typical structural changes of most of the alanates. The Al atom in $NaAlH_4$ is tetrahedrally coordinated, and the Al atom in the intermediate Na_3AlH_6 octahedrally coordinated. Pure aluminum metal is produced as a separated phase during the decomposition.

$NaAlH_4$ was first synthesized more than 50 years ago [95] as a substitute to the more expensive $LiAlH_4$ for reduction reactions in organic synthesis. The preparation was done by reaction of NaH and $AlBr_3$ in dimethyl ether. This complex hydride had not been initially considered as a hydrogen storage material, because it can effectively not be rehydrogenated unless very harsh conditions were used. However, in 1997, Bogdanovic and Schwickardi showed that $NaAlH_4$ can be a reversible system by doping it with titanium compounds, such as $Ti(OBu)_4$ [96]. The particular thermodynamic properties combined with the high hydrogen contents (7.5 and 10.6 wt.% for $NaAlH_4$ and $LiAlH_4$, respectively) are the reasons for the high interest in alanates nowadays.

The catalyst precursor that is frequently used is $TiCl_3$ and this compound is often added to $NaAlH_4$ by ball-milling [97, 98]. During the ball-milling, the dopant is highly dispersed within the hydride material and $TiCl_3$ is reduced to metallic titanium. The exact nature of the catalyst during the hydrogenation/dehydrogenation reaction remains unclear and highly disputed, owing to the difficult analytical problems. From extensive X-ray absorption studies of titanium-doped $NaAlH_4$, it was concluded that the formal titanium valence is zero and does not change during the cycling process [99, 100]. The titanium atoms are largely dispersed in the metallic aluminum phase in a state close to that of the $TiAl_3$ alloy. During the dehydrogenation reaction, the $TiAl_3$ alloy produced in the doping reaction is diluted with the generated aluminum.

The discovery that $NaAlH_4$ can be reversibly hydrogenated thanks to Ti-based catalysts was a real breakthrough. The rehydrogenation properties of doped $NaAlH_4$ are outstanding compared to the bulk un-doped material: Rehydrogenation times of few minutes can be achieved at 150°C under 100 bars of hydrogen. The Ti-doped $NaAlH_4$ material is experimentally able to store reversibly around 5 wt.% of hydrogen according to Eqs. 10.8 and 10.9 (cf. Fig. 10.20 with a 0.5 mol.% Ti-doping).

With the aim of increasing the hydrogen storage capacity, $LiAlH_4$ would be of course a promising hydrogen storage material with its 10.6 wt.% hydrogen content. The different decomposition steps of $LiAlH_4$ have been largely studied since 1965 [102]. It has been shown that the first decomposition step around 200°C leading to Li_3AlH_6 and Al (5.3 wt. H_2) is exothermic (−10 kJ·mol^{-1}) meaning inadequate for reversible hydrogen storage. In a further heating, Li_3AlH_6 is endothermically (+25 kJ·mol^{-1} H_2) decomposed into LiH and Al (2.6 wt.% H_2). The stability of $LiAlH_4$ at room temperature is caused by the slow kinetics of the solid-state transformation to the hexa-alanate and not due to thermodynamic reasons. Actually, when $LiAlH_4$ is ball-milled, especially in the presence of catalysts such as $TiCl_3$, $LiAlH_4$ is decomposed in few minutes into Li_3AlH_6 and Al [103].

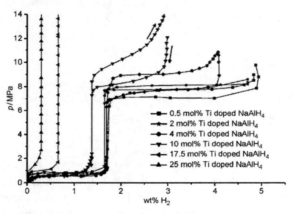

Figure 10.20 Pressure-composition-isotherms at 160°C of $NaAlH_4$ doped with different Ti contents. Reproduced from Ref. [101] with permission from the PCCP Owner Societies.

Other alanates that have high hydrogen contents are the alkaline earth compounds: $Mg(AlH_4)_2$ and $Ca(AlH_4)_2$. With respective total hydrogen contents of 9.3 and 7.9 wt.% for the Mg and Ca-based compounds, both materials are interesting for hydrogen storage. The synthesis of these materials is usually done by simple metathesis process between $NaAlH_4$ and $MgCl_2$ or $CaCl_2$ and can be carried out in solution [104] or in the solid state with the ball-milling method [105]. Unfortunately, from measurements of the decomposition enthalpies, it must be concluded that both alanates do not react reversibly under technologically relevant conditions. Mg(AlH$_4$)$_2$ is an unstable hydride, with a decomposition enthalpy of about 0 kJ·mol^{-1} H$_2$ [106, 107]. The decomposition occurs at about 130°C in one step without observation of an octahedral intermediate state, and it releases 7 wt.% of hydrogen according to this reaction:

$$Mg(AlH_4)_2 \rightarrow MgH_2 + 2\,Al + 3\,H_2 \tag{10.11}$$

Further heating above 300°C can lead to the decomposition of MgH$_2$ and to the formation of the Al$_3$Mg$_2$ alloy as shown by X-ray diffraction studies [107, 108]. The total hydrogen release is in this case around 9.0 wt.%.

In contrast to this pathway, calcium alanate Ca(AlH$_4$)$_2$ decomposes in a two-step mechanism with CaAlH$_5$ as an octahedral intermediate.

$$Ca(AlH_4)_2 \rightarrow CaAlH_5 + Al + 3/2\,H_2 \tag{10.12}$$

$$CaAlH_5 \rightarrow CaH_2 + Al + 3/2\,H_2 \tag{10.13}$$

About 2.9 wt.% H$_2$ is released in the first step, and additional 3.0 wt.% H$_2$ in the second step, thereby producing CaH$_2$ and aluminum. The first decomposition step is exothermic (-7.5 kJ·mol^{-1}) and therefore not reversible, but the second step is endothermic with a decomposition enthalpy of 32 kJ·mol^{-1}. This value lies in the right region for reversible hydrogen storage, but it is not yet clear whether the material is a reversible hydride compound or not, as no full rehydrogenation to a fully characterized CaAlH$_5$ has been achieved to date. In contrast to the isolated (AlH$_6$)$^{3-}$ units in the alkali metal alanates, the octahedral units in the calcium alanate intermediate are bridged over two corners, producing polyanionic chains with the empirical formula CaAlH$_5$ [109].

10.4.3 Silanides

The MSiH$_3$ compounds with heavy alkali metals (M = K, Rb, Cs) have been reported since the 1960s [110–113]. They are usually prepared by reaction between the alkali metal and silane SiH$_4$ in monoglyme at low temperatures. The direct hydrogenation of the corresponding MSi Zintl phases have been only very recently reported showing that reversible reactions are possible between MSi silicides and MSiH$_3$ silanides and thus that silanides can be considered as a new class of complex hydrides for reversible hydrogen storage. The MSi silicides are Zintl phases [114–116], meaning that silicon forms a polyanion that follows the Zintl–Klemm concept: Silicon adopts a structure which is similar to that of a neighboring isoelectronic element i.e., phosphorus. Namely, for the MSi silicides, silicon is present as [Si$_4$]$^{4-}$ tetrahedra identical to the P$_4$ molecules found in white phosphorus. NaSi crystallizes in a monoclinic unit cell (space group C2/C) [115] whereas, KSi, RbSi, and CsSi are found in cubic ones (space group P-43n) [116] (cf. Fig. 10.21).

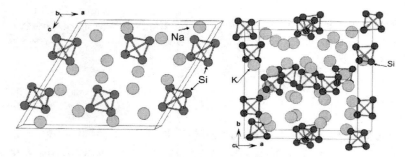

Figure 10.21 Crystal structures of the MSi Zintl phases. Monoclinic unit cell for NaSi and cubic unit cells for MSi with M = K, Rb, Cs. The [Si$_4$]$^{4-}$ tetrahedra are highlighted in green.

The reaction between KSi and KSiH$_3$ is indeed perfectly reversible allowing the storage of 4.3 wt.% of hydrogen [117] as shown in Fig. 10.22. The hydrogenation of KSi can be completed in few hours at 100°C under 50 bars H$_2$, whereas the desorption process requires a temperature of 200°C for the release of the whole hydrogen content in 2 h. From measurements of hydrogen equilibrium pressures at different temperatures, the thermodynamic data of the KSi/α-KSiH$_3$ equilibrium have been determined [117]. A desorption enthalpy of +23 kJ·mol^{-1} H$_2$ has been found, confirming that reversibility near ambient conditions is possible. More surprisingly, a very low entropy change of 55 J·K^{-1}·mol^{-1} H$_2$ was obtained, a much lower value than the entropy usually obtained for classical intermetallic hydrides (around 130 J·K^{-1}·mol^{-1} H$_2$ corresponding to the entropy change between hydrogen molecules in the gas phase and H atoms well localized in a rigid solid lattice [118]).

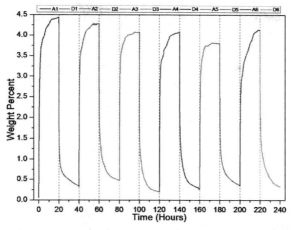

Figure 10.22 Reversible hydrogen storage based on the KSi/KSiH$_3$ equilibrium. Absorption at 100°C/50 bars and desorption at 200°C/< 1 bar.

This abnormally low loss of entropy upon hydrogenation could be related to the huge mobility of the (SiH$_3$)$^-$ anion in the α-KSiH$_3$ structure at room temperature and thus to the high entropy of this hydride. The SiH$_3^-$ mobility has been observed by Neutron Powder Diffraction on a deuterated sample with high B$_{iso}$ (isotropic displacement temperature) values for both the D atoms and the Si atoms. This has been similarly reported for the MGeH$_3$ phases with a rapid rotation of the (GeH$_3$)$^-$ anion around the three-fold axis at room temperature [119]. Upon cooling, α-KSiH$_3$ undergoes a structural transition around 270 K to orthorhombic β-KSiH$_3$ for which the SiH$_3^-$ anions are well fixed pyramidal entities [120].

In order to enhance the hydrogenation kinetics of KSi, ball-milling experiments with carbon have been performed. If the initial rate of hydrogen absorption can be effectively strongly increased, this leads to an absorption capacity limited to only 1.3 wt.% (0.9 H per KSi instead of 3 H when forming the KSiH$_3$ silanide) due to a disproportionation reaction leading to KH and the clathrate K$_8$Si$_{46}$ [121]. This disproportionation reaction is obviously not desired as KH is a very stable hydride and therefore the hydrogen cannot be easily desorbed. The origin of the modification of the reaction path (as illustrated on Fig. 10.23) could rely on the exothermicity of the hydrogenation reaction, as already explained for other intermetallics (for example, a slow hydrogenation of Zr_2Fe leads to the ternary Zr_2FeH_5 hydride, whereas a fast hydrogenation forms ZrH_2 and $ZrFe_2$ instead [122]). If the hydrogenation is fast, there is not enough time to dissipate the heat generated by the exothermicity of the reaction and there is locally a strong increase of the particles temperature, which could modify the reaction mechanism and thus favors more thermodynamically stable phases (i.e., KH).

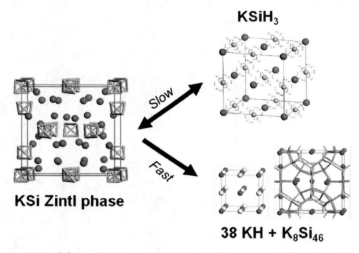

Figure 10.23 Schematic representation of the two possible routes upon hydrogenation of KSi. Slow absorption forms $KSiH_3$, whereas fast absorption leads to a disproportionation reaction.

If the $KSi/KSiH_3$ system is very promising with a good hydrogen storage capacity of 4.3 wt.% (volumetric capacity of 53 g $H_2 \cdot L^{-1}$) and suitable thermodynamic properties allowing reversibility near ambient conditions (1 bar equilibrium pressure at about 140°C), some efforts are still needed to manage the thermal issues and to get fast absorption/desorption kinetics without disproportionation reaction.

Similarly, the hydrogenation properties of Rbi and CsSi Zintl phases have been recently investigated. As for KSi, the formation of $MSiH_3$ silanides (M = Rb, Cs) is possible by direct hydrogenation at 100°C of the MSi silicides. No disproportionation has been noticed in the case of RbSi and CsSi, even on ball-milled powders showing fast hydrogen absorption/desorption kinetics. Both RbSiH$_3$ and CsSiH$_3$ can be reversibly desorbed allowing the storage of 2.6 and 1.85 wt.% H$_2$, respectively. The thermodynamics of the MSi/MSiH$_3$ equilibrium (M = Rb, Cs) have been investigated from experimentally determined van't Hoff plots and confronted to that of KSi/KSiH$_3$ (cf. Fig. 10.24a).

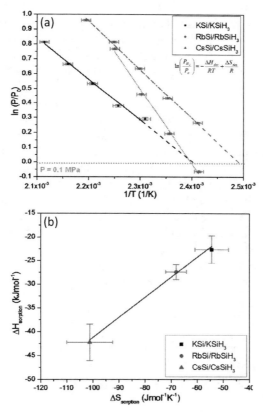

Figure 10.24 (a) Van't Hoff plots for the MSi/MSiH$_3$ equilibrium with M = K, Rb, Cs. (b) Linear enthalpy-entropy compensation effect.

It has been shown that larger is the alkali cation, higher is the enthalpy of hydrogenation (more exothermic) and therefore more is the MSiH$_3$ silanide stable. Following the same direction, the entropy change is also the more negative with the bigger alkali cation (101 J·K^{-1}·mol^{-1} H$_2$ in the case of CsSiH$_3$, a value much higher than that calculated for KSiH$_3$; 54 J·K^{-1}·mol^{-1} H$_2$). This shows that bigger is the alkali cation, less is the SiH$_3^-$ anion mobile in the crystal lattice. The exact origin of this huge modification of entropy change depending of the alkali cation is not yet fully understood. Nevertheless, as the enthalpy and entropy increase simultaneously and offset each other, a compensation effect is noticed for the MSi/MSiH$_3$ systems (M = K, Rb, Cs). A linear relationship can be even plotted as shown in Fig. 10.24b. As a result of this compensation effect, 1 bar equilibrium pressure is obtained at almost the same temperature of 140°C for all three systems.

Such enthalpy-entropy compensation effect is well established in the field of physisorption on zeolites. For example, in the case of the physisorption of linear alkanes on microporous zeolites [123], it has been shown that longer is the chain, higher is the physisorption enthalpy (stronger is the binding) and higher is the entropy loss (less is the linear alkane mobile). In the field of metal hydrides, if some entropy changes have been noticed due to mechanical stresses [124, 125], this is the first time that such a linear enthalpy-entropy compensation effect is reported for a family of complex hydrides, especially with such a large range of entropy modifications.

With the aim of increasing the 4.3 wt.% hydrogen capacity obtained with KSiH$_3$, the hydrogenation properties of the lighter alkali silicides, NaSi and LiSi have been also investigated. Unfortunately, no silanide are formed with these compounds: NaSi always disproportionate into NaH and Na$_8$Si$_{46}$ (whatever are the pressure-temperature conditions) [121] and LiSi forms LiH and Si upon hydrogenation at 300°C [126]. LiSiH$_3$ is still today a hypothetical compound that has never been reported, whereas NaSiH$_3$ has never been isolated as a solid-sate compound without solvent adducts [127].

10.5 Amides and Imides

As early as 1910, Dafert et al. reported a reaction between Li$_3$N and H$_2$ that forms a compound with a Li$_3$NH$_4$ composition [128]. These pioneering studies regain a lot of interest in 2002, when Chen et al. investigated the hydrogenation properties of lithium nitride [129]. It was emphasized that this compound could be a reversible hydrogen storage material. Indeed, lithium nitride is interesting, due to its lightweight, which leads to a hydrogen absorption capacity of 11.4 wt.% related to the formation of a mixture of 1 mole of $LiNH_2$ and 2 moles of LiH [129, 130]. The overall composition of the mixture is therefore Li_3NH_4 in agreement with the very early report made by Dafert et al. [128].

10.5.1 Hydrogenation of Li_3N

The pressure-composition-isotherms (PCI) at 528 K of Li_3N are presented in Fig. 10.25 [129]. Upon the first hydrogenation, about 3.8 H atoms are absorbed per Li_3N formula unit, which corresponds to 10.9 wt.% uptake. The desorption step of the first cycle cannot return to the origin, as only two hydrogen atoms are released. For the second cycle, the isotherm shows a good reversibility with a hydrogen capacity of about 5.2 wt.%. After full hydrogenation, the sample is composed of $LiNH_2$ and LiH phases in accordance with the following two-step reaction path:

$$Li_3N + H_2 \rightarrow Li_2NH + LiH \qquad (10.14)$$

$$Li_2NH + H_2 \rightarrow LiNH_2 + LiH \qquad (10.15)$$

Upon desorption, Li_3N is not re-formed, and the dehydrogenated sample consists of a mixture of Li_2NH and LiH phases, so that the reversible hydrogen storage is actually limited to the reaction between lithium imide (Li_2NH) and lithium amide ($LiNH_2$) (Eq. 10.15).

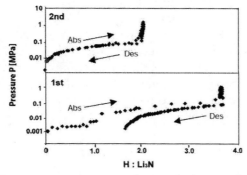

Figure 10.25 Pressure-composition-isotherms at 528 K of Li_3N: 1st and 2nd absorption-desorption cycles showing that only the reaction between Li_2NH and $LiNH_2$ is reversible. Reprinted from Ref. [129] by permission from Macmillan Publishers Ltd: Nature, copyright (2002).

The above results do not come as a surprise, since the enthalpies for reactions (10.14) and (10.15) are −165 and −65 kJ·mol^{-1} H$_2$, respectively. Reaction (10.15) is the only step viable for a practical hydrogen storage, which limits the capacity to 5.2 wt.%. In order to avoid the presence of inactive LiH formed during reaction (10.14) and to increase the reversible storage capacity, it is therefore relevant to use Li$_2$NH instead of Li$_3$N as the starting material. This allows the increase in hydrogen storage capacity up to 6.5 wt.%.

The exact mechanism for hydrogen absorption/desorption in the Li-N-H system has been further investigated. Pure lithium amide (LiNH$_2$) decomposes to lithium imide (Li$_2$NH) and ammonia around 650 K [131], whereas lithium hydride liberates hydrogen at temperatures above 773 K [132]. By thoroughly ball-milling LiNH$_2$ and LiH, hydrogen can be desorbed at a temperature as low as 523 K, with a very low ammonia release [133], as shown by TPD-MS experiments. Today, two possible mechanisms are still reported for the hydrogen desorption of the LiNH$_2$-LiH mixture: the ammonia-mediated mechanism and the direct solid–solid interaction mechanism. The ammonia-mediated mechanism was especially put forward by Hu et al. [134] and Ichikawa et al. [135]. The thermal decomposition of a material formed of two pellets stacked onto each other, one being LiH and the other one LiNH$_2$, was studied by a TPD-MS technique. When the vector gas (He) went through the LiNH$_2$ layer first and then through the LiH layer, only hydrogen was detected in the desorbing flow (cf. Fig. 10.26 right). In contrast, when the vector gas goes through the LiH layer first and then through the LiNH$_2$ layer, ammonia was detected (cf. Fig. 10.26 left) in the temperature range between 573 and 673 K. This suggests that ammonia formed by the LiNH$_2$ decomposition reacts with LiH to produce hydrogen. Therefore, it was deduced that the hydrogen desorption process for the LiH-LiNH$_2$ mixture follows a two-step reaction:

$$2\ LiNH_2 \rightarrow Li_2NH + NH_3 \quad\quad\quad\quad (10.16)$$

$$LiH + NH_3 \rightarrow LiNH_2 + H_2 \quad\quad\quad\quad (10.17)$$

The ultra-fast exothermic reaction of LiH and NH$_3$ prevents an ammonia contamination of the hydrogen desorbing flow. Moreover, this simply means that the hydrogen desorption kinetic of the

LiH-LiNH$_2$ mixture is controlled by the decomposition rate of LiNH$_2$ (reaction 10.16).

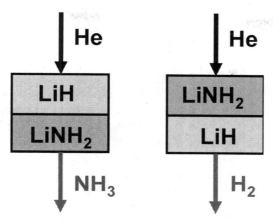

Figure 10.26 Illustration of the ammonia-mediated desorption mechanism for the Li-N-H system. The released gas depends of the stacking order of the LiH and LiNH$_2$ pellets.

On the contrary, Chen et al. claimed that the hydrogen desorption of the Li-N-H system occurs through a direct solid–solid interaction between LiH and LiNH$_2$ [130] and, especially, by the coupling between H$^{\delta-}$ in LiH (negatively charged hydrogen) and H$^{\delta+}$ in LiNH$_2$ (positively charged hydrogen). The hydrogen desorption rate is controlled by the LiH/LiNH$_2$ interface reaction in the early stage and by mass transport (Li and H atoms) through the Li$_2$NH imide layer formed between the LiH and LiNH$_2$ phases in the later stage of the reaction. David et al. demonstrated that the transformation between LiNH$_2$ and Li$_2$NH during hydrogen absorption/desorption cycling is a bulk reversible reaction that occurs in a non-stoichiometric manner within the cubic anti-fluorite structure [136], as observed by the structural refinement of synchrotron X-ray diffraction data. The existence of Li$_{2-x}$NH$_{1+x}$ phases is now quite well admitted.

Whatever is the hydrogen storage mechanism; the ammonia release should be avoided or minimized as much as possible for applications. This point is of critical importance. Hino et al. measured NH$_3$ and H$_2$ partial pressures in the desorbed gas from LiH + LiNH$_2$ + 1 mol.% TiCl$_3$ in a closed system by means of Infrared/Raman spectroscopies and gas chromatography

[137, 138]. The amount of desorbed NH_3 was drastically increased with temperature and the NH_3/H_2 ratio was of the order of 0.1% at 673 K.

10.5.2 The Li-Mg-N-H System

Substitution of Li by other metallic elements in the Li-N-H system could lead to materials with more suitable thermodynamic properties, namely having a 0.1 MPa hydrogen equilibrium pressure at a temperature lower than 553 K. Luo experimentally studied the desorption behavior of the 1:2 MgH_2-$LiNH_2$ mixture [139], for which plateau pressures were found much higher than those of 1:1 LiH-$LiNH_2$: This Li-Mg-N-H system has an equilibrium pressure of 3.4 MPa at 473 K, whereas the pressure is well below 0.1 MPa at the same temperature for the 1:1 LiH-$LiNH_2$ mixture. Using the van't Hoff equation, an enthalpy of about 45 kJ·mol^{-1} H_2 was found [139, 140], while the enthalpy is of the order of 65 kJ·mol^{-1} H_2 for 1:1 LiH-$LiNH_2$. Therefore, it can be concluded that the substitution of LiH by the less stable MgH_2 hydride significantly reduces the system stability. If the hydrogen equilibrium pressures are high at temperatures required for fast kinetics (473–513 K), the extrapolation of the van't Hoff plot reveals that a pressure around 0.1 MPa is expected at 363 K [141, 142] for the Li-Mg-N-H system, which would satisfy the pressure-temperature conditions required for mobile fuel-cells applications.

The desorption of the 1:2 MgH_2-$LiNH_2$ mixture leads to the formation of the $Li_2Mg(NH)_2$ ternary imide [143], which upon rehydrogenation does not go back to the starting phases, but to a mixture of $Mg(NH_2)_2$ and LiH instead. The following reaction summarizes the reversible hydrogen storage process, which possesses a promising 5.6 wt.% theoretical capacity:

$$Mg(NH_2)_2 + 2\ LiH \rightarrow Li_2Mg(NH)_2 + 2\ H_2 \qquad (10.18)$$

At 453 K, the full desorption of the 1:2 $Mg(NH_2)_2$-LiH mixture occurs in 180 min with a weight loss reaching 5.1%. The desorption rates are much faster at 473 and 493 K, with the achievement of the desorption process in 60 and 25 min, respectively. The activation energy of desorption process was estimated to be 107 kJ·mol^{-1} according to the Arrhenius law [144]. This high kinetic barrier puts at the present a restriction onto the low temperature applications of 1:2 $Mg(NH_2)_2$-LiH without the help of catalysts addition.

The experimental hydrogen storage capacity of the 1:2 $Mg(NH_2)_2$-LiH mixture remains around 5.0 wt.% at 473 K for at least 50 absorption/desorption cycles without significant ammonia contamination of the hydrogen desorbing flow (as checked by mass spectroscopy). In spite of the low density of the 1:2 $Mg(NH_2)_2$-LiH ball-milled powder (around 1.0 g·cm^{-3}), the volumetric capacity of the material is nevertheless rather high: around 50 g H_2·dm^{-3} vs. 70 g·dm^{-3} for liquid hydrogen at 20 K.

In order to investigate deeply the hydrogen storage mechanism of the Li-Mg-N-H system, accurate PCI curves were plotted for the 1:2 $Mg(NH_2)_2$-LiH mixture at 473 K [144]. The results have shown that a slope region is present at low hydrogen contents (below 1.5 wt.%) before the plateau region (cf. Fig. 10.27). This seemed to indicate that a two-step reaction occurs and Luo et al. proposed the reaction as follows [142]:

$$Li_2Mg(NH)_2 + 0.6\ H_2 \rightarrow Li_2MgN_2H_{3.2} \qquad (10.19)$$

$$Li_2MgN_2H_{3.2} + 1.4\ H_2 \rightarrow Mg(NH_2)_2 + 2\ LiH \qquad (10.20)$$

Although the $Li_2MgN_2H_{3.2}$ intermediate phase has not been identified by diffraction techniques, Aoki et al. performed PCI measurements for the dehydrogenation at 523 K of the 1:2 Mg $(NH_2)_2$-LiH mixture and observed the formation of an intermediate decomposition product [145]. By synchrotron X-ray diffraction, the phase composition was estimated to be $Li_{1.3}MgN_2H_{2.7}$ and this phase can be regarded as a complex of Mg imide and Li amide [145].

Xiong et al. pointed out that the TPD profiles of hydrogen desorption from 1:2 $Mg(NH_2)_2$-LiH and ammonia desorption from $Mg(NH_2)_2$ are completely different [146], indicating the unlikely involvement of NH_3 in the hydrogen desorption path of 1:2 $Mg(NH_2)_2$-LiH. Chen et al. measured the apparent activation energies of the desorption reactions to be 88 and 130 kJ·mol^{-1}, respectively [147]. It was concluded that the decomposition of $Mg(NH_2)_2$ was unlikely to be an elementary step in the chemical reaction of $Mg(NH_2)_2$ and LiH. As for the Li-N-H system, Chen et al. proposed a direct solid–solid reaction model where the reaction rate is controlled by mass transport through the Li-Mg ternary imide layer [147]. This mechanism accounts well for the formation of single-phase Li-Mg imides with different compositions and for the absence of NH_3 contamination in the hydrogen desorption flow.

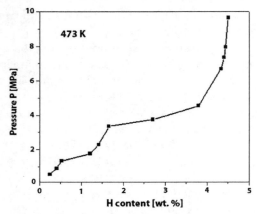

Figure 10.27 PCT desorption isotherm at 473 K for 1:2 $Mg(NH_2)_2$-LiH showing the existence of an intermediate phase before the formation of $Li_2Mg(NH)_2$.

10.5.3 Other Li-Metal-N-H Systems

In a similar approach as that pursued for the Li-Mg-N-H system, the desorption behaviors of the 1:2 $Ca(NH_2)_2$-LiH and 1:2 CaH_2-$LiNH_2$ mixtures were investigated to evaluate the feasibility of reversibly storing hydrogen in the Li-Ca-N-H system [143, 148]. In both cases, a strong hydrogen release occurs at 473–493 K with a negligible NH_3 contamination, which leads to the formation of the ternary imide $Li_2Ca(NH)_2$ at about 573 K [149, 150]. Surprisingly, the peak of hydrogen desorption of the 1:2 CaH_2-$LiNH_2$ mixture was detected at a lower temperature than that of 1:2 $Ca(NH_2)_2$-LiH (around 473 K versus 493 K) [148], whereas the decomposition temperature of $Ca(NH_2)_2$ is reported in the literature to be lower than that of $LiNH_2$ [151]. This observation was explained by the very high ionic mobility of H^- in CaH_2 [150], allowing a fast recombination with the positively charged hydrogen atoms of $LiNH_2$ and, therefore, a low temperature hydrogen release.

The experimental results showed that the 1:2 CaH_2-$LiNH_2$ mixture is able to release up to 4.5 wt.% of hydrogen, which is in agreement with the following equation:

$$CaH_2 + 2\ LiNH_2 \rightarrow Li_2Ca(NH)_2 + 2\ H_2 \tag{10.21}$$

The structure of the ternary imide $Li_2Ca(NH)_2$ was resolved by neutron powder diffraction on a deuterated sample [150] and by DFT calculations [152]. This structure is drastically different from that of $Li_2Mg(NH)_2$, the latter being regarded as vacancy-derivative from the Li_2NH anti-fluorite structure. Here, the $Li_2Ca(NH)_2$ structure consists of infinite layers of edge-shared $Ca(NH)_6$ octahedra separated by Li cations. Such a structural distinction is easily explained by comparing the ionic radii of Mg^{2+} and Ca^{2+} to that of Li^+. The ionic radii of Li^+ and Mg^{2+} are nearly identical (0.59 Å vs. 0.57 Å), which explains why the $Li_2Mg(NH)_2$ structure is closely related to that of Li_2NH. On the contrary, there is a huge discrepancy between Li^+ and Ca^{2+} ionic radii (0.59 Å vs. 1.00 Å). The large size of the Ca^{2+} cation leads to a preferential coordination number of +VI and, thus, the structure of $Li_2Ca(NH)_2$ consists of the stacking of $Ca(NH)_6$ octahedra and $Li(NH)_4$ tetrahedra along the c-axis direction.

Unfortunately, the reversibility of reaction (10.21) is not complete. The hydrogenation at 573 K of $Li_2Ca(NH)_2$ does not go back to the starting compounds. The absorption is limited to 2.3 wt.% and the rehydrogenated material consists of a mixture of $CaNH$, $LiNH_2$, and LiH [148] according to the following reaction:

$$Li_2Ca(NH)_2 + H_2 \rightarrow CaNH + LiNH_2 + LiH \tag{10.22}$$

The ternary $Li_2Ca(NH)_2$ imide behaves as if the CaNH layers were inert with hydrogen and, thus, the reversible process is the same as the classical reaction between Li amide and Li imide. This limitation reduces the storage capacity to 2.3 wt.% at 573 K and, therefore, the Li-Ca-N-H system cannot be considered as a promising hydrogen storage material.

Several groups [153, 154] showed that $LiNH_2$ can be strongly destabilized by mixing with $LiAlH_4$. For the 1:2 $LiNH_2$-$LiAlH_4$ mixture, a large amount of hydrogen is already desorbed upon ball-milling with the decomposition of $LiAlH_4$ into Li_3AlH_6 and Al. Heating under vacuum at 573 K of this ball-milled powder leads to an additional hydrogen release. The overall hydrogen desorption reaches almost 8.0 wt.%, which agrees fairly well with the following reaction:

$$\text{LiNH}_2 + 2\,\text{LiAlH}_4 \rightarrow 2\,\text{Al} + \text{Li}_2\text{NH} + \text{LiH} + 4\,\text{H}_2 \qquad (10.23)$$

Unfortunately, most of the desorbed hydrogen cannot be reloaded, probably due to thermodynamic reasons that are not fully elucidated yet.

The reactivity of Li_3AlN_2 with hydrogen was investigated by Chen et al. [155]. About 5.1 wt.% of hydrogen can be absorbed and desorbed by Li_3AlN_2: If the absorption can be achieved at temperatures below 623 K, the hydrogen release is much more difficult with a full hydrogen recovery at a temperature as high as 773 K. The fully hydrogenated sample consists of a mixture of LiNH_2, LiH and AlN as summarized in Eq. 10.24.

$$\text{Li}_3\text{AlN}_2 + 2\,\text{H}_2 \rightarrow \text{LiNH}_2 + 2\,\text{LiH} + \text{AlN} \qquad (10.24)$$

Indeed, as for the Li-Ca-N-H system, it is amazing to notice that this reaction is closely related to the Li_3N hydrogenation [129]. Here, Li_3AlN_2 can be simply considered as Li_3N + AlN, the latter nitride being inert during the hydrogenation process! The reversible step for this Li-Al-N-H system is equivalent to that in the Li-N-H system.

The investigation of the Li-Al-N-H system was also carried out through an original approach consisting first in the preparation of the $\text{LiAl(NH}_2)_4$ compound and, then, its mixture with LiH [156, 157]. $\text{LiAl(NH}_2)_4$ was prepared by reaction in liquid ammonia starting either from LiH and Al, or from LiAlH_4. The $\text{LiAl(NH}_2)_4$ ternary amide leads to a strong ammonia release at about 400 K (2 moles of NH_3 per mole of amide) leading to the formation of the LiAl(NH)_2 imide. Following a similar process as that demonstrated for the Li-N-H system, LiH was successfully added to $\text{LiAl(NH}_2)_4$ by ball-milling with the goal of trapping ammonia, due to the very fast reaction of LiH with NH_3. The following reaction with a theoretical hydrogen release of 6.2 wt.% was targeted:

$$\text{LiAl(NH}_2)_4 + 4\,\text{LiH} \rightarrow \text{LiAl(NH)}_2 + 2\,\text{Li}_2\text{NH} + 4\,\text{H}_2 \qquad (10.25)$$

The 1:4 $\text{LiAl(NH}_2)_4$-LiH ball-milled mixture is able to desorb up to 6.0 wt.% of hydrogen around 400 K, with no ammonia contamination as checked by mass spectrometry. This material is interesting as hydrogen is desorbed at a temperature about 70 K lower than the 1:2 $\text{Mg(NH}_2)_2$-LiH mixture. Unfortunately, the rehydrogenation at

403 K/10 MPa H_2 of the desorbed products is not successful. These poor performances are due to the low stability of the $LiAl(NH)_2$ imide formed upon desorption. This $LiAl(NH)_2$ compound is rapidly decomposed into a mixture of $LiNH_2$ and AlN [157], the latter phase being inert towards hydrogen at 403 K. Due to the lack of reactivity of AlN with hydrogen, the Li-Al-N-H system does not possess a reversible hydrogen storage process. Additional works are necessary to stabilize the $LiAl(NH)_2$ phase and other materials able to desorb large amounts of NH_3 at low temperatures are studied in the hope that their mixtures with appropriate amounts of LiH can give new materials able to store reversibly large amounts of hydrogen at low temperatures.

10.6 Ammonia-Borane

The amine–borane complexes have been identified as promising materials for hydrogen storage, due to their very high hydrogen contents: for instance, 24.5 wt.% for ammonium borohydride NH_4BH_4, 19.5 wt.% for ammonia-borane (NH_3BH_3) (also called borazane), and 17.8 wt.% for ammonia-triborane $NH_3B_3H_7$. Due to the interaction between $H^{\delta-}$ in the borane part and $H^{\delta+}$ in the amine part, the release of hydrogen is generally easy and can happen at low temperatures. Ammonia-borane (often noted AB) is one of the most studied compounds for hydrogen storage, due to its high capacities (19.5 wt.% and 146 g $H_2 \cdot dm^{-3}$) and its relative stability and non-toxicity.

Ammonia-borane is a white solid stable at room temperature with a melting point around 114°C. This compound is rather stable in aqueous solutions if the pH is kept neutral or basic. As soon as the pH decreases, ammonia-borane is quickly hydrolyzed and releases hydrogen [158]. The hydrolysis can be catalyzed by a large variety of metals. AB can be solubilized in large quantities in many solvents including ammonia (75 wt.%!), water (30 wt.%), THF (25 wt.%) and ethanol (6.5 wt.%). The analysis of the crystal structure of solid AB at room temperature (tetragonal, space group 14 mm) shows that the distance of the B-N bond is 1.56 Å [159, 160] and that the H-H distance between neighboring AB molecules is around 2.02 Å, smaller than the van der Waals distance, which is indicative of a strong hydrogen bonding. This hydrogen bonding is actually responsible for the solid state of AB at room temperature

[160]. Upon cooling of AB, a structural transition has been reported around 220 K leading to an orthorhombic unit cell (space group Pmn2$_1$) [161].

About the synthesis of ammonia-borane, it must be first indicated that the direct reaction between ammonia NH_3 and diborane B_2H_6 does not lead to pure AB, as the cleavage of B_2H_6 can be symmetric forming AB, but also asymmetric leading to the formation of an ionic isomer $[BH_2(NH_3)_2]^+[BH_4]^-$ called diammoniate of diborane (DADB) [162] (cf. Fig. 10.28). Although this synthesis method leads to a mixture of AB and DADB, the latter compound is insoluble in diethyl-ether and therefore pure AB can be obtained by simple filtration. Alternatively, AB can be prepared with reasonable yields by the reaction of ammonia gas with different borane complexes such as $BH_3\text{-}S(CH_3)_2$ [163].

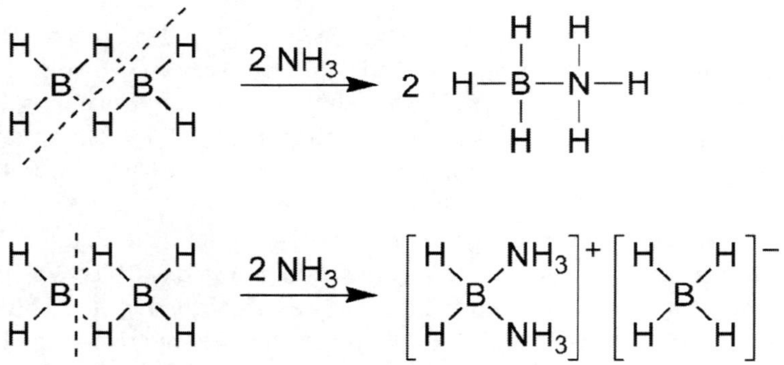

Figure 10.28 Symmetric and asymmetric cleavage of diborane by ammonia leading to the formation of ammoniaborane (AB) and diammoniate of diborane (DADB), respectively.

The more common method for the preparation of AB, especially at the laboratory scale, consists of the reaction between an ammonium salt and a metal borohydride to produce NH_4BH_4, this compound being quickly decomposed at room temperature to form ammonia-borane by release of 1 mole of hydrogen. This method has the great advantage of avoiding the use of gaseous diborane. The precursors used are often sodium borohydride and ammonium carbonate [164] and the synthesis is performed in THF according to the following reaction:

$$2\ NaBH_4 + (NH_4)_2CO_3 \rightarrow 2\ NH_3BH_3 + Na_2CO_3 + 2\ H_2 \quad (10.26)$$

This reaction has typically a yield of about 80%. Recently, it has been reported that yields exceeding 95% can be obtained by replacing ammonium carbonate by ammonium sulfate $(NH_4)_2SO_4$ or ammonium formate NH_4HCO_2 [165].

The thermal decomposition of ammonia-borane has been largely investigated, but the exact nature of the dehydrogenated products is still today subject to discussions. The dehydrogenation occurs in three main exothermic steps [166] as summarized in the following equations.

$$nNH_3BH_3 \rightarrow (NH_2BH_2)_n + nH_2 \quad (10.27)$$

$$(NH_2BH_2)_n \rightarrow (NHBH)_n + nH_2 \quad (10.28)$$

$$(NHBH)_n \rightarrow nBN + nH_2 \quad (10.29)$$

Generally, AB does not release hydrogen below its meting point (114°C): A long induction period (a few hours at 80°C) is needed before hydrogen desorption proceeds significantly. This induction period can be related to the formation of DADB, as shown by NMR spectroscopy experiments [167]. Upon heating at 80–90°C, DADB is formed by an AB molecule taking a hydride ion from another AB unit. The created DADB molecule reacts then with another AB unit to release hydrogen. Clearly, DADB appears to be less stable than AB, which implies a lower activation barrier for the hydrogen generation. The induction period seems therefore related to the time needed for the DADB nucleation and the consecutive reactions initiating the hydrogen release process, which leads in a first step to poly-aminoborane $(NH_2BH_2)_n$. If the heating rate of AB is fast (e.g., 10 K·min^{-1}), the first step of hydrogen release (1 mole of H_2) actually does not occur before the melting point.

The second step of hydrogen release (another mole of H_2) occurs around 150°C and leads to the formation of poly-iminoborane $(NHBH)_n$. At much higher temperatures (1400°C), the complete dehydrogenation of AB can be completed with the formation of boron nitride BN (Eq. 10.29). This temperature being too high for most of the applications, it is therefore considered that AB can release only 2 moles of hydrogen (13.0 wt.% H_2).

The characterization of the decompositions products of AB has revealed the complexity of the hydrogen release mechanism. The liberation of the first mole of H_2 leads to aminoborane NH_2BH_2, which is a monomer quickly polymerizing to form different $(NH_2BH_2)_x$ compounds (cf. Fig. 10.29). Cyclotriborazane $(NH_2BH_2)_3$ (CTB) is a solid analogous to cyclohexane, which is stable below 100°C. The pentamer $(NH_2BH_2)_5$ is another possible compound that has been detected by IR spectroscopy and X-ray diffraction. At slow heating rates, there is enough time for the unstable NH_2BH_2 to polymerize before it can escape the sample. On the contrary, at fast decomposition rates (10 K·min^{-1} or higher), a large amount of NH_2BH_2 can be release as a volatile product (up to 0.2 mol per mol of AB), which is highly problematic for many applications as fuel cells. The end solid product of the first decomposition step is poly-aminoborane $(NH_2BH_2)_n$ (PAB), for which the conformation (linear, branched, cyclic) is also strongly dependent of the heating rate [168].

The second decomposition step of AB (Eq. 10.28) leads to the formation of iminoborane NHBH, which is a monomer with a very low stability, quickly polymerizing to form $(NHBH)_x$ compounds. The cyclic borazine $(NHBH)_3$ analogous to benzene is a liquid at room temperature with a boiling point of 55°C and this compound can also contaminate the hydrogen desorption flow (contents up to 0.06 mol per mol of AB have been reported). At around 110°C, borazine decomposes to form hydrogen and poly-borazylene (PB), which consists of cross-linked borazine units (cf. Fig. 10.29). The end decomposition product of AB is therefore not only poly-iminoborane $(NHBH)_n$ but rather a mixture of PIB and PB, suggesting that hydrogen release exceeding 2 moles of H_2 could be achieved if no boron or nitrogen atoms are lost in volatile species during the desorption.

In order to promote the hydrogen release of ammonia-borane (i.e., to reduce the induction period and to suppress the release of volatiles species such as borazine), its nano-confinement into various porous hosts has been investigated [169, 170]. Autrey et al. have especially impregnated SBA-15 silica (pores size of 4 nm) with a solution of AB in ethanol [169]. After solvent removal, it has been shown that the resulting AB:SiO$_2$ composite with a 1:1 weight ratio is able to release hydrogen at a temperature about 15°C lower than that of neat AB (cf. Fig. 10.30). At 80°C, the time needed to complete the first decomposition step of neat AB is 5 h, whereas this process occurs in only 90 min at 50°C for the composite materials. In addition, thermodesorption and mass spectroscopy experiments have shown that the amount of borazine in the hydrogen desorption flow is drastically reduced for the nano-confined AB (cf. Fig. 10.30 bottom).

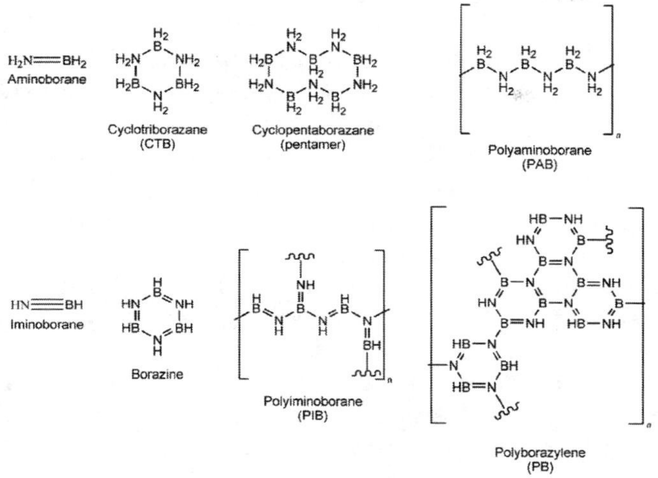

Figure 10.29 The possible decomposition products of ammonia-borane.

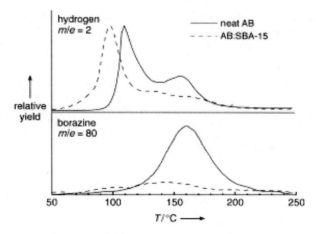

Figure 10.30 Hydrogen and borazine emissions of neat ammonia-borane (AB) and AB confined in a mesoporous SBA-15 silica. Reprinted from Ref. [169] with permission from John Wiley and Sons.

More recent studies on the nano-confinement of AB have shown that some thermodynamic modifications could be possible leading to a less exothermic desorption process. The enthalpy of desorption has been reduced from −21 kJ·mol^{-1} for neat AB to −1 kJ·mol^{-1} for AB confined in MCM-41 silica [171]. Moreover, it has been noticed that the structural transition observed at 220 K for neat AB is suppressed. The exact origin of these modifications is not yet fully understood and the exact nature of the interactions (bonding?) between AB and the matrix needs further characterizations.

In addition to these nano-confinement experiments, other studies have been devoted to the preparation of Metal-NH$_2$BH$_3$ compounds. Electron-donating substituent (as alkali metal) replacing one hydrogen atom at the nitrogen center lowers the protic character of the remaining hydrogen of the NH$_2$ groups and thus reduce the dihydrogen bond strength and leads to a lower decomposition temperature. By ball-milling of ammonia-borane with metal hydrides such as LiH or NaH, the LiNH$_2$BH$_3$ and NaNH$_2$BH$_3$ compounds can be easily obtained [172, 173] with the release of 1 mole of H$_2$ according to the following reaction:

$$MH + NH_3BH_3 \rightarrow MNH_2BH_3 + H_2 \quad (10.30)$$

LiNH$_2$BH$_3$ and NaNH$_2$BH$_3$ are interesting compounds, which are able to release about 2 moles of H$_2$ (11.0 and 7.5 wt.% H$_2$, respectively) at temperatures lower than 100°C. Moreover, the borazine contamination seems to be very low and is hardly detected by mass spectroscopy. Interestingly, the desorption processes of these MNH$_2$BH$_3$ compounds are mildly exothermic (around −5 kJ·mol^{-1} versus −36 kJ·mol^{-1} for AB when taking into account the two first decomposition steps for the latter). Nevertheless, the rehydrogenation of the desorbed products is still impossible and these compounds are not reversible hydrogen storage materials.

10.7 Conclusions

This chapter has described the different technologies and most of the materials available today for hydrogen storage. If a large variety of methods are known for storing hydrogen, many issues have still to be overcome before the large-scale implementation of hydrogen as an energy carrier becomes effective. Of course, the most suitable hydrogen storage method strongly depends of the application targeted and, in the future, probably several hydrogen storage methods, and not only one, will arrive on the market for each specific application.

Concerning the conventional hydrogen storages, namely compressed gas and cryogenic liquid, these methods are well-developed and mature. Liquid hydrogen at 20 K is the storage of choice for spacecrafts, but is hardly conceivable for smaller applications due to the need of a boil-off valve and the continual loss of hydrogen (around 1% per day). In addition, the liquefaction process of hydrogen is very energy consuming and this constitutes a severe drawback for the large development of liquid hydrogen as energy vector. The recent developments of composites tanks for compressed hydrogen (70 MPa) have allowed reaching reasonable ranges for different prototype fuel-cell vehicles, but all the infrastructure for high-pressure hydrogen has still to be developed and, more importantly, admitted by the citizens. The volumetric capacity of the composites tanks is quite low: 30 kg $H_2 \cdot m^{-3}$.

In spite of the recent discovery of the MOFs and the huge interest for these materials in the last few years, it appears that the physisorption of hydrogen is still very limited at room temperature (less than 1 wt.%) and thus irrelevant for technological applications. Large amounts of hydrogen are only adsorbed at low temperatures (typically that of liquid nitrogen, 77 K) with reproducible capacities of 5 and 7 wt.% obtained under 50 bars H_2, for activated carbons and MOFs, respectively. If these gravimetric capacities are high, it must be kept in mind that, due to the low density (less than 1 $g \cdot cm^{-3}$) of these high surface area materials, their volumetric capacity is generally lower than 50 kg $H_2 \cdot m^{-3}$.

On the contrary, metal hydrides have the main advantage of offering very high volumetric capacities which can exceed 100 $kg \cdot m^{-3}$, but their gravimetric capacities are typically of the order of 1–2 wt.%. The maximum capacity of the well-known LaNi$_5$-based alloys is around 1.2 wt.% with fast kinetics at room temperature. A significant increase of the hydrogen storage capacity of metal hydrides has been achieved with the development of the *bcc* alloys: Ti-V-Cr-Fe alloys have shown reversible capacities of about 2 wt.% with good thermodynamics and kinetics at room temperature. Because these alloys contain a large proportion of vanadium, which is expensive, its replacement by ferrovanadium is necessary before a large-scale commercialization occurs. The development of tanks with this type of metal hydrides has been pursued in the last decade.

A notable exception among metal hydrides is MgH_2. This hydride has a very high gravimetric capacity of 7.6 wt.%, but is unfortunately very stable with 1 bar hydrogen equilibrium pressure at 330°C. If strong kinetic improvements have been made possible through the combined use of mechanical-milling and addition of catalysts (as for instance Nb_2O_5), the thermodynamic of the Mg/MgH_2 reaction is unfavorable for reversibility near ambient conditions.

With the aim of increasing the gravimetric capacities, many complex hydrides have been studied since the pioneering work performed on $NaAlH_4$ at the end of the 1990s. It was shown that Ti-doping of sodium alanate allows a reversible hydrogen storage process with reasonable kinetics at 200°C. These Ti-doped materials offer a capacity of about 5.0 wt.% H_2. More recently, the studies of the equilibrium between silicides MSi and silanides $MSiH_3$ have demonstrated that reversible reactions are possible within these systems. The $KSi/KSiH_3$ system is able to store reversibly 4.3 wt.% H_2 at 100–150°C with suitable thermodynamic properties (1 bar equilibrium pressure around 130°C).

Much higher values of hydrogen release can be achieved with lithium borohydride and ammonia-borane (13.8 wt.% at about 350°C and 13.0 wt.% at 150°C, respectively), but these reactions are irreversible. The rehydrogenation of the desorbed products is only possible under very harsh conditions or needs a complicated chemistry involving a multistep process. In addition, the hydrogen desorption flow can be contaminated by gaseous species such as diborane or borazine in the cases of the borohydrides and ammonia-borane, respectively.

To conclude, the past two decades of research have resulted in the discovery of several promising hydrogen storage materials. Although the progress is constant, there are still challenging issues, such as thermodynamic modifications and kinetic improvements that need to be addressed. The urgent demands for applications involving hydrogen as an energy carrier require that a huge effort is still being devoted to this research and development area.

References

1. Nijkamp, M. G., Raaymakers, J., van Dillen, A. J., and de Jong, K. P. (2001). Hydrogen storage using physisorption—Materials demands, *Appl. Phys. A,* **72,** pp. 619–623.
2. Panella, B., Hirscher, M., and Roth, S. (2005). Hydrogen adsorption in different carbon nanostructures, *Carbon,* **43,** pp. 2209–2214.
3. Chahine, R., and Bose, T. K. (1994). Low-pressure adsorption storage of hydrogen, *Int. J. Hydrogen Energy,* **19,** pp. 161–164.
4. Texier-Mandoki, N., Dentzer, J., Piquero, T., Saadallah, S., David, P., and Vix-Guterl, C. (2004). Hydrogen storage in activated carbon materials: Role of the nanoporous texture, *Carbon,* **42,** pp. 2744–2747.
5. Thomas, K. M. (2007). Hydrogen adsorption and storage on porous materials, *Catal. Today*, **120,** pp. 389–398.
6. Yang, Z. X., Xia, Y. D., and Mokaya, R. (2007). Enhanced hydrogen storage capacity of high surface area zeolite-like carbon materials, *J. Am. Chem. Soc.,* **129,** pp. 1673–1679.
7. Yushin, G., Dash, R., Jagiello, J., Fischer, J. E. and Gogotsi, Y. (2006). Carbide-derived carbons: Effect of pore size on hydrogen uptake and heat of adsorption, *Adv. Funct. Mater.,* **16,** pp. 2288–2293.
8. Dillon, A. C., Jones, K. M., Bekkedahl, T. A., Kiang, C. H., Bethune, D. S., and Heban, M. J. (1997). Storage of hydrogen in single-walled carbon nanotubes, *Nature,* **386,** pp. 377–379.
9. Chambers, A., Park, C., Baker, R. T. K., and Rodriguez, N. M. (1998). Hydrogen storage in graphite nanofibers, *J. Phys. Chem. B,* **102,** pp. 4253–4256.
10. Tibbets, G. G., Meisner, G. P., and Olk, C. H. (2001). Hydrogen storage capacity of carbon nanotubes, filaments, and vapor-grown fibers, *Carbon,* **39,** pp. 2291–2301.
11. Hirscher, M., and Becker, M. (2003). Hydrogen storage in carbon nanotubes, *J. Nanosci. Nanotech.,* **3,** pp. 3–17.
12. Ritschel, M., Uhlemann, M., Gutfleisch, O., Leonhardt, A., Graff, A., Taschner, C., and Fink, J. (2002). Hydrogen storage in different carbon nanostructures, *Appl. Phys. Lett.,* **80,** pp. 2985–2987.
13. Schimmel, H. G., Nijkamp, G., Kearley, G. J., Rivera, A., de Jong, K. P., and Mulder, F. M. (2004). Hydrogen adsorption in carbon nanostructures compared, *Mater. Sci. Eng. B,* **108**, pp. 124–129.
14. Jorda-Beneyto, M., Suarez-Garcia, F., Lozano-Castella, D., Cazorla-Amoros, D., and Linares-Solano, A. (2007). Hydrogen storage on

chemically activated carbons and carbon nanomaterials at high pressure, *Carbon*, **45**, pp. 293–303.

15. Williams, K. A., Pradhan, B. K., Eklund, P. C., Kostov, M. K., and Cole, M. W. (2002). Raman spectroscopic investigation of H_2, HD, and D_2 physisorption on ropes of single-walled carbon nanotubes, *Phys. Rev. Lett.*, **88**, pp. 1655021–1655024.

16. Georgiev, P. A., Ross, D. K., De Monte, A., Montaretto-Marullo, U., Edwards, R. A. H., Ramirez-Cuesta, A. J., Adams, M. A., and Colognesi, D. (2005). In situ inelastic neutron scattering studies of the rotational and translational dynamics of molecular hydrogen adsorbed in single-wall carbon nanotubes (SWNTs), *Carbon*, **43**, pp. 895–906.

17. Panella, B., Hirscher, M., and Ludescher, B. (2007). Low-temperature thermal-desorption mass spectroscopy applied to investigate the hydrogen adsorption on porous materials, *Microporous Mesoporous Mater.*, **103**, pp. 230–234.

18. Harris, I. R., Langmi, H. W., Book, D., Wlaton, A., Johson, S. R., Al-Mamouri, M. M., Speight, J. D., Edwards, P. P., and Anderson, P. A. (2005). Hydrogen storage in ion-exchanged zeolites, *J. Alloys Compd.*, **404–406**, pp. 637–642.

19. Weitkamp, J., Fritz, M., and Ernst, S. (1995). Zeolites as media for hydrogen storage, *Int. J. Hydrogen Energy*, **20**, pp. 967–970.

20. Ramirez-Cuesta, A. J., Mitchell, P. C. H., Ross, D. K., Georgiev, P. A., Anderson, P. A., Langmi, H. W., and Book, D. (2007). Dihydrogen in cation-substituted zeolites X: An inelastic neutron scattering study, *J. Mater. Chem.*, **17**, pp. 2533–2539.

21. Palomino, G. T., Carayol, M. R. L., and Arean, C. O. (2006). Hydrogen adsorption on magnesium-exchanged zeolites, *J. Mater. Chem.*, **16**, pp. 2884–2885.

22. Rosi, N. L., Eckert, J., Eddaoudi, M., Vodak, D. T., Kim, J., O'Keeffe, M., and Yaghi, O. M. (2003). Hydrogen storage in microporous metal-organic frameworks, *Science*, **300**, pp. 1127–1129.

23. Dailly, A., Vajo, J. J., and Ahn, C. C. (2006). Saturation of hydrogen sorption in Zn-benzene-dicarboxylate and Zn-naphtalene-dicarboxylate, *J. Phys. Chem. B*, **110**, pp. 1099–1101.

24. Panella, B., Hirscher, M., Pütter, H., and Müller, U. (2006). Hydrogen adsorption in metal-organic-frameworks: Cu-MOFs and Zn-MOFs compared, *Adv. Funct. Mater.*, **16**, pp. 520–524.

25. Férey, G., Latroche, M., Serre, C., Millange, F., Loiseau, T., and Percheron-Guégan, A. (2003). Hydrogen adsorption in the nanoporous metal-

benzenedicarboxylate M(OH)(O$_2$C-C$_6$H$_4$-CO$_2$)(M = Al^{3+}, Cr^{3+}), MIL-53, *Chem. Commun.,* **24,** pp. 2976–2977.

26. Wong-Foy, A. G., Matzger, A. J., and Yaghi, O. M. (2006). Exceptional H$_2$ saturation uptake in microporous metal-organic-frameworks, *J. Am. Chem. Soc.,* **128,** pp. 3494–3495.

27. Furukawa, H., Miller, M. A., and Yaghi, O. M. (2007). Independent verification of the saturation hydrogen uptake in MOF-177 and establishment of a benchmark for hydrogen adsorption in metal-organic frameworks, *J. Mater. Chem.,* **17,** pp. 3197–3204.

28. Sculley, J., Yuan, D., and Zhou, H. C. (2011). The current status of hydrogen storage in metal–organic frameworks-updated, *Energy. Environ. Sci.,* **4,** pp. 2721–2735.

29. Roswell, J. L. C., Eckert, J., and Yaghi, O. M. (2005). Characterization of H$_2$ binding sites in prototypical metal-organic frameworks by inelastic neutron scattering, *J. Am. Chem. Soc.,* **127,** pp. 14904–14910.

30. Yildirim, T., and Hartman, M. R. (2005). Direct observation of hydrogen adsorption sites and nanocage formation in metal-organic frameworks, *Phys. Rev. Lett.,* **95,** pp. 2155041–2155044.

31. Roswell, J. L. C., and Yaghi, O. M. (2005). Strategies for hydrogen storage in metal-organic frameworks, *Angew. Chem. Int. Ed.,* **44,** pp. 4670–4679.

32. Jung, D. H., Kim, D., Lee, T. B., Choi, S. B., Yoon, J. H., Kim, J., Choi, K., and Choi, S. H. (2006). Grand canonical Monte Carlo simulation study on the catenation effect on hydrogen adsorption onto the interpenetrating metal-organic frameworks, *J. Phys. Chem. B,* **110,** pp. 22987–22990.

33. Schmitz, B., Müller, U., Trukhan, N., Schubert, M., Férey, G., and Hirscher, M. (2008). Heat of adsorption for hydrogen in microporous high-surface-area materials, *Chem. Phys. Chem.,* **9,** pp. 2181–2184.

34. Frost, H., and Snurr, R. Q. (2007). Design requirements for metal-organic frameworks as hydrogen storage materials, *J. Phys. Chem. C,* **111,** pp. 18794–18803.

35. Switendick, A. C. (1979). Band structure calculations for metal hydrogen systems, *Z. Phys. Chem. Neue Folge,* **117,** pp. 89–112.

36. Rao, B. K., and Jena, P. (1985). Switendick criterion for stable hydrides, *Phys. Rev. B,* **31,** pp. 6726–6730.

37. Westlake, D. G. (1983). A geometric model for the stoichiometry and interstitial site occupancy in hydrides (deuterides) of LaNi$_5$, LaNi$_4$Al and LaNi$_4$Mn, *J. Less Common Metals,* **91,** pp. 275–292.

38. Flanagan, T. B., and Clewley, J. D. (1982). Hysteresis in metal hydrides, *J. Less Common Metals*, **83**, pp. 127–141.
39. Flanagan, T. B., Park, C. N., and Oates, W. A. (1995). Hysteresis in solid state reactions, *Prog. Solid State Chem.*, **23**, pp. 291–363.
40. Luo, S., Luo, W., Clewley, J. D., Flanagan, T. B., and Wade, L. A. (1995). Thermodynamic studies of the LaNi$_{5-x}$Sn$_x$-H system from x = 0 to 0.5, *J. Alloys Compd.*, **231**, pp. 467–472.
41. Sandrock, G. (1999). A panoramic overview of hydrogen storage alloys from a gas reaction point of view, *J. Alloys Compounds*, **293–295**, pp. 877–888.
42. Ellinger, F. H., Holley, C. E., Mc Inteer, B. B., Pavone, D., Potter, R. M., Staritzky, E., and Zachariasen, W. H. (1955). The preparation and some properties of magnesium hydride, *J. Am. Chem. Soc.*, **77**, pp. 2647–2648.
43. Zaluska, A., Zaluski, L., and Ström-Olsen, J. O. (1999). Nanocrystalline magnesium for hydrogen storage, *J. Alloys Compounds*, **288**, pp. 217–225.
44. Huot, J., Liang, G., Boily, S., van Neste, A., and Schultz, R. (1999). Structural study and hydrogen sorption kinetics of ball-milled magnesium hydride, *J. Alloys Compounds*, **293–295**, pp. 495–500.
45. Oelerich, W., Klassen, T., and Bormann, R. (2001). Metal oxides as catalysts for improved hydrogen sorption in nanocrystalline Mg-based materials, *J. Alloys Compounds*, **315**, pp. 237–242.
46. Barkhordarian, G., Klassen, T., and Bormann, R. (2003). Fast hydrogen sorption kinetics of nanocrystalline Mg using Nb$_2$O$_5$ as catalyst, *Scrip. Mater.*, **49**, pp. 213–217.
47. Aguey-Zinsou, K. F., Ares Fernandez, J. R., Klassen, T., and Bormann, R. (2007). Effect of Nb$_2$O$_5$ on MgH$_2$ properties during mechanical milling, *Int. J. Hydrogen Energy*, **32**, pp. 2400–2407.
48. Porcu, M., Petford-Long, A. K., and Sykes, J. M. (2008). TEM studies of Nb$_2$O$_5$ catalyst in ball-milled MgH$_2$ for hydrogen storage, *J. Alloys Compounds*, **453**, pp. 341–346.
49. Schimmel, H. G., Huot, J., Chapon, L. C., Tichelaar, F. D., and Mulder, F. M. (2005). Hydrogen cycling of niobium and vanadium catalyzed nanostructured magnesium, *J. Am. Chem. Soc.*, **127**, pp. 14348–14354.
50. Kuijpers, F. A., and van Mal, H. H. (1971). Sorption hysteresis in the LaNi$_5$-H and SmCo$_5$-H systems, *J. Less Common Metals*, **23**, pp. 395–398.
51. Buschow, K. H., and van Mal, H. H. (1972). Phase relations and hydrogen absorption in the lanthanum-nickel system, *J. Less Common Metals*, **29**, pp. 203–210.

52. van Mal, H. H., Buschow, K. H., and Miedema, A. R. (1974). Hydrogen absorption in LaNi$_5$ and related compounds: Experimental observations and their explanation, *J. Less Common Metals*, **35**, pp. 65–76.

53. Shaltiel, D., Jacob, I., and Davidov, D. (1977). Hydrogen absorption and desorption properties of AB$_2$ Laves-phase pseudo-binary compounds, *J. Less Common Metals*, **53**, pp. 117–131.

54. Shoemaker, D. P., and Shoemaker, C. B. (1979). Concerning atomic sites and capacities for hydrogen absorption in the AB$_2$ Friauf-Laves phases, *J. Less Common Metals*, **68**, pp. 43–58.

55. Reilly, J. J., and Wiswall, R. H. (1974). Formation and properties of iron titanium hydride, *Inorg. Chem.*, **13**, pp. 218–222.

56. Reilly, J. J., and Wiswall, R. H. (1968). The reaction of hydrogen with alloys of magnesium and nickel and the formation of Mg$_2$NiH$_4$, *Inorg. Chem.*, **7**, pp. 2254–2256.

57. Zaluski, L., Zaluska, A., and Strom-Olsen, J. O. (1995). Hydrogen absorption in nanocrystalline Mg$_2$Ni formed by mechanical alloying, *J. Alloys Compounds*, **217**, pp. 245–249.

58. Orimo, S., and Fujii, H. (1997). Hydriding properties of the Mg$_2$Ni-H system synthesized by reactive mechanical grinding, *J. Alloys Compounds*, **232**, pp. L16–L19.

59. Janot, R., Rougier, A., Aymard, L., Nazri, G. A., and Tarascon, J. M. (2003). Enhancement of hydrogen storage in MgNi by Pd-coating, *J. Alloys Compounds*, **356–357**, pp. 438–441.

60. Janot, R., Darok, X., Rougier, A., Aymard, L., Nazri, G. A., and Tarascon, J. M. (2005). Hydrogen sorption properties for surface treated MgH$_2$ and Mg$_2$Ni alloys, *J. Alloys Compounds*, **404–406**, pp. 293–296.

61. Nohara, S., Inoue, H., Fukumoto, Y., and Iwakura, C. (1997). Effect of surface modification of an MgNi alloy with graphite by ball-milling on the rate of hydrogen absorption, *J. Alloys Compounds*, **252**, pp. L16–L18.

62. Bouaricha, S., Dodelet, J. P., Guay, C., Huot, J., and Schulz, R. (2001). Activation characteristics of graphite modified hydrogen absorbing materials, *J. Alloys Compounds*, **325**, pp. 245–251.

63. Akiba, E. (1999). Hydrogen-absorbing alloys, *Curr. Opin. Solid St. Mater.*, **4**, pp. 267–272.

64. Yan, Y., Chen, Y., Liang, H., Zhou, X., Wu, C., Tao, M., and Pang, L. (2008). Hydrogen storage properties of V–Ti–Cr–Fe alloys, *J. Alloys Compounds*, **454**, pp. 427–431.

65. Tamura, T., Kazumi, T., Kamegawa, A., Takamura, H., and Okada, M. (2003). Protium absorption properties and protide formations of Ti–Cr–V alloys, *J. Alloys Compounds*, **356–357**, pp. 505–509.
66. Challet, S., Latroche, M., and Heurtaux, F. (2007). Hydrogenation properties and crystal structure of the single BCC $(Ti_{0.355}V_{0.645})_{100-x}M_x$ alloys with M = Mn, Fe, Co, Ni (x = 7, 14 and 21), *J. Alloys Compounds*, **439**, pp. 294–301.
67. Huang, T., Wu, Z., Xia, B., Chen, J., Yu, X., Xu, N., Lu, C. and Yu, H. (2003). $TiCr_{1.2}(V-Fe)_{0.6}$ a novel hydrogen storage alloy with high capacity, *Sci. Tech. Adv. Mater.*, **4**, pp. 491–494.
68. Schlesinger, H. I., Brown, H. C., Abraham, B., Bond, A. C., Davidson, N., Finholt, A. E. et al. (1953). New developments in the chemistry of diborane and the borohydrides. I. General summary, *J. Am. Chem. Soc.*, **75**, pp. 186–190.
69. Nakamori, Y., Miwa, K., Ninomiya, A., Li, H., Ohba, N., Towata, S., Züttel, A., and Orimo, S. (2006). Correlation between thermodynamical stabilities of metal borohydrides and cation electronegativities: First-principles calculations and experiments, *Phys. Rev. B*, **74**, pp. 45126.
70. Brown, H. C., Choi, Y. M., and Narasimhan, S. (1981). Convenient procedure for the conversion of sodium borohydride into lithium borohydride in simple ether solvents, *Inorg. Chem.*, **20**, pp. 4454–4456.
71. Friedrichs, O., Buchter, F., Borgschulte, A., Remhof, A., Zwicky, C. N., Mauron, P., Bielmann, M., and Züttel, A. (2008). Direct synthesis of Li[BH$_4$] and Li[BD$_4$] from the elements, *Acta Mater.*, **56**, pp. 949–954.
72. Agresti, F., and Khandelwal, A. (2009). Evidence of formation of LiBH$_4$ by high-energy ball milling of LiH and B in a hydrogen atmosphere, *Scrip. Mater.*, **60**, pp. 753–755.
73. Filinchuk, Y., Chernyshov, D., and Dmitriev, V. (2008). Light metal borohydrides: Crystal structures and beyond, *Z. Krystallogr.*, **223**, pp. 649–659.
74. Soulié, J. P., Renaudin, G., Cerny, R., and Yvon, K. (2002). Lithium borohydride LiBH$_4$: I. Crystal structure, *J. Alloys Compounds*, **346**, pp. 200–205.
75. Matsuo, M., Nakamori, Y., Orimo, S., Maekawa, H., and Takamura, H. (2007). Lithium superionic conduction in lithium borohydride accompanied by structural transition, *Appl. Phys. Lett.*, **91**, pp. 224103.
76. Skripov, A. V., Soloninin, A. V., Filinchuk, Y., and Chernyshov, D. (2008). Nuclear Magnetic Resonance study of the rotational motion and the phase transition in LiBH$_4$, *J. Phys. Chem. C*, **112**, pp. 18701–18705.

77. Corey, R. L., Shane, D. T., Bowman, R. C., and Conradi, M. S. (2008). Atomic Motions in LiBH$_4$ by NMR, *J. Phys. Chem. C*, **112**, pp. 18706–18710.
78. Matsuo, M., Takamura, H., Maekawa, H., Li, H. W., and Orimo, S. (2009). Stabilization of lithium superionic conduction phase and enhancement of conductivity of LiBH$_4$ by LiCl addition, *Appl. Phys. Lett.*, **94**, pp. 84103.
79. Maekawa, H., Matsuo, M., Takamura, H., Ando, M., Noda, Y., Karahashi, T., and Orimo, S. (2009). Halide-stabilized LiBH$_4$, a room-temperature lithium fast-ion conductor, *J. Am. Chem. Soc.*, **131**, pp. 894–895.
80. Ohba, N., Miwa, K., Aoki, M., Noritake, T., Towata, S., Nakamori, Y., Orimo, S., and Züttel, A. (2006). First-principles study on the stability of intermediate compounds of LiBH$_4$, *Phys. Rev. B*, **74**, pp. 75110.
81. Züttel, A., Borgschulte, A., and Orimo, S. (2007). Tetrahydroborates as new hydrogen storage materials, *Scrip. Mater.*, **56**, pp. 823–828.
82. Her, J.-H., Yousufuddin, M., Zhou, W., Jalisatgi, S., Kulleck, J., Zan, J., Hwang, S.-J., Bowman, R., and Udovic, T. (2008). Crystal structure of Li$_2$B$_{12}$H$_{12}$: A possible intermediate species in the decomposition of LiBH$_4$, *Inorg. Chem.*, **47**, pp. 9757–9759.
83. Orimo, S., Nakamori, Y., Kitahara, G., Miwa, K., Ohba, N., Towata, S., and Züttel, A. (2005). Dehydriding and rehydriding reactions of LiBH$_4$, *J. Alloys Compounds*, **404–406**, pp. 427–430.
84. Au, M., and Jurgensen, A. (2006). Modified lithium borohydrides for reversible hydrogen storage, *J. Phys. Chem. B*, **110**, pp. 7062–7067.
85. Au, M., Spencer, W., Jurgensen, A., and Zeigler, K. (2008). Hydrogen storage properties of modified lithium borohydrides, *J. Alloys Compounds*, **462**, pp. 303–309.
86. Vajo, J., and Olson, G. (2007). Hydrogen storage in destabilized chemical systems, *Scrip. Mater.*, **56**, pp. 829–834.
87. Gross, A., Vajo, J., Van Atta, S., and Olson, G. (2008). Enhanced hydrogen storage kinetics of LiBH$_4$ in nanoporous carbon scaffolds, *J. Phys. Chem. C*, **112**, pp. 5651–5657.
88. Cahen, S., Eymery, J. B., Janot, R., and Tarascon, J. M. (2009). Improvement of the LiBH$_4$ hydrogen desorption by inclusion into mesoporous carbons, *J. Power Sources*, **189**, pp. 902–908.
89. Brun, N., Janot, R., Sanchez, C., Deleuze, H., Gervais, C., Morcrette, M., and Backov, R. (2010). Preparation of LiBH$_4$@carbon micro-macrocellular foams: Tuning hydrogen release through varying microporosity, *Energy Environ. Sci.*, **3**, pp. 824–830.

90. Cerny, R., Filinchuk, Y., Hagemann, H., and Yvon, K. (2007). Magnesium borohydride: Synthesis and crystal structure, *Angew. Chem. Int. Ed.*, **119**, pp. 5867–5870.

91. Her, J. H., Stephens, P. W., Gao, Y., Soloveichik, G. L., Rijssenbeek, J., Andrus, M., and Zhao, J. C. (2007). Structure of unsolvated magnesium borohydride Mg(BH$_4$)$_2$, *Acta Cryst. B*, **63**, pp. 561–568.

92. Chlopek, K., Frommen, C., Leon, A., Zabara, O., and Fichtner, M. (2007). Synthesis and properties of magnesium tetrahydroborate, *J. Mater. Chem.*, **17**, pp. 3496–3503.

93. Li, H. W., Kikuchi, K., Nakamori, Y., Ohba, N., Miwa, K., Towata, S., and Orimo, S. (2008). Dehydriding and rehydriding processes of well-crystallized Mg(BH$_4$)$_2$ accompanying with formation of intermediate compounds, *Acta Mater.*, **56**, pp. 1342–1347.

94. Matsunaga, T., Buchter, F., Mauron, P., Bielman, M., Nakamori, Y., Orimo, S., Ohba, N., Miwa, K., Towata, S., and Züttel, A. (2008). Hydrogen storage properties of Mg[BH$_4$]$_2$, *J. Alloys Compounds*, **459**, pp. 583–588.

95. Finholt, A. E., Barbaras, G. D., Barbaras, G. K., Urry, G., Wartik, T., and Schlesinger, H. I. (1955). The preparation of sodium and calcium aluminium hydrides, *J. Inorg. Nucl. Chem.*, **1**, pp. 317–325.

96. Bogdanovic, B., and Schwickardi, M. (1997). Ti-doped alkali metal aluminium hydrides as potential novel reversible hydrogen storage materials, *J. Alloys Compounds*, **253–254**, pp. 1–9.

97. Zaluski, L., Zaluska, A., and Strom-Olsen, J. O. (1999). Hydrogenation properties of complex alkali metal hydrides fabricated by mechano-chemical synthesis, *J. Alloys Compounds*, **290**, pp. 71–78.

98. Huot, J., Liang, G., and Schultz, R. (2001). Mechanically alloyed metal hydride systems, *Appl. Phys. A*, **72**, pp. 187–195.

99. Graetz, J., Reilly, J. J., Johnson, J., Ignatov. A., and Tyson, T. A. (2004). X-ray absorption study of Ti-activated sodium aluminium hydride, *Appl. Phys. Lett.*, **85**, pp. 500–502.

100. Leon, A., Kircher, O., Rothe, J., and Fichtner, M. (2004). Chemical state and local structure around titanium atoms in NaAlH$_4$ doped with TiCl$_3$ using X-ray absorption spectroscopy, *J. Phys. Chem. B*, **108**, pp. 16372–16376.

101. Streukens, G., Bogdanović, B., Felderhoff, M., and Schüth, F. (2006). Dependence of dissociation pressure upon doping level of Ti-doped sodium alanate-a possibility for "thermodynamic tailoring" of the system, *Phys. Chem. Chem. Phys.*, **8**, pp. 2889–2892.

102. Bloch, J., and Gray, A. P. (1965). The thermal decomposition of lithium aluminum hydride, *Inorg. Chem.*, **4**, pp. 304–305.

103. Balema, V. P., Wiench, J. W., Dennis, K. W., Pruski, M., and Pecharsky, V. K. (2001). Titanium catalyzed solid-state transformations in LiAlH$_4$ during high-energy ball-milling, *J. Alloys Compounds*, **329**, pp. 108–114.

104. Srivastava, S. C., and Ashby, E. C. (1971). Reaction of aluminum hydride with diethyl-magnesium in tetrahydrofuran. Characterization of new ethyl-substituted hydrido-magnesium aluminum hydrides, *Inorg. Chem.*, **10**, pp. 186–192.

105. Mamatha, M., Bogdanovic, B., Felderhoff, M., Pommerin, A., Schmidt, W., Schüth, F., and Weidenthaler, C. (2006). Mechano-chemical preparation and investigation of properties of magnesium, calcium and lithium–magnesium alanates, *J. Alloys Compounds*, **407**, pp. 78–86.

106. Claudy, P., Bonnetot, B., and Létoffé, J. M. (1979). Préparation et propriétés physico-chimiques de l'alanate de magnésium Mg(AlH$_4$)$_2$, *J. Thermal Anal.*, **15**, pp. 119–128.

107. Fichtner, M., Fuhr, O., and Kirchner, O. (2003). Magnesium alanate-a material for reversible hydrogen storage?, *J. Alloys Compounds*, **356–357**, pp. 418–422.

108. Fossal, A., Brinks, H. W., Fichtner, M., and Hauback, B. C. (2005). Thermal decomposition of Mg(AlH$_4$)$_2$ studied by in situ synchrotron X-ray diffraction, *J. Alloys Compounds*, **404–406**, pp. 752–756.

109. Weidenthaler, C., Frankcombe, T. J., and Felderhoff, M. (2006). First crystal structure studies of CaAlH$_5$, *Inorg. Chem.*, **45**, pp. 3849–3851.

110. Ring, M. A., and Ritter, D. M. (1961). Preparation and reactions of potassium silyl, *J. Am. Chem. Soc.*, **83**, pp. 802–805.

111. Ring, M. A., and Ritter, D. M. (1961). Crystal structure of potassium silyl, *J. Phys. Chem.*, **65**, pp. 182–183.

112. Amberger, E., Römer, R., and Layer, A. (1968). Reaktionen der hydrylanionen, V. Darstellung von SiH$_3$Na, SiH$_3$K. SiH$_3$Rb, SiH$_3$Cs und SnH$_3$K und ihre umsetzung mit CH$_3$I, *J. Organometal. Chem.*, **12**, pp. 417–423.

113. Weiss, E., Hencken, G., and Kühr, H. (1970). Kristallstrukturen und kernmagnetische Breitlinienresonanz der Alkalisilyle SiH$_3$M (M = K, Rb, Cs), *Chemische Berichte*, **103**, pp. 2868–2872.

114. Hohmann, E. (1948). Silicide und Germanide der Alkalimetalle, *Z. Anorg. Allg., Chem.*, **257**, pp. 113–126.

115. Goebel, T., Prots, Y., and Haarmann, F. (2008). Refinement of the crystal structure of tetrasodium tetrasilicide, Na$_4$Si$_4$, *Z. Krystallogr.*, **223**, pp. 187–188.

116. von Schnering, H. G., Schwarz, M. Chang, J. H., Peters, K., Peters, E. M., and Nesper, R. (2005). Refinement of the crystal structures of the tetrahedro-tetrasilicides K_4Si_4, Rb_4Si_4, and Cs_4Si_4, Z. Kristallogr.-New Cryst. Struct., **220**, pp. 525–527.

117. Chotard, J. N., Tang, W. S., Raybaud, P., and Janot, R. (2011). Potassium silanide ($KSiH_3$): A reversible hydrogen storage material, Chem. Eur. J., **17**, pp. 12302–12309.

118. Züttel, A. (2004). Hydrogen storage methods, Naturwissenschaften, **91**, pp. 157–172.

119. von Thirase, G., Weiss, E., Hennig, H. J., and Lechert, H. (1975). Präparative, röntgenographische und ^{1}H-Breitlinienresonanzuntersu chungen an Germylalkaliverbindungen, GeH_3M, Z. Anorg. Allg. Chem., **417**, pp. 221–228.

120. Mundt, O., Becker, G., Hartmann, H. M., and Schwarz, W. (1989). Metallderivate von Molekülverbindungen. II. Darstellung und Struktur des β-Kaliumsilanids, Z. Anorg. Allg. Chem., **572**, pp. 75–88.

121. Tang, W. S., Chotard, J.-N., Raybaud, P., and Janot, R. (2012). Hydrogenation properties of KSi and NaSi Zintl phases, Phys. Chem. Chem. Phys., **14**, pp. 13319–13324.

122. Janot, R., and Latroche, M. (2005). Development of a hydrogen absorbing layer in the outer shell of high pressure hydrogen tanks, Mater. Sci. Eng., B, **123**, pp. 187–193.

123. Eder, F., and Lercher, J. A. (2007). Alkane sorption in molecular sieves: The contribution of ordering, intermolecular interactions, and sorption on Bronsted acid sites, Zeolites, **18**, pp. 75–81.

124. Anastasopol, A., Pfeiffer, T. V., Middelkoop, J., Lafont, U., Canales-Perez, R. J., Schmidt-Ott, A., Mulder, F. M., and Eijt, S. W. H. (2013). Reduced enthalpy of metal hydride formation for Mg–Ti nanocomposites produced by spark discharge generation, J. Am. Chem. Soc., **135**, pp. 7891–7900.

125. Zhao-Karger, Z., Hu, J., Roth, A., Wang, D., Kübel, C., Lohstroh, W., and Fichtner, M. (2010). Altered thermodynamic and kinetic properties of MgH_2 infiltrated in microporous scaffold, Chem. Commun., **46**, pp. 8353–8355.

126. Tang, W. S., Chotard, J.-N., and Janot, R. (2013). Synthesis of single-phase LiSi by ball-milling: Electrochemical behavior and hydrogenation properties, J. Electrochem. Soc., **160**, pp. A1232–A1240.

127. Fehér, F., Krancher, M., and Fehér, M. (1991). Beiträge zur Chemie des Siliciums und Germaniums. XL. Die Bildung von Alkalimetallsilanylen,

MSi$_n$H$_{2n+1}$ (M = Na, K), - eine neue Methode zur Knüpfung von Silicium-Silicium-Bindungen ausgehend von SiH$_4$ und Natrium oder Kalium, *Z. Anorg. Allg, Chem.,* **606**, pp. 7-16.

128. Dafert, F. W., and Miklauz, R. (1910). New compounds of nitrogen and hydrogen with lithium, *Monatsh. Chem.,* **31**, pp. 981-993.

129. Chen, P., Xiong, Z., Luo, J., Lin, J., and Tan, K. L. (2002). Interaction of hydrogen with metal nitrides and imides, *Nature,* **420**, pp. 302-304.

130. Chen, P., Xiong, Z., Luo, J., Lin, J., and Tan, K. L. (2003). Interaction between lithium amide and lithium hydride, *J. Phys. Chem. B,* **107**, pp. 10967-10970.

131. Jacobs, H., and Juza, R. (1972). Neubestimmung der Kristallstruktur des Lithiumamids, *Z. Anorg. Allg. Chem.,* **391**, pp. 271-279.

132. Zaluska, A., Zaluski, L., and Ström-Olsen, J. O. (2002). Lithium-beryllium hydrides: The lightest reversible metal hydrides, *J. Alloys Compounds,* **307**, pp. 157-166.

133. Ichikawa, T., Isobe, S., Hanada, N., and Fujii, H. (2004). Lithium nitride for reversible hydrogen storage, *J. Alloys Compounds,* 365, pp. 271-276.

134. Hu, Y. H., and Ruckenstein, E. (2003). Ultrafast reaction between LiH and NH$_3$ during H$_2$ storage in Li$_3$N, *J. Phys. Chem. A,* **107**, pp. 9737-9739.

135. Ichikawa, T., Hanada, N., Isobe, S., Leng, H., and Fujii, H. (2004). Mechanism of novel reaction from LiNH$_2$ and LiH to Li$_2$NH and H$_2$ as a promising hydrogen storage system, *J. Phys. Chem. B,* **108**, pp. 7887-7892.

136. David, W. I., Jones, M. O., Gregory, D. H., Jewell, C. M., Johnson, S. R., Walton, A., and Edwards, P. P. (2007). A mechanism for non-stoichiometry in the lithium amide/lithium imide hydrogen storage reaction, *J. Am. Chem. Soc.,* **129**, pp. 1594-1601.

137. Hino, S., Ichikawa, T., Ogita, N., Udagawa, M., and Fujii, H. (2005). Quantitative estimation of NH$_3$ partial pressure in H$_2$ desorbed from the Li-N-H system by Raman spectroscopy, *Chem. Commun.,* **24**, pp. 3038-3040.

138. Hino, S., Ichikawa, T., Tokoyoda, K., Kojima, Y., and Fujii, H. (2007). Quantity of NH$_3$ desorption from the Li-N-H hydrogen storage system examined by Fourier transform infrared spectroscopy, *J. Alloys Compounds,* **446-447**, pp. 342-344.

139. Luo, W. (2004). (LiNH$_2$-MgH$_2$): A viable hydrogen storage system, *J. Alloys Compounds,* **381**, pp. 284-287.

140. Xiong, Z., Hu, J., Wu, G., and Chen, P. (2005). Hydrogen absorption and desorption in Mg–Na–N–H system, *J. Alloys Compounds*, **395**, pp. 209–212.

141. Luo, W., and Ronnebro, E. (2005). Towards a viable hydrogen storage system for transportation application, *J. Alloys Compounds*, **404–406**, pp. 392–395.

142. Luo, W., and Sickafoose, S. (2006). Thermodynamic and structural characterization of the Mg–Li–N–H hydrogen storage system, *J. Alloys Compounds*, **407**, pp. 274–281.

143. Xiong, Z., Wu, G., Hu, J., and Chen, P. (2004). Ternary imides for hydrogen storage, *Adv. Mater.*, **16**, pp. 1522–1525.

144. Janot, R., Eymery, J. B., and Tarascon, J. M. (2007). Investigation of the processes for reversible hydrogen storage in the Li–Mg–N–H system, *J. Power Sources*, **164**, pp. 496–502.

145. Aoki, M., Noritake, T., Kitahara, G., Nakamori, Y., Towata, S., and Orimo, S. (2007). Dehydriding reaction of $Mg(NH_2)_2$–LiH system under hydrogen pressure, *J. Alloys Compounds*, **428**, pp. 307–311.

146. Xiong, Z., Hu, J., Chen, P., Luo, W., Gross, K., and Wang, J. (2005). Thermodynamic and kinetic investigations of the hydrogen storage in the Li–Mg–N–H system, *J. Alloys Compounds*, **398**, pp. 235–239.

147. Chen, P., Xiong, Z., Yang, L., Wu, G., and Luo, W. (2006). Mechanistic investigations on the heterogeneous solid-state reaction of magnesium amides and lithium hydrides, *J. Phys. Chem. B*, **110**, pp. 14221–14225.

148. Tokoyoda, K., Hino, S., Ichikawa, T., Okamoto, K., and Fujii, H. (2007). Hydrogen desorption/absorption properties of Li–Ca–N–H system, *J. Alloys Compounds*, **439**, pp. 337–341.

149. Wu, G., Xiong, Z., Liu, T., Liu, Y., Hu, J., Chen, P., Feng, Y., and Wee, A. (2007). Synthesis and characterization of a new ternary imide $Li_2Ca(NH)_2$, *Inorg. Chem.*, **46**, pp. 517–521.

150. Wu, H. (2008). Structure of ternary imide $Li_2Ca(NH)_2$ and hydrogen storage mechanisms in amide–hydride system, *J. Am. Chem. Soc.*, **130**, pp. 6515–6522.

151. Hino, S., Ichikawa, T., Leng, H., and Fujii, H. (2005). Hydrogen desorption properties of the Ca–N–H system, *J. Alloys Compounds*, **398**, pp. 62–66.

152. Bhattacharya, S., Wu, G., Chen, P., Feng, Y., and Das, G. (2008). Lithium calcium imide $[Li_2Ca(NH)_2]$ for hydrogen storage: Structural and thermodynamic properties, *J. Phys. Chem. B*, **112**, pp. 11381–11384.

153. Lu, J., and Fang, Z. Z. (2005). Dehydrogenation of a combined LiAlH$_4$/LiNH$_2$ System, *J. Phys. Chem. B*, **109**, pp. 20830–20834.

154. Kojima, Y., Matsumoto, M., Kawai, S., Nakamori, Y., and Orimo, S. (2006). Hydrogen absorption and desorption by the Li–Al–N–H system, *J. Phys. Chem. B*, **110**, pp. 9632–9636.

155. Chen, P., Xiong, Z., Wu, G., Liu, Y., Hu, J., and Luo, W. (2007). Metal–N–H systems for the hydrogen storage, *Scrip. Mater.*, **56**, pp. 817–822.

156. Janot, R., Eymery, J. B., and Tarascon, J. M. (2007). Decomposition of LiAl(NH$_2$)$_4$ and reaction with LiH for a possible reversible hydrogen storage, *J. Phys. Chem. C*, **111**, pp. 2335–2340.

157. Eymery, J. B., Truflandier, L., Charpentier, T., Chotard, J. N., Tarascon, J. M., and Janot, R. (2010). Studies of covalent amides for hydrogen storage systems: Structures and bonding of the MAl(NH$_2$)$_4$ phases with M=Li, Na and K, *J. Alloys Compounds*, **503**, pp. 194–203.

158. Chandre, M., and Xu, Q. (2006). Dissociation and hydrolysis of ammonia-borane with solid acids and carbon dioxide: An efficient hydrogen generation system, *J. Power Sources*, **159**, pp. 855–860.

159. Matus, M. H., Anderson, K. D., Camaioni, D. M., Autrey, T., and Dixon, D. A. (2007). Reliable predictions of the thermochemistry of Boron–Nitrogen hydrogen storage compounds: B$_x$N$_x$H$_y$, x = 2, 3, *J. Phys. Chem. A*, **111**, pp. 4411–4421.

160. Shore, S. G., and Parry, R. W. (1955). The crystalline compound ammonia-borane, H$_3$NBH$_3$, *J. Am. Chem. Soc.*, **77**, pp. 6084–6085.

161. Hoon, C. F., and Reynhardt, E. C. (1983). Molecular dynamics and structures of amine boranes of the type R$_3$N–BH$_3$: X-ray investigations of H$_3$N–BH$_3$ at 295 K and 110 K, *J. Phys. C Solid State Phys.*, **16**, pp. 6129–6136.

162. Shore, S. G., and Boeddeker, K. W. (1964). Large scale synthesis of H$_2$B(NH$_3$)$_2^+$BH$_4^-$ and H$_3$NBH$_3$, *Inorg. Chem.*, **3**, pp. 914–915.

163. Adams, R. M., Beres, J., Dodds, A. and Morabito, A. J. (1971). Dimethyl sulfide-borane as a borane carrier, *Inorg. Chem.*, **10**, pp. 2072–2074.

164. Hu, M. G., Van Paasschen, J. M., and Geanangel, R. A. (1977). New synthetic approaches to ammonia-borane and its deuterated derivatives, *J. Inorg. Nucl. Chem.*, **39**, pp. 2147–2150.

165. Ramachandran, P. V., and Gagare, P. D. (2007). Preparation of ammonia borane in high yield and purity, methanolysis, and regeneration, *Inorg. Chem.*, **46**, pp. 7810–7817.

166. Baitalow, F., Baumann, J., Wolf, G., Jaenicke-Rößler, K., and Leitner, G. (2002). Thermal decomposition of B–N–H compounds investigated

by using combined thermoanalytical methods, *Thermochim. Acta,* **391**, pp. 159–168.

167. Stowe, A. C., Shaw, W. J., Linehan, J. C., Schmid, B., and Autrey, T. (2007). In situ solid state ^{11}B MAS-NMR studies of the thermal decomposition of ammonia borane: Mechanistic studies of the hydrogen release pathways from a solid state hydrogen storage material, *Phys. Chem. Chem. Phys.,* **9**, 1831–1836.

168. Stephens, F. H., Pons, V., and Baker, R. T. (2007). Ammonia–borane: The hydrogen source par excellence?, *Dalton Trans.,* pp. 2613–2626.

169. Gutowska, A., Li, L., Shin, Y., Wang, C. M., Li, X. S., Linehan, J. C., Smith, R. S., Kay, B. D., Schmid, B., Shaw, W., Gutowski, M., and Autrey, T. (2005). Nanoscaffold mediates hydrogen release and the reactivity of ammonia borane, *Angew. Chemie. Int. Ed.,* **44**, pp. 3578–3582.

170. Feaver, A. M., Sepehri, S., Shamberger, P. J., Stowe, A. C., Autrey, T., and Cao, G. (2007). Coherent carbon cryogel–ammonia borane nanocomposites for H_2 storage, *J. Phys. Chem. B,* **111**, pp. 7469–7472.

171. Paolone, A., Palumbo, O., Rispoli, P., Cantelli, R., Autrey, T., and Karkamkar, A. (2009). Absence of the structural phase transition in ammonia borane dispersed in mesoporous silica: Evidence of novel thermodynamic properties, *J. Phys. Chem. C,* **113**, pp. 10319–10321.

172. Kang, X., Fang, Z., Kong, L., Cheng, H., Yao, X., Lu, G., and Wang, P. (2008). Ammonia borane destabilized by lithium hydride: An advanced on-board hydrogen storage material, *Adv. Mater.,* **20**, pp. 2756–2759.

173. Xiong, Z., Yong, C. K., Wu, G., Chen, P., Shaw, W., Karkamkar, A., Autrey, T., Owen-Jones, M., Johnson, S. R., Edwards, P. P., and David, W. (2008). High-capacity hydrogen storage in lithium and sodium amidoboranes, *Nat. Mater.,* **7**, pp. 138–141.

PART 4
ENERGY TRANSMISSION AND CONSUMPTION

Chapter 11

Superconductors

Stuart C. Wimbush

*The Robinson Research Institute of Victoria University of Wellington,
PO Box 600, Wellington 6140, New Zealand*
stuart.wimbush@vuw.ac.nz

High-temperature superconducting materials have a major role to play in all aspects of sustainable energy applications right the way through from generation to consumption. While many severe technological hurdles have already had to be surmounted to extract practical utility from these materials, and a few still remain, they find themselves now, parallel with most other sustainable energy technologies, on the cusp of commercial viability. This chapter reviews the phenomenology of superconductivity as it relates to energy applications, the materials science challenges of presently available materials of particular interest to the sector, and the state of the art in energy applications of superconductors.

11.1   Introduction

The physical phenomenon of superconductivity, first observed in Leiden in 1911 [1], is most captivatingly and succinctly described

---

Materials for Sustainable Energy Applications: Conversion, Storage, Transmission, and Consumption
Edited by Xavier Moya and David Muñoz-Rojas
Copyright © 2016 Pan Stanford Publishing Pte. Ltd.
ISBN 978-981-4411-81-3 (Hardcover), 978-981-4411-82-0 (eBook)
www.panstanford.com

as the complete absence of resistance of a material to the flow of a direct electrical current. Although an incomplete description of the superconducting state, expressed in this manner it is instantly fathomable and the relevance of superconductors to sustainable energy applications is immediately apparent: Without resistance, there is no wasteful dissipation of energy and, in principle at least, the potential for entirely lossless energy transmission, conversion, and storage.

With the discovery of the high-temperature superconductors (HTS) in 1986 [2]—"high temperature" in this context referring to an operational temperature above the predicted and hitherto observed limit of around 30 K, and in particular above the 77 K boiling point of liquid nitrogen—the door was opened to widespread large-scale industrial application of technological superconductivity. Unlike the liquid helium typically used to attain lower temperatures, liquid nitrogen is cheap, easy to handle and readily available worldwide. Moreover, this temperature regime is well within the reach of electromechanical refrigeration systems, raising the prospect of liquid cryogen free cooling.

The subsequent 25 years of research and development of high-temperature superconductors have encompassed significant advances in the materials science involved in the production of these complex materials, the understanding of the fundamental physics of the superconducting state, and the engineering of the associated cooling technologies. While both the underlying theory and the production techniques continue steadily to evolve, today commercial partners stand ready to deliver the superconducting components needed to meet the global energy challenges of the mid-21st century, with successful demonstration projects already in place worldwide, large-scale installations in the advanced planning phase, and industrial scale-up of production capacity well under way.

11.2 Fundamental Phenomenology of Superconductivity

The superconducting state is a true thermodynamic phase of matter, similar in many respects to the ferromagnetic phase of magnetic materials. At a certain temperature termed the critical temperature T_c, well below the solidification temperature, the conduction electrons in a material pair up to form so-called Cooper pairs (named after Leon Cooper, one of the authors of the famous BCS theory of superconductivity [3], which explains the microscopic origin of this pairing). In a quantum mechanical description, the charge carrier formed by this pairing of electrons behaves as a boson (having integer spin) rather than a fermion (having the half-integer spin of a single electron), and in particular it need not obey the Pauli exclusion principle that applies to unpaired electrons. Instead, all of the Cooper pairs in the material "condense" into a single macroscopic ground state, the consequence of which is that all the Cooper pairs must acquire the same values for all physical quantities. To change one of those values, we must either change it for all the Cooper pairs in the material, or break the Cooper pair into its constituent (fermion) electrons.

11.2.1 Origin of Lossless Current Transport

This is the origin of the resistance-free direct current transport in a superconductor. When no current flows, each Cooper pair has a net momentum of zero (the equal positive and negative momenta of its constituent electrons canceling out). As a supercurrent is passed, this momentum takes on a value, but its value is of necessity the same for every Cooper pair in the material. Ordinarily, electrical resistance arises due to scattering of conduction electrons as they pass through the material, for example from impurity atoms or lattice defects, or even in the absence of these from thermally displaced host atoms. However, scattering is nothing other than a transfer of momentum between the electron and the lattice, and this is no longer permitted. Either we would have to change the momentum of all the Cooper pairs by an identical amount (which would constitute a variation in the macroscopic current, rather than a resistance), or we must supply sufficient momentum to a single Cooper pair to break it into its constituent electrons. Generally, there is insufficient momentum available from a single scattering event to do this, so the Cooper pair *may not* scatter, so it does not, and electrical resistance therefore does not arise.

11.2.2 Limitations on the Superconducting State

Nonetheless, there are thermodynamic limitations on the persistence of the superconducting state, as indicated in Fig. 11.1. As the temperature rises beyond T_c, the Cooper pairs will gain access to sufficient thermal energy to exceed the binding energy of their constituent electrons and the pairs will split. As the supercurrent is increased, the momentum of the Cooper pairs increases until at some point termed the depairing current density, J_d, their kinetic energy becomes sufficient to overcome their binding energy and break the pairing. Similarly, an applied magnetic field, through its influence on the intrinsic magnetic moment of the electrons, will tend to align their spin. Since the Cooper pairing relies on the coupling of electrons of opposing spin,[1]

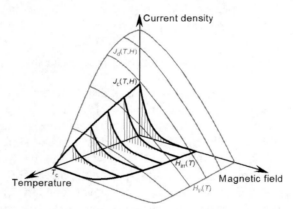

Figure 11.1 Limits on the extent of the superconducting state include the temperature, magnetic field and transport current. In addition to the thermodynamic limits of the critical temperature, T_c, the Pauli limiting field, H_P, and the depairing current density, J_d, there are practical limits determined by energy dissipation termed the irreversibility field, H_{irr}, and the critical current density, J_c, of greater relevance to application. Between these limits, the material remains superconducting, but can no longer carry a macroscopic supercurrent.

---

[1]Superconductors do exist that comprise paired electrons of parallel spin. However, those exotic "spin-triplet" superconductors that are known all have very low critical temperatures, making them irrelevant for application. Even here, an applied field that is not aligned to the spin direction will tend to split the pairs.

this will eventually, at a field termed the Pauli limiting field, H_P, disrupt the pairs, destroying the superconductivity. All three of these parameters, then, are seen to act against the binding energy of the Cooper pairs, attempting to split them into their constituent normal electrons and destroy the superconductivity. Together they create a critical surface within which superconductivity is able to persist.

In practice, neither of these latter two thermodynamic limits to the superconducting state is relevant for practical materials, but rather much lower limits determined by energy dissipation. For technological purposes, a superconducting material ceases to be useful when it can no longer carry a macroscopic supercurrent. Above these lower limits, termed the critical current density J_c and the irreversibility field H_{irr}, although the material may still be in the thermodynamic superconducting state, it cannot transport an electrical current without dissipation, for reasons explained below. These two parameters, therefore, provide the practical limitations to operation of the superconductor.

11.2.3 Flux Penetration and Flux Pinning

The second defining characteristic of the superconducting state, and the one that distinguishes superconductors from simple (hypothetical) perfect conductors, is its perfect diamagnetism, in other words its tendency to expel an applied magnetic field. It does this through the spontaneous creation of an internal supercurrent loop and associated magnetic field, exactly canceling the applied field. One class of superconductors, comprising most chemical elements, does this in a straightforward manner, fully excluding the magnetic field from the entire bulk of the superconductor (the Meissner–Ochsenfeld effect [4]). However, the consequence of this is that these type I superconductors break down and cease to superconduct at a rather low applied field H_c at which the energy cost of excluding the applied field (or establishing an opposing field to cancel it) exceeds the energy gain of allowing the Cooper pairs to settle into a lower ground state than their constituent electrons. Combined with their low transition temperatures, this effectively renders all the type I superconductors useless for application. Fortunately, a second class of superconductor, comprising the chemical elements vanadium, technetium and niobium as well as all known superconducting compounds including all the high-temperature superconductors, behaves slightly differently, as shown in Fig. 11.2. These type II superconductors find it energetically favorable to segregate into a mixed state at a lower critical field H_{c1}, allowing the superconductivity to break down in isolated regions of the material where the field is allowed to penetrate, while maintaining the superconducting state in other, field-free regions of the material. In this manner, overall superconductivity can be maintained up to a much greater applied field H_{c2}, termed the upper critical field, where the regions of field penetration encompass the entire material and superconductivity is once again destroyed.

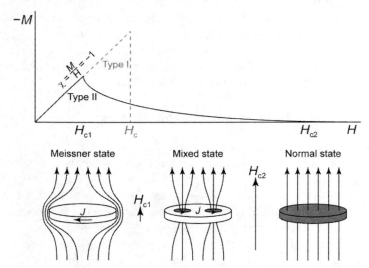

Figure 11.2 Different states of magnetic flux penetration in type I and type II superconductors. In type I superconductors, the Meissner state of complete flux expulsion is maintained until the critical field H_c is reached, at which point superconductivity is lost. In type II superconductors, partial flux penetration is permitted above a lower critical field H_{c1}, allowing overall superconductivity to persist up to a much higher critical field H_{c2}.

The catch in this is that the regions of material where the magnetic field penetrates form lines of flux threading through the material, and these flux lines are subject to a motive force analogous to the Lorentz force ($\mathbf{f} = \mathbf{J} \times \mathbf{B}$) when an electrical current is passed through the material. Unrestrained as, for example, in a perfect crystal, the flux lines will move under the influence of this force inducing an electric field ($\mathbf{E} = \mathbf{B} \times \mathbf{v}$) in the direction of the transport current. This mimics an Ohmic resistance ($\mathbf{E} \propto \mathbf{J}$), resulting in power dissipation. The only way to prevent this motion is by "pinning" the flux lines on defects within the material. Through their existing disruption of the superconducting state, such defects provide a preferred site for the flux lines to occupy, and the potential well created by this preferred site manifests itself as a pinning force opposing the motive force. However, at some point either the current density ($\mathbf{J}$) or the field ($\mathbf{B} = \mu_0 \mathbf{H}$) will reach a value at which the pinning force is overcome and dissipation will occur. This flux depinning at these two limiting values of current and field is the origin of the lower limits of operation of technological superconductors termed the critical current density J_c and the irreversibility field H_{irr}. As shown in Fig. 11.3, these are typically orders of magnitude lower than the corresponding thermodynamic limits, and it is an area of intensely active research to increase these values through improved flux pinning.

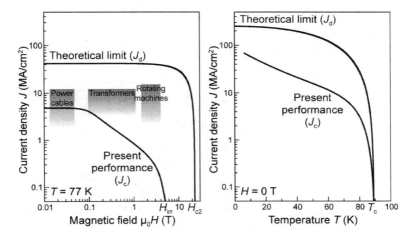

Figure 11.3 Present performance of RBCO-based superconductors compared with the theoretical limiting performance, and outlining field and current regimes of relevance to particular applications. After [5].

11.3 Superconducting Materials for Application

The timeline of superconducting materials development for application is driven by application requirements, manufacturing capability, economic considerations, and materials availability. The process of new materials discovery is unpredictable, and while the dream of room-temperature superconductivity persists, the consensus is that we must work with the materials we have until such times as they are supplanted by superior alternatives. The discovery of the first high-temperature superconducting material was rapidly followed by a plethora of further discoveries that yielded a range of promising materials to choose from. Ultimately, it was the capability to manufacture wires from the material that drove the selection of a particular family of chemical compounds for intensive development. As it became apparent that these materials would struggle to meet both the economic and the performance requirements of widespread application, a new choice was made proffering cheap, high-performance products, but only if an entirely new manufacturing paradigm could be mastered. And as scientific discovery continued in parallel with technology development, an unexpected contender arose to challenge the dominance of this unwieldy solution, at least in a subset of applications.

11.3.1 First Generation BSCCO Wires

The BSCCO ("bis-ko") series of superconducting materials named after its chemical formula $Bi_2Sr_2Ca_{n-1}Cu_nO_{2n+4+\delta}$ was discovered in 1988. The initial discovery was reported by Maeda et al. [7], although with an incorrect composition. Subsequent work the same year correctly identified $Bi_2Sr_2Ca_1Cu_2O_{8+\delta}$ (n = 2, Bi2212, T_c = 85 K) [8] and $Bi_2Sr_2Ca_2Cu_3O_{10+\delta}$ (n = 3, Bi2223, T_c = 110 K) [9] as high-temperature superconducting phases. Together, these two materials comprise the so-called first-generation (1G) high-temperature superconductors, and both have been and continue to be used to produce superconducting wires.

To form wires from the raw materials of 1G HTS, classical metallurgical processes of extrusion, drawing, rolling, and annealing are used. The fabrication process employs the powder-in-tube technique [10, 11] originally developed for the A-15 type low-temperature superconductor (LTS) materials such as Nb_3Sn [12]. In the case of BSCCO, the finely powdered material is packed into hollow tubes of silver that are bundled together and repeatedly drawn to mechanically sinter the powder before it undergoes heat treatment to form the superconducting phase. The resulting fine filamentary conductor provides the strength and flexibility required of the ceramic core material. At the same time, the silver matrix forms a normal metal stabilizer for the superconducting filaments embedded within it. This is required due to the high normal state resistance of the ceramic superconductor, much higher than a typical metal. Consequently, in the event of a localized or transient deficiency in the superconductor, the normal metal needs to be able to shunt the defect, otherwise the local Joule heating could be sufficient to melt the conductor, causing permanent damage. Silver is used because it is inert with respect to the core material while also lowering the formation temperature of Bi2223 and at the same time being permeable to the oxygen needed to achieve the optimal doping state (δ) of the superconductor embedded within. Alloys of silver may be used to improve the mechanical properties of the wire, but the silver content remains high.

Bi2212 is predominantly a candidate for the production of windings for high-field magnets due to its isotropic character and retention of high critical current density at low temperatures under high applied fields well above those achievable using LTS materials (Fig. 11.4). It also had the advantage that, until the discovery of MgB_2, it was the only HTS material able to be fabricated as a round wire that is the preferred form for coil winding applications (Fig. 11.5).

Of greater relevance to energy applications is the Bi2223 material that was developed into a conductor shortly afterwards. Here, the circular profile of the conductor had to be sacrificed to the need to apply an anisotropic (rolling) deformation (Fig. 11.6) in order to induce texturing of the superconductor material within the filaments, but in return it offered access to a much higher temperature operating range, particularly at low applied fields, thanks to its greatly increased T_c. This made Bi2223 the first viable candidate for the production of electrical power cables, motors, transformers and fault current limiters in spite of its diminished in-field performance and increased difficulty of handling.

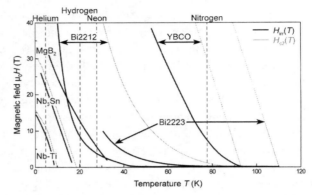

Figure 11.4 Field and temperature regimes of applicability of commercially available superconducting materials and liquid cryogens. Plotted for each material is the irreversibility field line $H_{\mathrm{irr}}(T)$ at which the critical current density J_c drops to zero, contrasted with the upper critical field line $H_{c2}(T)$ that represents the boundary of the superconducting state in the material. The illustration is schematic and assembled from a variety of sources, after [6] updated with more recent data for MgB_2 and Bi2223.

Figure 11.5 Stages in the production of BSCCO wires by the powder-in-tube technique, showing (left) bundling of powder-packed silver tubes into a silver outer sheath, (middle) first stage extrusion of the wire, (right) second stage drawing. The processed wires are cut obliquely to highlight the filamentary structure of the conductor comprising 55 individual superconducting filaments. Image courtesy of AMSC.

Figure 11.6 Cross-sectional microscope image of a Ag-sheathed Bi2223 wire with 85 superconducting filaments visible in black. The flattened shape of the conductor results from the final rolling deformation required to orient the superconductor material. RRI image.

BSCCO-based superconducting wires are available commercially from Oxford Instruments plc (Bi2212), Bruker HTS GmbH (Bi2223), Innova Superconductor Technology Co. Ltd. (Bi2223), and Sumitomo Electric Industries Ltd. (Pb-doped Bi2223). American Superconductor Corp. wound up its production of first-generation Bi2223 wires in 2006 to focus exclusively on second-generation RBCO tapes.

11.3.2 Second Generation RBCO Tapes

The first-generation HTS wires described above suffer from two primary deficiencies: They are intrinsically expensive due to the high fraction of silver content in the finished wire and they have poor in-field performance, requiring operation at very low temperatures if a magnetic field is to be withstood. The so-called second-generation (2G) HTS tapes were developed to address these issues, and feature the series of superconducting materials $RBa_2Cu_3O_{7-\delta}$ ($R123$), where R is a rare earth element (Sc, Y or a lanthanoid) or mixture thereof. The most common material to be used, and also the most intensively studied superconductor of all, is Y123 (T_c = 93 K) in spite of the fact that some members of the series have slightly higher critical temperatures (e.g., Nd123, T_c = 96 K) while others may have a slightly lower raw material cost. This has mainly historical reasons—Y123 was the first of this series to be discovered [13]—although there are also benefits in terms of phase purity: As the size of the rare earth ion is increased, it can begin to substitute onto the Ba site, disrupting the superconducting phase. Nonetheless, due to the interchangeability of the rare earth constituent, many manufacturers and researchers experiment with variable compositions.

It is interesting to observe that the RBCO materials were discovered *before* the BSCCO materials, and that their T_c is significantly *lower* than that of Bi2223. This emphasizes the dual point that often it is manufacturing capability that dictates which materials make it to market, while the critical temperature alone is not the determining factor for ultimate performance.

The primary challenge that had to be overcome in moving to this series of materials was the "grain boundary problem." It was demonstrated early on [14] following unremarkable attempts to prepare Y123 wires by the powder-in-tube process [15] that current flow across grain boundaries in R123 materials was problematic due to the extremely short coherence length of the Cooper pairs. This has the consequence that each grain boundary of a sufficiently large angle—just enough to disrupt the crystal structure over a distance of a few nanometers—acts as a barrier to the supercurrent, severely limiting the achievable critical current density. Whereas in BSCCO this problem was naturally circumvented by a percolating current flow in which the supercurrent effectively bypassed these barriers, in RBCO the microstructure necessary to facilitate this does not arise naturally, and instead the presence of these large angle grain boundaries had to be avoided. The solution to this problem (Fig. 11.7) was to adopt a thin film–based approach to forming the conductor,

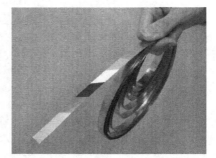

Figure 11.7 Photograph of a reel of RBCO coated conductor tape. The layers of the tape have been selectively etched or removed at the end to reveal (from left) the bare metal substrate, buffer layers, superconducting RBCO layer, silver capping layer, copper laminate, and polyimide insulation wrap. Image courtesy of Fujikura.

growing a superconducting film epitaxially on a textured substrate in such a way as to limit the range of grain boundary angles present to sufficiently small values that the supercurrent flow was not impeded. This "coated conductor" methodology constituted a major technological advance that has taken the better part of a decade to realize, and the various approaches to solving the problem are detailed in Section 11.4.

Second-generation RBCO tapes are presently commercially available from Bruker HTS GmbH, Superconductor Technologies Inc., Fujikura Ltd., American Superconductor Corp., and SuperPower Inc. (Furukawa Electric Co., Ltd.).

11.3.3   MgB$_2$ Wires

Magnesium diboride (MgB$_2$) was a latecomer to the party, its superconductivity with T_c of 39 K being discovered only in 2001 [16], which is all the more extraordinary given that this simple binary compound is a material known since the 1950s and available off-the-shelf from most chemistry stores. Unique at the time of its discovery for being the only non-cuprate superconductor believed to fall outside of the BCS framework (now joined by the iron arsenide series, mentioned below), much speculation can be entertained about how the landscape of superconductor research might have developed differently had it been identified prior to the cuprates.

The attractiveness of MgB$_2$, given its relatively low T_c, is in its "halfway house" position between expensive, underperforming HTS materials and difficult to handle, costly to run LTS materials for applications in moderate magnetic fields. With presently available materials, operation at liquid nitrogen temperature is not an option if fields greater than around 1 T are required (as is the case for applications such as motors and generators), and if a lower operating temperature is to be accepted in any case, then MgB$_2$ offers a low-cost alternative to HTS, while retaining a sufficiently high operating temperature that liquid cryogen free operation remains a possibility.

MgB$_2$ offers fabrication by a simple powder-in-tube metallurgical process (Fig. 11.8) similar to that employed for 1G HTS but without the requirement for a Ag stabilizing matrix since the constituents are less prone to contamination by the sheath material, and there is no requirement to achieve precise oxygenation. Care must be taken to protect the MgB_2 itself from immediate contact with contaminating factors, and most commonly a Ni or Ni-alloy material is used in direct contact. However, beyond this, the majority of the conductor matrix can be formed from any base metal such as Cu, Ni, Fe, or (stainless) steel depending on the mechanical and normal state electrical properties desired.

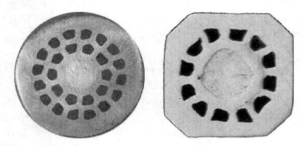

Figure 11.8 Round and square cross-section MgB_2 wires comprising superconductor filaments embedded in a Ni-alloy matrix, with an embedded Cu core for thermal and electrical stabilization. Image courtesy of Columbus Superconductors.

Two approaches to the formation of the MgB_2 phase are commonly employed: an ex situ reaction process that pre-forms the superconducting phase from its elemental constituents prior to fabrication of the conductor, and an in situ process that first prepares the conductor comprising elemental precursors that are subsequently reacted to form the superconducting compound. In the context of coil production, these are termed, respectively "react and wind" and "wind and react." The wind and react approach can offer advantages where mechanical stability of the reacted product is lacking, since it need not be further manipulated after reaction. The react and wind approach, as well as being more scalable, offers greater flexibility in terms of post-processing.

Magnesium diboride conductors offer a further unique advantage over the HTS materials described previously in that they allow the creation of a persistent current—the classic experiment of a closed superconducting loop in which an induced current will flow indefinitely. This is not available to the HTS materials due to the inability to create a resistance-free superconducting joint in the material, and also due to intrinsic thermal processes that lead to current leakage over time through flux creep. This makes MgB$_2$ wires of particular interest to magnetic resonance imaging (MRI) and nuclear magnetic resonance (NMR) applications, where field stability beyond that achievable in a driven mode is a distinct benefit.

Magnesium diboride wires prepared by ex situ reaction are presently commercially available from Columbus Superconductors SpA, while in situ reacted wires are available from Hyper Tech Research Inc.

11.3.4 New Materials on the Horizon

Research into superconducting materials continues apace, and new discoveries regularly present surprises. A recent such discovery was that of the iron arsenide superconductors $RO_{1-x}F_xFeAs$ [17] having T_c values as high as 52 K, the highest non-cuprate T_c to date. Subsequent discoveries of other iron arsenide compounds without oxygen confirmed this as a genuinely new family of high-temperature superconducting materials. Research into these materials is still at an early stage, and consequently it is hard to predict what impact on the field they might eventually have. Recent excitement was provoked by the observation that, unlike the other materials discussed here, these superconductors are almost isotropic, enabling the relatively easy production of bulk round wires with the promise of high critical fields [18]. However, J_c presently remains lower than ideal and the materials have their own processing challenges, not least of which is their arsenic content.

11.4 Coated Conductor Fabrication

The major technological achievement of high-temperature superconductor materials research over the last two decades has been the development of the coated conductor fabrication process for second-generation RBCO tapes. The challenge is that of producing kilometer lengths of homogeneous brittle ceramic material with tight crystallographic alignment ("texture") within a few degrees along the entire length of the conductor. The universal features of the several distinct solutions to this problem that have been developed are to employ a continuous thin film deposition procedure onto some form of flexible metallic substrate, relying on epitaxial growth of the superconducting layer to provide the necessary texture and its thin film nature to provide the requisite degree of flexibility and mechanical robustness to essential handling. In addition to the substrate tape and the superconducting film, a series of intermediate buffer layers are required to act as diffusion barriers and to provide crystal lattice matching to enable the epitaxial crystal growth, while a metallic passivating layer is also applied to provide environmental protection as well as to stabilize the current flow. Finally, encapsulation of the thin film stack within a metallic cladding is common to improve mechanical stability and solderability. The various processes that have been developed distinguish themselves in the method of achieving the initial texturing of the substrate upon which to grow the epitaxial layers and the method of thin film deposition itself.

The first fabrication method shown in Fig. 11.9 is that utilized by SuperPower in producing its coated conductors. Beginning with an off-the-shelf metallic alloy tape, the surface is first electropolished to produce a surface roughness less than the extremely thin (10 nm) buffer layers to be deposited, in order to ensure the integrity of those layers and enhance the degree of texturing ultimately achieved. An alumina buffer layer is then applied by sputtering to prevent oxidation of the metal surface and to limit diffusion of the metallic elements through the layers to the superconductor. An yttria layer is deposited as a nucleation layer for the magnesia layer in which texturing is introduced. This is done through a process termed ion beam assisted deposition (IBAD) [20] of magnesia [21] under which a preferred orientation of the crystalline layer is induced during nucleation using a directed ion beam aimed at the nucleation region. As soon as this texturing is achieved (within just a few nanometers of growth), the ion beam treatment is discontinued and a thicker layer of magnesia is grown homoepitaxially on top of the oriented layer in order to stabilize the texture. This is followed by the epitaxial growth of a layer of the perovskite LaMnO$_3$ (lattice parameter a = 0.40 nm) to provide lattice matching from the MgO (a = 0.42 nm) to the thick layer of RBCO (a = 0.39 nm) to follow, which is deposited by metal organic chemical vapor deposition (MOCVD) [22]. Postulated advantages of this approach include a small grain size of the resulting RBCO material and freedom of choice of the substrate material allowing for low cost, high strength and/or non-magnetic substrates. The throughput disadvantage of the relatively slow and localized IBAD process is overcome by the tiny amount of material required to be treated in this manner, but the expense associated with the requirement for multiple vacuum-deposited layers remains.

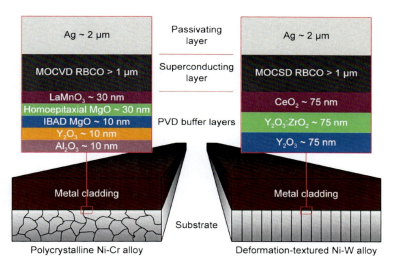

Figure 11.9 Comparison of the two primary methods of coated conductor fabrication employed today: (left) the IBAD-MOCVD process used by SuperPower, and (right) the RABiTS-MOCSD process used by American Superconductor. In each case, the first part of the name indicates the method used to achieve texturing, while the latter part describes the deposition process used for the RBCO layer. After [19]; not to scale.

The second approach, favored by American Superconductor, aims to be a lower-cost process. Utilizing a method of texturing the metallic substrate itself through a process of mechanical deformation and recrystallization termed in this context rolling-assisted biaxially textured substrates (RABiTS) [23], it is able to dispense with several of the buffer layers, retaining only those required to promote adherence to and prevent oxidation of the metallic tape surface (which would result in a loss of the preformed texture), limit diffusion of deleterious metallic elements through to the superconductor, and provide lattice matching from the Ni (lattice parameter a = 0.35 nm) through CeO_2 ($a/\sqrt{2}$ = 0.38 nm) to the RBCO (a = 0.39 nm). Also here, careful polishing of the textured tape is required prior to thin film deposition. The thick RBCO layer itself is deposited via a low-cost metal organic chemical solution deposition (MOCSD) [24] process. This involves slot die coating the buffered substrate with a solution formed typically from trifluoroacetate salts of the superconductor cations, followed by several heat treatment stages first to decompose the organic salts, then to react the precursor material to form the superconducting phase, and finally to oxygenate the superconductor.

In both cases, a final stage of the process involves applying a metallic capping layer, typically a thin film of silver for passivation and electrical stabilization, and to support the optional lamination or electroplating of the conductor to sandwich it between two metallic tapes for mechanical and thermal stability. Slitting the tape into narrower sections may also form part of the process as industrial scale-up has progressed from 4 mm-wide tapes (designed for drop-in compatibility with their 1G Bi2223 counterparts) to 10 mm wide, to 40 mm wide, and looking to the future, 100 mm wide or beyond.

Other manufacturers and research establishments apply a wide range of alternative deposition techniques to the fabrication of coated conductors, most notably IBAD-pulsed laser deposition (Fujikura Ltd. and Bruker HTS GmbH) and IBAD-reactive coevaporation (Superconductor Technologies Inc.). These alternative techniques aim to offer advantages such as a simplified buffer architecture, improved superconductor performance, lower cost precursor materials, increased throughput, or reduced capital equipment cost, but each introduces its own particular implementation challenges as well.

Currently active topics of research relate to the composition of the metallic alloy substrate to improve mechanical performance or reduce magnetization and associated hysteresis losses, surface treatment of the metallic substrate to encourage improved film growth, and the replacement of the vacuum buffer layer depositions with all-chemical processes to reduce cost and/or increase throughput. In the case of the MOCSD process, much work has been done on replacing the fluorine-based precursors with non-fluorinated alternatives due to a perceived health hazard of processing by-products. On an industrial level, the scale-up to ever increasing conductor lengths and production volumes and the processing challenges that entails are continuing endeavors.

11.5 Superconductors for Energy Applications

Superconductors both support and enable diverse energy applications ranging from individual components to complete systems. Whether through novel capabilities attainable only using superconducting technologies or through underlying support systems that enable the flexible energy infrastructure of the future or simply through the increased efficiency they offer, devices based on superconducting materials constitute a pivotal feature of the sustainable energy economy. This section outlines the many areas in which superconductors can contribute to sustainable energy applications, as well as detailing the major impediments to adoption of the technologies described, and the research that is under way to tackle these issues.

High-temperature superconductor–based technologies have enticing "green" credentials. Operating on electrical power, they are a natural development building upon the only practical energy source (aside from hydrogen) that is clean and free of greenhouse gases at the point of use, and that therefore forms the core of our future total energy infrastructure. The necessary cooling can be accomplished using non-flammable environmentally inert liquid nitrogen in contrast to conventional oil-based coolants that are both flammable and harmful to the environment. Even at lower temperatures, closed-cycle cooling systems utilizing helium or neon as the cryogen are equally benign. Superconducting heavy machinery is typically around a third the size and weight of equivalent conventional units, providing for both economic and operational advantages. These multifaceted benefits have resulted in the development of many new applications across the energy sector, of which the following is an overview.

11.5.1 Superconducting Power Cables

Perhaps the most evocative potential application of superconducting technology is the replacement of the swathes of power cables that festoon the furthest reaches of every developed country. The dream of transcontinental lossless power transmission has an undeniable appeal, and indeed this type of "global grid" may be the only feasible way of attaining another dream of the sustainable energy future: that of solar farms in remote desert locations or wind farms on the open seas supplying energy to distant population centers. However, this application is one of the most challenging and therefore least likely to be realized simply due to the scale involved. Superconductors require cooling, and consequently are best suited to localized, self-contained systems. By its very nature, a power grid is the exact converse of this ideal. Further, as outlined below, lossless transmission is not a reality for the alternating current networks that presently constitute the bulk of existing infrastructure. Nonetheless, the potential exists to reduce the 5–10% energy loss in transmission to around 0.5% using superconducting power cables, and even when the energy cost of the associated cooling is taken into account, a halving of the total loss remains feasible. For this reason, an open mind toward the potential of superconducting power cables remains prudent.

In spite of these severe barriers to adoption, early applications of this technology are to be found. The first on-grid installation of an HTS cable took place in Copenhagen, Denmark (NKT/NKT Cables, 30 m 1G, 70 MW) [25] in 2001, and test installations of both 1G and 2G HTS cables presently exist on the commercial grid at Kunming, Yunnan Province (Innova/Shanghai Cable Works, 33 m 1G, 28 MW) [26] in China, and at Albany, New York (SuperPower/Sumitomo, 320 m 1G plus 30 m 2G, 28 MW) [27], Columbus, Ohio (American Superconductor/Southwire-NKT Cables, 200 m 1G, 40 MW) [28], and Long Island, New York (American Superconductor/Nexans, 600 m 1G/2G, 330 MW) [29] in the US, with many more off-grid tests under way worldwide.

The first practical (rather than demonstration) installation was planned to link two substations in New Orleans, Louisiana (Undisclosed/Southwire, 1760 m, 28 MW) [30] in 2011, but was unfortunately canceled due to the economic downturn limiting growth in the area and therefore rendering the real-world installation unnecessary. An inner-city installation in Essen, Germany (Sumitomo/Nexans, 1000 m 1G, 40 MW) [31] is planned for 2013, with a medium-voltage 10 kV HTS cable replacing a conventional high-voltage 110 kV line between two transformer stations. The capability of superconducting cables to operate at lower transmission voltages than conventional cables thanks to their lower losses means that, in the future, these transformer stations themselves could become obsolete, reducing impact on residential and commercial centers.

A similar project is under development for New York City (American Superconductor/Southwire-NKT Cables, 300 m 2G, 55 MW) [32], and it is in densely populated cities like this that localized small-scale substitution of subterranean power cables with superconducting cables is a particularly attractive option, permitting an increase in the power capacity of existing conduits without the tremendous costs associated with providing new piping, thereby helping to ease the grid congestion that will only become worse as our energy demands and population densities increase. Both of these projects offer further benefits in terms of inherent fault-current limiting capabilities intended to increase the robustness of the grid infrastructure, helping to prevent costly and inconvenient extended power outages arising from initially minor faults (see Section 11.5.5).

In a similar vein, the privately funded Tres Amigas SuperStation [33] presently being developed in New Mexico, US, for commercial operation in 2015 will use 5 GW 2G HTS superconducting cables (American Superconductor/LS Cable) to link the three main grid networks of the US together (Fig. 11.10), enabling for the first time cross-continental transmission and redistribution of power, easing bottlenecks and providing capability for the effective integration of diverse, intermittent, renewable energy sources. Located at the intersection of the three existing networks, the facility will convert power obtained from each grid from ac to dc before passing it through the superconducting cables and finally converting synchronously back to ac for redelivery into the appropriate grid as required. It is in localized applications such as these that superconducting power cables are most likely to make an early commercial impact.

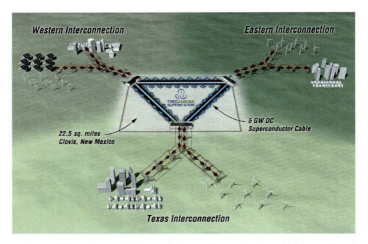

Figure 11.10 The planned Tres Amigas SuperStation linking the three existing major US electricity grids by 5 GW 2G HTS transmission cables for redistribution of power. Image courtesy of Tres Amigas LLC.

It is notable that in each of the projects listed above, the cable partner is as integral to the project if not even more so than the wire manufacturer. This is a clear indication of the progress of the technology from scientific novelty to engineering reality. The major engineering issues to be addressed in the manufacture of superconducting cables aside from the design of the cable itself are the provision of cooling, termination and splicing, and the minimization of losses. To understand how losses arise, it is first necessary to refine the concept of a cable. A cable comprises an assembly of conductors ("wires") connected in parallel. This is necessary in order to achieve the overall current capacity required. In the case of traditional copper cables, where a solid core is a possibility, separation into strands is nonetheless common both to afford flexibility to the cable and also to reduce eddy current losses. In the case of superconducting cables, we are (presently, at least) limited to around 200 A of current capacity for a single wire (at liquid nitrogen temperature). It is apparent, then, that production of a 2 kA cable, for example, requires at least 10 individual strands.

Continuing the analogy with a conventional cable, the strands comprising the cable are typically twisted, which is to say transposed with one another along the length of the cable. The purpose of this is—so far as possible given the finite strand diameter—to eliminate induction losses arising from the magnetic field generated by the current flow along individual strands. Without transposition, this "self field" resulting from the significant current being transported will produce a magnetic flux linking the strands, leading to ac losses as the current cycles. Striation alone is insufficient to prevent this, since ultimately the ends of the strands are connected at the termination of the cable, creating a current loop enclosing the flux. Only a complete transposition of strands such that each strand changes place with every other along the length of the cable will ensure that little to no flux is enclosed. Full transposition offers the additional advantage of current equalization, ensuring that the environment of each strand is the same averaged over the length of the conductor (such that, for example, no single strand is consistently better cooled, or subjected to a lesser magnetic field, or offers a lower induction current pathway) and that therefore there will be no imbalance in the currents being transported by different strands of the cable.

In summary, then, to create a workable cable, stranding and transposition are required. Most HTS power cables produced to date have used a variation of the Nuclotron-style design, depicted in Fig. 11.11, where individual strands spiral around a cylindrical former, which can conveniently be used to carry coolant. Various adaptations of this design are possible to produce single-phase cables or three phases coaxially (as shown) or simply bundled together within a single insulating jacket. Since the diameter of the cylindrical former must increase in proportion to the number of strands it supports, multiple cores or layering of strands upon a single core are commonly used to increase the current capacity. Simple layering of strands is possible, but breaks the requirement for transposition (inner layers are never transposed with outer layers) leading to increased losses. Cables are also needed for applications other than power transmission, where the requirement for low loss or a higher packing density may be more stringent. This is particularly the case in enclosed systems such as transformers, motors and generators, where close packing of ac coils is likely to be required.

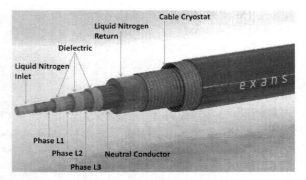

Figure 11.11 A three-phase concentric (triax) "cold dielectric" power cable assembled from superconductor tapes, featuring contraflow liquid cryogen pathways and an evacuated thermal insulation sleeve. Image courtesy of Nexans.

Various approaches to achieving cables having a higher packing density of fully transposed strands are illustrated in Fig. 11.12. The rope cable comprising multistage twisted strands is adopted from conventional cabling technology and finds some application in superconducting cable-in-conduit conductors thanks to its permeability to coolant flow. However, it can only be used with round-wire strands. The Rutherford cable (named after the Rutherford Laboratory in Oxfordshire, UK, where it was developed, and not after Rutherford himself) is a successful design used for low-temperature superconducting wires that effectively addresses the above issues while providing a high packing factor (resulting in high current density) regardless of the number of strands, good stacking possibilities and mechanical stability. Individual strands follow an elongated spiral while the cable as a whole is subjected to a mechanical deformation to produce structural and mechanical stability with low void fraction. Topologically, it can be considered as a flattened version of the Nuclotron cable.

The challenge of producing densely packed, low-loss cables from HTS tapes has been met by the Roebel cable approach [35], named after Ludwig Roebel, who in the early 20th century devised a method of producing low-loss high-power copper bars for power generators. His geometry and assembly method (Fig. 11.13) is applied essentially unchanged to HTS tapes in order to achieve transposition of the high aspect ratio conductor while retaining a high filling factor. This in turn can be seen as a sideways flattening of the Rutherford cable topology, and it overcomes the inability of HTS tapes to accommodate the tight bends required by Rutherford cabling, whilst also providing the advantage that the orientation of the superconductor remains fixed throughout transposition. The disadvantage of the Roebel method lies in its wastefulness of the tape material that must be punched to an appropriate meandering form before weaving into the cable.

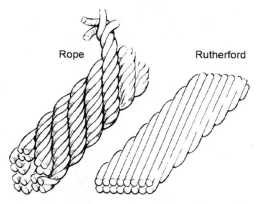

Figure 11.12 Approaches to forming a cable from an assembly of fully transposed strands. Images reproduced from [34] with permission.

Figure 11.13 Half transposition length of a 15/5 Roebel cable assembly formed from 15 strands of 5 mm lateral width cut from an original 12 mm wide tape. RRI image.

11.5.2 Superconducting Transformers

The transportation of electrical power from one location to another is accomplished by a series of stages of transformation between different voltage levels. In conventional power cables, high transmission voltages serve two purposes: increasing the power transmission (IV) while decreasing the resistive losses (I^2R) associated with increased currents. In superconducting cables, the latter are much less significant, and typically a cable will be operated close to its maximum current capacity, reducing the transmission voltage required for a given power. As previously described, this can help to reduce the number of transformer substations required, bringing particular benefits in dense urban areas. However, whether superconducting cables or conventional cables are used, voltage transformation is still required, and at each transformation stage, significant amounts of energy are lost to waste heat. Conventional oil-cooled transformers, in addition to being wasteful of energy, pose an environmental and a fire hazard. They also occupy significant amounts of space in locations where space for additional substations to handle increasing demand is limited and at a premium. Superconducting transformers (Fig. 11.14) address all of these issues. Even if used in conjunction with conventional power transmission technologies, high-temperature superconducting transformers are comparatively compact, offer increased efficiency and therefore less energy wastage, and are cooled by safe, abundant, and environmentally benign liquid nitrogen in place of oil. The necessity of cooling in this case even translates into an advantage since being actively cooled offers the opportunity to modulate that cooling in such a way as to operate at different loads in order to meet peak demand, thereby saving additional energy.

Figure 11.14 A three-phase, 1 MVA, 2G HTS distribution transformer under development at the Robinson Research Institute. This grid-connected prototype aims to demonstrate the utility of Roebel cable as an effective means of implementing a stable high-current winding with low ac loss, thus reducing the heat load on the cooling system. RRI image.

The most significant superconducting transformer project presently under way is led by Waukesha Electric Systems and aims to build, test, and install a 28 MVA three-phase device using SuperPower 2G HTS wire on Southern California Edison's Smart Grid by the end of 2013.

11.5.3 Superconducting Generators

Power cables and associated transmission facilities are just one particularly visible part of the ubiquitous electric power infrastructure. Throughout the power industry, every aspect of the infrastructure is struggling under a combination of aging technology, increasing demands on both quantity and quality of the power supply, and a general expectation of perpetual on-demand fault-free operation.

Superconducting power generators have a contribution to make to many of the smaller-scale energy generation devices associated with renewable energy technologies such as wind power and hydropower. In these areas, lower weight, higher efficiency, and improved reliability are critical performance requirements. The general inaccessibility of many of these generators, typically located offshore, in remote areas, or far above the ground, makes servicing challenging and costly, and therefore reliability is key. The environment in which they must operate may also be harsher than is typically encountered in conventional facilities, exposed to the elements and subjected to corrosive conditions at sea.

A 10 MW generator utilizing HTS technology has about one third the mass of a conventional generator. In the case of wind turbines, this reduction in mass allows for an increase in blade size and consequently greater power output (Fig. 11.15). The net effect is a doubling of the power capacity of a conventional system, and thereby a lowered cost for wind energy. More broadly, superconducting generators offer a wide range of benefits to the flexible, robust, smart grids of the future in terms of their high power density, high partial load efficiency, low reactance, low harmonic generation, low noise level, and high grid stability. In challenging applications such as those demanded by renewable energy sources, direct drive systems enabled by compact, lightweight superconducting generators provide for flexible and robust mechanical design impervious to deflections coupled with high reliability and lower maintenance requirements, as well as increased safety for service or installation personnel through the ability to "turn off" the rotor coils, a feature unavailable to permanent magnet designs.

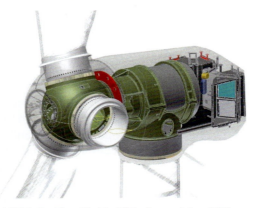

Figure 11.15 AMSC SeaTitan™ 10 MW direct-drive HTS generator wind turbine. The increased power density of the superconducting generator enables a nacelle mass equivalent to a 5 MW conventional system. Image courtesy of AMSC.

11.5.4 Superconducting Energy Storage Devices

The transition to renewable energy sources brings further complications due to intermittency. The sun does not always shine where needed, the wind does not always blow, and even an optimal mixture of sources cannot be *guaranteed* to provide the required energy at the required time, in line with expectation. Even considering conventional power sources, the ever increasing complexity of the grid means that instabilities are arising with increasing frequency and a temporary shortfall in available energy may occur, a situation society finds intolerable.

Here again, superconductor-based technologies provide a solution to the requirement for short-term energy storage and release on the scale required and with the necessary flexibility at an achievable cost and an acceptable level of loss. Small-scale superconducting magnetic energy storage (SMES) systems, based on low-temperature superconductors, are a proven technology [36] in widespread use for many years now. At the point of distribution, they enhance the capacity and reliability of stability-constrained grids and at the point of consumption, they are employed by large industrial sites to guarantee reliability and improve power quality. Large-scale systems based on high-temperature superconductors such as those required to compensate temporary shortfalls in the power output of natural energy sources are the present focus of development (Fig. 11.16).

The principle behind SMES is simple: A superconducting coil is energized when power is available, and the energy stored in the magnetic field generated by the coil. The superconducting nature of the coil means that this energy is not dissipated significantly over time as in a conventional electromagnetic coil, so that when the energy is required, it can simply be extracted from the SMES device by reversing its connections and employing it as a source. Since the energy is stored in the form required (as a circulating current), it can be drawn from the device almost instantaneously. It is this fast response that constitutes the SMES' primary benefit to energy applications, and enables a range of capabilities that go beyond what would traditionally be considered as pure energy storage functionality. Its lack of moving parts and virtually infinite cycling capability provide extremely high reliability and lifetime. The suggested potential capability of large-scale SMES units is for storage of 5 GWh of electricity with a return efficiency (storage plus retrieval cost) of greater than 95%—favorable compared to the 70–80% ascribed to pumped-storage hydroelectricity, the most efficient alternative—combined with a rapid response time on the scale of milliseconds (less than a single power line cycle) [37]. Other proposed benefits include almost instantaneous tracking of load changes allowing conventional generators to operate at a more efficient and cost-effective fixed output, stabilization of system frequency and voltage, provision of reserve capacity allowing generators to operate close to full load, reducing the number of generators required, and energy management (smart grid) functionality. SMES systems are consequently expected to be able to store energy more effectively than any conventional energy storage system such as chemical batteries, flywheels, or hydro-storage.

Superconductors can also be employed as bearings in flywheel energy storage devices. A flywheel transforms electrical energy into kinetic energy, which is stored in the rotating flywheel. This stored energy is then converted back into electrical energy as needed.

By levitating the flywheel in a vacuum enclosure on frictionless superconductor bearings, frictional losses are virtually eliminated. While conventional flywheels suffer energy losses of 3–5% per hour, HTS-based flywheels have been shown to operate at <0.1% energy loss per hour [39].

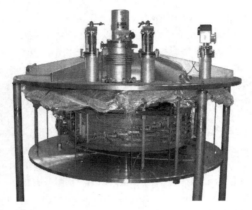

Figure 11.16 An 800 kJ solenoidal HTS SMES device comprising a 1G Bi2212 coil operating at 20 K using conduction cooling from electrical cryocoolers. The coil is shown exposed; in operation, the device is contained within a vacuum enclosure. Reproduced from [38] with permission.

11.5.5 Superconducting Fault Current Limiters

As grids expand, becoming more complex with increased numbers of sources, and ultimately with each sink also acting as a source in the feed-in/demand-side participation networks of the future, the likelihood of unacceptable power surges occurring increases, and becomes more difficult to guard against. Such surges can be initiated by diverse events including poor weather, falling trees, traffic accidents involving infrastructure, animal interference and other unpredictable occurrences. An increase in the cumulative summation of these fault current levels increases the risk that they will exceed the ratings of protection devices and grid components, exposing grids to costly damage and downtime for repairs. Superconductors offer a new solution to this problem in the form of superconducting fault current limiters; essentially smart, passive, self-recovering circuit breakers. These sit in the current path, virtually transparent to the network, until a fault condition results in an abnormally high current flow that exceeds their critical current. At this point, they undergo the transition from the superconducting to the normal state with an accompanying large increase in resistance that acts to reduce the current flow to safe levels. This switching can happen extremely quickly, fast enough to prevent any damage downstream, and as the fault condition is removed and the current flow returns to normal, the fault current limiter reverts to the superconducting state, and normal operation is resumed without manual intervention [40].

The technological challenge behind the development of superconducting fault current limiters is in ensuring that the superconductor does not suffer damage during the current-limiting "fault" phase, and thus that it can resume normal operation after a prolonged fault period. Various refinements to the basic design described are implemented to achieve this, and superconducting fault current limiters today comprise one of the closest to market energy applications of superconductors. Technological progress continues, however, and Nexans recently implemented the first real-world 2G HTS fault current limiter (Fig. 11.17) in an upgrade of an earlier 1G version [41]. The device has a current rating of 560 A at 12 kV, and offers 90% lower losses than the original version, while reacting faster in the event of a fault condition occurring.

Figure 11.17 Nexans 2G HTS fault current limiter unit installed at a power plant in Saxony, Germany. The device has a current rating of 560 A at 12 kV. Image courtesy of Nexans.

11.6 Superconductors for Transportation Applications

Transportation is a major consumer of energy, accounting for fully one quarter of global energy demand. Almost two thirds of the oil presently consumed goes to power transportation, a proportion that has been growing over time. Although often considered distinct from other energy usage (for example, that of electricity), in terms of sustainable energy applications it is necessary to incorporate transportation into these considerations, since as oil becomes increasingly inaccessible (both literally and through steadily rising cost), the sustainable energy sources are going to have to accept this burden also, and in practice this will mean electric vehicles powered via the grid, dramatically increasing the demand and the load on the infrastructure.

Superconductors both piggyback on the existing trend toward electric vehicles and also contribute to other major forms of transportation in terms of a new swathe of technologies including advanced ship propulsion systems, magnetically levitated trains, and railway traction transformers. The incorporation of superconducting technologies into transportation system design can increase flexibility, improve efficiency and performance, reduce weight and consequently fuel consumption, and extend the range of all forms of transportation.

11.6.1 Superconducting Motors

Through the greater magnetic flux density that can be generated by a superconducting coil or bulk magnet compared to a conventional electromagnetic coil or ferromagnet, rotating machines based on high-temperature superconductors offer a higher power density or torque than conventional machines, resulting in a smaller size and weight for equivalent performance. Prototype superconducting motors and generators are one third the size and weight of their conventional copper-wound counterparts (Fig. 11.18). They are quieter, due to the absence of cyclic magnetostriction of the ferromagnetic core, and they are more efficient—even when the power consumption necessary for cooling is taken into account—as a result of the practical elimination of Joule heating in the coils as well as of the hysteresis and eddy current losses associated with a conventional ferromagnetic core. The higher torque available can also eliminate the requirement for gearing and transmission drives, which increases flexibility and can further reduce losses. All of these features translate directly into potential cost savings that make superconducting motors highly attractive to a range of large-scale transportation applications.

Figure 11.18 Schematic comparison of the relative sizes of conventional and superconducting 40 MW class motors/generators. Image courtesy of Superconductor Technologies Inc.

11.6.2 Marine Propulsion Systems

In many senses, the marine sector is ahead of the game, with the early adoption of electric propulsion systems characterized as the most important change in ship design since the adoption of the diesel engine in the 1920s. Almost all new large commercial ocean-going vessels are electrically propelled, and in 2000, the US Navy announced that it was migrating to an all-electric fleet. More recently, since 2009 an EU-funded project aims to extend the applicability of efficient electric propulsion systems to a wider range of vessels in order to reduce the environmental impact of the combined European commercial shipping fleet. Historically, the large size and weight of conventional electrical motors and generators has been a barrier to the adoption of electric propulsion systems, which otherwise offer significant benefits. For these reasons, high-temperature superconductors offer important advantages along the lines outlined above (Fig. 11.19). Superconducting motors are particularly suited to marine propulsion since the elimination of rotor losses results in far higher efficiency, particularly under partial-load conditions, where many ships operate for most of the time. This improved efficiency translates directly into a longer cruising range and greater fuel economy, while smaller motors enable electric ships to use shallower ports. In common with other sectors, electric propulsion also enables new, more flexible system design by decoupling the rigid connection between engine and motor. The same power plant can be used for propulsion as for other shipboard requirements, leaving more space for passengers and cargo, or in the case of military applications, for more powerful weapon systems.

Figure 11.19 A 36.5 MW superconducting ship propulsion motor designed and manufactured by AMSC for the US Navy. The motor is one third the size and weight of a conventional 36.5 MW marine motor and less than half the size of a power-dense naval motor. Image courtesy of AMSC.

11.6.3 Magnetically Levitated Trains

Magnetically levitated ("maglev") trains are the sustainable alternative to intracontinental air travel. In Japan, where a high-speed, highly efficient rail service is the norm, rail travel is already a preferred alternative to air for many travelers. The ever-increasing inconvenience and frequent disruption associated with air travel is a driving factor for consumer preference, quite aside from environmental considerations. Maglev trains utilize attractive and repulsive interactions between magnets located within the guide track and in the carriage to both propel the vehicle using a linear motor arrangement and to levitate the carriage within the track, enabling faster speeds and greater reliability due to decreased wear. Permanent magnets, electromagnets, and superconducting magnets have all been trialed in various designs.

In Yamanashi prefecture in Japan, an 18 km superconducting maglev train test track has been operational since 1996 (Fig. 11.20), and manned test runs have exceeded 500 km/h. This compares favorably with current commercial high speed trains based on conventional technologies, which run at a maximum speed of around 300 km/h. The train also holds the world speed record for trains of just over 580 km/h, set in 2003. The test track is now being extended to 42 km in preparation for its incorporation into a commercial maglev line approved for construction by the Japanese government in 2011 [42]. The Chūō Shinkansen line will have a cruising speed of 505 km/h, linking Tokyo and Nagoya (290 km) in 40 min. The plan is for this to be completed by 2027, and extended to Osaka (a total of 550 km) by 2045. For optimal journeys such as these, the actual travel time can be reduced to about half the typical flight time, while energy consumption at high speed is reduced to around a third of that associated with air travel. More importantly, however, the system is electrically powered, meaning it can benefit from all the advances in electrical power generation that are anticipated to occur over the timescale of its development.

While the existing design utilizes LTS Nb-Ti magnet coils, research is ongoing into the use of 2G HTS magnet coils instead. A further notable test installation exists at Southwest Jiaotong University in China, where a design operating on superconducting *permanent* magnets (rather than magnet coils) formed from bulk RBCO was demonstrated in late 2000, becoming the first crewed HTS maglev system test [43]. Similar small-scale demonstrations have subsequently been built elsewhere, but as yet no operational system exists.

Other forms of maglev technology exist that do not rely on superconductors, most notably the Shanghai Transrapid maglev train, the only commercially operated high-speed maglev train, which utilizes conventional electromagnets. Low-speed maglev systems are also in use for local transportation.

Figure 11.20 The experimental MLX01 superconducting magnetically levitated train runs on the Yamanashi test track in Japan. Image credit: Yosemite/Wikimedia Commons/CC-BY-SA-3.0.

11.6.4 Electric Aircraft

Looking well beyond what has been realized to date, superconductor-based aircraft are a strongly feasible future possibility currently under active investigation by NASA [44] (Fig. 11.21) and EADS [45] for realization around 2030–2035, offering many of the advantages already described of a transition away from fossil fuel combustion engines reducing emissions, flexible systems design enabling a reduction in the noise impact of aircraft operation and the option of operating at reduced power for much of the flight, increased efficiency and improved reliability. Given the weight considerations of primary importance to aviation, a particularly attractive possibility is the combination of both cooling system and fuel source in a single liquid hydrogen-based design. Hydrogen-powered aircraft are not new, and also independently of superconducting technologies, liquid hydrogen fuelled aircraft are under consideration [46], making the combination with superconductor technology an obvious next step.

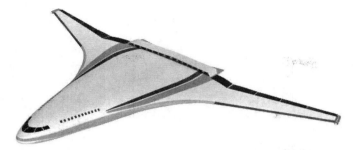

Figure 11.21 NASA's N3-X concept is a 300-passenger hybrid wing-body aircraft with turboelectric distributed propulsion incorporating two superconducting generators mounted in the wingtips that power 15 superconducting motor-driven propulsors embedded in the fuselage via superconducting transmission lines. NASA image.

11.6.5 Personal Electric Vehicles

The most challenging area of technology is consumer technology, and here as well the breakthrough of superconductors is some way off. With the transition to conventional electric vehicles now well under way, it may be seen as only a matter of time until developments in the technology of superconductors trickle through to the consumer level (Fig. 11.22). Enabling breakthroughs for this to occur would include the discovery of higher-temperature superconducting materials, significant advances in cooling technology, or the establishment of the hydrogen economy.

Figure 11.22 A demonstration electric vehicle modified by Sumitomo Electric Industries to feature a liquid nitrogen-cooled 30 kW, 120 Nm superconducting motor formed from their 1G BSCCO wire [47]. Image courtesy of Sumitomo Electric Industries Ltd.

11.7 Paradigm-Shifting Energy Technologies

There are a number of technologies on the horizon that would completely change the game in regards to energy supply and usage. Each of these is worthy of a full treatment in its own right, but here the briefest of overviews will be given, focusing on the relevance of superconductivity in each case.

11.7.1 Fusion Power

The long-term solution to the problem of the ever-increasing energy demand of a globally developed society is presently considered to rely on fusion power—the release of energy from the fusing together of atomic nuclei in a process analogous to that undergone in the sun. While the other sustainable energy sources discussed in this book may provide more immediate, stop-gap measures, given the self-evident difficulty in supplying a country such as the United States with sufficient energy from renewable sources, it is hard to see how these same sources alone could also adequately supply the people of India and China, for example, each with four times the population, at a similar, and similarly increasing, standard of living.

The proposed fusion reaction consumes deuterium and tritium to produce helium. Deuterium (heavy hydrogen) is present in sufficiently useful quantities in water, while tritium will initially be sourced elsewhere but could ultimately be produced within the reactor itself from lithium. A working reactor would require about a kilogram of fuel per day. To achieve fusion, a high-energy plasma of fuel must attain a temperature of 150 million K while confined by strong magnetic fields, since no material containment can withstand such a temperature. These magnetic fields rely on superconducting magnets for their production.

China, the European Union, India, Japan, Korea, Russia, and the United States are presently engaged in a joint project to construct the €13bn International Thermonuclear Experimental Reactor (ITER), with initial operation expected in 2020 and energy production planned for 2027. ITER is intended to demonstrate the feasibility of fusion power by generating 500 MW of power from an input of 50 MW over a limited period of time (several minutes). However, the energy it produces will not be retained. It will be followed by a commercial demonstration reactor, DEMO, intended to supply 2–4 GW of electrical power to the grid from the same 50 MW input as early as 2040. Finally, there is already a plan for the prototype commercial fusion reactor, PROTO, expected to be built post-2050.

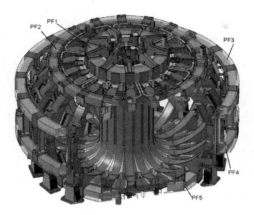

Figure 11.23 Cut-away schematic diagram of the ITER reactor design, highlighting the 18 Nb_3Sn toroidal field coils in gold. Five of six Nb-Ti poloidal field coils are labeled PF1 to PF5. Additionally, there is a central Nb_3Sn solenoid comprising six coil packs stacked vertically and 18 Nb-Ti correction coils positioned between the toroidal and poloidal field coils. Image credit: © ITER Organization, http://www.iter.org/.

ITER makes use of LTS cable-in-conduit materials for its superconducting magnets (Fig. 11.23), requiring 80,000 km (400 tons) of Nb_3Sn superconducting wire for its toroidal field coils. Before ITER, worldwide production of Nb_3Sn was just 20 tons per year. Additional to this is a further 250 tons of (cheaper) Nb-Ti wire for the (lower field) poloidal coils. One third of the total cost of the reactor is the cost of the superconducting magnets and associated cooling system. The debate is still open as to whether DEMO will progress to HTS alternatives, enabling operation at either higher fields or higher temperatures, but they presently appear promising candidates, offering more compact, simpler reactor designs with reduced cooling costs [48], and the potential for significant performance improvements to be attained in the intervening time period.

11.7.2 Hydrogen Economy

The use of hydrogen as a medium for the storage of energy is discussed at length elsewhere in this volume, and hydrogen-powered vehicles are a particular driver for research into this technology. Hydrogen has the unique advantage as a fuel source that, like electricity, it is clean at the point of use, producing no harmful by-products of combustion, merely water. Furthermore, electrolysis is a viable method of hydrogen production, while the generation of electricity from hydrogen in a fuel cell is an equally viable process. To some extent, then, the two are mutable, and their combination to form the basic energy infrastructure thus seems sensible. However, it is possible to take this vision of the future energy infrastructure one step further, to incorporate superconductivity. If hydrogen were to be stored and transported in liquid form then it becomes feasible to use it not only as a fuel but also as a coolant for a ubiquitous high-temperature superconducting infrastructure lying alongside it. Liquid hydrogen pipelines could incorporate superconducting power cables for distribution (a so-called SuperCable [49]), and vehicles powered by liquid hydrogen would naturally make use of superconducting motors cooled by the fuel tank. This vision of the future has been termed the SuperGrid (Fig. 11.24). The realization of such an infrastructure would inevitably lead to a sharp increase in the uptake of HTS technologies as efficient cooling became widely and readily accessible [50].

11.7.3 Room-Temperature Superconductivity

It is often supposed that the discovery of a new superconducting material having a critical temperature above room temperature would trigger a revolution in the applicability of superconducting technologies. But would it? In fact, this seemingly self-evident aspiration may be nothing more than an unhelpful impediment to the necessary effort involved in bringing any new technology out of the laboratory and into the real world. First, let's examine the possibility of the existence of a room-temperature superconductor. The BCS theory of superconductivity originally predicted an upper limit to T_c of around 30 K. Various tweaks and modifications are able to plausibly stretch this value to the 39 K of MgB_2, but it still

Figure 11.24 Artist's impression of a hypothetical SuperGrid energy pipe sharing a tunnel with a high-speed, long-distance train line. A liquid hydrogen pipeline forms the cooling core of a superconducting power cable for electrical power distribution. Image reproduced from [51] with permission of the Electric Power Research Institute.

falls far short of the 133 K T_c of $HgBa_2Ca_2Cu_3O_{8+\delta}$ (Hg1223), or the highest T_c ever observed of 164 K for the same material under pressure, although adaptations can be postulated to encompass these too [52]. Although there is presently no accepted theory of high-temperature superconductivity to supplant the BCS theory, theoretical suggestions have been made that the 160–170 K values already observed are in fact the limit for cuprate superconductors [53]. Even then, the latest discoveries (Section 11.3.4) have shown that high-temperature superconductivity is not restricted to cuprates, and there is no reason to suppose, therefore, that higher temperature materials cannot be found. Without a working theory of their operation, however, there is no strategy to guide discovery, as reflected in the fact that every discovery of a new family of high-temperature superconductors to date has been purely serendipitous.

In any event, this discussion overlooks the fact that T_c is not the whole story, nor even a particularly important part of it. Given a workable cooling system, the T_c of the material is almost entirely irrelevant to application. We have already seen this in the example of Bi2223 (T_c = 110 K) being supplanted by Y123 (T_c = 93 K) in the transition from first-generation to second-generation HTS, and the reason for it is clear from Fig. 11.4. Parameters such as the in-field performance and the deviation of practical operational limits from fundamental ones are far more critical to the viability of a particular material than just its T_c. A simple material akin to LTS but with a transition temperate around 100 K would offer far greater utility than a complex room-temperature superconductor with extrapolated properties similar to those of the cuprates. Furthermore, to be *operated* in the vicinity of room temperature (say, water-cooled), the material would need to have a T_c of perhaps 100–200°C, three to four times the maximum known value. Any less and its performance would be limited to the point of uselessness.

The prospect of room-temperature superconductivity is attractive and enticing [54], and its discovery would certainly provide exhibition pieces for schools and museums everywhere. But it is unlikely to prove a panacea to the valuable technological developments required and presently under way to extract utility from existing HTS materials.

11.8 Other Applications of Superconductors

The effort expended on developing these new technologies for energy applications will yield further urgently required capabilities not touched on here. As with any new technology, it is difficult to predict all the applications that will eventually arise. In medicine, industry, research and communications technology, superconducting machines, instruments and devices have the potential to be as revolutionary as in the energy sector. Some of these technologies are obvious offshoots of those described above, while others are unique to their particular field of application. Broadly speaking, all existing electrical devices stand to be revitalized by superconductivity, bringing new or improved capabilities to bear. Powerful magnets and advanced sensors drive medical and industrial advances, while superconducting electronics power advanced communications, information technology and ultimately quantum computers or high-speed computing based on fluxonics rather than electronics to surpass the fundamental limits of silicon-based technology.

The most widespread application of superconductors today is in the MRI machines used in medicine for radiation-free functional imaging of the body and brain. Superconducting sensors can also be applied to enhance the sensitivity of other medical techniques such as magnetoencephalography (MEG) and magnetocardiography (MCG). In industry, superconductors are applied to provide higher strength, more effective magnetic separation capabilities and higher efficiency induction heating for metal processing, cutting energy demand and processing costs. The superconducting bearings developed for flywheels have other applications where a frictionless bearing may be required, and small-scale versions of the levitating linear motors used in maglev vehicles may find other applications, for example in cleanroom environments. Plug-and-play HTS filter units are deployed in mobile phone networks as a cost-effective alternative to additional base stations for increasing bandwidth.

In scientific research, the drive to understand superconductivity as a fundamental state of matter pushes back the boundaries of human knowledge. Applications of superconductors in cutting-edge research abound for the capabilities they offer that are not available by other means, for example high-power magnets in particle accelerators such as the LHC at CERN, ultra-precise sensors in the form of SQUIDs (superconducting quantum interference devices) able to detect magnetic fields at the quantum level, and the equivalent of MRI applied to chemistry in the form of NMR for proteomics, drug discovery, and materials investigation. In the field of space exploration, where available power is at a premium, superconducting devices are particularly attractive for long-term operation. Even the base electronic unit of the volt is today defined using a Josephson junction formed from a superconductor.

Many of these research technologies have parallels within the broader energy sector. The superconducting magnets used for particle accelerators also find application in the fusion reactors of tomorrow, while superconducting sensors are employed in the ever more challenging search for new sources of fossil fuels.

11.9 Cooling

It would be remiss to write an entire chapter on super-conducting technologies without at least mentioning cooling. Indeed, knowledge of the existence of superconductivity is now so widespread that upon mentioning it, one is almost certain to receive an immediate objection along the lines that cooling is a problem. It is important, therefore, to also recognize the present status of cryogenic technology. Reliable cost-effective cooling is a critical aspect of HTS technology, and significant technological advances have been made in this area too.

Low-temperature superconductors have critical temperatures ranging up to around 23 K, and are generally operated in a liquid helium bath at 4.2 K or below (achieved through reducing the pressure in the vessel). This is the basis of operation of all present-day commercial MRI machines. Liquid helium is expensive, difficult to handle, store and transport, and most importantly from the viewpoint of sustainable energy, it is a limited resource. In fact, helium is less renewable even than coal and oil since once released as a gas it escapes the atmosphere and is gone forever. Supplies of liquid helium are most economically obtained as a by-product of mining natural gas, with two thirds or more of global consumption for a long time being satisfied by the government-mandated depletion of the US reserve built up in the first half of the 20th century, intended to be complete by 2015. As this date draws close, and demand for helium *increases* as supplies decrease, the scarcity of helium is rising along with its long artificially deflated price, doubling from 2002–2007. Already, shortages of supply regularly affect research projects and indeed it was a (possibly apocryphal) failure of the supply of liquid helium rather than of the superconductor itself that is said to have ended the longest-running experiment aimed at detecting a diminishment of the supercurrent in a closed persistent superconducting loop (leaving the conclusion that there was no detectable reduction in current after 2.5 years).

High-temperature superconductors are, however, a different story. With their critical temperatures above the 77 K boiling point of liquid nitrogen, they offer numerous other possibilities for cooling. As a liquid cryogen, liquid nitrogen is cheap, abundant and easy to handle. If a lower temperature is required for increased performance, a small number of other liquid cryogens are available (cf. Fig. 11.4): Neon (27 K) is favored for its inert nature and lower overall cost than other low-temperature cryogens (resulting from significantly higher cooling capacity than the alternatives, not simply raw material cost) while hydrogen (20 K) is of particular interest for its potential to be tightly integrated into a new energy economy combining coolant and fuel source in a single, clean fluid.

At temperatures above 20 K or so, the use of efficient, self-starting *closed cycle* refrigeration is also feasible. This is refrigeration of a more familiar type, whereby a working medium (in this case, commonly a small volume of helium gas) is continuously cycled in a closed loop around an electrically powered refrigeration system to keep essential components cool. In this field also, many technological advances in reliability, efficiency, and capability have been made and are continuing. It is this type of cryocooler (Fig. 11.25) that powers existing mainstream superconductor-based technologies such as superconducting filters in mobile phone base stations.

Figure 11.25 A Stirling cycle electrical cryocooler. The copper cold head seen at the left has a cooling capacity of 5 W at 77 K. This can be used to conduction cool superconducting components, or alternatively to drive a closed-cycle gas cooling system. Image provided by Superconductor Technologies Inc. All rights reserved, 2012.

11.10 Cost

We should not conclude without a discussion of costs, although to do so is to immediately date the text given the rapid rate of technological progress in the newer materials. The standardized unit of cost for superconducting wire is the US$/kA-m, in words, the price per meter of wire normalized to the current that wire can carry. On top of this come the running costs associated with cooling, and the measure must be specified for a defined operating temperature and magnetic field, since both of these affect the achievable current. For comparison purposes, the cost of a Cu cable operating at room temperature is around $30–60/kA-m, while a conventional (LTS) superconducting cable is around $1/kA-m at 4.2 K, 2 T (Nb-Ti), or $10/kA-m at 4.2K, 10 T ($Nb_3Sn$). (The seemingly low cost of these LTS conductors is offset by the extremely high cost of cooling, emphasizing the previous point. For a fairer comparison, it is necessary to consider the performance of Cu at such low temperatures, which is 30–200 times better, depending on purity, than at room temperature. It then becomes apparent that it is only because of their unique properties, such as the high magnetic fields they can generate, that LTS materials find use.)

1G HTS presently sells for around $200/kA-m at 77 K, 0 T. If one were to operate this wire at 4.2 K, due to the increase in current capacity at the lower temperature, this would scale to around $20/kA-m, and one can immediately see (unsurprisingly) that conventional LTS cables are a better choice at these temperatures, offering greater field capability at lower cost. Reductions in the cost of 1G HTS are limited by the high Ag content, and it is for this reason that the focus has shifted to 2G HTS for the potential of cost reduction.

2G HTS presently sells for more than 1G HTS, but is predicted to come down in cost to around $100/kA-m at 77 K, 0 T over the next 2–3 years as a result of both processing cost reduction and performance improvements, with the cost continuing to fall beyond that timeframe [55]. Initial cost expectations (or desires, perhaps) of $10/kA-m at 77 K, 0 T now no longer seem likely to be realized, but the real cost advantage of 2G HTS occurs in-field, at a targeted operating condition such as 65 K, 3 T, where due to its superior in-field performance relative to 1G, the cost-performance remains unchanged. This opens up new areas of application that were previously inaccessible to 1G HTS.

The potential of MgB_2 under a restricted range of fields and temperatures becomes apparent if we consider the present estimate for eventual costs of $1/kA-m at 25 K, 1 T [56], 20 times lower than 1G HTS at the same temperature and field and strongly competitive with predictions for the eventual cost of 2G HTS. In these targeted regimes, therefore, the robustness and ease of handling of MgB_2 may make it the preferred choice, offering all the advantages of an LTS material at a more conveniently accessible (and therefore cheaper to achieve) temperature.

11.11 Summary

Superconductors have a broad role to play across the gamut of sustainable energy applications, from a supporting role for the plethora of renewable energy generation technologies presently vying for superiority to an enabling role in new technologies for energy storage and transmission to a revolutionary role across the variety of ways we consume energy. Superconductors can provide a major contribution to lessening CO_2 emissions and other environmental impacts, while reducing the societal impact of the underlying energy infrastructure at the same time as increasing its robustness and capacity. Looking further to the future, superconductors are central to paradigm-shifting technologies such as fusion power or the transition from an oil-based to a hydrogen-based economy. We can expect that the progress of science will steadily provide us with improved materials capabilities as the gap between present performance and fundamental limitations is closed, while advances in manufacturing will lead to more reliable and cheaper production upon the basis of which superconductor-based technologies will become ever more beneficial and prevalent, perhaps ultimately to the point where a room-temperature superconductor will make them as commonplace as copper is today. In the meantime, the materials we have are already at the stage where serious consideration must be given to their price-competitive alternative to conventional technologies, with often only the inherent conservatism of the energy industry acting as the biggest barrier to adoption. The widespread, fast-paced, needs-driven introduction of a range of new energy technologies to tackle the energy crisis provides an opportunity for the simultaneous incorporation and "road-testing" of superconducting technologies with a view to making them the future standard.

References

1. Kamerlingh Onnes, H. (1911). Further experiments with liquid helium. G. On the electrical resistance of pure metals, etc. VI. On the sudden

change in the rate at which the resistance of mercury disappears, *Commun. Phys. Lab. Univ. Leiden* 124c, and earlier works of the same year cited therein. Republished (1912) in *Proc. KNAW*, **14**, 818–821.

2. Bednorz, G., and Müller, K. A. (1986). Possible high T_c superconductivity in the Ba–La–Cu–O system, *Z. Phys. B*, **64**, 189–193.

3. Bardeen, J., Cooper, L., and Schrieffer, J. (1957). Microscopic theory of superconductivity, *Phys. Rev.*, **106**, 162–164; Bardeen, J., Cooper, L. N., and Schrieffer, J. R. (1957). Theory of superconductivity, *Phys. Rev.*, **108**, 1175–1204.

4. Meissner, W., and Ochsenfeld, R. (1933). Ein neuer Effekt bei Eintritt der Supraleitfähigkeit, *Die Naturwissenschaften*, **21**, 787. Reprinted with English translation in Forrest, A. M. (1983). Meissner and Ochsenfeld revisited, *Eur. J. Phys.*, **4**, 117–120.

5. Sarrao, J. (2006). *Basic Research Needs for Superconductivity: Report of the Basic Energy Sciences Workshop on Superconductivity* (US Department of Energy, USA) p. 109, http://science.energy.gov/~/media/bes/pdf/reports/files/sc_rpt.pdf.

6. Larbalestier, D., Gurevich, A., Feldmann, D. M., and Polyanskii, A. (2001). High-T_c superconducting materials for electric power applications, *Nature*, **414**, 368–377.

7. Maeda, H., Tanaka, Y., Fukutomi M., and Asano, T. (1988). A new high-T_c oxide superconductor without a rare earth element, *Jpn. J. Appl. Phys.*, **27**, L209–L210.

8. Subramanian, M. A., Torardi, C. C., Calabrese, J. C., Gopalakrishnan, J., Morrissey, K. J., Askew, T. R., Flippen, R. B., Chowdhry, U., and Sleight, A. W. (1988). A new high-temperature superconductor: $Bi_2Sr_{3-x}Ca_xCu_2O_{8+y}$, *Science*, **239**, 1015–1017.

9. Tallon, J. L., Buckley, R. G., Gilberd, P. W., Presland, M. R., Brown, I. W. M., Bowden, M. E., Christian, L. A., and Goguel, R. (1988). High-T_c superconducting phases in the series $Bi_{2.1}(Ca, Sr)_{n+1}Cu_nO_{2n+4+\delta}$, *Nature*, **333**, 153–156.

10. Heine, K., Tenbrink, J. and Thoner, M. (1989). High-field critical current densities in $Bi_2Sr_2Ca_1Cu_2O_{8+x}$/Ag wires, *Appl. Phys. Lett.*, **55**, 2441–2443.

11. Hikata, T., Sato, K., and Hitotsuyanagi, H. (1989). Ag-sheathed Bi-Pb-Sr-Ca-Cu-O superconducting wires with high critical current density, *Jpn. J. Appl. Phys.*, **28**, 82–84.

12. van Beijnen, C., and Elen, J. (1975). Potential fabrication method of superconducting multifilament wires of the A-15 type, *IEEE Trans. Magn.*, **11**, 243–246.

13. Wu, M. K., Ashburn, J. R., Torng, C. J., Hor, P. H., Meng, R. L., Gao, L., Huang, Z. J., Wang, Y. Q., and Chu, C. W. (1987). Superconductivity at 93 K in a new mixed-phase Y-Ba-Cu-O compound system at ambient pressure, *Phys. Rev. Lett.*, **58**, 908–910.
14. Dimos, D., Chaudhari, P., and Mannhart, J. (1990). Superconducting transport properties of grain boundaries in $YBa_2Cu_3O_7$ bicrystals, *Phys. Rev. B*, **41**, 4038–4049.
15. Jin, S., Sherwood, R. C., van Dover, R. B., Tiefel, T. H., and Johnson, D. W. (1987). High T_c superconductors—composite wire fabrication, *Appl. Phys. Lett.*, **51**, 203–204.
16. Nagamatsu, J., Nakagawa, N., Muranaka, T., Zenitani, Y., and Akimitsu, J. (2001). Superconductivity at 39 K in magnesium diboride, *Nature*, **410**, 63–64.
17. Kamihara, Y., Watanabe, T., Hirano, M., and Hosono, H. (2008). Iron-based layered superconductor $La[O_{1-x}F_x]FeAs$ (x = 0.05–0.12) with T_c = 26 K, *J. Am. Chem. Soc.*, **130**, 3296–3297.
18. Weiss, J. D., Tarantini, C., Jiang, J., Kametani, F., Polyanskii, A. A., Larbalestier, D. C., and Hellstrom, E. E. (2012). High intergrain critical current density in fine-grain $(Ba_{0.6}K_{0.4})Fe_2As_2$ wires and bulks, *Nat. Mater.*, **11**, 682–685.
19. Foltyn, S. R., Civale, L., MacManus-Driscoll, J. L., Jia, Q. X., Maiorov, B., Wang, H., and Maley, M. (2007). Materials science challenges for high-temperature superconducting wire, *Nat. Mater.*, **6**, 631–642.
20. Iijima, Y., Tanabe, N., Kohno, O., and Ikeno, Y. (1992). In-plane aligned $YBa_2Cu_3O_{7-x}$ thin films deposited on polycrystalline metallic substrates, *Appl. Phys. Lett.*, **60**, 769–771.
21. Wang, C. P., Do, K. B., Beasley, M. R., Geballe, T. H., and Hammond, R. H. (1997). Deposition of in-plane textured MgO on amorphous Si_3N_4 substrates by ion-beam-assisted deposition and comparisons with ion-beam-assisted deposited yttria-stabilized-zirconia, *Appl. Phys. Lett.*, **71**, 2955–2957.
22. Selvamanickam, V., Carota, G., Funk, M., Vo, N., Haldar, P., Balachandran, U., Chudzik, M., Arendt, P., Groves, J. R., DePaula, R., and Newnam, B. (2001). High-current Y-Ba-Cu-O coated conductor using metal organic chemical-vapor deposition and ion-beam-assisted deposition, *IEEE Trans. Appl. Supercond.*, **11**, 3379–3381.
23. Goyal, A., Norton, D. P., Budai, J. D., Paranthaman, M., Specht, E. D., Kroeger, D. M., Christen, D. K., He, Q., Saffian, B., List, F. A., Lee, D. F., Martin, P. M., Klabunde, C. E., Hartfield, E., and Sikka, V. K. (1996). High critical current density superconducting tapes by epitaxial

deposition of $YBa_2Cu_3O_x$ thick films on biaxially textured metals, *Appl. Phys. Lett.*, **69**, 1795–1797.

24. Malozemoff, A. P., Annavarapu, S., Fritzemeier, L., Li, Q., Prunier, V., Rupich, M., Thieme, C., Zhang, W., Goyal, A., Paranthaman, M., and Lee, D. F. (2000). Low-cost YBCO coated conductor technology, *Supercond. Sci. Technol.*, **13**, 473–476.

25. Willén, D., Hansen, F., Däumling, M., Rasmussen, C. N., Østergaard, J., Træholt, C., Veje, E., Tønnesen, O., Jensen, K.-H., Olsen, S. K., Rasmussen, C., Hansen, E., Schuppach, O., Visler, T., Kvorning, S., Schuzster, J., Mortensen, J., Christiansen, J., and Mikkelsen, S. D. (2002). First operation experiences from a 30 kV, 104 MVA HTS power cable installed in a utility substation, *Phys. C*, **372–376**, 1571–1579.

26. Xin, Y., Hou, B., Bi, Y., Cao, K., Zhang, Y., Wu, S., Ding, H., Wang, G., Liu, Q., and Han, Z. (2004). China's 30 m, 35 kV/2 kA ac HTS power cable project, *Supercond. Sci. Technol.*, **17**, S332–S335.

27. US Department of Energy Project Fact Sheet (2008). Albany HTS Power Cable, http://energy.gov/sites/prod/files/oeprod/Documentsand-Media/Albany_03_05_08.pdf.

28. US Department of Energy Project Fact Sheet (2008). Columbus HTS Power Cable, http://energy.gov/sites/prod/files/oeprod/Documents-andMedia/columbus_03.05.08.pdf.

29. US Department of Energy Project Fact Sheet (2008). Long Island HTS Power Cable, http://energy.gov/sites/prod/files/oeprod/Documents-andMedia/LIPA_5_16_08.pdf.

30. Southwire Press Release (2007). Superconducting technology from Southwire to power New Orleans-area electric circuit, http://www.htstriax.com/news/08282007_NO_pressrelease.pdf.

31. Nexans Press Release (2012). RWE Deutschland, Nexans and KIT launch "AmpaCity" project, http://www.nexans.com/Corporate/2012/Nexans_Ampacity_GB_final.pdf.

32. Southwire Project Profile (2009). Manhattan New York HTS Triax Project, http://www.htstriax.com/docs/htsTriax_ManhattanProfile.pdf.

33. Tres Amigas LLC (2012). Uniting North America's Power Grid, http://www.tresamigasllc.com/docs/TRSAMGS_CAPBRO_0710_FNL_web.pdf.

34. Wilson, M. (1987). *Superconducting Magnets* (Oxford University Press, USA) ISBN 0198548109, p. 308.

35. Goldacker, W., Nast, R., Kotzyba, G., Schlachter, S. I., Frank, A., Ringsdorf, B., Schmidt, C., and Komarek, P. (2006). High current DyBCO-ROEBEL assembled coated conductor (RACC), *J. Phys. Conf. Ser.*, **43**, 901–904.
36. Boom, R. W., and Peterson, H. A. (1972). Superconductive energy storage for power systems, *IEEE Trans. Magn.*, **8**, 701–703.
37. Ali, M. H., Wu, B., and Dougal, R. A. (2010). An overview of SMES applications in power and energy systems, *IEEE Trans. Sustainable Energy*, **1**, 38–47.
38. Tixador, P. (2008). Superconducting magnetic energy storage: Status and perspective, IEEE/CSC & ESAS European Superconductivity News Forum, 3, CR5, http://snf.ieeecsc.org/sites/ieeecsc.org/files/CR5_Final3_012008.pdf.
39. Higasa, H., Ishikawa, F., Shibayama, M., Ono, T., Yokoyama, S., Nakamura, S., Yamada, T., and Yoshida, Y. (1994). A feasibility study of an 8-MWh flywheel-type power storage system using oxide superconductors, *Elect. Eng. Jpn.*, **114**, 20–31.
40. Noe, M., and Steurer, M. (2007). High-temperature superconductor fault current limiters: Concepts, applications, and development status, *Supercond. Sci. Technol.*, **20**, R15–R29.
41. Nexans Press Release (2012). Nexans goes live on grid with world's first fault current limiter based on second-generation superconductors, http://www.nexans.com/Corporate/2012/Nexans_SCFCL_Final_GB.pdf.
42. Central Japan Railway Company Annual Report (2011). *Promoting the Chuo shinkansen using the superconducting maglev system*, pp. 24–25, http://english.jr-central.co.jp/company/ir/annualreport/_pdf/annualreport2011.pdf.
43. Wang, J., Wang, S., and Zheng, J. (2009). Recent development of high temperature superconducting maglev system in China, *IEEE Trans. Appl. Supercond.*, **19**, 2142–2147.
44. Masson, P. J., Brown, G. V., Soban, D. S., and Luongo, C. A. (2007). HTS machines as enabling technology for all-electric airborne vehicles, *Supercond. Sci. Technol.*, **20** 748–756.
45. EADS Press Release (2011). EADS showcases VoltAir all-electric propulsion system concept, http://www.airbusgroup.com/int/en/news-media/press-releases/Airbus-Group/Financial_Communication/2011/06/20110622_eads_voltair.html.
46. EC Project GRD1-1999-10014 (2003). Liquid hydrogen fuelled aircraft—system analysis (CRYOPLANE), *Final Technical Report*.

47. Sumitomo Electric Industries Press Release (2008). Sumitomo Electric prototyped world's first superconducting electric car, available from http://web.archive.org/web/20120222073004/http://global-sei.com/news/press/08/08_12.html.
48. Fietz, W. H., Heller, R., Schlachter, S. I., and Goldacker, W. (2011). Application of high temperature superconductors for fusion, *Fusion Eng. Des.*, **86**, 1365–1368.
49. Grant, P. M. (2005). The SuperCable: Dual delivery of chemical and electric power, *IEEE Trans. Appl. Supercond.*, **15**, 1810–1813.
50. Mikheenko, P. (2011). Superconductivity for hydrogen economy, *J. Phys. Conf. Ser.*, **286**, art. 012014.
51. Electric Power Research Institute White Paper 1013089 (2006). The hydrogen-electric energy SuperGrid.
52. Tallon, J. L., Storey, J. G., and Mallett, B. (2012). The design of high-T_c superconductors—room-temperature superconductivity?, *Phys. C*, **482**, 45–49.
53. Kresin, V. Z., Wolf, S. A., and Ovchinnikov, Y. N. (1997). Exotic normal and superconducting properties of the high-T_c oxides, *Phys. Rep.*, **288**, 347–354.
54. Mourachkine, A. (2004). *Room-Temperature Superconductivity* (Cambridge International Science Publishing, UK) ISBN 1904602274.
55. Lehner, T. (2011). Development of 2G HTS wire for demanding electric power applications, presented at Enermat—new materials for energy, Spain, June 20–21, 2011, http://www.superpower-inc.com/system/files/2011_0620+ENERMAT+Spain_TL+Web.pdf.
56. Grant, P. M. (2001). Potential electric power applications for magnesium diboride, *MRS Proc.*, **689**, art. E1.1.

Chapter 12

Solid-State Lighting: An Approach to Energy-Efficient Illumination

Mariano Perálvarez, Jorge Higuera, Wim Hertog, Óscar Motto, and Josep Carreras

*Catalonia Institute for Energy Research (IREC), Lighting Group,
Jardins de les Dones de Negre, 1. PL2, 08930, Sant Adrià de Besòs, Barcelona, Spain*
mperalvarez@irec.cat

12.1 Properties of Light

12.1.1 Introduction

Light can be defined as waves of radiation whose wavelength lies in the spectral range from 380 nm (i.e. wavelengths longer than ultraviolet) to 750 nm (wavelengths shorter than the infrared). This spectral range corresponds to the radiation to which the human eye is most sensitive, and for this reason it is commonly called the *visible range* (see Fig. 12.1).

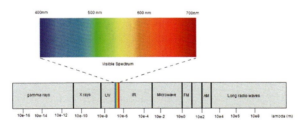

Figure 12.1 The electromagnetic spectrum.

Materials for Sustainable Energy Applications: Conversion, Storage, Transmission, and Consumption
Edited by Xavier Moya and David Muñoz-Rojas
Copyright © 2016 Pan Stanford Publishing Pte. Ltd.
ISBN 978-981-4411-81-3 (Hardcover), 978-981-4411-82-0 (eBook)
www.panstanford.com

This radiation can be described as the propagation of two mutually coupled vector waves, an electric field and magnetic field, which travel perpendicular to each other and to the propagation direction. This is known as the *electromagnetic* formalism of light, and within this framework light waves can be defined in terms of three parameters:

- wavelength
- intensity
- polarization state

Light, as with all electromagnetic waves, propagates in a vacuum at a constant speed, c, of 299792458 kms^{-1}. However, when electromagnetic waves propagate through a material of a given refractive index n, this speed, v, varies according to the following expression:

$$v = \frac{c}{n}$$

This interpretation is an oversimplification of the interaction light and matter; the decrease in the propagation velocity when waves propagate through a material is derived from more complex phenomena whose discussion is out of the scope of this book.

It is worth noting that even if the electromagnetic formalism provides the broadest treatment of the behaviour of light within the limits of classical optics, it fails to cover certain optical phenomena that can only be explained on the basis of a quantum nature of light. In 1905, Albert Einstein extended the quantum theory of thermal radiation proposed by Max Planck in 1900 to cover not only vibrations of the source of radiation, but also vibrations of the radiation itself. He thus suggested that light and other forms of electromagnetic radiation travel as tiny bundles of energy called *photons*. Some years later, Louis de Broglie provided a new perspective of light showing that in some cases it appears to act as a beam of particles and in other cases as a wave. De Broglie demonstrated a complementarity between the particle and wave nature of light, which is known as *particle-wave duality*.

According to the quantum formalism, light and other electromagnetic radiation is composed of a set of zero-rest-mass particles called photons. These particles carry electromagnetic energy and momentum, and their angular momentum (spin) is related to the polarization of the propagating radiation. Photons also have a wave-like character that defines their location in time and space and the rules by which they exhibit wave-like behaviour such as interference and diffraction. The energy of a single photon, E, is related to the frequency, ν, (i.e., the colour in the case of light waves in the visible range) at which the electromagnetic radiation is oscillating by the following expression:

$$E = h \cdot \nu = h \cdot \frac{c}{\lambda},$$

where c is the speed of light, λ is the wavelength of the electromagnetic radiation and h is the Planck Constant, whose value is 6.626×10^{-34} Js.

By way of illustration, we consider the case of monochromatic radiation of frequency ν_m. The total energy that this radiation possesses, E_T, corresponds to an integer number, n_m, of photons:

$$E_T = n_m \cdot h \cdot \nu_m$$

12.1.2 The Visual System

The perception of light depends on the response of our visual system to a received luminous stimulus. In this process, the translation of light into a signal that is recognizable by our brain is performed by the *retina*. The retina is a layered structure composed of several layers of interconnected nerve cells. Only three types of nerve cells (known as *photoreceptors*) in this neural network are sensitive to light: rods, cones and intrinsically photosensitive retinal ganglion cells (ipRGCs). When light is incident on our eye, the cornea and lens project an image onto these photoreceptors. Depending on the luminance level of the light falling on the retina, either the rods or cones, or a combination of both, are active. Luminance is the luminous intensity per unit area of light travelling in a certain direction (SI unit: cdm^{-2}). At luminance levels higher than 100 cdm^{-2}, three different kinds of cones are active, which detect three different light wavelength ranges. These long-wavelength (L), middle-wavelength (M) and short-wavelength (S) cone responses are summed to give a single achromatic response value, known as the photopic vision response. This response sets limits on which wavelengths can be seen by the eye, and can be displayed as the eye's luminosity function, $V(\lambda)$. The $V(\lambda)$ curve is a standard function established by the Commission Internationale d'Eclairage (CIE) (Fig. 12.2). Wavelengths below 380 nm and above 750 nm are not visible, whilst the range in between contains the full range of colours perceived by the human visual system. It is important to note that the extremes of the visible range are not precisely defined and may depend on the intensity or optical power of the stimulus.

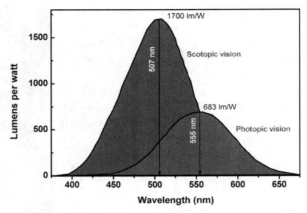

Figure 12.2 Variation of the photopic and scotopic luminosity functions with illumination wavelength.

Nerve cells of the retina encode these stimuli into two colour opponent signals depending on the wavelength (L, M or S) of the incident light: a red-green channel (L-M + S) and a blue-yellow channel (L + M – S). Together with the achromatic $V(\lambda)$ channel, which provides the luminance information, this nerve system carries the data needed to see our world in colour. In order to perceive a stimulus as achromatic white, both the neural responses of the red-green channel and the blue-yellow channel must be in balance, and the response of the $V(\lambda)$ channel has to be high enough for the light to be perceived as white instead of gray or black.

Rods, in contrast to cones, are active at luminance levels below 0.003 cdm^{-2} and therefore are responsible for what it is called scotopic vision. Their absorption peak is at 498 nm (blue), but in vivo, if the lens absorption and macular pigments are taken into account, the effective maximum sensitivity of the rod when integrated into the eye is shifted to 507 nm, which the peak position of the scotopic curve is shown Fig. 12.2. It should be noted that whereas photopic vision provides us with highly detailed colour sight, scotopic vision is achromatic and low resolution. In addition, as the highest sensitivity scotopic vision is at shorter wavelengths than that of photopic vision, human vision in dark conditions is more sensitive to the blue range and almost blind in the red range. A third wavelength range called the mesopic region, which is related to luminance levels in between those defined for photopic and scotopic visions, may also be used where both cones and rods contribute to vision, enabling us to see colours in semi-dark conditions.

Finally, ipRGCs have not yet been considered. This is because these photoreceptors do not affect image formation and processing. With their sensitive response lying in a narrow range between 460 and 480 nm, the ipRGCs are involved in the regulation of our sleep-wake cycle. When blue light in the 460–480 nm range reaches the retina, these photoreceptors are stimulated, thereby triggering the brain to release hormones responsible for preparing the body for sleep.

12.1.3 The Chromaticity Diagram

The range of visible colours may be represented in a so-called *chromaticity diagram*. While there are many variations of such diagrams, each one with its advantages and disadvantages, a long-surviving standard is the 1931 CIE chromaticity diagram shown in Fig. 12.3. The diagram contains monochromatic colours ranging from blue to red, with achromatic white in the middle, although achromatic white is not constrained to one unique point. Taking the change of sunlight throughout the day as an example, bluish morning light and reddish evening light are perceived as white light, despite both having a different spectrum and very different chromaticity coordinates. This natural change in chromaticity of white light closely follows the black body curve, represented here as a black line. The lines perpendicular to the black body locus set the limits for white light as defined by the CIE [1]. Non-spectral purple colours are found on the axis connecting the red end with the blue end of the spectral locus. It is important to keep in mind that the 1931 CIE chromaticity diagram is highly non-uniform. In practice this means that a colour difference between two colour coordinates located in one region of the diagram does not translate into the same visible colour difference between two colours at the same distance from each other but located elsewhere in the 1931 CIE chromaticity diagram. More modern colour diagrams aim to solve this issue. Colour diagrams and colour spaces are used extensively in the world of colour management, where different input and output devices such as photographic cameras and printers are standardized to ensure a colour-correct (re)production environment.

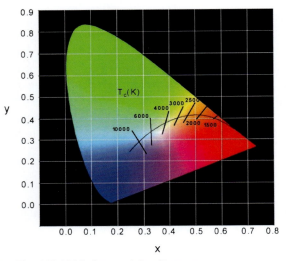

Figure 12.3 The CIE 1931 chromaticity diagram.

12.1.4 Luminous Efficacy of Radiation

The luminous efficacy of radiation (LER) defines the ratio of luminous flux (in lumens) of a source to its optical power (in watts) or, in other words, the ratio of luminous flux to radiant flux. Therefore, the LER indicates how well electromagnetic radiation generates visible light, or how bright 1 watt of radiation appears to the human eye.

Luminous efficacy itself is expressed in SI units as lumens per watt (lm/W). The lumen is defined to be unity for a radiant energy of 1/683 watt at 555 nm (the maximum of the photopic luminosity function).

It should be noted that, in addition to the LER parameter, the lighting industry extensively uses an alternative parameter known as the *luminous efficacy* of a source, which is the ratio of luminous flux to electrical power. It measures how efficiently an electrical watt generates a luminous flux.

12.1.5 Colour Temperature

The *colour temperature*, or CT (in Kelvin), is a spectral parameter of a light source that describes the temperature of an ideal black body that radiates light with a hue similar to the light source. In the case of thermal emitters such as filament light bulbs, the colour temperature is the actual temperature of the filament inside the light bulb. For other light sources such as gas discharge lamps or light-emitting diodes (LEDs), the actual temperature of the light-emitting area is different from the colour temperature, and we instead speak of the *correlated colour temperature* or CCT. A low CT or CCT (2500 to 3500 K) is associated with warm, yellowish white light, while a higher CT or CCT correlates to neutral to bluish white light (4000 to 6500 K). Overcast daylight in northern latitudes has a CCT around 6500 K. The CCT and CT are important parameters to define white light used for general illumination.

12.1.6 Colour Rendering Index (CRI)

An important quality of white light is the ability to render the colours of an object illuminated by that light. The colour rendering qualities of a light source depend on a complicated interaction between the emission spectrum of the light source, the reflectance spectrum of the object, and the visual system of the observer.

Colour fidelity describes the spectral qualities of the light source in comparison to a reference light source. The well-known General Colour Rendering Index (CRI), or R_a, is a colour fidelity indicator defined by the CIE where eight reference reflectance spectra, in this case a subset of the Munsell Book of Colours [2], are illuminated by both the test light source and the reference source. The general CRI was first proposed in 1964 and updated in 1995 [3]. The colour rendering index only measures the fidelity of a light source in comparison to the sun and other incandescent bodies, but does not consider the colour preference of an observer. The larger the colour difference between both reflected spectra, the lower the colour rendering of that particular sample. The arithmetic mean of all individual colour rendering indices gives the general colour rendering index of the light source. When the differences between the colour rendering by the test source and reference source of those eight samples is 0, the colour rendering index reaches a value of 100.

The full method for calculating the CRI is given below:

- Find the chromaticity coordinates and correlated colour temperature of the light source under test.
- The CRI is only valid when the distance of the chromaticity coordinates from the Planckian black body locus, Δ_{uv}, is less than 5.4×10^{-3} in the 1960 CIE (u, v) colour space.
- If CCT < 5000 K, use a black body radiator as a reference source. If the CCT > 5000 K, the reference must be CIE standard illuminant D. Ensure that both sources have the same CCT.
- Illuminate the first eight Munsell samples with the reference source.
- Illuminate the first eight Munsell samples with the test source.
- Use the von Kries chromatic adaptation for each sample.
- Calculate the Euclidian distance (ΔE_i) between the pair of colour coordinates for each sample.
- The CRI of each sample is found by using the following formula: $R_i = 100 - 4.6 \, \Delta E_i$
- The General CRI or R_a is found by calculating the mean of all R_is.

Colour preference indicators, such as the memory colour rendering index (MCRI) [4], rely on qualitative comparisons of colour rendering instead of direct measurements. In situations where an accurate reference spectrum is not needed, colour preference indicators might provide a better indication of the colour rendering qualities of a light source.

12.1.7 Spectrum and Quality of Light

The perceived colour of an object depends on its reflectance spectrum multiplied by the spectrum of the light source with which that object is illuminated. The colour is not an exclusive property of an observed material but the reflectance spectrum is unique for each material.

Having understood the generic procedure for calculating the CRI, some logical questions arise. Commercial manufacturers of light sources always optimize the luminaires to obtain a CRI (R_a) as high as possible, with huge development efforts on increasing the value obtained by this indicator.

This gives a privileged role to eight Munsell colour samples instead of creating a spectrum optimized for the reflection spectra found in daily life. Figure 12.4 shows three different spectra with identical CRI (R_a) and very similar CCT values.

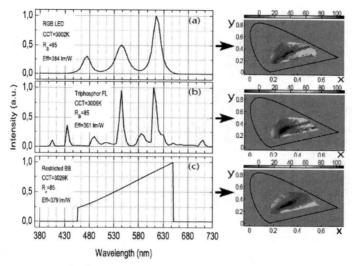

Figure 12.4 Comparison for three different light spectra according their CRI.

When instead of using only eight Munsell samples as defined by the CIE for the calculation of the CRI, the full set of 1269 Munsell samples is used, large differences between the three spectra in terms of colour rendering ability are observed. The 1931 chromaticity diagram in the right-hand column shows the result of using the CRI calculation method using 1269 Munsell samples.

Consequently, it is possible to have three light sources with the same CRI which reproduce the colours of a painting in a museum in three completely different ways.

All of this demonstrates the oversimplification of the CRI as a metric to transform spectral information into a single number. If we take into account other factors to determine the colorimetric characteristics of a light source, such as the correlated colour temperature (CCT), the distance from the chromaticity coordinates to the black body locus (Δ_{uv}), and other important additional factors such as the LER.

12.2 Light Sources

12.2.1 Introduction

Since the discovery of fire by *Homo erectus*, light generation has been achieved in many different ways. Camp fires, torches, wax candles and oil lamps are some examples of how humankind has created artificial light sources. However, after the invention of the electricity and subsequently the incandescent light bulb, general lighting underwent a turning point. Electric lamps were a straightforward method of obtaining brighter, more efficient and safer light than combustible sources. Incandescent bulbs have an outer glass cover containing a filament that is heated to a temperature between 2600 and 2900 K, thereby producing a typically warm white glow. The incandescent bulb has enjoyed a very high level of adoption in residential lighting due to the warmness of the light produced, its adaptability to different power supplies and low cost. This is despite the fact that less than 10% of the electrical power supplied to the bulb is converted into visible light. The remaining energy is mostly lost as heat, and incandescent lamp lifetimes are typically 1000 to 1500 h. At present, governments around the world have passed measures to phase out incandescent light bulbs for general lighting in favour of more energy-efficient alternatives. Phase-out regulations effectively ban the manufacture, importation or sale of incandescent light bulbs for general lighting. These regulations allow the sale of future versions of incandescent bulbs if they are sufficiently energy efficient. This is the case for tungsten-halogen lamps which have been introduced into the market in an attempt to improve the light output and reliability of traditional incandescent lamps. The main characteristic of these lamps is the addition of a halogen gas to the glass envelope. This strategy is beneficial in two ways: First, the tungsten on the filament is continuously recycled due to the halogen cycle. The filament particles which are evaporated during high temperature operation react with the halogen gas, thereby inducing tungsten re-deposition on the filament. With this simple step, bulb lifetimes increase up to 2000 to 4000 h. Second, the halogen cycle allows a higher filament operating temperature, resulting in a higher colour temperature (2900 to 3100 K) and a whiter appearance than light produced by standard incandescent bulbs. This blue shift in emission wavelength means that a larger amount of the emitted light falls within the visible range, which translates into a net improvement of lamp efficacy.

There are other commercial approaches than incandescent technology, such as fluorescent lighting. These lamps are partially evacuated glass tubes containing a small amount of mercury. Plasma, formed by electrons emitted by two barium-coated electrodes, ionizes the mercury, causing it to emit ultraviolet (UV) radiation. This radiation is then absorbed by a phosphorescent coating on the inside of the glass envelope. The absorbed energy is subsequently re-emitted (i.e., down-converted) in the form of visible light. These phosphors, in combination with the emission lines of mercury, determine the final spectrum of the light source. Fluorescent tubes operate at much cooler temperatures than incandescent lamps and have an operational lifetime up to 20000 h. Another commercial technology is the so-called high intensity discharge (HID) lamp. These types of lamps include metal halide, sodium vapour or xenon lamps, and can have typical lifespans ranging from 10000 to 15000 h, depending on the technology. In this case, light generation relies on the direct emission from a carrier gas and metal salts in the discharge arc. In the case of light-emitting diodes (LEDs), in contrast, luminescence is achieved in a very different way: a semiconductor diode based on direct-gap III-V elements is forward-biased, thereby inducing the radiative recombination of electron and holes within the semiconductor. LEDs have typical lifespans up to 50000 h and are highly efficient light sources, exhibiting internal quantum efficiencies, in many cases, close to 100. By way of illustration, Fig. 12.5 and Table 12.1 show the typical efficacy and power consumption of different lighting technologies. An incandescent bulb consuming 60 W produces 830 lumens, which equates to only 14 lm/W.[1] Fluorescent lamps have a higher efficacy. A standard T8 fluorescent tube delivers 2700 lumens at 36 W (including the energy consumption

---

[1]This quantity refers to luminous efficacy of a source. This concept is accurately defined in Section 12.1.2.

of the lamp driving circuit) giving an efficacy value of 75 lm/W. Compact Fluorescent Lamps (CFLs) obtain an efficacy of 60 lm/W. A typical CFL lamp has a wall-plug efficiency of 7% to 10%, versus 1.5% to 2.5% for an incandescent bulb. White LEDs, in contrast, have a significantly higher luminous efficacy than any other light source. Values up to 180 lm/W are commercially available, and record efficiencies of 303 lm/W, as reported by CREE in 2014 [5], are achieved in laboratory devices. At lamp level, the efficacy drops slightly to 100 to 120 lm/W because of losses in LED driver electronics and optical components. With such a tremendous increase in efficacy over other light sources combined with the very long lifetime, high reliability and flexibility regarding emission spectra, it is clear that the light-emitting diode is on its way to be the light source of the future.

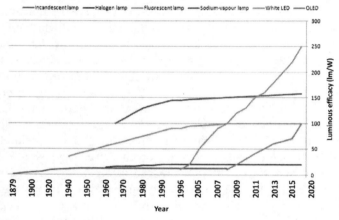

Figure 12.5 Yearly progression of luminous efficacy for different lighting technologies.

Table 12.1 Comparison of common lighting technologies

Brightness (lumens)	Power consumption (W)		
	Incandescent	CFL bulb	LED luminaire
450	40	9–11	5–8
800	60	13–20	10–13
1100	75	18–25	13–18
1600	100	23–30	20–26

Despite having excellent properties, the adoption of LEDs is rather slow. According to the US Department of Energy, the current market penetration of LED lighting is around 2% to 3% (in unit sales). Thanks to the continuous efficacy improvement and to the subsequent reduction of the cost per lumen (Haitz's law [6]), it is expected that the LED market share will account for 52% of worldwide lamp and luminaire shipment volumes by 2020.

12.3 LED Physics

In the previous section, we discussed how efficient LED-based lamps are in comparison to other mainstream technologies. Such efficiency is intimately related to how the emission is generated within these tiny devices.

A diode is a two-terminal device that allows current to flow in only one direction (asymmetric conductance). It is composed of two connected crystals of semiconducting material, one with an excess of positive charge and the other with an excess of negative charge. When electricity is passed through, the excess electrons in the negatively charged semiconductor crystal move into the positively charged crystal, where they combine with positively charged ions in the crystal, resulting in the release of energy in the form of light. The colour of this luminescence will depend on the materials and fabrication procedures that make up the device. This emission mechanism is very different from those observed in any other standard light source. Unlike incandescent or fluorescent lamps, which create light with filaments and gases encased in glass bulbs, the luminescence is obtained from a semiconductor that converts electrical energy into light energy. This is why LED lighting is also commonly referred as *solid-state lighting*.

12.3.1 Semiconductors

Semiconductors, as indicated by their name, are materials with electrical conductance in between those of conductors and insulators, so are particularly useful for devices where careful control of conductance is required. Conductors have typical conductivities ranging from 1 to 10^8 $\Omega^{-1}m^{-1}$. Insulators cover the range from 10^{-9} to 10^{-18} $\Omega^{-1}m^{-1}$, and semiconductors have conductivities within the range 1 to 10^{-9} $\Omega^{-1}m^{-1}$. Apart from this difference in conductance, semiconductors differ from insulators because the electrical conductivity of a semiconductor increases with temperature. This is to say, above Absolute Zero, there is a finite probability that an electron in the lattice is mobile, leaving behind a region of positive charge called a *hole*. This relationship between temperature and semiconductor conductivity contrasts with the behaviour of conductors, where an increase of temperature reduces conductivity.

If a voltage is applied to both ends of a semiconductor crystal, the current that flows consists of both mobile electrons and holes. That is, mobile electrons can move through the material, and the positively charged regions of crystal lattice (holes) that are left behind are then filled with other electrons, thereby contributing to the conduction. This second mechanism is commonly called *hole conduction* due to the fact that this mechanism is equivalent to holes migrating across the material in the direction opposite to the electron movement.

The number of electrons and holes available for conduction is determined by the band theory of solids. The electrons in solids, in contrast with single atoms, do not have discrete energies. The available energy states for electrons are structured in bands. In Fig. 12.6, we have a rough representation of the energy bands in insulators, semiconductors and conductors.

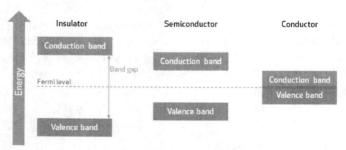

Figure 12.6 Schematic drawing of band structure in solids.

Conduction occurs when electrons move from the *valence band* in the *conduction band*. In insulators, the electrons in the valence band (VB) are well separated from the conduction band (CB) by a large energy gap (i.e., the *band gap*). In semiconductors, this energy band gap is relatively small, and the CB and VB may be so close that thermal or other excitations can allow electrons to bridge the band gap and allow current to flow. Finally in conductors, the CB and VB overlap. A crucial parameter in determining the final conducting properties is the Fermi energy, defined as energy of the highest occupied electron state at Absolute Zero or, in other words, as the highest in energy of the available electron energy levels.

12.3.1.1 Doping

Owing to the small band gap, the conductivity of a semiconductor can be dramatically increased by the addition of impurity atoms which donate or accept electronic charge, known as a *dopant* atom. A dopant is a chemical element that enriches the semiconductor with electrons (donor) or holes (acceptor). According to this concept, three different semiconductor crystal types may be distinguished: intrinsic, n-type (electrons in excess) and p-type (holes in excess). Intrinsic semiconductors are not particularly useful for devices because have properties in between those of insulators and conductors. N-type semiconductors have electron donor energy levels within the semiconductor band gap, close to the conduction band. This now allows the Fermi level to lie halfway between the donor levels and the conduction band (see Fig. 12.7). In this case, the conduction is via electrons (known as *majority carriers* when in n-type material) that are now more easily promoted to the CB. In p-type materials the behaviour is similar. The addition of acceptor levels from a p-type dopant creates hole energy levels appear within the band gap close to the VB allows the Fermi level to move in between these levels, resulting in conduction carried out by holes. With this procedure, the conductivity of an intrinsic semiconductor can be increased more than a factor of 10^6.

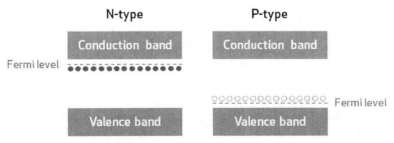

Figure 12.7 Fermi level modification by action of n-type (left) and p-type (right) doping.

12.3.1.2 p-n junctions

One of the most important mechanisms in semiconductor physics occurs when n-type and p-type materials are placed in contact on the same device, known as a *p-n junction*. When such a junction is formed, some of the free electrons in the n-type region diffuse, in a transient process, to the p-type side and recombine with holes there to form negatively charged ions (see Fig. 12.8a). Consequently, in the n-type side of the structure, diffused electrons leave free donor sites behind, thereby creating positively charged ions. The transient ends when the electric field (built-up field, E_{sp}) generated by the space charge on both sides of the junction is high enough to block any further electron transfer. At the steady state, the region around the p-n junction has a high built-up electric field and is depleted of free carriers, and is known as the *depletion zone*.

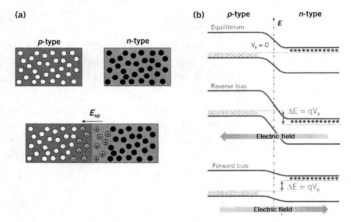

Figure 12.8 Formation of the depletion zone around the p-n junction, (a). Junction under different bias, V_a, conditions, (b).

The principal characteristic of a p–n junction is that current cannot pass in both directions through it, which arises from the nature of the charge transport process in the two types of materials. According to the criteria fixed in Fig. 12.8b, when a positive bias is applied to the p-side of the junction (reverse bias), the established electric field pushes electrons on the n-side and holes in the p-side out of the junction, thereby hindering the electronic flow across it. When a negative bias is applied to the p-side, in contrast, if the applied electric field is higher than the built-up potential, then conduction through the junction is enabled. Then electrons in the n-side are attracted to the p-side, where they combine with holes.

P–N junctions are the key component of diode devices. The charge carriers may release energy by a variety of mechanisms. When they recombine, one of them being *radiative recombination*, that is, the emission of a photon. When this mechanism takes place efficiently and dominates over non-radiative carrier recombination pathways, then the diode is a working LED.

12.3.1.3 Direct and indirect band gaps

In contrast to Figs. 12.6 to 12.8, actual energy bands in solids are not flat. In real solids, these bands are bent and exhibit a set of relative minima and maxima resulting from the crystalline structure of the semiconductor. Thus, the energy band gap is defined as the minimum energy difference between the top of the valence band and the bottom of the conduction band. However, the top of the valence band and the bottom of the conduction band are not generally at the same value of the crystal momentum of the electron, as represented in Fig. 12.9a,b. When the top of the valence band and the bottom of the conduction band occur at the same value of momentum, the material is called a *direct band gap* semiconductor. If the maximum energy of the valence band occurs at a different value of momentum to the minimum in the conduction band energy, then the material is called an *indirect band gap* semiconductor.

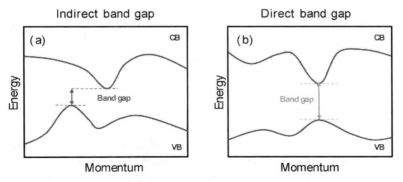

Figure 12.9 Schematic drawing of indirect (a) and direct (b) band gaps.

The difference between both kinds of semiconductors is especially relevant when dealing with optical devices. Interactions among electrons, holes, phonons, photons, and other particles are required to satisfy conservation of energy and crystal momentum. When, in a direct band gap semiconductor, an electron loses energy on falling from the CB to the VB, the process usually takes place by the emission of one photon. In other words, since the top of the VB and the bottom of the CB share the same momentum coordinates, only the generation of one photon is needed to conserve energy and momentum. In an indirect band gap material, in contrast, the electron can only fall if the process involves a change in the electron momentum, that is, if it is assisted by a phonon (quantum of solid lattice vibrations). In this case, light emission is a two-particle process, generating a photon and a phonon, which is a very unlikely (and slow) phenomenon in comparison to direct band gap transitions. Indirect gap semiconductor diodes therefore do not emit light when under forward bias. These differences represent the basis of why direct band gap semiconductors are the materials of choice in the implementation of light-emitting diodes. In fact, most materials employed for fabricating LEDs are based on III-V materials which exhibit direct band gaps.

12.3.1.4 LED architecture

An LED is a forward-biased p-n junction fabricated from direct gap semiconductor alloys that emits light via injection electroluminescence. As discussed above, the materials for LEDs are III-V compound semiconductors. These materials show high internal quantum efficiencies, in some cases close to 100%, and very short carrier recombination lifetimes between 1 and 100 ns, suitable for fast modulation rates. Furthermore, the band gaps of III-V alloys can be engineered by alloying with other group III elements to give a range of LED emission wavelengths.

The efficiency of an LED depends on two factors: the efficiency of light generation in the junction, and the efficiency of light extraction from the junction. In this section we will briefly discuss the former case; the latter case will be treated in detail in Section 12.8. Although modern LEDs have typical efficiencies close to 100%, the efficiencies of early LEDs were quite low. The first devices were based on simple p-n junctions of the same semiconductor type (known as a *homojunction*), where electron–

hole radiative recombination was inefficient due to a relatively high leakage current and carrier migration phenomena. In addition, emitted photons in such a structure can be absorbed by the crystal since the photon energy is similar to the band gap. As a consequence, photons emitted towards the substrate are mostly absorbed. This performance was improved by the introduction of a novel concept: a local variation of the band gap on either side of the junction to increase carrier confinement. To do this, a p-n junction is constructed out of two semiconductors with different band gap energies (known as a *heterojunction*). The difference in band gap creates a one-way path for carriers. Electrons are attracted from the material of higher band gap (which is typically n-type) to that of the lower band gap (which is typically p-type). Therefore, the holes are confined to the p-side of the heterojunction and electrons are injected from the n-side. In the p-side, the radiative recombination time is very short due to the high concentration of holes, which translates into very high internal quantum efficiencies. Furthermore, in such a system, emitted photons will have now energies lower than the band gap at the n-side (substrate). These photons can be recovered, e.g. through mirrors created by cleaving the semiconductor crystal facets, thus improving the device efficiency. This mechanism is further improved with the incorporation of a double heterojunction (DH) where the narrow band gap material is now sandwiched between two wide band gap layers. The double heterojunction forms a barrier which restricts the region of electron–hole recombination to within the lower band gap material, thereby leading to more efficient confinement of the carriers in the active region, increasing the chances that they will meet and recombine radiatively, rather than being lost to non-radiative processes such as trapping by crystalline defects.

    An important breakthrough in LED development occurred due to research into quantum confinement, where extreme physical confinement of carriers was applied to light-emitting heterostructures. An improvement over double heterostructures is to use quantum-well (QW) structures, which contain a very thin active layer (smaller than the de Broglie wavelength of the electron) within which the carriers and photons are confined, and where quantum effects play an important role. The additional carrier confinement and increase in the local carrier concentration enhances the internal quantum efficiency and reduces the carrier recombination lifetime. The use of a number of quantum wells separated by higher band gap barrier layers (known as a multiple quantum well, or MQW) will increase the carrier capture cross section when the device is driven in forward bias, allowing a higher emission intensity. However, the addition of increasing numbers of quantum wells will result in the formation of crystalline defects which are non-radiative pathways for carriers, leading to decreased efficiencies. Commercial LEDs therefore tend to have fewer than ten quantum wells. Recent increases in power efficiency of up to one order of magnitude have been reported [7], and it is likely that developments in MQW architecture will result in higher-brightness LEDs in the near future.

Another important aspect regarding LED performance is *efficiency droop*, which occurs when LEDs are driven at high currents. This is a phenomenon that has represented a key obstacle to LED lighting development, since overhead lights in homes and offices require high powers and brightness, and currently III-V LEDs suffer from a loss in efficiency in these conditions. The physical mechanism causing efficiency droop has been highly debated, and addressing droop is a major research drive in the field of white lighting technology.

12.3.1.5 Manufacturing processes

Heterostructures shown in the previous sections have been effectively used in semiconductor devices for many years. In such structures, two or more layers of different materials are grown one on another. To do this, different techniques have been developed:

Molecular beam epitaxy (MBE): this is a type of ultra-high vacuum evaporation, in which the atoms or molecules of the material are directed as precise beams from effusion cells to a heated substrate. The molecules land on the surface of the substrate, condense, and build up very slowly and systematically in ultra-thin layers. This system allows growth with high precision and purity (>99.99999%) and it is possible to grow layers with changes in composition between a few monolayers. However, due to the high vacuum conditions and very slow growth rate, this technique is not used commercially.

Metalorganic chemical vapour deposition (MOCVD): Metalorganics, i.e., carbon-based compounds containing metal ions, containing group III elements are mixed with hydrides of group V elements using hydrogen as the carrier gas, and are thermally activated to react and produce III-V binary or mixed compounds, which form a layer on the substrate. In contrast to MBE, the growth takes place by chemical reaction and not physical deposition. In addition, the process does not need high vacuum.

Chemical beam epitaxy: Combines the distinctive feature of molecular beams in MBE with gaseous sources of MOCVD. The metalorganic compounds and the hydrides are directed to the substrate by forming beams at relatively low pressures.

12.4 Light Emitting Diodes Based on III-V Junctions

The story of LEDs started in 1907 when H. J. Round, one of the Marconi's assistants in the UK, discovered electroluminescence (EL) by applying more than 100 V to a crystalline material used as a sandpaper abrasive. LED operation was established during the 1920s by Russian scientist, O. V. Losev. In his studies on ZnO and SiC LEDs, he related LED emission to diode action, established the threshold voltage for the onset of light emission, and measured the current–voltage characteristics of the device in detail. Unfortunately, his prolific career prematurely ended during the Second World War when he died during the blockade of Leningrad. In the 1930s, the French physicist G. Destriau discovered electroluminescence from zinc oxide, and is often cited as the true discoverer of the electroluminescence effect. In the early 1950s, the discovery of the transistor and the advances in semiconductor physics that went hand-in-hand with it made possible to accurately explain the origin of LED light emission. Ten years later, the first commercial red GaAsP LEDs were available on the market. In this progression that has finally resulted in the massive adoption of LED technology in recent years, it is clear that Losev, as the first to understand the potential of semiconductor light sources, together with H. J. Round and G. Destriau, deserves a significant place in the history of optoelectronics technology.

12.4.1 Gallium Arsenide

GaAs-based devices were developed during the early 1960s as infrared emitters. The LED emission peaks at 950 nm and is completely invisible to the human eye. Efficiencies in the range of 1% were achieved. This success undoubtedly boosted research in the field, but the applications of these devices was limited due to the lack of visible light emission.

12.4.2 Gallium Arsenide Phosphide and Gallium Phosphide

Soon after the discovery of GaAs LEDs, researchers discovered the possibility of shifting the emission wavelength towards the visible range by growing a ternary alloy, a mixture of GaP and GaAs [8]. The exact wavelength of the emission was controlled by accurately monitoring the ternary alloy composition. In 1968 the first commercial GaAsP device was introduced to the market. The device, composed of an active layer of GaAsP on a GaAs substrate, exhibited emission in the red range with wall-plug efficiencies of ~0.1 to 0.3%. Shortly thereafter, more efficient devices were introduced based on GaP doped with Zn and O. These devices, fabricated on GaP substrates, emitted in the red region with efficiencies peaking at 2%. Another interesting alternative for early LEDs was the discovery of isoelectronic impurities such as nitrogen, which can act as efficient emitting centres when incorporated into GaP and GaAsP alloys. These GaAsP:N- and GaP:N-based LEDs emitted in the yellow-green, yellow and orange range and exhibited wall-plug efficiencies ranging from 0.1% to 1%. This relatively low efficiency was caused by the GaAsP/GaAs lattice mismatch, resulting in a high density of lattice strain-induced misfit dislocations in the GaAsP epitaxial films. Further development resulted in the discovery of the possibility to alloy GaAs and InP, two compounds with perfect lattice match, to form a quaternary compound: InGaAsP. This material is typically used for fabricating LEDs in the near infrared range, including the optical fibre telecommunication window between 1330 and 1550 nm. The exact emission wavelength is defined by the GaAs-to-InP ratio.

12.4.3 Aluminium Gallium Arsenide

An important breakthrough in LED performance was the development of the AlGaAs/GaAs single heterojunction LED emitting in the red range. These devices introduced a novel concept: local variation of the energy band gap resulting in carrier confinement. As explained in Section 12.3.1.4, the use of semiconductors with different energy band gaps results in carrier confinement and, subsequently, in improved internal quantum efficiencies. This approach was further improved in 1985 with the introduction of the first commercial an AlGaAs/GaAs double heterojunction (DH). This structure is formed by sandwiching a narrow band gap material between two layers of wide band gap semiconductor. This provides more efficient confinement of the carriers in the active region, leading to more efficient light emission. This commercial device also emitted in the red range, with an efficiency of around 5%. It should be noted that despite these noticeable improvements, the AlGaAs/GaAs LEDs presented an obvious disadvantage compared to early GaP-based diodes: unlike GaP, GaAs substrates are not transparent to visible light (known as an *absorbing substrate*, AS), which led to low output powers. Fortunately, this issue was rapidly solved with the introduction of transparent AlGaAs substrates that prevented optical absorption, enhancing the emission efficiency by a factor of two. The first commercial transparent substrate (TS) LED [9] was introduced in 1987 and showed an efficiency of about 10%.

In the light of the promising results provided by heterostructure devices, the parallel advances in quantum well physics[2] attracted many experts in the field to assess the feasibility of incorporating such structures into LEDs. The subsequent introduction of MQW heterostructures into semiconductor devices boosted the development of high-brightness LEDs.

12.4.4 Aluminium Gallium Indium Phosphide

A common problem observed in all the structures discussed above is the limited tunability of the alloy composition. For low

---

[2]This development could not be attained without the appearance of new fabrication methods with precision monitoring of thickness and composition.

P-fractions, GaAsP has a direct band gap and is almost lattice-matched to GaAs substrates. As the P concentration is increased in order to extend the emission to shorter wavelengths, the lattice mismatch and therefore the defect density increases appreciably. This leads to rather low emission efficiencies since carriers are lost at those non-radiative recombination centres. Furthermore, when the P content is too high the band-gap becomes indirect, which also contributes to the reduction of the emission efficiency. In a similar way, tunability of AlGaAs-based compounds towards the visible range is restricted to a certain range of Al concentrations. Emission wavelengths shorter than 660 nm cannot be achieved since the compound becomes indirect at high Al fractions. At low concentrations, in contrast, the emission shifts to the infrared range. The problem was solved with the development of one of the so-called *high-brightness* alloys: AlInGaP. With the AlInGaP quaternary material system, tunability from the yellow-green range to the red is enabled whilst maintaining lattice matching on GaAs. Within this range, the device efficiency increases with the emission wavelength, reaching its maximum value in the red region. Record wall-plug efficiencies of over 60% at 609 nm have been achieved [10].

12.4.5 Gallium Nitride and Indium Gallium Nitride

In the early 1960s, green and red emission had been already achieved through the usage of GaP:N and GaAsP respectively. All that was needed to obtain white light was an efficient blue-emitter. Until then, only LEDs made of SiC had exhibited blue light emission but with efficiencies between 0.05 to 0.1%. From that moment on, much effort was devoted to the research of GaN which at the time was expected to provide bright blue emission. In this process, the first GaN single crystal was fabricated in 1968 and, surprisingly, it seemed to have intrinsic n-type behaviour. The following aim was to find an efficient *p*-type dopant to enable the creation of a GaN p–n junction. After several attempts, the difficulties in achieving efficient p-doping in III-N alloys were solved in 1989 when p-conductivity was generated by activating Mg dopant atoms by electron beam irradiation. Finally, Nichia Chemicals developed a more straightforward method for activating these dopants by subjecting the doped nitride films to high temperature annealing. This resulted, in 1994, in the first commercial blue LED on the market. This device consisted in a GaN crystal deposited onto a sapphire substrate, since GaN substrates were not commercially available at the time. Figure 12.10 shows an example of such a device. One interesting advantage of GaN-based alloys is the possibility of tuning the emission wavelength over a wide range, from 362 nm (3.4 eV) to 615 nm (2 eV), by altering the indium concentration in the InGaN alloy. Alternatively to sapphire substrates, which are electrically insulating and exhibit a sizeable lattice mismatch with GaN, InGaN can also be grown on SiC, a substrate that is conductive and has a closer lattice match with GaN, with only 0.5 mismatch %. This technology was developed by Cree Research and resulted in the *superbright* GaN/SiC LED chip introduced in 1995. At present, Cree still is at the forefront in this area through its SC^3 Technology Platform with commercially available wall-plug efficiencies over 55%.[5] GaN and InGaN-based LEDs are used commercially in white lighting in combination with Ce:YAG yellow phosphor powder, which down-converts the blue light to white light. They are being used in combination with AlGaInP to develop white LED emission by combining red, blue and green LEDs.

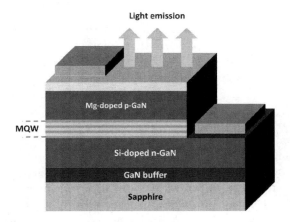

Figure 12.10 Typical GaN LED device structure.

12.4.6 ZnSe

ZnSe LEDs are white light-emitting devices that do not rely on phosphor conversion to generate white light. These diodes, based on II-IV compound materials emit blue-green light from the active p-n junction. A part of this emission is absorbed by the conductive substrate which emits broadband light peaking at 585 nm by photoluminescence. The combination of both emission bands results in white light with a CCT of 3400 K and typical CRI of 68. The efficiency of ZnSe LEDs barely reaches 4%. This relatively low efficiency combined with lifetimes of less than 1000 h lead to a very low market penetration.

12.4.7 Materials for UV LEDs

The fabrication of violet and ultraviolet LEDs (<400 nm) on sapphire substrates, commonly used for blue and green InGaN LEDs, results very high defect densities due to lattice mismatch with the substrate. UV InGaN LEDs down to 365 nm are commercially available, but their efficiency and lifetimes are a fraction of those figures for blue LEDs based on the same technology.

The use of AlGaN (where the binary compound AlN has a band gap in the deep UV range) to produce UV devices results in a significant increase in both efficiency and LED lifetime. Recently developed 280 nm LEDs with external quantum efficiencies of 5% are now available. Further improvements in efficiency and lifetime will allow deep UV LEDs in applications to sterilize medical equipment and water, two markets that are currently dominated by heavy and fragile mercury discharge lamps, which are difficult to dispose of safely.

12.5 Organic Light Emitting Diodes

Organic light emitting diodes (OLEDs) are light-emitting diodes made from semiconducting organic (carbon-based) materials. They are composed of at least one undoped organic layer between two electrodes. The emission takes place when, by device biasing, electrons and holes from opposite electrodes are injected into the active material, forming excitons which decay radiatively to give visible emission. The organic compound may be a molecule with a relatively small number of atoms, in crystalline form (small-molecule OLEDs, SMOLEDs) or a polymer (PLEDs). It should be noted that OLEDs typically refers to small-molecule devices.

The first work dealing with efficient emission from SMOLEDs was reported in 1987 by Tang and van Slyke [11], who introduced a multilayered organic semiconductor structure that induced balanced electron/hole densities, thereby leading to much more efficient devices. SMOLEDs are based on the usage of small carbon-based compounds as opposed to large polymer chains. The organic layers are deposited by vacuum thermal evaporation onto a transparent electrode (which is in turn deposited on a glass substrate), typically made of indium tin oxide (ITO). This fabrication method makes SMOLED fabrication an expensive process which is of limited use for large-area devices in comparison to other processing techniques. However, contrary to polymer-based devices, layers fabricated by vacuum deposition are highly homogeneous, a basic requirement for constructing multi-layer structures. Figure 12.11a shows a schematic drawing of a generic OLED.

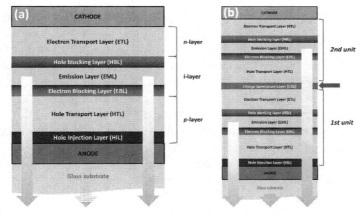

Figure 12.11 Schematic drawing of a single unit p-i-n type OLED (a) and a tandem structure (b).

Undoped layers (responsible for exciton formation and emission) are sandwiched between a *p*-type layer for transporting holes, and an n-type layer for transporting electrons, both of them designed to facilitate carrier transport towards the emitting layer. The *p*-layer is overgrown with a hole-injection layer, responsible for improving carrier injection from the anode. The undoped part consists of the emission layer (EML), where radiative recombination takes place, an electron blocking layer (EBL) and a hole blocking layer (HBL). The blocking layers ensure charge confinement within the EML, improving recombination efficiency. One important advantage of organic systems is the possibility of combining several single unit p-i-n architectures to construct a stacked system (see Fig. 12.11b). To do so, the emissive units are separated by a charge generation layer (CGL) that serves as an anode for one unit and as a cathode for a second unit. Stacked systems contribute to maintaining the optical power level at low drive currents, which benefits device reliability.

The first evidence of light emission from a polymer-based light-emitting device was recorded in 1990 when R.H. Friend's group at Cambridge University demonstrated electroluminescence from a structure based on poly (p-phenylene vinylene) (PPV). Major research and development efforts are being made for PLED devices, as their compatibility with large area manufacturing techniques may be the key to future cost-effective general illumination devices.

PLEDs are based on conjugated polymers that provide the devices with highly conductive layers. Polymers are directly deposited on the transparent electrode (generally ITO on glass) and, in contrast to SMOLEDs, the architecture is simple since neither transport layers nor blocking layers are needed. Another important difference with respect to SMOLEDs is that, due to the very high molecular weight, polymer deposition is incompatible with vacuum thermal evaporation. The polymers used are soluble in common organic solvents so large-area, uniform, defect-free thin films can be cast from solutions at room temperature. Deposition techniques such as spin coating, drop-casting, ink-jet printing or roll-to-toll web coating are then available, all of them very cheaply, especially when compared to vacuum deposition. In addition, because of the high elasticity of polymers, they are flexible and can be easily fabricated on to rigid or flexible substrates in flat or curved shapes. These characteristics make PLEDs a cheaper approach than SMOLEDs.

Generally speaking, OLEDs (both SMOLEDs and PLEDs) have many advantages over conventional LEDs. A key advantage is the potential for exact spectral tuning within one OLED device due to the variety of emitting polymers across the visible spectrum. Either by mixing these elements or by stacking units with different emission wavelengths (in the case of SMOLEDs) a myriad of emission spectra can be created.

However, OLEDs are not as bright as inorganic LEDs, but their large emission surface provides a soft and pleasant luminance distribution. Moreover, the light source itself is touchable and easy to handle as it produces little or no concentrated heat, which means luminaires can be mounted close to the user without causing any discomfort.

However, despite the attractive features provided by OLED technology, their performance is still far behind conventional LEDs. To illustrate, Table 12.2 summarizes the typical features of state-of-the-art OLED panels, as reported by different manufacturers in 2013 and 2014. Despite considerable efforts in recent years by companies such as CDT, DuPont and Merck, there is still a wide performance gap.

Table 12.2 Performance indicators from different OLEDs manufacturers at a luminance of 3000 cd/m^2 [12] Marcador no definido

Metric	LG Chem[1]	Panasonic[2]	CDT/Sumitomo[3]
Electrical efficiency (%)	80	75	46
Internal quantum efficiency (%)	75	85	72
Extraction efficiency (%)	42	50	46
Spectral efficiency (%)	90	85	89
Panel efficiency (%)	23	27	14
Panel efficacy (lm/W)	82	98	48

[1] A hybrid triple stack with fluorescent blue emitters and phosphorescent red and green.
[2] A double stack with all phosphorescent emitters.
[3] A single stack with polymer/oligomer emitter.

12.6 White Light with LEDs

Creating white light using LEDs is not a trivial task given that one LED is a highly monochromatic emitter, while white light is, by definition, polychromatic.

12.6.1 Wavelength Converters

Wavelength converters are organic or inorganic materials that absorb radiation in a certain wavelength range and re-emit light in another wavelength range. Wavelength converter materials include fluorescent dyes, semiconductors and phosphors. Phosphors are widely used in fluorescent lighting systems where they convert the ultraviolet light emitted by a mercury discharge lamp into a range of visible wavelengths. Their emission colour and spectral width depends on the phosphorescent materials and dopants in the phosphor, and covers the full visible spectrum plus the near-infrared and near-ultraviolet region.

An important characteristic of a fluorescent or phosphorescent converter is the efficiency, η, given by the following equation:

$$\eta = \eta_{ext} \cdot \left(\frac{\lambda_1}{\lambda_2}\right),$$

where

$$\eta_{ext} = \left(\frac{\#\text{photons per sec}@\lambda_1}{\#\text{photons per sec}@\lambda_2}\right)$$

In LED-based applications, phosphors can be coupled either directly or remotely to an ultraviolet, violet or blue LED, creating a single emitter light source with the possibility of a very high CRI, gamut area or efficiency. Compared to the multi-emitter systems described in the next section, phosphor-converted LEDs have the advantage of being significantly easier and cheaper to manufacture and operate. There is no need for complex electronics to provide a specific current to each LED die, and no feedback mechanism to keep chromaticity points under control.

12.6.1.1 Phosphors for LEDs

Immediately after its invention in the early 1990s, the blue LED was combined with a yellow $Y_3Al_5O_{12}:Ce^{3+}$ (YAG) phosphor to create white LED emission. These dichromatic LEDs still dominate the white LED market today.

The YAG phosphor very efficiently absorbs blue radiation between 440 and 460 nm, and re-emits a broadband spectrum with a half-width of 130 nm, centred at 550 nm. The result is a conversion system with an internal quantum efficiency (IQE), that is, the efficiency of conversion of charge carriers into photons, over 90% and relatively good colour rendering properties due to the broad emission spectrum of the YAG phosphor. The addition of Ga shifts the spectrum of the YAG phosphor to shorter wavelengths while the addition of Gd shifts the spectrum to longer wavelengths. This spectral tunability allows manufacturers to create white LEDs with different CCTs down to 3000 K. The thickness of the phosphor layer determines the exact location of the chromaticity coordinates of the white LED. A thicker phosphor layer will absorb more blue light, moving the chromaticity coordinates towards the yellow region of the CIE diagram.

Additional materials can be incorporated into the phosphor mix to improve colour rendering qualities or achieve lower CCTs.

Aluminate phosphors provide blue-green to pure green light with a very wide spectral range, allowing the creation of highly efficient white LEDs with colour rendering indices exceeding 95.

Silicate phosphors provide green to orange light with a relatively narrow spectral width, making them ideal for use in display backlighting. The narrow emission spectrum of a silicate phosphor allows the creation of weaker colour filters in LCD displays, creating a larger colour gamut and a more efficient display.

Nitride phosphors are red-emitting converters primarily used to create high colour rendering, warm white LED sources.

Not all white LEDs are based on phosphors with incident blue InGaN radiation (known as the *pump*). A different approach uses violet or near-UV LEDs to excite a mixture of three or more phosphors to create white light. The huge advantage of this method is that the radiation of the pump does not contribute significantly to the final output spectrum. This allows manufacturers to create LEDs with exceptional colour rendering properties and very low unit-to-unit spectral variation. The downside is that the large difference between excitation and emission wavelengths results in a relatively low down-conversion efficiency. InGaN UV-based white LEDs have a significantly lower efficiency than InGaN blue/YAG-based emitters.

12.6.1.2 Phosphor application methods

The way phosphor particles are distributed inside the LED package influences the colour distribution and efficiency of the output beam.

Early LED packages used the so-called proximate phosphor distribution where the phosphor was mixed with the encapsulating material. The thickness and concentration of the phosphor layer determined the final chromaticity coordinates of the LED. The main problem with this method is that the length of the light path through the phosphor varies over the location of the individual semiconductor device, known as a *die*. This results in an inhomogeneous colour distribution in the output beam pattern.

A conformal coating covers the whole semiconductor wafer with a very evenly controlled layer of phosphor particles. This allows manufacturers to greatly enhance the beam uniformity of the resulting LED and also reduces phosphor material consumption while reducing manufacturing variation between devices with regard to colour differences and luminous flux.

Very-high-power devices or luminaires have another problem: As the quantum efficiency of the phosphor is less than 100 percent, part of the absorbed radiation is converted into heat. This heat has a detrimental effect on the efficiency of the phosphor itself and lowers the efficiency of the LED die. Placing the phosphor layer at a relatively large distance from the die (several mm to cm) reduces these effects considerably. A remote phosphor application also reduces reflection losses between the phosphor and the LED die, resulting in an overall increase in system efficiency, colour stability and lifetime.

12.6.2 Multichromatic LED Sources

Multichromatic LED sources use a combination of monochromatic emitters to create a certain spectrum. While this increases, sometimes considerably, the complexity of the electronics and optical system, it offers many advantages regarding colour or spectral tuning. While many phosphor-based white LED systems are in fact multichromatic light sources, this paragraph focuses on LED sources using a combination of monochromatic emitters.

12.6.2.1 Dichromatic LEDs

While the combination of two complementary wavelengths can provide a chromaticity point close to the black-body locus, the colour rendering of objects is heavily distorted. Narrow spectrum dichromatic light sources are not useable for general illumination.

12.6.2.2 Trichromatic LEDs

Trichromatic LEDs are widely used in lighting and display applications. Their most famous incarnation, the RGB LED, consists of three emitters packaged together to form a compact semiconductor light source with a tunable chromaticity point. Usually the peak wavelengths of the three LEDs are chosen to maximize the colour gamut whilst keeping a relatively high efficacy. This results in emission peaks at around 455, 530 and 610 nm. RGB LEDs are primarily used in the display industry as direct view LED screens for large outdoor events and efficient, wide colour gamut backlights for high-end LCD panels. At the same time the entertainment industry is abandoning electricity-hungry discharge lamps with different colour filter combinations in favour of much more energy friendly RGB LEDs.

In colour-critical applications where a stable white point must not be affected by temperature variations or the aging of semiconductor material, a closed loop feedback system is necessary. A tristimulus sensor tuned to the peak wavelengths of the red, green and blue emitters provides a feedback signal to the LED driver electronics, allowing careful adjustment of the LED drive current when any change in chromaticity is detected.

12.6.2.3 Polychromatic LEDs

A new trend in general illumination systems is combining high quality light with an interactive control where the user can select an ideal illumination depending on a specific location, the time of the day or the mood of that person. While RGB-based lighting solutions are very flexible, their spiky spectra (see Fig. 12.12) rate poorly when rendering subtle hues of daily objects. When

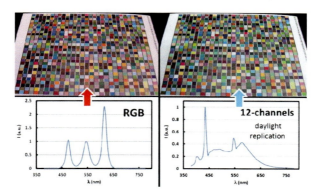

Figure 12.12 Colour rendering capabilities comparison between a RGB-based lamp and a daylight spectrum replicated by a 12-channel LED engine. The colours on the right are clearly more vivid and accurate.

combining a sufficient number of narrow-band-emitting LEDs, a tailored lighting spectrum without significant gaps or valleys can be created. Such polychromatic light engines can have both excellent colour fidelity and high efficiency. The challenge in this kind of source is finding effective and efficient methods of combining the output of those monochromatic emitters to form a single homogeneous beam of light.

12.6.2.4 Spectral characteristics of multichromatic LED sources

The colour rendering qualities of a white light source created by monochromatic spectra are highly dependent on the reflectance spectra of the object being illuminated.

Trichromatic light sources offer better colour fidelity then *dichromatic* light sources. However, objects with reflection spectra not matching the emission spectrum of the light source will be rendered incorrectly. This does not necessarily mean that the object illuminated by that spectrum will look highly undersaturated or with low colour contrast. Depending on the interaction between the source spectrum and the object's reflection spectrum, a scene can look more saturated with better colour contrast under trichromatic illumination compared to a broadband illumination source with high colour fidelity.

Polychromatic light sources can have very high colour fidelity values. Depending on the number of monochromatic emitters in the source, CRI values of 95 to 100 are achievable. In extreme cases, even a complete spectral match to a reference spectrum is possible.

In the case of a monochromatic light source emitting at 555 nm, the LER is maximized at 683 lm/W. Of course, this pure green light can hardly be called a white light source since its colour rendering properties are non-existent. LED light sources containing individual emitters can be carefully optimized to provide a balance between LER, gamut area coverage and colour rendering. Depending on the application, one of these criteria might be more important than another. By choosing the LED peak wavelengths closer to the 555 nm emission wavelength, the LER is maximized. LEDs with a narrow emission spectrum located near the edges of the chromaticity diagram favour a very wide colour gamut, a useful feature in certain display applications.

12.7 New Approaches

12.7.1 Silicon-Based Emitters

Silicon is an appealing material for potentially implementing low-cost solid-state lighting (SSL) sources. Its availability as a raw material and compatibility with mainstream CMOS processes are two characteristics that have caught the attention of many researchers in the field. Up to now, silicon has demonstrated its versatility in a wide variety of photonics applications: waveguides, modulators and optical sensors are some of the functionalities already covered by silicon-based materials. Nevertheless, the achievement of an efficient LED source entirely based on silicon compounds still remains elusive. Silicon is an indirect semiconductor where the radiative recombination of electron–hole (e–h) pairs can only occur with the assistance of a further process to conserve the momentum, typically the intervention of phonons. This three-particle process is very unlikely, which results in longer radiative lifetimes (of the order of microseconds). In these conditions, fast non-radiative interactions (of the order of nanoseconds) are favoured, quenching most of the emission and leading to very low internal quantum efficiencies, about 10^{-4}%. In addition, there are two more phenomena that also contribute to limit the emission capabilities of silicon. The first one is the Auger process in which the e–h pair recombines non-radiatively, transferring its energy to a third carrier (typically an electron or hole). The probability of Auger recombination is proportional to the square of the number of excited states and therefore dominates the recombination processes for high carrier injection rates. The second process is light absorption by excited carriers, known as *free carrier absorption*. This can deplete the population of excited carriers and at the same time increase the optical losses of the generated signal. This process is mostly observed when the density of free carriers is relatively high, that is, under high (electrical and optical) pump powers. Despite these drawbacks, the potential manufacturing benefits of an eventual SSL source based on silicon are huge. Some research efforts focus on bulk Si structures [13–16] while others have been working on bolder strategies:

Porous silicon: In 1990, after repeated reports of inefficient emission from bulk Si, the first article concerning porous silicon

(PS) was published [17]. In that work Canham et al. reported broad spectrum emission in the red-infrared range from anodized Si layers. Porous silicon is fabricated by accurate electrochemical etching of Si layers resulting in a relatively high degree of porosity consisting of an irregular array of nanometre-scale porous structures. Such small structures promote quantum confinement that increases electron–hole wave function overlap, resulting in increased light emission efficiency and a shift in the emission peak to higher energies. The magnitude of the peak shift can be controlled by the size of the nanostructures [17, 18]. However, despite promising results, this material was abandoned due to the high reactivity of the sponge-like porous structure that causes rapid aging of the device and uncontrollable variations in LED performance, and also due to the incompatibility of the Si anodization process with mainstream CMOS technology.

Si nanocrystals: In comparison to porous silicon, Si nanocrystals (Si-nc) embedded in transparent dielectric matrices represent a better alternative for light-emitting device fabrication. Protected from exposure to air, the silicon nanoparticles have high chemical and thermal stability which leads to more efficient and reliable devices. Moreover, the fabrication of Si-nc-based composites is fully compatible with current CMOS technology. However, this approach appears to be unsuitable due to the inherent electrical limitations of using a dielectric material with a limited number of emitting centres as an active layer [19–23].

Rare earth doping: This is a similar strategy to that of Si-nanocrystals. Rare earth ions are randomly distributed across a dielectric host matrix. Despite this approach providing a better control of emission wavelengths by control of the concentration of introduced emitting species, it fails for the same reasons as Si-nanocrystal-based emission [24, 25].

12.7.2 Quantum Dots

In recent years, quantum dots (QDs) have arisen as an alternative emissive material in SSL devices. With sizes ranging from 3 to 12 nm in diameter, the electronic structure of QDs is dominated by quantum size effects, providing them with their signature narrow-band emission that can be spectrally positioned by controlling the nanocrystal size during synthesis.

Quantum dots can be categorized by their synthesis method as either colloidal or self-assembled (epitaxial). In the first case, QDs are prepared by wet chemical procedures whereas in the second case the nanoparticles are created from relatively high energy dry methods of epitaxial growth from the vapour phase. At present, colloidal QDs of inorganic semiconductor materials are the preferred trend because, apart from the aforementioned narrow emission and extraordinary colour tuning versatility, they involve fast synthesis protocols. Nanoparticles are synthesized from organometallic precursors and they retain a passivating layer of ligands, which confers solubility in a wide variety of solvents. Such capability enables the use of low-cost deposition techniques such as spin-coating, inkjet printing and micro-contact printing, which represents a clear advantage over epitaxially grown nanoparticles. In addition, a wide variety of ligands exist which increase the versatility of the QDs, making them soluble in aqueous solutions. Because nanoparticles are made from inorganic semiconductor materials, they are often more resistant to photo-degradation than organic dyes.

A typical quantum dot LED (QLED) consists of a layer of quantum dots sandwiched between two layers: the first layer transports electrons while the second layer transports holes. When an electric field is applied to the outer layers, electrons and holes move into the active layer and are captured by the quantum dots where they recombine and generate light.

QDs employed for efficient QLEDs are mostly based on chalcogenide semiconductors like CdSe or CdS; however, their application in electronic devices is no longer feasible due to environmental regulations. New approaches are being investigated to overcome this issue. One example is the silicon QD, previously discussed in this chapter.

12.8　LED Packaging

One of the most important criteria for commercial LEDs is the brightness. Under this premise, packaging must be designed to maximize the ratio of the extracted light over the total emission at the junction. To do this, light has to be redirected to the output surface of the device by accurately designing the die shape and the primary optics included in the LED package.

A second important function of the LED package is thermal management. LED dies generate a considerable amount of heat, and lifetime and efficacy are direct functions of the junction temperature. For this reason an efficient thermal package is crucial to improve LED performance. The method of attaching the die to the package, and the material of the LED package itself, determine the final thermal performance of the LED device.

Finally the LED package provides mechanical protection from the environment as the LED die is sensitive to mechanical damage cause by adverse handling.

12.8.1 Low-Power LED Packaging

Conventional low-power LEDs have been used in electronics for years, usually as indicators lights, signals and backlights. Compared to high-power LEDs, the principal difference is the smaller semiconductor die area and absence of relevant mechanisms for heat dissipation. Low current operation limits heat generation at the junction which facilitates, to some extent, the design of LED packaging. All the LEDs possess two electrical leads. In the case of low-power LEDs, the die is glued to a cup reflector, conceived to increase light extraction, which in turn is bonded to one of the two power leads. The top contact is then connected to the second lead by a bond wire. The whole module is encapsulated in an epoxy resin which provides the LED with chemical and mechanical stability and at the same time reduces the refractive index contrast between the semiconductor and air, thereby reducing total internal reflection effects and enlarging the light escape cone. Typically the encapsulant is hemispherically shaped, which generates a Lambertian emission pattern. This shape can be changed to flat-topped or more focused radiation patterns depending on the required application.

12.8.2 Mid- and High-Power LED Packaging

Mid- and high-power LEDs differ from their low-power cousins mainly by the size of the semiconductor die. While low-power LEDs usually use die sizes around 0.1 mm^2 (and corresponding low forward currents), high-power LEDs boast die sizes of at least 1 mm^2. This tenfold increase in die size allows the use of much higher forward currents (typically in the range from 350 to 700 mA). In addition, whereas in low-power LEDs, usually employed for indicator purposes, the packaging is quite standardized, high-power LEDs feature a wide variety of shapes and sizes. At present, a variety of possible structures which achieve high light output with minimum material usage and optimized heat extraction are fabricated by LED manufacturers, despite the fact that room for further improvements in efficacy is shrinking.

12.8.3 Thermal Management

Thermal management becomes a relevant subject in LEDs when high operating currents are considered. Under these conditions the LED junction of mid- and high-power LEDs delivers higher output powers but at the cost of generating a significant amount of heat. In order to avoid the effect of temperature on device performance (Fig. 12.13) and reliability, it is crucial to dissipate most of this heat towards heat sinks or printed circuit boards (PCBs) designed for efficient thermal management.

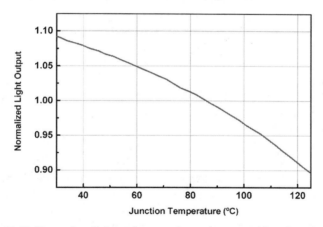

Figure 12.13 Example light output dependency with the junction temperature. The luminescence is normalized for temperature at the junction of 85°C.

All LED dies for high-power purposes are directly attached to a thermally conductive path that acts as a heat-spreader. At present there is huge variety of LED structures, but they have a number of common design features. As indicated above, the die is soldered by a metal-based solder to a ceramic, copper or aluminium heat sink slug with a high thermal conductivity, as shown in Fig. 12.14.

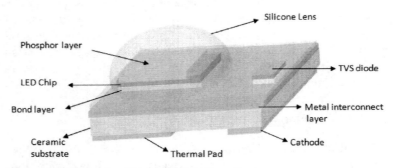

Figure 12.14 Example of high-power LED packaging.

However, even if the packaging design is very effective in reducing the junction temperature, the generated heat is often still too high to assure high performance and long lifetime. In this case, the solution is to attach the LED to a thermally conductive surface which removes heat from the LED package. Care has to be taken to ensure electrical isolation between the die and surrounding metal contacts. This is usually accomplished by using thermally conductive but electrically isolating substrates such as ceramics or a metal-core printed circuit board. MCPBs have a layered structure consisting of a top structure containing the conductive circuit traces, and an aluminium base separated by a thin dielectric that provides electrical isolation between the conductive traces and the base material. In cases where dissipated power is extremely high, MCPCBs are not advised since their thermal conductivity is not high enough to maintain the system within a safe temperature range. In those cases, ceramic circuit boards provide better LED operating conditions and nowadays their usage is widespread among the different manufacturers in the field.

It is clear that thermal management is a key issue when designing a LED-based system. A meticulous analysis of the LED system to be produced can avoid unnecessary reliability problems, thereby assuring performance and lifetime. The current practice is to simulate the device performance by computational fluid dynamics (CFD) analysis. This method, especially useful in earlier stages of the design process, allows prediction of the temperature of every component of the LED system.

12.8.4 Some Effects Related to Excessive Junction Temperature

As stated in the previous section, LED performance is very sensitive to variations in operating temperature. When the heat generated at the junction is not effectively dissipated, the operating temperature increases. On a nanometre scale, this translates into an increase in the amplitude of lattice vibrations within the LED die and the surrounding materials (solders, electrodes, etc.). The increase in thermal agitation in turn gives rise to higher electrical resistance, and subsequently to lower light output powers. At the same time, the inter-atomic spacing increases by the action of these stronger atomic vibrations, which leads to an effective LED die enlargement (by thermal expansion), quantified by the linear expansion coefficient of the material. The larger inter-atomic spacing implies a decrease in the potential seen by the electrons in the material, and therefore a reduction in the energy band gap occurs. In other words, an increase in the device temperature correlates to an emission peak shift. Therefore, besides the necessity of efficient heat dissipation strategies, it is also important to implement a stable and reliable driving current system. Variations in driving current can induce temperature shifts that result in brightness and chromatic oscillations.

However, elevated temperatures have many other effects on device performance. Operating under these conditions induces thermal stress in packaging components due to the mismatch in coefficients of thermal expansion between the adjacent parts of the die. The active layer of the die is very sensitive to mechanical stresses, and high temperatures can lead to cracks, or delamination of thin films. In the case of phosphor-converted LEDs, phosphors, typically deposited on the LED die, can be also easily affected. Heat fosters non-radiative mechanisms and induces peak broadening, thereby leading to a drop in phosphor conversion efficiency and a drift in emission colour. Moreover, the mechanical stress generated between die and the silicone encapsulant can induce morphological variations of the phosphor layer, such as roughening or delamination. In the first case, it is possible that roughening may cause the mean free path through the phosphor layer to increase, causing a shift towards the warm colour temperature side of the spectrum. In the second case, the high temperatures can cause the phosphor layer on the edges of the chip to curl up, resulting in a net increase in blue light coming from the LED chip, leading to cooler colour temperature light emission.

Concerning moulding compounds used as encapsulants, high temperature operation also plays an important role. Most modern encapsulants are based on polysiloxane materials [26] because of their excellent optical and thermal behaviour (transparency and stability). When such polymers are exposed to high temperatures they can decompose, forming thermo-oxidative cross-links leading to encapsulant discolouration. The result of this phenomenon is that the material's transparency decreases, with a subsequent drop in light output.

Additional heat dissipation may be the result of moisture penetration, atmospheric pollutants, and reductions in LED driver efficiency, reliability and current stability that may impact LED performance. The combination of temperature, moisture, and voltage bias can cause metal migration failures, which may be followed by catastrophic failure of the device.

12.8.5 The Role of the Packaging on Light Extraction

Apart from its relevance to thermal issues, packaging plays an important role in optical performance. As well as being responsible for maximizing the extraction of the light generated at the junction, which is complicated due to the high refractive index contrast between the junction and the surroundings, it plays a role in shaping the emission pattern of the output beam.

Despite recent efficacy improvement in LEDs, optical powers of current LEDs are still far from their theoretical limit. Internal quantum efficiencies may be close to 100%; however, only a small percentage of the light generated at the junction emerges from the device. This is due to different optical phenomena acting within the semiconductor: total internal reflection, dispersion and absorption. Of these, the major contribution is from total internal reflection. As depicted in Fig. 12.15, all the rays that emerge at the output surface of the LED at angles greater than the critical angle, θ_C, are reflected and confined within the semiconductor.

The critical angle is defined in terms of the refractive indices of the material where the light is generated, n_m (typically in the range of 2.9 to 3.6 for III-V materials), and the surrounding media, n_0 (for air, $n_0 = 1$), according to Snell's Law:

$$n_m \sin \theta_i = n_0 \sin \theta_t,$$

where θ_i and θ_t are the angles of the incident and transmitted rays normal to the surface. The critical angle will therefore be that at which the transmitted rays are parallel to the interface, $\theta_t = 90°$:

$$\theta_c = \sin^{-1}\left(\frac{n_0}{n_m}\right),$$

where θ_c defines a cone in three dimensions, known as the *escape cone*, that represents the fraction of emitted light that can escape from the diode. To a lesser extent, Fresnel reflection of the light back into the device from the die surface also contributes to the total internal reflection. By way of illustration, devices with an absorbing substrate (AS-LEDs) exhibit typical extraction efficiencies of around 1% to 2%, since most of light transmitted towards sideways and downwards is absorbed by the substrate. In contrast, transparent substrate LEDs (TS-LEDs) exhibit typical extraction efficiencies up to 8%, since the transparency of the substrate facilitate emission from the device edges. Besides the use of transparent substrates, many others strategies are applied to enhance the percentage of extracted light. Some of them are as follows:

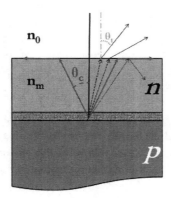

Figure 12.15 Schematic drawing of total internal reflection effects within an LED die.

Device shaping: The basic idea is to fabricate devices with materials and geometries that optimize the light extraction from the active layers. Chips may be shaped into cones, spheres or pyramids, making use of elementary concepts such as total internal reflection to significantly increase the extraction efficiency. A classic example is the shaping of TS AlGaInP LEDs into truncated inverted pyramids (TIP LEDs). A more recent approach is CREE's new generation of silicon carbide-based Direct Attach (DA) LED chips [27]. As shown in Fig. 12.16, these have a bevel-cut surface that increases light extraction efficiency and results in light emission at a wide angle than planar devices.

Figure 12.16 3D schematic drawing of a shaped die inspired in direct-attach LED technology from CREE.

Surface roughening: Despite the promising light extraction provided by device shaping, the semiconductor manufacturing process is not suited for making dies in any shape other than cubes because it uses straight saw cuts to remove dies from the wafer. Consequently, manufacturing shaped dies requires an additional manufacturing step, which can become expensive. Therefore a more cost-effective and practical alternative is to roughen the upper surface of the die before the individual dies are cut from the wafer. By roughening, patterning (for example, forming photonic crystals) or micro structuring the upper surface, the angles of reflected light are randomized which gives a significant improvement in output power.

Mirrors: The placement of mirrors in between the absorbing substrate and the active layers is a suitable technique to increase light extraction. The principle is very simple: the mirror reduces the effects of substrate absorption and redirects the light to the top surface, increasing the fraction of emitted light that reaches the air. Two kinds of mirrors are typically used: distributed Bragg reflectors (DBRs) and metallic reflectors. In this second case, the buried metallic mirror is fabricated by bonding the LED active layer onto the substrate at chip- or wafer-level with an intermediate metal layer.

Resonant cavity: Resonant-cavity LEDs (RCLEDs) are devices with the active layer inside an optical cavity that is approximately tuned to the emission wavelength. The spontaneous emission properties from the light-emitting region inside the cavity are enhanced by the resonant cavity effect. The light intensity emitted along the cavity axis is higher compared than for conventional LEDs. Moreover, the signal has a higher spectral purity and the emission pattern is more directed along the axis of the cavity. The cavity is typically fabricated by two DBRs, but metals are also used, especially as top contacts.

Specific lenses: Another approach to improve extraction efficiencies is by designing specific lenses to decrease the probability of total internal reflection of the light escaping from the external surface. The basic premise is to reduce the incident angle and therefore increase the chances of light being directly emitted without reabsorption or multiple reflections. In addition, such designs must provide radiation patterns with specific shapes in order to comply with the requirements of different illumination applications.

12.9 LED Drivers

The steep current–voltage characteristic of LED requires a current-controlled power source. In the lighting market the term *driver* is usually applied to a device that can accept an input voltage and produce the required current. An important factor that affects the choice of a particular driver is the way in which multiple LEDs are actually wired together.

Since most commercial applications for mid- to high-power LEDs for general lighting require multiple LEDs to be connected together to produce the required amount of luminous flux, some considerations must be given as to how to connect them. The three general wiring configurations are

- **Series**: LEDs are connected to form a chain from the cathode to anode.
- **Parallel**: All of the cathodes are connected together and all of the anodes are connected together.
- **Series/Parallel**: LEDs are connected in series to form a chain, and then the chains are wired in parallel.

12.9.1 Linear Constant-Current Drivers

Linear drivers rely on transistors to deliver a constant current to the LED. A generic block diagram of a constant-current LED driver is shown below in Fig. 12.17. The output current is determined by the net resistance (R_L) between the input voltage (VCC) and the emitter of a p-n-p-transistor, where $I_{LED} = I_{RL} = (V_{ZENER} - V_{EB})/R_L$.

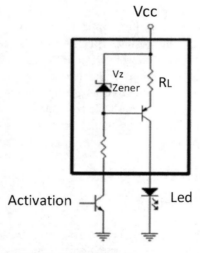

Figure 12.17 Linear constant current LED driver.

An external transistor may be added to provide an ON/OFF activation pulse.

A linear constant current driver includes a small number of components in the circuit without any switching element it means that noise is kept to a minimum. However, the waste heat can be more high if the average load or the difference between the input and output voltages are high. In addition, the efficiency decreases when the difference between input and output voltages is high. It should be noted that when the supply voltage is only marginally higher than the LED voltage, the efficiency of a circuit using a linear regulator may be higher than a more complicated switching regulator design.

Linear LED drivers have the disadvantage of their large size, because a step-down transformer is almost always required, and a 50 or 60 Hz mains transformer is bulky and heavy. Also, smoothing capacitors added to the power supply to stabilize the voltage input are also very bulky. The main limitation of a linear supply is that the LED forward voltage must be lower than the supply voltage.

12.9.2 Switching Constant-Current LED Drivers

Switching DC-to-DC converters (buck, boost, buck-boost or SEPIC (Single-ended primary-inductor converter) converters) are widely used as a constant current source for clusters of LEDs (see Fig. 12.18). The output voltage will vary depending on the forward voltage drop across the LEDs selected. Typically, the LED current is set with an external sense resistor (R_S) and is directly regulated by a feedback pin (F_B) that regulates the voltage across the sense resistor. LED brightness can often be regulated by applying a PWM signal to the appropriate PWM input pin on the regulator.

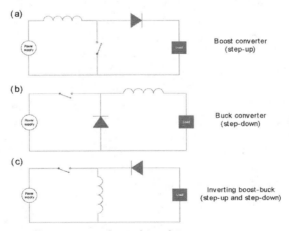

Figure 12.18 Different types of switching drivers.

12.9.2.1 Buck converters

The buck converter is the simplest of the switching drivers, and is a step-down converter for applications where the load voltage does not exceed the supply voltage. In a buck converter circuit, a power MOSFET (metal-oxide-semiconductor field-effect transistor) is used to switch the supply voltage across an inductor and an LED cluster connected in series. The inductor is used to store energy when the MOSFET is turned on. This energy is then used to provide current for the LED when the MOSFET is turned off. A diode across the LED and inductor circuit provides a return path for the current during the MOSFET's off time.

Buck converters can achieve efficiency values higher than 95% and are therefore very popular in battery-powered applications. A peak-current-controlled buck converter can give reasonable LED current variation over a wide range of input and LED voltages, and needs no design effort in feedback control design.

12.9.2.2 Boost converter

Boost converters are ideal for LED driver applications where the voltage for the LEDs is greater than the input voltage. Normally, a boost converter would only be used when the output voltage is at least 1.5 times the input voltage.

- The converter can easily be designed to operate at efficiencies greater than 90%.
- Both the switching control and load (LEDs) are connected to a common ground. This simplifies load current sensing, unlike the buck converter where either a switching amplifier or a current sensing measurement must be chosen at the side of the LEDs. In the current sensing method, a sense resistor is connected in series with the ground path.
- The inductor is the main energy storage device in a switching regulator. The energy is stored in the inductor and transferred to the load. The size of the inductor, the voltage across it, and the time of the switching determines the inductor current ripple. The current ripple need be filtered to meet the required electromagnetic interference (EMI) standards.

Boost converters have some disadvantages, especially when used as LED drivers, due to the low dynamic impedance of the LED string.

- The output current of the boost converter is a pulsed waveform. Thus, a large output capacitor is required to reduce the ripple in the LED current.
- The large output capacitor makes PWM dimming more challenging. Turning the boost converter on and off to achieve PWM dimming means that the capacitor must be charged and discharged every PWM dimming cycle. This increases the rise and fall time of the LED current.
- Open loop feedback must be avoided due the impact caused by the temperature variation and fluctuation in current that need be sensed and considered in the controller. A closed-loop feedback circuit is required to stabilize the converter. A close loop feedback allows sensing the current to change the voltage or the duty cycle.
- There is no control over the output current during a short circuit condition at the time of load fault unless the output dc power of the converter can be disconnected from the load within several hundred microseconds at the instant of short-circuit fault.

12.9.2.3 Boost-buck converter

A boost-buck converter is a single-switch converter, which combines the principles of the buck converter and the boost converter in a single circuit.

The converter has many advantages:

- The converter can both boost and buck the input voltage. Thus, it is ideal for cases where the output LED string voltage can be either above or below the input voltage during operation. This condition is most common in automotive applications, or when a single driver design is needed to cover a wide range of voltage supply and load conditions.
- The converter has inductors on both the input and output sides. Operating both stages in continuous conduction mode (CCM) will enable continuous currents in both inductors with low current ripple, which would greatly reduce the filter capacitor requirements at both input and output. Continuous input current would also help greatly in meeting conducted EMI standards at the input.

- All the switching nodes in the circuit are isolated between the two inductors. The input and output nodes are relatively quiet. This will minimize the radiated EMI from the converter. With proper layout and design, the converter can easily meet radiated EMI standards.
- One of the advantages of the boost-buck converter is the capacitive isolation. Any failure of the switching transistor will short the input and not affect the output, thus protecting the LEDs from failure of the MOSFET.
- Both inductors can be coupled together on one core. When coupled on a single core, the ripple in the inductor current from one side can be transferred completely to another side (ripple cancellation technique). This would allow, for example, the input ripple to be transferred completely to the output side, making it very easy for the converter to meet conducted EMI standards.

12.10 Lighting Control Systems and Applications

This section provides detailed information on modern LED lighting control systems, covering heterogeneous sensors and actuators, LED drivers, communications technologies and selected applications for smart cities, all of them conceived to increase comfort and minimize energy consumption.

12.10.1 Smart Lighting Control Systems

Smart lighting is a development in the functionality of LED and OLED luminaires, with the possibility of integrating a wide set of electronic capabilities as well as providing light. The most important features include intelligent functions such as spectral tunability and adaptive dimming according to the requirements of the illuminated space, and taking into account the natural lighting available.

A typical smart LED lighting control system receives information from numerous sensors and actuators and autonomously decides when to turn on, dim, or shut off the light system based on decision-making control logic. Additional features include the adjustment of artificial light levels based on the available daylight, environmental needs or user preference, thereby improving energy consumption and visual comfort. Typical sensors incorporated in smart lighting systems are passive infrared sensors (PIRs), illuminance and colour sensors, and mini-spectrometers.

12.10.2 Occupancy Sensors

There are four basic sensor technologies for occupancy lighting sensors as shown in Fig. 12.19. These are photocell, infrared, ultrasonic, acoustic or dual-technology. Infrared (IR) technology senses body heat and is required to be in the line of sight of people in the room in order to properly operate. IR technology is used mainly in offices. Passive infrared sensors (PIRs) are triggered when a warm object moves either into or out of the line of sight defined for the sensor.

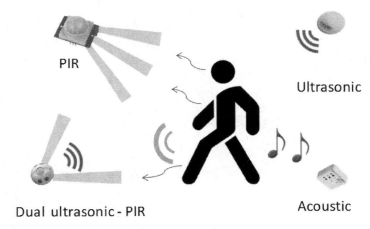

Figure 12.19 Occupancy sensors for smart lighting systems.

Ultrasonic (US) technology emits a high-frequency sound that reflects off room surfaces. Ultrasonic occupancy sensors emit sound pulses at frequencies well above both human and animal hearing capabilities. Ultrasonic sensors utilize the Doppler Effect to sense changes in the direct environment of the sensor. The most suitable location of occupancy devices is a critical design decision. They suit irregularly shaped spaces and room obstructions such as medium to high partitions, large furniture or structural columns.

Acoustic or audible sensors rely on voice activation. Background noise, such as a constant hum, and low-level noise are ignored. This technology works well in areas with high partitions and obstructions, or high air movement within the space during unoccupied periods.

Dual or triple technology sensors are intelligent sensors that self-adjust to occupancy data collected in a prescribed learning period. These sensors also reduce false-on and -off conditions. However, they must still be properly located, adjusted and calibrated in commissioning, and regularly maintained.

12.10.2.1 Illuminance sensors

Generally, two types of sensors are used for daylight harvesting and adaptive compensation: Light-dependent resistors (LDRs), usually cadmium sulphide (CdS) photoresistors, and digital ambient illuminance sensors. Both can measure natural and artificial lighting. The LDR photocell contains a resistor whose resistance decreases with increasing incident light intensity, whereas digital ambient light sensors contain internal digital photodiodes. The photodiodes are photosensors that generate a current when a p-n junction in the semiconductor is irradiated by light.

The spectral response of digital ambient light sensors is close to the peak of human visual sensitivity, as shown in Fig. 12.20. These sensors are intended to respond to light under a wide variety of lighting conditions.

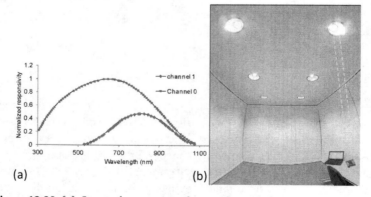

Figure 12.20 (a) Spectral response of an ambient light sensor containing two photodiodes. (b) Interaction of an ambient light sensor in a tunable LED lighting system (IREC showroom).

12.10.2.2 Colour sensors

A colour sensor device provides the detection of the red (λ = 615 nm), green (λ = 540 nm) and blue (λ = 465 nm) (RGB) regions of the visible spectrum, and also allows detection of light intensity under a variety of lighting conditions and through a variety of attenuation materials (Fig. 12.21).

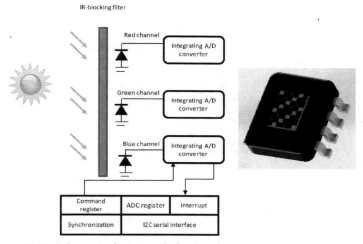

Figure 12.21 Schematic drawing of a basic colour sensor.

These devices are intended for precise measurement, determination, and discrimination of colour. They contain a digital interface which may be connected directly to a microprocessor.

12.10.2.3 Spectral sensors

Ultra-compact mini-spectrometers known as micro-opto-electro-mechanical systems (MOEMS) combine nanotechnology and nano-electromechanical systems (NEMS) technology with image sensors [28]. An example can be seen in Fig. 12.22. These spectrometers contain compact polychromators to disperse light into different directions in order to isolate different components of the spectrum of the light. Modern compact mini-spectrometers contain a nano-imprint grating, a CCD image sensor and a driver circuit, which are connected together in a low-cost compact device. The mini-spectrometers use a transmission matrix grating, fabricated by a holographic process, as a wavelength dispersive element. The grating separates the light into a spectrum and the CMOS image sensor converts the spectrum of light focused by the mirror or lens into an electrical signal.

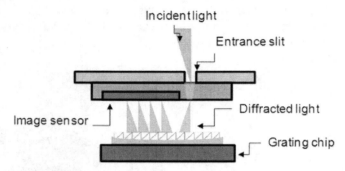

Figure 12.22 Example of a portable mini-spectrophotometer architecture.

The typical full-width-at-half-maximum (defined as the spectral width at 50% of the peak power value) spectral resolution of mini-spectrometers ranges from 1 to 10 nm. Applications range from analytical measurement tools to portable light characterization instruments.

12.10.3 Sustainable Energy-Efficient Applications for Smart Cities

Emergent lighting systems are evolving rapidly to increase visual comfort and reduce overall energy consumption. In addition, LED lighting systems may be used as both lighting sources and to transmit optical data. The following examples explore various lighting applications for indoor and outdoor environments.

12.10.3.1 Smart outdoor urban lighting

A considerable amount of energy used for lighting is devoted to street lighting. The evolution of LED technology and the incorporation of sophisticated intelligent drivers can improve energy savings, safety and comfort as a whole. LED-based street lighting not only complies with requirements in energy-efficient urban planning, but also provides safer environments for vehicles and pedestrians. An intelligent street light module adjusts the light levels automatically to the actual traffic density, and changes the spectral power distribution according to the light level or weather conditions.

12.10.3.2 Visible light communications

Visible light communications (VLC) refers to a data communication medium by means of electromagnetic radiation in the 380 to 750 nm range (visible range). Conceptually, a VLC system is quite simple, consisting of on a transmitter and a receiver, as show in Fig. 12.23. The transmitter is in its most simple form an LED which can be modulated by an LED driver according to the data to be transmitted. The receiver consists on a photo-detector and a means for demodulating the received signal. The data to be transmitted is converted into an electrical signal using digital modulation. This electrical signal is used to modulate an LED that is powered by a driver which supplies the high current density needed for high-power LEDs. The transmitted optical signal is detected by a photodiode which converts the light into current. Silicon-based photodiodes are used as they can detect visible light, although avalanche photodiodes (APDs) have high sensitivities, PIN photodiodes are the most used due to lower levels of noise, parasitic capacitance and cost.

APDs differ from PIN photodiodes in that APDs have gain so that with the correct circuitry better sensitivity can be achieved with an APD. However, a higher operating voltage may be required. In addition, an avalanche photodiode produces a much higher level of noise than a PIN photodiode. A typical PIN photodiode has lower capacitance and thus a higher detection bandwidth from visible spectrum region until near infrared. The faster PIN photodiodes have bandwidths of the order of tens of gigahertz.

Finally, the received signal is amplified by a trans-impedance amplifier and, then, demodulated to recover the original binary data.

Despite such a priori simplicity, VLC is a very promising and powerful technology. Unlike radio frequency (RF)-based communications, it uses a part of the electromagnetic spectrum that is not regulated. In addition, it makes use of previously installed lighting systems which clearly results in reduced cost. Moreover, VLC offers other advantages over RF communications such as lack of electromagnetic fields which affect other electronic appliances, which is critical in environments with carefully controlled electronic equipment such as hospitals. Also, no radiation is generated besides harmless light which does not have any negative effects on the human body, and which can also be dimmed to be more comfortable for the eyes.

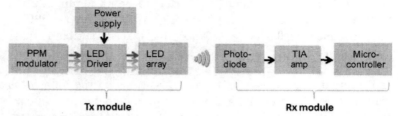

Figure 12.23 Block diagram of a basic visible light communication (VLC) system.

Aircraft cabin lighting can have a significant impact on aircraft weight and construction costs, and may be designed to aid the biological clocks and circadian rhythms of passengers. The circadian cycle is a series of physical, mental and behavioural changes that follow a roughly 24 h cycle, responding primarily to light and darkness in an organism's environment. Interior aircraft lighting can provide secure electronic communication, since the light will remain in the enclosed cabin, thus preventing any external antenna from intercepting any communication signals.

Furthermore, VLC can make use of different wavelength LEDs and optical filters to create multiple-input multiple-output (MIMO) systems, resulting in a dramatic increase in data rates. A MIMO system employs an array of LEDs and spatial light modulators (SLMs) with quadrature amplitude modulation and optical frequency division multiplexing (OFDM) modulation. VLC may soon act as a communication technology in vehicle traffic control infrastructures, for example as vehicle-to-infrastructure or vehicle-to-vehicle communication using traffic lights or car lamps, thus providing an intelligent accident prevention system.

12.10.3.3 Adaptive LED lighting

Smart LED lighting engines with the ability to transmit and receive data also offer huge possibilities, not only for energy savings but also for the functional selection of light to create a healthier environment, for example by mimicking the natural illumination conditions that govern human circadian rhythms, or by selecting particular wavelengths that promote a desired effect on cognitive abilities and behaviour, such as sustained concentration. This can be done very accurately by spectral selection or even through CCT adaptation by using the most efficient combination of phosphor-converted white LEDs, coloured LEDs, and OLEDs as the building blocks of the light source, whilst satisfying different constraints on energy efficiency.

12.10.3.4 Indoor spectrally tunable LED luminaires

Spectrally tunable LED light engines are luminaires which can be dynamically adapted to the needs of an indoor space in real time, through the establishment of algorithmic strategies and feedback systems.

Intelligent LED light engines are particularly well suited to spaces that demand complicated requirements on lighting. This is the case for museum lighting, where careful selection of spectral content may achieve an optimal balance between energy efficiency, quality of light and art conservation.

In addition these systems can be deployed as networks of intelligent LED lighting devices, which can communicate through intelligent protocols and interact with people in real time. These solutions are only a few examples of intelligent LED light control systems that will be incorporated into future smart cities.

References

1. CIE 1960, "Uniform Chromaticity Scale".
2. N. Ohta, A. Robertson, *Colourimetry Fundamentals and Applications*, John Wiley & Sons England (2005).
3. CIE 13.3 (1995), "Method of Measuring and Specifying Colour Rendering Properties of Light Sources".
4. C. J. BARTLESON, Memory Colors of Familiar Objects, *JOSA*, **50**(1) (1960), 73–77.
5. www.cree.com/News-and-Events/Cree-News/Press-Releases/2014/March/300LPW-LED-barrier.
6. Haitz's law, *Nature Photonics*, **1** (2007), 23.
7. E. Fred Schubert, *Light-Emitting Diodes*, Cambridge University Press (2006).

8. G. Craford. LEDs Challenge the incandescents, *IEEE Circ. Devices Mag.*, **8** (1992), 24–29.

9. J. P Dakin, R Brown. *Handbook of Electronics*, vol. I. CRC Press Taylor & Francis Group (2006).

10. Laboratory record: red LED breaks through the 200 lm/W barrier. *Optik & Photonik*, **6**, p. 16, 2011.

11. C. W. Tang and S. A. Van Slyke, Organic electroluminescent diodes, *Appl. Phys. Lett.*, **51** (1987), 913.

12. U.S. Department of Energy (DOE), Solid-State Lighting Research and Development, "Multi-Year Program Plan" (2014).

13. J. Kramer, P. Sertz, E. F. Steigmeier, H. Auderset, B. Delley, and H. Baltes, Light-emitting devices in industrial CMOS technology, *Sens. Actuators A*, **37–38** (1993), 527–533.

14. W. L. Ng, M. A. Lourenco, R. M. Gwilliam, S. Ledain, G. Shao, and K. P. Homewood, An efficient room-temperature silicon-based light-emitting diode, *Nature*, **410** (2001), 192–194.

15. M. A. Green, J. Zhao, A. Wang, P. J. Reece, and M. Gal, Efficient silicon light-emitting diodes, *Nature*, **412** (2001), 805–808.

16. S. Pillai, K. R. Catchpole, T. Trupke, G. Zhang, J. Zhao, and M. A. Green, Enhanced emission from Si-based light-emitting diodes using surface plasmons, *Appl. Phys. Lett.*, **88** (2006), 161102-1-3.

17. L. T. Canham, Silicon quantum wire array fabrication by electrochemical and chemical dissolution of wafers, *Appl. Phys. Lett.*, **57** (1990), 1046–1048.

18. L. Pavesi and D. Lockwood, *Silicon Photonics*, vol. 94 of Topics in Applied Physics, Springer, Berlin Germany (2004).

19. P. F. Trwoga, A. J. Kenyon, and C. W. Pitt, DC electroluminescence from PECVD grown thin films of silicon-rich silica, *Electron. Lett.*, **32** (1996), 1703–1704.

20. S. Fujita and N. Sugiyama, Visible light-emitting devices with Schottky contacts on an ultra-thin amorphous silicon layer containing silicon nanocrystals, *Appl. Phys. Lett.*, **74**, 308–310.

21. L. Dal Negro, R. L. Li, J. Warga, S. Yerci, S. N. Basu, S. Hamel, and G. Galli, *Light Emission from Silicon-Rich Nitride Nanostructures*, World Scientific Publishing Co. Pte. Ltd., Singapore (2009).

22. R. J. Walters, G. I. Bourianoff, and H. A. Atwater, Field-effect electroluminescence in silicon nanocrystals, *Nat. Mater.*, **4** (2005), 143–146.

23. M. Perálvarez, J. Barreto, Josep Carreras, A. Morales, D. Navarro-Urrios, Y. Lebour, C. Dominguez, and B. Garrido, Si nanocrystal-based LEDs fabricated by ion implantation and plasma-enhanced chemical vapour deposition, *Nanotechnology*, **20** (2009), 405201.
24. L. Rebohle, J. Lehmann, A. Kanjilal, S. Prucnal, A. Nazarov, I. Tyagulskii, W. Skorupa, and M. Helm. The correlation between electroluminescence properties and the microstructure of Europium-implanted MOS light emitting diodes, *Nuclear Instrum. Methods Phys. Res. Section B: Beam Interact. Mater. Atoms*, **267** (2009), 1324.
25. L. Rebohle, C. Cherkouk, S. Prucnal, M. Helm, and W. Skorupa, Rare-earth implanted Si-based light emitters and their use for smart biosensor applications, *Vacuum*, **83** (2009), S24–S28.
26. J.-S. Kim, S. C. Yang, and B.-S. Bae, Thermally stable transparent sol–gel based siloxane hybrid material with high refractive index for light emitting diode (LED) encapsulation, *Chem. Mater*, **22**(11) (2010), 3549–3555.
27. CREE. Direct Attach DA1000 LEDs and DA2432. Data-Sheet. 2014. http://www.cree.com/LED-Chips-and-Materials/Chips/Chips/Direct-Attach/DA.
28. www.hamamatsu.com. "Mini-spectrometer MS Series". 2014.

Chapter 13

Solid-State Refrigeration Based on Caloric Effects

Seda Aksoy

Department of Physics Engineering, Istanbul Technical University, Maslak, Istanbul 34469, Turkey

eaksoy@itu.edu.tr

Today, most cooling systems around room temperature are based on gas-compression technology, and they still use greenhouse gases, i.e., CFCs (chlorofluorocarbons) and HCFCs (hydrochlorofluorocarbons). These gases are believed to be responsible for ozone layer depletion and global climate change. Increasing living standards and extensive use of refrigeration in industry play an important role on energy consumption, consumption of non-renewable energy resources and other environmental issues such as noise, hazardous materials, and recyclability. There is a great deal of interest to develop a more sustainable technological solution based on an alternative cooling technology both in the ambient temperature and in cryogenic temperatures. The solid-state cooling technology based on caloric effects can rapidly become competitive with the conventional one. The caloric effects are observed in materials as an adiabatic temperature change or an isothermal entropy change under an

---

Materials for Sustainable Energy Applications: Conversion, Storage, Transmission, and Consumption
Edited by Xavier Moya and David Muñoz-Rojas
Copyright © 2016 Pan Stanford Publishing Pte. Ltd.
ISBN 978-981-4411-81-3 (Hardcover), 978-981-4411-82-0 (eBook)
www.panstanford.com

external field. According to the type of external field, it is called with a specific name such as the magnetocaloric, the mechanocaloric and the electrocaloric effect. In this chapter, caloric effects and caloric materials will be discussed. The general overview will be given about the cooling technology and the prototype refrigerators. In this research field, room-temperature solid-state cooling is the main interest due to the impact on energy saving and environmental concerns.

13.1 Magnetocaloric Effect

The magnetocaloric effect (MCE) is defined as heating or cooling of a magnetic material due to the external magnetic field. When a magnetic material is exposed to a magnetic field, randomly oriented magnetic spins align to the magnetic field direction, and the magnetic entropy of the system reduces. If the process is performed in adiabatic conditions, the total entropy of the system must remain constant, and the reduction in the magnetic entropy causes an increase in the lattice entropy of the material. Consequently, the atomic vibrations increase in the material, and the temperature of the material increases. If the applied magnetic field is removed under adiabatic conditions, all spins go back to its randomized state by thermal agitation energy. This causes a decrease in the magnetic entropy and in the temperature of the material. The MCE is therefore represented by either the variation of magnetic entropy of the magnetic subsystem ΔS, or the temperature variation under the effect of a magnetic field ΔT_{ad}.

13.1.1 Theory of the MCE

The total entropy (S_T) of a magnetic material at constant pressure is the sum of the magnetic entropy of the material (S_m), lattice entropy caused by the vibration of crystal lattice (S_l), and the electronic entropy of the free electrons of the material (S_e). The S_l and the S_e are independent from the magnetic field and only depend on temperature. However, the S_m strongly depends on both the temperature and the magnetic field. So that, the total entropy change under an external magnetic field is equal to the magnetic entropy change, $\Delta S = \Delta S_m$.

When a material is exposed to a magnetic field, the adiabatic temperature change (ΔT_{ad}) and the isothermal entropy change (ΔS) are called magnetocaloric potentials. These potentials have high values around phase transition region. Phase transitions are classified simply by the order of the derivative of the free energy with respect to temperature or other external parameters such as pressure and magnetic field. In a first-order phase transition, a first derivative of free energy, which gives the values of physical properties of the system such as volume, entropy, and magnetic moment, becomes discontinuous at the transition temperature. One of the characteristics is a sudden jump in magnetization or in entropy. In a second-order phase transition, the first derivative of the free energy is a continuous function, but the second derivative of free energy, which gives properties such as specific heat, compressibility, and magnetic susceptibility of the system, becomes discontinuous at the transition temperature. Therefore, there is no jump in magnetization or entropy in a second-order phase transition.

According to the principle of thermodynamics, the internal energy of a ferromagnetic system under a magnetic field can be expressed as, $dU = TdS - pdV + \mu_0 HdM$, where H is the intensity of the magnetic field, p is the pressure, V is the volume of the sample, and μ_0 is the magnetic permeability of the vacuum. If there is no volume change, dU can be expressed as $dU = TdS + \mu_0 HdM$. The full differential of the total entropy in a closed system depends on the temperature T, the pressure p, and the external field H. At constant pressure and temperature, the total entropy changes only with magnetic field so that

$$dS(H)_{p,T} = \mu_0 \left(\frac{\partial S}{\partial H}\right) dH. \tag{13.1}$$

The relation between the temperature derivative of the magnetization and the field derivative of the entropy is given by the Maxwell relation:

$$\mu_0 \left(\frac{\partial M}{\partial T}\right)_H = \left(\frac{\partial S}{\partial H}\right)_T. \tag{13.2}$$

The integral of Eq. 13.2 for isothermal (and isobaric) entropy change will be

$$\Delta S = \Delta S_m = \mu_0 \int_{H_1}^{H_2} \left(\frac{\partial M}{\partial T}\right)_H dH. \qquad (13.3)$$

For a magnetic material that has a field-induced first-order transition at a certain value of magnetic field H, the Gibbs energies of the magnetic phases should be equal, so that the magnetic entropy change can be calculated using the magnetic Clausius–Clapeyron equation as

$$\left(\frac{\mu_0 \Delta H}{\Delta T}\right) = -\left(\frac{\Delta S}{\Delta M}\right) \qquad (13.4)$$

Under adiabatic and isobaric conditions ($dS = dp = 0$), the heat capacity is written as $C = dQ/dT = T(dS/dT)$, and the temperature change will be

$$dT = -\mu_0 \left(\frac{T}{C}\right)\left(\frac{dM}{dT}\right)_T dH. \qquad (13.5)$$

By integrating Eq. 13.5, the adiabatic temperature change ($\Delta T_{ad} = T_1 - T_0$) can be calculated using the following relation:

$$\Delta T_{ad} = -\mu_0 \int_{H_1}^{H_2} \left(\frac{T}{C}\right)\left(\frac{dM}{dT}\right)_T dH \qquad (13.6)$$

Schematic diagrams of temperature dependence of the entropy for a magnetic system are given in Fig. 13.1. For a second-order phase transition (Fig. 13.1a), the entropy shows a continuous function. In the magnetic system, the magnetic spins order with increasing the external magnetic field and the magnetic entropy is decreases. When the temperature is kept constant, the isothermal magnetic entropy change is $\Delta S_m = S_1 - S_0$, and when the entropy is kept constant, the adiabatic temperature change is $\Delta T_{ad} = T_1 - T_0$.

In general, in first-order phase transitions, the entropy is not a continuous function. There is a jump at the transition temperature in the entropy, because of the presence of latent heat resulting from transition enthalpy change [1, 2]. Figure 13.1b shows schematically entropy as a function of temperature $S(T)$ for a first-order transition under $H_0 = 0$, and a larger value of H_1, where an application of a magnetic field shifts the transition temperature to higher values. T_0 and T_1 are indicated as the transition temperatures at H_0 and H_1, respectively. The entropy change associated with the first-order transition brings an extra contribution to MCE and this effect is called giant magnetocaloric effect (GMCE). As a result, the GMCE shows much larger magnetic entropy change compared with those of the conventional MCE. When the temperature shift is relatively small with respect to the jump in the entropy, one can obtain a giant ΔS, but a small ΔT, and the other way around [3].

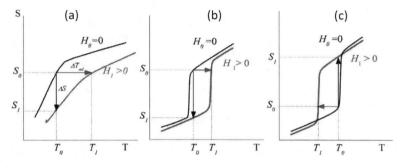

Figure 13.1 Schematic presentation of temperature dependence of the total entropy of a ferromagnetic material in zero-field and under an applied field for (a) the second-order transition, (b) the first-order transition and (c) the first-order transitions associated with the inverse MCE.

Figure 13.1c shows schematically $S(T)$ for H_0 and H_1, where applying a magnetic field shifts the first-order transformation temperature to lower temperatures, which gives rise to the inverse MCE. T_0 and T_1 are indicated as the transition temperatures at H_0 and H_1. A similar $S(T)$ curve is observed in the martensitic transformation. When a magnetic field is applied at T_0, entropy increases from S_0 to S_1. On the other hand, when a magnetic field is applied adiabatically at T_0, temperature decreases from T_0 to T_1. This means that the magnetic field can cause the material to release heat so that $\Delta S > 0$ and $\Delta T_{ad} < 0$. This is known as the inverse MCE.

The MCE can be determined experimentally by direct or indirect methods. One is the direct measurement of initial (T_0) and final temperatures (T_1) of the material, when the external magnetic field is changed from an initial H_0 value to a final value of H_1. Then, the adiabatic temperature change is simply found by

$\Delta T_{ad} = T_1 - T_0$. This measurement can be performed in an adiabatic calorimeter or simply using thermal sensors that are directly connected to the material [4]. The magnetic field change should be rapid, so the sample moves in and out of a constant magnetic field, or the field is accomplished by a pulsed or ramped magnetic field on fixed samples [5]. Permanent magnets (up to 2 T) or superconducting magnets (up to 10 T) are used. Considering the error of the reading of temperature sensors or the quality of the insulation of the sample the accuracy of the direct methods is claimed to be within the 5–10% range [6–8].

Indirect methods are based on determination of both ΔS and ΔT from magnetization and heat capacity measurements. This method is quite often used to investigate MCEs. The magnetization is normally measured as a function of H and at constant T. This allows to obtain ΔS by numerical integration of Eq. 13.3. The accuracy of the calculated ΔS measurements depends on the accuracy of the measurements of the magnetic moment, temperature T and magnetic field H. The error lies within the range of 3–10% [3, 9]. Equation 13.3 can only be used to calculate the ΔS of a system that has a second-order magnetic phase transition. Indeed, for a first-order transition, the derivative $\partial M/\partial T$ is infinite at the transition temperature. However, in real material, these values are usually finite, and even for a first-order transition Eq. 13.3 can be used to calculate for the MCE. It is necessary to know that in such first-order transition system, the calculated ΔS is usually higher than the real value. In a transition region, which the applied field increases ($\Delta H > 0$), the MCE sign is given by the sign of $\partial M/\partial T$. In direct MCE, when $\partial M/\partial T < 0$, the resulting $\Delta S < 0$ and $\Delta T > 0$, and in the inverse MCE, when $\partial M/\partial T > 0$, then $\Delta S > 0$ and $\Delta T < 0$. In general, the change in magnetization with temperature $|\partial M/\partial T|$ is large around the transition and this causes a large MCE. It has been reported that the upper limit of ΔT cannot exceed ~18 K under an applied field of 1 T [10].

The MCE can also be calculated indirectly from heat capacity as a function of temperature at constant magnetic field and pressure $C(T)_{p,H}$. The advantage of this indirect method is that it allows determining various parameters that are required to complete the characterization of the magnetocaloric properties in a magnetic material, such as C, S, ΔS and ΔT. The total entropy $S(T, H)$ of a material can be calculated from heat capacity $C(T)$:

$$S(T, H) = \int_0^T \left(\frac{C(T)}{T}\right) dT + S_0, \tag{13.7}$$

where S_0 is the entropy at 0 K, and it is usually assumed to be zero. Hence, if $S(T, H_0)$ and $S(T, H_1)$ are known, both ΔT_{ad} and ΔS can be obtained. It is also possible to calculate ΔT_{ad} from the heat capacity and magnetization measurements using Eq. 13.6. The accuracy of the ΔT_{ad} and ΔS is related to the accuracy of the heat capacity measurements, which is assumed not to be field-dependent, and the relative error decreases in both values for higher field changes ΔH. More detailed information about the MCE can be obtained from reference [11].

13.1.2 Magnetocaloric Materials

13.1.2.1 Rare earth (lanthanide) elements and their compounds

Early research on the MCE was carried out in paramagnetic salts to reach ultra-low temperature. Giauque won the 1949 Nobel prize in chemistry with the work on paramagnetic gadolinium sulfate $Gd_2(SO_4)_3 \cdot 8H_2O$, and reached a temperature of 0.25 K from the initial temperature of 1.5 K [12]. The low-temperature MCE was reported also for other paramagnetic salts, including ferric ammonium alum, chromic potassium alum, and cerous magnesium nitrate [6]. Paramagnetic intermetallic compounds were used in adiabatic demagnetization application and $PrNi_5$ is still one of the most studied materials in nuclear adiabatic demagnetization devices. Soft ferromagnetic materials are suitable for application of helium and nitrogen liquefaction in temperatures from 4 to 77 K. The largest MCE is observed in a range of 10 K–80 K temperature range in lanthanide metals [6]. In Fig. 13.2a, $\Delta T/\Delta H(T)$ is shown for selected compounds. At temperatures lower than 100 K, Dy and Gd-based compounds are one of the best magnetic refrigerant materials however; this temperature range is not suitable for many applications [13, 14]. Figure 13.2b shows $-\Delta S/\Delta H(T)$ for rare earth-based materials as a function of transition temperature.

Gd is one of the most popular magnetocaloric materials, and it has been used in most of the magnetic refrigerator prototypes. Around the ferromagnetic-paramagnetic transition temperature T_C = 294 K, the adiabatic temperature change is 6 K (2 K) under a magnetic field change of 5 T (2 T) [7, 13]. It is reported that Gd should have very high purity, because impurities (e.g., carbon, nitrogen and oxygen) lower the MCE nearly 2.5 times. Binary compounds of Gd-R (R is a lanthanide metal such as Tb, Dy, and Er) were investigated to improve magnetocaloric properties of Gd; however, substitutions only changed the T_C slightly without any improvement in its MCE. Only Gd_5Si_4 compound displayed a MCE as large as that of Gd at T_C = 335 K. A summary of magnetocaloric rare earth materials is given by Gschneidner et al. [24].

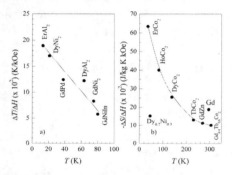

Figure 13.2 (a) The adiabatic temperature change per an applied field change (ΔH) as a function of transition temperature for intermetallic materials in the temperatures 10 K < T < 80 K. (b) The entropy change per an applied field change as a function of transition temperature for rare earth-based compounds. Solid lines are drawn to guide the eye. The data is taken from Refs. [6, 15–23].

In 1997, a GMCE was found in $Gd_5(Si_xGe_{1-x})_4$ alloys with $0 \leq x \leq 0.5$ [25]. It displays larger ΔS than those of Gd as a result of coupled second-order magnetic transition and first-order structural transition from paramagnetic monoclinic $Gd_5Si_2Ge_2$-type structure to ferromagnetic orthorhombic Gd_5Si_4-type structure (or from antiferromagnetic Sm_5Ge_4-type structure to the ferromagnetic Gd_5Si_4-type structure). The alloys show a number of other features such as a colossal magnetostriction and a giant magnetoresistance. The first-order magnetic phase transition temperature in $Gd_5(Si_xGe_{1-x})_4$ strongly depends on the molar ratio of Si and Ge. At T = 276 K, ΔS = –18.5 J/kg K under a 5 T magnetic field change, and it is possible to tune the transition temperature by changing Si concentration x from 20 K to room temperature.

In the low-temperature range (T = 40 K), ΔS = −26 J/kg K, and in the intermediate temperature range (T = 145 K), ΔS = −68 J/kg K [26]. The adiabatic temperature change for a 5 T magnetic field change was reported at T = 276 K to be ΔT_{ad} = 15.2 K and at T = 70 K, ΔT_{ad} = 15 K [27]. The disadvantages of this system are the existence of thermal hysteresis and high cost of Gd. When Gd is exchanged with other rare earth metals $R_5(Si_xGe_{1-x})_4$ (R = lanthanides), a first-order transition still exists, but the transition temperature is observed below room temperature [28–30].

Heavy lanthanide metals and their compounds, with high magnetic moments, are also potential candidates for magnetocaloric applications. It was found that $La(Fe_xSi_{1-x})_{13}$ series shows a first-order field-induced transition about $T \sim$ 200 K [31, 32]. In order to shift the transition temperature to room-temperature Co or H are introduced, and a GMCE is maintained. Co addition significantly increases T_C, but decreases ΔS. By hydrogenation ($La(Fe_xSi_{1-x})_{13}H_y$), it was successfully achieved the tuning of T_C to room temperature with its GMCE character [33–35]. In addition, the thermal conductivity of $La(Fe_xSi_{1-x})_{13}H_y$ alloy is higher than other magnetocaloric material.

13.1.2.2 3d-transition metal compounds and manganites

MnAs is another magnetocaloric system that exhibits a first-order magnetostructural phase transition from a ferromagnetic hexagonal NiAs-type structure to a paramagnetic orthorhombic MnP-type structure at 318 K. A giant magnetocaloric effect is observed at the transition temperature, with ΔS = −30 J/kg K and ΔT_{ad} = 13 K for 5 T magnetic field change. Nevertheless, MnAs is not suitable for application due to the large thermal hysteresis associated with the transition [36]. To solve this problem, Sb is substituted for As in $Mn(As_xSb_{1-x})$. It makes T_C tunable from 317 to 225 K without changing magnetocaloric properties around room temperature [37]. The addition of Fe and P in $MnFe(P_{1-x}As_x)$ alloys keeps also the magnetocaloric properties near room temperature. These transition metal–based alloys are cheaper than rare earth–based alloys but have a serious disadvantage due to the high toxicity of As [38–40]. Therefore, various substitutions have been done to replace As by other elements such as Ge. In $Mn_{1.1}Fe_{0.9}(P_{1-x}Ge_x)$, a GMCE has been reported with a ΔS = −74 J/kg K for 5 T magnetic field [41].

The intermetallic binary Fe-Rh is one of the first materials shows a negative GMCE in first-order transition from ferromagnetic to antiferromagnetic phase at T_N = 311 K. A negative GMCE was observed with ΔS = 12.4 J/kg K and ΔT_{ad} = −11 K in a 5 T magnetic field change. The disadvantage of Fe-Rh for any kind of application is the high cost of Rh that makes this alloy an unsuitable material [42].

The Ni-Mn–based Heusler alloys Ni-Mn-Z (Z = Ga, Sn, In, Sb) are ferromagnetic alloys that undergo a reversible first-order martensitic transition from a cubic austenite phase to a martensite phase. The martensite phase can be a non-modulated tetragonal structure or various modulated structures such as monoclinic (10 M and 14 M) and orthorhombic (10O and 4O). The structure strongly depends on the valance electron concentration and when it increases the structure develops most of time in a sequence of 10 M, 14 M and tetragonal $L1_0$ [43]. The investigations in Ni_2MnGa led to the discovery of a large MCE that has a positive ΔS with a value of 107 J/kg K [44]. A largest MCE ΔS = 20.7 J/kg K is reported in $Ni_{2.18}Mn_{0.82}Ga$ at a magnetic field change of 2 T [45]. It is suggested to be a good magnetic refrigerant material between 300 K and 350 K temperature range. In particular, $Ni_{50}Mn_{34}In_{16}$ [46–48], $Ni_{50}Mn_{50-x}Sn_x$ (x = 15, 16) [49], $Ni_{45}Co_5Mn_{37}In_{13}$ [50] and $Ni_{50}Mn_{36}Sb_{14}$ [51–53] undergo a martensitic transformation around 250 K and also have a large inverse MCE below room temperature. The reason of the inverse MCE is the appearance of strong AFM correlations in the martensite phase [54]. However, in addition to the inverse MCE, the conventional MCE is also observed at the T_C of the austenite phase, and these two features make these alloys interesting for multifunctional applications. Today, ΔT_{ad} is still small in Heusler alloys compared to the other prototype materials.

Another class of materials, manganites, also displays the MCE. They are perovskite-type $LaMnO_3$ materials whose general formula is $A_{1-x}B_xMnO_3$ (where A is for rare earth elements such as La, Pr, Nd, Sm, Eu, Gd, Ho, Tb, and B is for Na, K, Ag, Sr, Ca, Ba, and Pb). These manganites have the advantages of low cost, good chemical stability, large electrical resistivity. However, manganites show much smaller ΔT than Gd.

Since the magnetic moment in rare earths is larger than in $3d$ transition materials, a large MCE has been searched extensively in rare earth metals or lanthanides ($4f$ metals) and their alloys rather than $3d$ transition metals. Most of the research has been focused on materials showing a higher MCE near room temperature. However, there are a number of other factors to be considered from the practical point of view [24, 55], as materials should have some properties before they are used in a commercial magnetic refrigerator. These properties can be listed as follows: (a) having a Curie temperature near room temperature, (b) a large ΔT_{ad} in the vicinity of phase transition, (c) negligible thermal or magnetic hysteresis to enable high operation frequency, (d) large cooling power, (e) low specific heat and high thermal conductivity, (f) high electrical resistance to avoid eddy current loss, (g) non-toxic, (h) resistance to corrosion, (i) good mechanical properties, (j) low cost, (k) a wide operating temperature range, and (l) simple sample preparation process for mass production [56].

Table 13.1 Selected giant magnetocaloric materials with values of ΔS and ΔT_{ad} in magnetic fields change of ΔH

Material	T (K)	ΔS (J/kg K)	ΔT_{ad} (K)	ΔH (kOe)	Refs.
$Fe_{0.49}Rh_{0.51}$	313	22	−13	20	[57, 58]
$Gd_5Ge_2Si_2$	276	−18.5	15	50	[25]
$Tb_5Ge_2Si_2$	110	−21.8	5.4	50	[24, 59]
$ErCo_2$	36	−31.7	7.2	50	[11]
$LaFe_{11.7}Si_{1.3}$	188	−29	4	14	[60]
$LaFe_{11.57}Si_{1.43}H_{1.3}$	291	−24	6.9	20	[33]
$La_{0.6}Ca_{0.4}MnO_3$	260	−5	2.1	30	[61]
MnAs	312	−32	13	50	[36]
$MnAs_{0.9}Sb_{0.1}$	283	−30	—	50	[36]
$MnFeP_{0.45}As_{0.55}$	308	−18	—	50	[62]
$Ni_{52.6}Mn_{23.1}Ga_{24.3}$	297	−18	—	50	[63]
$Ni_{50}Mn_{34}In_{16}$	190	12	−2	50	[46, 64]
$Ni_{50}Mn_{37}Sn_{13}$	300	20	—	50	[49, 65]
$Ni_{45}Co_5Mn_{36.6}In_{13.6}$	292	15.2	—	20	[50]
$Ni_{45.2}Co_{5.1}Mn_{36.7}In_{13}$	317	19	6.2	20	[66]
$Ni_2Mn_{0.75}Cu_{0.25}Ga$	308	−64	—	50	[67]
$Mn_{1.96}Cr_{0.05}Sb$	198	7	—	50	[62]
$MnCoGeB_{0.02}$	287	−47.2	—	50	[68]

13.1.3 Refrigeration Technology

In 1881, irreversible heating under magnetic field was discovered below the Curie point of Fe [69]. The magnetocaloric effect was reported for the first time in the vicinity of the Curie temperature of Ni [70]. However, adiabatic demagnetization has been used for many years in research to reach low temperatures in the mK range. It is a magnetic cooling technology, and it is based on MCE. The fundamental principle of the phenomena was introduced by Debye (1926) [71] and Giauque (1927) [72]. The first experimental demonstration was carried out in 1933 by Giauque and MacDougall [12]. In 1976, Brown reported a prototype of a room-temperature magnetic refrigerator [73]. In 2001, another room-temperature magnetic refrigerator has been also built for commercial purposes. The current interest in refrigeration technology is to use magnetic materials with large MCE and to obtain an efficient method of cooling from room temperature. The magnetic cooling offers an alternative energy-efficient and environmentally friendly technology to the conventional refrigeration technology.

A magnetic refrigerator (MR) contains a magnetic working material, a magnet system to magnetize/demagnetize; hot/cold heat exchangers and a heat transfer fluid (e.g., water, antifreeze liquid, water-alcohol mixture, ferrofluids or some gases). The magnetic working material, so-called refrigerant, is a magnetocaloric material and its temperature changes under the magnetic field. The working principle of the refrigerator is that the working material absorbs heat from the low-temperature load (CHEX) and releases heat at the high-temperature sink (HHEX).

Figure 13.3 illustrates schematically a comparison of a simple conventional refrigeration cycle and a simple magnetic refrigeration cycle. The four basic stages are shown for a conventional gas-compression refrigerator (GCR) system. These are compression of a gas (+P), extraction of heat (–Q), expansion of the gas (–P) and absorption of heat (+Q). For magnetic refrigeration (MR) stages, instead of compression of a gas, a magnetocaloric material is moved into a magnetic field (+H) and instead of expansion it is moved out of the field ($H = 0$).

In stage 1, in GCR, the refrigerant gas is compressed adiabatically in the compressor, and the gas heats up. This stage is called adiabatic magnetization in MR, and the application of external magnetic field (+H) causes alignment of spins. The magnetocaloric material heats up ($T + \Delta T$).

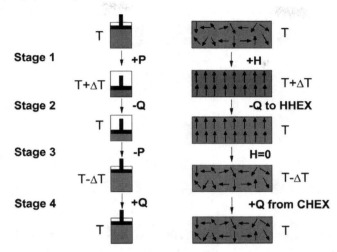

Figure 13.3 Schematic presentation of the gas-compression conventional refrigeration (left) and the magnetic refrigeration (right).

In stage 2, the heat is removed during isothermal compression in GCR. In MR, this is called isofield enthalpic transfer; the heat is removed by a coolant fluid flown from the CHEX to the HHEX.

In the next stage, called adiabatic demagnetization, the magnetic field is removed in adiabatic conditions so the total entropy remains constant. The spins become disoriented and the material cools down ($T - \Delta T$). In this case, the adiabatic temperature change can be defined as ΔT (in GCR it is called adiabatic expansion).

Isofield entropic transfer is the last stage of the MR cycle. The magnetic field is kept zero ($H = 0$). Since the material is colder than the refrigerated environment, heat (Q) is absorbed from the cold heat exchanger (CHEX). Once the material and the environment are in thermal equilibrium the cycle begins from the stage 1. Stage 4 is called, in GCR, isothermal expansion. Heat is absorbed in the evaporator.

One of the possible solid-state refrigeration cycles is the Carnot cycle. The cyclical repetition of the process makes the load cool. The magnetic cooling can also follow various thermodynamic cycles as shown in Fig. 13.4, such as Brayton cycle and Ericsson cycle [74].

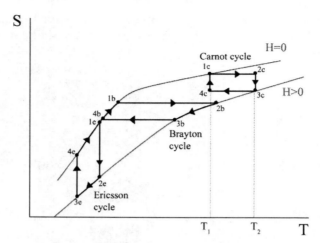

Figure 13.4 Schematic representation of total entropy as a function of temperature with three possible refrigeration cycles.

The Carnot cycle is a reference cycle consisting of two adiabatic and two isothermal processes. The refrigerant is partially magnetized (1c–2c), and its temperature increases from T_1 to T_2 adiabatically. Then, the intensity of the magnetic field increases isothermally. The heat transfer fluid absorbs the heat that is generated in the refrigerant due to the magnetization (2c–3c). In the process 3c–4c, the magnetic field decreases partially in adiabatic conditions and the temperature decreases from T_2 to T_1. Then the refrigerant is demagnetized completely (4c–1c), and it absorbs heat from the fluid to recover the energy lost during demagnetization. The ideal coefficient of performance (COP) can be calculated for Carnot cycle COP = $T_1/T_2 - T_1$.

The most common thermodynamic cycle for a MR is the Brayton cycle. This cycle consists of four processes; two adiabatic and two isofield (this cycle was described in the previous page). The temperature of the working material increases when the magnetic field is applied adiabatically (1b–2b). At constant magnetic field, the transfer fluid absorbs the heat from the material and its temperature drops (2b–3b). Then, the material is adiabatically demagnetized, and its temperature further decreases (3b–4b). When the material absorbs heat from the fluid, its temperature increases (4b–1b). The Ericsson cycle is similar to the Brayton cycle, and it consists of two isothermal and two isofield processes.

The material is isothermally magnetized/demagnetize (1e–2e and 3e–4e), and the material rejects/absorbs heat from the fluid (2e–3e and 4e–1e). Isofield processes in both of these cycles require heat regeneration to increase the temperature difference of the refrigerator. Thermal energy in the first regenerative cycles (2b–3b and 2e–3e) is absorbed by the working material during the fluid flow from CHEX to HHEX, and the absorbed thermal energy is returned to the fluid during the flow from HHEX to CHEX in the second regeneration cycles (4b–1b and 4e–1e). In the regeneration processes, heat exchange is directly related to the temperature difference between CHEX and HHEX. If the difference is very small, the efficiency of the cycle decreases dramatically.

There are several other types of thermodynamic cycles that can be applied to change the efficiency of an MR, to decrease the total cost, or to get the simplest MR design [74]. However, most research has focused on the active magnetic regenerator (AMR) cycle. Regenerator is used to transfer heat between different parts of the cycle. An AMR operates between two isofield and two adiabatic processes, as at the Brayton cycle. In Fig. 13.5, an AMR refrigeration cycle is illustrated. Firstly, the refrigerant (or regenerator) is adiabatically magnetized, and this causes an increase in temperature (Fig. 13.5a). A displacer provides heat transfer fluid flow back and forth through the regenerator. At constant magnetic field, the fluid is forced by displacer to flow through the regenerator from CHEX to HHEX. The rejected heat by the regenerator is absorbed by the fluid and it is released in the HHEX (Fig. 13.5b). The regenerator is adiabatically demagnetized, and its temperature decreases to a lower temperature below that of the CHEX (Fig. 13.5c). Finally, the fluid is forced by the displacer to flow from HHEX to CHEX. The hot fluid increases the temperature of the regenerator to the initial temperature while its temperature lowers than the temperature of CHEX. The fluid absorbs heat passing through the CHEX (Fig. 13.5d). The cycling causes cooling at CHEX, which is related to the refrigeration capacity and warming at HHEX. In AMR refrigerators, the magnetic working material not only acts as refrigerant providing temperature gradient under a magnetic field but also as a regenerator for heat transfer fluid. There are various designs with respect to improving efficiency such as a solid block with tube channels, parallel/perpendicular arranged plates to the heat flow direction or a packed bed of spherical particles [76].

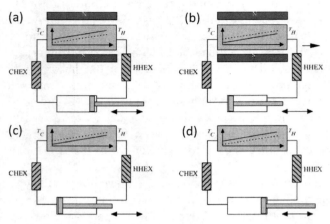

Figure 13.5 Schematic presentation of an AMR cycle. The dashed lines and the solid lines represent the initial and the final temperature profiles through the regenerator in each process.

The first MR prototype was built using Gd plate and a 7 T magnetic field by Brown [73]. In the mid 1990s, an AMR magnetic refrigerator was developed by Gschneidner et al [76]. In this system, the magnetic material rotates in and out of the strong magnetic field. Several prototypes of refrigerators with AMR cycle have been developed by various research groups since then [77]. Research is still continued with the aim of increasing the efficiency either in the magnetocaloric material, or in the design of the AMR or in the magnet system. For more detail on the current refrigeration system, see Gschneidner et al. [78].

13.2 Mechanocaloric Effect

Mechanocaloric effect is the mechanical analogue of the magnetocaloric effect. The entropy or the temperature of a solid can be changed by application of an external mechanical field. When it is induced by uniaxial stress (or a compressive stress), it is called the elastocaloric effect (eCE), additionally, when it is induced by a hydrostatic pressure, it is called the barocaloric effect (BCE). In eCE, the mechanical field is stress σ, and the corresponding strain (elongation along the field direction) is ε. Through the application of pressure, materials gain elastic energy, entropy decreases due to the elastic deformation, and materials heat up. Adiabatic release of the pressure leads to cool down.

The isothermal entropy change can be obtained from Maxwell equations, $(\partial S/\partial \sigma)_T = V(\partial \varepsilon /\partial T)_\sigma$, where V is the volume. The total entropy change, $\Delta S = S_2(T, H, p_2) - S_1(T, H, p_1)$, can be calculated as

$$\Delta S_{eCE} = V \int_{\sigma_1}^{\sigma_2} \left(\frac{\partial \varepsilon}{\partial T}\right) d\sigma. \tag{13.8}$$

The adiabatic temperature change can then be found as (assuming heat capacity to be stress independent):

$$dT_{eCE} = -\frac{T}{C} \Delta S_{eCE}. \tag{13.9}$$

In the BCE, the mechanical field is an external pressure p, and from the Maxwell relations $(\partial S/\partial p = -\partial V/\partial T)$, the entropy change ΔS_{BCE} is calculated as

$$\Delta S_{BCE} = -\int_{p_1}^{p_2} \left(\frac{\partial V}{\partial T}\right)_{p,H} dp \tag{13.10}$$

and ΔT_{ad}^{bar} can be found using Eq. 13.9 or from the direct measurement of the temperature under applied pressure, $\Delta T_{ad}^{bar} = T_{p_2} - T_{p_1}$.

13.2.1 Mechanocaloric Materials

The mechanocaloric effect has been reported in various rare earth–based compounds. For instance, in the metallic Kondo single crystal $Ce_3Pb_{20}Ge_6$, an applied uniaxial pressure change of $p = 0.3$ GPa along [111] direction causes nearly 0.75 K cooling at 4.4 K [79]. In CeSb, the adiabatic temperature change ΔT is -1.2 K for $p = 0.26$ GPa at 21 K. In $EuNi_2(Si_{0.15}Ge_{0.85})_2$, ΔT is -0.5 K for $p = 0.48$ GPa at 60 K [80]. However, in these alloys the applied pressure/stress does not yield changes in the magnetic part of the entropy, so that the observed ΔT_{ad} is relatively small. For some materials, it is possible to reach a large value of caloric effects, where magnetic and structural phase transitions (first-order phase transition) coexist, or where the pressure/stress can shift the transition temperature. $Pr_{1-x}La_xNiO_3$ exhibits a first-order structural phase transition from a high-temperature rhombohedral to a low-temperature orthorhombic phase. When $x = 0.34$, the transition occurs at around 360 K and the transition temperature shifts to lower temperatures under applied hydrostatic pressure by 5 K/kbar. At 300 K, the adiabatic temperature change is $\Delta T = 2$ K for $p = 15$ kbar [81]. For $La_{0.7}Pb_{0.3}MnO_3$ single crystal, which undergoes a ferromagnetic phase transition at 338 K, the barocaloric potentials increase linearly with application of pressure, and the rate of ΔT and ΔS are 0.25 K/kbar, and 4.3 J/kg K·kbar, respectively [82].

RCo_2 (R = Er, Ho, and Dy), MnAs, $La(Fe,Si)_{13}H_x$ and $R_5Si_2Ge_2$ (R = Gd and Tb) alloys and Heusler alloys exhibit substantial changes in the vicinity of phase transitions [83–89]. In Fig. 13.6, the comparison of the elastocaloric and the magnetocaloric effects is given for $Ni_{52.6}Mn_{21.9}Ga_{24.2}Fe_{1.3}$ Heusler alloy in the vicinity of martensitic transition temperature $T = 324$ K. Application of stress increases transition temperature, and the absolute value of ΔS_{eCE} increases with a rate of 0.5 J/kg K MPa, which is very similar to the increase of ΔS under magnetic field [90]. In $Ni_{46-x}Cu_xMn_{43}Sn_{11}$ Heusler alloy, the maximum value of ΔS around the martensitic transformation is -1.7 J/kg K under an applied stress of 5.24 MPa for an external magnetic field of 0.5 T [91].

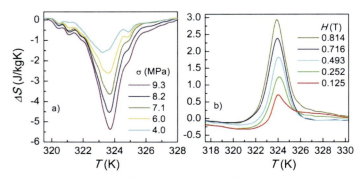

Figure 13.6 (a) Stress and (b) magnetic field–induced entropy changes as a function of temperature for $Ni_{52.6}Mn_{21.9}Ga_{24.2}Fe_{1.3}$. Reproduced with permission from ref. [90]. Copyright 2010, AIP Publishing LLC.

A giant barocaloric effect was reported in $Ni_{49.26}Mn_{36.08}In_{14.66}$ near the martensitic transformation temperature. The pressure-induced entropy change was found as ΔS = 24.4 J/kg K for p = 2.6 kbar [89]. The elastocaloric effect has been investigated in Cu-Zn-Al single crystal around the martensitic transformation. The stress-induced maximum entropy change remained almost constant at −1.2 J/mol K from 105 to 143 MPa. Besides, the strain-induced maximum entropy change is −1.2 J/mol K for $\Delta\varepsilon$ = 0.08 at 308 K [92]. In Fe-Rh alloy, a temperature change of $\Delta T \approx$ −5 K is obtained under a tensile stress of 526 MN/m^2 [93].

La-Fe-Co-Si shows a paramagnetic to ferromagnetic transition around room temperature accompanied by conventional magnetocaloric properties. However, a giant inverse barocaloric effect is observed for this material, as the sample warms up when pressure is adiabatically released, and the entropy decreases under isothermal conditions when pressure is applied. The pressure-induced entropy change is ΔS = 8.6 J/kg K for p = 2.1 kbar, and the adiabatic temperature change is ΔT = 1.5 K for p = 1 kbar at 240 K [94].

Materials whose transition temperature can be shifted by application of an external pressure are good candidates to study magnetocaloric and barocaloric effects upon variation of magnetic field and applied pressure. The giant magnetocaloric effect has been studied under hydrostatic pressure in $Gd_5Ge_2Si_2$. In Fig. 13.7, the entropy change is given as a function of temperature under different applied pressures. The rate of increase in T_C is around 6 K/kbar. The entropy change increases with increasing pressure, and the first-order transition disappears above 6 kbar applied pressure [87]. When the BCE is studied below 3 kbar applied pressure, the maximum pressure-induced entropy change is 13 J/kg K, and ΔT = −1 K (Δp = 2 kbar) [95]. However, ΔT = −6 K for a 2 kbar pressure variation was calculated by Oliveira et al. [96].

13.2.2 Mechanocaloric Refrigeration

Currently, most solid-state refrigeration prototypes are based on the MCE. However, a prototype of the pressure-induced solid-state refrigerator was illustrated in 2011 [97]. A barocaloric magnetic material is placed in a pressure cell, which is connected to the CHEX and HHEX through tubes where the heat exchange fluid flows. A fixed magnetic field perpendicular to the pressure direction is generated by a permanent magnet.

The material was chosen from a ferromagnetic material. Its magnetization around the magnetic ordering temperature decreases when the external pressure is applied. The refrigerator is working in a thermodynamic cycle, including two isobaric and two adiabatic processes, which is very similar to that of the working principle of a magnetic refrigerator. When an external pressure is applied adiabatically, the magnetization of the material, and consequently the temperature, decreases. Then, the heat exchange fluid passes through the pressure cell, absorbs heat and reduces the temperature in isobaric condition. When the pressure is adiabatically removed, the magnetization increases, and the temperature increases. The heat exchange fluid passes through the pressure cell and releases the heat to the magnetic material and the temperature increases to the initial temperature. The more suitable material for room-temperature cooling under hydrostatic pressure was proposed as MnAs-based alloys and Heusler alloys.

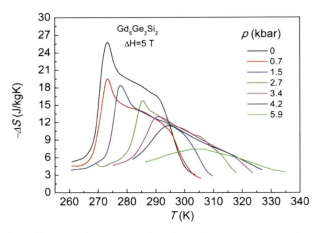

Figure 13.7 Entropy change as a function of temperature under various applied pressure for $Gd_5Ge_2Si_2$ at constant magnetic field of 5 T. Reproduced with permission from ref. [87]. Copyright 2005, AIP Publishing LLC.

13.3 Electrocaloric Effect

Electrocaloric effect is the electrical analogue of the magnetocaloric effect. When an electric field is applied to a dielectric (or ferroelectric) material, a change in the temperature of the material can be induced due to the change in the polarization. The electric field-induced temperature and entropy changes in a polarizable material are called the electrocaloric (EC) effect. In polarizable materials, the net polarization increases under an applied electric field. This causes a decrease in entropy of the system and, consequently, an increase in temperature. When the external field is removed, the entropy increases, and the temperature decreases.

In general, the EC entropy change can be derived from the pyroelectric coefficient $((\partial D/\partial T)_E)$ under a constant field through the Maxwell relation $((\partial D/\partial T)_E = (\partial S/\partial E)_T)$ as a function of the electric field:

$$\Delta S = -\int_{E_1}^{E_2} \left(\frac{\partial D}{\partial T}\right)_T dE, \tag{13.11}$$

where D is the electric displacement, $D = \varepsilon_0 E + P$, P is the polarization and ε_0 is the vacuum dielectric permittivity ($\varepsilon_0 = 8.85 \times 10^{-12}$ F/m). The adiabatic temperature change for a dielectric (and ferroelectric) material can be written as

$$\Delta T_{ad} = -\int_{E_1}^{E_2} \frac{T}{C_E} \left(\frac{\partial D}{\partial T}\right)_E dE, \tag{13.12}$$

where C_E is the specific heat under constant electric field. The EC effect can be written in terms of polarization as

$$\left(\frac{dT}{dE}\right)_S = -\frac{T}{\rho C_E} \left(\frac{\partial P}{\partial T}\right)_E, \tag{13.13}$$

where ρ is the density. Accordingly, to obtain a large ΔS and ΔT, the EC materials should have a large pyroelectric coefficient. For ferroelectric material, a large pyroelectric effect is observed near the ferroelectric-paraelectric transition temperature and above this transition temperature a large effect can be induced by the application of electric field. A large ΔT can be found with a large C_E, in addition, the EC material should exhibit high dielectric strength to resist the high electric fields. An upper limit of ΔT is suggested by Pirc et al. ($\Delta T = kT\ln\Omega/3\varepsilon_0\Theta C_E$ where k is the Boltzmann constant, Ω is the number of discrete equilibrium orientations of dipole entities, Θ is Curie constant), and most of the published EC materials show a ΔT, which is lower than this limit. However, for materials having giant EC effect the limit value is so small with respect to the experimental results [98].

The EC effect can be measured directly from the adiabatic temperature change ΔT, or indirectly from the change of the polarization with temperature at constant electric fields. The experimental calculation of the EC effect is very similar to the calculations of the MCE. The polarization hysteresis at different temperatures within a specific electric field range are numerically differentiated to obtain ΔS. For ΔT, the temperature dependence of heat capacity should be known. In direct methods, the temperature of the material can be measured with a temperature sensor or an adiabatic calorimeter on application/removal of the electric field [99].

13.3.1   Electrocaloric Materials

The EC effect was studied firstly on Rochelle salt in 1930; however, the temperature change was too small, ΔT = 0.0036°C under an electric field of 1.4 kV/cm [100]. After ferroelectric materials with a large polarization were discovered, many studies were reported on several bulk ceramics and single crystal materials with a small ΔT. In 2006, Mischenko et al. reported a giant adiabatic temperature change of 12°C (under 48 MV/m) in $Pb(Zr_{0.95}Ti_{0.05})O_3$ (PZT) ceramic thin film near the ferroelectric-paraelectric transition temperature (~222°C) [101]. After the discovery of the giant EC effect, research interest on suitable EC materials exponentially increased. In general, EC properties can be observed in bulk crystals or ceramics, ceramic thick and thin films and polymer films. In these materials when a higher electric field is applied, it is possible to observe higher ΔT. However, the maximum applied electric field is limited by the breakdown field of the material. In thin films and ferroelectric polymers (with large dielectric strength), the breakdown field is high enough to withstand the electric field. Therefore, ΔT increases with increasing applied electric field. For a PMN-PT ($0.65PbMg_{1/3}Nb_{2/3}O_3$-$0.35PbTiO_3$) thin film, ΔT = 31 K has been reported. This system has the highest temperature change strength, $\Delta T/\Delta E$ = 0.41 × 10^{-6} mK/V, in comparison to the other oxide thin films [102]. In single-crystal $BaTiO_3$, ΔT = 0.87°C was measured at paraelectric to ferroelectric transition temperature (T_C = 397 K) and a giant electrocaloric strength has been reported as $\Delta T/\Delta E$ = 2.2 × 10^{-6} mK/V [103]. On polycrystalline La-doped Pb(ZrTi)O_3 (PLZT) relaxor ceramic thin films with a thickness of 450 nm, ΔT was measured above 40°C and it is almost constant around the ferroelectric phase transition under an applied electric field change of 125 MV/m [104]. In lead-free thin films such as SrBi$_2$Ta$_2$O$_9$ the EC effect is reported as a 0.08 × 10^{-6} mK/V with a $\Delta T \sim$ 5 K.

EC effect has been investigated on β-formed polyvinylidene-fluoride-based copolymers (PVDF), which have a spontaneous polarization due to the dipole moment alignment. The temperature change of 21 K and the entropy change of 56 J/kgK has been reported under the electric field change of ΔE > 300 MV/m [105]. However, when extremely high fields are applied to polymers, the induced ΔT is similar the one induced in oxide thin films.

Recently, a negative electrocaloric effect, in which the sign of ΔT is reversed, has been measured in relaxor Na$_{0.5}$Bi$_{0.5}$TiO$_3$-BaTiO$_3$ (BNT-BT) ceramics [106]. The coexistence of positive and negative EC effect was measured in PMN-PT (70/30) single crystal [107]. Several thin films and ceramic materials exhibit EC properties around their Curie temperature, which is above room temperature. These materials can be used in heat pumping or power generation technology. However, the transition temperature can be shifted to room temperature by changing the composition of the material. Some of EC materials are listed in Table 13.2. For more information about electrocaloric effects and the EC materials, see Correira et al. and Moya et al. [121, 122].

13.3.2 Electrocaloric Refrigeration

EC-based cooling is an environmentally friendly alternative to the current gas-compression refrigeration. The main advantage over magnetocaloric cooling is that the generation of the high electric field, which is required for the cooling cycle, is easier and cheaper than the generation of the magnetic field, which is required in magnetic refrigeration. The principle of a solid-state EC refrigeration cycle goes through the same stages as the magnetocaloric refrigeration cycle given in the Fig. 13.4. The only difference is that the external field is an electric field E, instead of a magnetic field H. The thermal contacts between the cold and hot heat exchangers are provided either moving the working body or circulating the heat exchange fluid.

- Stage 1 is adiabatic polarization. When an external electric field is applied adiabatically, it causes alignment of dipole moments, and the working material heats up to $T + \Delta T$.
- Stage 2 is isofield enthalpic transfer, where the electric field is kept constant and the heat is removed.
- Stage 3 is adiabatic depolarization. The electric field is removed in this stage adiabatically, and the dipole moments orient randomly. Consequently, the working material temperature decreases.
- In the last stage, isofield entropic transfer, under the absence of electric field, the heat flows from the cold heat exchanger into the material.

Table 13.2 The maximum electrocaloric temperature change ΔT_{ad} at T for selected materials in an electric field change of ΔE

Material	T(°C)	ΔT_{ad}(°C)	ΔE (MV/m)	Ref.
Ceramic				
PMN-PT 85/15	18	1.71	1.6	[108]
$Pb_{0.99}Nb_{0.02}(Zr_{0.75}Sn_{0.20}Ti_{0.05})O_3$	163	2.6	2.5	[109]
Single Crystal				
PMN-PT 72/28	130	2.7	1.2	[110]
KH_2PO_4	−150	0.08	0.07	[111, 112]
$(NH_2CH_2COOH)_3 \cdot H_2SeO_4$	22	0.18	0.15	[113]
Thick film				
$(Co,Sb)-Pb(Sc_{0.5}Ta_{0.5})O_3$	18	3.5	13.5	[114]
$BaTiO_3$	80	1.8	17.6	[115]
PMN-PT 70/30	160	2.8	9	[116]
Thin film				
$PbZrO_3$	235	11.4	40	[117]
$Pb_{0.88}La_{0.08}(Zr_{0.65}Ti_{0.35})O_3$	45	40	125	[118]
PMN-PT 65/35	140	31	74.7	[102]
PMN-PT 67/33	145	14.5	60	[119]
$SrBi_2Ta_2O_3$	288	5	60	[120]
P(VDF-TrFE) 70/30	117	21	300	[105]

In Fig. 13.8, a thermodynamic refrigeration cycle, a Carnot cycle, based on EC effect is shown. A partially applied electric field (1c–2c) causes the temperature change due to the polarization of the material. Then, the heat is rejected to the HHEX while the entropy of the material decreases in the isothermal conditions (2c–3c). When the electric field is reduced, the polar ordering is lost in the material, and the temperature of the material decreases (3c–4c). To complete the cycle, the material absorbs heat from the cold sink (4c–1c).

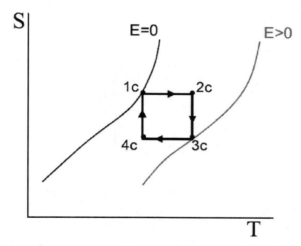

Figure 13.8 Schematic representation of a refrigeration cycle based on the EC effect.

Electrocaloric refrigeration represents a new alternative cooling technology, and it is promising likely the magnetocaloric cooling technology. In the design of the EC refrigerator (ECR), prototypes use thermal diodes (i.e., a thermal switch) or an active electrocaloric regenerator (AER). In the ECR with thermal diodes, a material or a cascade system of EC materials lies between diodes and the heat exchange fluid flows continuously through the micro heat exchangers [123]. Particularly, in new developed prototypes, AERs are preferred, and they have almost the same working principle as AMR (see Fig. 13.5 and consider an external field generated by electrodes instead of a magnet). Recently, a new ECR prototype has been reported that uses two AERs, and each of them operates inversely. The electric field is supplied by two voltage sources for each AER. When one is polarized (heats up), the other one is depolarized (cools down). The heat is rejected through the polarized AER by the working fluid from the CHEX to the HHEX. Simultaneously, the flowing fluid through the HHEX to the CHEX transfers the heat that is required to increase the temperature of the depolarized AER [124]. This kind of twin AER system increases the efficiency of the ECR.

13.4 Conclusion

In this chapter, the theory and thermodynamics of caloric effects, as well as the properties of caloric materials for cooling technology have been described briefly. Research reveals that the electrocaloric refrigerators can lead to similar or even better performance than magnetic refrigerators. Despite the promising results, electrocaloric refrigeration still requires more research to reach the level of magnetic refrigeration. Construction of energy efficient, environmentally friendly elastocaloric/barocaloric-based refrigeration is only at the first stage.

The most efficient commercial cooling units operate with 60% of Carnot efficiency, and almost no further improvement related to increase the efficiency is possible in the market. However, in the caloric effect–based solid-state refrigerators, it is possible to reach a 60–70% Carnot efficiency without using a compressor unit. On the other hand, prototypes based on caloric materials have not shown large efficiencies for large temperature ranges. Also, magnetocaloric cooling technology requires large magnetic fields and consequently large magnets. This problem can be solved when an electric field is considered as an external field for refrigeration.

When the giant effects (large values of ΔS and ΔT) are obtained in caloric materials that undergo a first-order phase transition, some disadvantages appears due to the nature of the phase transition, such as hysteresis losses. The materials should have no hysteresis to enable high operating frequencies and large cooling power. The main refrigerant materials used in room-temperature MR prototypes are so far Gd, Gd-based alloys or La(Fe,Si)$_{13}$H$_x$ alloys. One of the largest obstacles in magnetic refrigerators is the price of Gd. Therefore, further research in new alternative caloric materials is still desirable.

From the practical point of view, materials should have a large ΔT, low toxicity, sufficient thermal conductivity, long-term stability, low hysteresis in the vicinity of the transition temperature, etc. In most electrocaloric materials, it is important to improve dielectric strength, which results in higher EC properties. Research is focused on environmentally friendly materials, and the lowest possible total cost of the entire refrigeration system. On the other hand, the engineering design of refrigerators should be optimized to increase the efficiency.

References

1. Pecharsky, V. K., Gschneidner Jr., K. A., Pecharsky, A. O., and Tishin, A. M. (2001). Thermodynamics of the magnetocaloric effect, *Phys. Rev. B*, **64**, pp. 144406–144418.
2. Spichkin, Y. I., and Tishin, A. M. (2005). Magnetocaloric effect at the first-order magnetic phase transitions, *J. Alloy. Comp.*, **403**, pp. 38–44.
3. Pecharsky, V. K., and Gschneidner, Jr., K. A. (1999). Magnetocaloric effect from indirect measurements: Magnetization and heat capacity, *J. Appl. Phys.*, **86**, pp. 565–575.
4. Kurth, C., Schittny, T., and Bärner, K. (1985). Magnetic B–T phase diagram of anion substituted MnAs. Magnetocaloric experiments, *Phys. Stat. Sol. (a)*, **91**, pp. 105–113.
5. Ponomarev, B. K. (1986). Magnetic properties of gadolinium in the region of paraprocess, *J. Magn. Magn. Mater*, **61**, pp. 129–138.
6. Pecharsky, V. K., and Gschneidner, Jr. K. A. (1999). Magnetocaloric effect and magnetic refrigeration, *J. Magn. Magn. Mater.*, **200**, pp. 44–56.
7. Dan'kov, S. Y., Tishin, A. M., Pecharsky, V. K., and Gschneidner, Jr., K. A. (1997). Experimental device for studying the magnetocaloric effect in pulse magnetic field, *Rev. Sci. Instrum.*, **68**, pp. 2432–2437.
8. Gopal, B. R. Chahine, R., and Bose, T. K. (1997). A sample translatory type insert for automated magnetocaloric effect measurements, *Rev. Sci. Instrum.*, **68**, pp. 1818–1822.
9. Földeàki, M., Chahine, R., and Bose, T. K. (1995). Magnetic measurements: A powerful tool in magnetic refrigerator design, *J. Appl. Phys.*, **77**, pp. 3528–3537.

10. Zverev, V. I., Tishin, A. M., and Kuz'min, M. D. (2010). The maximum possible magnetocaloric ΔT effect, *J. Appl. Phys.*, **107**, p. 043907(3).
11. Tishin, A. M., and Spichkin, Y. I. (2003). *The Magnetocaloric Effect and Its Application* (Institute of Physics Publishing, UK).
12. Giauque, W. F., and MacDougall, D. P. (1933). Attainment of temperature below 1° absolute by demagnetization of $Gd_2(SO_4)_3 \cdot 8H_2O$, *Phys. Rev.*, **43**, p. 768.
13. Tishin, A. M., Gschneidner, Jr., K. A., and Pecharsky, V. K. (1999). Magnetocaloric effect and heat capacity in the phase transition region, *Phys. Rev. B*, **59**, pp. 503–511.
14. Benford, S. M. (1979). The magnetocaloric effect in dysprosium, *J. Appl. Phys.*, **50**, pp. 1868–1870.
15. Duc, N. H., Kim Anh, D. T., and Brommer, P. E. (2002). Metamagnetism, giant magnetoresistance and magnetocaloric effects in RCo_2-based compounds in the vicinity of the Curie temperature, *Phys. B Condens. Matter*, **319**, pp. 1–8.
16. Ranke, P. J. von, Nóbrega, E. P., Oliveira, I. G. de, Gomes, A. M., and Sarthour, R. S. (2001). Influence of the crystalline electrical field on the magnetocaloric effect in the series RNi_2 (R = Pr, Nd, Gd, Tb, Ho, Er), *Phys. Rev. B*, **63**, p. 184406.
17. Canepa, F., Napoletano, M., Palenzona, A., Merlo, F., and Cirafici, S. (1999). Magnetocaloric properties of GdNiGa and GdNiIn intermetallic compounds, *J. Phys. D:Appl. Phys.*, **32**, pp. 2721–2725.
18. Foldeaki, M., Giguère, A., Gopal, B. R., Chahine, R., Bose, T. K., Liu, X. Y., and Barclay, J. A. (1997). Composition dependence of magnetic properties in amorphous rare-earth-metal-based alloys, *J. Magn. Magn. Mater.*, **174**, pp. 295–308.
19. Zimm, C. B., Ludeman, E. M., Severson, M.C., and Henning, T. A. (1992). Materials for regenerative magnetic cooling spanning 20K to 80K, *Adv. Cryog. Eng.*, **37B**, p. 883.
20. Gschneidner, Jr. K. A., Pecharsky, V. K., Gailloux, M. J., and Takeya, H. (1996). Utilization of the magnetic entropy in active magnetic regenerator materials, *Adv. Cryog. Eng.*, **42**, p. 465.
21. Gschneidner, Jr. K. A., Pecharsky, V. K., and Malik, S. K. (1996). The $(Dy_{1-x}Er_x)Al_2$ alloys as active magnetic regenerators for magnetic refrigeration, *Adv. Cryog. Eng.*, **42**, p. 475.
22. Ranke, P. J. von, Pecharsky, V. K., and Gschneidner, Jr. K. A. (1998). Influence of the crystalline electrical field on the magnetocaloric effect of $DyAl_2$, $ErAl_2$ and $DyNi_2$, *Phys. Rev. B*, **58**, pp. 12110–12116.

23. Pecharsky, V. K., and Gschneidner, Jr. K. A. (1999). Gd-Zn alloys as active magnetic regenerator materials for magnetic refrigeration, *Cryocoolers 10*, p.629 (Kluwer Academic/Plenum Publishers).
24. Gschneidner, Jr., K. A., Pecharsky, V. K. and Tsokol, A. O. (2005). Recent developments in magnetocaloric materials, *Rep. Prog. Phys.*, **68**, pp. 1479–1539.
25. Pecharsky, V. K., and Gschneidner, Jr., K. A. (1997). Giant magnetocaloric effect in $Gd_5(Si_2Ge_2)$, *Phys. Rev. Lett.*, **78**, pp. 4494–4497.
26. Pecharsky, V. K., and Gschneidner, Jr., K. A. (1997). Tunable magnetic regenerator alloys with a giant magnetocaloric effect for magnetic refrigeration from 20 to 290 K, *Appl. Phys. Lett.*, **70**, pp. 3299–3301.
27. Pecharsky, V. K., and Gschneidner, Jr., K. A. (1998). The giant magnetocaloric effect in $Gd_5(Si_xGe_{1-x})_4$ materials for magnetic refrigeration, *Adv. Cryog. Eng.*, **43**, pp. 1729–1736.
28. Gschneidner, Jr., K. A., Pecharsky, V. K., Pecharsky, A. O., Ivtchenko, V. V., and Levin, E. M. (2000). The nonpareil $R_5(Si_xGe_{1-x})_4$ phases, *J. Alloys Comp.*, **303**, pp. 214–222.
29. Morellon, L., Magen, C., Algarabel, P. A., Ibarra, M. R., and Ritter, C. (2001). Magnetocaloric effect in $Tb_5(Si_xGe_{1-x})_4$, *Appl. Phys. Lett.*, **79**, pp. 1318–1320.
30. Ivtchenko, V. V., Pecharsky, V. K., and Gschneidner, Jr., V. K. (2000). Magnetothermal properties of $Dy_5(Si_xGe_{1-x})_4$ alloys, *Adv. Cryog. Eng.*, **46A**, pp. 405–412.
31. Fujita, A., Akamatsu, Y., and Fukamichi, K. (1999). Itinerat electron metamagnetic transition in $La(Fe_xSi_{1-x})_{13}$ intermetallic compounds, *J. Appl. Phys.*, **85**, pp. 4756–4758.
32. Fujita, A., Fujieda, S., Fukamichi, K., Mitamura, H., and Goto, T. (2001). Itinerant-electron metamagnetic transition and large magnetovolume effects in $La(Fe_xSi_{1-x})_{13}$ compounds, *Phys. Rev. B*, **65**, p. 014410(6).
33. Fujieda, S., Fujida, A., and Fukamichi, K. (2002). Large magnetocaloric effect in $La(Fe_xSi_{1-x})_{13}$ itinerant electron metamagnetic compounds, *Appl. Phys. Lett.*, **81**, pp. 1276–1278.
34. Fujita, A., Fujieda, S., Hasegawa, Y., and Fukamichi, K. (2003). Itinerant-electron metamagnetic transition and large magnetocaloric effects in $La(Fe_xSi_{1-x})_{13}$ compounds and their hydrides, *Phys. Rev. B*, **67**, p. 104416(12).
35. Hu, F. X., Shen, B. G., Sun, J. R., Wang, G. J., and Cheng, Z. H. (2002). Very large magnetic entropy change near room temperature in $LaFe_{11.2}Co_{0.7}Si_{1.1}$, *Appl. Phys. Lett.*, **80**, pp. 826–828.

36. Wada, H., and Tanabe, Y. (2001). Giant magnetocaloric effect of MnAs$_{1-x}$Sb$_x$, *Appl. Phys. Lett.*, **79**, pp. 3302–3304.

37. Wada, H., Morikawa, T., Taniguchi, K., Shibata, T., Yamada, Y., and Akishige, Y. (2003). Giant magnetocaloric effect of MnAs$_{1-x}$Sb$_x$ in the vicinity of first-order magnetic transition, *Phys. B Condens. Matt.*, **328**, pp. 114–116.

38. Zach, R., Guillot, M., and Fruchart, R. (1990). The influence of high magnetic fields on the first order magnetoelastic transition in MnFe(P$_{1-y}$As$_y$) systems, *J. Magn. Magn. Mater.*, **89**, pp. 221–228.

39. Bacmann, M., Soubeyroux, J. L., Barret, R., Fruchart, D., Zach, R., Niziol, S., and Fruchart, R. (1994). Magnetoelastic transition and antiferro-ferromagnetic ordering in the system MnFeP$_{1-y}$As$_y$, *J. Magn. Magn. Mater.*, **134**, pp. 59–67.

40. Tegus, O., Brück, E., Buschow, K. H. J., and de Boer, F. R. (2002). Transition-metal-based magnetic refrigerants for room-temperature applications, *Nature*, **415**, pp. 150–152.

41. Liu, D., Yue, M., Zhang, J., McQueen, T. M., Lynn, J. W., Wang, X., Chen, Y., Li, J., Cava, J., Liu, X., Altounian, Z., and Huang, Q. (2009). Origin and tuning of the magnetocaloric effect in the magnetic refrigerant Mn$_{1.1}$Fe$_{0.9}$(P$_{0.8}$Ge$_{0.2}$), *Phys. Rev. B.*, **79**, p. 014435(7).

42. Annaorazov, M. P., Nikitin, S. A., Tyurin, A. L., Asatryan, K. A., and Devletov, A. K. (1996). Anomalously high entropy change in FeRh alloy, *J. Appl. Phys.*, **79**, pp. 1689–1695.

43. Acet, M., Mañosa, L., and Planes, A. (2011). Magnetic field induced effects in martensitic Heusler-based magnetic shape memory alloys, *Handbook Magn. Mater.*, **19**, pp. 231–290.

44. Marcos, J., Planes, A., Mañosa, L. L., Casanova, F., Battle, X., Labarta, A., and Martinez, B. (2002). Magnetic field induced entropy change and magnetoelasticity in Ni-Mn-Ga alloys, *Phys. Rev. B*, **66**, p. 224413(6).

45. Cherechukin, A. A., Takagi, T., Matsumoto, M., and Buchelnikov, V. D. (2004). Magnetocaloric effect in Ni$_{2+x}$Mn$_{1-x}$Ga Heusler alloys, *Phys. Lett. A*, **326**, pp. 146–151.

46. Krenke, T., Duman, E., Acet, M., Wassermann, E. F., Moya, X., Mañosa, L. L., Planes, A., Suard, E., and Ouladdiaf, B. (2007). Magnetic superelasticity and inverse magnetocaloric effect in Ni-Mn-In, *Phys. Rev. B*, **75**, p. 104414(6).

47. Sharma, V. K., Chattopadhyay, M. K., and Roy, S. B. (2007). Large inverse magnetocaloric effect in Ni$_{50}$Mn$_{34}$In$_{16}$, *J. Phys. D: Appl. Phys.*, **40**, pp. 1869–1873.

48. Han, Z. D., Wang, D. H., Zhang, C. L., Tang, S. L., Gu, B. X., and Du Y. W. (2006). Large magnetic entropy changes in the $Ni_{45.4}Mn_{41.5}In_{13.1}$ ferromagnetic shape memory alloy, *Appl. Phys. Lett.*, **89**, p. 182507(3).

49. Krenke, T., Duman, E., Acet, M., Wassermann, E. F., Moya, X., Mañosa, L. L., and Planes, A. (2005). Inverse magnetocaloric effect in ferromagnetic Ni-Mn-Sn alloys, *Nat. Mater.*, **4**, pp. 450–454.

50. Kainuma, R., Imano, Y., Ito, W., Sutou, Y., Morito, H., Okamoto, S., Kitakami, O., Oikawa, K., Fujita, A., Kanomata, T., and Ishida, K. (2006). Magnetic-field-induced shape recovery by reverse phase transformation, *Nature*, **439**, pp. 957–960.

51. Du, J., Zheng, Q., Ren, W. J., Feng, W. J., Liu, X. G., and Zhang, Z. D. (2007). Magnetocaloric effect and magnetic-field-induced shape recovery effect at room temperature in ferromagnetic Heusler alloy Ni-Mn-Sb, *J. Phys. D: Appl. Phys.*, **40**, pp. 5523–5526.

52. Khan, M., Ali, N., and Stadler, S. (2007). Inverse magnetocaloric effect in ferromagnetic $Ni_{50}Mn_{37+x}Sb_{13-x}$ Heusler alloys, *J. Appl. Phys.*, **101**, p. 053919(3).

53. Aksoy, S., Acet, M., Eberhard, E. F., Krenke, T., Moya, X., Mañosa, L. L., Planes, A., and Deen P. P. (2009). Structural properties and magnetic interactions in martensitic Ni-Mn-Sb alloys, *Philos. Mag.*, **89**, pp. 2093–2109.

54. Aksoy, S., Acet, M., Deen, P. P., Mañosa, L. L., and Planes A. (2009). Magnetic correlations in martensitic Ni-Mn-based Heusler shape-memory alloys: Neutron polarization analysis, *Phys. Rev. B*, **79**, p. 212401(4).

55. Yu, B. F., Gao, Q., Zhang, B., Meng, X. Z., and Chen, Z. (2003). Review on research of room temperature magnetic refrigeration, *Int. J. Refrigeration*, **26**, pp. 622–636.

56. Romero Gómez J., Ferreiro Garcia R., De Miguel Catoina A., and Romero Gómez (2013). Magnetocaloric effect: A review of the thermodynamic cycles in magnetic refrigeration, *Renew. Sustain. Energy Rev.*, **17**, pp. 74–82.

57. Annaorazov, M. P., Asatryan, K. A., Myalikgulyev, G., Nikitin, S. A., Tishin, A. M., and Tyurin, A. L. (1992). Alloys of the Fe-Rh system as a new class of working material for magnetic refrigerators, *Cryogenics*, **32**, pp. 867–872.

58. Nikitin, S. A., Myalikgulyev, G., Tishin, A. M., Annaorazov, M. P., Astaryan, K. A., and Tyurin, A. L. (1990). The magnetocaloric effect in $Fe_{49}Rh_{51}$ compounds, *Phys. Lett. A*, **148**, pp. 363–366.

59. Morellon, L., Magen, C., Algarabel, P. A., Ibarra, M. R., and Ritter, C. (2001). Magnetocaloric effect in $Tb_5(Si_xGe_{1-x})_4$, *Appl. Phys. Lett.*, **79**, pp. 1318–1320.

60. Hu, F. X., Ilyn, M., Tishin, A. M., Sun, J. R., Wang, G. J., Chen, Y. F., Wang, F., Cheng, Z. H., and Shen, B. G. (2003). Direct measurements of magnetocaloric effect in the first-order system $LaFe_{11.7}Si_{1.3}$, *J. Appl. Phys.*, **93**, pp. 5503–5506.

61. Bohigas, X., Tejada, J., Marínez-Sarrión, M. L., Tripp, S., and Black, R. (2000). Magnetic and calorimetric measurements on the magnetocaloric effect in $La_{0.6}Ca_{0.4}MnO_3$, *J. Magn. Magn. Mater.*, **208**, pp. 85–92.

62. Tegus, O., Brück, E., Zhang, L., Dagula, Buschow, K. H. J., and de Boer, F. R. (2002). Magnetic-phase transitions and magnetocaloric effects, *Phys. B*, **319**, pp. 174–192.

63. Hu, F. X., Shen, B. G., Sun, J. R., and Wu, G. H. (2001). Large magnetic entropy change in a Heusler alloy $Ni_{52.6}Mn_{23.1}Ga_{24.3}$ single crystal, *Phys. Rev. B*, **64**, p. 132412(4).

64. Aksoy, S., Krenke, T., Acet, M., Wassermann, E. F., Moya, X., Mañosa, L. L., and Planes, A. (2007). Tailoring magnetic and magnetocaloric properties of martensitic transitions in ferromagnetic Heusler alloys, *Appl. Phys. Lett.*, **91**, p. 241916(3).

65. Planes, A., Mañosa, L. L., and Acet, M. (2009). Magnetocaloric effect and its relation to shape-memory properties in ferromagnetic Heusler alloys, *J. Phys. Condens. Matter.*, **21**, p. 233201(29).

66. Liu, J., Gottschall, T., Skokov, K. P., Moore, J. D., and Gutfleisch, O. (2012). Giant magnetocaloric effect driven by structural transitions, *Nat. Mater.*, **11**, pp. 620–626.

67. Khan, M., Dubenko, I., Stadler, S., and Ali, N. (2007). Phase transitions and corresponding magnetic entropy changes in $Ni_2Mn_{0.75}Cu_{0.25-x}Co_xGa$ Heusler alloys, *J. Appl. Phys.*, **102**, p. 023901(5).

68. Trung, N. T., Zhang, L., Caron, L., Buschow, K. H. J., and Brück, E. (2010). Giant magnetocaloric effects by tailoring the phase transitions, *Appl. Phys. Lett.*, **96**, p. 172504(3).

69. Warburg, E. (1881). Magnetische untersuchungen, *Ann. Phys. Chem.*, **13**, pp. 141–164 (in German).

70. Weiss, M. M. P., and Piccord A. (1918). Sur un nouveau phenomena magnétocalorique, *Compt. Rend.*, **166**, pp. 352–354 (in French).

71. Debye, P. (1926). Einige bemerkungen zur magnetisierung bei tiefer temperatur, *Ann. Phys.*, **386**, pp. 1154–1160 (in German).

72. Giauque, W. F. (1927). A thermodynamic treatment of certain magnetic effects. A proposed method of producing temperatures considerably below 1° absolute, *J. Amer. Chem. Soc.*, **49**, pp. 1864–1870.
73. Brown, G. V. (1976). Magnetic heat pumping near room temperature, *J. Appl. Phys.*, **47**, pp. 3673–3680.
74. Kitanovski, A., Plaznik, U., Tušek, J., and Poredoš, A. (2014). New thermodynamic cycles for magnetic refrigeration, *Int. J. Refrigeration*, **37**, pp. 28–35.
75. Hall, J. L., Reid, C. E., Spearing, I. G., and Barclay, J. A. (1996). Thermodynamic considerations for the design of active magnetic regenerative refrigerators, *Adv. Cryogenic Eng.*, **41B**, pp. 1653–1663.
76. Gschneidner, Jr., K. A., and Pecharsky, V. K. (1999). Magnetic refrigeration materials, *J. Appl. Phys.*, **85**, pp. 5365–5368.
77. Yu, B., Liu, M., Egolf, P. W., and Kitanovsky, A. (2010). A review of magnetic refrigerator and heat pump prototypes built before the year 2010, *Int. J. Refrigeration*, **33**, pp. 1029–1060.
78. Gschneidner, Jr., K. A., and Pecharsky, V. K. (2008). Thirty years of near room temperature magnetic cooling: Where we are today and future prospects, *Int. J. Refrigeration*, **31**, pp. 945–961.
79. Strässle, T., Furrer, A., Dönni, A., and Komatsubara, T. (2002). Barocaloric effect: The use of pressure for magnetic cooling in $Ce_3Pd_{20}Ge_6$, *J. Appl. Phys.*, **91**, pp. 8543–8545.
80. Strässle, T., Furrer, A., Hossain, Z., and Geibel, Ch. (2003). Magnetic cooling by the application of external pressure in rare-earth compounds, *Phys. Rev. B*, **67**, p. 054407(12).
81. Müller, K. A., Fauth, F., Fischer, S., Koch, M., Furrer, A., and Lacorre, P. (1998). Cooling by adiabatic pressure application in $Pr_{1-x}La_xNiO_3$, *Appl. Phys. Lett.*, **73**, pp. 1056–1058.
82. Kartashev, A. V., Mikhaleva, E. A., Gorev, M. V., Bogdanov, E. V., Cherepakhin, A. V., Sablina, K. A., Mikhashonok, N. V., Flerov, I. N., and Volkov, N. V. (2013). Thermal properties, magneto- and baro-caloric effects in $La_{0.7}Pb_{0.3}MnO_3$ single crystal, *J. Appl. Phys.*, **113**, p. 073901(6).
83. Syschenko, O., Fujita, T., Sechovsky, V., Diviš, M., and Fujii, H. (2001). Magnetism in $RECo_2$ compounds under high pressure, *J. Magn. Magn. Mater.*, **226–230**, pp. 1062–1067.
84. Gama, S., Coelho, A. A., de Campos A., Carvalho, A. M., Gandra, F. C. G., von Ranke, P. J., and de Oliveira, N. A. (2004). Pressure-induced colossal magnetocaloric effect in MnAs, *Phys. Rev. Lett.*, **93**, p. 237202(4).

85. Morellon, L., Arnold, Z., Magen, C., Ritter, C., Prokhnenko, O., Skorokhod, Y., Algarabel, P. A., Ibarra, M. R., and Kamarad, J. (2004). Pressure enhancement of the giant magnetocaloric effect in $Tb_5Si_2Ge_2$, *Phys. Rev. Lett.*, **93**, p. 137201(4).

86. Morellon, L., Arnold, Z., Algarabel, P. A., Magen, C., Ibarra, M. R., and Skorokhod, Y. (2004). Pressure effect in the giant magnetocaloric compounds $Gd_5(Si_xGe_{1-x})_4$, *J. Phys. Condens. Matter.*, **16**, pp. 1623–1630.

87. Carvalho, A. M. G., Alves, C. S., de Campos, A., Coelho, A. A., Gama, S., Gandra, F. C. G., von Ranke, P. J., and de Oliveira N. A. (2005). The magnetic and magnetocaloric properties of $Gd_5Ge_2Si_2$ compound under hydrostatic pressure, *J. Appl. Phys.*, **97**, p. 10M320(3).

88. Lyubina, J., Nenkov, K., Schultz, L., and Gutfleisch, O. (2008). Multiple metamagnetic transition in the magnetic refrigerant $La(Fe,Si)_{13}H_x$, *Phys. Rev. Lett.*, **101**, p. 177203(4).

89. Mañosa, L., González-Alonso, D., Planes A., Bonnot, E., Barrio, M., Tamarit, J. L., Aksoy, S., and Acet, M. (2010). Giant solid-state barocaloric effect in the Ni-Mn-In magnetic shape-memory alloy, *Nat. Mater.*, **9**, pp. 478–481.

90. Soto-Parra, D. E., Vives, E., González-Alonso, D., Mañosa, L., Planes, A., Romero, R., Matutes-Aquino, J. A., Ochoa-Gamboa, R. A., and Flores-Zúñiga, H. (2010). Stress- and magnetic field-induced entropy changes in Fe-doped Ni-Mn-Ga shape-memory alloys, *Appl. Phys. Lett.*, **96**, p. 071912(3).

91. Castillo-Villa, P. O., Mañosa, L., Planes, A., Soto-Parra, D. E., Sánchez-Llamazares, J. L., Flores-Zúñiga, H., and Frontera, C. (2013). Elastocaloric and magnetocaloric effects in Ni-Mn-Sn(Cu) shape memory alloy, *J. Appl. Phys.*, **113**, p. 053506(6).

92. Bonnot, E., Romero, R., Mañosa, L., Vives, E., and Planes, A. (2008). Elastocaloric effect associated with the martensitic transition in shape-memory alloys, *Phys. Rev. Lett.*, **100**, p. 125901(4).

93. Nikitin, S. A., Myalikgulyev, G., Annaorazov, M. P., Tyurin, A. L., Myndyev, R. W., and Akopyan, S. A. (1992). Giant elastocaloric effect in FeRh alloy, *Phys. Lett. A*, **171**, pp. 234–236.

94. Mañosa, L., González-Alonso, D., Planes, A., Barrio, M., Tamarit, J., Titov, I. S., Acet, M., Bhattacharyya, A., and Majumdar, S. (2011). Inverse barocaloric effect in the giant magnetocaloric La-Fe-Si-Co compound, *Nat. Commun.*, **2**, p. 595(5).

95. Yuce, S., Barrio, M., Emre, B., Stern-Taulats, E., Planes, A., Tamarit, J., Mudryk, Y., Gschneidner, Jr., K. A., Pecharsky, V. K., and Mañosa, L.

(2012). Barocaloric effect in the magnetocaloric prototype $Gd_5Si_2Ge_2$, *Appl. Phys. Lett.*, **101**, p. 071906(4).

96. Oliveira, N. A. de (2013). Giant magnetocaloric and barocaloric effects in $R_5Si_2Ge_2$ (R=Tb, Gd), *J. Appl. Phys.*, **113**, p. 033910(5).
97. Oliveira, N. A. de (2011). Barocaloric effect and the pressure induced solid state refrigerator, *J. Appl. Phys.*, **109**, p. 053515(3).
98. Pirc, R., Kutnjak, Z., Blinc, R., and Zhang, Q. M. (2011). Upper bounds on the electrocaloric effect in polar solids, *Appl. Phys. Lett.*, **98**, p. 021909(3).
99. Valant, M. (2012). Electrocaloric materials for future solid-state refrigeration technologies, *Prog. Mater. Sci.*, **57**, pp. 980–1009.
100. Kobeko, P., and Kurtschatov, J. (1930). Dielectric characteristics of seignette's salts, *Z. Phys.*, **66**, pp. 192–205.
101. Mischenko, A. S., Zhang, Q., Scott, J. F., Whatmore, R. W., and Mathur, N. D. (2006). Giant electrocaloric effect in thin-film $PbZr_{0.95}Ti_{0.05}O_3$, *Science*, **311**, pp. 1270–1271.
102. Saranaya, D., Chaudhuri, A. R., Parui, J., and Krupanidhi, S. B. (2009). Electrocaloric effect of PMN-PT thin films near morphotropic phase boundary, *Bull. Mater. Sci.*, **32**, pp. 259–262.
103. Moya, X., Stern-Taulats, E., Crossley, S., González-Alonso, D., Kar-Narayan, S., Planes, A., Mañosa, L., and Mathur, N. D. (2013). Giant electrocaloric strength in single crystal $BaTiO_3$, *Adv. Mater.*, **25**, pp.1360–1365.
104. Lu, S. G., Rožič, B., Zhang, Q. M., Kutnjak, Z., Li, X., Furman, E., Gorny, L. J., Lin, M., Malič, B., Kosec, M., Blinc, R., and Pirc, R. (2010). Organic and inorganic relaxor ferroelectrics with giant electrocaloric effect, *Appl. Phys. Lett.*, **97**, p. 162904(3).
105. Liu, P. F., Wang, J. L., Meng, X. J., Yang, J., Dkhil, B., and Chu, J. H. (2010). Huge electrocaloric effect in Langmuir-Blodgett ferroelectric polymer thin films, *New J. Phys.*, **12**, p. 023035(8).
106. Bai, Y., Zheng, G., and Shi, S. (2011). Abnormal electrocaloric effect of $Na_{0.5}Bi_{0.5}TiO_3$-$BaTiO_3$ lead-free ferroelectric ceramics above room temperature, *Mater. Res. Bull.*, **46**, pp. 1866–1869.
107. Li, B., Wang, J. B., Zhong, X. L., Wang, F., Zeng, Y. K., and Zhou, Y. C. (2013). The coexistence of the negative and positive electrocaloric effect in ferroelectric thin films for solid-state refrigeration, *Eur. Phys. Lett.*, **102**, p. 47004.
108. Shaobo, L., and Yanqiu, L. (2004). Research on the electronic effect of PMN/PT solid solution for ferroelectrics MEMS microcoolers, *Mater. Sci. Eng. B*, **113**, pp. 46–49.

109. Fuith, A., Kabelka, H., Birks, E., Shebanovs, L., and Sternberg, A. (2000). Thermodynamic properties at the phase transition of Pb(Zr, Sn, T,)O_3, *Ferroelectrics*, **237**, pp. 153–159.

110. Chukka, R., Cheah, J. W., Chen, Z., Yang, P., Shannigrahi, S., Wang, J., and Wang, L. (2011). Enhanced cooling capacities of ferroelectric materials at morphotropic phase boundary, *Appl. Phys. Lett.*, **98**, p. 242902(3).

111. Weisman, G. G. (1969). Electrocaloric effect in potassium dihydrogen phosphate. *IEEE Trans. Electron. Dev.*, **16**, pp. 588–593.

112. Benepe, J. N., and Reese, W. (1971). Electronic studies of KH_2PO_4, *Phys. Rev. B*, **3**, pp. 3032–3039.

113. Strukov, B. A., Taraskin, S. A., and Varikash, V. M. (1968). Thermal and electrocaloric properties of ferroelectric triglycine selenite near the Currie point, *Fizika Tverdogo Tela*, **10**, pp. 1836–1842 (in Russian).

114. Shebanovs, L., Borman, K., Lawless, W. N., and Kalvane, A. (2002). Electrocaloric effect in some perovskite ferroelectric ceramics and multilayer capacitors, *Ferroelectrics*, **273**, pp. 137–142.

115. Bai, Y., Zheng, G., and Shi, S. (2010). Direct measurement of giant electrocaloric effect in $BaTiO_3$ multilayer thick film structure, *Appl. Phys. Lett.*, **96**, p. 192902(3).

116. Rožič, B., Kosec, M., Uršič, H., Holc, J., Malič, B., Zhang, Q. M., Blinc, R., Pirc, R., and Kutnjak, Z. (2011). Influence of the critical point on the electrocaloric response of relaxor ferroelectrics, *J. Appl. Phys.*, **110**, p. 064118(5).

117. Parui, J., and Krupanidhi, S. B. (2008). Electrocaloric effect in antiferroelectric $PbZrO_3$ thin films. *Phys. Status Solidi (RRL)*, **2**, pp. 230–232.

118. Lu, S. G., Rožič, B., Zhang, Q. M., Kutnjak, Z., Pirc, R., Lin, M., Li, X., and Gory, L. (2010). Comparison of directly and indirectly measured electrocaloric effect in relaxor ferroelectric polymer, *Appl. Phys. Lett.*, **97**, p. 202901(3).

119. Feng, Z., Shi, D., and Duo, S. (2011). Large electrocaloric effect in highly (001)-oriented $0.67PbMg_{1/3}Nb_{2/3}O_3$-$0.33PbTiO_3$ thin films, *Solid State Commun.*, **151**, pp. 123–126.

120. Chen, H., Ren, T. L., Wu, X. M., Yang, Y., and Liu, L. T. (2009). Giant electrocaloric effect in lead-free thin film of strontium bismuth tantalite, *Appl. Phys. Lett.*, **94**, p. 182902(3).

121. Correia, T., and Zhang, Q. (2014). *Electrocaloric Materials: New Generation of Cooler* (Engineering Materials, Springer).

122. Moya, X., Kar-Narayan, S., and Mathur, N. D. (2014). Caloric materials near ferroic phase transition, *Nat. Mater.*, **13**, p. 439.
123. Es'kov, A. V., Karmanenko, S. F., Pakhomov, O. V., and Starkov, A. S. (2009). Simulation of a solid-state cooler with electrocaloric element, *Phys. Solid State*, **51**, pp. 1574–1577.
124. Ožbolt, M., Kitanovski, A., Tušek, J., and Poredoš, A. (2014). Electrocaloric refrigeration: Thermodynamics, state of the art and future perspectives, *Int. J. Refrigeration*, **40**, pp. 174–188.

Index

AB, *see* ammonia-borane
AB_2 alloys   590–592
AB_5 alloys   589–591
absorbing materials   28, 35, 44, 63–64, 541
ACs, *see* activated carbons
activated carbons (ACs)   375–382, 387, 411, 568, 577–579, 623
activation, chemical   377–378, 382
activation energy   37, 298, 517–519, 524, 577, 612–613
active electrocaloric regenerator (AER)   777–778
actuators   19, 249, 742
adiabatic conditions   754, 765–766
adiabatic temperature change   753, 755–757, 759–761, 765, 769–771, 773–774
AEC, *see* alkaline electrolysis cell
AER, *see* active electrocaloric regenerator
AFC, *see* alkaline fuel cell
ALD, *see* atomic layer deposition
alkaline electrolysis cell (AEC)   515–517, 521–522, 524–525, 530–531
alkaline electrolyzers   517, 522–523, 525, 536
alkaline fuel cell (AFC)   284, 286, 304–305
alkaline water electrolysis   522

alloying reactions   321–322, 336
alloys   194, 197, 306, 321, 331, 589, 591, 595, 603, 623, 710, 716, 760–762, 769–770, 778
ammonia-borane (AB)   617–622, 624
APD, *see* avalanche photodiode
aqueous electrolytes   316–317, 319, 352, 359, 379, 381–382, 385, 401, 403, 406, 413, 418, 422–427, 431, 437
  acidic   407, 409, 431
  neutral   425, 427, 445
artificial photosynthesis   16, 22, 493, 540–541, 543, 545, 547, 549, 551, 553, 556
asymmetric capacitors   413–414, 421
asymmetric cells   410, 418–419, 421, 438, 443
atomic layer deposition (ALD)   27, 75–77
avalanche photodiode (APD)   747

barocaloric effect (BCE)   768–771
BCE, *see* barocaloric effect
biofuels   10, 14, 21
biomass   5–6, 9, 21, 291, 376–377, 407

BSCCO 648
  Bi2212 649
  Bi2223 649

CA, *see* carbon aerogel
CAES, *see* compressed air energy storage
caloric effects, solid-state refrigeration based on 753–778
caloric materials 754, 778
capacitors
  conventional 352–354, 357–359, 361, 375
  hybrid 372
carbon aerogel (CA) 375, 380–381, 383, 405, 408, 431–432
carbon dioxide 500–501, 505, 536–537, 540, 551–552
carbon dioxide electrolysis 536–538
carbon dioxide photoreduction 551–552
carbon dioxide reduction 540, 552–554
carbon electrodes 282, 376, 422, 429
carbon fibers 403, 572
carbon nanostructures 408
carbon nanotube (CNT) 68–69, 73–74, 382–384, 405–410, 421, 428, 431–432, 579
  single-walled 375
carbon paste electrode (CPE) 367, 369
carbonates, molten 288, 291
CCT, *see* correlated color temperature

cells
  photoelectrochemical 540, 548–549, 554–555
  supercapacitor 371, 390, 430, 440
charge collection efficiency 98, 104–105, 126, 136
charge storage mechanisms 360–361, 363, 365, 367, 369, 372, 396, 434
chemical energy 278–279, 290–291, 494, 496, 498, 500, 502–503, 506, 539, 551
chemical energy storage 495, 497, 499, 540, 555
chemical storage media (CSM) 498–499, 505
chemical vapor deposition (CVD) 75, 385–386
climate change 7
closed cycle refrigeration 685
CNT, *see* carbon nanotube
compressed air energy storage (CAES) 495–496
conducting polymer (CP) 370, 372, 388–389, 391–392, 394, 400, 404–405, 408–410, 413, 418, 429
conventional hydrogen storage 570–571, 573, 575, 623
correlated color temperature (CCT) 699–700, 702, 718, 723
CP, *see* conducting polymer
CPE, *see* carbon paste electrode
cryocooler, *see* closed cycle refrigeration
CSM, *see* chemical storage media
Curie temperature 211–212, 763, 775
CV, *see* cyclic voltammetry

Index | 793

CVD, see chemical vapor deposition
cyclic voltammetry (CV) 325, 367, 369, 381, 412, 418, 430, 434–435, 437–441, 443
CZTS 53–54

DADB, see diammoniate of diborane
DCFCs, see direct carbon fuel cells
DH, see double heterojunction
diammoniate of diborane (DADB) 618–619
dielectric material 260, 357–359, 728
direct carbon fuel cells (DCFCs) 290–291
double heterojunction (DH) 292, 300, 498, 504, 506, 512, 514, 711, 715, 755–756, 758–760, 763
double layer capacitance 365, 394, 404, 436
double synchronized switch harvesting (DSSH) 230–231
DSSC, see dye-sensitized solar cell
DSSH, see double synchronized switch harvesting
dye-sensitized solar cell (DSSC) 48–49, 57, 62, 64, 71, 93, 95–96, 101–109, 111–121, 123, 125, 127–129, 131, 133, 135–138
  principle 101, 103, 105, 107, 109, 111, 113, 115, 117, 119, 121, 123, 125, 127, 129

eCE, see elastocaloric effect
EDLC, see electrochemical double-layer capacitor
EIS, see electrochemical impedance spectroscopy
elastocaloric effect (eCE) 768–769, 771
electroactive film electrodes 430–431
electroactive materials 314, 316, 318, 370–371, 426, 428, 431–433, 443
electrocaloric effect 754, 772–773, 775, 777
electrocaloric materials 774
electrocaloric refrigeration 775, 777–778
electrochemical capacitors 351, 359–361, 365
electrochemical cell 292, 296–297, 313–314, 316, 325, 496, 511
electrochemical devices 278, 291, 495, 502, 504
electrochemical double-layer capacitor (EDLC) 372–375, 387–388, 394, 401, 410–411, 414, 445–447
electrochemical energy storage 14, 495
electrochemical impedance spectroscopy (EIS) 37, 137, 325, 430, 518–519
electrodes
  carbon paste 367, 369
  interdigitated 239, 241, 257–260
electrolyzers 303–304, 505, 512–513, 515, 518–521, 523, 530–532, 543, 554, 556
  solid oxide 521, 529, 537
electromechanical coupling 207, 218, 224, 258–259

electronic devices, low-power 208–209, 232, 248
electronics, low-power 208, 232, 239, 263
electrons, photogenerated 543–544, 547
electroplating 321, 658
energy conditioning circuitry 224–231
energy harvesting 206, 236–238, 242, 252–253, 259, 262–264
  nonlinear piezoelectric 237, 250
  vibration 207, 217, 232, 238, 252–253
energy harvesting circuitry 206, 226, 230, 233
energy harvesting systems 205, 228, 232, 239
  multi-source 253–255
energy harvesting technologies 16, 207, 236
energy storage
  flywheel 669
  magnetic 668
EQE, *see* external quantum efficiency
equivalent series resistance (ESR) 425–427, 429–432, 441–443
ESR, *see* equivalent series resistance
external quantum efficiency (EQE) 34–35

faradaic efficiency 294, 300, 520, 522, 527, 555
FC, *see* fuel cell

feed-in grid 670
first-generation (1G) HTS, *see* BSCCO
fluorine-doped tin oxide (FTO) 67, 70–71
FTO, *see* fluorine-doped tin oxide
fuel cell (FC) 16, 22, 277–292, 294–302, 304–306, 353, 356, 500, 502–505, 556, 620, 680
fuel cell efficiency 294–301
fuel cell technologies 280, 282, 284, 288, 302, 304–305
fuel cells
  alkaline 284, 286, 304
  high-temperature 282, 284, 288, 290, 298, 300, 302, 306, 539
  polymer electrolyte membrane 284, 286, 305
fusion power 678
  ITER 678

galvanostatic cycling (GC) 325, 381, 412–414, 417, 421, 430, 440, 442
GC, *see* galvanostatic cycling
giant magnetocaloric effect (GMCE) 757, 760–761
Gibbs energies 292, 295, 756
GMCE, *see* giant magnetocaloric effect
graphene 68–69, 73–74, 375, 381, 383–387, 403, 406, 409, 411–412, 414
graphite 333–336, 339, 383–385, 392

Index | 795

heat engines 279–280, 502–503
helium reserve 684
high-temperature superconductor (HTS) 642, 646, 648, 653, 669, 672–673, 681, 684, 686
  discovery 642
  first-generation (1G), see BSCCO
  Hg1223 681
  second-generation (2G), see RBCO
  theory of 681
high voltage direct current (HVDC) 20–21
hole transporting materials 63–64, 106–107
HTS, see high-temperature superconductor
HTS materials 649, 654, 682
HVDC, see high voltage direct current
hybrid materials 12, 58, 96, 370, 372, 400, 404–409
hydrides 331, 587, 591, 593, 605–606, 624, 713
hydrocarbons 280, 306, 382, 505, 551, 553, 568
hydrogen adsorption 581
hydrogen economy 680
  SuperGrid 680
hydrogen evolution reaction 518, 528–529, 544
hydrogen physisorption 576–577, 579, 581, 583, 623
hydrogen-powered aircraft 676
hydrogen production 495, 499, 508, 521–523, 530, 538, 551, 680

hydrogen storage 567–570, 572, 574, 576–584, 586–588, 590, 592, 594–596, 598, 600, 602, 604, 606, 616–618, 622
  reversible 588, 592, 602–605, 609
hydrogen storage materials 579, 596, 601–602, 608, 615, 624

IBAD, see ion beam assisted deposition
ILs, see ionic liquids
internal quantum efficiency (IQE) 35, 712, 722
ion beam assisted deposition (IBAD) 656
ionic liquids (ILs) 136, 340, 412, 422–423, 426–427, 528
IQE, see internal quantum efficiency
iron arsenides 655

lead-acid batteries 319, 325–327, 329
lead zirconate titanate (PZT) 205, 210, 212–214, 241, 256, 260, 262–263, 774
LED, see light-emitting diode
Li-ion batteries 329, 333, 339, 342–343, 388
light, properties of 693–701
light-emitting diode (LED) 13–14, 699, 703–705, 710,

712–713, 715–716, 718, 721–723, 725–726, 730–731, 734, 738–740, 742, 748
liquid electrolytes 58, 423, 512, 517–518, 522–523, 525, 536
liquid hydrogen 302, 499, 569, 573–575, 595, 613, 623, 676, 680
lithium batteries, rechargeable 332
lossless transmission 660
low-temperature superconductor (LTS) 649, 668, 682, 684, 686
LTS, see low-temperature superconductor

maglev 674
magnetic refrigeration 764–765, 775, 778
magnetic refrigerator 759, 763–768, 772, 778
magnetocaloric effect (MCE) 754–755, 757–765, 767–768, 770–772, 774
magnetocaloric materials 759, 761, 764, 768
manganese oxides 396–397, 399, 413–414, 418
MCE, see magnetocaloric effect
MCFCs, see molten carbonate fuel cells
MEMS harvesters 240–242
metal hydride 283, 329, 331–332, 576, 584–587, 589, 591, 593, 595–596, 608, 622–624

metal organic chemical solution deposition (MOCSD) 658
metal organic chemical vapor deposition (MOCVD) 657, 713
metal-organic frameworks (MOFs) 577, 581, 583–584, 623
MgB_2 653
  in situ and ex situ 654
  persistent joint 654
MOCSD, see metal organic chemical solution deposition 658
MOCVD, see metal organic chemical vapor deposition
modal electromechanical coupling coefficient 219, 222–223
MOFs, see metal-organic frameworks
molten carbonate fuel cells (MCFCs) 284–285, 288–289, 306, 535
MOSFET 740, 742

n-type materials 28, 31–32, 40, 198, 707
nanocomposite materials 375, 405–406, 409–410, 413–414, 418, 445
nanocomposites 372, 383, 400, 404–410
nanostructured carbons 381
NHE, see normal hydrogen electrode
normal hydrogen electrode (NHE) 420, 552

OCV, *see* open circuit voltage
OER, *see* oxygen evolution reaction
OLED, *see* organic light emitting diode
OMCs, *see* ordered mesoporous carbons
open circuit voltage (OCV) 32–33, 66, 121, 281–282, 292–293, 296–298, 318, 440, 511, 514, 521
OPV, *see* organic photovoltaics
ordered mesoporous carbons (OMCs) 379, 403
organic electrolytes 317, 335, 359, 375–376, 386–387, 393, 418, 422–423, 427, 431, 446
organic light emitting diode (OLED) 14, 718–721, 749
organic photovoltaics (OPV) 95–97, 100–101
  basics of 95–99
organic semiconductors 49–50, 96
oxygen evolution reaction (OER) 340–341, 516, 518, 528–529, 544

p-type materials 28–29, 31, 551, 707–708
PAFCs, *see* phosphoric acid fuel cells
PCE, *see* power conversion efficiency
PCI, *see* pressure-composition-isotherms
PECVD, *see* plasma-enhanced chemical vapor deposition

PEM, *see* polymer electrolyte membrane
PEMEC, *see* proton exchange membrane electrolysis cells
PEMFCs, *see* polymer electrolyte membrane fuel cells
PGEC, *see* phonon- glass electron-crystal
phonon-glass electron-crystal (PGEC) 195–196, 198
phosphoric acid fuel cells (PAFCs) 284–285, 288, 305
photo-conversion efficiency 64, 71
photoanode 101, 104, 110, 548, 550, 554
photocatalyst 540, 546–548, 552
photocathode 548–550
photoreceptors 695, 697
photovoltaic devices 62, 65
photovoltaics 12, 16–18, 20, 66, 68, 77, 255
piezoelectric devices 207, 214–215, 229–230, 236, 247
piezoelectric energy harvesters 206, 216–217, 224, 247
piezoelectric energy harvesting 205–207, 209, 214, 216–217, 224, 226, 232–237, 239, 241, 243, 245, 247–248, 256, 263
piezoelectric materials 205–206, 208, 210–212, 214–215, 217, 238, 242, 252–254, 256, 258, 263
piezoelectric transduction 205, 208–223, 242
  principles of 210–223
PIN photodiodes 747

plasma-enhanced chemical vapor deposition (PECVD) 41, 47
PLEDs 718, 720
PLEs, see protected lithium electrodes
polyaniline 409
polymer electrolyte membrane (PEM) 283, 521, 527, 568
polymer electrolyte membrane fuel cells (PEMFCs) 284–287, 305–306, 540
polymer electrolytes 283, 339, 423
polyoxometalates 405
polypyrrole 408
polyvinylidene fluoride (PVDF) 212–214, 243, 259, 261, 426–427, 775
porous carbon materials 372, 577
powder-in-tube technique 648
power conversion efficiency (PCE) 32–33, 41, 73, 94, 100–102, 105, 113–115, 120, 127, 136–137
pressure-composition-isotherms (PCI) 585, 602, 609
protected lithium electrodes (PLEs) 340–341
proton exchange membrane electrolysis cells (PEMEC) 515–517, 521–522, 525, 527–531
pseudocapacitance 354–355, 359–360, 365–366, 368–373, 387–388, 396, 398, 401, 404, 434, 445
pseudocapacitive materials 383, 399, 405, 413, 418, 421, 446

pseudocapacitive processes 368, 388, 411
pseudocapacitors 370, 372, 387–388, 394, 397–398
PSH, see pumped-storage hydroelectricity
pumped-storage hydroelectricity (PSH) 495, 555, 669
PVDF, see polyvinylidene fluoride
PZT, see lead zirconate titanate
PZT cantilevers 241, 243, 247

QDs, see quantum dots
QDSSCs, see quantum dot-sensitized solar cells
quantum dot-sensitized solar cells (QDSSCs) 64–65, 106, 121
quantum dots (QDs) 36, 56–57, 64, 67, 121–123, 728–729

RBCO 651
redox flow batteries 343–345
rolling assisted, biaxially textured substrates (RABiTS) 657

Sabatier reaction 500–501
SALD, see spatial atomic layer deposition
second-generation (2G) HTS, see RBCO
scanning electron microscopy (SEM) 430

SECE, *see* synchronous electric charge extraction
SEI, *see* solid electrolyte interphase
SEM, *see* scanning electron microscopy
semiclassical theory of thermoelectricity in solids 168–191
sensing systems 206, 233
 self-powered 232–233, 235
sensitizers 108–109, 112–114, 120–122
silicon 12, 28, 30, 40, 42, 47–48, 94, 197, 290, 336, 502, 568, 604, 727, 736
single-walled carbon nanotubes (SWCNTs) 375, 381, 383, 386–387, 408, 579
small-molecule OLEDs (SMOLEDs) 718–720
SMES, *see* superconducting magnetic energy storage
SMOLEDs, *see* small-molecule OLEDs
SOECs, *see* solid oxide electrolysis cells
SOFCs, *see* solid oxide fuel cells
solar cell, efficiency 69–71
solar cells 27–78, 542
 excitonic 49
 hybrid 38–39, 49, 64–65
 organic 49–50, 100
 perovskite 58, 102
 physics of 27–37
 tandem 65, 67
solid electrolyte interphase (SEI) 129, 131, 335, 338
solid electrolytes 282, 285, 304–305, 386, 418, 422, 424, 597
solid oxide cells 531, 539

solid oxide electrolysis cells (SOECs) 501, 515–517, 521–522, 529, 531–539
solid oxide fuel cells (SOFCs) 282, 284–285, 289–291, 305–307, 504, 530–531, 533–535, 539
solid-state lighting (SSL) 693–728
spatial atomic layer deposition (SALD) 27, 75–79
SSHI, *see* synchronized switch harvesting on inductor
SSL, *see* solid-state lighting
storage capacitors 235, 259–260
supercapacitors 15–16, 351–352, 355, 370–371, 393, 421
 asymmetric 420
 high-performance environmentally friendly 425–429
 hybrid 400
superconducting
 cable
  Nuclotron 663
  Roebel 664
  Rutherford 664
 electric aircraft 676
 energy storage devices 668
 fault current limiters 670
 generators 667
 magnetic energy storage (SMES) 495, 668
 magnetically levitated trains 674
 marine propulsion systems 673
 materials 648
 motors 672
 personal electric vehicles 677

power cables   660
quantum interference device (SQUID)   683
transformers   665
superconducting magnetic energy storage   495
superconducting materials   645, 648, 651, 655, 659
superconducting power cables   660–661, 680–681
superconductivity   13, 200, 641–643, 645–647, 653, 678, 680, 682–683
  room-temperature   648, 682
superconductor
  BCS theory   643, 681
  coated conductor   655
  cooling   683
  cost   685
  critical current density J_c 645, 647
  critical field H_c 645
  critical temperature T_c 643–644
  depairing current density J_d 644
  flux pinning   645
  grain boundary problem   652
  high-temperature, see high-temperature superconductors
  irreversibility field H_{irr} 645, 647
  lossless current transport   643
  Meissner–Ochsenfeld effect, perfect diamagnetism   645
  mixed state   646
  Pauli limiting field H_P 645
  persistent current   684
  pinning force   647
  room-temperature   680
  type I and II   645–646
SWCNTs, see single-walled carbon nanotubes
synchronized switch harvesting on inductor (SSHI)   229–230
synchronous electric charge extraction (SECE)   228–229
syngas   280–281, 494, 500–501, 533, 536, 539, 568

tandem cells   66–67
TBP, see tert-butylpyridine
TCMs, see transparent conductive materials
TCOs, see transparent conductive oxides
tert-butylpyridine (TBP)   124, 131, 134–135
thermoelectric materials   159–162, 166–168, 189, 193, 195, 197, 199, 202
thermoelectric power   157–159, 164, 177–178, 180–186, 188, 191–193
thermoelectricity, applications of   157–167
thermoelectrics   155–156, 158, 160, 162–164, 166, 168, 170, 172, 174, 176, 178, 180, 182, 194, 200–202
transition metal oxides   200, 372, 388, 394, 399
transparent conductive materials (TCMs)   27, 67–71, 73–74
transparent conductive oxides (TCOs)   41–42, 63, 67–70, 74, 136

transparent electrodes 67, 71, 73–74, 96, 719–720
transportation 672
Tres Amigas SuperStation 661

water electrolysis 278, 508–509, 512–513, 520, 536, 568
water splitting 540, 543, 550–552
wind turbines

offshore 667
superconducting 667

YSZ, *see* yttria-stabilised zirconia
yttria-stabilised zirconia (YSZ) 289, 307, 533

Z-scheme water splitting 545–546
zeolites 501, 577–580, 608